# Ebook Set up and Use

Use of the ebook is subject to the single user licence at the back of this book.

Electronic access to your price book is now provided as an ebook on the VitalSource® Bookshelf platform. You can access it online or offline on your PC/Mac, smartphone or tablet. You can browse and search the content across all the books you've got, make notes and highlights and share these notes with other users.

## Setting up

**1. Create a VitalSource Bookshelf account** at https://online.vitalsource.com/user/new (or log into your existing account if you already have one).

**2. Retrieve the code** by scratching off the security-protected label inside the front cover of this book. Log in to Bookshelf and click the **Redeem** menu at the top right of the screen. and Enter the code in the **Redeem code** box and press **Redeem**. Once the code has been redeemed your Spon's Price Book will download and appear in your **library**. N.B. the code in the scratch-off panel can only be used once, and has to be redeemed before end December 2017.

When you have created a Bookshelf account and redeemed the code you will be able to access the ebook online or offline on your smartphone, tablet or PC/Mac. Your notes and highlights will automatically stay in sync no matter where you make them.

## Use ONLINE

1. Log in to your Bookshelf account at https://online.vitalsource.com).
2. Double-click on the title in your **library** to open the ebook.

## Use OFFLINE

Download BookShelf to your PC, Mac, iOS device, Android device or Kindle Fire, and log in to your Bookshelf account to access your ebook, as follows:

*On your PC/Mac*
Go to https://support.vitalsource.com/hc/en-us/ and follow the instructions to download the free VitalSource Bookshelf app to your PC or Mac. Double-click the VitalSource Bookshelf icon that appears on your desktop and log into your Bookshelf account. Select **All Titles** from the menu on the left – you should see your price book on your Bookshelf. If your Price Book does not appear, select **Update Booklist** from the **Account** menu. Double-click the price book to open it.

*On your iPhone/iPod Touch/iPad*
Download the free VitalSource Bookshelf App available via the iTunes App Store. Open the Bookshelf app and log into your Bookshelf account. Select **All Titles** - you should see your price book on your Bookshelf. Select the price book to open it. You can find more information at https://support.vitalsource.com/hc/en-us/categories/200134217-Bookshelf-for-iOS

*On your Android™ smartphone or tablet*
Download the free VitalSource Bookshelf App available via Google Play. Open the Bookshelf app and log into your Bookshelf account. You should see your price book on your Bookshelf. Select the price book to open it. You can find more information at https://support.vitalsource.com/hc/en-us/categories/200139976-Bookshelf-for-Android-and-Kindle-Fire

*On your Kindle Fire*
Download the free VitalSource Bookshelf App from Amazon. Open the Bookshelf app and log into your Bookshelf account. Select All Titles – you should see your price book on your Bookshelf. Select the price book to open it. You can find more information at https://support.vitalsource.com/hc/en-us/categories/200139976-Bookshelf-for-Android-and-Kindle-Fire

## Support

If you have any questions about downloading Bookshelf, creating your account, or accessing and using your ebook edition, please visit http://support.vitalsource.com/

For questions or comments on content, please contact us on sponsonline@tandf.co.uk

---

All materials in *Spon's Online*, which is supplied with *Spon's External Works & Landscape Price Book 2017*, are copyright protected. No such materials may be used, displayed, modified, adapted, distributed, transmitted, transferred, published or otherwise reproduced in any form or by any means now or hereafter developed other than strictly in accordance with the terms of the Electronic Version Single-User Licence Agreement at the end of this book. The ebook may be used by more than one user if additional access codes are purchased from Electronic Sales, Spon Press, 3 Park Square, Milton Park, Abingdon, Oxon OX14 4RN Tel: +44 (0) 20 7017 6000

© 2017 COPYRIGHT ALL RIGHTS RESERVED

# Free Updates

with three easy steps…

1. Register today on www.pricebooks.co.uk/updates

2. We'll alert you by email when new updates are posted on our website

3. Then go to www.pricebooks.co.uk/updates
   and download the update.

All four Spon Price Books – *Architects' and Builders'*, *Civil Engineering and Highway Works*, *External Works and Landscape* and *Mechanical and Electrical Services* – are supported by an updating service. Two or three updates are loaded on our website during the year, typically in November, February and May. Each gives details of changes in prices of materials, wage rates and other significant items, with regional price level adjustments for Northern Ireland, Scotland and Wales and regions of England. The updates terminate with the publication of the next annual edition.

As a purchaser of a Spon Price Book you are entitled to this updating service for this 2017 edition – free of charge. Simply register via the website www.pricebooks.co.uk/updates and we will send you an email when each update becomes available.

If you haven't got internet access or if you've some questions about the updates please write to us at Spon Price Book Updates, Spon Press Marketing Department, 3 Park Square, Milton Park, Abingdon, Oxfordshire, OX14 4RN.

**Find out more about Spon books**
Visit www.pricebooks.co.uk for more details.

# Spon's External Works and Landscape Price Book

# 2017

# Spon's External Works and Landscape Price Book

*Edited by*
**AECOM**

*in association with*
LandPro Ltd
Landscape Surveyors

**2017**

*Thirty-sixth edition*

First edition 1978
Thirty-sixth edition published 2017
by CRC Press
2 Park Square, Milton Park, Abingdon, Oxon, OX14 4RN

and by CRC Press
Taylor & Francis, 6000 Broken Sound Parkway, NW, Suite 300, Boca Raton, FL 33487

*CRC Press is an imprint of the Taylor & Francis Group, an informa business*

© 2017 Taylor & Francis

The right of AECOM Ltd to be identified as the Author of this Work has been asserted by them in accordance with the Copyright, Designs and Patents Act 1988

All rights reserved. No part of this book may be reprinted or reproduced or utilized in any form or by any electronic, mechanical, or other means, now known or hereafter invented, including photocopying and recording, or in any information storage or retrieval system, without permission in writing from the publishers.

*Trademark notice:* Product or corporate names may be trademarks or registered trademarks, and are used only for identification and explanation without intent to infringe.

The publisher makes no representation, express or implied, with regard to the accuracy of the information contained in this book and cannot accept legal responsibility or liability for any errors or omissions that may be made.

*British Library Cataloguing in Publication Data*
A catalogue record for this book is available from the British Library

ISBN13: 978-1-4987-8615-7
Ebook: 978-1-4987-8647-8
ISSN: 0267-4181

Typeset in Arial by Taylor & Francis Books

Printed and bound in Great Britain by
TJ International Ltd, Padstow, Cornwall

# Contents

| | |
|---|---|
| *Preface to the Thirty-Sixth Edition* | vii |
| *Acknowledgements* | xi |
| *How to Use this Book* | xxiii |
| *How this Book is Compiled* | xxix |

## PART 1: GENERAL — 1

| | |
|---|---|
| Labour Rates Used in this Edition | 3 |
| Computation of Labour Rates Used in this Edition | 5 |
| Notes on the New Rules of Measurement (NRM) | 11 |
| Computation of the Cost of Materials | 13 |
| The Landfill Tax | 15 |
| The Aggregates Levy | 18 |
| Cost Indices | 23 |
| Regional Variations | 25 |

## PART 2: RATES OF WAGES – BUILDING INDUSTRY — 27

| | |
|---|---|
| Building Industry Wages | 29 |

## PART 3: APPROXIMATE ESTIMATING RATES — 33

| | | | | |
|---|---|---|---|---|
| Cost Models | 35 | 8.3.2 External Planting | 82 |
| Preliminaries | 40 | 8.3.3 Irrigation Systems | 96 |
| 8.1.1 Site Clearance | 42 | 8.4.1 Fencing and Railings | 98 |
| 8.1.2 Preparatory Groundworks | 45 | 8.4.2 Walls and Screens | 102 |
| 8.2.1 Roads, Paths, Pavings and Surfacings | 52 | 8.4.3 Retaining Walls | 110 |
| 8.2.2 Special Surfacings and Pavings | 74 | 8.5 External Fixtures | 113 |
| 8.3.1 Soft Landscaping, Seeding and Turfing | 75 | 8.6 External Drainage | 115 |

## PART 4: PRICES FOR MEASURED WORKS — 123

| | | | | |
|---|---|---|---|---|
| Introduction | 123 | 29 Decoration | 228 |
| New Items 2017 | 124 | 33 Drainage Above Ground | 230 |
| 1 Preliminaries | 141 | 34 Drainage Below Ground | 231 |
| 2 Off site Manufactured Materials, Components and Buildings | 153 | 35 Site Works – Bases | 265 |
| | | Hardcore bases | 265 |
| 3 Demolitions | 155 | Type 1 bases | 265 |
| 4 Alterations, Repairs and Conservation | 160 | Cellular confined bases | 266 |
| 5 Excavating and Filling | 161 | Surface treatments to granular bases | 266 |
| 6 Ground Remediation and Soil Stabilization | 171 | 35 Site Works – Kerbs and Edgings | 267 |
| 10 Crib Walls, Gabions and Reinforced Earth | 180 | Foundations to kerbs | 267 |
| 11 In Situ Concrete Works | 186 | Precast concrete kerbs | 268 |
| 13 Precast Concrete | 194 | Precast concrete edgings | 270 |
| 14 Masonry | 195 | Stone kerbs | 270 |
| 19 Waterproofing | 211 | Stone edgings | 271 |
| 21 Cladding and Covering | 213 | Brick edgings | 273 |
| 28 Floor, Wall, Ceiling and Roof Finishings | 214 | Channels | 274 |

| | | | |
|---|---|---|---|
| Timber edgings | 275 | Cycle holders/cycle shelters | 321 |
| Metal edgings | 275 | Flagpoles | 322 |
| 35 Site Works – Pavings | 277 | Generally | 323 |
| Specialized mortars for pavings | 277 | Pergolas | 323 |
| Excavation of pavings | 277 | Picnic tables/sundry furniture | 324 |
| Filling to make up levels for pavings | | Plant containers | 324 |
| and surfaces | 278 | Seating | 326 |
| Brick pavings | 278 | Signage | 330 |
| Cellular confined granular pavings | 282 | Smoking shelters | 331 |
| Grass concrete | 283 | Stones and boulders | 331 |
| Cobble pavings | 285 | Tree grilles/tree protection | 332 |
| Granular pavings | 285 | 35 Site Works – Sports, Play Equipment, Surfaces | 333 |
| Precast concrete block pavings | 287 | Market prices of sports and play | |
| SUDS permeable block paving | 289 | surfacing materials | 333 |
| Precast concrete slab pavings | 291 | Sports equipment | 333 |
| Deterrent paving | 295 | Playground equipment | 334 |
| Stone pavings | 296 | Natural sports pitches | 338 |
| Stone setts | 298 | Play surfaces | 340 |
| Reconstituted stone pavings | 298 | Tennis courts | 341 |
| Resin bound surfacing | 299 | Sports surfaces | 342 |
| Resin bonded surfacing | 300 | 35 Site Works – Water and Water Features | 343 |
| Macadam surfacing | 301 | Clarification notes on labour costs in | |
| Marking of car parks | 304 | this section | 343 |
| Suspended paving | 304 | Boreholes | 343 |
| Formwork for concrete pavings | 304 | Cold water | 343 |
| Reinforcement for concrete pavings | 305 | Rainwater harvesting | 344 |
| In situ concrete pavings | 305 | Irrigation | 346 |
| Expansion joints in pavings | 307 | Fountains and water features | 349 |
| Woodchip sufaces | 307 | Internal water features | 350 |
| Timber decking | 307 | Water and river bank remediation/ecology | 351 |
| Steps | 310 | Lakes and ponds | 354 |
| 35 Site Works – Street Furniture | 313 | Swimming pools | 359 |
| Barriers | 313 | 36 Fencing | 362 |
| Bases for street furniture | 313 | 37 Soft Landscaping | 381 |
| Bins: litter/waste/dog waste/salt and grit | 315 | 39 Electrical Services | 480 |
| Bollards | 318 | | |

**PART 5: TABLES AND MEMORANDA**     **489**

**INDEX**     **583**

# Preface to the Thirty-Sixth Edition

### Market conditions

As in previous years the landscape market continues to be buoyant. We perceive that there is less pressure on post tender negotiations and that the main requirement of employers is programme and quality rather than minor savings at the price point. This has led to a softening of landscape prices in the tender sector. The major landscape contractors continue to scoop up the prime projects but the second and third tier contractors who carry out specialist works have been able to raise their prices in competitive tender and still win good projects.

### Key influences

Manufacturers and suppliers have advised us to keep our database prices for their products the same in many cases for this publication.

The fuel price is lower than last year's price which has influenced only delivery rates.

The labour market has however risen strongly. Costs of labour and especially management are the major factor in the higher tender prices submitted by prospective contractors who wish to ensure quality finishes and management on their projects.

### Recent Construction Activity Trends

Headline UK output data published by the Office for National Statistics (ONS) suggests that a weaker rate of construction growth is apparent. This is a continuing trend seen since the second half of 2015. Between January and March 2016, output in the construction industry was estimated to have decreased by 1.1% compared with 2015Q4. Downward pressure on the quarter came from all new work which decreased by 0.6% and repair and maintenance (R&M) which decreased by 1.9%.

Between 2015Q1 and 2016Q1 output was estimated to have decreased by 1.9%.

### Materials

Construction materials experiencing the largest price changes in the 12 months to March 2016 in the UK are shown in the table below:

| Construction materials | Change between March 2015 and March 2016 |
|---|---|
| Insulating materials | +4.8% |
| Concrete blocks, bricks, tiles and flagstones | +4.1% |
| Sanitary ware | +3% |
| Precast concrete products | +2.9% |
| Builders woodwork | +2.6% |
| Sand and gravel | +2.6% |
| Imported sawn or planed wood | −6.5% |
| Fabricated structural steel | −10% |
| Imported plywood | −13.2% |
| Concrete reinforcing bars | −16.5% |

From March 2015 to March 2016 the average rise for the BIS recorded materials show a slight fall of 0.7%. This figure does however hide some dramatic differences for individual materials.

## Measurement changes – New Rules of Measurement (NRM)

The NRM is a suite of documents to provide a standard set of measurement rules that are understandable by all those involved in a construction project.

This year's edition maintains the refinement as part of the move towards compliance with NRM1 and NRM2. There will be further refinement over future editions as NRM becomes more embedded in general practice and workflows. SMM7 measurement is largely omitted from this publication

There are three parts to the NRM suite:

**NRM 1: Order of cost estimating and cost planning for building projects.**

**NRM 2: Detailed measurement for building works: provides detailed rules for the measurement and description of building works for the purpose of obtaining tender prices.**

**NRM 3: Order of cost estimating and cost planning for building maintenance works. NB NOT covered in this book.**

### Preliminaries

There are continuing signs that preliminaries costs are beginning to harden, but they still typically range from 10% to 15%, sometimes lower, sometimes higher. Procurement route, location and a host of other things effect the preliminaries allowances. We advise that for prediction purposes on projects that preliminaries should be estimated at 13% of the contract sum.

**Prices included within this edition do not include for VAT, professional fees etc. which must be added if appropriate.**

### Supplier prices

Similarly to last year, there have been many suppliers who have only reflected minimal increases or even no increase in their prices especially for material supplies. The larger suppliers have increased their list costs by a standard 5%–6% but we perceive that there is room to negotiate on material cost rates.

### Labour rates

This year we have increased the price of the base cost of labour by £1.00 (Approximately £1800.00 per annum).

This is due to our perception of a mostly stable labour market at lower levels but a shortage at the skill levels. Contract and project managers are in very short supply thus pushing up the average rate as per our calculations in the front of the book. Readers may make adjustments to these rates in accordance with the instructions and the indices tables published in the general section following this preface.

The standard rate for this year's calculations for Major works is £23.75 per hour, an increase of 4.4% driven mostly by the demand for experienced labour and increased management costs. There is a separate rate used for bricklayers of 30.00 per hour due to shortages and demand for this skill.

### Plant rates

The hired in heavy plant suppliers referenced in this publication have advised us during our update to keep their prices the same. The price of red diesel has decreased slightly and this would have some effect on the hourly rates of the plant used. Our supplier of prices to this publication comments that the large national cross-hiring companies are driving hire prices down.

**Profit and overhead**

*Spon's External Works and Landscape* prices do not allow for profit or site overhead. Readers should evaluate the market conditions in the sector or environment in which they operate and apply a percentage to these rates. Company overhead is allowed for within the labour rates used. Please refer to 'Part 1 General' for an explanation of the build-up of this year's labour rates.

**Prices for suppliers and services**

We acknowledge the support afforded to us by the suppliers of products and services who issue us with the base information used. Their contact details are published in the acknowledgements section at the front of this book. We advise that readers wishing to evaluate the cost of a product or service should approach these suppliers directly should they wish to confirm prices prior to submission of tenders or quotations.

Whilst every effort is made to ensure the accuracy of the information given in this publication, neither the editors nor the publishers in any way accept liability for loss of any kind resulting from the use made by any person of such information.

We remind readers that we would be grateful to hear from them during the course of the year with suggestions or comments on any aspect of the contents of this year's book and suggestions for improvements. Our contact details are shown below.

AECOM Ltd
Aldgate Tower
2 Leman Street
London
E1 8FA

SAM HASSALL
LANDPRO LTD
Landscape Surveyors
14 Upper Bourne Lane
Farnham
Surrey
GU10 4RQ

Tel: 020 7645 2000
e-mail: spons.europe@aecom.com

Tel: 01252 795030
e-mail: info@landpro.co.uk

# SPON'S PRICEBOOKS 2017 From AECOM

## Spon's Architects' and Builders' Price Book 2017

Editor: AECOM

To NRM1 and NRM2, and with more plant prices separated out in the measured works section. With further developments:

- The in situ concrete section and plasterboard linings and partitions sections have been heavily revised and developed.
- A laboratory, a car park and an updated London Office have been added as new Cost Models
- And new items are added: Foamglas insulation, Carlite waterproof concrete, and an expanded range of stone flooring.

Hbk & VitalSource® ebook 824pp approx.
978-1-4987-8611-9 £160
VitalSource® ebook
978-1-4987-8643-0 £160
(inc. sales tax where appropriate)

## Spon's Civil Engineering and Highway Works Price Book 2017

Editor:AECOM

This year some item descriptions have been revised to reflect prices in the global market. Plus a revised structure for manhole construction; new prices for pre-cast manholes with monolithic precast concrete bases; updated rail supply and installation rates and a few more added items; revised piling rates and descriptions; new highway pipe rates; reduced steel prices.

Hbk & VitalSource® ebook 688pp approx.
978-1-4987-8612-6 £180
VitalSource® ebook
978-1-4987-8646-1 £180
(inc. sales tax where appropriate)

## Spon's External Works and Landscape Price Book 2017

Editor:AECOM

This NRM edition includes a number of new and significantly developed items: acoustic fencing, architectural metalwork, block and beam systems, culverts, deep concrete manholes to 3 m, headwalls, pre-cast steps and new retaining walls. Green roofs and sportsfields have now been broken down into details, rather than presented as a lump sum.

Hbk & VitalSource® ebook 624pp approx.
978-1-4987-8615-7 £150
VitalSource® ebook
978-1-4987-8648-5 £150
(inc. sales tax where appropriate)

## Spon's Mechanical and Electrical Services Price Book 2017

Editor:AECOM

This NRM edition includes a number of new and significantly developed items: WiFi equipment, FP 600 cable, wireless fire alarms; and a much enhanced clock section.

Hbk & VitalSource® ebook 840pp approx.
978-1-4987-8616-4 £160
VitalSource® ebook
978-1-4987-8650-8 £160
(inc. sales tax where appropriate)

> Receive our VitalSource® ebook free when you order any hard copy Spon 2017 Price Book
>
> Visit www.pricebooks.co.uk

To order:
Tel: 01235 400524 Fax: 01235 400525
Post: Taylor & Francis Customer Services, Bookpoint Ltd, 200 Milton Park, Abingdon, Oxon, OX14 4SB, UK
Email: book.orders@tandf.co.uk
A complete listing of all our books is on www.crcpress.com

**CRC Press**
Taylor & Francis Group

# Acknowledgements

This list has been compiled from the latest information available but as firms frequently change their names, addresses and telephone numbers due to reorganization, users are advised to check this information before placing orders.

ACO Technologies Plc
ACO Business Park
Hitchin Road
Shefford
Bedfordshire SG17 5TE
Tel: 01462 816666
Website: www.aco.co.uk
E-mail: technologies@aco.co.uk
**Linear drainage**

Addagrip Terraco Ltd
Addagrip House
Bell Lane Industrial Estate
Uckfield
East Sussex TN22 1QL
Tel: 01825 761333
Website: www.addagrip.co.uk
E-mail: sales@addagrip.co.uk
**Resin bound and bonded surfacing**

Agripower Ltd
Broomfield Farm
Rignall Road
Great Missenden
Buckinghamshire HP16 9PE
Tel: 01494 866776
Website: www.agripower.co.uk
E-mail: info@agripower.co.uk
**Land drainage and sports grounds**

Amber Precast Ltd
West Way
Somercotes
Derbyshire DE55 4QJ
Tel: 01773 540707
Website: www.amberprecast.co.uk
Email: info@amberprecast.co.uk
**Precast concrete**

Amenity Land Solutions Ltd (ALS)
Allscott Park
Allscott
Telford
Shropshire TF6 5DY
Tel: 01902 255036
Website: www.amenity.co.uk
E-mail: sales@amenity.co.uk
**Horticultural products supplier**

Anderton Concrete Products Ltd
Units 1 & 2 Cosgrove Business Park
Soot Hill
Anderton
Northwich
Cheshire CW9 6AA
Tel: 01606 535300
Website: www.andertonconcrete.co.uk
E-mail: civils@andertonconcrete.co.uk
**Concrete retaining wall systems**

Architectural Heritage Ltd
Taddington Manor
Cutsdean
Cheltenham
Gloucestershire GL54 5RY
Tel: 01386 584414
Website: www.architectural-heritage.co.uk
E-mail: puddy@architectural-heritage.co.uk
**Ornamental stone buildings and water features**

Arundel Stone Sussex Ltd
62 Aldwick Road
Bognor Regis
West Sussex PO21 2PE
Tel: 01243 829151
Website: www.arundelstone.co.uk
Email: info@arundelstone.co.uk
**Reconstituted stone**

AVS Fencing Supplies Ltd
Unit 1 AVS Trading Park
Chapel Lane
Milford
Surrey GU8 5HU
Tel: 01483 410960
Website: www.avsfencing.co.uk
E-mail: sales@avsfencing.co.uk
**Fencing**

BBS Green Roofing Ltd
Vitis House
50 Dickens Street
Battersea
London SW8 3EQ
Tel: 0207 622 6225
Website: http://www.green-roofing.co.uk/
Email: bbs@green-roofing.co.uk
**Specialist Green Roof consultants/Installers**

Bituchem Group
Forest Vale Industrial Estate
Cinderford
Gloucestershire GL14 2YH
Tel: 01594 826768
Website: www.bituchem.com
E-mail: info@bituchem.com
**Coloured hard landscaping materials**

Breedon Special Aggregates
(A division of Breedon Aggregates England Ltd)
Breedon Quarry, Main Street
Breedon-on-the-Hill
Derby
DE73 8AP
Tel: 01332 694001
Website: www.breedon-special-aggregates.co.uk
E-mail: specialaggs@breedonaggregates.com
**Specialist gravels**

British Sugar Topsoil
Sugar Way
Peterborough PE2 9AY
Tel: 0870 240 2314
Website: www.bstopsoil.co.uk
E-mail: topsoil@britishsugar.com
**Topsoil**

Broxap Street Furniture
Rowhurst Industrial Estate
Chesterton
Newcastle-under-Lyme
Staffordshire ST5 6BD
Tel: 0844 800 4085
Website: www.broxap.com
E-mail: sales@broxap.com
**Street furniture**

Buildbase Civils and Lintels
Bloomfield Road
Farnworth
Bolton BL4 9LP
Tel: 01204 862222
Website: www.civilsandlintels.co.uk
E-mail: farnworthspecials@civilsandlintels.co.uk
**Wall drainage and erosion control**

Butek Landline
1 Bluebridge Industrial Estate
Halstead
Essex CO9 2EX
Tel: 01787 476699
Website: www.landline.co.uk
E-mail: sales@landline.co.uk
**Pond and lake installation**

Capital Garden Products Ltd
Gibbs Reed Barn
Pashley Road
Ticehurst
East Sussex TN5 7HE
Tel: 01580 201092
Website: www.capital-garden.com
E-mail: sales@capital-garden.com
**Plant containers**

CED Ltd
728 London Road
West Thurrock
Grays
Essex RM20 3LU
Tel: 01708 867237
Website: www.ced.ltd.uk
E-mail: sales@ced.ltd.uk
**Natural stone**

Cedar Nursery
Horsley Road
Cobham
Surrey KT11 3JX
Tel: 01932 862473
Website: www.landscaping.co.uk
E-mail: sales@landscaping.co.uk
**Aggregate containment grids**

Charcon Hard Landscaping
Hulland Ward
Ashbourne
Derbyshire DE6 3ET
Tel: 01335 372222
Website: www.aggregate.com
E-mail: ukenquiries@aggregate.com
**Paving and street furniture**

Cleartrack (EvL) Ltd
Ploughmans Barn
Station Farm
Denton Road
Horton
Northants NN7 2BG
Tel: 01604 871360
Website: www.cleartrack.co.uk
E-mail: info@cleartrack.co.uk
**Ground clearance**

CU Phosco Ltd
Charles House
Lower Road
Great Amwell
Ware
Hertfordshire SG12 9TA
Tel: 01920 860600
Website: www.cuphosco.co.uk
E-mail: sales@cuphosco.co.uk
**Lighting**

DLF Seeds Ltd
Thorn Farm
Evesham Road
Inkberrow
Worcestershire WR7 4LJ
Tel: 01386 791102
Website: www.dlf.co.uk
E-mail: amenity@dlf.co.uk
**Grass and wildflower seed**

Earth Anchors Ltd
15 Campbell Road
Croydon
Surrey CR0 2SQ
Tel: 020 8684 9601
Website: www.earth-anchors.com
E-mail: sales@earth-anchors.com
**Anchors for site furniture**

Elliott Hire
Manor Drive
Peterborough
Cambridgeshire PE4 7 AP
Tel: 01733 298700
Website: www.elliottuk.com
Email: hirediv@elliottuk.com
**Site offices**

English Woodlands
Burrow Nursery
Cross in Hand
Heathfield
East Sussex TN21 0UG
Tel: 01435 862992
Website: www.ewburrownursery.co.uk
E-mail: sales@ewburrownursery.co.uk
**Plant protection**

Eve Trakway
Bramley Vale
Chesterfield
Derbyshire S44 5GA
Tel: 08700 767676
Website: www.evetrakway.co.uk
E-mail: marketing@evetrakway.co.uk
**Portable roads**

Everris Limited
Epsilon House
West Road
Ipswich
Suffolk IP3 9FJ
Tel: 01473 237100
Website: www.everris.com
E-mail: prof.sales@everris.com
**Landscape chemicals**

Exterior Solutions Ltd
Unit 8
Penn Street Works
Penn Street
Amersham
Buckinghamshire HP7 0PX
Tel: 01494 722204
Website: www.exteriordecking.co.uk
E-mail: design@exteriordecking.co.uk
**Decking**

Fairwater Ltd
Lodge Farm
Malthouse Lane
Ashington
West Sussex RH20 3BU
Tel: 01903 892228
Website: www.fairwater.co.uk
E-mail: info@fairwater.co.uk
**Water garden specialists**

FP McCann Ltd
Whitehill Road
Coalville
Leicester LE67 1ET
Tel: 01530 240000
Website: www.fpmccann.co.uk
E-mail: info@fpmccann.co.uk
**Precast concrete pipes**

Furnitubes Street Furniture
Meridian House
Royal Hill
Greenwich
London SE10 8RT
Tel: 020 8378 3200
Website: www.furnitubes.com
E-mail: spons@furnitubes.com
**Street furniture**

Gatic Slotdrain
Poulton Close
Dover
Kent CT17 0UF
Tel: 01304 203545
Website: www.gatic.com
E-mail: info@gatic.com
**Linear drainage**

Germinal GB
Camp Road
Witham St Hughs
Lincoln
Lincolnshire LN6 9QJ
Tel: 01522 868714
Website: www.bshamenity.com
E-mail: seeds@bshlincoln.co.uk
**Grass and wildflower seed**

Grass Concrete Ltd
Duncan House
142 Thornes Lane
Thornes
West Yorkshire WF2 7RE
Tel: 01924 379443
Website: www.grasscrete.com
E-mail: info@grasscrete.com
**Grass block paving**

Greenblue Urban Ltd
Northpoint
Compass Park
Junction Road
Bodiam
East Sussex TN32 5BS
Tel: 01424 717797
E-mail: enquiries@greenblueurban.com
**Root directors**

Greenfix Soil Stabilization and Erosion Control
Allens West
Durham Lane
Eaglescliffe
Stockton on Tees
County Durham TS16 0RW
Tel: 01642 888693
Website: www.greenfix.co.uk
E-mail: stockton@greenfix.co.uk
**Seeded erosion control mats**

Greentech Ltd
Sweethills Park
Nun Monkton
York
North Yorkshire YO26 8ET
Tel: 01423 332100
E-mail: sales@green-tech.co.uk
Web: www.green-tech.co.uk
**Landscaping suppliers**

Griffin Nurseries
New Barn Farm
Rake road
Milland
Liphook
Hampshire GU30 7JU
Tel: 01428 741655
Website: www.griffinnurseries.co.uk
E-mail: enquiries@griffinnurseries.co.uk
**Hedging**

Grundon Waste Management Ltd
Goulds Grove
Ewelme
Wallingford
Oxfordshire OX10 6PJ
Tel: 0870 4438278
Website www.grundon.com
E-mail: sales@grundon.com
**Gravel**

Guncast Swimming Pools Ltd
Unit L
The Old Bakery
Golden Square
Petworth
West Sussex GU28 0AP
Tel: 0179 8343725
Website: www.guncast.com
E-mail: info@guncast.com
**Swimming pools**

Haddonstone Ltd
The Forge House
Church Lane
East Haddon
Northampton
Northamptonshire NN6 8DB
Tel: 01604 770711
Website: www.haddonstone.com
E-mail: info@haddonstone.co.uk
**Landscape ornaments and architectural cast stone**

Hags SMP Ltd
Ten Acre Lane
Thorpe
Egham
Surrey TW20 8RJ
Tel: 01784 489100
Website: www.smp.co.uk
E-mail: sales@smp.co.uk
**Playground equipment**

Havells Sylvania Fixtures UK Ltd
Avis Way
Newhaven
East Sussex BN9 0ED
Tel: 0870 6062030
Website: www.havells-sylvania.com
E-mail: info.concord@havells-sylvania.com
**Lighting**

Headland Amenity Ltd
1010 Cambourne Business Park
Cambourne
Cambridgeshire CB23 6DP
Tel: 01223 597834
Website: www.headlandamenity.com
E-mail: info@headlandamenity.com
**Chemicals**

Heicom UK
4 Frog Lane
Tunbridge Wells
Kent TN1 1YT
Tel: 01892 522360
Website: www.treesand.co.uk
E-mail: mike.q@treesand.co.uk
**Tree sand**

Hepworth Wavin Plc
Hazelhead
Crow Edge
Sheffield
South Yorkshire S36 4HG
Tel: 0870 4436000
Website: www.hepworthdrainage.co.uk
E-mail: info@hepworthdrainage.co.uk
**Drainage**

Hillier Nurseries
Ampfield House
Ampfield
Romsey
Hampshire SO51 9PA
Tel: 01794 368733
Website: www.hilliertrees.co.uk
E-mail: trees@hillier.co.uk
**Trees**

HSS Hire
Unit 3
14 Wates Way
Mitcham
Surrey CR4 4HR
Tel: 020 8685 9500
Website: www.hss.com
E-mail: hire@hss.com
**Tool and plant hire**

Inturf
The Chestnuts
Wilberfoss
York
North Yorkshire YO41 5NT
Tel: 01759 321000
Website: www.inturf.com
E-mail: info@inturf.co.uk
**Turf**

J Toms Ltd
7 Marley Farm
Headcorn Road
Smarden
Kent TN27 8PJ
Tel: 01233 770066
Website: www.jtoms.co.uk
E-mail: jtoms@btopenworld.com
**Plant protection**

Jacksons Fencing
Stowting Common
Ashford
Kent TN25 6BN
Tel: 01233 750393
Website: www.jacksons-fencing.co.uk
E-mail: sales@jacksons-fencing.co.uk
**Fencing**

James Coles & Sons (Nurseries) Ltd
The Nurseries
Uppingham Road
Thurnby
Leicestershire LE7 9QB
Tel: 01162 412115
Website: www.colesnurseries.co.uk
E-mail: sales@colesnurseries.co.uk
**Nurseries**

Johnsons Wellfield Quarries Ltd
Crosland Hill
Huddersfield
West Yorkshire HD4 7AB
Tel: 01484 652311
Website: www.johnsons-wellfield.co.uk
E-mail: sales@johnsons-wellfield.co.uk
**Natural Yorkstone pavings**

Jones of Oswestry
Whittington Road
Oswestry
Shropshire SY11 1HZ
Tel: 01691 653251
Website: www.jonesofoswestry.com
E-mail: sales@jonesofoswestry.com
**Channels, gulleys, manhole covers**

Kinley Systems
Haywood Way
Hastings
East Sussex TN35 4PL
Tel: 01424 201111
Website www.kinleysystems.com
E-mail: sales@exceledge.co.uk
**Aluminium edgings**

Kompan Ltd
20 Denbigh Hall
Bletchley
Milton Keynes
Bucks MK3 7QT
Tel: 01908 642466
Website: www.kompan.com
E-mail: kompan.uk@kompan.com
**Play equipment**

Landscape Plus
Unit 1
Kemble Business Park
Malmesbury
Wiltshire SN16 9SH
Tel: 01666 577577
Website: www.landscapeplus.com
E-mail: info@landscapeplus.com
**Lighting**

Lappset UK Ltd
Lappset House
Henson Way
Kettering
Northants NN16 8PX
Tel: 01536 412612
Website: www.lappset.com
E-mail: customerservicesuk@lappset.com
**Play equipment**

LDC Limited
The Charcoal House
Blacksmith Lane
Guildford
Surrey GU4 8NQ
Tel: 01483 573817
Website: www.ldclandscape.co.uk
E-mail: info@ldc.co.uk
**Willow walling**

Leaky Pipe Systems Ltd
Frith Farm
Dean Street
East Farleigh
Maidstone
Kent ME15 0PR
Tel: 01622 746495
Website: www.leakypipe.co.uk
E-mail: sales@leakypipe.co.uk
**Irrigation systems**

Leca UK
Regus House
Herons Way
Chester
Cheshire CH4 9QR
Tel: 08443 351770
Website: www.leca.co.uk
Email: sales@leca.co.uk
**Light aggregates**

Lindum Turf
West Grange
Thorganby
York
North Yorkshire YO19 6DJ
Tel: 01904 448675
Website: www.turf.co.uk
Email: lindum@turf.co.uk
**Meadow and specialized turf**

Lorenz von Ehren
Maldfeldstrasse 4
D-21077 Hamburg
Germany
Tel: 0049 40 761080
Website: www.lve.de
E-mail: sales@lve.de
**Topiary**

Marshalls Plc
Landscape House
Premier Way
Lowfields Business Park
Elland
West Yorkshire HX5 9HT
Tel: 01422 312000
Website: www.marshalls.co.uk
**Hard landscape materials, street furniture, linear drainage**

Melcourt Industries Ltd
Boldridge Brake
Long Newnton
Tetbury
Gloucestershire GL8 8RT
Tel: 01666 502711
Website: www.melcourt.co.uk
E-mail: mail@melcourt.co.uk
**Mulch and compost**

Meon UK
Railside
Northarbour Spur
Portsmouth
Hampshire PO6 3TU
Tel: 023 9220 0606
Website: www.dekorgrip.com
E-mail: support@dekorgrip.com
**Resin surfaces and safety surfaces**

Milton Precast
Cooks Lane
Milton Regis
Sittingbourne
Kent ME10 2QF
Tel: 01795 425191
Website: www.miltonprecast.com
E-mail: sales@miltonprecast.com
**Soakaway rings**

Mobilane UK
PO Box 449
Stoke-on-Trent
Staffordshire ST6 9 AE
Tel: 07711 895261
Website: www.mobilane.co.uk
E-mail: sales@mobilane.co.uk
**Green screens**

Neptune Outdoor Furniture Ltd
Thompsons Lane
Marwell
Winchester
Hampshire SO21 1JH
Tel: 01962 777799
Website: www.nofl.co.uk
E-mail: info@nofl.co.uk
**Street furniture**

Nomix Enviro
A Division of Frontier Agriculture Ltd
The Grain Silos
Weyhill Road
Andover
Hampshire SP10 3NT
Tel: 01264 388050
Website: www.nomixenviro.co.uk
E-mail: nomixenviro@frontierag.co.uk
**Herbicides**

Oakover Nurseries Ltd
Maidstone Road
Hothfield
Ashford
Kent TN26 1 AR
Tel: 01233 713016
Website: www.oakovernurseries.co.uk
E-mail: enquiries@oakovernurseries.co.uk
**Native plant nurseries**

Outdoor Design
Block A Unit 1
Ford Airfield Industrial Estate
Arundel
West Sussex BN18 0HY
Tel: 01903 716960
Website: www.boothengineering.co.uk
E-mail: info@boothengineering.co.uk
**Steel fabricator**

PBA Solutions
Bryn
Rake Road
Liss
Hampshire GU33 7HB
Tel: 01730 893460
Website: www.pba-solutions.com
E-mail: info@pba-solutions.com
**Japanese knotweed consultants**

PC Landscapes Ltd
Abbott House
Hale Road
Farnham
Surrey GU9 9QH
Tel: 01252 891150
Website: www.pclandscapes.co.uk
Email: info@pclandscapes.co.uk
**Landscaping services**

Phi Group
Harcourt House
13 Royal Crescent
Cheltenham
Gloucestershire GL50 3DA
Tel: 01242 707600
Website: www.phigroup.co.uk
E-mail: southern@phigroup.co.uk
**Retaining wall specialists**

Platipus Anchors Ltd
Kingsfield Business Centre
Philanthropic Road
Redhill
Surrey RH1 4DP
Tel: 01737 762300
Website: www.platipus-anchors.com
E-mail: info@platipus-anchors.com
**Tree anchors**

Rainwater Harvesting Ltd
Southgate Way
Peterborough
Cambridgeshire PE2 6YQ
Tel: 01733 405111
Website: www.rainwater-harvesting.co.uk
E-mail: info@rainwaterharvesting.co.uk
**Rain water harvesting**

Readyhedge Ltd
Court Gate Nursery
Station Road
Eckington
Pershore
Worcestershire WR10 3BB
Tel: 01386 750585
Website: www.readyhedgeltd.com
E-mail: info@readyhedgeltd.com
**Pre-grown hedging**

Rigby Taylor Ltd
The Riverway Estate
Portsmouth Road
Peasmarsh
Guildford
Surrey GU3 1LZ
Tel: 01483 446900
Website: www.rigbytaylor.com
E-mail: sales@rigbytaylor.com
**Horticultural supply**

RIW Ltd
Arc House
Terrace Road South
Binfield
Bracknell
Berkshire RG42 4PZ
Tel: 01344 397777
Website: www.riw.co.uk
E-mail: enquiries@riw.co.uk
**Waterproofing products**

Road Equipment Ltd
28–34 Feltham Road
Ashford
Middlesex TW15 1DL
Tel: 01784 256565
E-mail: roadequipment@aol.com
**Plant hire**

Rolawn Ltd
York Road
Elvington
North Yorkshire YO41 4XR
Tel: 01904 608661
Website: www.rolawn.co.uk
E-mail: info@rolawn.co.uk
**Industrial turf**

Saint-Gobain PAM UK Ltd
Lows Lane
Stanton-by-Dale
Ilkeston
Derbyshire DE7 4QU
Tel: 0115 930 5000
Website: www.saint-gobain-pam.co.uk
E-mail: sales.uk.pam@saint-gobain.com
**Cast iron drainage**

Salix River & Wetland Services Ltd
Blackhills Nurseries
Blackhills Lane
Gower
Swansea SA2 7JN
Tel: 0870 350 1851
Website: www.salixrw.com
Email: info@salixrw.com
**River and wetland services**

Scotscape Ltd
Unit 1
Wandle Technology Park
Mill Green Road
Mitcham Junction
Surrey CR4 4HZ
Tel: 020 8254 5000
Website: www.scotscape.net
E-mail: sales@scotscape.net
**Living walls**

Southern Conveyors
Unit 2, Denton Slipways Site
Wharf Road
Gravesend
Kent DA12 2RU
Tel: 01474 564145
Website: www.southernconveyors.co.uk
E-mail: sales@southernconveyors.co.uk
**Conveyors**

Spadeoak (Duracourt)
Town Lane
Wooburn Green
High Wycombe
Buckinghamshire HP10 0PD
Tel: 01628 529421
Website: www.duracourt.co.uk
E-mail: info@duracourt.co.uk
**Tennis courts**

Spadeoak
Town Lane
Wooburn Green
High Wycombe
Buckinghamshire HP10 0PD
Tel: 01628 529421
Website: www.spadeoak.co.uk
E-mail: email@spadeoak.co.uk
**Macadam contractors**

Steelway Brickhouse
Brickhouse Lane
West Bromwich
West Midlands B70 0DY
Tel: 01215 214500
Website: www.steelway.co.uk
E-mail: sales@steelwaybrickhouse.co.uk
**Access covers**

Steelway Fensecure
Queensgate Works
Bilston Road
Wolverhampton
West Midlands WV2 2NJ
Tel: 01902 490919
Website: www.steelway.co.uk
E-mail: sales@steelway.co.uk
**Fencing**

SteinTec
728 London Road
West Thurrock
Essex RM20 3LU
Tel: 01708 860049
Website: www.steintec.co.uk
E-mail: info@steintec.co.uk
**Specialized mortars**

Stonebank Ironcraft Ltd
Parkway Farm
Fossebridge
Northleach
Cheltenham
Gloucestershire GL54 3JL
Tel: 01285 720 737
Website: www.stonebank-ironcraft.co.uk
E-mail: info@stonebank-ironcraft.co.uk
**Solid steel products**

SureSet UK Limited
32 Deverill Road
Trading Estate
Sutton Veny
Warminster
Wiltshire BA12 7BZ
Tel: 01985 841873
Website: www.sureset.co.uk
E-mail: mail@sureset.co.uk
**Solid steel products**

Tensar International
Cunningham Court
Shadsworth Business Park
Blackburn
Lancashire BB2 4PJ
Tel: 01254 262431
Website: www.tensar-international.com
E-mail: sales@tensar.co.uk
**Erosion control, soil stabilization**

Terranova Lifting Ltd
Terranova House
Bennett Road
Reading
Berkshire RG2 0QX
Tel: 0118 931 2345
Website: www.terranova-lifting.co.uk
E-mail: cranes@terranovagroup.co.uk
**Crane hire**

The Garden Trellis Company
Unit 1, Brunel Road
Gorse Lane Industrial Estate
Great Clacton
Essex CO15 4LU
Tel: 01255 688361
Website: www.gardentrellis.co.uk
Email: info@gardentrellis.co.uk
**Garden joinery experts**

TK Solutions Ltd
122 South Avenue
Southend-on-Sea
Essex SS2 4HU
Tel: 01702 308612
Email: tksukltdnew@gmail.com
Skype Name: tomer.kniznik
**Irrigation supplies**

Townscape Products Ltd
Fulwood Road South
Sutton-in-Ashfield
Nottinghamshire NG17 2JZ
Tel: 01623 513355
Website: www.townscape-products.co.uk
E-mail: sales@townscape-products.co.uk
**Hard landscaping**

Trulawn Ltd
Unit 7E Vulcan Way
Sandhurst
Berkshire GU47 9DB
Tel: 01252 819695
Website: www.trulawn.co.uk
E-mail: sales@trulawn.co.uk
**Artificial grass**

Wade International Ltd
Third Avenue
Halstead
Essex CO9 2SX
Tel: 01787 475151
Website: www.wadedrainage.co.uk
E-mail: sales@wade.eu
**Stainless steel drainage**

Wavin Plastics Ltd
Parsonage Way
Chippenham
Wiltshire SN15 5PN
Tel: 01249 766600
Website: www.wavin.co.uk
E-mail: info@wavin.co.uk
**Drainage products**

Wicksteed Leisure Ltd
Digby Street
Kettering
Northamptonshire NN16 8YJ
Tel: 01536 517028
Website: www.wicksteed.co.uk
E-mail: sales@wicksteed.co.uk
**Play equipment**

Willowbank Services
Curload
Stoke St Gregory
Taunton
Somerset TA3 6JD
Tel: 01823 690113
Website: www.willowbankservices.co.uk
E-mail: info@willowbankservices.co.uk
**Waterside bank stabilization**

Woodscape Ltd
Church Works
Church Street
Church
Lancashire BB5 4JT
Tel: 01254 383322
Website: www.woodscape.co.uk
E-mail: sales@woodscape.co.uk
**Street furniture**

Wybone Ltd
Mason Way
Platts Common Industrial Estate
Barnsley
South Yorkshire S74 9TF
Tel: 01226 744010
Website: www.wybone.co.uk
E-mail: sales@wybone.co.uk
**Street furniture**

Yeoman Aggregates Ltd
Stone Terminal
Horn Lane
Acton
London W3 9EH
Tel: 020 8896 6820
Website: www.foster-yeoman.co.uk
E-mail: sales@yeoman-aggregates.co.uk
**Aggregates**

Zinco Green Roof Systems Ltd
St. John's Innovation Centre
Cowley Road
Cambridge CB4 0WS
Tel: 01223 853843
Website: www.zinco-greenroof.co.uk
E-mail: office@zinco-greenroof.co.uk
**Green roof systems**

# ESSENTIAL READING FROM TAYLOR AND FRANCIS

## Spon's Asia Pacific Construction Costs Handbook, Fifth Edition

LANGDON & SEAH

In the last few years, the global economic outlook has continued to be shrouded in uncertainty and volatility following the financial crisis in the Euro zone. While the US and Europe are going through a difficult period, investors are focusing more keenly on Asia. This fifth edition provides overarching construction cost data for 16 countries: Brunei, Cambodia, China, Hong Kong, India, Indonesia, Japan, Malaysia, Myanmar, Philippines, Singapore, South Korea, Sri Lanka, Taiwan, Thailand and Vietnam.

May 2015: 234X156 mm: 452 pp
Hb: 978-1-4822-4358-1: £160.00

To Order: Tel: +44 (0) 1235 400524  Fax: +44 (0) 1235 400525
or Post: Taylor and Francis Customer Services,
Bookpoint Ltd, Unit T1, 200 Milton Park, Abingdon, Oxon, OX14 4TA UK
Email: book.orders@tandf.co.uk

For a complete listing of all our titles visit:
www.tandf.co.uk

| Item<br>Excluding site overheads and profit | Unit | Total rate £ |
|---|---|---|

## HOW TO USE THIS BOOK

### How this book is presented

**All prices in this book exclude overhead and profit**
The price information sections of this book are split into two subsections as follows:
   Approximate Estimating Rates – This is presented in a 3 column format. This section combines prices from the Measured Works items to form a composite rate for the work described.
   Measured Works – This is presented in an 8 column format and is a detailed analysis of single tasks commonly employed throughout the external works and landscape sector.

### Column formats

**Description of the column formats used in this book**
Approximate Estimating Rates
   Only the 3 column format is used; users of the book cannot see the components which make up the prices in the book format. An electronic version is supplied free with this book. Users of the electronic system can see more detail in the electronic format.
Measured Works
   Where an 8 column analysis is shown these are tasks carried out directly by the external works or landscape contractor and are shown at cost.
   Where prices are shown in the 8 column section without analysis, prices have been supplied by specialist subcontractors and are assumed to have the specialist subcontractor's selling price of the item included.

**Explanation of the columns used in the Measured Works analysis**
Prime Cost
   Commonly known as the PC. Prime Cost is the actual price of the material item being addressed such as paving, shrubs, bollards or turves, as sold by the supplier. Prime Cost is given 'per square metre', 'per 100 bags' or 'each' according to the way the supplier sells his product. In researching the material prices for the book we requested that the suppliers price for full loads of their product delivered to a site close to the M25 in London. Spon's rates do not include VAT. Some companies may be able to obtain greater discounts on the list prices than those shown in the book. Prime Cost prices for those products and plants which have a wide cost range will be found under the heading of Market Prices in the main sections of this book, so that the user may select the product most closely related to their specification.

| Item<br>Excluding site overheads and profit | Unit | Total rate £ |
|---|---|---|
| **Column formats – cont**<br><br>**Explanation of the columns used in the Measured Works analysis – cont**<br>Labour Hours<br>   This is the total amount of man hours required to carry out the quantity and unit of the measured works item which is being analysed.<br>Labour £<br>   This is the value of the labour used. Please refer to the labour rates used in this edition which are detailed preceding the Approximate Estimating Rates section.<br>Plant<br>   This covers the cost of machinery used to work out the cost of the item addressed. Plant can range from heavy excavators to light electrical breakers. All plant is assumed as hired in except for concrete mixers and normal site tools which a contractor would be expected to carry in his or her company vehicle (hand drills, grinders, etc). Plant includes for fuel and down time. The rates used are shown in the section of this book Computation of Mechanical Plant Costs which follows this section.<br>   Delivery and removal to site costs are not included in any rates.<br>Material<br>   The PC material plus the additional materials required to fix the PC material. Every job needs materials for its completion besides the product bought from the supplier. Paving needs sand for bedding, expansion joint strips and cement pointing; fencing needs concrete for post setting and nails or bolts; tree planting needs manure or fertilizer in the pit, tree stakes, guards and ties. If these items were to be priced out separately, *Spon's External Works and Landscape Price Book* (and the Bill of Quantities) would be impossibly unwieldy, so they are put together under the heading of Materials.<br>Delivery Prices<br>   These are not generally included. Exceptions would be aggregates or loose materials which are delivered by the load. Our specification to suppliers of information for this book is that all materials are delivered in full loads to a site close to the M25 in North London.<br>Unit<br>   The unit being used to measure the item.<br>Total Rate £<br>   The total price 'At Cost' – without profit of the item being addressed.<br>Where only a figure is shown in the columns<br>   This is generally a subcontract rate which has been received from a specialist subcontractor.<br>   This rate would generally include the subcontractor's overhead and profit but not the main contractors oncosts.<br>Where a figure is shown in the Material column only<br>   This is generally the market price of a single material only. | | |

# How to Use this Book

| Item<br>Excluding site overheads and profit | Unit | Total rate £ |
|---|---|---|
| **Worked examples using data in this edition**<br>The following operations describe the trench excavation, pipe laying and backfilling operations as contained in section NRM Section 34 of this edition. | | |
| **Example 1** | | |
| **Pipe laying 100 m** | | |
| Remove topsoil 150 × 300 mm wide | 100 m | 33.50 |
| Excavate for drain 150 × 450 mm; excavated material to spoil heaps on site by machine | 100 m | 1125.00 |
| Lay flexible plastic pipe 100 mm diameter wide | 100 m | 185.00 |
| Backfilling with gravel rejects, blinding with sand and topping with 150 mm topsoil | 100 m | 390.00 |
| TOTAL FOR ALL OF THE ABOVE COMBINED | 100 m | 1725.00 |
| This total above item is a composite of all of the above Measured Works items combined and is typical of an entry in the Approximate Estimating Rates section of this book. | | |
| **Example 2** | | |
| The following individual items are shown as components of a tree planting specification for one tree. These individual items are all extracted from the paving section of this book. | | |
|     The PC of 900 × 600 x 50 mm precast concrete flags is shown in the PC (Prime cost) column as: | nr | 4.15 |
|     This equates to the rate for the PC material of: | m² | 7.70 |
|     Costs of bedding and pointing are added to give a total Materials cost of: | m² | 5.80 |
|     Further costs for labour and mechanical plant give a resultant price of: | m² | 23.00 |
| If a quotation of PC £9.00/m² was received from a supplier of the flags the resultant price would be calculated as: | | |
|     Original price of the flags: | m² | 7.70 |
|     Gives an extra over of: | m² | 1.30 |
|     Less the original cost PC (£7.50) plus the revised cost (£9.00) to give: | m² | 24.00 |
| **Example 3** | | |
| The following samples are extracted from the tree planting section of this edition. | | |
|     The prime cost of an Acer platanoides 8–10 cm bare root tree is given as: | nr | 17.50 |
|     To which the cost of handling and planting only to tree pits excavated separately is added: | nr | 9.50 |
|     Giving an overall cost of: | nr | 27.00 |
| The following items are also shown as component prices in the Measured Works section | | |
|     labour and mechanical plant for mechanical excavation of a 600 × 600 x 600 mm pit | nr | 4.20 |
|     a single tree stake | nr | 6.30 |
|     importing Topgrow compost in 75 litre bags | nr | 2.60 |
|     backfilling with imported topsoil | nr | 10.30 |
|     transporting the excavated material to spoil heaps on site 25 m distant | nr | 0.90 |
|     and disposing of the excavated material | nr | 8.30 |

| Item<br>Excluding site overheads and profit | Unit | Total rate £ |
|---|---|---|
| **Worked examples using data in this edition – cont** | | |
| **The following samples are extracted from the tree planting section of this edition – cont** | | |
| The following price which is a composite of all the above might be shown in the Approximate Estimating Rates section: | | |
|    Tree planting; supply and plant Acer platanoides 8–10 cm in 600 × 600 mm tree pit excavated mechanically; allow for a single tree stake and Topgrow compost and backfilling with imported topsoil; all excavated material removed from site | nr | 62.00 |
| **Example 4 – Composite labour rates** | | |
| The worked example below shows the build-up for Keyblock paviours as used elsewhere in this book. Readers will note that there appears to be an anomaly between the labour time and the labour rate. This however is due to the useage of a composite labour rate. In this case a 3 man team which is 3 x the hourly rate shown in the labour rates section. 2017 = £23.75 / hour | | |
| **Precast concrete paving blocks; Keyblok Marshalls Plc; on prepared base (not included); on 50 mm compacted sharp sand bed; blocks laid in 7 mm loose sand and vibrated; joints filled with sharp sand and vibrated; level and to falls only**<br>Herringbone bond | | |
|    200 × 100 × 60 mm; natural grey | m² | 33.50 |
| The labour value is shown as £19.96 but the labour hours are 0.2801 hrs. This should equate to £6.65 but you will see a labour value of £19.96. A three man team is used. This equates to 3 x £6.65 = £19.95. Please ignore minor rounding calculations. Readers should extrapolate in these instances. | | |
| **ADJUSTMENT AND VARIATION OF THE RATES IN THIS BOOK** | | |
| It will be appreciated that a variation in any one item in any group will affect the final Measured Works price. Any cost variation must be weighed against the total cost of the contract. A small variation in Prime Cost where the items are ordered in thousands may have more effect on the total cost than a large variation on a few items. A change in design that necessitates the use of earth moving equipment which must be brought to the site for that one job will cause a dramatic rise in the contract cost. Similarly, a small saving on multiple items will provide a useful reserve to cover unforeseen extras. Using the tree planting example above, if the tree size was to be increased from an 8-10 cm tree to a 10–12 cm tree of the same variety, this might necessitate an increase in tree pit size, more compost, a larger tree stake and more excavated material being disposed of off site. | | |
| The resultant variation would again be referenced from the tree planting section as follows: | | |
|    The revised PC (prime cost) of an Acer platanoides 10–12 cm bare root tree is given as: | nr | 19.50 |
|    To which the cost of handling and planting only to tree pits excavated separately is added: | nr | 13.80 |
|    Giving an overall cost of: | nr | 33.00 |
| Add to this the following variations as described: | | |
|    Tree pit increased in size from 600 × 600 x 600 mm to 900 × 900 x 600 mm | nr | 9.40 |
|    Additional compost | nr | 8.60 |
|    Increased staking requirements to two stakes | nr | 8.60 |

| Item<br>Excluding site overheads and profit | Unit | Total rate £ |
|---|---|---|
| Increased volume of imported topsoil | nr | 23.00 |
| increased disposal volume | nr | 19.20 |
| Giving a resultant price for the increased tree size as | nr | 94.00 |
| The variation therefore from the original specification for the 8–10 cm tree to the revised 10–12 cm tree is calculated as an additional sum per tree of | nr | 32.00 |
| **Labour Rates Used in this Edition** | | |
| **General landscape construction** | | |
| Specialist labour rate for brickwork | hr | 30.00 |
| General labour rate for all other works | hr | 24.00 |
| **Landscape maintenance** | | |
| General labour rate for all works | hr | 19.00 |
| **Computation of Mechanical Plant Costs** | | |
| **Hired plant** | | |
| Figures in brackets reflect the actual working hours per week of the hired machinery | | |
| Dumper 3 tonne and operator composite | hr | 31.00 |
| Dumper 3 tonne Thwaites self drive (32 hours/week) | hr | 35.50 |
| Dumper 5 tonne Thwaites self drive (32 hours/week) | hr | 33.00 |
| Dumper 6 tonne Thwaites self drive (32 hours/week) | hr | 36.00 |
| Excavator 360 Tracked 21 tonne operated (32 hours/week) | hr | 58.00 |
| Excavator 360 Tracked 8 tonne self drive (32 hours/week) | hr | 46.50 |
| Excavator Tracked 5 tonne self drive (32 hours/week) | hr | 39.00 |
| Fork lift telehandler self drive (24 hours/week) | hr | 33.00 |
| JCB 3CX 4 x 4 Sitemaster + breaker operated (32 hours/week) | hr | 49.00 |
| JCB 3CX 4 x 4 Sitemaster operated (32 hours/week) | hr | 46.00 |
| Mini Excavator 1.5 tonne self drive (32 hours/week) | hr | 6.25 |
| Mini Excavator JCB 803 Rubber tracks self drive (32 hours/week) | hr | 7.80 |
| Mini Excavator JCB 803 Steel tracks self drive (32 hours/week) | hr | 7.80 |
| Skip loader 1 tonne (32 hours/week) | hr | 4.70 |
| Fuel charge Red diesel (32 hours/week) | l | 0.80 |
| HSS Ltd | | |
| Access tower; alloy; 5.2 m | day | 31.00 |
| Cultivator; 110 kg | hr | 3.00 |
| Diamond blade consumable; 450 mm; concrete | mm | 54.00 |
| Heavy-duty breaker; 110 V; 2200 W | hr | 5.10 |
| Heavy-duty breaker; petrol; 5 hrs/day; single tool | hr | 6.05 |
| Mesh fence; temporary security fencing; 2.85 × 2.0 m high | 1m/ wk | 2.25 |
| Oxy-acetylene cutting kit | hr | 5.10 |
| Petrol masonry saw bench; 350 mm | hr | 3.50 |
| Petrol poker vibrator + 50 mm head | hr | 2.65 |
| Post hole borer; 1 man; weekly rate | hr | 4.10 |
| Vibrating plate compactor | hr | 2.70 |
| Vibrating roller; 136 kg; 10.1 kN; 4 hrs/day | hr | 4.10 |
| Vibration damped breaker (light); 1600 w; 110 v | hr | 3.00 |

# ESSENTIAL READING FROM TAYLOR AND FRANCIS

## Professional Practice for Landscape Architects, 3rd Edition
### TENNANT, GARMORY & WINSCH

Professional Practice for Landscape Architects third edition deals with the practical issues of being a successful landscape architect professional.

Endorsed by the Landscape Institute, this book is an indispensable guide for licentiate members of the Institute on their Pathway to Chartership. It follows the revised 2013 syllabus covering all aspects of professional judgement, ethics and values, the legal system, organisation and management, legislation and the planning system, environmental policy and control, procurement and implementation. It also serves as a reminder and reference for fully qualified professionals in their everyday practice and for landscape students.

Valuable information is presented in an easy to follow manner with diagrams and schedules, key acts, professional documents and contracts clearly explained and made easy to understand.

A handy list of questions are included to aid with P2C revision, answers of which are found within the text.

September 2015; 234 x 156 mm; 466 pp
Pb: 978-1-138-78597-7; £26.99

---

To Order: Tel: +44 (0) 1235 400524  Fax: +44 (0) 1235 400525
or Post: Taylor and Francis Customer Services,
Bookpoint Ltd, Unit T1, 200 Milton Park, Abingdon, Oxon, OX14 4TA UK
Email: book.orders@tandf.co.uk

For a complete listing of all our titles visit:
www.tandf.co.uk

# How this Book is Compiled

## Introduction

First-time users of *Spon's External Works and Landscape Price Book* and others who may not be familiar with the way in which prices are compiled may find it helpful to read this section before starting to calculate the costs of landscape works.

Please also refer to the How to Use this Book notes on page xxiii
The cost of an item of landscape construction or planting is made up of many components:

- the cost of the product
- the labour and additional materials needed to carry out the job
- the cost of running the contractor's business

These are described more fully below.

**Important notes on the profit element of rates in this book**

The rates shown in the Approximate Estimating Rates and Measured Works sections of this book do not contain a profit element unless the rate has been provided by a subcontractor. Prices are shown at cost. This is the cost including the costs of company overhead required to perform this task or project. For reference please see the tables on pages 6–10.

*Analyzed rates versus subcontractor rates*

As a general rule if a rate is shown as an analysed rate in the Measured Works section, i.e. it has figures shown in the columns other than the Unit and Total Rate columns, it can be assumed that this has no profit or overhead element and that this calculation is shown as a direct labour/material supply/plant task being performed by a contractor.

On the other hand, if a rate is shown as a Total Rate only, this would normally be a subcontractor's rate and would contain the profit and overhead for the subcontractor.

The foregoing applies for the most part to the Approximate Estimating Rates section, however in some items there may be an element of subcontractor rates within direct works build-ups.

As an example of this, to excavate, lay a base and place a macadam surface, a general landscape contractor would normally perform the earthworks and base installation. The macadam surfacing would be performed by a specialist subcontractor to the landscape contractor.

The Approximate Estimating Rate for this item uses rates from the Measured Works section to combine the excavation, disposal, base material supply and installation with all associated plant. There is no profit included on these elements. The cost of the surfacing, however, as supplied to us in the course of our annual price enquiry from the macadam subcontractor, would include the subcontractor's profit element.

The landscape contractor would add this to his section of the work but would normally apply a mark up at a lower percentage to the subcontract element of the composite task. Users of this book should therefore allow for the profit element at the prevailing rates. Please see the worked examples below and the notes on overheads for further clarification.

## Introductory notes on pricing calculations used in this book

There are two pricing sections to this book:

**Approximate Estimating Rates** – Average contract value £100,000.00–300,000.00

**Measured Works** – Average contract value £100,000.00–300,000.00

### Typical Project Profile Used in this Book

| | |
|---|---|
| Contract value | £100,000.00–300,000.00 |
| Labour rate (see page 3) | £23.75 per hour |
| Labour rate for maintenance contracts | £19.00 per hour |
| Number of site staff | 35 |
| Project area | 6000 m$^2$ |
| Project location | Outer London |
| Project components | 50% hard landscape<br>50% soft landscape and planting |
| Access to works areas | Very good |
| Contract | Main contract |
| Delivery of materials | Full loads |

As explained in more detail later, the prices are generally based on wage rates and material costs current at June 2016. They do not allow for preliminary items, which are dealt with in the Preliminaries section of this book, or for any Value Added Tax (VAT) which may be payable.

Adjustments should be made to standard rates for time, location, local conditions, site constraints and any other factors likely to affect the costs of a specific scheme.

Term contracts for general maintenance of large areas should be executed at rates somewhat lower than those given in this section.

There is now a facility available to readers that enables a comparison to be made between the level of prices given and those for projects carried out in regions other than outer London; this is dealt with on page 25.

## Units of measurement

The units of measurement have been varied to suit the type of work and care should be taken when using any prices to ascertain the basis of measurement adopted.

The prices per unit of area for executing various mechanical operations are for work in the following areas and under the following conditions:

Prices per m$^2$ relate to areas not exceeding 100 m$^2$

    (any plan configuration)

Prices per 100 m$^2$ relate to areas exceeding 100 m$^2$ but not exceeding ¼ ha

    (generally clear areas but with some subdivision)

Prices per ha relate to areas over ¼ ha

    (clear areas suitable for the use of tractors and tractor-operated equipment)

The prices per unit area for executing various operations by hand generally vary in direct proportion to the change in unit area.

## Approximate estimating

These are combined Measured Work prices which give an approximate cost for a complete section of landscape work. For example, the construction of a car park comprises excavation, levelling, road-base, surfacing and marking parking bays. Each of these jobs is priced separately in the Measured Works section, but a comprehensive price is given in the Approximate Estimating Rates section, which is intended to provide a quick guide to the cost of the job. It will be seen that the more items that go to make up an approximate estimate, the more possibilities there are for variations in the PC prices and the user should ensure that any PC price included in the estimate corresponds to his specification.

The figures given in *Spon's External Works and Landscape Price Book* are intended for general guidance, and if a significant variation appears in any one of the cost groups, the Price for Measured Work should be recalculated.

## Measured works

*Prime Cost:* Commonly known as the PC. Prime Cost is the actual price of the material item being addressed such as paving, shrubs, bollards or turves, as sold by the supplier. Prime Cost is given 'per square metre', 'per 100 bags' or 'each' according to the way the supplier sells his product. In researching the material prices for the book we requested that the suppliers price for full loads of their product delivered to a site close to the M25 in London. Spon's rates do not include VAT. Some companies may be able to obtain greater discounts on the list prices than those shown in the book. Prime Cost prices for those products and plants which have a wide cost range will be found under the heading of Market Prices in the main sections of this book, so that the user may select the product most closely related to his specification.

*Materials:* The PC material plus the additional materials required to fix the PC material. Every job needs materials for its completion besides the product bought from the supplier. Paving needs sand for bedding, expansion joint strips and cement pointing; fencing needs concrete for post setting and nails or bolts; tree planting needs manure or fertilizer in the pit, tree stakes, guards and ties. If these items were to be priced out separately, Spon's External Works and Landscape Price Book (and the Bill of Quantities) would be impossibly unwieldy, so they are put together under the heading of Materials.

*Labour:* This figure covers the cost of planting shrubs or trees, laying paving, erecting fencing etc. and is calculated on the wage rate (skilled or unskilled) and the time needed for the job. Extras such as highly skilled craft work, difficult access, intermittent working and the need for labourers to back up the craftsman all add to the cost. Large regular areas of planting or paving are cheaper to install than smaller intricate areas, since less labour time is wasted moving from one area to another.

*Plant, consumable stores and services*: This rather impressive heading covers all the work required to carry out the job which cannot be attributed exactly to any one item. It covers the use of machinery ranging from JCBs to shovels and compactors, fuel, static plant, water supply (which is metered on a construction site), electricity and rubbish disposal. The cost of transport to site is deemed to be included elsewhere and should be allowed for elsewhere or as a preliminary item. Hired plant is calculated on an average of 36 hours working time per week.

*Subcontract rates*: Where there is no analysis against an item, this is deemed to be a rate supplied by a subcontractor. In most cases these are specialist items where most external works contractors would not have the expertise or the equipment to carry out the task described. It should be assumed that subcontractor rates include for the subcontractor's profit. An example of this may be found for the tennis court rates in section 35.

### Labour rates used in this edition

The rates for labour used in this edition have been based on surveys carried out on a cross section of external works contractors. These rates include for company overheads such as employee administration, transport insurance, and oncosts such as National Insurance. Please see the calculation of these rates shown on pages 6–7. The rates do not include for profit.

*Overheads*: An allowance for this is included in the labour rates which are described above. The general overheads of the contract such as insurance, site huts, security, temporary roads and the statutory health and welfare of the labour force are not directly assignable to each item, so they are distributed as a percentage on each, or as a separate preliminary cost item. The contractor's and subcontractor's profits are not included in this group of costs. Site overheads, which will vary from contract to contract according to the difficulties of the site, labour shortages, inclement weather or involvement with other contractors, have not been taken into account in the build-up of these rates, while overhead (or profit) may have to take into account losses on other jobs and the cost to the contractor of unsuccessful tendering.

In many instances a composite team rate is used in the calculations.

### Adjustment and variation of the rates in this book

It will be appreciated that a variation in any one item in any group will affect the final Measured Works price. Any cost variation must be weighed against the total cost of the contract. A small variation in Prime Cost where the items are ordered in thousands may have more effect on the total cost than a large variation on a few items. A change in design that necessitates the use of earth moving equipment which must be brought to the site for that one job will cause a dramatic rise in the contract cost. Similarly, a small saving on multiple items will provide a useful reserve to cover unforeseen extras.

### Worked examples

These are shown on pages xxv–xxvi.

### How this book is updated each year

The basis for this book is a database of Material, Labour, Plant and Subcontractor resources each with its own area of the database.

### Material, plant and subcontractor resources

Each year the suppliers of each material, plant or subcontract item are approached and asked to update their prices to those that will prevail in September of that year.

These resource prices are individually updated in the database. Each resource is then linked to one or many tasks. The tasks in the task library section of the database is automatically updated by changes in the resource library. A quantity of the resource is calculated against the task. The calculation is generally performed once and the links remain in the database.

On occasions where new information or method or technology are discovered or suggested, these calculations would be revisited. A further source of information is simple time and production observations made during the course of the last year.

*Labour resource update*

Most tasks, except those shown as subcontractor rates (see above), employ an element of labour. The Data Department at Aecom conducts ongoing research into the costs of labour in various parts of the country.

Tasks or entire sections are then re-examined and recalculated. Comments on the rates published in this book are welcomed and may be submitted to the contact addresses shown in the Preface.

It is one thing to
imagine a better world.
**It's another to deliver it.**

**Understanding change,
unlocking potential, creating
brilliant new communities.**

The Tate Modern extension
takes an iconic building and
adds to it. Cost management
provided by AECOM.

**aecom.com**

# SPON'S PRICEBOOKS 2017 From AECOM

## Spon's Architects' and Builders' Price Book 2017

Editor: AECOM

To NRM1 and NRM2, and with more plant prices separated out in the measured works section. With further developments:

- The in situ concrete section and plasterboard linings and partitions sections have been heavily revised and developed.
- A laboratory, a car park and an updated London Office have been added as new Cost Models
- And new items are added: Foamglas insulation, Carlite waterproof concrete, and an expanded range of stone flooring.

Hbk & VitalSource® ebook 824pp approx.
978-1-4987-8611-9 £160
VitalSource® ebook
978-1-4987-8643-0 £160
(inc. sales tax where appropriate)

## Spon's Civil Engineering and Highway Works Price Book 2017

Editor: AECOM

This year some item descriptions have been revised to reflect prices in the global market. Plus a revised structure for manhole construction; new prices for pre-cast manholes with monolithic precast concrete bases; updated rail supply and installation rates and a few more added items; revised piling rates and descriptions; new highway pipe rates; reduced steel prices.

Hbk & VitalSource® ebook 688pp approx.
978-1-4987-8612-6 £180
VitalSource® ebook
978-1-4987-8646-1 £180
(inc. sales tax where appropriate)

## Spon's External Works and Landscape Price Book 2017

Editor: AECOM

This NRM edition includes a number of new and significantly developed items: acoustic fencing, architectural metalwork, block and beam systems, culverts, deep concrete manholes to 3 m, headwalls, pre-cast steps and new retaining walls. Green roofs and sportsfields have now been broken down into details, rather than presented as a lump sum.

Hbk & VitalSource® ebook 624pp approx.
978-1-4987-8615-7 £150
VitalSource® ebook
978-1-4987-8648-5 £150
(inc. sales tax where appropriate)

## Spon's Mechanical and Electrical Services Price Book 2017

Editor: AECOM

This NRM edition includes a number of new and significantly developed items: WiFi equipment, FP 600 cable, wireless fire alarms; and a much enhanced clock section.

Hbk & VitalSource® ebook 840pp approx.
978-1-4987-8616-4 £160
VitalSource® ebook
978-1-4987-8650-8 £160
(inc. sales tax where appropriate)

---

Receive our VitalSource® ebook free when you order any hard copy Spon 2017 Price Book

Visit www.pricebooks.co.uk

---

To order:
Tel: 01235 400524 Fax: 01235 400525
Post: Taylor & Francis Customer Services, Bookpoint Ltd, 200 Milton Park, Abingdon, Oxon, OX14 4SB, UK
Email: book.orders@tandf.co.uk
A complete listing of all our books is on www.crcpress.com

**CRC Press**
Taylor & Francis Group

# PART 1

# General

This part of the book contains the following sections:

|  | page |
|---|---|
| Labour Rates Used in this Edition | 3 |
| Computation of Labour Rates Used in this Edition | 5 |
| Notes on the New Rules of Measurement (NRM) | 11 |
| Computation of the Cost of Materials | 13 |
| The Landfill Tax | 15 |
| The Aggregates Levy | 18 |
| Cost Indices | 23 |
| Regional Variations | 25 |

# ESSENTIAL READING FROM TAYLOR AND FRANCIS

# NRM1 Cost Management Handbook

David P Benge

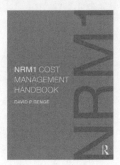

The definitive guide to measurement and estimating using NRM1, written by the author of NRM1

The 'RICS New rules of measurement: Order of cost estimating and cost planning of capital building works' (referred to as NRM1) is the cornerstone of good cost management of capital building works projects - enabling more effective and accurate cost advice to be given to clients and other project team members, while facilitating better cost control.

The NRM1 Cost Management Handbook is the essential guide to how to successfully interpret and apply these rules, including explanations of how to:

- quantify building works and prepare order of cost estimates and cost plans
- use the rules as a toolkit for risk management and procurement
- analyse actual costs for the purpose of collecting benchmark data and preparing cost analyses
- capture historical cost data for future order of cost estimates and elemental cost plans
- employ the rules to aid communication
- manage the complete 'cost management cycle'
- use the elemental breakdown and cost structures, together with the coding system developed for NRM1, to effectively integrate cost management with Building Information Modelling (BIM).

March 2014: 246 x 174: 640pp
Pb: 978-0-415-72077-9: £41.99

To Order: Tel: +44 (0) 1235 400524  Fax: +44 (0) 1235 400525
or Post: Taylor and Francis Customer Services,
Bookpoint Ltd, Unit T1, 200 Milton Park, Abingdon, Oxon, OX14 4TA UK
Email: book.orders@tandf.co.uk

For a complete listing of all our titles visit:
www.tandf.co.uk

# Labour Rates Used in this Edition

Based on surveys carried out on a cross section of external works contractors, the following rates for labour have been used in this edition.

These rates include for company overheads such as employee administration, transport insurance, and oncosts such as National Insurance.

The rates used do not include for profit.

The rates used may include a subcontractor's profit where a subcontractor's rates are used to build up the costs of an item in the book.

The rates for all labour used in this edition are as follows:

| | |
|---|---|
| General contracting | £23.75/man-hour |
| Brickwork 1+1 man gang | £53.75/team-hour |
| Brickwork 1+2 man gang | £83.75/team-hour |
| Maintenance contracting | £19.00/man-hour |

# ESSENTIAL READING FROM TAYLOR AND FRANCIS

## Estimator's Pocket Book

Duncan Cartlidge

The Estimator's Pocket Book is a concise and practical reference covering the main pricing approaches, as well as useful information such as how to process sub-contractor quotations, tender settlement and adjudication. It is fully up-to-date with NRM2 throughout, features a look ahead to NRM3 and describes the implications of BIM for estimators.

It includes instructions on how to handle:

- the NRM order of cost estimate;
- unit-rate pricing for different trades;
- pro rata pricing and dayworks
- builders' quantities;
- approximate quantities.

Worked examples show how each of these techniques should be carried out in clear, easy-to-follow steps. This is the indispensible estimating reference for all quantity surveyors, cost managers, project managers and anybody else with estimating responsibilities. Particular attention is given to NRM2, but the overall focus is on the core estimating skills needed in practice.

May 2013 186x123: 310pp
Pb: 978-0-415-52711-8: £21.99

To Order: Tel: +44 (0) 1235 400524  Fax: +44 (0) 1235 400525
or Post: Taylor and Francis Customer Services,
Bookpoint Ltd, Unit T1, 200 Milton Park, Abingdon, Oxon, OX14 4TA UK
Email: book.orders@tandf.co.uk

For a complete listing of all our titles visit:
www.tandf.co.uk

# Computation of Labour Rates Used in this Edition

Different organizations will have varying views on rates and costs, which will in any event be affected by the type of job, availability of labour and the extent to which mechanical plant can be used. However this information should assist the reader to:

(1)   compare the prices to those used in his own organization
(2)   calculate the effect of changes in wage rates or prices of materials
(3)   calculate prices for work similar to the examples given

From June 2016 – assumed basic weekly rates of pay for craft and general operatives are £457.34 and £343.83 respectively; to these rates have been added allowances for the items below in accordance with the recommended procedure of the Chartered Institute of Building in its Code of Estimating Practice. The resultant hourly rates are £14.91 and £11.16 for craft operatives and general operatives respectively.

The items for which allowances have been made are:

- Lost time
- Construction Industry Training Board Levy
- Holidays with pay
- Accidental injury, retirement and death benefits scheme
- Sick pay
- National Insurance
- Severance pay and sundry costs
- Employer's liability and third party insurance

The tables that follow illustrate how the above hourly rates have been calculated. Productive time has been based on a total of 1801.8 hours worked per year for daywork calculations above.

**How the labour rate has been calculated**

A survey of typical landscape/external works companies indicates that they are currently paying above average wages for multi-skilled operatives regardless of specialist or supervisory capability. In our current overhaul of labour constants and costs we have departed from our previous policy of labour rates based on national awards and used instead a gross rate per hour for all rate calculations. Estimators can readily adjust the rate if they feel it is inappropriate for their work.

The productive labour of any organization must return their salary plus the cost of the administration which supports the labour force.

The labour rate used in this edition is calculated as follows:

**Team size**

A three man team with annual salaries as shown and allowances for National Insurance, uniforms, site tools and a company vehicle, returns a basic labour rate of £19.57 per hour to which company overhead is added as per the following tables.

Basic labour rate calculation for Spon's 2017

Standard Working Hours per year (2016/7)
= 1802

Less allowance for sick leave and down time
= -39

Actual Working Hours per year (2016/7)
=1763

| Working Hours per year 2015/2016 | | 1763 | Allows sick and down time of 39 hours | | | | |
|---|---|---|---|---|---|---|---|
| | Number of | Labour team | NI | Net cost | Clothing | Site tools | TOTAL 3 Man Team |
| Foreman | 1 | 35000.00 | 4480.00 | 39480.00 | 200.00 | 150.00 | |
| Craftsman | 1 | 24000.00 | 3072.00 | 27072.00 | 200.00 | 150.00 | |
| Labourer | 1 | 19500.00 | 2496.00 | 21996.00 | 200.00 | 150.00 | |
| | 3.00 | 78500.00 | 10048.00 | 88548.00 | 600.00 | 450.00 | **£ 89598.00** |
| **Overtime Hours** | | | | | | | |
| nominal hours | 40.00 | 29.78 | 3.81 | 1,343.62 | | | |
| /annum | 40.00 | 20.42 | 2.61 | 921.34 | | | |
| at 1.5 times normal | 40.00 | 16.59 | 2.12 | 748.59 | | | |
| standard rate | | | | | | | |
| **Total staff in Team** | 3.00 | | | 86312.79 | 600.00 | 450.00 | 87362.79 |
| **Vehicle Costs** inclusive of fuel insurances etc. | Working days | £/Day | | | | | |
| | 240.00 | 45.50 | | | | | £ 10920.00 |
| | | | | | | TOTAL | £ 103531.55 |
| **Net Average Cost per Man Hour** | | | 19.57 | | | | |

The basic average labour rate excluding overhead costs (below) is

$$\frac{£\ 103531.55}{\text{Total staff in team (3) x Working hours per year (1763)}} = £19.57 \text{ per hour}$$

Add to the above the overhead costs as per the table on the next page.

## Overhead costs

| COST CENTRE | Number of | Cost | Total |
|---|---|---|---|
| MD: Salary only: excludes profits + NI + Vehicle | 1 | 65000.00 | 65000.00 |
| Senior contracts managers + NI + Vehicle (48K basic) | 1 | 57000.00 | 57000.00 |
| Other contracts managers + NI + Vehicle £32K basic | 1 | 40000.00 | 40000.00 |
| Business administrator/ secretary | 1 | 23000.00 | 23000.00 |
| Bookkeeper | 1 | 20000.00 | 20000.00 |
| Rental | 12 | 1200.00 | 14400.00 |
| Insurances | 1 | 5000.00 | 5000.00 |
| Telephone and Mobiles | 12 | 250.00 | 3000.00 |
| Office Equipment | 1 | 1000.00 | 1000.00 |
| Stationary | 12 | 50.00 | 600.00 |
| Advertising | 12 | 300.00 | 3600.00 |
| Other vehicles not allocated to contract teams | 1 | 9000.00 | 9000.00 |
| Other consultants | 1 | 6000.00 | 6000.00 |
| Accountancy | 1 | 2500.00 | 2500.00 |
| Lights heating water | 12 | 200.00 | 2400.00 |
| Other expenses | 1 | 8000.00 | 8000.00 |
| **TOTAL OFFICE OVERHEAD** | | | 260500.00 |

Add to the above basic rate the company overhead costs for a small company as per the table below. These costs are absorbed by the number of working men multiplied by the number of working hours supplied in the table below. This then generates an hourly overhead rate which is added to the net cost rate above.

## Illustrative overhead for company employing 12–50 landscape operatives

The final labour rate used is then generated as follows:

$$\text{Net labour rate} = \frac{\text{Total Office Overhead (£260500.00)}}{\text{Working hours per year (1763)} \times \text{Labour resources employed}}$$

| Total nr of site staff | Admin Cost per hour | Total Rate per Man Hour |
|---|---|---|
| 12 | 12.31 | 31.89 |
| 15 | 9.85 | 29.43 |
| 18 | 8.21 | 27.78 |
| 20 | 7.39 | 26.96 |
| 25 | 5.91 | 25.49 |
| 35 | 4.22 | 23.80* |
| 40 | 3.69 | 23.27 |
| 45 | 3.28 | 22.86 |
| 50 | 2.96 | 22.53 |
| 55 | 2.69 | 22.26 |

Hourly labour rates used for Spon's External Works and Landscape 2017

This year's rates are calculated on the rounded value of a 40 man working team rounded to the nearest £0.25

| | |
|---|---|
| General contracting | £23.75/man-hour |
| Brickwork 1+1 man gang | £53.75/team-hour |
| Brickwork 1+2 man gang | £83.75/team-hour |
| Maintenance contracting | £19.00/man-hour |

## Smaller organizations

Smaller organizations should adjust the rates shown in this book to suit a labour rate which is generally slightly higher due to the balance of company overhead and number of employees. The costs for smaller organizations are no longer shown in this edition but users may adjust the rates using the resultant labour rates shown below. The calculation of this labour rate is as follows:

- A typical smaller works labour calculation is based on the above but the administration cost of the organization is £115640.00 per annum
- The owner of the business does not carry out site works and is assumed to be paid a salary of £58000.00 including National insurance. The balance of his income is taken in dividend and profit share figure constitutes part of the £115640.00 shown
- There are between 6 and 15 site operatives

**Illustrative overhead for small company employing 6–15 landscape operatives**

| COST CENTRE | Number of | Cost | Total |
|---|---|---|---|
| MD: Salary only: excludes profits + NI + Vehicle | 1 | 58000.00 | 58000.00 |
| Secretary | 1 | 19000.00 | 19000.00 |
| Bookkeeper | 1 | 9000.00 | 9000.00 |
| Rental | 12 | 325.00 | 3900.00 |
| Insurances | 1 | 4000.00 | 4000.00 |
| Telephone and Mobiles | 12 | 200.00 | 2400.00 |
| Office Equipment | 1 | 1000.00 | 1000.00 |
| Stationary | 12 | 20.00 | 240.00 |
| Advertising | 12 | 200.00 | 2400.00 |
| Other vehicles not allocated to contract teams | 1 | 4000.00 | 4000.00 |
| Other consultants | 1 | 1000.00 | 1000.00 |
| Accountancy | 1 | 1500.00 | 1500.00 |
| Lights heating water | 12 | 100.00 | 1200.00 |
| Other expenses | 1 | 8000.00 | 8000.00 |
| **TOTAL OFFICE OVERHEAD** | | | 115640.00 |

The final labour rate used is then generated as follows:

$$\text{Net labour rate} = \frac{\text{Total Office Overhead (£115640.00)}}{\text{Working hours per year (1763)} \times \text{Labour resources employed}}$$

| Total nr of site staff | Admin Cost per hour | Total Rate per Man Hour |
|---|---|---|
| 3 | 21.86 | 41.44 |
| 6 | 10.93 | 30.51 |
| 9 | 7.29 | 26.86 |
| 12 | 5.47 | 25.04 |
| 15 | 4.37 | 23.95 |

Hourly labour rates for a typical small company working in London for 2017 could be calculated on the rounded value of a 9 man working team (£27.00).

| General and Maintenance Contracting (Smaller organizations) | £27.00 |

**Calculation of Annual Hours Worked**

| Normal hours worked | | |
|---|---|---|
| Normal hours worked per week | | 39 hours |
| Normal hours worked per year | 39 hrs × 52 wks | 2028 hours |
| | | |
| Less holidays | | |
| Annual holidays | 39 hrs × 4.2 wks | −163.8 hours |
| Public holidays | 8 days @7.8 hours | −62.4 hours |
| | | |
| Less sick and downtime | 39 hrs × 1 wk | −39 hours |
| Actual hours worked per annum | | 1763 hours |

# Notes on the New Rules of Measurement (NRM) Used in this Edition

**New Rules of Measurement (NRM)**

The NRM suite covers the life cycle of cost management and means that, at any point in a building's life, quantity surveyors will have a set of rules for measuring and capturing cost data. In addition, the *BCIS Standard Form of Cost Analysis* (SFCA) 4th edition has been updated so that it is aligned with the NRM suite.
The three volumes of the NRM are as follows.

- **NRM1** – Order of cost estimating and cost planning for capital building works.
- **NRM2** – Detailed measurement for building works.
- **NRM3** – Order of cost estimating and cost planning for building maintenance works.

## NRM1

In May 2009 the New Rules of Measurement (NRM1) Order of cost estimating and elemental cost planning were introduced by the RICS. The NRM have been written to provide a standard set of measurement rules that are understandable by all those involved in a construction project, including the employer; thereby aiding communication between the project/design team and the employer. In addition, the NRM should assist the quantity surveyor / cost manager in providing effective and accurate cost advice to the employer and the rest of the project team.

## NRM2

Developed to provide a detailed set of measurement rules to support the procurement of construction works. These rules will ultimately be a direct replacement for the *Standard Method of Measurement for Building Works (SMM7)*.

This edition of Spon's is a hybrid of the NRM2 and SMM7 in order to help the user transfer between the two formats. Headings are in line with NRM2 but sub-headings and content reflect SMM7 more closely. This will continue to be developed over several editions of the book.

The NRM suite is published by the RICS and readers are strongly advised to read the documents.

# ESSENTIAL READING FROM TAYLOR AND FRANCIS

## Metric Handbook Planning and Design Data, 5th Edition

Pamela Buxton

The Metric Handbook is the major handbook of planning and design information for architects and architecture students. Covering basic design data for all the major building types, it is the ideal starting point for any project. For each building type, the book gives the basic design requirements and all the principal dimensional data, and succinct guidance on how to use the information and what regulations the designer needs to be aware of.

As well as building types, the Metric Handbook deals with broader aspects of design such as materials, acoustics and lighting, and general design data on human dimensions and space requirements. The Metric Handbook provides an invaluable resource for solving everyday design and planning problems.

Mar 2015: 203 x 292: 858 pp
Pb: 978-0-415-72542-2 £39.99

To Order: Tel: +44 (0) 1235 400524  Fax: +44 (0) 1235 400525
or Post: Taylor and Francis Customer Services,
Bookpoint Ltd, Unit T1, 200 Milton Park, Abingdon, Oxon, OX14 4TA UK
Email: book.orders@tandf.co.uk

For a complete listing of all our titles visit:
www.tandf.co.uk

# Computation of the Cost of Materials

Percentages of default waste are placed against material resources within the supplier database. These range from 2.5% (for bricks) to 20% (for topsoil). An allowance for the cost of unloading, stacking etc. should be added to the cost of materials.

The following are typical hours of labour for unloading and stacking some of the more common building materials.

| Material | Unit | Labourer hours |
| --- | --- | --- |
| Cement | tonne | 0.67 |
| Lime | tonne | 0.67 |
| Common bricks | 1000 | 1.70 |
| Light engineering bricks | 1000 | 2.00 |
| Heavy engineering bricks | 1000 | 2.40 |

ESSENTIAL READING FROM TAYLOR AND FRANCIS

# JCT Contract Administration Pocket Book

Andy Atkinson

Successfully managing your JCT contracts is a must, and this handy reference is the swiftest way to doing just that. Making reference to best practice throughout, the JCT Standard Building Contracts SBC/Q, DB and MW are used as examples to take you through all the essential contract administration tasks, including:

- Procurement
- Payment
- Final accounts
- Progress, completion and delay
- Subcontracting
- Defects and quality control

In addition to the day to day tasks, this also gives you an overview of what to expect from common sorts of dispute resolution under the JCT, as well as a look at how to administer contracts for BIM-compliant projects. This is an essential starting point for all students of construction contract administration, as well as practitioners needing a handy reference to working with JCT contracts.

June 2015 186x123: 144pp
Pb: 978-1-138-78192-4: £24.99

To Order: Tel: +44 (0) 1235 400524  Fax: +44 (0) 1235 400525
or Post: Taylor and Francis Customer Services,
Bookpoint Ltd, Unit T1, 200 Milton Park, Abingdon, Oxon, OX14 4TA UK
Email: book.orders@tandf.co.uk

For a complete listing of all our titles visit:
www.tandf.co.uk

# The Landfill Tax

## The Tax

The Landfill Tax came into operation on 1 October 1996. It is levied on operators of licensed landfill sites in England, Wales and Northern Ireland at the following rates with effect from 1 April 2016:

- Inactive or inert wastes [Lower Rate]   £2.65 per tonne   Included are soil, stones, brick, plain and reinforced concrete, plaster and glass – lower rate

- All other taxable wastes [Higher Rate]   £84.40 per tonne   Included are timber, paint and other organic wastes generally found in demolition work and builders skips – standard rate

The standard and lower rates of Landfill Tax were due to increase in line with the RPI, rounded to the nearest 5 pence, from 1 April 2016. However, legislation introduced in the Finance Bill 2016 provided for new rates of Landfill Tax until 31 March 2019 as follows:

- Lower Rate       From 1 April 2016    £2.65/tonne
- Lower Rate       From 1 April 2017    £2.70/tonne
- Lower Rate       From 1 April 2018    £2.80/tonne
- Standard Rate    From 1 April 2016    £84.40/tonne
- Standard Rate    From 1 April 2017    £86.10/tonne
- Standard Rate    From 1 April 2018    £88.95/tonne

Following industry engagement to address compliance, on 1 April 2015 the government introduced a loss on ignition testing regime on fines (residual waste from waste processing) from waste transfer stations by April 2015. Only fines below a 10% threshold will be considered eligible for the lower rate, though there is a 12 month transitional period where the threshold will be 15%. This transitional period came to an end on 31 March 2016. The government intends to provide further longer term certainty about the future level of landfill tax rates once the consultation process on testing regime has concluded, but in the meantime is committed to ensuring that the rates are not eroded in real terms (Finance Bill 2014).

The Landfill Tax (Qualifying Material) Order 2011 came into force on 1 April 2011. This has amended the qualifying criteria that are eligible for the lower rate of Landfill Tax. The revisions introduced arose primarily from the need to reflect changes in wider environmental policy and legislation since 1996, such as the implementation of the European Landfill Directive.

A waste will be lower rated for Landfill Tax only if it is listed as a qualifying material in the Landfill Tax (Qualifying Material) Order 2011.

The principal changes to qualifying material are:

- Rocks and subsoils that are currently lower rated will remain so
- Topsoil and peat will be removed from the lower rate, as these are natural resources that can always be recycled/reused
- Used foundry sand, which has in practice been lower rated by extra-statutory concession since the tax's introduction in 1996, will now be included in the lower rate Order
- Definitions of qualifying ash arising from the burning of coal and petroleum coke (including when burnt with biomass) will be clarified
- The residue from titanium dioxide manufacture will qualify, rather than titanium dioxide itself, reflecting industry views
- Minor changes will be made to the wording of the calcium sulphate group of wastes to reflect the implementation of the Landfill Directive since 2001
- Water will be removed from the lower rate – water is now banned from landfill so its inclusion in the list of lower rated wastes is unnecessary; where water is used as a waste carrier the water is not waste and therefore not taxable.

**Calculating the weight of waste**

There are two options:

- If licensed sites have a weighbridge tax will be levied on the actual weight of waste.
- If licensed sites do not have a weighbridge tax will be levied on the permitted weight of the lorry based on an alternative method of calculation based on volume to weight factors for various categories of waste.

**Effect on prices**

The tax is paid by landfill site operators only. Tipping charges reflect this additional cost.

As an example, *Spon's A & B* rates for mechanical disposal will be affected as follows:

| | | |
|---|---|---|
| • Inactive waste | Spon's A & B 2017 net rate | £21.69 per m$^3$ |
| | Tax, 1.9 tonne per m$^3$ (un-bulked) @ £2.65 | £5.04 per m$^3$ |
| | Spon's rate including tax | £26.63 per m$^3$ |
| • Active waste | Active waste will normally be disposed of by skip and will probably be mixed with inactive waste. The tax levied will depend on the weight of materials in the skip which can vary significantly. | |

**Exemptions**

The following disposals are exempt from Landfill Tax subject to meeting certain conditions:

- dredgings which arise from the maintenance of inland waterways and harbours
- naturally occurring materials arising from mining or quarrying operations
- pet cemeteries
- material from the reclamation of contaminated land (see below)
- inert waste used to restore landfill sites and to fill working and old quarries where a planning condition or obligation is in existence
- waste from visiting NATO forces

The exemption for waste from contaminated land has been phased out completely from 1 April 2012 and no new applications for Landfill Tax exemption are now accepted.

## Devolution of landfill tax to Scotland from 1 April 2015

The Scotland Act 2012 provided for landfill tax to be devolved to Scotland. From 1 April 2015, operators of landfill sites in Scotland are no longer liable to pay UK Landfill Tax for waste disposed at their Scottish sites. Instead, they will be liable to register and account for Scottish Landfill Tax (SLfT).

Operators of landfill sites only in Scotland were deregistered from UK Landfill Tax with effect from 31 March 2015.

The Standard and Lower rates of the Scottish Landfill Tax will mirror the rates applied to the current UK Landfill Tax for the time being.

## Devolution of Landfill Tax to Wales from April 2018

The Wales Act 2014 provides for Landfill Tax to be devolved to Wales. This is expected to take effect in April 2018. Further information will be provided on developments by HM Revenue & Customs at a later date.

For further information contact the HMRC VAT and Excise Helpline, telephone 0300 200 3700.

# The Aggregates Levy

The Aggregates Levy came into operation on 1 April 2002 in the UK, except for Northern Ireland where it has been phased in over five years from 2003.

It was introduced to ensure that the external costs associated with the exploitation of aggregates are reflected in the price of aggregate, and to encourage the use of recycled aggregate. There continues to be strong evidence that the levy is achieving its environmental objectives, with sales of primary aggregate down and production of recycled aggregate up. The Government expects that the rates of the levy will at least keep pace with inflation over time, although it accepts that the levy is still bedding in.

The rate of the levy remains at £2.00 per tonne from 1 April 2014 and is levied on anyone considered to be responsible for commercially exploiting 'virgin' aggregates in the UK and should naturally be passed by price increase to the ultimate user.

All materials falling within the definition of 'Aggregates' are subject to the levy unless specifically exempted.

It does not apply to clay, soil, vegetable or other organic matter.

The intention is that it will:

- Encourage the use of alternative materials that would otherwise be disposed of to landfill sites
- Promote development of new recycling processes, such as using waste tyres and glass
- Promote greater efficiency in the use of virgin aggregates
- Reduce noise and vibration, dust and other emissions to air, visual intrusion, loss of amenity and damage to wildlife habitats

### Definitions

'Aggregates' means any rock, gravel or sand which is extracted or dredged in the UK for aggregates use. It includes whatever substances are for the time being incorporated in it or naturally occur mixed with it.

'Exploitation' is defined as involving any one or a combination of any of the following:

- Being removed from its original site, a connected site which is registered under the same name as the originating site or a site where it had been intended to apply an exempt process to it, but this process was not applied
- Becoming subject to a contract or other agreement to supply to any person
- Being used for construction purposes
- Being mixed with any material or substance other than water, except in permitted circumstances

The definition of 'aggregate being used for construction purposes' is when it is:

- Used as material or support in the construction or improvement of any structure
- Mixed with anything as part of a process of producing mortar, concrete, tarmacadam, coated roadstone or any similar construction material

### Incidence

It is a tax on primary aggregates production – i.e. 'virgin' aggregates won from a source and used in a location within the UK territorial boundaries (land or sea). The tax is not levied on aggregates which are exported or on aggregates imported from outside the UK territorial boundaries.

It is levied at the point of sale.

## Exemption from tax

An 'aggregate' is exempt from the levy if it is:

- Material which has previously been used for construction purposes
- Aggregate that has already been subject to a charge to the Aggregates Levy
- Aggregate which was previously removed from its originating site before the start date of the levy
- Aggregate which is moved between sites under the same Aggregates Levy Registration
- Aggregate which is removed to a registered site to have an exempt process applied to it
- Aggregate which is removed to any premises where china clay or ball clay will be extracted from the aggregate
- Aggregate which is being returned to the land from which it was won provided that it is not mixed with any material other than water
- Aggregate won from a farm land or forest where used on that farm or forest
- Rock which has not been subjected to an industrial crushing process
- Aggregate won by being removed from the ground on the site of any building or proposed building in the course of excavations carried out in connection with the modification or erection of the building and exclusively for the purpose of laying foundations or of laying any pipe or cable
- Aggregate won by being removed from the bed of any river, canal or watercourse or channel in or approach to any port or harbour (natural or artificial), in the course of carrying out any dredging exclusively for the purpose of creating, restoring, improving or maintaining that body of water
- Aggregate won by being removed from the ground along the line of any highway or proposed highway in the course of excavations for improving, maintaining or constructing the highway otherwise than purely to extract the aggregate
- Drill cuttings from petroleum operations on land and on the seabed
- Aggregate resulting from works carried out in exercise of powers under the New Road and Street Works Act 1991, the Roads (Northern Ireland) Order 1993 or the Street Works (Northern Ireland) Order 1995
- Aggregate removed for the purpose of cutting of rock to produce dimension stone, or the production of lime or cement from limestone
- Aggregate arising as a waste material during the processing of the following industrial minerals:
    - anhydrite
    - ball clay
    - barytes
    - calcite
    - china clay
    - clay, coal, lignite and slate
    - feldspar
    - flint
    - fluorspar
    - fuller's earth
    - gems and semi-precious stones
    - gypsum
    - any metal or the ore of any metal
    - muscovite
    - perlite
    - potash
    - pumice
    - rock phosphates
    - sodium chloride
    - talc
    - vermiculite
    - spoil from the separation of the above industrial minerals from other rock after extraction
    - material that is mainly but not wholly the spoil, waste or other by-product of any industrial combustion process or the smelting or refining of metal.

Anything that consists 'wholly or mainly' of the following is exempt from the levy (note that 'wholly' is defined as 100% but 'mainly' as more than 50%, thus exempting any contained aggregates amounting to less than 50% of the original volumes:

- clay, soil, vegetable or other organic matter
- drill cuttings from oil exploration in UK waters
- material arising from utility works, if carried out under the New Roads and Street Works Act 1991

However, when ground that is more than half clay is mixed with any substance (for example, cement or lime) for the purpose of creating a firm base for construction, the clay becomes liable to Aggregates Levy because it has been mixed with another substance for the purpose of construction.

Anything that consists completely of the following substances is exempt from the levy:

- Spoil, waste or other by-products from any industrial combustion process or the smelting or refining of metal – for example, industrial slag, pulverized fuel ash and used foundry sand. If the material consists completely of these substances at the time it is produced it is exempt from the levy, regardless of any subsequent mixing
- Aggregate necessarily arising from the footprint of any building for the purpose of laying its foundations, pipes or cables. It must be lawfully extracted within the terms of any planning consent
- Aggregate necessarily arising from navigation dredging
- Aggregate necessarily arising from the ground in the course of excavations to improve, maintain or construct a highway or a proposed highway.
- Aggregate necessarily arising from the ground in the course of excavations to improve, maintain or construct a railway, monorail or tramway.

Relief from the levy either in the form of credit or repayment is obtainable where:

- it is subsequently exported from the UK in the form of aggregate
- it is used in an exempt process
- where it is used in a prescribed industrial or agricultural process
- it is waste aggregate disposed of by dumping or otherwise, e.g. sent to landfill or returned to the originating site

The Aggregates Levy Credit Scheme (ALCS) for Northern Ireland was suspended with effect from 1 December 2010 following a ruling by the European General Court.

An exemption for aggregate obtained as a by-product of railway, tramway and monorail improvement, maintenance and construction was introduced in 2007.

Exemptions to the levy were suspended on 1 April 2014, following an investigation by the European Commission into whether they were lawful under State aid rules. The Commission announced its decision on 27 March 2015 that all but part of one exemption (for shale) were lawful and all the exemptions have, therefore, been reinstated apart from the exemption for shale. The effective date of reinstatement is 1 April 2014, which means that businesses which paid Aggregates Levy on materials for which the exemption has been confirmed as lawful may claim back the tax they paid while the exemption was suspended.

However, under EU law the UK government is required to recover unlawful State aid with interest from businesses that benefited from it. HM Revenue & Customs therefore initiated a process in 2015 of clawing back the levy, plus compound interest, for the deliberate extraction of shale aggregate for commercial exploitation from businesses it believes may have benefited between 1 April 2002 and 31 March 2014.

## Discounts

From 1 July 2005 the standard added water percentage discounts listed below can be used. Alternatively a more exact percentage can be agreed and this must be done for dust dampening of aggregates:

- washed sand 7%
- washed gravel 3.5%
- washed rock/aggregate 4%

## Impact

The British Aggregates Association suggested that the additional cost imposed by quarries is more likely to be in the order of £3.40 per tonne on mainstream products, applying an above average rate on these in order that by-products and low grade waste products can be held at competitive rates, as well as making some allowance for administration and increased finance charges.

With many gravel aggregates costing in the region of £20.00 per tonne, there is a significant impact on construction costs.

## Avoidance

An alternative to using new aggregates in filling operations is to crush and screen rubble which may become available during the process of demolition and site clearance as well as removal of obstacles during the excavation processes.

Example: Assuming that the material would be suitable for fill material under buildings or roads, a simple cost comparison would be as follows (note that for the purpose of the exercise, the material is taken to be 1.80 tonne per m³ and the total quantity involved less than 1,000 m³):

| | £/m³ | £/tonne |
|---|---|---|
| **Importing fill material:** | | |
| Cost of 'new' aggregates delivered to site | 37.10 | 20.16 |
| Addition for Aggregates Tax | 3.60 | 2.00 |
| **Total cost of importing fill materials** | **40.70** | **22.61** |
| **Disposing of site material:** | | |
| Cost of removing materials from site | 26.63 | 14.79 |
| **Crushing site materials:** | | |
| Transportation of material from excavations or demolition to stockpiles | 0.88 | 0.49 |
| Transportation of material from temporary stockpiles to the crushing plant | 2.36 | 1.31 |
| Establishing plant and equipment on site; removing on completion | 2.36 | 1.31 |
| Maintain and operate plant | 10.62 | 5.90 |
| Crushing hard materials on site | 15.34 | 8.52 |
| Screening material on site | 2.36 | 1.31 |
| **Total cost of crushing site materials** | **33.92** | **18.84** |

From the above it can be seen that potentially there is a great benefit in crushing site materials for filling rather than importing fill materials.

Setting the cost of crushing against the import price would produce a saving of £6.78 per m³. If the site materials were otherwise intended to be removed from the site, then the cost benefit increases by the saved disposal cost to £33.41 per m³.

Even if there is no call for any or all of the crushed material on site, it ought to be regarded as a useful asset and either sold on in crushed form or else sold with the prospects of crushing elsewhere.

| Specimen Unit rates | Unit[3] | £ |
|---|---|---|
| Establishing plant and equipment on site; removing on completion | | |
|     Crushing plant | trip | 1,400.00 |
|     Screening plant | trip | 700.00 |
| Maintain and operate plant | | |
|     Crushing plant | week | 8,500.00 |
|     Screening plant | week | 2,100.00 |
| Transportation of material from excavations or demolition places to temporary stockpiles | $m^3$ | 3.50 |
| Transportation of material from temporary stockpiles to the crushing plant | $m^3$ | 2.80 |
| Breaking up material on site using impact breakers | | |
|     mass concrete | $m^3$ | 16.50 |
|     reinforced concrete | $m^3$ | 19.00 |
|     brickwork | $m^3$ | 7.00 |
| Crushing material on site | | |
|     mass concrete not exceeding 1000 $m^3$ | $m^3$ | 15.00 |
|     mass concrete 1000–5000 $m^3$ | $m^3$ | 14.00 |
|     mass concrete over 5000 $m^3$ | $m^3$ | 13.00 |
|     reinforced concrete not exceeding 1000 $m^3$ | $m^3$ | 18.00 |
|     reinforced concrete 1000–5000 $m^3$ | $m^3$ | 16.00 |
|     reinforced concrete over 5000 $m^3$ | $m^3$ | 15.00 |
|     brickwork not exceeding 1000 $m^3$ | $m^3$ | 14.00 |
|     brickwork 1000–5000 $m^3$ | $m^3$ | 13.00 |
|     brickwork over 5000 $m^3$ | $m^3$ | 12.00 |
| Screening material on site | $m^3$ | 2.50 |

More detailed information can be found on the HMRC website (www.hmrc.gov.uk) in Notice AGL 1 Aggregates Levy published August 2015.

# Cost Indices

The purpose of this section is to show changes in the cost of carrying out landscape work (hard surfacing and planting) since 2010. It is important to distinguish between costs and tender prices: the following table reflects the change in cost to contractors but does not necessarily reflect changes in tender prices. In addition to changes in labour and material costs, which are reflected in the indices given below, tender prices are also affected by factors such as the degree of competition at the time of tender and in the particular area where the work is to be carried out, the availability of labour and materials, and the general economic situation. This can mean that in a period when work is scarce, tender prices may fall despite the fact that costs are rising, and when there is plenty of work available, tender prices may increase at a faster rate than costs.

## The Constructed Cost Index

A Constructed Cost Index has been developed based on NEDO Price Adjustment Formula Indices (PAFI) (Building Series 2). Cost indices for the various building trades are published monthly by BCIS and are available via a subscription service.

The indices comprise 49 Building Work indices plus seven "Appendices" and other specialist indices. The Building Work indices are compiled by monitoring the cost of labour and materials for each category and applying a weighting to these to calculate a single index.

Although the PAFI indices are prepared for use with price-adjustment formulae for calculating reimbursement of increased costs during the course of a contract, they also present a time series of cost indices for the main components of landscaping projects. They can therefore be used as the basis of an index for landscaping costs.

The method used here is to construct a composite index by allocating weightings to the indices representing the usual work categories found in a landscaping contract, the weightings being established from an analysis of actual projects. These weightings totalled 100 in 1976 and the composite index is calculated by applying the appropriate weightings to the appropriate PAFI indices on a monthly basis, which is then compiled into a quarterly index and rebased to 1976 = 100.

**Constructed Landscaping Cost Index 1976 = 100**

| Year | First Quarter | Second Quarter | Third Quarter | Fourth Quarter | Annual Average |
|---|---|---|---|---|---|
| 2000 | 493 | 496 | 513 | 514 | 504 |
| 2001 | 512 | 514 | 530 | 530 | 522 |
| 2002 | 531 | 540 | 573 | 574 | 555 |
| 2003 | 577 | 582 | 603 | 602 | 591 |
| 2004 | 602 | 610 | 639 | 639 | 623 |
| 2005 | 641 | 648 | 684 | 683 | 664 |
| 2006 | 688 | 692 | 710 | 710 | 710 |
| 2007 | 714 | 718 | 737 | 742 | 728 |
| 2008 | 752 | 771 | 803 | 794 | 780 |
| 2009 | 789 | 792 | 796 | 802 | 795 |
| 2010 | 804 | 810 | 814 | 817 | 811 |
| 2011 | 825 | 835 | 845 | 852 | 839 |
| 2012 | 858 | 858 | 860 | 862 | 860 |
| 2013 | 874 | 873 | 874 | 874 | 874 |
| 2014 | 878 | 880 | 890 | 886 | 884 |
| 2015 | 881 | 889 | 897 | 894 | 890 |
| 2016 | 890 | 987 | | | 893 |

This index is updated every quarter in *Spon's Price Book Update*. The updating service is available, free of charge, to all purchasers of Spon's Price Books (complete the reply-paid card enclosed).

# Regional Variations

Prices in *Spon's External Works and Landscape Price Book* are based upon conditions prevailing for a competitive tender in the Outer London area. For the benefit of readers, this edition includes regional variation adjustment factors which can be used for an assessment of price levels in other regions.

Special further adjustment may be necessary when considering city centre or very isolated locations.

| Region | Adjustment Factor |
| --- | --- |
| Outer London (Spon's 2017) | 1.00 |
| Inner London | 1.06 |
| South East | 0.98 |
| South West | 0.91 |
| East Midlands | 0.90 |
| West Midlands | 0.88 |
| East of England | 0.93 |
| Yorkshire & Humberside | 0.88 |
| North East | 0.86 |
| North West | 0.89 |
| Scotland | 0.90 |
| Wales | 0.88 |
| Northern Ireland | 0.76 |

The following example illustrates the adjustment of prices for regions other than Outer London, by use of regional variation adjustment factors.

| | | |
| --- | --- | --- |
| A. | Value of items priced using *Spon's External Works and Landscape Price Book* for Outer London | £100,000 |
| B. | Adjustment to value of A. to reflect North West region price level 100,000 × 0.89 | £ 89,000 |

# ESSENTIAL READING FROM TAYLOR AND FRANCIS

## Construction Project Manager's Pocket Book

Duncan Cartlidge

Construction project management requires a broad range of skills, from technical expertise to leadership, negotiation, team building and communication. This no–nonsense guide covers all of the essentials of the role, including:

- Pre-construction activities
- Design management and BIM
- Procurement
- Feasibility studies
- Environmental management systems
- People skills
- Recommended document formats
- Occupancy activities

Construction project management activities are tackled in the order they occur on real projects, with reference made to the RIBA Plan of Work and OGC Gateway process throughout. This is the ideal concise reference which no project manager, construction manager, or quantity surveyor should be without.

February 2015: 186x123: 284pp
Pb: 978-0-415-73239-0: £20.99

---

To Order: Tel: +44 (0) 1235 400524  Fax: +44 (0) 1235 400525
or Post: Taylor and Francis Customer Services,
Bookpoint Ltd, Unit T1, 200 Milton Park, Abingdon, Oxon, OX14 4TA UK
Email: book.orders@tandf.co.uk

For a complete listing of all our titles visit:
www.tandf.co.uk

PART 2

# Rates of Wages – Building Industry

# ESSENTIAL READING FROM TAYLOR AND FRANCIS

## Landscape Architect's Pocket Book

Siobhan Vernon, Rachel Tennant & Nicola Garmory

An indispensable tool for all landscape architects, this time-saving guide answers the most frequently asked questions in one pocket-sized volume. It is a concise, easy-to-read reference that gives instant access to a wide range of information needed on a daily basis, both out on site and in the office.

Covering all the major topics, including hard landscaping, soft landscaping as well as planning and legislation, the pocket book also includes a handy glossary of important terms, useful calculations and helpful contacts. Not only an essential tool for everyday queries on British standards and procedures, this is a first point of reference for those seeking more extensive, supplementary sources of information, including websites and further publications.

This new edition incorporates updates and revisions from key planning and environmental legislation, guidelines and national standards.

April 2013: 186x123: 340pp
Pb: 978-0-415-63084-9: £19.99

To Order: Tel: +44 (0) 1235 400524  Fax: +44 (0) 1235 400525
or Post: Taylor and Francis Customer Services,
Bookpoint Ltd, Unit T1, 200 Milton Park, Abingdon, Oxon, OX14 4TA UK
Email: book.orders@tandf.co.uk

For a complete listing of all our titles visit:
www.tandf.co.uk

## BUILDING INDUSTRY WAGES

Note

The following data is published for information only and is not used for within the calculation of rates for this publication. The exception is the calculation shown for the working hours per year.

The Working Rule Agreement includes a pay structure with general operative and additional skilled rates of pay as well as craft rate. Plus rates and additional payments will be consolidated into basic pay to provide the following rates (for a normal 39 hour week) which will come into effect from the following dates:

**AUTHORS' NOTE: AT THE TIME OF COMPILING, THE FOLLOWING CIJC RATES AGREEMENT HAD NOT BEEN ACHIEVED FOR 2016/2017 WAGE LEVELS. WE HAVE ASSUMED AN INCREASE OF 3.5%. WE WILL NOTIFY READERS OF THE ACTUAL RATES IN OUR UPDATE IN NOVEMBER/DECEMBER 2016**

The following basic rates of pay are assumed effective from 27 June 2016

|  | Rate per 39-hour week (£) | Rate per hour (£) |
| --- | --- | --- |
| **Craft Rate** | 457.34 | 11.73 |
| **Skill Rate 1** | 435.72 | 11.17 |
| **Skill Rate 2** | 419.09 | 10.75 |
| **Skill Rate 3** | 392.06 | 10.05 |
| **Skill Rate 4** | 370.44 | 9.50 |
| **General operative** | 343.83 | 8.82 |

## BUILDING CRAFT AND GENERAL OPERATIVES

### Effective from 27 June 2016

|  |  |  |  | Craft operatives | | General operatives | |
|---|---|---|---|---|---|---|---|
|  |  |  |  | £ | £ | £ | £ |
| Wages at standard basic rate |  |  |  |  |  |  |  |
| Productive time | 44.30 | wks |  | 457.34 | 20,259.96 | 343.83 | 15,231.81 |
| Lost time allowance, say | 0.90 | wks |  | 457.34 | 411.60 | 343.83 | 309.45 |
| Overtime, say | 0.00 | wks |  | 0.00 | 0.00 | 0.00 | 0.00 |
|  |  |  |  |  | 20,671.56 |  | 15,541.26 |
| Holiday Pay | 5.80 | wks |  | 457.34 | 2,652.55 | 343.83 | 1,994.23 |
| CITB Levy | 0.50% | year |  |  | 116.62 |  | 87.68 |
|  |  |  |  |  | 23,440.73 |  | 17,623.17 |
| Employer's contributions to: |  |  |  |  |  |  |  |
| EasyBuild Stakeholder Pension(Death and accident cover is provided free) | 52.00 | wks |  | 7.50 | 390.00 | 7.50 | 390.00 |
| National Insurance (average weekly payment) | 52.00 | wks |  | 40.68 | 2,115.36 | 25.24 | 1,312.54 |
|  |  |  |  |  | 25,946.09 |  | 19,325.71 |
| Severance pay and sundry costs |  |  |  | 1.50% | 389.19 |  | 289.89 |
|  |  |  |  |  | 26,335.28 |  | 19,615.59 |
| Employer's Liability and Third Party Insurance |  |  |  | 2.00% | 526.71 |  | 392.31 |
| **Total cost per annum** |  |  |  |  | **26,861.99** |  | **20,007.90** |
| **Total cost per hour** |  |  |  |  | **15.55** |  | **11.58** |
| **Total cost per week** |  |  |  |  | **606.37** |  | **451.65** |

## DAYWORK RATES

Example calculation of Price Cost of Labour in Daywork

**Effective from 29 June 2016**

|  |  |  | Craft operative | | General operative | |
|---|---|---|---|---|---|---|
|  |  |  | Rate £ | Annual cost £ | Rate £ | Annual cost £ |
| Basic Wages | 52 wks | 441.87 | 22,977.24 | 332.21 | 17,274.71 |
| CITB levy (0.5% of payroll) | 0.5% |  | 114.89 |  | 86.37 |
| Pension & Welfare Benefit | 52 wks | 12.86 | 668.72 | 12.86 | 668.72 |
| Employer's National Insurance contribution |  |  |  |  |  |
| (13.8% after the first £155 per week) | 46.2 wks |  | 2,195.39 |  | 1,408.44 |
| **Annual labour cost** |  |  | **25,956.24** |  | **19,438.24** |
| **Hourly base rates as defined** |  |  | **14.41** |  | **10.79** |

This is the prime cost of employment per person, which the employer has to meet even if there is no work for the employee to do.

NOTES:

1. Calculated following Definition of Prime Cost of Daywork carried out under a Building Contract, published by the Royal Institution of Chartered Surveyors and the Construction Confederation.
2. Standard basic rates effective from 27 June 2016.
3. Standard working hours per annum calculated as follows:

   | 52 weeks @ 39 hours | = | 2028 |
   |---|---|---|
   | Less |  |  |
   | 4.2 weeks holiday @ 39 hours | = | 163.8 |
   | 8 days public holidays @ 7.8 hours | = | 62.4 |
   |  |  | 226.2 |
   |  |  | 1801.8 |

4. All labour costs incurred by the contractor in his capacity as an employer, other than those contained in the hourly base rate, are to be taken into account under Section 6.
5. The above example is for guidance only and does not form part of the Definition; all the basic costs are subject to re-examination according to the time when and in the area where the daywork is executed.
6. N.I. payments are at not-contracted out rates applicable from April 2016
7. Basic rate and GMB number of weeks = 52 weeks – 4.2 weeks annual holiday – 1.6 weeks public holiday = 46.2 weeks

## ESSENTIAL READING FROM TAYLOR AND FRANCIS

## Landscape Architecture and Digital Technologies
### WALLISS & RAHMANN

Landscape Architecture and Digital Technologies explores how digital technologies are reshaping design and making in landscape architecture. While the potentials of digital technologies are well documented within landscape planning and visualisation, their application within design practice is far less understood. This book highlights the role of the digital model in encouraging a new design logic that moves from the privileging of the visual to a focus on processes of formation, bridging the interface of the conceptual and material, the virtual and the physical.

Drawing on interviews and projects from a range of international designers -including , Snøhetta, Arup, Gustafson Porter, ASPECT Studios, Grant Associates, Catherine Mosbach, Philippe Rahm, PARKKIM, LAAC and PEG office of landscape + architecture among others, the authors explore the influence of parametric modelling, scripting, real-time data, simulation, prototyping, fabrication, and Building Information Modelling on the design and construction of contemporary landscapes. This engagement with practice is expanded through critical reflection from academics involved in landscape architecture programs around the world that are reshaping their research and pedagogy to reflect an expanded digital realm.

February 2016; 246 x 174 mm; 296 pp
Pb: 978-0-415-74585-7; £36.99

To Order: Tel: +44 (0) 1235 400524  Fax: +44 (0) 1235 400525
or Post: Taylor and Francis Customer Services,
Bookpoint Ltd, Unit T1, 200 Milton Park, Abingdon, Oxon, OX14 4TA UK
Email: book.orders@tandf.co.uk

For a complete listing of all our titles visit:
www.tandf.co.uk

# PART 3

# Approximate Estimating Rates

Prices in this section are based upon the Prices for Measured Works, but allow for incidentals which would normally be measured separately in a Bill of Quantities. They do not include for Preliminaries which are priced elsewhere in this book.

Items shown as subcontract or specialist rates would normally include the specialist's overhead and profit. All other items which could fall within the scope of works of general landscape and external works contractors would not include profit.

Based on current commercial rates, profits of 10% to 35% may be added to these rates to indicate the likely 'with profit' values of the tasks below. The variation quoted above is dependent on the sector in which the works are taking place – domestic, public or commercial.

ESSENTIAL READING FROM TAYLOR AND FRANCIS

# Architect's Legal Pocket Book
# Second Edition

Matthew Cousins

A little book that's big on information. The Architect's Legal Pocket Book is the definite reference on legal issues for architects and architectural students. This handy pocket guide covers key legal principles which will help you to quickly understand the law and where to go for further information.

Now in a fully updated new edition, this bestselling book covers a wide range of subjects focused on the UK including building legislation and the Localism Bill, negligence, liability, planning policy and development, listed buildings, party wall legislation, and rights of light. This edition also contains greater coverage of contracts including the RIBA contracts, dispute resolution and legal issues in professional practice.

Illustrated with clear diagrams and featuring key cases, this is an invaluable source of practical information and a comprehensive guide of the current law for architects. It is a book no architect should be without.

August 2015 186x123: 402pp
Pb: 978-1-138-82144-6: £24.99

To Order: Tel: +44 (0) 1235 400524  Fax: +44 (0) 1235 400525
or Post: Taylor and Francis Customer Services,
Bookpoint Ltd, Unit T1, 200 Milton Park, Abingdon, Oxon, OX14 4TA UK
Email: book.orders@tandf.co.uk

For a complete listing of all our titles visit:
www.tandf.co.uk

## COST MODELS

| Item<br>Excluding site overheads and profit | Unit | Range £ |
|---|---|---|
| **HOUSING AND HOUSE BUILDER** | | |
| Cost Models for areas as described below. These models below exclude items such as demolitions, bulk groundworks, drainage unless specifically mentioned as included; Site or project preliminaries of 7.5%–12% may also be added to the rates; These items below are based on areas of 1000 m$^2$ ; Users should add O & P to the rates below; We advise 15% for illustration purposes | | |
| Housing cost models meeting minimum planning requirements or to maximize sales potential at economic cost | | |
| Soft landscaped space only space; planting:turf ratios as shown; planting sizes and densities as indicated; 1 tree of 18–20 cm girth at 1 tree per 100 m$^2$ overall external area; Mulch and watering after planting | | |
|     Landscaped area; Turf and trees only; imported topsoil to turf areas at 50 thick | 100 m$^2$ | 1075.00 |
|     Landscaped area 70:30 turf to planted area with imported topsoil to turf areas at 50 thick | 100 m$^2$ | 2000.00 |
|     Landscaped area 50:50 turf to planted area; imported topsoil to turf areas at 50 thick | 100 m$^2$ | 2325.00 |
|     Landscaped area; Planting and trees only; Planting of shrubs 100% at 400 mm centres | 100 m$^2$ | 3200.00 |
|     Landscaped area; Planting and trees only; Planting of shrubs and groundcovers 100% Shrubs: Groundcovers 70:30 at 400:250 mm centres | 100 m$^2$ | 3650.00 |
|     Landscaped area; Planting and trees only; Planting of shrubs and groundcovers 100% Shrubs: Groundcovers 30:70 at 400:250 mm centres | 100 m$^2$ | 4150.00 |
| Mixed Hard and Soft landscaped Space with parking; with 1 tree of 18–20 cm every 100 m$^2$; imported topsoil to turf areas at 50 thick | | |
|     Landscaped area where 50% of the external area is hard surface of macadam parking area; with surround of PC kerbs and 50% soft landscape of 70% turf and 30% planted area at 50 thick | 100 m$^2$ | 4400.00 |
|     Landscaped area where 50% of the external area is hard surface of block paved parking area; surround of PC kerbs 50% soft landscape of 70% turf and 30% planted area | 100 m$^2$ | 5200.00 |
| External area where the area is 50% parking and 50% soft landscape. The soft landscape is 50% turf and 50% planting; planting at 400 mm ccs – 2–3 litre plants per m$^2$; Trees at 1 nr 16–18 cm girth per 100 m$^2$ throughout the parking and landscape area | | |
|     macadam parking area | 100 m$^2$ | 4700.00 |
|     concrete block paved parking area | 100 m$^2$ | 5800.00 |
| External area where the area is 50% parking and 50% soft landscape. The soft landscape comprises 100% planting; planting at 400 mm ccs – 2–3 litre plants per m$^2$; Trees at 1 nr 16–18 cm girth per 100 m$^2$ throughout the parking and landscape area | | |
|     macadam parking area | 100 m$^2$ | 3150.00 |
|     concrete block paved parking area | 100 m$^2$ | 4100.00 |
| External area where the area is 50% parking and 50% soft landscape. The soft landscape comprises 100% planting; planting at 400 mm ccs – 2–3 litre plants per m$^2$; Trees at 1 nr 16–18 cm girth per 100 m$^2$ throughout the parking and landscape area; imported topsoil to planting areas 400 mm thick inclusive of removal of existing material and disposal offsite | | |
|     macadam parking area | 100 m$^2$ | 5000.00 |
|     concrete block paved parking area | 100 m$^2$ | 6000.00 |

## COST MODELS

| Item<br>Excluding site overheads and profit | Unit | Range £ |
|---|---|---|
| **HOUSING AND HOUSE BUILDER – cont** | | |
| | | |
| **Green infrastructure to new housing development; Comprising public greenspace with footpaths, play and amenity areas, mitigation of new internal spine roads; Landscaping to individual properties not included; Prices shown relate to the total site area** | | |
| Rural locations or not in major metropolitan areas; average property size 350 m²; Landscape impact to suit development for first time buyers; amenity areas catering for families with young children | | |
|    Rural location with 100 properties; public space, green buffers and green mitigation comprises 25% of the total area; Total site area 58000 m²; amenity space buffers and accesses 23000 m²; Green infrastructure space – 14600 m²; 800 m² of fenced equipped play area | 100 m² | 335.00 |
|    Economic housing – first time buyers; amenity catering for young children; Total site area 13500 m²; public space and accesses; 5100 m² ; Green space including footways – 4500 m²; fenced and gated green area with benches and litterbins with LEAP compliant play equipment | 100 m² | 500.00 |
|    Economic housing – first time buyers; amenity catering for young children; Total site area 13500 m²; public space and accesses; 5100 m² ; Green space including footways – 4500 m²; fenced and gated green area with benches and litterbins with no play equipment but fenced, seeded play area | 100 m² | 380.00 |
| | | |
| **Green infrastructure and access; fenced isolated site of 25000 m²; comprising 35 properties; prices shown include for creation of roadways 5.5 m wide with 500 mm roadway buildup inclusive of kerbs and soft verge one side; Allowance for development is 16250 m²; Roadways 2300 m²; footways 2 m wide 415 m (430 m²); All costs to sales related development areas excluded; services excluded** | | |
| No allowance for community shared space; Seeded footway 2.00 m wide one side only; Trees 14–16 at 1 tree per 15 lm | | |
|    greenfield site | 1000 m² | 10000.00 |
|    brownfield site | 1000 m² | 11000.00 |
| Public open space internally comprising 5% of the site; with graded and levelled bow fenced and gated play area for informal ball games; grass cutting for 3 years at 22 times per year | | |
|    greenfield site | 1000 m² | 12000.00 |
|    brownfield site | 1000 m² | 13000.00 |
| | | |
| **SCIENCE AND OFFICE PARKS** | | |
| | | |
| **Business or Science park models. Users should add O & P to the rates below; We advise 15% for illustration purposes** | | |
| Generally areas of more dense landscape incorporating relaxation areas and larger planting schemes; planting is generally more mature at implementation with larger trees; Planting schemes often incorporate larger swathes of groundcovers to give graded views over landscaped areas; planting density and sizes are generally of higher standard than housing | | |

## COST MODELS

| Item<br>Excluding site overheads and profit | Unit | Range £ |
|---|---|---|
| Soft landscaped space only space; planting:turf ratios as shown; planting sizes and densities as indicated; 1 tree of 25–30 cm girth at 1 tree per 100 m² overall external area; Mulch and watering after planting | | |
|    Landscaped area; Turf and trees only; imported topsoil to turf areas at 50 thick | 100 m² | 1275.00 |
|    Landscaped area 70:30 turf to planted area with imported topsoil to turf areas at 50 thick | 100 m² | 2225.00 |
|    Landscaped area 50:50 turf to planted area; imported topsoil to turf areas at 50 thick | 100 m² | 2500.00 |
|    Landscaped area; Planting and trees only; Planting of shrubs 100% at 400 mm centres | 100 m² | 3450.00 |
|    Landscaped area; Planting and trees only; Planting of shrubs and groundcovers 100% Shrubs: Groundcovers 70:30 at 400:250 mm centres | 100 m² | 3900.00 |
|    Landscaped area; Planting and trees only; Planting of shrubs and groundcovers 100% Shrubs: Groundcovers 30:70 at 400:250 mm centres | 100 m² | 4400.00 |
| **Mixed hard and soft landscaped space with parking; 15% of the external space are pathways or building surrounds paved with good quality textured concrete pavings. The remaining area of 85% of the external area is divided into 50:50 Soft landscape: Parking; 1 tree 25–30 cm per 100 m² of external area; Parking area treated as described** | | |
| External area where 50% of the remaining external area is parking area; with surround of PC kerbs and 50% soft landscape of 70% turf and 30% planted area; imported topsoil to turf areas at 50 thick | | |
|    macadam parking area | 100 m² | 6400.00 |
|    concrete block paved parking area | 100 m² | 7100.00 |
| The soft landscape is 50% turf and 50% planting; planting at 400 mm ccs – 2–3 litre plants per m²; | | |
|    macadam parking area | 100 m² | 6700.00 |
|    concrete block paved parking area | 100 m² | 7600.00 |
| The soft landscape comprises 100% planting; planting at 400 mm ccs – 2–3 litre plants per m²; | | |
|    macadam parking area | 100 m² | 6400.00 |
|    concrete block paved parking area | 100 m² | 7300.00 |
| External area where the area is 50% parking and 50% soft landscape. The soft landscape comprises 100% planting; planting at 400 mm ccs – 2–3 litre plants per m²; imported topsoil to planting areas 400 mm thick inclusive of removal of existing material and disposal offsite | | |
|    macadam parking area | 100 m² | 7800.00 |
|    concrete block paved parking area | 100 m² | 8700.00 |
| **CIVIC ENVIRONMENTS AND STREETSCAPE** | | |
| Civic and Public areas in the current climate encompass higher quality surfaces for public access. Planting is generally larger. There is a need for furniture and lighting; Users should add O & P to the rates below; We advise 15% for illustration purposes | | |

## COST MODELS

| Item<br>Excluding site overheads and profit | Unit | Range £ |
|---|---|---|
| **CIVIC ENVIRONMENTS AND STREETSCAPE – cont** | | |
| **Civic town centre area of paved open space; model area 500 m²** | | |
| General open area; one 30–35 cm tree with root cell planting system every 400 m²; Metal tree grid; One bench per 200 m² and one bin per 500 m²; Granite kerb surround | | |
|    Concrete slab paved PC £28.30/m² | 100 m² | 9400.00 |
|    Granite paved | 100 m² | 16000.00 |
|    Granite paved with stepped terraces. 3 steps of 10 m long each in an area of 500 m² | 100 m² | 19000.00 |
|    Diamond sawn Yorkstone paved | 100 m² | 19000.00 |
|    Granite paved with stepped terraces. 3 steps of 10 m long each in an area of 500 m² | 100 m² | 22000.00 |
| General open area; one 30–35 cm tree with root cell planting system every 200 m²; Metal tree grid; One bench per 200 m² and one bin per 500 m²; Granite kerb surround | | |
|    Granite paved | 100 m² | 17000.00 |
|    Granite paved with stepped terraces. 3 steps of 10 m long each in an area of 500 m² | 100 m² | 20000.00 |
|    Diamond sawn Yorkstone paved | 100 m² | 20000.00 |
|    Granite paved with stepped terraces. 3 steps of 10 m long each in an area of 500 m² | 100 m² | 23000.00 |
| **Hard landscaped civic streetscape space of 1000 m² plan area with paved amenity areas comprising kerb surrounded paved area with street furniture. Raised beds filled with planting; trees of 30–35 cm at 100 m² centres** | | |
| Raised seating walls surrounding planted area with bollard lights at 100 m² centres; planting of low box hedges and shrubs comprising 10% of the total area | | |
|    Reconstituted stone slab paved | 100 m² | 15500.00 |
|    Granite paved | 100 m² | 21000.00 |
|    Yorkstone paved | 100 m² | 24000.00 |
| Sunken area with long bullnosed steps with bollard and in-ground lighting; Planting area comprising 15% of the available area with box hedging and annual bedding plants and associated maintenance for 1 year | | |
|    Reconstituted stone slab paved | 100 m² | 22000.00 |
|    Granite paved | 100 m² | 27000.00 |
|    Yorkstone paved | 100 m² | 28000.00 |
| **Hard landscaped civic streetscape space of 1000 m² plan area with paved amenity areas comprising kerb surrounded paved area with street furniture. Raised water feature of 25 m²; lighting bollards; Trees of 30–35 cm at 10 m centres to the perimeter** | | |
| Sunken area with long bullnosed steps with bollard lighting; Planting area comprising 15% of the available area with box hedging and annual bedding plants and associated maintenance for 1 year. | | |
|    Reconstituted stone slab paved | 100 m² | 24000.00 |
|    Granite paved | 100 m² | 28000.00 |
|    Yorkstone paved | 100 m² | 29500.00 |

## COST MODELS

| Item<br>Excluding site overheads and profit | Unit | Range £ |
|---|---|---|
| **PLAYGROUNDS** | | |
| **Urban play area comprising fenced play equipment for children to 12 years; Cushioned fall play equipment as specified by equipment manufacturer. Bow fenced all sides with benches and litter bins; appropriate signage; Users should add O & P to the rates below; We advise 15% for illustration purposes** Traditional equipment swings, slides and spring equipment; six activities; Safety surfacing below and surrounding equipment; three benches and litter bin; turfed where not surfaced | | |
| Local parish traditional equipment | 1200 m² | 61000.00 |
| City and urban medium sized play equipment; Modern design climbing, zipwire net climbing and medium sized theme multiplay units Hags-SMP | 1200 m² | 83000.00 |
| City and urban but with larger multiplay themed play equipment | 1500 m² | 125000.00 |
| Mixed themed and traditional play area with fitness elements and adventure climbing equipment | 3000 m² | 150000.00 |
| **LEAP Play areas for compliance with housing development; 35–100 units; Kompan Ltd** Local Equipped Area for Play (LEAP) is a play area equipped for children of early school age (mainly 4-8 year olds). The activity zone should have a minimum area of 400m², with grass playing space and at least five types of play equipment with appropriate safety surfacing. There should also be seating for accompanying adults. Contains at least five types of play equipment; Contains seating for parents and/or carers; Contains a litter bin signage fencing | | |
| minimum requirement | nr | 21500.00 |
| minimum requirement but with green seeded kickabout area | nr | 22000.00 |
| minimum requirement brownfield site | nr | 25500.00 |
| **NEAP Play area for compliance for housing developers; Kompan Ltd** A Neighbourhood Equipped Area for Play (NEAP) is a play area for 8–14 year olds which should include a grassed kickabout area, a hard surfaced area for a ball games or wheeled activities, eight types of play equipment appropriate to children in this age group and seating, including a youth shelter. This requires an activity zone of at least 1000m² | | |
| minimum requirement | nr | 135000.00 |
| minimum requirement brownfield site | nr | 150000.00 |
| **DOMESTIC COST MODELS** | | |
| **Suburban new or refurbished landscape comprising; Driveways and parking; Recreational pavings; Access pavings; Perimeter fencing; Trees; Planting beds and turf; Measured at per 1000 m² of total site area; Users should add O & P to the rates below; We advise 15% for illustration purposes** Total property size; 3300 m² of which building footprint comprises 280 m² (8.5%) Access and parking 330 m² (10%); Closeboard fencing to perimeter; Balance of landscape (2700 m² +/- 80%) area as follows | | |
| Hard recreation and access areas/Soft landscaped areas; standard size planting | 1000 m² | 86000.00 |
| As above but with enhanced planting and larger trees | 1000 m² | 110000.00 |
| As above but with brick entrance walls gates and piers | 1000 m² | 115000.00 |
| As immediately above but with a lighting scheme | 1000 m² | 120000.00 |
| As immediately above but with a stone paved steps to the patio 5 steps high × 2.00 m wide | 1000 m² | 120000.00 |
| As immediately above but with timber pergolas and water feature | 1000 m² | 130000.00 |

## PRELIMINARIES

| Item<br>Excluding site overheads and profit | Unit | Range £ |
|---|---|---|
| **CONTRACT ADMINISTRATION AND MANAGEMENT** | | |
| **Prepare tender bid for external works or landscape project; measured works contract; inclusive of bid preparation, provisional programme and method statements** | | |
| Measured works bid; measured bills provided by the employer; project value | | |
| £30,000.00 | nr | 375.00 |
| £50,000.00 | nr | 600.00 |
| £100,000.00 | nr | 1275.00 |
| £200,000.00 | nr | 1975.00 |
| £500,000.00 | nr | 2500.00 |
| £1,000,000.00 | nr | 3500.00 |
| Lump sum bid; scope document, drawings and specification provided by the employer; project value | | |
| £30,000.00 | nr | 740.00 |
| £50,000.00 | nr | 960.00 |
| £100,000.00 | nr | 1850.00 |
| £200,000.00 | nr | 3150.00 |
| £500,000.00 | nr | 4250.00 |
| £1,000,000.00 | nr | 6400.00 |
| Prepare and maintain Health and Safety file; prepare risk and COSHH assessments, method statements, works programmes before the start of works on site; project value | | |
| £35,000.00 | nr | 280.00 |
| £75,000.00 | nr | 420.00 |
| £100,000.00 | nr | 560.00 |
| £200,000.00–£500,000.00 | nr | 1400.00 |
| **Contract management** | | |
| Site management by site surveyor carrying out supervision and administration duties only; inclusive of oncosts | | |
| full time management | week | 1200.00 |
| managing one other contract | week | 600.00 |
| **Parking** | | |
| Parking expenses where vehicles do not park on the site area; per vehicle | | |
| metropolitan area; city centre | week | 200.00 |
| metropolitan area; outer areas | week | 160.00 |
| suburban restricted parking areas | week | 40.00 |
| **Congestion charging** | | |
| London only | week | 58.00 |

*Approximate Estimating Rates*

## PRELIMINARIES

| Item<br>Excluding site overheads and profit | Unit | Range £ |
|---|---|---|
| **SITE SETUP AND SITE ESTABLISHMENT** | | |
| **Site fencing; supply and erect temporary protective fencing and remove at completion of works** | | |
| Cleft chestnut paling | | |
| 1.20 m high 75 mm larch posts at 3 m centres | 100 m | 1725.00 |
| 1.50 m high 75 mm larch posts at 3 m centres | 100 m | 1900.00 |
| Heras fencing | | |
| 2 week hire period | 100 m | 990.00 |
| 4 week hire period | 100 m | 1450.00 |
| 8 week hire period | 100 m | 2350.00 |
| 12 week hire period | 100 m | 3250.00 |
| 16 week hire period | 100 m | 4150.00 |
| 24 week hire period | 100 m | 6000.00 |
| **Site compounds** | | |
| Establish site administration area on reduced compacted Type 1 base; 150 mm thick; allow for initial setup of site office and welfare facilities for site manager and maximum eight site operatives | | |
| temporary base or hard standing; 20 × 10 m | 200 m$^2$ | 3250.00 |
| **To hardstanding area above; erect office and welfare accommodation; allow for furniture, telephone connection, and power; restrict access using temporary safety fencing to the perimeter of the site administration area** | | |
| Establishment; delivery and collection costs only | | |
| site office and chemical toilet; 20 m of site fencing | nr | 990.00 |
| site office and chemical toilet; 40 m of site fencing | nr | 1075.00 |
| site office and chemical toilet; 60 m of site fencing | nr | 1150.00 |
| **Weekly hire rates of site compound equipment** | | |
| site office 2.4 × 3.6 m; chemical toilet; 20 m of site fencing | week | 105.00 |
| site office 4.8 × 2.4 m; chemical toilet; 20 m of site fencing; armoured store 2.4 × 3.6 m | week | 120.00 |
| site office 2.4 × 3.6 m; chemical toilet; 40 m of site fencing | week | 150.00 |
| site office 2.4 × 3.6 m; chemical toilet; 60 m of site fencing | week | 195.00 |
| **Removal of site compound** | | |
| Remove base; fill with topsoil and reinstate to turf on completion of site works | 200 m$^2$ | 1975.00 |
| **Surveying and setting out** | | |
| Note: In all instances a quotation should be obtained from a surveyor. The following cost projections are based on the costs of a surveyor using EDM (electronic distance measuring equipment) where an average of 400 different points on site are recorded in a day. The surveyed points are drawn in a CAD application and the assumption used that there is one day of drawing up for each day of on site surveying. | | |
| Survey existing site using laser survey equipment and produce plans of existing site conditions | | |
| site 1000 to 10,000 m$^2$ with sparse detail and features; boundary and level information only; average five surveying stations | nr | 750.00 |
| site up to 1000 m$^2$ with dense detail such as buildings, paths, walls and existing vegetation; average five survey stations | nr | 1500.00 |
| site up to 4000 m$^2$ with dense detail such as buildings, paths, walls and existing vegetation; average 10 survey stations | nr | 2250.00 |

## 8.1.1 SITE CLEARANCE

| Item<br>Excluding site overheads and profit | Unit | Range £ |
|---|---|---|
| **CLEARING VEGETATION** | | |
| **Clear existing scrub vegetation including shrubs and hedges and stack on site** | | |
| By machine | | |
|   light scrub and grasses | 100 m² | 21.50 |
|   light scrub and grasses but remove vegetation to licensed tip | 100 m² | 58.00 |
|   undergrowth brambles and heavy weed growth | 100 m² | 50.00 |
|   small shrubs | 100 m² | 50.00 |
| By hand | | |
|   light scrub and grasses to spoil heaps 25 m distance | 100 m² | 76.00 |
|   light scrub and grasses to spoil heap; subsequently disposed off site | 100 m² | 96.00 |
|   light scrub and grasses to tip | 100 m² | 60.00 |
| **Fell trees on site; grub up roots; all by machine; remove debris to licensed tip** | | |
| trees 600 mm girth | each | 345.00 |
| trees 1.5–3.00 m girth | each | 1075.00 |
| trees over 3.00 m girth | each | 1725.00 |
| **Chip existing vegetation on site; remove to stockpile 100 m** | | |
| light scrub and bramble; average height 500 mm | 100 m² | 22.00 |
| dense scrub and bramble; average height 1.00 m | 100 m² | 45.00 |
| mixed scrub and shrubs average 1.50 m high | 100 m² | 76.00 |
| dense overgrown shrubs and bramble; average 2.00 m high | 100 m² | 100.00 |
| **Turf stripping for preservation** | | |
| Cut and strip existing turf area by machine; turves 50 mm thick; turf to be lifted for preservation | | |
|   load to dumper or palette by hand; transport on site not exceeding 100 m travel to new location | 100 m² | 89.00 |
|   all by hand; including barrowing to stacking locally 10 m distance | 100 m² | 550.00 |
| **Turf stripping for disposal; spray turf with glyphosate and return to site after 3 weeks; strip dead grass; load to stockpile on site; rotavate to prepare area for new surface treatments to within 50 mm of final levels** | | |
| By knapsack sprayer; excavator and dumper; pedestrian rotavator | | |
|   5 tonne excavator and 6 tonne dumper; stock pile distance 25 m | 100 m² | 58.00 |
|   5 tonne excavator and 3 tonne dumper; stock pile distance 50 m | 100 m² | 59.00 |
|   5 tonne excavator and 3 tonne dumper; stock pile distance 100 m | 100 m² | 63.00 |
| By tractor drawn equipment | | |
|   spraying with glyphosate; 2 passes | 100 m² | 560.00 |

## 8.1.1 SITE CLEARANCE

| Item<br>Excluding site overheads and profit | Unit | Range £ |
|---|---|---|
| **CLEARING FENCES AND GATES AND LIGHT STRUCTURES** | | |
| **Clear site; demolish existing closeboard fence; clear shrubs 50% in shrub beds and 50% free-standing; spray turf and return to clear ground of sprayed vegetation; chip vegetation on site and transport to spoil heaps 50 m; works by mechanical excavator** | | |
| Rectangular site with 70% turf and 30% shrub area | | |
| 1000 m² | nr | 1900.00 |
| 2000 m² | nr | 3450.00 |
| 4000 m² | nr | 6500.00 |
| Cleartrack (Evl) Ltd; as above but clearance by mechanical spraying and in situ mulcher; vegetation only to spoil heap; turf mulched in | | |
| 5000 m² | nr | 2150.00 |
| **Clear site; demolish existing chain-link fence; clear shrubs 50% in shrub beds and 50% free-standing; spray turf and return to clear ground of sprayed vegetation; chip vegetation on site and transport to spoil heaps 50 m; works by high capacity mechanical mulcher** | | |
| Cleartrack (Evl) Ltd | | |
| overgrown areas of weeds and brambles with maximum stem diameter of 100 mm | ha | 3500.00 |
| overgrown areas of weeds and brambles with maximum stem diameter of 150 mm | ha | 6900.00 |
| Rectangular site with 60% turf grass/scrub area and 40% shrubs and overgrowth | | |
| site area; 1000 m² | nr | 2175.00 |
| site area; 2000 m² | nr | 4100.00 |
| site area; 4000 m² | nr | 7600.00 |
| **Demolish fencing and gates** | | |
| Demolish fencing; break out posts, remove fencing stack and load mechanically for removal to tip | | |
| Chain link fencing; vegetation not present on angle iron stakes driven in | 100 m | 250.00 |
| Chain link fencing; overgrown with vegetation on angle iron stakes driven in | 100 m | 740.00 |
| **DEMOLITIONS OF EXTERNAL STRUCTURES AND SURFACES** | | |
| **Demolish existing free standing walls; grub out foundations; remove arisings to tip** | | |
| Brick wall; 112 mm thick | | |
| 300 mm high | m | 18.10 |
| 500 mm high | m | 28.00 |
| Brick wall; 225 mm thick | | |
| 300 mm high | m | 23.50 |
| 500 mm high | m | 31.00 |
| 1.00 m high | m | 50.00 |
| 1.20 m high | m | 53.00 |
| 1.50 m high | m | 66.00 |
| 1.80 m high | m | 74.00 |

## 8.1.1 SITE CLEARANCE

| Item<br>Excluding site overheads and profit | Unit | Range £ |
|---|---|---|
| **DEMOLITIONS OF EXTERNAL STRUCTURES AND SURFACES – cont** | | |
| **Demolish existing structures** | | |
| Timber sheds on concrete base 150 mm thick inclusive of disposal and backfilling with excavated material; plan area of building | | |
| 10 m² | nr | 300.00 |
| 15 m² | nr | 430.00 |
| Building of cavity wall construction; insulated with tiled roof; inclusive of foundations; allow for disconnection of electrical water and other services; grub out and remove 10 m of electrical cable, water and waste pipes | | |
| 10 m² | nr | 1850.00 |
| 20 m² | nr | 2450.00 |
| **CLEARING SURFACES** | | |
| **Demolish existing surfaces** | | |
| Break up plain concrete slab; remove to licensed tip | | |
| 150 mm thick | m² | 13.20 |
| 200 mm thick | m² | 17.60 |
| 300 mm thick | m² | 26.00 |
| Break up plain concrete slab and preserve arisings for hardcore; break down to maximum size of 200 × 200 mm and transport to location; maximum distance 50 m | | |
| 150 mm thick | m² | 7.50 |
| 200 mm thick | m² | 10.10 |
| 300 mm thick | m² | 15.10 |
| Break up reinforced concrete slab and remove to licensed tip | | |
| 150 mm thick | m² | 18.30 |
| 200 mm thick | m² | 24.50 |
| 300 mm thick | m² | 36.50 |
| Break out existing surface and associated 150 mm thick granular base load to 20 tonne muck away vehicle for removal off site | | |
| macadam surface; 70 mm thick | m² | 13.30 |
| macadam surface; 150 mm thick | m² | 18.80 |
| block paving; 50 mm thick | m² | 12.30 |
| block paving; 80 mm thick | m² | 13.50 |

## 8.1.2 PREPARATORY GROUNDWORKS

| Item<br>Excluding site overheads and profit | Unit | Range £ |
|---|---|---|
| **EXCAVATION OF BANKS** | | |
| **Excavate existing slope to vertical to form flat area; grade bank behind to 45°; no allowance for removal or backfilling of excavated materials** | | |
| Existing slope angle 10°; excavation by 7 tonne machine; excavated material to location on site 50 m distant; linear measurement of flat area at the base of the slope | | |
|    2.00 m wide; height 350 mm | m | 3.10 |
|    4.00 m wide; height 860 mm | m | 12.30 |
|    6.00 m wide; height 1.28 m | m | 28.00 |
|    8.00 m wide; height 1.71 m | m | 49.00 |
|    10.00 m wide; height 2.14 m | m | 77.00 |
| Existing slope angle 20°; excavation by 7 tonne machine; excavated material to loctaion on site 50 m distant; linear measurement of flat area at the base of the slope | | |
|    2.00 m wide; height 1.14 m | m | 8.30 |
|    4.00 m wide; height 2.29 m | m | 33.00 |
|    6.00 m wide; height 3.43 m | m | 74.00 |
|    8.00 m wide; height 4.58 m | m | 130.00 |
|    10.00 m wide; height 5.72 m | m | 200.00 |
| **Excavate and grade banks to grade to receive stabilization treatments below; removal of excavated material not included** | | |
| By 21 tonne excavator | m$^3$ | 0.40 |
| By 7 tonne excavator | m$^3$ | 2.30 |
| By 5 tonne excavator | m$^3$ | 2.40 |
| By 3 tonne excavator | m$^3$ | 3.85 |
| **Excavate sloped bank to vertical to receive retaining or stabilization treatment (priced separately); allow 1.00 m working space at top of bank; backfill working space and remove balance of arisings off site; measured as length along the top of the bank** | | |
| Bank height 1.00 m; excavation by 5 tonne excavator | | |
|    10° | m | 105.00 |
|    30° | m | 33.00 |
|    45° | m | 18.80 |
|    60° | m | 10.90 |
| Bank height 1.00 m; excavation by 5 tonne excavator; arisings moved to stockpile on site | | |
|    10° | m | 12.30 |
|    30° | m | 5.95 |
|    45° | m | 3.40 |
|    60° | m | 2.00 |
| Bank height 2.00 m; excavation by 5 tonne excavator | | |
|    10° | m | 425.00 |
|    30° | m | 130.00 |
|    45° | m | 75.00 |
|    60° | m | 43.00 |

## 8.1.2 PREPARATORY GROUNDWORKS

| Item<br>Excluding site overheads and profit | Unit | Range £ |
|---|---|---|
| **EXCAVATION OF BANKS – cont** | | |
| **Excavate sloped bank to vertical to receive retaining or stabilization treatment (priced separately) – cont** | | |
| Bank height 2.00 m; excavation by 5 tonne excavator but arisings moved to stockpile on site | | |
| 10° | m | 77.00 |
| 30° | m | 23.00 |
| 45° | m | 13.70 |
| 60° | m | 7.85 |
| Bank height 3.00 m; excavation by 21 tonne excavator | | |
| 20° | m | 430.00 |
| 30° | m | 270.00 |
| 45° | m | 150.00 |
| 60° | m | 82.00 |
| As above but arisings moved to stockpile on site | | |
| 20° | m | 48.00 |
| 30° | m | 30.00 |
| 45° | m | 12.50 |
| 60° | m | 6.75 |
| **EXCAVATING GENERALLY** | | |
| **Prices for excavating and reducing to levels are for work in light or medium soils; multiplying factors for other soils are as follows** | | |
| Clay | 1.5 | – |
| Compact gravel | 1.2 | – |
| Soft chalk | 2.0 | – |
| Hard rock | 3.0 | – |
| **Excavating topsoil for preservation** | | |
| Remove topsoil average depth 300 mm; deposit in spoil heaps not exceeding 100 m travel to heap; turn heap once during storage | | |
| by machine | m³ | 10.10 |
| by hand | m³ | 145.00 |
| **Excavate to reduce levels; remove spoil to stockpile not exceeding 100 m travel; all by machine** | | |
| Average 0.25 m deep | m² | 2.20 |
| Average 0.25 m deep | m³ | 8.70 |
| Average 0.30 m deep | m² | 2.60 |
| Average 0.30 m deep | m³ | 7.95 |
| Average 1.0 m deep; using 21 tonne 360 tracked excavator | m³ | 4.40 |
| Average 1.0 m deep; using JCB | m³ | 7.95 |
| **Excavate to reduce levels; remove spoil to stockpile not exceeding 100 m travel; all by hand** | | |
| Average 0.10 m deep | m² | 12.80 |
| Average 0.20 m deep | m² | 25.50 |
| Average 0.30 m deep | m² | 38.00 |

## 8.1.2 PREPARATORY GROUNDWORKS

| Item<br>Excluding site overheads and profit | Unit | Range £ |
|---|---|---|
| Average 100–300 mm deep | m³ | 130.00 |
| Average 1.0 m deep | m³ | 145.00 |
| Extra for carting spoil to licensed tip off site (20 tonnes) | m³ | 39.00 |
| Extra for carting spoil to licensed tip off site (by skip; machine loaded) | m³ | 60.00 |
| Extra for carting spoil to licensed tip off site (by skip; hand loaded) | m³ | 130.00 |
| **Spread excavated material to levels in layers not exceeding 150 mm; using scraper blade or bucket** | | |
| Average thickness 100 mm | m² | 0.50 |
| Average thickness 100 mm but with imported topsoil | m² | 4.70 |
| Average thickness 200 mm | m² | 1.00 |
| Average thickness 200 mm but with imported topsoil | m² | 9.40 |
| Average thickness 250 mm | m² | 1.30 |
| Average thickness 200–250 mm | m³ | 5.20 |
| Average thickness 250 mm but with imported topsoil | m² | 11.80 |
| Average thickness 250 mm but with imported topsoil | m³ | 47.00 |
| Extra for work to banks exceeding 30° slope | 30% | – |
| **Rip subsoil using approved subsoiling machine to a depth of 600 mm below topsoil at 600 mm centres; all tree roots and debris over 150 × 150 mm to be removed; cultivate ground where shown on drawings to depths as shown** | | |
| 100 mm | 100 m² | 10.80 |
| 200 mm | 100 m² | 11.20 |
| 300 mm | 100 m² | 11.80 |
| 400 mm | 100 m² | 17.00 |
| **Form landform from imported material; mounds to falls and grades to receive turf or seeding treatments; material dumped by 20 tonne vehicle to location; volume 100 m³ spread over area; thickness varies from 150–500 mm; prices to not include for surface preparations**<br>Imported standard grade topsoil | | |
|     by large excavator | m³ | 40.00 |
|     by 5 tonne excavator | m³ | 42.00 |
|     by 3 tonne excavator | m³ | 43.50 |
| **Form landform from excavated material; mounds to falls and grades to receive turf or seeding treatments; material moved by dumper to location maximum 50 m; volume 100 m³ spread over a 200 m² area; thickness varies from 150–500 mm; prices to not include for surface preparations**<br>Imported recycled topsoil | | |
|     by large excavator | m³ | 3.30 |
|     by 5 tonne excavator | m³ | 5.30 |
|     by 3 tonne excavator | m³ | 6.70 |

## 8.1.2 PREPARATORY GROUNDWORKS

| Item<br>Excluding site overheads and profit | Unit | Range £ |
|---|---|---|
| **EXCAVATION OF TRENCHES** | | |
| Excavate trenches for foundations; trenches 225 mm deeper than specified foundation thickness; pour plain concrete foundations GEN 1 10 N/mm and to thickness as described; disposal of excavated material off site; dimensions of concrete foundations | | |
| 250 mm wide × 250 mm deep | m | 13.00 |
| 300 mm wide × 300 mm deep | m | 17.70 |
| 400 mm wide × 600 mm deep | m | 42.00 |
| 600 mm wide × 400 mm deep | m | 45.00 |
| Excavate trenches for foundations; trenches 225 mm deeper than specified foundation thickness; pour plain concrete foundations GEN 1 10 N/mm and to thickness as described; disposal of excavated material off site; dimensions of concrete foundations | | |
| 250 mm wide × 250 mm deep | m³ | 255.00 |
| 300 mm wide × 300 mm deep | m³ | 195.00 |
| 400 mm wide × 600 mm deep | m³ | 160.00 |
| 600 mm wide × 400 mm deep | m³ | 185.00 |
| Excavate trenches for services; grade bottoms of excavations to required falls; remove 20% of excavated material off site | | |
| Trenches 300 mm wide | | |
| 600 mm deep | m | 9.00 |
| 750 mm deep | m | 11.50 |
| 900 mm deep | m | 13.90 |
| 1.00 m deep | m | 15.20 |
| **Ha-ha**<br>Excavate ditch 1200 mm deep × 900 mm wide at bottom battered one side 45° slope; excavate for foundation to wall and place GEN 1 concrete foundation 150 × 500 mm; construct one and a half brick wall (brick PC £300.00/1000) battered 10° from vertical; laid in cement:lime:sand (1:1:6) mortar in English garden wall bond; precast concrete coping weathered and throated set 150 mm above ground on high side; rake bottom and sides of ditch and seed with low maintenance grass at 35 g/m² | | |
| wall 600 mm high | m | 320.00 |
| wall 900 mm high | m | 410.00 |
| wall 1200 mm high | m | 520.00 |
| Excavate ditch 1200 mm deep × 900 mm wide at bottom battered one side 45° slope; place deer or cattle fence 1.20 m high at centre of excavated trench; rake bottom and sides of ditch and seed with low maintenance grass at 35 g/m² | | |
| fence 1200 mm high | m | 24.00 |

## 8.1.2 PREPARATORY GROUNDWORKS

| Item<br>Excluding site overheads and profit | Unit | Range £ |
|---|---|---|
| **SOIL STABILIZATION** | | |
| Please refer to slope excavation above for preparation of banks | | |
| **REVETMENTS** | | |
| **Gabion mattress revetments**<br>Construct revetment of Reno mattress gabions laid on firm level ground; tightly packed with broken stone or concrete and securely wired; all in accordance with manufacturer's instructions | | |
| gabions 6 × 2 × 0.17 m; one course | m$^2$ | 39.50 |
| gabions 6 × 2 × 0.17 m; two courses | m$^2$ | 65.00 |
| **Grass concrete; bank revetments; excluding bulk earthworks**<br>Bring bank to final grade by machine; lay regulating layer of Type 1; lay filter membrane; lay 100 mm drainage layer of broken stone or approved hardcore 28–10 mm size; blind with 25 mm sharp sand; lay grass concrete surface (price does not include edgings or toe beams); fill with approved topsoil and fertilizer at 35 g/m$^2$; seed with dwarf rye based grass seed mix | | |
| Grasscrete in situ reinforced concrete surfacing GC 1; 100 mm thick | m$^2$ | 57.00 |
| Grasscrete in situ reinforced concrete surfacing GC 2; 150 mm thick | m$^2$ | 69.00 |
| Grassblock 103 mm open matrix blocks; 406 × 406 × 103 mm | m$^2$ | 68.00 |
| **Grade slope to even grade; Stabilize river bank edge with woven willow walling; lay bank stabilization mat pinned to graded embankment 5 m long (measured per m of river bank edge)**<br>Slope edge to river bank 5 m long seeded with Greenfix stabilization matting | | |
| Brushwood faggott and pre-planted coir revetment edge | m | 170.00 |
| Brushwood faggott revetment edge | m | 150.00 |
| **StableGRASS grass reinforcement: cellular slope retention; clear existing bank of light scrub, shrubs and grasses; spray with glyphosate; grade surface only to even grade by machine and final hand grade; lay soil retaining system**<br>Backfilled with excavated material | | |
| 38 mm deep | m$^2$ | 22.00 |
| Backfilled with imported topsoil and seeded | | |
| 100 mm deep | m$^2$ | 1.95 |

## 8.1.2 PREPARATORY GROUNDWORKS

| Item<br>Excluding site overheads and profit | Unit | Range £ |
|---|---|---|
| **SOIL REINFORCEMENT** | | |
| **Geogrid soil reinforcement (Note: measured per metre run at the top of the bank)** | | |
| **Backfill excavated bank; lay Tensar geogrid; cover with excavated material as work proceeds** | | |
| 2.00 m high bank; geogrid at 1.00 m vertical lifts | | |
|     10° slope | m | 710.00 |
|     20° slope | m | 395.00 |
|     30° slope | m | 285.00 |
| 2.00 m high bank; geogrid at 0.50 m vertical lifts | | |
|     45° slope | m | 67.00 |
|     60° slope | m | 53.00 |
| 3.00 m high bank | | |
|     10° slope | m | 710.00 |
|     20° slope | m | 395.00 |
|     30° slope | m | 285.00 |
|     45° slope | m | 610.00 |

## 8.1.2 PREPARATORY GROUNDWORKS

| Item<br>Excluding site overheads and profit | Unit | Range £ |
|---|---|---|
| **EROSION CONTROL** | | |
| **Bank stabilization; erosion control mats** | | |
| Excavate fixing trench for erosion control mat 300 × 300 mm; backfill after placing mat selected below | m | 7.40 |
| **To anchor trench above; place mat to slope to be retained; allow extra over to the slope for the area of matting required to the anchor trench** | | |
| By machine; final grade only of excavated bank by excavator; lay erosion control solution; sow with low maintenance grass at 35 g/m²; spread imported topsoil 25 mm thick incorporating medium grade sedge peat at 3 kg/m² and fertilizer at 30 g/m²; water lightly | | |
|     unseeded Eromat Light | m² | 2.95 |
|     seeded Covamat Standard | m² | 3.60 |
|     lay Greenfix biodegradable pre-seeded erosion control mat; fix with 6 × 300 mm steel pegs at 1.0 m centres | m² | 4.10 |
|     Tensar open textured erosion mat | m² | 5.60 |
| By hand; final grade only of excavated bank by hand; lay erosion control solution; sow with low maintenance grass at 35 g/m²; spread imported topsoil 25 mm thick incorporating medium grade sedge peat at 3 kg/m² and fertilizer at 30 g/m²; water lightly | | |
|     unseeded Eromat Light | m² | 3.50 |
|     seeded Covamat Standard | m² | 4.15 |
|     lay Greenfix biodegradable pre-seeded erosion control mat; fix with 6 × 300 mm steel pegs at 1.0 m centres | m² | 4.70 |
|     Tensar open textured erosion mat | m² | 6.20 |
| **Cellular retaining systems; Greenfix Ltd; final grade to excavated bank by machine and by hand; lay bank retaining system** | | |
| Geoweb; filling with angular aggregate | m | 25.00 |
| **Haul road for site enablement** | | |
| Reduce existing soil to stockpile on site; Lay Tensar Biaxial geogrid on 200 mm compacted hardcore base; backfill with 200 mm Type 1 granular fill and compact | m² | 18.50 |
| Reduce existing soil to stockpile on site; Lay 200 mm compacted hardcore base blinded with 50 mm sharp sand; Lay Cellweb load support | m² | 33.50 |

## 8.2.1 ROADS, PATHS, PAVINGS AND SURFACINGS

| Item<br>Excluding site overheads and profit | Unit | Range £ |
|---|---|---|
| **EXCAVATION KERBS AND EDGINGS TO PAVED SURFACES** | | |
| **Excavate for pathways or roadbed to receive surface 100 mm thick priced separately; grade; 100 mm hardcore; kerbs PC 125 × 255 mm on both sides; including foundations haunched** | | |
| Excavation 350 deep; Type 1 base 150 thick | | |
| 1.50 m wide | m | 110.00 |
| 2.00 m wide | m | 120.00 |
| 3.00 m wide | m | 150.00 |
| 4.00 m wide | m | 175.00 |
| 5.00 m wide | m | 200.00 |
| 6.00 m wide | m | 225.00 |
| 7.00 m wide | m | 250.00 |
| Excavation 450 mm deep; Type 1 base 250 mm thick | | |
| 1.50 m wide | m | 210.00 |
| 2.00 m wide | m | 255.00 |
| 3.00 m wide | m | 345.00 |
| 4.00 m wide | m | 435.00 |
| 5.00 m wide | m | 530.00 |
| 6.00 m wide | m | 620.00 |
| 7.00 m wide | m | 710.00 |
| **EDGINGS TO PATHS AND PAVINGS** | | |
| **PRECAST CONCRETE KERBS** | | |
| **Note: excavation is by machine unless otherwise mentioned** | | |
| **Excavate trench and construct concrete foundation 300 mm wide × 300 mm deep; lay precast concrete kerb units bedded in semi-dry concrete; slump 35 mm maximum; haunching one side; disposal of arisings off site** | | |
| Kerbs laid straight; site mixed concrete bases | | |
| 125 mm high × 125 mm thick; bullnosed; type BN | m | 32.00 |
| 125 × 255 mm; ref HB2; SP | m | 35.00 |
| 150 × 305 mm; ref HB1 | m | 42.50 |
| Kerbs laid straight; ready mixed concrete bases | | |
| 125 mm high × 125 mm thick; bullnosed; type BN | m | 30.00 |
| 125 × 255 mm; ref HB2; SP | m | 33.00 |
| 150 × 305 mm; ref HB1 | m | 40.50 |
| Straight kerbs 125 mm × 255 mm laid to radius ready mixed concrete bases | | |
| 4.5 m radius | m | 42.00 |
| 6.1 m radius | m | 41.50 |
| 9.15 m radius | m | 41.00 |
| 12.20 m radius | m | 40.00 |
| Dropper kerbs | | |
| 125 mm × 255–150 mm; left or right handed | m | 37.00 |
| Quadrants | | |
| 305 mm radius | nr | 42.00 |
| 455 mm radius | nr | 44.00 |

## 8.2.1 ROADS, PATHS, PAVINGS AND SURFACINGS

| Item<br>Excluding site overheads and profit | Unit | Range £ |
|---|---|---|
| **Excavate trench and construct concrete foundation 300 mm wide × 150 mm deep; lay precast concrete kerb units bedded in semi-dry concrete; slump 35 mm maximum; haunching one side; disposal of arisings off site** | | |
| Conservation kerbs laid straight | | |
| 225 mm wide × 150 mm high; | m | 59.00 |
| 155 mm wide × 255 mm high | m | 58.00 |
| Conservation kerbs laid to radius; 145 mm × 255 mm | | |
| radius 3.25 m | m | 74.00 |
| radius 6.50 m | m | 67.00 |
| radius 9.80 m | m | 66.00 |
| solid quadrant; 305 × 305 mm | nr | 67.00 |
| **STONE KERBS** | | |
| **Excavate and construct concrete foundation 450 mm wide × 150 mm deep; lay kerb units bedded in semi-dry concrete; slump 35 mm maximum; haunching one side** | | |
| Granite kerbs | | |
| straight; 125 × 250 mm | m | 62.00 |
| curved; 125 × 250 mm | m | 72.00 |
| **Precast concrete edging on concrete foundation 100 × 150 mm deep and haunching one side 1:2:4 including all necessary excavation disposal and formwork** | | |
| Rectangular, chamfered or bullnosed; to one side of straight path | | |
| 50 × 150 mm | m | 29.00 |
| 50 × 200 mm | m | 37.00 |
| 50 × 250 mm | m | 39.00 |
| Rectangular, chamfered or bullnosed as above but to both sides of straight paths | | |
| 50 × 150 mm | m | 64.00 |
| 50 × 200 mm | m | 74.00 |
| 50 × 250 mm | m | 78.00 |
| **Timber edgings; softwood** | | |
| Straight | | |
| 150 × 38 mm | m | 7.40 |
| 150 × 50 mm | m | 9.90 |
| Curved; over 5 m radius | | |
| 150 × 38 mm | m | 9.50 |
| 150 × 50 mm | m | 12.80 |
| Curved; 4–5 m radius | | |
| 150 × 38 mm | m | 10.30 |
| 150 × 50 mm | m | 13.70 |

## 8.2.1 ROADS, PATHS, PAVINGS AND SURFACINGS

| Item<br>Excluding site overheads and profit | Unit | Range £ |
|---|---|---|
| **STONE KERBS – cont** | | |
| **Timber edgings – cont** | | |
| Curved; 3–4 m radius | | |
|    150 × 38 mm | m | 11.60 |
|    150 × 50 mm | m | 15.30 |
| Curved; 1–3 m radius | | |
|    150 × 38 mm | m | 13.60 |
|    150 × 50 mm | m | 17.40 |
| **Brick or concrete block edge restraint; excavate for foundation; lay concrete 1:2:4 250 mm wide × 150 mm deep; on 50 mm thick sharp sand bed; lay blocks or bricks** | | |
| Blocks 200 × 100 × 60 mm; PC £10.51/m²; butt jointed | | |
|    header course | m | 34.50 |
|    stretcher course | m | 30.50 |
| Bricks 215 × 112.5 × 50 mm; PC £300.00/1000; with mortar joints | | |
|    header course | m | 35.50 |
|    header course; brick on edge | m | 45.00 |
|    stretcher course; flat or brick on edge | m | 35.00 |
|    stretcher course; 2 rows brick on edge | m | 48.00 |
| Bricks 215 × 112.5 × 50 mm; PC £600.00/1000; with mortar joints | | |
|    header course | m | 38.50 |
|    header course; brick on edge | m | 48.00 |
|    stretcher course; flat or on edge | m | 27.00 |
|    stretcher course; 2 rows brick on edge | m | 68.00 |
| **STONE EDGINGS** | | |
| **Sawn Yorkstone edgings; excavate for foundations; lay concrete 1:2:4 150 mm deep × 33.3% wider than the edging; on 35 mm thick mortar bed** | | |
| Yorkstone 50 mm thick | | |
|    100 mm wide × random lengths | m | 48.00 |
|    100 × 100 mm | m | 48.00 |
|    100 mm wide × 200 mm long | m | 57.00 |
|    250 mm wide × random lengths | m | 57.00 |
|    500 mm wide × random lengths | m | 88.00 |
| **Granite edgings; excavate for footings; lay concrete 1:2:4 150 mm deep × 33.3% wider than the edging; on 35 mm thick mortar bed** | | |
| Granite 50 mm thick | | |
|    100 mm wide × 200 mm long | m | 41.50 |
|    100 mm wide × random lengths | m | 49.00 |
|    250 mm wide × random lengths | m | 51.00 |
|    setts; 100 × 100 mm | m | 61.00 |
|    300 mm wide × random lengths | m | 57.00 |

## 8.2.1 ROADS, PATHS, PAVINGS AND SURFACINGS

| Item<br>Excluding site overheads and profit | Unit | Range £ |
|---|---|---|
| **BASES FOR PAVING** | | |
| **Excavate ground and reduce levels to receive 38 mm thick slab and 25 mm mortar bed; dispose of excavated material off site** | | |
| Lay granular fill Type 1; limestone aggregate 150 mm thick laid to falls and compacted | | |
| all by machine | m² | 19.00 |
| all by hand except disposal by grab | m² | 56.00 |
| Lay granular fill Type 1; recycled material; 150 mm thick laid to falls and compacted | | |
| all by machine | m² | 16.20 |
| all by hand except disposal by grab | m² | 53.00 |
| Lay granular fill Type 1; limestone aggregate 200 mm thick laid to falls and compacted | | |
| all by machine | m² | 25.50 |
| all by hand except disposal by grab | m² | 71.00 |
| Lay granular fill Type 1; recycled material; 200 mm thick laid to falls and compacted | | |
| all by machine | m² | 21.50 |
| all by hand except disposal by grab | m² | 68.00 |
| Lay 1:2:4 concrete base 150 mm thick laid to falls | | |
| all by machine | m² | 28.00 |
| all by hand except disposal by grab | m² | 69.00 |
| 150 mm hardcore base with concrete base 150 mm deep | | |
| concrete 1:2:4 site mixed | m² | 40.00 |
| concrete PAV 1 35 ready mixed | m² | 40.00 |
| concrete PAV 1 35 ready mixed; but for 50 mm thick slab | m² | 41.50 |
| Hardcore base 150 mm deep; concrete reinforced with A142 mesh | | |
| site mixed concrete 1:2:4; 150 mm deep | m² | 60.00 |
| site mixed concrete 1:2:4; 250 mm deep | m² | 82.00 |
| concrete PAV 1 35 N/mm² ready mixed; 150 mm deep | m² | 49.50 |
| concrete PAV 1 35 N/mm² ready mixed; 250 mm deep | m² | 72.00 |
| **CHANNELS** | | |
| **Excavate and construct concrete foundation 450 mm wide × 150 mm deep; lay precast concrete kerb units bedded in semi-dry concrete; slump 35 mm maximum; haunching one side; lay channel units bedded in 1:3 sand mortar; jointed in cement:sand mortar 1:3** | | |
| Kerbs; 125 mm high × 125 mm thick; bullnosed; type BN | | |
| dished channel; 125 × 225 mm; Ref CS | m | 59.00 |
| square channel; 125 × 150 mm; Ref CS2 | m | 57.00 |
| bullnosed channel; 305 × 150 mm; Ref CBN | m | 70.00 |
| **Brick channel; class B engineering bricks** | | |
| Excavate and construct concrete foundation 600 mm wide × 200 mm deep; lay channel to depths and falls bricks to be laid as headers along the channel; bedded in 1:3 sand mortar bricks close jointed in cement:sand mortar 1:3 | | |
| 3 courses wide | m | 83.00 |
| **3 coursed granite setts; dished** | | |
| Excavate and construct concrete foundation 400 mm wide × 200 mm deep; lay channel to depths and falls; bedded in 1:3 sand mortar | | |
| 3 courses wide | m | 79.00 |

## 8.2.1 ROADS, PATHS, PAVINGS AND SURFACINGS

| Item<br>Excluding site overheads and profit | Unit | Range £ |
|---|---|---|
| **CONCRETE ROADS** | | |
| | | |
| **Excavate weak points of excavation by hand and fill with well rammed Type 1 granular material** | | |
| 100 mm thick | m² | 13.00 |
| 200 mm thick | m² | 16.50 |
| | | |
| **Reinforced concrete roadbed; excavation and disposal off site; reinforcing mesh A142; expansion joints at 50 m centres; on 150 mm hardcore blinded sand; kerbs 155 × 255 mm to both sides; haunched one side** | | |
| Concrete 150 mm thick | | |
|    4.00 m wide | m | 240.00 |
|    5.00 m wide | m | 265.00 |
|    6.00 m wide | m | 360.00 |
|    7.00 m wide | m | 380.00 |
| Concrete 250 mm thick | | |
|    4.00 m wide | m | 305.00 |
|    5.00 m wide | m | 355.00 |
|    6.00 m wide | m | 420.00 |
|    7.00 m wide | m | 480.00 |
| | | |
| **FOOTPATHS** | | |
| | | |
| Note: For pathway surfaces; select appropriate finish from the Measured Works section | | |
| | | |
| **Excavate 250 mm; disposal off site; grade to levels and falls; lay 150 mm Type 1 granular material; all disposal off site; Lay or fix edge as described** | | |
| Lay Permaloc Asphalt edge 100 mm metal edgings; path width | | |
|    1.50 m wide | m | 62.00 |
|    3.00 m wide | m | 86.00 |
| Lay timber edging 150 × 38 mm; path width | | |
|    1.50 m wide | m | 35.00 |
|    3.00 m wide | m | 59.00 |
| Lay precast edgings kerbs 50 × 150 mm on both sides; including foundations haunched one side in 11.5 N/mm² concrete with all necessary formwork; path width | | |
|    1.50 m wide | m | 85.00 |
|    3.00 m wide | m | 110.00 |
| Lay concrete edging Keyblock header edging; path width | | |
|    1.50 m wide | m | 55.00 |
|    3.00 m wide | m | 79.00 |
| Lay concrete edging Keyblock stretcher edging; path width | | |
|    1.50 m wide | m | 58.00 |
|    3.00 m wide | m | 100.00 |
| Lay brick on flat header edging; brick £300.00/1000; path width | | |
|    1.50 m wide | m | 85.00 |
|    3.00 m wide | m | 110.00 |

## 8.2.1 ROADS, PATHS, PAVINGS AND SURFACINGS

| Item<br>Excluding site overheads and profit | Unit | Range £ |
|---|---|---|
| Lay brick on flat header edging; brick £600.00/1000; path width | | |
|   1.50 m wide | m | 91.00 |
|   3.00 m wide | m | 115.00 |
| Lay brick on edge header edging; brick £300.00/1000; path width | | |
|   1.50 m wide | m | 100.00 |
|   3.00 m wide | m | 125.00 |
| Lay brick on edge header edging; brick £600.00/1000; path width | | |
|   1.50 m wide | m | 110.00 |
|   3.00 m wide | m | 130.00 |
| Lay brick on flat or brick on edge stretcher edging; brick £300.00/1000; path width | | |
|   1.50 m wide | m | 74.00 |
|   3.00 m wide | m | 98.00 |
| Lay brick on edge strecher edging; brick £600.00/1000; path width | | |
|   1.50 m wide | m | 77.00 |
|   3.00 m wide | m | 100.00 |
| Lay new granite setts 100 × 100 × 100 mm single row; path width | | |
|   1.50 m wide | m | 130.00 |
|   3.00 m wide | m | 160.00 |
| Lay new granite setts 200 × 100 × 100 mm single row; path width | | |
|   1.50 m wide | m | 105.00 |
|   3.00 m wide | m | 130.00 |
| **BRICK PAVING** | | |
| **WORKS BY MACHINE** | | |
| **Excavate and lay base Type 1 150 mm thick; remove arisings; all by machine; lay clay brick paving** | | |
| 200 × 100 × 50 mm thick; butt jointed on 50 mm sharp sand bed | | |
|   PC £300.00/1000 | m² | 78.00 |
|   PC £600.00/1000 | m² | 93.00 |
| 200 × 100 × 50 mm thick; 10 mm mortar joints on 35 mm mortar bed | | |
|   PC £300.00/1000 | m² | 98.00 |
|   PC £600.00/1000 | m² | 110.00 |
| **Excavate and lay base Type 1 250 mm thick all by machine; remove arisings; lay clay brick paving** | | |
| 200 × 100 × 50 mm thick; 10 mm mortar joints on 35 mm mortar bed | | |
|   PC £300.00/1000 | m² | 110.00 |
| **Excavate and lay 150 mm readymix concrete base reinforced with A393 mesh; all by machine; remove arisings; lay clay brick paving** | | |
| 200 × 100 × 50 mm thick; 10 mm mortar joints on 35 mm mortar bed; running or stretcher bond | | |
|   PC £300.00/1000 | m² | 115.00 |
|   PC £600.00/1000 | m² | 130.00 |
| 200 × 100 × 50 mm thick; 10 mm mortar joints on 35 mm mortar bed; butt jointed; herringbone bond | | |
|   PC £300.00/1000 | m² | 97.00 |
|   PC £600.00/1000 | m² | 110.00 |

## 8.2.1 ROADS, PATHS, PAVINGS AND SURFACINGS

| Item<br>Excluding site overheads and profit | Unit | Range £ |
|---|---|---|
| **BRICK PAVING – cont** | | |
| **Excavate and lay base readymix concrete base 150 mm thick reinforced with A393 mesh; all by machine; remove arisings; lay clay brick paving** 215 × 102.5 × 50 mm thick; 10 mm mortar joints on 35 mm mortar bed | | |
| PC £300.00/1000; herringbone | m² | 120.00 |
| PC £600.00/1000; herringbone | m² | 130.00 |
| **PATHS** | | |
| **Brick paved paths; bricks 215 × 112.5 × 65 mm with mortar joints; prices are inclusive of excavation, disposal off site, 100 mm hardcore, 100 mm 1:2:4 concrete bed; edgings of brick 215 mm wide, haunched; all jointed in cement:lime:sand mortar (1:1:6)** | | |
| Path 1015 mm wide laid stretcher bond; edging course of headers | | |
| rough stocks PC £450.00/1000 | m | 105.00 |
| engineering brick | m | 100.00 |
| Path 1015 mm wide laid stack bond | | |
| rough stocks PC £450.00/1000 | m | 110.00 |
| engineering brick | m | 110.00 |
| Path 1115 mm wide laid header bond | | |
| rough stocks PC £450.00/1000 | m | 115.00 |
| engineering brick | m | 110.00 |
| Path 1330 mm wide laid basketweave bond | | |
| rough stocks PC £450.00/1000 | m | 135.00 |
| engineering brick | m | 130.00 |
| Path 1790 mm wide laid basketweave bond | | |
| rough stocks PC £450.00/1000 | m | 180.00 |
| engineering brick | m | 130.00 |
| **Brick paved paths; brick paviors 200 × 100 × 50 mm chamfered edge with butt joints; prices are inclusive of excavation, 100 mm Type 1 and 50 mm sharp sand; jointing in kiln dried sand brushed in; exclusive of edge restraints** | | |
| Path 1000 mm wide laid stretcher bond; edging course of headers | | |
| rough stocks PC £450.00/1000 | m | 95.00 |
| engineering brick | m | 105.00 |
| Path 1330 mm wide laid basketweave bond | | |
| rough stocks PC £450.00/1000 | m | 140.00 |
| engineering brick | m | 130.00 |
| Path 1790 mm wide laid basketweave bond | | |
| rough stocks PC £450.00/1000 | m | 180.00 |
| engineering brick | m | 135.00 |

## 8.2.1 ROADS, PATHS, PAVINGS AND SURFACINGS

| Item<br>Excluding site overheads and profit | Unit | Range £ |
|---|---|---|
| **WORKS BY HAND** | | |
| Excavate and lay base Type 1 150 mm thick by hand; arisings barrowed to spoil heap maximum distance 25 m and removal off site by grab; lay clay brick paving | | |
| 200 × 100 × 50 mm thick; butt jointed on 50 mm sharp sand bed | | |
|   PC £300.00/1000 | m² | 115.00 |
|   PC £600.00/1000 | m² | 150.00 |
| Excavate and lay 150 mm concrete base; 1:3:6 site mixed concrete reinforced with A393 mesh; remove arisings to stockpile and then off site by grab; lay clay brick paving | | |
| 215 × 102.5 × 50 mm thick; 10 mm mortar joints on 35 mm mortar bed | | |
|   PC £300.00/1000 | m² | 110.00 |
|   PC £600.00/1000 | m² | 145.00 |
| **BLOCK PAVING** | | |
| **CONCRETE SETT PAVING** | | |
| Excavate ground and bring to levels; supply and lay 150 mm granular fill Type 1 laid to falls and compacted; supply and lay setts bedded in 50 mm sand and vibrated; joints filled with dry sand and vibrated; all excluding edgings or kerbs measured separately | | |
| Marshalls Mono; Tegula precast concrete setts | | |
|   random sizes 60 mm thick | m² | 61.00 |
|   single size 60 mm thick | m² | 58.00 |
|   random size 80 mm thick | m² | 66.00 |
|   single size 80 mm thick | m² | 80.00 |
| **CAR PARKS** | | |
| Excavate 450 mm and prepare base of 100 hardcore and 250 mm Type 1; Work to falls, crossfalls and cambers not exceeding 15° mark out car parking bays 5.0 × 2.4 m with thermoplastic road paint; surfaces all mechanically laid; edgings not included | | |
| Marshalls Keyblock concrete blocks 200 × 100 × 80 mm; grey | | |
|   per bay; 5.0 × 2.4 m | each | 950.00 |
|   gangway | m² | 81.00 |
| Marshalls Keyblock concrete blocks 200 × 100 × 80 mm; colours | | |
|   per bay; 5.0 × 2.4 m | each | 960.00 |
|   gangway | m² | 81.00 |
| Marshalls Keyblock concrete blocks 200 × 100 × 60 mm; grey | | |
|   per bay; 5.0 × 2.4 m | each | 930.00 |
|   gangway | m² | 79.00 |
| Marshalls Keyblock concrete blocks 200 × 100 × 60 mm; colours | | |
|   per bay; 5.0 × 2.4 m | each | 930.00 |
|   gangway | m² | 80.00 |

## 8.2.1 ROADS, PATHS, PAVINGS AND SURFACINGS

| Item<br>Excluding site overheads and profit | Unit | Range £ |
|---|---|---|
| **BLOCK PAVING – cont** | | |
| **Excavate 450 mm and prepare base of 100 hardcore and 250 mm Type 1 – cont** | | |
| Marshalls Tegula concrete blocks 200 × 100 × 80 mm; random sizes | | |
| per bay; 5.0 × 2.4 m | each | 1075.00 |
| gangway | m² | 90.00 |
| Marshalls Tegula concrete blocks 200 × 100 × 80 mm; single size | | |
| per bay; 5.0 × 2.4 m | each | 1025.00 |
| gangway | m² | 88.00 |
| **BLOCK PAVING EDGINGS** | | |
| **Edge restraint to block paving; excavate for groundbeam; lay concrete 1:2:4 200 mm wide × 150 mm deep; on 50 mm thick sharp sand bed; inclusive of haunching one side** | | |
| Blocks 200 × 100 × 60 mm; PC £10.51/m²; butt jointed | | |
| header course | m | 19.30 |
| stretcher course | m | 15.20 |
| Bricks 215 × 112.5 × 50 mm; PC £300.00/1000; with mortar joints | | |
| header course | m | 20.00 |
| stretcher course | m | 19.70 |
| **Excavate ground; supply and lay granular fill Type 1 150 mm thick laid to falls and compacted; supply and lay block pavers; laid on 50 mm compacted sharp sand; vibrated; joints filled with loose sand excluding edgings or kerbs measured separately** | | |
| Concrete blocks | | |
| 200 × 100 × 60 mm | m² | 54.00 |
| 200 × 100 × 80 mm | m² | 57.00 |
| **Reduce levels; lay 150 mm granular material Type 1; lay vehicular block paving to 90° herringbone pattern; on 50 mm compacted sand bed; vibrated; jointed in sand and vibrated; excavate and lay precast concrete edging 50 × 150 mm; on concrete foundation 1:2:4** | | |
| Blocks 200 × 100 × 60 mm | | |
| 1.0 m wide clear width between edgings | m | 110.00 |
| 1.5 m wide clear width between edgings | m | 140.00 |
| 2.0 m wide clear width between edgings | m | 170.00 |
| Blocks 200 × 100 × 60 mm but blocks laid 45° herringbone pattern including cutting edging blocks | | |
| 1.0 m wide clear width between edgings | m | 120.00 |
| 1.5 m wide clear width between edgings | m | 150.00 |
| 2.0 m wide clear width between edgings | m | 175.00 |
| Blocks 200 × 100 × 60 mm but all excavation by hand disposal off site by grab | | |
| 1.0 m wide clear width between edgings | m | 260.00 |
| 1.5 m wide clear width between edgings | m | 300.00 |
| 2.0 m wide clear width between edgings | m | 340.00 |

## 8.2.1 ROADS, PATHS, PAVINGS AND SURFACINGS

| Item<br>Excluding site overheads and profit | Unit | Range £ |
|---|---|---|
| **MACADAM SURFACES** | | |
| **Macadam roadway over 1000 m$^2$** | | |
| Excavate 350 mm for pathways or roadbed; bring to grade; lay 100 mm well-rolled hardcore; lay 150 mm Type 1 granular material; lay precast edgings kerbs 50 × 150 mm on both sides; including foundations haunched one side in 11.5 N/mm$^2$ concrete with all necessary formwork; machine lay surface of 90 mm macadam 60 mm base course and 30 mm wearing course; all disposal off site | | |
| 1.50 m wide | m | 135.00 |
| 2.00 m wide | m | 145.00 |
| 3.00 m wide | m | 205.00 |
| 4.00 m wide | m | 250.00 |
| 5.00 m wide | m | 300.00 |
| 6.00 m wide | m | 350.00 |
| 7.00 m wide | m | 400.00 |
| **Macadam roadway 400–1000 m$^2$** | | |
| Excavate 350 mm for pathways or roadbed; bring to grade; lay 100 mm well-rolled hardcore; lay 150 mm Type 1 granular material; lay precast edgings kerbs 50 × 150 mm on both sides; including foundations haunched one side in 11.5 N/mm$^2$ concrete with all necessary formwork; machine lay surface of 90 mm macadam 60 mm base course and 30 mm wearing course; all disposal off site | | |
| 1.50 m wide | m | 140.00 |
| 2.00 m wide | m | 165.00 |
| 3.00 m wide | m | 220.00 |
| 4.00 m wide | m | 270.00 |
| 5.00 m wide | m | 325.00 |
| 6.00 m wide | m | 380.00 |
| 7.00 m wide | m | 430.00 |
| **MACADAM FOOTPATH** | | |
| **Excavate footpath to reduce level; remove arisings to tip on site maximum distance 25 m; lay Type 1 granular fill 100 mm thick; lay base course of 28 mm size dense bitumen macadam 50 mm thick; wearing course of 10 mm size dense bitumen macadam 20 mm thick; timber edge 150 × 38 mm** | | |
| Areas over 1000 m$^2$ | | |
| 1.0 m wide | m | 41.00 |
| 1.5 m wide | m | 57.00 |
| 2.0 m wide | m | 73.00 |
| Areas 400–1000 m$^2$ | | |
| 1.0 m wide | m | 43.00 |
| 1.5 m wide | m | 60.00 |
| 2.0 m wide | m | 95.00 |

## 8.2.1 ROADS, PATHS, PAVINGS AND SURFACINGS

| Item<br>Excluding site overheads and profit | Unit | Range £ |
|---|---|---|
| **MACADAM SURFACES – cont** | | |
| **CAR PARKS** | | |
| Excavate 450 mm and prepare base of 100 hardcore and 250 mm Type 1; Work to falls, crossfalls and cambers not exceeding 15° mark out car parking bays 5.0 × 2.4 m with thermoplastic road paint; surfaces all mechanically laid; edgings not included | | |
| Macadam base course 60 mm thick and wearing course 40 mm thick | | |
|    per bay; 5.0 × 2.4 m | each | 820.00 |
|    gangway | m² | 68.00 |
| Bay for disabled people; Macadam base course 60 mm thick and wearing course 40 mm thick | | |
|    per bay; 5.80 × 3.25 ambulant | each | 1275.00 |
|    per bay; 6.55 × 3.80 wheelchair | each | 1650.00 |
| **Excavate 280 mm and prepare base of 150 mm Type 1** | | |
| Surface treatments as shown | | |
|    Macadam 60:40 | m² | 49.50 |
|    Tegula setts 80 mm; random sizes | m² | 72.00 |
|    Tegula setts 80 mm; single size | m² | 69.00 |
|    Keyblock 80 mm; Colours | m² | 62.00 |
|    Keyblock 80 mm; Grey | m² | 62.00 |
| **IN SITU CONCRETE PAVINGS AND SURFACES** | | |
| **CONCRETE PAVING** | | |
| **Pedestrian areas** | | |
| Excavate to reduce levels; lay 100 mm Type 1; lay PAV 1 air entrained concrete; joints at maximum width of 6.0 m cut out and sealed with sealant; inclusive of all formwork and stripping | | |
|    100 mm thick | m² | 55.00 |
| **Trafficked areas** | | |
| Excavate to reduce levels; lay 150 mm Type 1 lay PAV 2 40 N/mm² air entrained concrete; reinforced with steel mesh 200 × 200 mm square at 2.22 kg/m²; joints at max width of 6.0 m cut out and sealed with sealant | | |
|    150 mm thick | m² | 76.00 |
| **Reinforced concrete paths;** excavate path and dispose of excavated material off site; bring to grade; lay reinforcing mesh A142 lapped and joined; lay in situ reinforced concrete PAV1 35 N/mm²; with 15 impregnated fibreboard expansion joints sealant at 50 m centres; on 100 mm hardcore blinded sand; edgings 150 × 50 mm to both sides; including foundations haunched one side in 11.5 N/mm² concrete with all necessary formwork | | |
| Concrete 150 mm thick | | |
|    1.00 m wide | m | 73.00 |
|    2.00 m wide | m | 115.00 |
|    3.00 m wide | m | 160.00 |

## 8.2.1 ROADS, PATHS, PAVINGS AND SURFACINGS

| Item<br>Excluding site overheads and profit | Unit | Range £ |
|---|---|---|
| **WORKS BY HAND; Ready mixed concrete – mixed on site** | | |
| Reinforced concrete paths; excavate path by hand barrow to spoil heap 25 m and dispose by grab off site; bring to grade; lay reinforcing mesh A142 lapped and joined; lay in situ reinforced ready mixed concrete mixed on site (1:2:4:); with 15 impregnated fibreboard expansion joints sealant at 50 m centres; on 100 mm hardcore blinded sand; edgings 150 × 50 mm to both sides; including foundations haunched one side in 11.5 N/mm² concrete with all necessary formwork | | |
| Concrete 150 mm thick | | |
|    1.00 m wide | m | 100.00 |
|    2.00 m wide | m | 190.00 |
|    3.00 m wide | m | 270.00 |
| **GRASS CONCRETE** | | |
| **IN SITU GRASS CONCRETE TO CAR PARKS** | | |
| Grass Concrete Grasscrete in situ continuously reinforced cellular surfacing; fill with topsoil and peat (5:1) and fertilizer at 35 g/m²; seed with dwarf rye grass at 35 g/m²; Excludes marking | | |
|    GC1; 100 mm thick for cars and light traffic; per bay; 5.0 × 2.4 m | each | 930.00 |
|    GC1; 100 mm thick for cars and light traffic; gangway | m² | 76.00 |
| GC2; 150 mm thick for HGV traffic including dust carts | | |
|    per bay; 5.0 × 2.4 m | each | 1075.00 |
|    gangway | m² | 90.00 |
| Car park as above but with interlocking concrete blocks 200 × 100 × 60 mm; grey; but mark out bays | | |
|    per bay; 5.80 × 3.25 ambulant | each | 630.00 |
|    per bay; 6.55 × 3.80 wheelchair | each | 830.00 |
|    gangway | m² | 33.50 |
| **MODULAR GRASS CONCRETE PAVING** | | |
| Reduce levels; lay 150 mm Type 1 granular fill compacted; supply and lay precast grass concrete blocks on 20 sand and level by hand; fill blocks with 3 mm sifted topsoil and pre-seeding fertilizer at 50 g/m²; sow with perennial ryegrass/chewings fescue seed at 35 g/m² | | |
| Grass concrete | | |
|    GB103 406 × 406 × 103 mm | m² | 58.00 |
|    GB83 406 × 406 × 83 mm | m² | 56.00 |
|    extra for geotextile fabric underlayer | m² | 0.80 |
| **Firepaths** | | |
| Excavate to reduce levels; lay 300 mm well rammed hardcore; blinded with 100 mm type 1; supply and lay Marshalls Mono Grassguard 180 precast grass concrete blocks on 50 mm sand and level by hand; fill blocks with sifted topsoil and pre-seeding fertilizer at 50 g/m² | | |
|    firepath 3.8 m wide | m | 335.00 |
|    firepath 4.4 m wide | m | 390.00 |
|    firepath 5.0 m wide | m | 440.00 |
|    turning areas | m² | 89.00 |
| Charcon Hard Landscaping Grassgrid; 366 × 274 × 100 mm | m² | 43.00 |

## 8.2.1 ROADS, PATHS, PAVINGS AND SURFACINGS

| Item<br>Excluding site overheads and profit | Unit | Range £ |
|---|---|---|
| **RESIN BOUND AND RESIN BONDED SURFACES** | | |
| **Resin bonded surfaces; excavate 230 mm; bring to grade; lay 150 mm Type 1 granular material; hand lay base course of 60 mm macadam and resin bonded wearing course; all disposal off site; edgings excluded** | | |
| Natratex resin bonded macadam wearing course | | |
|    general paved area | $m^2$ | 61.00 |
|    pathway 1.50 m wide | m | 98.00 |
|    pathway 3.00 m wide | m | 195.00 |
| Addagrip; resin bonded aggregate; 1–3 mm golden pea gravel with buff adhesive | | |
|    general paved area | $m^2$ | 68.00 |
|    pathway 1.50 m wide | m | 110.00 |
|    pathway 3.00 m wide | m | 215.00 |
| Addagrip; resin bonded aggregate; 6 mm chocolate gravel 18 mm depth | | |
|    general paved area | $m^2$ | 98.00 |
|    pathway 1.50 m wide | m | 150.00 |
|    pathway 3.00 m wide | m | 305.00 |
| Addagrip; resin bonded aggregate; Cobalt blue recycled glass 15 mm depth | | |
|    general paved area | $m^2$ | 82.00 |
|    pathway 1.50 m wide | m | 130.00 |
|    pathway 3.00 m wide | m | 260.00 |
| **Resin bound surfaces; excavate 230 mm; bring to grade; lay 150 mm Type 1 granular material; hand lay base course of 60 mm macadam and resin bound wearing course; all disposal off site; edgings excluded** | | |
| Sureset; resin bound permeable surfacing; natural gravel 6 mm aggregate 18 mm thick | | |
|    paved area | $m^2$ | 81.00 |
|    pathway 1.50 m wide | m | 130.00 |
|    pathway 3.00 m wide | m | 255.00 |
| Sureset; resin bound permeable surfacing; crushed rock 6 mm aggregate 18 mm thick | | |
|    paved area | $m^2$ | 110.00 |
|    pathway 1.50 m wide | m | 130.00 |
|    pathway 3.00 m wide | m | 420.00 |
| Sureset; resin bound permeable surfacing; marble 6 mm aggregate 18 mm thick | | |
|    paved area | $m^2$ | 87.00 |
|    pathway 1.50 m wide | m | 135.00 |
|    pathway 3.00 m wide | m | 270.00 |
| Sureset; resin bound permeable surfacing; recycled glass 6 mm aggregate 18 mm thick | | |
|    paved area | m | 100.00 |
|    pathway 1.50 m wide | m | 160.00 |
|    pathway 3.00 m wide | m | 315.00 |

*Approximate Estimating Rates*

## 8.2.1 ROADS, PATHS, PAVINGS AND SURFACINGS

| Item<br>Excluding site overheads and profit | Unit | Range £ |
|---|---|---|
| **GRAVEL PATHWAYS** | | |
| Reduce levels and remove spoil to dump on site; lay 150 mm hardcore well rolled; lay geofabric; fix timber edge 150 × 38 mm to both sides of straight paths | | |
| Lay Cedec gravel 50 mm thick; watered and rolled | | |
|    1.0 m wide | m | 33.00 |
|    1.5 m wide | m | 43.00 |
|    2.0 m wide | m | 60.00 |
| Lay Breedon gravel 50 mm thick; watered and rolled | | |
|    1.0 m wide | m | 29.00 |
|    1.5 m wide | m | 38.00 |
|    2.0 m wide | m | 47.50 |
| Reduce levels and remove spoil to dump on site; lay 150 mm Type 1 well rolled; lay geofabric; fix timber edge 150 × 38 mm to both sides of straight paths | | |
| Lay Cedec gravel 50 mm thick; watered and rolled | | |
|    1.0 m wide | m | 38.00 |
|    1.5 m wide | m | 51.00 |
|    2.0 m wide | m | 70.00 |
| Lay Breedon gravel 50 mm thick; watered and rolled | | |
|    1.0 m wide | m | 34.00 |
|    1.5 m wide | m | 46.00 |
|    2.0 m wide | m | 58.00 |
| Lay 20 mm shingle; 30 mm thick | | |
|    1.0 m wide | m | 26.50 |
|    1.5 m wide | m | 33.50 |
|    2.0 m wide | m | 47.00 |
| **CELLULAR CONFINED SURFACES** | | |
| Excavate ground and reduce levels to receive cellular confined paving; dispose of excavated material off site | | |
| Lay granular fill Type 1; limestone aggregate 150 mm thick laid to falls and compacted; lay confined cellular surface | | |
|    Geoweb 100 thick | m² | 44.00 |
|    StablePATH honeycomb cellular gravel grid; 25 mm deep for foot traffic and bicycles filled with shingle | m² | 35.00 |
|    StablePATH honeycomb cellular gravel grid; 25 mm deep for foot traffic and bicycles filled with decorative aggregate | m² | 36.50 |
|    StablePATH honeycomb cellular gravel grid 40 mm deep for vehicular traffic; filled with shingle | m² | 37.00 |
|    StablePATH honeycomb cellular gravel grid 40 mm deep for vehicular traffic; filled with decorative aggregate | m² | 38.50 |

## 8.2.1 ROADS, PATHS, PAVINGS AND SURFACINGS

| Item<br>Excluding site overheads and profit | Unit | Range £ |
|---|---|---|
| **SLAB PAVINGS** | | |
| **SLAB/BRICK PATHS** | | |
| **Stepping stone path inclusive of hand excavation and mechanical disposal, 100 mm Type 1 and sand blinding; slabs laid 100 mm apart to existing turf** | | |
| 600 mm wide; 600 × 600 × 50 mm slabs | | |
|     natural finish | m | 59.00 |
|     coloured | m | 59.00 |
|     exposed aggregate | m | 68.00 |
| 900 mm wide; 600 × 900 × 50 mm slabs | | |
|     natural finish | m | 64.00 |
|     coloured | m | 67.00 |
|     exposed aggregate | m | 80.00 |
| **Pathway inclusive of mechanical excavation and disposal; 100 mm Type 1 and sand blinding; slabs close butted** | | |
| Straight butted path 900 mm wide; 600 × 900 × 50 mm slabs | | |
|     natural finish | m | 28.00 |
|     coloured | m | 29.50 |
|     exposed aggregate | m | 38.50 |
| Straight butted path 1200 mm wide; double row; 600 × 900 × 50 mm slabs; laid stretcher bond | | |
|     natural finish | m | 47.50 |
|     coloured | m | 51.00 |
|     exposed aggregate | m | 71.00 |
| Straight butted path 1200 mm wide; one row of 600 × 600 × 50 mm; one row 600 × 900 × 50 mm slabs | | |
|     natural finish | m | 46.00 |
|     coloured | m | 50.00 |
|     exposed aggregate | m | 79.00 |
| Straight butted path 1500 mm wide; slabs of 600 × 900 × 50 mm and 600 × 600 × 50 mm; laid to bond | | |
|     natural finish | m | 57.00 |
|     coloured | m | 61.00 |
|     exposed aggregate | m | 79.00 |
| Straight butted path 1800 mm wide; two rows of 600 × 900 × 50 mm slabs; laid bonded | | |
|     natural finish | m | 63.00 |
|     coloured | m | 77.00 |
|     exposed aggregate | m | 105.00 |
| Straight butted path 1800 mm wide; three rows of 600 × 900 × 50 mm slabs; laid stretcher bond | | |
|     natural finish | m | 69.00 |
|     coloured | m | 82.00 |
|     exposed aggregate | m | 110.00 |

## 8.2.1 ROADS, PATHS, PAVINGS AND SURFACINGS

| Item<br>Excluding site overheads and profit | Unit | Range £ |
|---|---|---|
| **FLAG PAVING TO PEDESTRIAN AREAS** | | |
| Prices are inclusive of all mechanical excavation and disposal. | | |
| **Supply and lay precast concrete flags; excavate ground and reduce levels; treat substrate with total herbicide; supply and lay granular fill Type 1 150 mm thick laid to falls and compacted** | | |
| Standard precast concrete flags bedded and jointed in lime:sand mortar (1:3) | | |
|    450 × 450 × 50 mm; chamfered | m² | 52.00 |
|    600 × 300 × 50 mm | m² | 49.00 |
|    600 × 450 × 50 mm | m² | 48.00 |
|    600 × 600 × 50 mm | m² | 44.00 |
|    750 × 600 × 50 mm | m² | 43.00 |
|    900 × 600 × 50 mm | m² | 42.00 |
| Coloured flags bedded and jointed in lime:sand mortar (1:3) | | |
|    600 × 600 × 50 mm | m² | 47.00 |
|    400 × 400 × 65 mm | m² | 55.00 |
|    750 × 600 × 50 mm | m² | 48.00 |
|    900 × 600 × 50 mm | m² | 46.00 |
| Marshalls Saxon; textured concrete flags; reconstituted Yorkstone in colours; butt jointed bedded in lime:sand mortar (1:3) | | |
|    300 × 300 × 35 mm | m² | 88.00 |
|    450 × 450 × 50 mm | m² | 79.00 |
|    600 × 300 × 35 mm | m² | 70.00 |
|    600 × 600 × 50 mm | m² | 66.00 |
| Tactile flags; Marshalls blister tactile pavings; red or buff; for blind pedestrian guidance laid to designed pattern | | |
|    450 × 450 mm | m² | 67.00 |
|    400 × 400 mm | m² | 73.00 |
| **PEDESTRIAN DETERRENT PAVING** | | |
| **Excavate ground and bring to levels; treat substrate with total herbicide; supply and lay granular fill Type 1 150 mm thick laid to falls and compacted; supply and lay precast deterrent paving units bedded in lime:sand mortar (1:3) and jointed in lime:sand mortar (1:3)** | | |
| Marshalls Mono | | |
|    Contour deterent paving; 600 × 600 × 75 mm | m² | 64.00 |
| Townscape | | |
|    Abbey square cobble pattern pavings; reinforced; 600 × 600 × 60 mm | m² | 140.00 |
|    Geoset chamfered studs; 600 × 600 × 60 mm | m² | 140.00 |

## 8.2.1 ROADS, PATHS, PAVINGS AND SURFACINGS

| Item<br>Excluding site overheads and profit | Unit | Range £ |
|---|---|---|
| **BARK PAVINGS** | | |
| Excavate to reduce levels; remove all topsoil to dump on site; treat area with herbicide; lay 150 mm clean hardcore; blind with sand; lay 0.7 mm geotextile filter fabric water flow 50 l/m²/sec; supply and fix treated softwood edging boards 50 × 150 mm to bark area on hardcore base extended 150 mm beyond the bark area; boards fixed with galvanized nails to treated softwood posts 750 × 50 × 75 mm driven into firm ground at 1.0 m centres; edging boards to finish 25 mm above finished bark surface; tops of posts to be flush with edging boards and once weathered | | |
| Supply and lay 100 mm Melcourt conifer walk chips 10–40 mm size | | |
|   1.00 m wide | m | 30.00 |
|   2.00 m wide | m | 45.00 |
|   3.00 m wide | m | 60.00 |
|   4.00 m wide | m | 72.00 |
| Extra over for wood fibre | m² | -1.55 |
| Extra over for hardwood chips | m² | -1.37 |
| **COBBLE PAVINGS** | | |
| **BEACH COBBLE PAVING** | | |
| Excavate ground and bring to levels and fill with compacted Type 1 fill 100 mm thick; lay GEN 1 concrete base 100 mm thick; supply and lay cobbles individually laid by hand bedded in cement:sand mortar (1:3) 25 mm thick minimum; dry grout with 1:3 cement:sand grout; brush off surplus grout and water in; sponge off cobbles as work proceeds; all excluding formwork edgings or kerbs measured separately | | |
| Scottish beach cobbles 50–75 mm | m² | 140.00 |
| Scottish beach cobbles 100–200 mm | m² | 165.00 |
| Kidney flint cobbles 75–100 mm | m² | 130.00 |
| **NATURAL STONE PAVING** | | |
| **GRANITE SETT PAVING – PEDESTRIAN** | | |
| Excavate and grade; 100 mm hardcore; Type 1 50 mm thick; 100 mm concrete 1:2:4; supply and lay granite setts 100 × 100 × 100 mm bedded in cement:sand mortar (1:3) 25 mm thick minimum; jointed in mortar 1:3 | | |
| Setts laid to bonded pattern and jointed; bedded in cement:sand mortar 1:3 25 mm thick | | |
|   new setts 100 × 100 × 100 mm | m² | 140.00 |
|   second-hand cleaned 100 × 100 × 100 mm | m² | 130.00 |
| Setts laid to bonded pattern and jointed; bedded in SteinTec mortar 25 mm thick | | |
|   new setts 100 × 100 × 100 mm | m² | 140.00 |
|   second-hand cleaned 100 × 100 × 100 mm | m² | 130.00 |
| Setts laid in curved pattern | | |
|   new setts 100 × 100 × 100 mm | m² | 150.00 |
|   second-hand cleaned 100 × 100 × 100 mm | m² | 140.00 |

## 8.2.1 ROADS, PATHS, PAVINGS AND SURFACINGS

| Item<br>Excluding site overheads and profit | Unit | Range £ |
|---|---|---|
| **GRANITE SETT PAVING – TRAFFICKED AREAS** | | |
| **Excavate to reduce levels; lay 100 mm hardcore; compact; Type 150 mm thick; lay 150 mm site mixed concrete 1:2:4 reinforced with steel fabric; granite setts 100 × 100 × 100 mm bedded in cement:sand mortar (1:3) 25 mm thick minimum; jointed in mortar 1:3; all excluding edgings or kerbs measured separately** | | |
| Site mixed concrete | | |
| new setts 100 × 100 × 100 mm | m² | 150.00 |
| second-hand cleaned 100 × 100 × 100 mm | m² | 140.00 |
| Ready mixed concrete | | |
| new setts 100 × 100 × 100 mm | m² | 150.00 |
| second-hand cleaned 100 × 100 × 100 mm | m² | 140.00 |
| **NATURAL STONE SLAB PAVING** | | |
| **WORKS BY MACHINE** | | |
| **Excavate ground by machine and reduce levels; to receive 65 mm thick slab and 35 mm mortar bed; dispose of excavated material off site; treat substrate with total herbicide; lay granular fill Type 1 150 mm thick laid to falls and compacted; lay to random rectangular pattern on 35 mm mortar bed** | | |
| New riven slabs | | |
| laid random rectangular on 35 mm 1:3 mortar bed | m² | 170.00 |
| laid random rectangular on 30 mm thick Steintec mortar bed | m² | 180.00 |
| New riven slabs; but to 150 mm plain concrete base | | |
| laid random rectangular on 35 mm 1:3 mortar bed | m² | 180.00 |
| laid random rectangular on 30 mm thick Steintec mortar bed | m² | 190.00 |
| New riven slabs; disposal by grab | | |
| laid random rectangular | m² | 170.00 |
| Reclaimed Cathedral grade riven slabs | | |
| laid random rectangular | m² | 200.00 |
| Reclaimed Cathedral grade riven slabs; but to 150 mm plain concrete base | | |
| laid random rectangular | m² | 210.00 |
| Reclaimed Cathedral grade riven slabs; disposal by grab | | |
| laid random rectangular | m² | 200.00 |
| New slabs sawn 6 sides | | |
| laid random rectangular | m² | 140.00 |
| three sizes; laid to coursed pattern | m² | 140.00 |
| New slabs sawn 6 sides; but to 150 mm plain concrete base | | |
| laid random rectangular | m² | 150.00 |
| three sizes; laid to coursed pattern | m² | 150.00 |
| New slabs sawn 6 sides; disposal by grab | | |
| laid random rectangular | m² | 140.00 |
| 3 sizes, laid to coursed pattern | m² | 135.00 |

## 8.2.1 ROADS, PATHS, PAVINGS AND SURFACINGS

| Item<br>Excluding site overheads and profit | Unit | Range £ |
|---|---|---|
| **NATURAL STONE PAVING – cont** | | |
| **WORKS BY HAND** | | |
| Excavate ground by hand and reduce levels, to receive 65 mm thick slab and 35 mm mortar bed; barrow all materials and arisings 25 m; dispose of excavated material off site by grab; treat substrate with total herbicide; lay granular fill Type 1 150 mm thick laid to falls and compacted; lay to random rectangular pattern on 35 mm mortar bed | | |
| New riven slabs | | |
|   laid random rectangular | m² | 210.00 |
| New riven slabs laid random rectangular; but to 150 mm plain concrete base | | |
|   laid random rectangular | m² | 220.00 |
| New riven slabs; but disposal to skip | | |
|   laid random rectangular | m² | 230.00 |
| Reclaimed Cathedral grade riven slabs | | |
|   laid random rectangular | m² | 240.00 |
| Reclaimed Cathedral grade riven slabs; but to 150 mm plain concrete base | | |
|   laid random rectangular | m² | 255.00 |
| Reclaimed Cathedral grade riven slabs; but disposal to skip | | |
|   laid random rectangular | m² | 265.00 |
| New slabs sawn 6 sides | | |
|   laid random rectangular | m² | 175.00 |
|   three sizes; sawn 6 sides laid to coursed pattern | m² | 175.00 |
| New slabs sawn 6 sides; but to 150 mm plain concrete base | | |
|   laid random rectangular | m² | 190.00 |
|   three sizes; sawn 6 sides laid to coursed pattern | m² | 190.00 |
| **TIMBER DECKING** | | |
| **HARDWOOD DECKS** | | |
| Timber decking; Exterior Decking Limited; timber decking to roof gardens; cranage or movement of materials to roof top not included; coated one coat with Seasonite | | |
| Joist size 47 × 47 mm | | |
|   Ipe; pre-drilled and countersunk; screw fixed | m² | 170.00 |
|   Ipe; Exterpark; 21 mm invisibly fixed | m² | 185.00 |
|   Ipe; Exterpark; 28 mm invisibly fixed | m² | 250.00 |
|   Ipe; Exterpark; 35 mm invisibly fixed | m² | 320.00 |
|   Rustic Teak; 21 mm invisibly fixed | m² | 200.00 |
|   Merbau; 21mm invisibly fixed | m² | 165.00 |
|   Merbau; 28 mm invisibly fixed | m² | 195.00 |
| Timber decking to commercial applications; double beams at 300 mm centres; coated one coat with Seasonite | | |
| Joist size 47 × 150 mm | | |
|   Ipe; pre-drilled and countersunk; screw fixed | m² | 210.00 |
|   Ipe; Exterpark; 21 mm invisibly fixed | m² | 230.00 |
|   Ipe; Exterpark; 28 mm invisibly fixed | m² | 290.00 |

## 8.2.1 ROADS, PATHS, PAVINGS AND SURFACINGS

| Item<br>Excluding site overheads and profit | Unit | Range £ |
|---|---|---|
| Ipe; Exterpark; 35 mm invisibly fixed | m² | 360.00 |
| Rustic Teak; 21 mm invisibly fixed | m² | 245.00 |
| Merbau; 21mm invisibly fixed | m² | 210.00 |
| Merbau; 28 mm invisibly fixed | m² | 240.00 |
| **SOFTWOOD DECKS** | | |
| **Timber decking; AVS Ltd support structure of timber joists for decking laid at 400 mm centres on blinded base (measured separately)** | | |
| Yellow balau | | |
|   joists 38 × 88 mm | 10 m² | 150.00 |
|   joists 47 × 100 mm | 10 m² | 140.00 |
|   joists 47 × 150 mm | 10 m² | 150.00 |
| As above but boards 141 × 28 mm thick | | |
|   joists 47 × 100 mm | 10 m² | 115.00 |
|   joists 47 × 150 mm | 10 m² | 120.00 |
| As above but decking boards in red cedar 150 mm wide × 32 mm thick | | |
|   joists 38 × 88 mm | 10 m² | 120.00 |
|   joists 47 × 100 mm | 10 m² | 120.00 |
|   joists 47 × 150 mm | 10 m² | 120.00 |
| Add to all of the above for handrails fixed to posts 100 × 100 × 1370 mm high | | |
|   square balusters at 100 mm centres | m | 87.00 |
|   square balusters at 300 mm centres | m | 67.00 |
|   turned balusters at 100 mm centres | m | 120.00 |
|   turned balusters at 300 mm centres | m | 77.00 |
| **Steps** | | |
| **Excavate for new steps; grade to correct levels; remove excavated material from site and lay concrete base; Construct formwork; supply and lay site mixed concrete; Fix treads and risers to steps** | | |
| Stone clad steps; 1.20 m wide; tread depth 225 mm; Granite cut off site | | |
|   1 riser | nr | 295.00 |
|   2 risers | nr | 495.00 |
|   3 risers | nr | 700.00 |
|   4 risers | nr | 950.00 |
|   6 risers | nr | 1350.00 |
| Stone clad steps; 1.20 m wide; tread depth 225 mm; Riven Yorkstone cut on site | | |
|   1 riser | nr | 290.00 |
|   2 risers | nr | 480.00 |
|   3 risers | nr | 680.00 |
|   4 risers | nr | 920.00 |
|   6 risers | nr | 1300.00 |
| Stone clad steps; 2.40 m wide; tread depth 225 mm; Granite cut off site | | |
|   1 riser | nr | 560.00 |
|   2 risers | nr | 950.00 |
|   3 risers | nr | 1350.00 |
|   4 risers | nr | 1875.00 |
|   6 risers | nr | 2700.00 |

## 8.2.1 ROADS, PATHS, PAVINGS AND SURFACINGS

| Item<br>Excluding site overheads and profit | Unit | Range £ |
|---|---|---|
| **PLAY AREA FOR BALL GAMES** | | |
| Excavate for playground; bring to grade; lay 225 mm consolidated hardcore; blind with 100 mm type 1; lay macadam base of 40 mm size dense bitumen macadam to 75 mm thick; lay wearing course 30 mm thick | | |
| Over 1000 m² | 100 m² | 5700.00 |
| 400–1000 m² | 100 m² | 5900.00 |
| Excavate for playground; bring to grade; lay 100 mm consolidated hardcore; blind with 100 mm type 1; lay macadam base of dense bitumen macadam to 50 mm thick; lay wearing course 20 mm thick | | |
| Over 1000 m² | 100 m² | 4250.00 |
| 400–1000 m² | 100 m² | 5100.00 |
| Excavate for playground; bring to grade; lay 100 mm consolidated hardcore; blind with 100 mm type 1; lay macadam base of dense bitumen macadam to 50 mm thick; lay wearing course of Addagrip resin coated aggregate 3 mm diameter × 6 mm thick | | |
| Over 1000 m² | 100 m² | 6800.00 |
| 400–1000 m² | 100 m² | 7700.00 |
| **SAFETY SURFACED PLAY AREAS** | | |
| Excavate ground and reduce levels to receive safety surface; dispose of excavated material off site; treat substrate with total herbicide; lay granular fill Type 1 150 mm thick laid to falls and compacted; lay macadam base 40 mm thick; supply and lay Rubaflex wet pour safety system to thicknesses as specified | | |
| All by machine except macadam by hand | | |
|   black | | |
|     15 mm thick | m² | 65.00 |
|     35 mm thick | m² | 71.00 |
|     60 mm thick | m² | 78.00 |
|   coloured | | |
|     15 mm thick | m² | 72.00 |
|     35 mm thick | m² | 64.00 |
|     60 mm thick | m² | 86.00 |
| All by hand except disposal by grab | | |
|   black | | |
|     15 mm thick | m² | 92.00 |
|     35 mm thick | m² | 98.00 |
|     60 mm thick | m² | 105.00 |
|   coloured | | |
|     15 mm thick | m² | 96.00 |
|     35 mm thick | m² | 105.00 |
|     60 mm thick | m² | 115.00 |
| Excavate for playground; bring to grade; lay with 150 mm type 1; lay macadam base of dense bitumen macadam to 50 mm thick; lay safety surface | | |
|   Wetpour coloured; 0.50 m critical fall height | m | 74.00 |
|   Wetpour black; 0.50 m critical fall height | m | 66.00 |
|   Wetpour coloured; 1.50 m critical fall height | m | 84.00 |
|   Wetpour black; 1.50 m critical fall height | m | 76.00 |

## 8.2.1 ROADS, PATHS, PAVINGS AND SURFACINGS

| Item<br>Excluding site overheads and profit | Unit | Range £ |
|---|---|---|
| Excavate for playground; bring to grade; lay with 150 mm type 1; lay safety surface<br>Meon UK | | |
|    Black; 10 m$^2$–50 m$^2$ | m | 73.00 |
|    Black; over 50 m$^2$ | m | 70.00 |
|    Colours; 10 m$^2$–50 m$^2$ | m | 94.00 |
|    Colours; over 50 m$^2$ | m | 88.00 |
| **BARK PLAY AREA** | | |
| **Excavate playground area to 450 mm depth; lay 150 mm broken stone or clean hardcore; lay filter membrane; lay bark surface**<br>Melcourt Industries | | |
|    Playbark 10/50; 300 mm thick | m$^2$ | 44.50 |
|    Playbark 8/25; 300 mm thick | m$^2$ | 51.00 |

## 8.2.2 SPECIAL SURFACINGS AND PAVINGS

| Item<br>Excluding site overheads and profit | Unit | Range £ |
|---|---|---|
| **SPORTS FIELD CONSTRUCTION** | | |
| Strip existing vegetation; grade to levels using laser guided equipment; cultivate ground; apply pre-seeding fertilizer at 900 kg/ha; apply pre-seeding selective weedkiller; seed in two operations with sports pitch type grass seed at 350 kg/ha; harrow and roll lightly; including initial cut; size; mark out and paint lines; erect goal posts | | |
| New ground construction of existing field; greenfield or brownfield; existing lightweight or loam soils | | |
|     association football; senior 114 × 72 m | each | 31000.00 |
|     association football; junior 106 × 58 m | each | 23000.00 |
| New ground construction of existing field; greenfield or brownfield; existing clay soil; replacement of rootzone 150 mm | | |
|     association football; senior 114 × 72 m | each | 110000.00 |
|     association football; junior 106 × 58 m | each | 85000.00 |
| Plain sports pitches; site clearance, grading and drainage not included | | |
| Agripower Ltd; cultivate ground and grade to levels; apply pre-seeding fertilizer at 900 kg/ha; apply pre-seeding selective weedkiller; seed in two operations with sports pitch type grass seed at 350 kg/ha; harrow and roll lightly; including initial cut; size | | |
| Existing level and graded ground previously used for playing surface | | |
|     association football; senior 114 × 72 m | each | 3500.00 |
|     association football; junior 106 × 58 m | each | 2600.00 |
|     rugby union pitch; 156 × 81 m | each | 5400.00 |
|     rugby league pitch; 134 × 60 m | each | 3400.00 |
|     hockey pitch; 95 × 60 m | each | 2425.00 |
|     shinty pitch; 186 × 96 m | each | 7700.00 |
|     men's lacrosse pitch; 100 × 55 m | each | 2650.00 |
|     women's lacrosse pitch; 110 × 73 m | each | 3400.00 |
|     target archery ground; 150 × 50 m | each | 3100.00 |
|     cricket outfield; 160 × 142 m | each | 3300.00 |
|     cycle track outfield; 160 × 80 m | each | 5400.00 |
|     polo ground; 330 × 220 m | each | 31000.00 |
| Agripower Ltd; cricket square; excavate to depth of 100 mm; lay imported marl or clay loam; bring to accurate levels; apply pre-seeding fertilizer at 50 g/m; apply selective weedkiller; seed with cricket square type g grass seed at 50 g/m²; rake in and roll lightly; erect and remove temporary protective chestnut fencing; allow for initial cut and watering three times 22.8 × 22.8 m | each | 15000.00 |
| **JOGGING TRACK** | | |
| Excavate track 250 mm deep; lay filter membrane; lay 100 mm depth gravel waste or similar; lay 100 mm compacted gravel | 100 m² | 9200.00 |
| Extra for treated softwood edging 50 × 150 mm on 50 × 50 mm posts | | |
|   both sides | m | 10.30 |

## 8.3.1 SOFT LANDSCAPING, SEEDING AND TURFING

| Item<br>Excluding site overheads and profit | Unit | Range £ |
|---|---|---|
| **8.3.1 Soft landscaping, seeding and turfing** | | |
| **Sports ground for amenity sport** | | |
| **To existing field; Spray to remove existing turf. Scarify top 25 mm; Grade and level using ripper power harrow cultivator and stone burier combination inclusive of grass seed and fertilizers** | | |
| Light to medium soils | | |
|    Existing area to 1:100 existing grade | 6000 m$^2$ | 3000.00 |
| Clay soils | | |
|    Existing area to 1:100 existing grade | 6000 m$^2$ | 4000.00 |
| **To existing badly drained field; Spray to remove existing turf. Scarify top 200 mm of clay surface; Grade and level using ripper power harrow cultivator and stone burier combination; Install land drainage with outfall to soakaway of 4 m$^3$; backfil with imported loam 200 mm; seed with grass seed and fertilizers.** | | |
| Clay soils | | |
|    Existing area to 1:100 existing grade | 6000 m$^2$ | – |

## 8.3.1 SOFT LANDSCAPING, SEEDING AND TURFING

| Item<br>Excluding site overheads and profit | Unit | Range £ |
|---|---|---|
| **SURFACE PREPARATION FOR SEEDING AND TURFING** | | |
| **BY MACHINE** | | |
| **Mechanized preparation; Stone burier and cultivator; Treat area with systemic herbicide in advance of operations; rip subsoil using subsoiling machine to a depth of 500 mm at 1.20 m centres in light to medium soils; rotavate to 200 mm deep in two passes** | | |
| Larger areas | | |
|   light to meduim soils | 100 m² | 34.00 |
|   heavier soils | 100 m² | 57.00 |
| **Rip area by tractor and subsoiler; soil preparation by pedestrian operated rotavator; clearance and raking by hand; herbicide application by knapsack sprayer** | | |
| light to medium soils | 100 m² | 87.00 |
| heavier soils | 100 m² | 88.00 |
| **Spread and lightly consolidate topsoil brought from spoil heap not exceeding 100 m; in layers not exceeding 150 mm; grade to specified levels; remove stones over 25 mm; all by machine** | | |
| 100 mm thick | 100 m² | 100.00 |
| **Extra to the above for imported topsoil** | | |
| topsoil PC £51.00 m³ allowing for 20% settlement | | |
|   100 mm thick | 100 m² | 395.00 |
| topsoil to BS3882 PC £51.00 m³ allowing for 20% settlement | | |
|   100 mm thick | 100 m² | 510.00 |
| **Excavate existing topsoil and remove off site; fill with new good quality topsoil to BS3882; add 50 mm mushroom compost (delivered in 65 m³ loads) and cultivate in; grade to level removing stones** | | |
| For turfing | | |
|   200 mm deep | m² | 22.00 |
|   300 mm deep | m² | 31.00 |
| **SURFACE PREPARATION BY HAND** | | |
| **To area previously cleared and sprayed; cultivate existing topsoil to receive new ornamental planting; rotavate using pedestrian rotavator; dig over by hand 1 spit deep; rake and grade to level to receive new planting** | | |
| For areas to be seeded or turfed | m² | 1.00 |
| **Spread only and lightly consolidate topsoil brought from spoil heap in layers not exceeding 150 mm; grade to specified levels; remove stones over 25 mm; all by hand** | | |
| 100 mm thick | 100 m² | 475.00 |
| As above but inclusive of loading to location by barrow maximum distance 25 m; finished topsoil depth | | |
|   100 mm thick | 100 m² | 1050.00 |

## 8.3.1 SOFT LANDSCAPING, SEEDING AND TURFING

| Item<br>Excluding site overheads and profit | Unit | Range £ |
|---|---|---|
| As above but loading to location by barrow maximum distance 100 m; finished topsoil depth | | |
| 100 mm thick | 100 m² | 760.00 |
| Extra to above for imported topsoil PC £51.00 m³ allowing for 20% settlement | | |
| 100 mm thick | 100 m² | 395.00 |
| **SEEDING** | | |
| Spray with systemic herbicide; Cultivate topsoil 200 mm a fine tilth by tractor; remove stones over 25 mm by mechanical stone rake and bring to final tilth by harrow; apply pre-seeding fertilizer at 50 g/m² and work into top 50 mm during final cultivation; seed with certified grass seed in two operations; roll seedbed lightly after sowing | | |
| General amenity grass areas; 35 g/m²; BSH A3 | | |
|    general grass seeded areas | 100 m² | 64.00 |
|    lawn areas for public or domestic general use | 100 m² | 86.00 |
|    high grade lawn areas | 100 m² | 105.00 |
| General amenity grass at 35 g/m²; DLF Seeds Ltd Promaster 10 | | |
|    general turfed areas | 100 m² | 77.00 |
|    lawn areas for public or domestic general use | 100 m² | 97.00 |
|    high grade lawn areas | 100 m² | 115.00 |
| Shaded areas; BSH A6 at 50 g/m² | 100 m² | 73.00 |
| Motorway and road verges; DLF Seeds Ltd Promaster 120 at 25–35 g/m² | 100 m² | 93.00 |
| Spray with systemic herbicide; Cultivate topsoil 200 mm a fine tilth by tractor; remove stones over 25 mm by mechanical stone rake and bring to final tilth by harrow; Add 50 mm of imported topsoil over surface of area to be seeded; apply pre-seeding fertilizer at 50 g/m² and work into top 50 mm during final cultivation; seed with certified grass seed in two operations; roll seedbed lightly after sowing | | |
| General amenity grass areas; 35 g/m²; BSH A3 | | |
|    general turfed areas | 100 m² | 290.00 |
|    lawn areas for public or domestic general use | 100 m² | 310.00 |
|    high grade lawn areas | 100 m² | 330.00 |
| Spray with systemic herbicide; Cultivate topsoil 200 mm a fine tilth by tractor; remove stones over 25 mm by mechanical stone rake and bring to final tilth by harrow; Add 50 mm of topsoil from site spoil heaps maximum distance 100 m over surface of area to be seeded; apply pre-seeding fertilizer at 50 g/m² and work into top 50 mm during final cultivation; seed with certified grass seed in two operations; roll seedbed lightly after sowing | | |
| General amenity grass areas; 35 g/m²; BSH A3 | | |
|    general turfed areas | 100 m² | 110.00 |
|    lawn areas for public or domestic general use | 100 m² | 135.00 |
|    high grade lawn areas | 100 m² | 150.00 |

## 8.3.1 SOFT LANDSCAPING, SEEDING AND TURFING

| Item<br>Excluding site overheads and profit | Unit | Range £ |
|---|---|---|
| **SEEDING – cont** | | |
| **Bring top 200 mm of topsoil to a fine tilth using pedestrian operated rotavator; remove stones over 25 mm; apply pre-seeding fertilizer at 50 g/m² and work into top 50 mm during final hand cultivation; seed with certified grass seed in two operations; rake and roll seedbed lightly after sowing** | | |
| General amenity grass areas; 35 g/m²; BSH A3 | | |
| general grass seeded areas | 100 m² | 140.00 |
| lawn areas for public or domestic general use | 100 m² | 155.00 |
| high grade lawn areas | 100 m² | 170.00 |
| Extra to above for using imported topsoil spread by machine | | |
| 100 mm minimum depth | 100 m² | 440.00 |
| 150 mm minimum depth | 100 m² | 660.00 |
| extra for slopes over 30° | 50% | – |
| Extra to above for using imported topsoil spread by hand; maximum distance for transporting soil 100 m | | |
| 100 mm minimum depth | m² | 870.00 |
| 150 mm minimum depth | m² | 1300.00 |
| extra for slopes over 30° | 50% | – |
| Extra for mechanically screening top 25 mm of topsoil through 6 mm screen and spreading on seedbed; debris carted to dump on site not exceeding 100 m | m³ | 6.95 |
| **Bring top 200 mm of topsoil to a fine tilth; remove stones over 50 mm; apply pre-seeding fertilizer at 50 g/m² and work into top 50 mm of topsoil during final cultivation; seed with certified grass seed in two operations; harrow and roll seedbed lightly after sowing** | | |
| Areas inclusive of fine levelling by specialist machinery | | |
| outfield grass at 350 kg/ha; DLF Seeds Ltd Promaster 40 | ha | 9900.00 |
| outfield grass at 350 kg/ha; DLF Seeds Ltd Promaster 70 | ha | 9600.00 |
| sportsfield grass at 300 kg/ha; DLF Seeds Ltd J Pitch | ha | 9500.00 |
| Seeded areas prepared by chain harrow | | |
| low maintenance grass at 350 kg/ha | ha | 10500.00 |
| verge mixture grass at 150 kg/ha | ha | 10000.00 |
| Extra for wild flora mixture at 30 kg/ha | | |
| BSH WSF 75 kg/ha | ha | 3750.00 |
| extra for slopes over 30° | 50% | – |
| Cut existing turf to 1.0 × 1.0 × 0.5 m turves; roll up and move to stack not exceeding 100 m | | |
| by pedestrian operated machine; roll up and stack by hand | 100 m² | 84.00 |
| all works by hand | 100 m² | 260.00 |
| Extra for boxing and cutting turves | 100 m² | 4.60 |
| **TURFING** | | |
| **Cultivate topsoil in two passes; remove stones over 25 mm and bring to final tilth; apply pre-seeding fertilizer at 50 g/m² and work into top 50 mm during final cultivation; roll turf bed lightly** | | |
| Using tractor drawn implements and mechanical stone rake | 100 m² | 77.00 |
| Cultivation by pedestrian rotavator; all other operations by hand | | |
| light or medium soils | m² | 1.20 |
| heavy wet or clay soils | m² | 1.35 |

## 8.3.1 SOFT LANDSCAPING, SEEDING AND TURFING

| Item<br>Excluding site overheads and profit | Unit | Range £ |
|---|---|---|
| Cultivate topsoil in two passes; remove stones over 25 mm and bring to final tilth; apply pre-seeding fertilizer at 50 g/m² and work into top 50 mm during final cultivation; lay imported turves maximum distance from turf stack 25 m | | |
|    using tractor drawn implements and mechanical stone rake | m² | 5.35 |
|    cultivation by pedestrian rotavator; all other operations by hand | m² | 5.20 |
| Cultivate topsoil in two passes; remove stones over 25 mm and bring to final tilth; apply pre-seeding fertilizer at 50 g/m² and work into top 50 mm during final cultivation; roll turf bed supply and lay imported turf; Rolawn Medallion | | |
|    using tractor drawn implements and mechanical stone rake | m² | 4.85 |
|    cultivation by pedestrian rotavator; all other operations by hand | m² | 4.60 |
| Turfed areas; watering on two occasions and carrying out initial cut | | |
|    by ride-on triple mower | 100 m² | 1.40 |
|    by pedestrian mower | 100 m² | 5.95 |
|    by pedestrian mower; box cutting | 100 m² | 6.65 |
| Extra for using imported topsoil spread and graded by machine | | |
|    25 mm minimum depth | m² | 1.20 |
|    75 mm minimum depth | m² | 3.30 |
|    100 mm minimum depth | m² | 4.35 |
|    150 mm minimum depth | m² | 6.55 |
| Extra for using imported topsoil spread and graded by hand; distance of barrow run 25 m | | |
|    25 mm minimum depth | m² | 2.30 |
|    75 mm minimum depth | m² | 5.90 |
|    100 mm minimum depth | m² | 7.50 |
|    150 mm minimum depth | m² | 11.30 |
| Extra for work on slopes over 30° including pegging with 200 galvanized wire pins | m² | 1.95 |
| **INTURF BIG ROLL** | | |
| **Supply, deliver in one consignment, fully prepare the area and install in Big Roll format Inturf 553, a turfgrass comprising dwarf perennial ryegrass, smooth stalked meadowgrass and fescues; installation by tracked machine** | | |
| Preparation by tractor drawn rotavator | m² | 6.45 |
| Cultivation by pedestrian rotavator; all other operations by hand | m² | 6.35 |
| **Erosion control** | | |
| On ground previously cultivated, bring area to level, treat with herbicide; lay 20 mm thick open texture erosion control mat with 100 mm laps, fixed with 8 × 400 mm steel pegs at 1.0 m centres; sow with low maintenance grass suitable for erosion control on slopes at 35 g/m²; spread imported topsoil 25 mm thick and fertilizer at 35 g/m²; water lightly using sprinklers on two occasions | m² | 7.80 |
|    as above but hand-watering by hose pipe maximum distance from mains supply 50 m | m² | 7.90 |

## 8.3.1 SOFT LANDSCAPING, SEEDING AND TURFING

| Item<br>Excluding site overheads and profit | Unit | Range £ |
|---|---|---|
| **ARTIFICIAL TURF** | | |
| **Excavate ground and reduce levels to receive 20 mm thick artificial grass surface on a compacted 150 thick Type 1 base; dispose of excavated material off site** | | |
| Lay artificial turf; by machine | | |
|    Trulawn Value; 16 mm | m² | 53.00 |
|    Trulawn Play; 20 mm; entry level; lawn and play areas | m² | 55.00 |
|    Trulawn Continental; 20 mm; realistic appearance; lawns and patios | m² | 51.00 |
|    Trulawn Optimum; 26 mm; softest pile; lush green; lawns and patios | m² | 63.00 |
|    Trulawn Luxury; 30 mm; deep pile; realistic lawn | m² | 65.00 |
| All by hand except disposal by grab | | |
|    Trulawn Value; 16 mm | m² | 84.00 |
|    Trulawn Play; 20 mm; entry level; lawn and play areas | m² | 86.00 |
|    Trulawn Continental; 20 mm; realistic appearance; lawns and patios | m² | 82.00 |
|    Trulawn Optimum; 26 mm; softest pile; lush green; lawns and patios | m² | 94.00 |
|    Trulawn Luxury; 30 mm; deep pile; realistic lawn | m² | 97.00 |
| **MAINTENANCE OF TURFED/SEEDED AREAS** | | |
| **Maintenance executed as part of a landscape construction contract** | | |
| **Grass cutting** | | |
| Grass cutting; fine turf; using pedestrian guided machinery; arisings boxed and disposed of off site | | |
|    per occasion | 100 m² | 6.00 |
|    per annum 26 cuts | 100 m² | 155.00 |
|    per annum 18 cuts | 100 m² | 110.00 |
| Grass cutting; standard turf; using self propelled three gang machinery | | |
|    per occasion | 100 m² | 0.45 |
|    per annum 26 cuts | 100 m² | 12.20 |
|    per annum 18 cuts | 100 m² | 8.45 |
| Grass cutting for one year; recreation areas, parks, amenity grass areas; using tractor drawn machinery | | |
|    per occasion | ha | 33.00 |
|    per annum 26 cuts | ha | 870.00 |
|    per annum 18 cuts | ha | 600.00 |
| **Maintain grass area inclusive of grass cutting; fertilization twice per annum; leaf and litter clearance** | | |
| Using pedestrian guided machinery; arisings boxed and disposed of off site | | |
|    26 cuts; public areas | 100 m² | 190.00 |
|    26 cuts; private areas | 100 m² | 170.00 |
|    18 cuts; public areas | 100 m² | 140.00 |
|    18 cuts; private areas | 100 m² | 125.00 |

## 8.3.1 SOFT LANDSCAPING, SEEDING AND TURFING

| Item<br>Excluding site overheads and profit | Unit | Range £ |
|---|---|---|
| **Maintain grass area inclusive of grass cutting; fertilization twice per annum; leaf and litter clearance** | | |
| Using ride-on three cylinder machinery; arisings left on site | | |
|     26 cuts; public areas | 100 m² | 43.00 |
|     26 cuts; private areas | 100 m² | 24.00 |
|     18 cuts; public areas | 100 m² | 40.50 |
|     18 cuts; private areas | 100 m² | 21.00 |
| **Aeration of turfed areas** | | |
| Aerate ground with spiked aerator; apply spring/summer fertilizer once; apply autumn/winter fertilizer once; cut grass, 16 cuts; sweep up leaves twice | | |
|     as part of a landscape contract, defects liability | ha | 2125.00 |
|     as part of a long-term maintenance contract | ha | 1000.00 |

## 8.3.2 EXTERNAL PLANTING

| Item<br>Excluding site overheads and profit | Unit | Range £ |
|---|---|---|
| **SURFACE PEPARATIONS FOR PLANTING** | | |
| **SURFACE PREPARATION BY MACHINE** | | |
| Treat area with systemic non selective herbicide one month before starting cultivation operations; rip up subsoil using subsoiling machine to a depth of 250 mm below topsoil at 1.20 m centres in light to medium soils; rotavate to 200 mm deep in two passes; cultivate with chain harrow; roll lightly; clear stones over 50 mm | | |
| By tractor | 100 m² | 43.00 |
| As above but carrying out operations in clay or compacted gravel | 100 m² | 45.00 |
| As above but ripping by tractor rotavation by pedestrian operated rotavator; clearance and raking by hand; herbicide application by knapsack sprayer | 100 m² | 65.00 |
| As above but carrying out operations in clay or compacted gravel | 100 m² | 66.00 |
| Spread and lightly consolidate topsoil brought from spoil heap not exceeding 100 m; in layers not exceeding 150 mm; grade to specified levels; remove stones over 25 mm; all by machine | | |
| 100 mm thick | m² | 100.00 |
| 150 mm thick | m² | 150.00 |
| 300 mm thick | m² | 285.00 |
| 450 mm thick | m² | 425.00 |
| **Extra to the above for imported topsoil** | | |
| Topsoil PC £51.00 m³ allowing for 20% settlement | | |
| 100 mm thick | 100 m² | 395.00 |
| 150 mm thick | 100 m² | 590.00 |
| 300 mm thick | 100 m² | 1175.00 |
| 450 mm thick | 100 m² | 1775.00 |
| 500 mm thick | 100 m² | 1975.00 |
| 600 mm thick | 100 m² | 2375.00 |
| 750 mm thick | 100 m² | 3000.00 |
| 1.00 m thick | 100 m² | 3950.00 |
| Topsoil to BS3882 PC £51.00 m³ allowing for 20% settlement | | |
| 100 mm thick | 100 m² | 510.00 |
| 150 mm thick | 100 m² | 770.00 |
| 300 mm thick | 100 m² | 1525.00 |
| 450 mm thick | 100 m² | 2275.00 |
| 500 mm thick | 100 m² | 2550.00 |
| 600 mm thick | 100 m² | 3050.00 |
| 750 mm thick | 100 m² | 3800.00 |
| 1.00 m thick | 100 m² | 5100.00 |
| Excavate existing topsoil and remove off site; fill with new good quality topsoil to BS3882; add 50 mm mushroom compost (delivered in 65 m³ loads) and cultivate in; grade to level removing stones | | |
| For planting | | |
| 300 mm deep | m² | 30.00 |
| 400 mm deep | m² | 42.00 |
| 500 mm deep | m² | 51.00 |

## 8.3.2 EXTERNAL PLANTING

| Item<br>Excluding site overheads and profit | Unit | Range £ |
|---|---|---|
| Extra for incorporating mushroom compost at 50 mm/m² into the top 150 mm of topsoil (compost delivered in 20 m³) loads | | |
|    manually spread; mechanically rotavated | 100 m² | 190.00 |
|    mechanically spread and rotavated | 100 m² | 140.00 |
| Extra for incorporating manure at 50 mm/m² into the top 150 mm of topsoil loads | | |
|    manually spread; mechanically rotavated; 20 m³ loads | 100 m² | 290.00 |
|    mechanically spread and rotavated; 60 m³ loads | 100 m² | 180.00 |
| **SURFACE PREPARATION BY HAND** | | |
| **To area previously cleared and sprayed; cultivate existing topsoil to receive new ornamental planting; rotavate using pedestrian rotavator; dig over by hand 1 spit deep; rake and grade to level to receive new planting** | | |
| For ornamental planting beds | m² | 0.60 |
| **Spread only and lightly consolidate topsoil brought from spoil heap in layers not exceeding 150 mm; grade to specified levels; remove stones over 25 mm; all by hand** | | |
| 100 mm thick | 100 m² | 475.00 |
| 150 mm thick | 100 m² | 710.00 |
| 300 mm thick | 100 m² | 1425.00 |
| 450 mm thick | 100 m² | 2125.00 |
| As above but inclusive of loading to location by barrow maximum distance 25 m; finished topsoil depth | | |
|    100 mm thick | 100 m² | 1050.00 |
|    150 mm thick | 100 m² | 1575.00 |
|    300 mm thick | 100 m² | 3100.00 |
|    450 mm thick | 100 m² | 4700.00 |
|    500 mm thick | 100 m² | 5200.00 |
|    600 mm thick | 100 m² | 6300.00 |
| As above but loading to location by barrow maximum distance 100 m; finished topsoil depth | | |
|    100 mm thick | 100 m² | 760.00 |
|    150 mm thick | 100 m² | 1150.00 |
|    300 mm thick | 100 m² | 2275.00 |
|    450 mm thick | 100 m² | 3400.00 |
|    500 mm thick | 100 m² | 3800.00 |
|    600 mm thick | 100 m² | 4550.00 |
| Extra to the above for incorporating mushroom compost at 50 mm/m² into the top 150 mm of topsoil; by hand | 100 m² | 225.00 |
| Extra to above for imported topsoil PC £51.00 m³ allowing for 20% settlement | | |
|    100 mm thick | 100 m² | 395.00 |
|    150 mm thick | 100 m² | 590.00 |
|    300 mm thick | 100 m² | 1175.00 |
|    450 mm thick | 100 m² | 1775.00 |
|    500 mm thick | 100 m² | 1975.00 |
|    600 mm thick | 100 m² | 2375.00 |
|    750 mm thick | 100 m² | 3000.00 |
|    1.00 m thick | 100 m² | 3950.00 |

## 8.3.2 EXTERNAL PLANTING

| Item<br>Excluding site overheads and profit | Unit | Range £ |
|---|---|---|
| **TREE PLANTING** | | |
| Specification Corresponding Tree Girth Size | | |
| Light Standard (LS) 6–8 cm | | |
| Standard (S) 8–10 cm | | |
| Select Standard (SS) 10–12 cm | | |
| Heavy Standard (HS) 12–14 cm | | |
| Extra Heavy Standard (EHS) 14–16 cm | | |
| Advanced Heavy Standard (AHS) 16–18 cm | | |
| Semi-mature 18–20 cm + | | |
| **Excavate tree pit by hand; fork over bottom of pit; plant tree with roots well spread out; backfill with excavated material incorporating treeplanting compost at 1 m³ per 3 m³ of soil; one tree stake and two ties; tree pits square in sizes shown** | | |
| Standard bare root tree in pit; PC £22.75 | | |
| 600 × 600 mm deep | each | 60.00 |
| 900 × 600 mm deep | each | 70.00 |
| Standard rootballed tree in pit; PC £31.25 | | |
| 600 × 600 mm deep | each | 70.00 |
| 900 × 600 mm deep | each | 81.00 |
| 1.00 × 1.00 m deep | each | 110.00 |
| Selected standard bare root tree in pit; PC £39.25 | | |
| 900 × 900 mm deep | each | 84.00 |
| 1.00 × 1.00 m deep | each | 120.00 |
| Selected standard rootball tree in pit; PC £51.75 | | |
| 900 × 900 mm deep | each | 120.00 |
| 1.00 × 1.00 m deep | each | 175.00 |
| Heavy standard bare root tree in pit; PC £70.00 | | |
| 900 × 900 mm deep | each | 140.00 |
| 1.00 × 1.00 m deep | each | 155.00 |
| 1.50 m × 750 mm deep | each | 200.00 |
| Heavy standard rootball tree in pit; PC £85.50 | | |
| 900 × 900 mm deep | each | 155.00 |
| 1.00 × 1.00 m deep | each | 175.00 |
| 1.50 m × 750 mm deep | each | 220.00 |
| Extra heavy standard bare root tree in pit; PC £105.00 | | |
| 1.00 × 1.00 m deep | each | 200.00 |
| 1.20 × 1.00 m deep | each | 200.00 |
| 1.50 m × 750 mm deep | each | 255.00 |
| Extra heavy standard rootball tree in pit; PC £122.15 | | |
| 1.00 × 1.00 m deep | each | 220.00 |
| 1.50 m × 750 mm deep | each | 265.00 |
| 1.50 × 1.00 m deep | each | 305.00 |

## 8.3.2 EXTERNAL PLANTING

| Item<br>Excluding site overheads and profit | Unit | Range £ |
|---|---|---|
| **SEMI-MATURE TREE PLANTING** | | |
| **Trees in soft landscape environments** | | |
| Excavate tree pit deep by machine; fork over bottom of pit; plant rootballed tree Acer platanoides Emerald Queen using telehandler where necessary; backfill with excavated material, incorporating Melcourt Topgrow bark/manure mixture at 1 m³ per 3 m³ of soil; Platipus underground guying system; tree pits 1500 × 1500 × 1500 mm deep inclusive of Platimats; excavated material not backfilled to tree pit spread to surrounding area | | |
| 16–18 cm girth; PC £85.00 | each | 290.00 |
| 18–20 cm girth; PC £105.00 | each | 315.00 |
| 20–25 cm girth; PC £130.00 | each | 420.00 |
| 25–30 cm girth; PC £170.00 | each | 530.00 |
| 30–35 cm girth; PC £310.00 | each | 800.00 |
| As above but tree pits 2.00 × 2.00 × 1.5 m deep | | |
| 40–45 cm girth; PC £600.00 | each | 1250.00 |
| 45–50 cm girth; PC £850.00 | each | 1500.00 |
| 55–60 cm girth; PC £1,200.00 | each | 1925.00 |
| 67–70 cm girth; PC £2,200.00 | each | 3200.00 |
| 75–80 cm girth; PC £4,800.00 | each | 5800.00 |
| Extra to the above for imported topsoil moved 25 m from tipping area and disposal off site of excavated material | | |
| Tree pits 1500 × 1500 × 1500 mm deep | | |
| 16–18 cm girth | each | 300.00 |
| 18–20 cm girth | each | 290.00 |
| 20–25 cm girth | each | 275.00 |
| 25–30 cm girth | each | 260.00 |
| 30–35 cm girth | each | 250.00 |
| **Tree pit in Trafficked environments** | | |
| Excavate tree pit; remove excavated material; break up bottom of pit; line pit with root barrier; Lay drainage layer 200 mm thick; backfill tree pit with cellular load supporting cells to 7.5 tonne traffic weight limit; Plant tree with underground guying system; inclusive of root director. Add rootball irrigation system; backfill over rootball with topsoil enriched with composts and fertilizers | | |
| Tree pit  tree size | | |
| 2.00 × 2.00 × 1.00 m deep; 30–35 cm tree | nr | 1475.00 |
| 2.50 × 2.50 × 1.00 m deep; 40–45 cm tree | nr | 2300.00 |
| Excavate tree pit; remove excavated material; break up bottom of pit; line pit with root barrier; Lay drainage layer 200 mm thick; backfill tree pit with cellular load supporting cells to 40 tonne traffic weight limit; Plant tree with underground guying system; inclusive of root director. Add rootball irrigation system; backfill over rootball with topsoil enriched with composts and fertilizers | | |
| Tree pit  tree size | | |
| 2.00 × 2.00 × 1.00deep; 30–35 cm tree | nr | 1975.00 |
| 2.50 × 2.50 × 1.00deep; 40 – 45 cm tree | nr | 3100.00 |

## 8.3.2 EXTERNAL PLANTING

| Item<br>Excluding site overheads and profit | Unit | Range £ |
|---|---|---|
| **TREE PLANTING – cont** | | |
| | | |
| **TREE PLANTING WITH MOBILE CRANES** | | |
| | | |
| **Excavate tree pit 1.50 × 1.50 × 1.00 m deep; supply and plant semi-mature trees delivered in full loads; trees lifted by crane; inclusive of backfilling tree pit with imported topsoil, compost, fertilizers and underground guying using Platipus anchors** | | |
| Self managed lift; local authority applications, health and safety, traffic management or road closures not included; tree size and distance of lift; 35 tonne crane | | |
|   25–30 cm; maximum 25 m distance | each | 860.00 |
|   30–35 cm; maximum 25 m distance | each | 1050.00 |
|   35–40 cm; maximum 25 m distance | each | 1250.00 |
|   55–60 cm; maximum 15 m distance | each | 1975.00 |
|   80–90 cm; maximum 10 m distance | each | 7100.00 |
| Managed lift; inclusive of all local authority applications, health and safety, traffic management or road closures all by crane hire company; tree size and distance of lift; 35 tonne crane | | |
|   25–30 cm; maximum 25 m distance | each | 870.00 |
|   30–35 cm; maximum 25 m distance | each | 1075.00 |
|   35–40 cm; maximum 25 m distance | each | 1275.00 |
|   55–60 cm; maximum 15 m distance | each | 2025.00 |
|   80–90 cm; maximum 10 m distance | each | 7200.00 |
| Self managed lift; local authority applications, health and safety, traffic management or road closures not included; tree size and distance of lift; 80 tonne crane | | |
|   25–30 cm; maximum 40 m distance | each | 860.00 |
|   30–35 cm; maximum 40 m distance | each | 1050.00 |
|   35–40 cm; maximum 40 m distance | each | 1275.00 |
|   55–60 cm; maximum 33 m distance | each | 1975.00 |
|   80–90 cm; maximum 23 m distance | each | 7200.00 |
| Managed lift; inclusive of all local authority applications, health and safety, traffic management or road closures all by crane hire company; tree size and distance of lift; 35 tonne crane | | |
|   25–30 cm; maximum 40 m distance | each | 870.00 |
|   30–35 cm; maximum 40 m distance | each | 1075.00 |
|   35–40 cm; maximum 40 m distance | each | 1275.00 |
|   55–60 cm; maximum 33 m distance | each | 2050.00 |
|   80–90 cm; maximum 23 m distance | each | 7400.00 |

## 8.3.2 EXTERNAL PLANTING

| Item<br>Excluding site overheads and profit | Unit | Range £ |
|---|---|---|
| **PLANTING SHRUBS AND GROUNDCOVERS** | | |
| **SHRUBS** | | |
| Excavate planting holes 250 × 250 × 300 mm deep to area previously ripped and rotavated; excavated material left alongside planting hole | | |
| By mechanical auger | | |
| 250 mm centres (16 plants per m$^2$) | m$^2$ | 13.40 |
| 300 mm centres (11.11 plants per m$^2$) | m$^2$ | 9.35 |
| 400 mm centres (6.26 plants per m$^2$) | m$^2$ | 5.25 |
| 450 mm centres (4.93 plants per m$^2$) | m$^2$ | 4.10 |
| 500 mm centres (4 plants per m$^2$) | m$^2$ | 3.35 |
| 600 mm centres (2.77 plants per m$^2$) | m$^2$ | 2.30 |
| 750 mm centres (1.77 plants per m$^2$) | m$^2$ | 1.50 |
| 900 mm centres (1.23 plants per m$^2$) | m$^2$ | 1.00 |
| 1.00 m centres (1 plant per m$^2$) | m$^2$ | 0.80 |
| 1.50 m centres (0.44 plants per m$^2$) | m$^2$ | 0.40 |
| As above but excavation by hand | | |
| 250 mm centres (16 plants per m$^2$) | m$^2$ | 13.40 |
| 300 mm centres (11.11 plants per m$^2$) | m$^2$ | 9.35 |
| 400 mm centres (6.26 plants per m$^2$) | m$^2$ | 5.25 |
| 450 mm centres (4.93 plants per m$^2$) | m$^2$ | 4.10 |
| 500 mm centres (4 plants per m$^2$) | m$^2$ | 3.35 |
| 600 mm centres (2.77 plants per m$^2$) | m$^2$ | 2.30 |
| 750 mm centres (1.77 plants per m$^2$) | m$^2$ | 1.50 |
| 900 mm centres (1.23 plants per m$^2$) | m$^2$ | 1.00 |
| 1.00 m centres (1 plant per m$^2$) | m$^2$ | 0.80 |
| 1.50 m centres (0.44 plants per m$^2$) | m$^2$ | 0.40 |
| Clear light vegetation from planting area and remove to dump on site; dig planting holes; plant whips with roots well spread out; backfill with excavated topsoil; including one 38 × 38 mm treated softwood stake, two tree ties and mesh guard 1.20 m high; planting matrix 1.5 × 1.5 m; allow for beating up once at 10% of original planting, cleaning and weeding round whips once, applying fertilizer once at 35 gm/m² ; using the following mix of whips, bare rooted | | |
| Plant bare root plants average PC £0.42 each to a required matrix plant mix as above | 100 m² | 290.00 |
| Plant bare root plants average PC £0.75 each to a required matrix plant mix as above | 100 m² | 320.00 |

## 8.3.2 EXTERNAL PLANTING

| Item<br>Excluding site overheads and profit | Unit | Range £ |
|---|---|---|
| **PLANTING SHRUBS AND GROUNDCOVERS – cont** | | |
| Cultivate and grade shrub bed; bring top 300 mm of topsoil to a fine tilth, incorporating mushroom compost at 50 mm and Enmag slow release fertilizer; rake and bring to given levels; remove all stones and debris over 50 mm; dig planting holes average 300 × 300 × 300 mm deep; supply and plant specified shrubs in quantities as shown below; backfill with excavated material as above; water to field capacity and mulch 50 mm bark chips 20–40 mm size; water and weed regularly for 12 months and replace failed plants | | |
| Shrubs 3 l PC £3.40; ground covers 9 cm PC £1.50 | | |
| 100% shrub area | | |
| 300 mm centres | 100 m² | 5500.00 |
| 400 mm centres | 100 m² | 3350.00 |
| 500 mm centres | 100 m² | 2350.00 |
| 600 mm centres | 100 m² | 1825.00 |
| 100% groundcovers | | |
| 200 mm centres | 100 m² | 5900.00 |
| 300 mm centres | 100 m² | 2950.00 |
| 400 mm centres | 100 m² | 1925.00 |
| 500 mm centres | 100 m² | 1450.00 |
| groundcover 30%/shrubs 70% at the distances shown below | | |
| 200/300 mm | 100 m² | 5500.00 |
| 300/400 mm | 100 m² | 3200.00 |
| 300/500 mm | 100 m² | 2500.00 |
| 400/500 mm | 100 m² | 2175.00 |
| groundcover 50%/shrubs 50% at the distances shown below | | |
| 200/300 mm | 100 m² | 5700.00 |
| 300/400 mm | 100 m² | 3100.00 |
| 300/500 mm | 100 m² | 2650.00 |
| 400/500 mm | 100 m² | 2125.00 |
| **BULB PLANTING** | | |
| Cultivate ground by machine and rake to level; plant bulbs as shown; bulbs PC £25.00/100 | | |
| 15 bulbs per m² | 100 m² | 730.00 |
| 25 bulbs per m² | 100 m² | 1175.00 |
| 50 bulbs per m² | 100 m² | 2350.00 |
| Cultivate ground by machine and rake to level; plant bulbs as shown; bulbs PC £13.00/100 | | |
| 15 bulbs per m² | 100 m² | 520.00 |
| 25 bulbs per m² | 100 m² | 850.00 |
| 50 bulbs per m² | 100 m² | 1675.00 |
| Form holes in grass areas and plant bulbs using bulb planter, backfill with organic manure and turf plug; bulbs PC £13.00/100 | | |
| 15 bulbs per m² | 100 m² | 790.00 |
| 25 bulbs per m² | 100 m² | 1325.00 |
| 50 bulbs per m² | 100 m² | 2600.00 |

## 8.3.2 EXTERNAL PLANTING

| Item<br>Excluding site overheads and profit | Unit | Range £ |
|---|---|---|
| **HEDGE PLANTING** | | |
| **Works by machine; excavate trench for hedge 300 mm wide × 300 mm deep; deposit spoil alongside and plant hedging plants in single row at 200 mm centres; backfill with excavated material** | | |
| Trench 300 mm × 300 mm; bare root hedging plants PC £0.48 per plant; single row | | |
| 200 mm centres | m | 6.50 |
| 300 mm centres | m | 5.50 |
| 400 mm centres | m | 4.75 |
| 600 mm centres | m | 4.00 |
| 800 mm centres | m | 3.65 |
| Trench 600 mm × 300 mm deep; bare root hedging plants PC £0.48 per plant; double staggered row | | |
| 300 mm centres | m | 10.30 |
| 400 mm centres | m | 3.80 |
| 500 mm centres | m | 7.35 |
| 600 mm centres | m | 7.00 |
| 800 mm centres | m | 6.00 |
| 1.00 m centres | m | 5.60 |
| **Works by hand; excavate trench for hedge 300 mm wide × 450 mm deep; deposit spoil alongside and plant hedging plants in single row at 200 mm centres; backfill with excavated material incorporating organic manure at 1 m³ per 5 m³; carry out initial cut; including delivery of plants from nursery** | | |
| Trench 300 mm × 300 mm; bare root hedging plants PC £0.48 per plant; single row | | |
| 200 mm centres | m | 9.90 |
| 300 mm centres | m | 8.90 |
| 400 mm centres | m | 8.20 |
| 600 mm centres | m | 7.40 |
| 800 mm centres | m | 7.10 |
| Trench 600 mm × 300 mm deep; bare root hedging plants PC £0.48 per plant; double staggered row | | |
| 300 mm centres | m | 18.40 |
| 400 mm centres | m | 16.40 |
| 500 mm centres | m | 15.50 |
| 600 mm centres | m | 15.10 |
| 800 mm centres | m | 14.10 |
| 1.00 m centres | m | 13.70 |
| **Readyhedge Ltd; Fully preformed hedge planting; excavate trench by machine 700 mm wide × 500 mm deep; add compost at 100 mm/m² mixed to excavated material; plant mature hedge plants; backfill with excavated material and compost; allow for disposal of 50% of excavated material to spoil heaps 50 m distant** | | |
| Beech or hornbeam hedging | | |
| 1.50 m × 500 mm wide at 500 mm centres | m | 240.00 |
| 1.50 m × 500 mm wide at 750 mm centres | m | 150.00 |
| 2.00 m × 500 mm wide at 500 mm centres | m | 335.00 |
| 2.00 m × 500 mm wide at 750 mm centres | m | 240.00 |
| 1.50 m × 500 mm wide at 500 mm centres | m | 240.00 |

## 8.3.2 EXTERNAL PLANTING

| Item<br>Excluding site overheads and profit | Unit | Range £ |
|---|---|---|
| **HEDGE PLANTING – cont** | | |
| | | |
| **Readyhedge Ltd – cont** | | |
| Yew (Taxus) hedging | | |
| 1.50 m × 500 mm wide at 500 mm centres | m | 350.00 |
| 1.50 m × 500 mm wide at 750 mm centres | m | 225.00 |
| 1.50 m × 500 mm wide at 1.00 m centres | m | 180.00 |
| 1.75 m × 500 mm wide at 500 mm centres | m | 395.00 |
| 1.75 m × 500 mm wide at 750 mm centres | m | 255.00 |
| 1.75 m × 500 mm wide at 1.00 m centres | m | 200.00 |
| 2.00 m × 500 mm wide at 500 mm centres | m | 460.00 |
| 2.00 m × 500 mm wide at 750 mm centres | m | 320.00 |
| 2.00 m × 500 mm wide at 1.00 m centres | m | 235.00 |
| Laurel (Prunus) hedging | | |
| 2.00 m × 500 mm wide at 1.00 m centres | m | 10.80 |
| 1.40/1.60 m high × 400 mm wide | m | 170.00 |
| 1.60/1.80 m high × 500 mm wide | m | 190.00 |
| 1.80/2.00 m high × 500 mm wide | m | 175.00 |
| Box (Buxus) hedging | | |
| 800 mm/1.00 m high × 300 mm wide | m | 140.00 |
| 1.00 m/1.20 m high × 300 mm wide | m | 170.00 |
| | | |
| **Readyhedge Ltd; pre-clipped and preformed hedge planting; excavate trench by machine 500 mm wide × 500 mm deep; add compost at 100 mm/m² mixed to excavated material; plant mature hedge plants; backfill with excavated material and compost; allow for disposal of 50% of excavated material to spoil heaps 50 m distant** | | |
| Beech or Hornbeam hedging | | |
| 1.75 m high × 300 mm wide at 400 mm centres | m | 97.00 |
| 1.75 m high × 300 mm wide at 500 mm centres | m | 79.00 |
| 1.75 m high × 300 mm wide at 750 mm centres | m | 59.00 |
| 1.75 m high × 300 mm wide at 900 mm centres | m | 47.50 |
| 2.00 m high × 300 mm wide at 400 mm centres | m | 170.00 |
| 2.00 m high × 300 mm wide at 500 mm centres | m | 135.00 |
| | | |
| **Readyhedge; strip hedge planting; excavate trench 700 mm wide × 500 mm deep by machine; add compost at 100 mm/m² mixed to excavated material; plant instant strip hedging; backfill with excavated material and compost; allow for disposal of 50% of excavated material to spoil heaps 50 m distant** | | |
| Beech hedging | | |
| 1.40/1.60 m high × 400 mm wide | m | 125.00 |
| 1.60/1.80 m high × 500 mm wide | m | 130.00 |
| Yew (Taxus) hedging | | |
| 1.40/1.60 m high × 400 mm wide | m | 100.00 |
| 1.60/1.80 m high × 500 mm wide | m | 240.00 |
| 1.80/2.00 m high × 500 mm wide | m | 105.00 |
| Laurel (Prunus) hedging | | |
| 1.40/1.60 m high × 400 mm wide | m | 170.00 |
| 1.60/1.80 m high × 500 mm wide | m | 190.00 |
| 1.80/2.00 m high × 500 mm wide | m | 175.00 |

## 8.3.2 EXTERNAL PLANTING

| Item<br>Excluding site overheads and profit | Unit | Range £ |
|---|---|---|
| Box (Buxus) hedging | | |
|    800 mm/1.00 m high × 300 mm wide | m | 140.00 |
|    1.00/1.20 m high × 300 mm wide | m | 170.00 |
| Griffin Nurseries; individual feathered hedge planting; excavate trench by machine 500 mm wide × 500 mm deep; add compost at 100 mm/m² mixed to excavated material; plant mature hedge plants; backfill with excavated material and compost; allow for disposal of 50% of excavated material to spoil heaps 50 m distant | | |
| Yew hedging | | |
|    1.0–1.20 m high × 300 mm wide at 2/m | m | 54.00 |
|    1.25–1.50 m high × 300 mm wide at 3/m | m | 23.00 |
|    1.50–1.75 m high at 450 mm centres | m | 170.00 |
|    1.50–1.75 m high at 600 mm centres | m | 130.00 |
|    2.00–2.25 m high at 600 mm centres | m | 255.00 |
|    2.00–2.25 m high at 900 mm centres | m | 180.00 |
| Griffin Nurseries; Works by machine; excavate trench for hedge 300 mm wide × 300 mm deep; deposit spoil alongside and plant hedging plants in single row; backfill with excavated material | | |
| Box hedging; rootballed | | |
|    400–500 mm high; 200 mm centres | m | 41.50 |
|    400–500 mm high; 300 mm centres | m | 29.00 |
|    400–500 mm high; 500 mm centres | m | 18.50 |
|    500–600 mm mm high; 200 mm centres | m | 55.00 |
|    500–600 mm mm high; 300 mm centres | m | 38.00 |
|    500–600 mm mm high; 600 mm centres | m | 24.50 |
|    600–800 mm high; 250 mm centres | m | 60.00 |
|    600–800 mm high; 400 mm centres | m | 39.00 |
|    800 mm–1.0 mm high; 300 mm centres | m | 77.00 |
|    800 mm–1.0 m high; 500 mm centres | m | 48.00 |
|    1.0–1.25 m high; 600 mm centres | m | 60.00 |
|    1.0–1.25 m high; 900 mm centres | m | 41.50 |
| **ANNUAL BEDDING** | | |
| **BEDDING** | | |
| Spray surface with glyphosate; lift and dispose of turf when herbicide action is complete; cultivate new area for bedding plants to 400 mm deep; spread compost 100 mm deep and chemical fertilizer Enmag and rake to fine tilth to receive new bedding plants; remove all arisings | | |
| Existing turf area | | |
|    disposal to skip | m² | 6.20 |
|    disposal to compost area on site; distance 25 m | m² | 6.40 |

## 8.3.2 EXTERNAL PLANTING

| Item<br>Excluding site overheads and profit | Unit | Range £ |
|---|---|---|
| **ANNUAL BEDDING – cont** | | |
| **Plant bedding to existing planting area; bedding planting PC £0.33 each**<br>Clear existing bedding; cultivate soil to 230 mm deep; incorporate compost 75 mm and rake to fine tilth; collect bedding from nursery and plant at 100 mm centres; irrigate on completion; maintain weekly for 12 weeks | | |
|     mass planted; 100 mm centres | m$^2$ | 35.00 |
|     to patterns; 100 mm centres | m$^2$ | 39.00 |
|     mass planted; 150 mm centres | m$^2$ | 21.00 |
|     to patterns; 150 mm centres | m$^2$ | 25.00 |
|     mass planted; 200 mm centres | m$^2$ | 21.00 |
|     to patterns; 200 mm centres | m$^2$ | 25.00 |
| **Watering of bedding by hand held hose pipe; The prices shown here are per occasion; Bedding may require up to 40 visits during a dry season** | | |
| Flow rate 25 litres/minute; per occasion | | |
|     10 litres/m$^2$ | 100 m$^2$ | 0.10 |
|     15 litres/m$^2$ | 100 m$^2$ | 0.20 |
|     20 litres/m$^2$ | 100 m$^2$ | 0.30 |
|     25 litres/m$^2$ | 100 m$^2$ | 0.35 |
| Flow rate 40 litres/minute | | |
|     10 litres/m$^2$ | 100 m$^2$ | 0.10 |
|     15 litres/m$^2$ | 100 m$^2$ | 0.10 |
|     20 litres/m$^2$ | 100 m$^2$ | 0.20 |
|     25 litres/m$^2$ | 100 m$^2$ | 0.20 |
| **LIVING WALL** | | |
| **Living wall; Scotscape Ltd; Design and installation of planted modules with automatic irrigation systems** | | |
| Fabric based systems; indicative area rates as shown | | |
|     Wall 20 m$^2$ | m$^2$ | 570.00 |
|     Wall 50 m$^2$ | m$^2$ | 440.00 |
|     Wall 100 m$^2$ | m$^2$ | 410.00 |
|     Wall 150 m$^2$ | m$^2$ | 410.00 |
| Annual maintenance costs for living walls | | |
|     Walls over 30–70 m$^2$ | item | 3500.00 |
|     Walls under over 150 m$^2$ | item | 7500.00 |
| **PLANTERS** | | |
| **To brick planter; coat insides with two coats RIW liquid asphaltic composition; fill with 50 mm shingle and cover with geofabric; fill with screened topsoil incorporating 25% Topgrow compost and Enmag** | | |
| Planters 1.00 m deep | | |
|     1.00 × 1.00 m | each | 125.00 |
|     1.00 × 2.00 m | each | 210.00 |
|     1.00 × 3.00 m | each | 300.00 |

## 8.3.2 EXTERNAL PLANTING

| Item<br>Excluding site overheads and profit | Unit | Range £ |
|---|---|---|
| **Planters 1.50 m deep** | | |
|    1.00 × 1.00 m | each | 185.00 |
|    1.00 × 2.00 m | each | 280.00 |
|    1.00 × 3.00 m | each | 450.00 |
| **Container planting; fill with 50 mm shingle and cover with geofabric; fill with screened topsoil incorporating 25% Topgrow compost and Enmag** | | |
| Planters 1.00 m deep | | |
|    400 × 400 × 400 mm deep | each | 9.80 |
|    400 × 400 × 600 mm deep | each | 12.20 |
|    1.00 m × 400 mm wide × 400 mm deep | each | 21.00 |
|    1.00 m × 600 mm wide × 600 mm deep | each | 31.50 |
|    1.00 m × 100 mm wide × 400 mm deep | each | 33.00 |
|    1.00 m × 100 mm wide × 600 mm deep | each | 48.00 |
|    1.00 m × 100 mm wide × 1.00 m deep | each | 78.00 |
|    1.00 m diameter × 400 mm deep | each | 26.00 |
|    1.00 m diameter × 1.00 m deep | each | 62.00 |
|    2.00 m diameter × 1.00 m deep | each | 240.00 |
| **Street planters**<br>Supply and locate in position precast concrete planters; fill with topsoil placed over 50 mm shingle and terram; plant with assorted 5 litre and 3 litre shrubs to provide instant effect | | |
|    700 mm diameter × 470 mm high; white exposed aggregate finish; Marshalls Boulevard | each | 810.00 |
| **LANDSCAPE MAINTENANCE** | | |
| **As part of a term maintenance contract** | | |
| **MAINTENANCE OF ORNAMENTAL SHRUB BEDS** | | |
| **Ornamental shrub beds; private gardens or business parks and resticted commercial environments** | | |
| **Hand weed ornamental shrub bed during the growing season; clear leaves and litter; edge beds when necessary; prune shrubs or hedges and remove; replace mulch to 50 mm at end of each year; (mulched areas only) planting less than 2 years old** | | |
| Mulched beds; weekly visits; 12 month period planting centres shown; public areas; shopping areas | | |
|    600 mm centres | 100 m² | 160.00 |
|    400 mm centres | 100 m² | 140.00 |
|    300 mm centres | 100 m² | 125.00 |
|    groundcovers | 100 m² | 110.00 |
| Mulched beds; weekly visits; 12 month period planting centres shown; private areas ; business or office parks or similar | | |
|    600 mm centres | 100 m² | 140.00 |
|    400 mm centres | 100 m² | 115.00 |
|    300 mm centres | 100 m² | 98.00 |
|    groundcovers | 100 m² | 79.00 |

## 8.3.2 EXTERNAL PLANTING

| Item<br>Excluding site overheads and profit | Unit | Range £ |
|---|---|---|
| **LANDSCAPE MAINTENANCE – cont** | | |
| **Hand weed ornamental shrub bed during the growing season – cont** | | |
| Non-mulched beds; weekly visits; planting centres | | |
| 600 mm centres | 100 m$^2$ | 130.00 |
| 400 mm centres | 100 m$^2$ | 125.00 |
| 300 mm centres | 100 m$^2$ | 120.00 |
| groundcovers | 100 m$^2$ | 115.00 |
| Non-mulched beds; monthly visits; planting centres | | |
| 600 mm centres | 100 m$^2$ | 105.00 |
| 400 mm centres | 100 m$^2$ | 98.00 |
| 300 mm centres | 100 m$^2$ | 94.00 |
| groundcovers | 100 m$^2$ | 90.00 |
| **Maintain rose bed** | | |
| Weed around base of plants; dead head regularly; fertilize once per annum with Enmag; prune once per annum; edge with half moon edging tool; add compost to 50 mm once per annum; clear of litter and leaves | | |
| private or restricted access beds | 100 m$^2$ | 280.00 |
| public areas | 100 m$^2$ | 310.00 |
| **Remulch planting bed at the start of the planting season; top up mulch 25 mm thick; Melcourt Ltd** | | |
| Larger areas maximum distance 25 m; 80 m$^3$ loads | | |
| Ornamental bark mulch | 100 m$^2$ | 160.00 |
| Melcourt Bark Nuggets | 100 m$^2$ | 150.00 |
| Amenity Bark | 100 m$^2$ | 115.00 |
| Forest biomulch | 100 m$^2$ | 110.00 |
| Smaller areas; maximum distance 25 m; 25 m$^3$ loads | | |
| Ornamental bark mulch | 100 m$^2$ | 195.00 |
| Melcourt Bark Nuggets | 100 m$^2$ | 185.00 |
| Amenity Bark | 100 m$^2$ | 150.00 |
| Forest biomulch | 100 m$^2$ | 140.00 |
| **Shrub beds in ornamental gardens** | | |
| Maintain mature shrub bed; shrubs average 1.00 m–3.00 m high with occasional mature trees in bed or in proximity; prune one third of all planting by 30% every year in rotation; fertilize; add compost; annually; add fertilizers; clear leaves; edge bed with half moon tool; water with hand held hose on 6 occasions | | |
| massed plants; 1.00 high | m$^2$ | 430.00 |
| plants up to 2 m high; | m$^2$ | 485.00 |
| plants 2 m–3 m high; | m$^2$ | 630.00 |

## 8.3.2 EXTERNAL PLANTING

| Item<br>Excluding site overheads and profit | Unit | Range £ |
|---|---|---|
| **VEGETATION CONTROL** | | |
| **Native planting; roadside, railway or forestry planted areas.** | | |
| Post-planting maintenance; control of weeds and grass; herbicide spray applications; maintain weed free circles 1.00 m diameter to planting less than 5 years old in roadside, rail or forestry planting environments and the like; strim grass to 50–75 mm; prices per occasion (three applications of each operation normally required) | | |
| Knapsack spray application; glyphosate; planting at | | |
|    1.50 m centres | ha | 1175.00 |
|    1.75 m centres | ha | 750.00 |
|    2.00 m centres | ha | 810.00 |
| Maintain planted areas; control of weeds and grass; maintain weed free circles 1.00 m diameter to planting less than 5 years old in roadside, rail or forestry planting environments and the like; strim surrounding grass to 50–75 mm; prices per occasion (three applications of each operation normally required) | | |
| Herbicide spray applications; CDA (controlled droplet application) glyphosate and strimming; plants planted at the following centres | | |
|    1.50 m centres | ha | 650.00 |
|    1.75 m centres | ha | 750.00 |
|    2.00 m centres | ha | 810.00 |
| Post-planting maintenance; control of weeds and grass; herbicide spray applications; CDA (controlled droplet application); Xanadu glyphosate/ diuron; maintain weed free circles 1.00 m diameter to planting less than 5 years old in roadside, rail or forestry planting environments and the like; strim grass to 50–75 mm; prices per occasion (1.5 applications of herbicide and three strim operations normally required) | | |
| Plants planted at the following centres | | |
|    1.50 m centres | ha | 700.00 |
|    1.75 m centres | ha | 780.00 |
|    2.00 m centres | ha | 840.00 |
| **MAINTENANCE AS DEFECTS LIABILITY AS PART OF A LANDSCAPE CONSTRUCTION PROJECT** | | |
| Clean and weed plant beds; fork over as necessary; water with full liability for the duration of the first growing season using moveable sprinklers; clear leaf fall; top up mulch at the end of the 12 month period | | |
| 1st year after planting | | |
|    weekly visits to high profile areas, new planting areas | 100 m² | 390.00 |
|    monthly visits to new planting areas | 100 m² | 340.00 |

## 8.3.3 IRRIGATION SYSTEMS

| Item<br>Excluding site overheads and profit | Unit | Range £ |
|---|---|---:|
| **IRRIGATION SYSTEMS** | | |
| **LANDSCAPE IRRIGATION** | | |
| **Automatic irrigation; TK Solutions Ltd** | | |
| Large garden consisting of 24 stations; 7000 m² irrigated area | | |
|    turf only | nr | 11000.00 |
|    70/30% turf/shrub beds | nr | 14000.00 |
| Medium sized garden consisting of 12 stations; 3500 m² irrigated area | | |
|    turf only | nr | 7700.00 |
|    70/30% turf/shrub beds | nr | 11000.00 |
| Medium sized garden consisting of 6 stations of irrigated area; 1000 m² | | |
|    turf only | nr | 5200.00 |
|    70/30% turf/shrub beds | nr | 5800.00 |
|    50/50% turf/shrub beds | nr | 6300.00 |
| **Leaky pipe irrigation; Leaky Pipe Ltd** | | |
| Works by machine; main supply and connection to laterals; excavate trench for main or ring main 450 mm deep; supply and lay pipe; backfill and lightly compact trench | | |
|    20 mm LDPE | 100 m | 250.00 |
|    16 mm LDPE | 100 m | 240.00 |
| Works by hand; main supply and connection to laterals; excavate trench for main or ring main 450 mm deep; supply and lay pipe; backfill and lightly compact trench | | |
|    20 mm LDPE | 100 m | 1625.00 |
|    16 mm LDPE | 100 m | 1600.00 |
| Turf area irrigation; laterals to mains; to cultivated soil; excavate trench 150 mm deep using hoe or mattock; lay moisture leaking pipe laid 150 mm below ground at centres of 350 mm | | |
|    low leak | 100 m² | 500.00 |
|    high leak | 100 m² | 465.00 |
| Landscape area irrigation; laterals to mains; moisture leaking pipe laid to the surface of irrigated areas at 600 mm centres | | |
|    low leak | 100 m² | 245.00 |
|    high leak | 100 m² | 220.00 |
| Landscape area irrigation; laterals to mains; moisture leaking pipe laid to the surface of irrigated areas at 900 mm centres | | |
|    low leak | 100 m² | 165.00 |
|    high leak | 100 m² | 150.00 |
|    multistation controller | each | 410.00 |
|    solenoid valves connected to automatic controller | each | 610.00 |

### 8.3.3 IRRIGATION SYSTEMS

| Item<br>Excluding site overheads and profit | Unit | Range £ |
|---|---|---|
| **RAINWATER HARVESTING** | | |
| **Rainwater Harvesting Ltd; rainwater tanks; soft landscape areas installed below ground** | | |
| Excavate in soft landscape area by machine to install rainwater harvesting tank; grade base and lay compacted granular base to level; connect filters to rainwater down pipe of building 10 m distance and lay 110 mm uPVC pipe in trench. Install rainwater tank complete with filters for irrigation and submersible pump; cover with geofabric and backfill with excavated material compacting lightly; remove surplus material off site | | |
| 3700 litre | nr | 3700.00 |
| 4500 litre | nr | 3850.00 |
| 6500 litre | nr | 4600.00 |
| 9000 litre | nr | 5600.00 |
| 13000 litre; 2 nr 6500 tanks | nr | 7300.00 |
| Excavate in soft landscape area by hand to install rainwater harvesting tank; grade base and lay compacted granular base to level; connect filters to rainwater down pipe of building 10 m distance and lay 110 mm uPVC pipe in trench. Install low profile rainwater tank complete with filters for irrigation and submersible pump; cover with geofabric and backfill with excavated material compacting lightly; remove surplus material off site | | |
| Flat tank 1500 litre | nr | 3900.00 |
| Flat tank 3000 litre | nr | 5300.00 |

## 8.4.1 FENCING AND RAILINGS

| Item<br>Excluding site overheads and profit | Unit | Range £ |
|---|---|---|
| **TEMPORARY FENCING** | | |
| Site fencing; supply and erect temporary protective fencing and remove at completion of works | | |
| Cleft chestnut paling | | |
|   1.20 m high 75 mm larch posts at 3 m centres | 100 m | 1725.00 |
|   1.50 m high 75 mm larch posts at 3 m centres | 100 m | 1900.00 |
| Heras fencing | | |
|   2 week hire period | 100 m | 990.00 |
|   4 week hire period | 100 m | 1450.00 |
|   8 week hire period | 100 m | 2350.00 |
|   12 week hire period | 100 m | 3250.00 |
|   16 week hire period | 100 m | 4150.00 |
|   24 week hire period | 100 m | 6000.00 |
| **CHAIN LINK AND WIRE FENCING** | | |
| Chain link fencing; supply and erect chain link fencing; form post holes and erect concrete posts and straining posts with struts at 50 m centres all set in 1:3:6 concrete; fix line wires | | |
| 3 mm galvanized wire 50 mm chainlink fencing | | |
|   1800 mm high | m | 43.00 |
| Plastic coated 3.15 gauge galvanized wire mesh | | |
|   900 mm high | m | 29.00 |
|   1200 mm high | m | 31.00 |
|   1800 mm high | m | 42.00 |
| Extra for additional concrete straining posts with 1 strut set in concrete | | |
|   900 mm high | each | 140.00 |
|   1200 mm high | each | 175.00 |
|   1800 mm high | each | 205.00 |
| Extra for additional concrete straining posts with 2 struts set in concrete | | |
|   900 mm high | each | 240.00 |
|   1200 mm high | each | 300.00 |
|   1800 mm high | each | 350.00 |
| Extra for additional angle iron straining posts with 2 struts set in concrete | | |
|   900 mm high | each | 88.00 |
|   1200 mm high | each | 100.00 |
|   1400 mm high | each | 120.00 |
|   1800 mm high | each | 120.00 |
| **AGRICULTURAL FENCING** | | |
| **RABBIT FENCING** | | |
| Construct rabbit-stop fencing; erect galvanized wire netting; mesh 31 mm; 900 mm above ground, 150 mm below ground turned out and buried; on 75 mm diameter treated timber posts 1.8 m long driven 700 mm into firm ground at 4.0 m centres; netting clipped to top and bottom straining wires 2.63 mm diameter; straining post 150 mm diameter × 2.3 m long driven into firm ground at 50 m intervals | | |
|   turned in 150 mm | 100 m | 1350.00 |
|   buried 150 mm | 100 m | 1475.00 |

## 8.4.1 FENCING AND RAILINGS

| Item<br>Excluding site overheads and profit | Unit | Range £ |
|---|---|---|
| **Deer fence**<br>Construct deer-stop fencing; erect five 4 mm diameter plain galvanized wires and five 2 ply galvanized barbed wires at 150 mm spacing on 45 × 45 × 5 mm angle iron posts 2.4 m long driven into firm ground at 3.0 m centres, driven into firm ground at 3.0 m centres, with timber droppers 25 × 38 mm × 1.0 m long at 1.5 m centres | 100 m | 1650.00 |
| **Forestry fencing**<br>Supply and erect forestry fencing of three lines of 3 mm plain galvanized wire tied to 1700 × 65 mm diameter angle iron posts at 2750 m centres with 1850 × 100 mm diameter straining posts and 1600 × 80 mm diameter struts at 50.0 m centres driven into firm ground | | |
| 1800 mm high; three wires | 100 m | 1175.00 |
| 1800 mm high; three wires including cattle fencing | 100 m | 1525.00 |
| **TIMBER FENCING** | | |
| Clear fence line of existing evergreen shrubs 3 m high average; grub out roots by machine chip on site and remove off site; erect close boarded timber fence in treated softwood, pales 100 × 22 mm lapped, 150 × 22 mm gravel boards | | |
| Concrete posts 100 × 100 mm at 3.0 m centres set into ground in 1:3:6 concrete | | |
| 1500 mm high | m | 93.00 |
| 1800 mm high | m | 100.00 |
| As above but with softwood posts; three arris rails | | |
| 1350 mm high | m | 84.00 |
| 1650 mm high | m | 93.00 |
| 1800 mm high | m | 95.00 |
| Erect chestnut pale fencing; cleft chestnut pales; two lines galvanized wire, galvanized tying wire, treated softwood posts at 3.0 m centres and straining posts and struts at 50 m centres driven into firm ground | | |
| 900 mm high; posts 75 mm diameter × 1200 mm long | m | 14.20 |
| 1200 mm high; posts 75 mm diameter × 1500 mm long | m | 16.90 |
| Construct timber rail; horizontal hit and miss type; rails 150 × 25 mm; posts 100 × 100 mm at 1.8 m centres; twice stained with coloured wood preservative; including excavation for posts and concreting into ground (C7P) | | |
| In treated softwood | | |
| 1800 mm high | m | 64.00 |
| Construct cleft oak rail fence with rails 300 mm minimum girth tennoned both ends; 125 × 100 mm treated softwood posts double mortised for rails; corner posts 125 × 125 mm, driven into firm ground at 2.5 m centres | | |
| Three rails | m | 19.50 |
| Four rails | m | 24.00 |
| **Birdsmouth fencing; timber** | | |
| 600 mm high | m | 25.00 |
| 900 mm high | m | 26.50 |

## 8.4.1 FENCING AND RAILINGS

| Item<br>Excluding site overheads and profit | Unit | Range £ |
|---|---|---|
| **SECURITY FENCING** | | |
| **Supply and erect chainlink fence; 51 × 3 mm mesh; with line wires and stretcher bars bolted to concrete posts at 3.0 m centres and straining posts at 10 m centres; posts set in concrete 450 × 450 mm × 33% of height of post deep; fit straight extension arms of 45 × 45 × 5 mm steel angle with three lines of barbed wire and droppers; all metalwork to be factory hot-dip galvanized for painting on site** | | |
| Galvanized 3 mm mesh | | |
|   1800 mm high | m | 60.00 |
| As above but with PVC coated 3.15 mm mesh (diameter of wire 2.5 mm) | | |
|   900 mm high | m | 38.50 |
|   1200 mm high | m | 45.00 |
|   1800 mm high | m | 59.00 |
| Add to fences above for base of fence to be fixed with hairpin staples cast into concrete ground beam 1:3:6 site mixed concrete; mechanical excavation disposal to on site spoil heaps | | |
|   125 × 225 mm deep | m | 6.00 |
| Add to fences above for straight extension arms of 45 × 45 × 5 mm steel angle with three lines of barbed wire and droppers | m | 2.05 |
| **Supply and erect palisade security fence Jacksons Barbican 2500 mm high with rectangular hollow section steel pales at 150 mm centres on three 50 × 50 × 6 mm rails; rails bolted to 80 × 60 mm posts set in concrete 450 × 450 × 750 mm deep at 2750 mm centres; tops of pales to points and set at 45° angle; all metalwork to be hot-dip factory galvanized for painting on site** | m | 74.00 |
| Supply and erect single gate to match above complete with welded hinges and lock | | |
|   1.0 m wide | each | 500.00 |
|   4.0 m wide | each | 590.00 |
|   8.0 m wide | pair | 1275.00 |
| **Supply and erect Orsogril proprietary welded steel mesh panel fencing on steel posts set 750 mm deep in concrete foundations 600 × 600 mm; supply and erect proprietary single gate 2.0 m wide to match fencing** | | |
| 930 mm high | 100 m | 7000.00 |
| 1326 mm high | 100 m | 8900.00 |
| 1722 mm high | 100 m | 11000.00 |

## 8.4.1 FENCING AND RAILINGS

| Item<br>Excluding site overheads and profit | Unit | Range £ |
|---|---|---|
| **BALL STOP FENCING** | | |
| Supply and erect plastic coated 30 × 30 mm netting fixed to 60.3 mm diameter 12 mm solid bar lattice galvanized dual posts; top, middle and bottom rails with three horizontal rails on 60.3 mm diameter nylon coated tubular steel posts at 3.0 m centres and 60.3 mm diameter straining posts with struts at 50 m centres set 750 mm into FND2 concrete footings 300 × 300 × 600 mm deep; include framed chain link gate 900 × 1800 mm high to match complete with hinges and locking latch | | |
| 4500 mm high | 100 m | 11000.00 |
| 5000 mm high | 100 m | 14000.00 |
| 6000 mm high | 100 m | 15000.00 |
| **RAILINGS** | | |
| Conservation of historic railings; Eura Conservation Ltd | | |
| Remove railings to workshop off site; shotblast and repair mildly damaged railings; remove rust and paint with three coats; transport back to site and re-erect | | |
|    railings with finials 1.80 m high | m | 590.00 |
|    railings ornate cast or wrought iron | m | 930.00 |
| Supply and erect mild steel bar railing of 19 mm balusters at 115 mm centres welded to mild steel top and bottom rails 40 × 10 mm; bays 2.0 m long bolted to 51 × 51 mm ms hollow section posts set in C15P concrete; all metal work galvanized after fabrication | | |
| 900 mm high | m | 82.00 |
| 1200 mm high | m | 115.00 |
| 1500 mm high | m | 130.00 |
| **GATES** | | |
| **STOCK GATE** | | |
| Erect stock gate, ms tubular field gate, diamond braced 1.8 m high hung on tubular steel posts set in concrete (C7P); complete with ironmongery; all galvanized | | |
| Width 3.00 m | each | 370.00 |
| Width 4.20 m | each | 400.00 |

## 8.4.2 WALLS AND SCREENS

| Item<br>Excluding site overheads and profit | Unit | Range £ |
|---|---|---|
| **FOUNDATIONS FOR WALLS** | | |
| **IN SITU CONCRETE FOUNDATIONS** | | |
| Excavate foundation trench mechanically; remove spoil off site; lay 1:3:6 site mixed concrete foundations; distance from mixer 25 m; depth of trench to be 225 mm deeper than foundation to allow for three underground brick courses priced separately | | |
| Concrete poured to blinded exposed ground | | |
| 200 mm deep × 400 mm wide | m | 27.50 |
| 300 mm deep × 500 mm wide | m | 45.00 |
| 400 mm deep × 400 mm wide | m | 44.00 |
| 400 mm deep × 600 mm wide | m | 66.00 |
| 600 mm deep × 600 mm wide | m | 92.00 |
| Concrete poured to formwork | | |
| 200 mm deep × 400 mm wide | m | 56.00 |
| 300 mm deep × 500 mm wide | m | 75.00 |
| 300 mm deep × 600 mm wide | m | 80.00 |
| 300 mm deep × 800 mm wide | m | 100.00 |
| 400 mm deep × 400 mm wide | m | 72.00 |
| 400 mm deep × 600 mm wide | m | 110.00 |
| 600 mm deep × 600 mm wide | m | 140.00 |
| Excavate foundation trench by hand; remove spoil off site; lay 1:3:6 site mixed concrete foundations; distance from mixer 25 m; depth of trench to be 225 mm deeper than foundation to allow for three underground brick courses priced separately | | |
| Disposal to spoil heap 25 m by barrow and off site by grab vehicle | | |
| 200 mm deep × 400 mm wide | m | 72.00 |
| 300 mm deep × 500 mm wide | m | 100.00 |
| 400 mm deep × 400 mm wide | m | 100.00 |
| 400 mm deep × 600 mm wide | m | 150.00 |
| 600 mm deep × 600 mm wide | m | 200.00 |
| Excavate foundation trench mechanically; remove spoil off site; lay ready mixed concrete GEN1 discharged directly from delivery vehicle to location; depth of trench to be 225 mm deeper than foundation to allow for three underground brick courses priced separately; foundation size | | |
| 200 mm deep × 400 mm wide | m | 21.00 |
| 300 mm deep × 500 mm wide | m | 33.00 |
| 400 mm deep × 400 mm wide | m | 32.00 |
| 400 mm deep × 600 mm wide | m | 48.00 |
| 400 mm deep × 800 mm wide | m | 59.00 |
| 400 mm deep × 1000 mm wide | m | 74.00 |
| 400 mm deep × 1200 mm wide | m | 89.00 |
| 600 mm deep × 600 mm wide | m | 68.00 |
| **Reinforced concrete wall to foundations above (site mixed concrete)** | | |
| Up to 1.00 m high × 200 mm thick | m² | 79.00 |
| Up to 1.00 m high × 300 mm thick | m² | 79.00 |

## 8.4.2 WALLS AND SCREENS

| Item<br>Excluding site overheads and profit | Unit | Range £ |
|---|---|---|
| **Reinforced concrete wall to foundations above (ready mix concrete RC35)** | | |
| 1.00 m high × 200 mm thick | m² | 115.00 |
| 1.00 m high × 300 mm thick | m² | 130.00 |
| | | |
| **Grading; excavate to reduce levels for concrete slab; lay waterproof membrane, 100 mm hardcore and form concrete slab reinforced with A142 mesh in 1:2:4 site mixed concrete to thickness; remove excavated material off site** | | |
| Concrete 1:2:4 site mixed | | |
| 100 mm thick | m² | 56.00 |
| 150 mm thick | m² | 66.00 |
| 250 mm thick | m² | 84.00 |
| 300 mm thick | m² | 93.00 |
| Concrete ready mixed GEN 2 | | |
| 100 mm thick | m² | 53.00 |
| 150 mm thick | m² | 60.00 |
| 250 mm thick | m² | 75.00 |
| 300 mm thick | m² | 82.00 |
| | | |
| **BLOCK WALLING** | | |
| | | |
| Notes: Measurements allow for works above ground only | | |
| | | |
| **Concrete block walls; including excavation of foundation trench 450 mm deep; remove spoil off site; lay GEN 1 concrete foundations 600 × 300 mm thick; 1 block below ground** | | |
| Solid blocks 7 N/mm²; 100 mm thick | | |
| 500 mm high | m² | 65.00 |
| 750 mm high | m² | 75.00 |
| 1.00 m high | m² | 86.00 |
| 1.25 m high | m² | 96.00 |
| 1.50 m high | m² | 105.00 |
| 1.80 m high | m² | 120.00 |
| 140 mm thick | | |
| generally | m² | 86.00 |
| 100 mm blocks; laid on flat | | |
| 500 mm high | m² | 98.00 |
| 750 mm high | m² | 125.00 |
| 100 mm blocks laid flat; 2 courses underground | | |
| 1.00 m high | m² | 160.00 |
| 1.20 m high | m² | 180.00 |
| 1.50 m high | m² | 210.00 |
| 1.80 m high | m² | 240.00 |
| Hollow blocks filled with concrete; 440 × 215 × 215 mm | | |
| 500 mm high | m² | 95.00 |
| 750 mm high | m² | 115.00 |
| 1.00 m high | m² | 135.00 |
| 1.25 m high | m² | 150.00 |
| 1.50 m high | m² | 180.00 |
| 1.80 m high | m² | 200.00 |

## 8.4.2 WALLS AND SCREENS

| Item<br>Excluding site overheads and profit | Unit | Range £ |
|---|---|---|
| **BLOCK WALLING – cont** | | |
| **Concrete block retaining walls; 2 courses underground; including excavation of foundation trench 450 mm deep; remove spoil off site; lay GEN 1 concrete foundations 1200 × 400 mm thick** | | |
| Solid blocks 7 N/mm$^2$; 100 mm thick | | |
|    1.00 m high | m | 165.00 |
|    1.50 m high | m | 195.00 |
|    1.80 m high | m | 210.00 |
| 100 mm blocks; laid on flat | | |
|    1.00 m high | m$^2$ | 250.00 |
|    1.50 m high | m$^2$ | 300.00 |
|    1.80 m high | m$^2$ | 330.00 |
| Hollow blocks filled with concrete; 215 mm thick | | |
|    1.00 m high | m$^2$ | 260.00 |
|    1.50 m high | m$^2$ | 320.00 |
|    1.80 m high | m$^2$ | 345.00 |
| Hollow blocks with steel bar cast into the foundation | | |
|    1.00 m high | m$^2$ | 290.00 |
|    1.50 m high | m$^2$ | 355.00 |
|    1.80 m high | m$^2$ | 380.00 |
| **Concrete block wall with brick face 112.5 mm thick to stretcher bond; including excavation of foundation trench 450 mm deep; remove spoil off site; lay GEN 1 concrete foundations 600 × 300 mm thick; place stainless steel ties at 4 nr/m$^2$ of wall face** | | |
| Solid blocks 7 N/mm$^2$ with stretcher bond brick face; reclaimed bricks PC £800.00/1000 | | |
|    100 mm thick | m$^2$ | 200.00 |
|    140 mm thick | m$^2$ | 205.00 |
| **BRICK WALLING** | | |
| **One brick thick wall (225 mm thick); excavation 400 mm deep; remove arisings off site; lay GEN 1 concrete foundations 450 mm wide × 250 mm thick; all in English Garden Wall bond; laid in cement:lime:sand (1:1:6) mortar with flush joints, fair face one side; DPC two courses engineering brick in cement:sand (1:3) mortar; brick on edge coping** | | |
| Wall 900 mm high above DPC | | |
|    engineering brick (class B) | m | 210.00 |
|    brick PC £400.00/1000 | m | 215.00 |
|    brick PC £800.00/1000 | m | 260.00 |
| Curved wall; 6.00 m radius 900 mm high above DPC | | |
|    engineering brick (class B) | m | 250.00 |
|    brick PC £400.00/1000 | m | 320.00 |
|    brick PC £800.00/1000 | m | 340.00 |
| Wall 1200 mm high above DPC | | |
|    engineering brick (class B) | m | 270.00 |
|    brick PC £400.00/1000 | m | 265.00 |
|    brick PC £800.00/1000 | m | 325.00 |

## 8.4.2 WALLS AND SCREENS

| Item<br>Excluding site overheads and profit | Unit | Range £ |
|---|---|---|
| Curved wall; 6.00 m radius; 1200 mm high above DPC | | |
|    engineering brick (class B) | m | 325.00 |
|    brick PC £400.00/1000 | m | 420.00 |
|    brick PC £800.00/1000 | m | 630.00 |
| Wall 1800 mm high above DPC | | |
|    engineering brick (class B) | m | 370.00 |
|    brick PC £400.00/1000 | m | 375.00 |
|    brick PC £800.00/1000 | m | 470.00 |
| Curved wall; 6.00 m radius; 1800 mm high above DPC | | |
|    engineering brick (class B) | m | 465.00 |
|    brick PC £400.00/1000 | m | 590.00 |
|    brick PC £800.00/1000 | m | 630.00 |
| | | |
| One and a half brick wall; excavate 450 mm deep; remove arisings off site; lay GEN 1 concrete foundations 600 × 300 mm thick; two thick brick piers at 3.0 m centres; all in English Garden Wall bond; laid in cement:lime:sand (1:1:6) mortar with flush joints; fair face one side; DPC two courses engineering brick in cement:sand (1:3) mortar; coping of headers on edge | | |
| Wall 900 mm high above DPC | | |
|    engineering brick (class B) | m | 335.00 |
|    brick PC £400.00/1000 | m | 320.00 |
| Wall 1200 mm high above DPC | | |
|    engineering brick (class B) | m | 470.00 |
|    brick PC £400.00/1000 | m | 270.00 |
| Wall 1800 mm high above DPC | | |
|    engineering brick (class B) | m | 580.00 |
|    brick PC £400.00/1000 | m | 420.00 |
|    brick PC £800.00/1000 | m | 1275.00 |
| | | |
| **BRICK WALLING WITH PIERS** | | |
| | | |
| Half brick thick wall; 102.5 mm thick; with one brick piers at 2.00 m centres; excavate foundation trench 500 mm deep; remove spoil off site; lay site mixed concrete foundations 1:3:6 350 × 150 mm thick; laid in cement:lime:sand (1:1:6) mortar with flush joints; fair face one side; DPC two courses underground; engineering brick in cement:sand (1:3) mortar; coping of headers on end | | |
| Wall 900 mm high above DPC | | |
|    engineering brick (class B) | m | 240.00 |
|    brick PC £300.00/1000 | m | 140.00 |
|    brick PC £800.00/1000 | m | 190.00 |

## 8.4.2 WALLS AND SCREENS

| Item<br>Excluding site overheads and profit | Unit | Range £ |
|---|---|---|
| **BRICK WALLING – cont** | | |
| **One brick thick wall; 225 mm thick; with one and a half brick piers at 3.0 m centres; excavation 400 mm deep; remove arisings off site; lay GEN 1 concrete foundations 450 mm wide × 250 mm thick; all in English Garden Wall bond; laid in cement:lime:sand (1:1:6) mortar with flush joints; fair face one side; DPC two courses engineering brick in cement:sand (1:3) mortar; precast concrete coping 152 × 75 mm** | | |
| Wall 900 mm high above DPC | | |
| engineering brick (class B) | m | 410.00 |
| brick PC £400.00/1000 | m | 210.00 |
| brick PC £800.00/1000 | m | 580.00 |
| Wall 1200 mm high above DPC | | |
| engineering brick (class B) | m | 460.00 |
| brick PC £400.00/1000 | m | 260.00 |
| brick PC £800.00/1000 | m | 3400.00 |
| Wall 1800 mm high above DPC | | |
| engineering brick (class B) | m | 670.00 |
| brick PC £400.00/1000 | m | 375.00 |
| brick PC £800.00/1000 | m | 4850.00 |
| **Faced block walling** | | |
| **Concrete block wall with brick face 112.5 mm thick to bond pattern using snap headers; including excavation of foundation trench 450 mm deep; remove spoil off site; lay GEN 1 concrete foundations 600 × 300 mm thick; place stainless steel ties at 4 nr/m² of wall face; lay coping of brick on edge in engineering brick** | | |
| Solid blocks 7 N/mm² with snap header brick face; bricks PC £800.00/1000 | | |
| 500 mm high | m² | 145.00 |
| 750 mm high | m² | 180.00 |
| 1.00 m high | m² | 210.00 |
| 1.50 m high | m² | 290.00 |
| 1.80 m high | m² | 350.00 |
| **PIERS** | | |
| **BLOCK PIERS** | | |
| **Block piers; blocks 440 × 215 × 100 mm thick laid on flat piers on concrete footings 600 × 600 × 250 mm thick; inclusive of excavations and disposal off site; coping of engineering brick on edge; two courses underground** | | |
| Pier 450 × 450 mm; to receive cladding treatment priced separately | | |
| 500 mm high | nr | 55.00 |
| 750 mm high | nr | 70.00 |
| 1.00 m high | nr | 92.00 |
| 1.25 m high | nr | 105.00 |
| 1.50 m high | nr | 120.00 |
| 1.80 m high | nr | 150.00 |

*Approximate Estimating Rates*

## 8.4.2 WALLS AND SCREENS

| Item<br>Excluding site overheads and profit | Unit | Range £ |
|---|---|---|
| **BRICK PIERS** | | |
| **Brick piers on concrete footings 500 × 500 × 250 mm thick; inclusive of excavations and disposal off site; coping of engineering brick on edge; two courses underground** | | |
| Pier 215 × 215 mm; one brick thick; brick PC £400/1000 | | |
|   500 mm high | nr | 62.00 |
|   750 mm high | nr | 77.00 |
|   1.00 m high | nr | 92.00 |
|   1.25 m high | nr | 105.00 |
|   1.50 m high | nr | 120.00 |
|   1.80 m high | nr | 140.00 |
| Pier 215 × 215 mm; one brick thick; brick PC £800/1000 | | |
|   500 mm high | nr | 69.00 |
|   750 mm high | nr | 87.00 |
|   1.00 m high | nr | 105.00 |
|   1.25 m high | nr | 120.00 |
|   1.50 m high | nr | 140.00 |
|   1.80 m high | nr | 160.00 |
| Pier 337.5 × 337.5 mm; one and a half brick thick; brick PC £400/1000 | | |
|   500 mm high | nr | 105.00 |
|   750 mm high | nr | 135.00 |
|   1.00 m high | nr | 170.00 |
|   1.25 m high | nr | 200.00 |
|   1.50 m high | nr | 230.00 |
|   1.80 m high | nr | 270.00 |
| Pier 337.5 × 337.5 mm; one and a half brick thick; brick PC £800/1000 | | |
|   500 mm high | nr | 300.00 |
|   750 mm high | nr | 400.00 |
|   1.00 m high | nr | 510.00 |
|   1.25 m high | nr | 610.00 |
|   1.50 m high | nr | 720.00 |
|   1.80 m high | nr | 840.00 |
| **Brick piers on concrete footings 800 × 800 × 600 mm thick; inclusive of excavations and disposal off site; coping of Haddonstone S150C classically moulded corbelled piercap PC £120.00; 120 mm thick overall** | | |
| Pier 450 × 450 mm; two bricks thick; brick PC £400/1000; height of brickwork above ground | | |
|   500 mm high | nr | 340.00 |
|   750 mm high | nr | 390.00 |
|   1.00 m high | nr | 435.00 |
|   1.25 m high | nr | 480.00 |
|   1.50 m high | nr | 530.00 |
|   1.80 m high | nr | 590.00 |
|   2.00 m high | nr | 630.00 |

## 8.4.2 WALLS AND SCREENS

| Item<br>Excluding site overheads and profit | Unit | Range £ |
|---|---|---|
| **PIERS – cont** | | |
| **Brick piers on concrete footings 800 × 800 × 600 mm thick – cont** | | |
| Pier 450 × 450 mm; two bricks thick; brick PC £800/1000; height of brickwork above ground | | |
| 500 mm high | nr | 370.00 |
| 750 mm high | nr | 430.00 |
| 1.00 m high | nr | 490.00 |
| 1.25 m high | nr | 550.00 |
| 1.50 m high | nr | 610.00 |
| 1.80 m high | nr | 680.00 |
| 2.00 m high | nr | 720.00 |
| **Brick piers on concrete footings 1000 × 1000 × 600 mm thick; inclusive of excavations and disposal off site; coping Haddonstone S215C classically moulded corbelled pyramidal piercap PC £226.00** | | |
| Pier 600 × 600 mm; three bricks thick; brick PC £800/1000; height of brickwork above ground | | |
| 1.50 m high | nr | 1175.00 |
| 1.80 m high | nr | 1325.00 |
| 2.00 m high | nr | 1425.00 |
| **IN SITU CONCRETE WALLS** | | |
| Excavate foundation trench mechanically; remove spoil off site; fix reinforcement starter bars 12 mm at 200 mm centres; lay 1:3:6 site mixed concrete foundations; distance from mixer 25 m; depth of trench to be 225 mm deeper than foundation; cast in situ concrete walls inclusive of bar reinforcement 12 mm; footings cast to blinded exposed ground | | |
| Wall height 500 mm above ground; on foundation 400 mm × 200 mm thick | | |
| 150 thick wall; site mixed concrete 20 tonne aggregate loads | m | 99.00 |
| 150 thick wall; ready mixed concrete ST2 | m | 97.00 |
| 250 thick wall; site mixed concrete; 20 tonne aggregate loads | m | 140.00 |
| 250 thick wall; ready mixed concrete ST2 | m | 150.00 |
| Wall 1.00 m high on foundation 500 mm wide × 300 mm deep | | |
| 150 thick wall; site mixed concrete; 20 tonne aggregate loads | m | 160.00 |
| 150 thick wall; ready mixed concrete ST2 | m | 150.00 |
| 250 thick wall; site mixed concrete; 20 tonne aggregate loads | m | 180.00 |
| 250 thick wall; ready mixed concrete ST2 | m | 165.00 |
| Wall 1.50 m high on foundation 500 mm wide × 300 mm deep | | |
| 250 thick wall; site mixed concrete; 20 tonne aggregate loads | m | 345.00 |
| 250 thick wall; ready mixed concrete ST2 | m | 320.00 |
| Wall 1.80 m high on foundation 600 mm wide × 300 mm deep | | |
| 250 thick wall; site mixed concrete; 20 tonne aggregate loads | m | 440.00 |
| 250 thick wall; ready mixed concrete ST2 | m | 410.00 |
| Wall 1.80 m high on foundation 800 mm wide × 300 mm deep | | |
| 250 thick wall; site mixed concrete; 20 tonne aggregate loads | m | 450.00 |
| 250 thick wall; ready mixed concrete ST2 | m | 420.00 |

## 8.4.2 WALLS AND SCREENS

| Item<br>Excluding site overheads and profit | Unit | Range £ |
|---|---|---|
| **Curved walls; 6.00 m radius** | | |
| Wall 1.50 m high on foundation 500 mm wide × 300 mm deep | | |
|    250 thick wall; site mixed concrete; 20 tonne aggregate loads | m | 370.00 |
|    250 thick wall; ready mixed concrete ST2 | m | 345.00 |
| Wall 1.80 m high on foundation 600 mm wide × 300 mm deep | | |
|    250 thick wall; site mixed concrete; 20 tonne aggregate loads | m | 470.00 |
|    250 thick wall; ready mixed concrete ST2 | m | 440.00 |
| **Curved walls; 3.00 m radius** | | |
| Wall 1.50 m high on foundation 500 mm wide × 300 mm deep | | |
|    250 thick wall; site mixed concrete; 20 tonne aggregate loads | m | 410.00 |
|    250 thick wall; ready mixed concrete ST2 | m | 385.00 |
| Wall 1.80 m high on foundation 600 mm wide × 300 mm deep | | |
|    250 thick wall; site mixed concrete; 20 tonne aggregate loads | m | 510.00 |
|    250 thick wall; ready mixed concrete ST2 | m | 485.00 |
| **SITE MIXED CONCRETE** | | |
| **Mix concrete on site; aggregates delivered in 20 tonne loads; deliver mixed concrete to location by mechanical dumper distance 25 m** | | |
|    1:3:6 | $m^3$ | 100.00 |
|    1:2:4 | $m^3$ | 120.00 |
| As above but ready mixed concrete | | |
|    10 N/mm² | $m^3$ | 105.00 |
|    15 N/mm² | $m^3$ | 110.00 |
| **Mix concrete on site; aggregates delivered in 20 tonne loads; deliver mixed concrete to location by barrow distance 25 m** | | |
| Aggregates delivered in 20 tonne loads to site | | |
|    1:3:6 | $m^3$ | 160.00 |
|    1:2:4 | $m^3$ | 170.00 |
| Aggregates delivered in 1 tonne bags | | |
|    1:3:6 | $m^3$ | 200.00 |
|    1:2:4 | $m^3$ | 210.00 |
| Ready mixed concrete | | |
|    10 N/mm² | $m^3$ | 160.00 |
|    15 N/mm² | $m^3$ | 165.00 |

## 8.4.3 RETAINING WALLS

| Item<br>Excluding site overheads and profit | Unit | Range £ |
|---|---|---|
| **FOOTINGS FOR RETAINING WALLS** | | |
| **Excavate trench for retaining wall; Cart material off site; pour concrete footings with L shaped T20 starter bars; total length 2.00 m at 200 mm centres; Concrete poured to 225 mm below ground level** | | |
| Ready mixed concrete | | |
| 600 mm wide × 300 mm deep | m | 54.00 |
| 800 mm wide × 300 mm deep | m | 65.00 |
| 1.00 m wide × 300 mm deep | m | 76.00 |
| 1.20 m wide × 300 mm deep | m | 97.00 |
| 1.00 m wide × 400 mm deep | m | 100.00 |
| 1.50 m wide × 300 mm deep | m | 110.00 |
| 1.50 m wide × 400 mm deep | m | 160.00 |
| **BLOCK RETAINING WALLS** | | |
| **Note: The following models do not allow for excavation to the existing bank in preparation of the retaining system. Please amalgamate with the Grading and preparation section above.** | | |
| **Concrete block retaining walls; Stepoc** | | |
| Excavate trench 750 mm deep and lay concrete foundation 750 mm wide × 400 mm deep; construct Forticrete precast hollow concrete block wall with 450 mm below ground laid all in accordance with manufacturer's instructions; fix reinforcing bar 12 mm as work proceeds; fill blocks with ready mixed concrete; Allow for two coats of bitumen based waterproofing to back of wall; backfilling against face of wall with drainage trench perforated pipe filled with gravel rejects 600 mm deep | | |
| Walls 1.00 m high | | |
| type 200; 400 × 225 × 200 mm | m | 205.00 |
| type 256; 400 × 225 × 256 mm | m | 230.00 |
| As above but 1.50 m high | | |
| type 200; 400 × 225 × 200 mm | m | 250.00 |
| type 256; 400 × 225 × 256 mm | m | 280.00 |
| Walls 1.50 m high as above but with foundation 1.80 m wide × 450 mm deep | | |
| type 200; 400 × 225 × 200 mm | m | 370.00 |
| type 256; 400 × 225 × 256 mm | m | 400.00 |
| Walls 1.80 m high | | |
| type 200; 400 × 225 × 200 mm | m | 395.00 |
| type 256; 400 × 225 × 256 mm | m | 435.00 |
| **Hollow concrete block retaining wall** | | |
| Excavate trench 750 mm deep and lay concrete foundation 750 mm wide × 400 mm deep; construct hollow concrete block wall with 450 mm below ground laid all in accordance with manufacturer's instructions; fix reinforcing bar 12 mm as work proceeds; fill blocks with concrete 1:3:6 as work proceeds; allow for two coats of bitumen based waterproofing to back of wall; backfilling against face of wall with drainage trench perforated pipe filled with gravel rejects 600 mm deep | | |
| Walls 1.00 m high | m | 185.00 |
| Walls 1.25 m high | m | 200.00 |
| Walls 1.50 m high | m | 220.00 |

## 8.4.3 RETAINING WALLS

| Item<br>Excluding site overheads and profit | Unit | Range £ |
|---|---|---|
| Excavate trench 750 mm deep and lay concrete foundation 1.80 m wide × 400 mm deep; construct hollow concrete block wall with 225 mm below ground; fix reinforcing bar 12 mm as work proceeds; fill blocks with concrete 1:3:6 as work proceeds; Allow for two coats of bitumen based waterproofing to back of wall; backfilling against face of wall with drainage trench perforated pipe filled with gravel rejects 600 mm deep | | |
| Walls 1.50 m high | m | 345.00 |
| Walls 1.80 m high | m | 370.00 |
| **PRECAST CONCRETE RETAINING WALLS** | | |
| Excavate trench and lay concrete foundation 600 × 300; 1:3:6: plain concrete, supply and install Milton Precast Concrete precast concrete L shaped units, constructed all in accordance with manufacturer's instructions; backfill with approved excavated material compacted as the work proceeds | | |
| 1500 mm high × 1000 mm wide | m | 250.00 |
| 2500 mm high × 1000 mm wide | m | 490.00 |
| 3000 mm high × 1000 mm wide | m | 580.00 |
| **GABION WALLS** | | |
| Gabion walls; excavation costs excluded; see Grading and preparation of banks above | | |
| Gabions 500 mm high; allowance of 1.00 m working space at top of bank; backfill working space and remove balance of arisings off site; lay concrete footing; 200 mm deep × 1.50 m wide | | |
| Height retained | | |
| 500 mm | m | 110.00 |
| 1.00 m | m | 170.00 |
| 1.50 m | m | 235.00 |
| 2.00 m | m | 295.00 |
| **TIMBER RETAINING WALLS** | | |
| Timber log retaining walls | | |
| Excavate trench 300 mm wide to one third of the finished height of the retaining walls below; lay 100 mm hardcore; fix machine rounded logs set in concrete 1:3:6; remove excavated material from site; fix geofabric to rear of timber logs; backfill with previously excavated material set aside in position; all works by machine | | |
| 100 mm diameter logs | | |
| 500 mm high (constructed from 1.80 m lengths) | m | 81.00 |
| 1.20 m high (constructed from 1.80 m lengths) | m | 125.00 |
| 1.60 m high (constructed from 2.40 m lengths) | m | 180.00 |
| 2.00 m high (constructed from 3.00 m lengths) | m | 230.00 |

### 8.4.3 RETAINING WALLS

| Item<br>Excluding site overheads and profit | Unit | Range £ |
|---|---|---|
| **TIMBER RETAINING WALLS – cont** | | |
| | | |
| **Excavate trench 300 mm wide to one third of the finished height of the retaining walls below – cont** | | |
| 150 mm diameter logs | | |
|    1.20 m high (constructed from 1.80 m lengths) | m | 220.00 |
|    1.60 m high (constructed from 2.40 m lengths) | m | 255.00 |
| 200 mm diameter logs | | |
|    2.00 m high (constructed from 3.00 m lengths) | m | 290.00 |
| | | |
| **Timber crib wall** | | |
| Excavate trench to receive foundation 300 mm deep; place plain concrete foundation 150 mm thick in 11.50 N/mm² concrete (sulphate-resisting cement); construct timber crib retaining wall and backfill with excavated spoil behind units | | |
|    timber crib wall system; average 5.0 m high | m | 1275.00 |
|    timber crib wall system; average 4.0 m high | m | 950.00 |
|    timber crib wall system; average 2.0 m high | m | 425.00 |
|    timber crib wall system; average 1.00 m high | m | 215.00 |
| | | |
| **Railway sleeper wall; Excavate and remove spoil off site; Pour concrete footing 500 × 250 deep; lay sleeper retaining wall** | | |
| Grade 1 hardwood sleeper 2 450 m × 250 mm wide × 150 mm high; Sleepers set to galvanized pins set in the concrete | | |
|    150 mm high | m | 42.50 |
|    300 mm high | m | 69.00 |
|    450 mm high | m | 95.00 |
|    600 mm high | m | 120.00 |
| Grade 1 hardwood; 2 galvanized angle iron stakes per sleeper length set into concrete | | |
|    750 mm high | m | 170.00 |
|    900 mm high | m | 190.00 |

## 8.5 EXTERNAL FIXTURES

| Item<br>Excluding site overheads and profit | Unit | Range £ |
|---|---|---|
| **TRIP RAILS** | | |
| **Steel trip rail**<br>Erect trip rail of galvanized steel tube 38 mm internal diameter with sleeved joint fixed to 38 mm diameter steel posts as above 700 mm long; set in C7P concrete at 1.20 m centres; metalwork primed and painted two coats metal preservative paint | m | 180.00 |
| **8.5.1 SITE/STREET FURNITURE AND EQUIPMENT** | | |
| **Bollards and access restriction**<br>Supply and install 10 nr cast iron Doric bollards 920 mm high above ground × 170 mm diameter bedded in concrete base 400 mm diameter × 400 mm deep | 10 nr | 2050.00 |
| **Benches and seating**<br>In grassed area excavate for base 2500 × 1575 mm and lay 100 mm hardcore, 100 mm concrete, brick pavers in stack bond bedded in 25 mm cement:lime:sand mortar; supply and fix where shown on drawing proprietary seat, hardwood slats on black powder coated steel frame, bolted down with 4 nr 24 × 90 mm recessed hex-head stainless steel anchor bolts set into concrete | set | 1275.00 |
| **Cycle stand**<br>Supply and fix cycle stand 1250 m × 550 mm of 60.3 mm black powder coated hollow steel sections, one-piece with rounded top corners; set 250 mm into paving | each | 410.00 |
| **8.5.2 ORNAMENTAL WATER FEATURES** | | |
| **FAIRWATER LTD** | | |
| **Excavate for small pond or lake maximum depth 1.00 m; remove arisings off site; grade and trim to shape; lay 75 mm sharp sand; line with 0.75 mm butyl liner 75 mm sharp sand and geofabric; cover over with sifted topsoil; anchor liner to anchor trench; install balancing tank and automatic top-up system<br>Pond or lake of organic shape | | |
|    100 m²; perimeter 50 m | each | 7100.00 |
|    250 m²; perimeter 90 m | each | 11000.00 |
|    500 m²; perimeter 130 m | each | 27000.00 |
|    1000 m²; perimeter 175 m | each | 55000.00 |
| **Excavate for lake average depth 1.0 m, allow for bringing to specified levels; reserve topsoil; remove spoil to approved dump on site; remove all stones and debris over 75 mm; lay polythene sheet including welding all joints and seams by specialist; screen and replace topsoil 200 mm thick**<br>Prices are for lakes of regular shape | | |
|    500 micron sheet | 1000 m² | 16000.00 |
|    1000 micron sheet | 1000 m² | 20000.00 |
| Extra for removing spoil to tip | m³ | 39.00 |
| Extra for 25 mm sand blinding to lake bed | 100 m² | 140.00 |
| Extra for screening topsoil | m² | 1.30 |
| Extra for spreading imported topsoil | 100 m² | 910.00 |

## 8.5 EXTERNAL FIXTURES

| Item<br>Excluding site overheads and profit | Unit | Range £ |
|---|---|---|
| **8.5.2 ORNAMENTAL WATER FEATURES – cont** | | |
| | | |
| **Plant aquatic plants in lake topsoil** | | |
| | | |
| Representative plant prices are shown below and users should add these to to the labour rates shown | | |
|     Apongeton distrachyium | nr | 3.20 |
|     Acorus calamus | nr | 2.20 |
|     Butomus umbellatus | nr | 2.20 |
|     Typha latifolia | nr | 2.20 |
|     Nymphaea PC | nr | 8.50 |
| Planting aquatics labours only | 100 | 95.00 |
| | | |
| **Formal water features** | | |
| | | |
| Excavate and construct water feature of regular shape; lay 100 mm hardcore base and 150 mm concrete 1:2:4 site mixed; line base and vertical face with butyl liner 0.75 micron and construct vertical sides of reinforced blockwork; rendering two coats; anchor the liner behind blockwork; install pumps, balancing tanks and all connections to mains supply | | |
| 1.00 × 1.00 m × 1.00 m deep | nr | 2600.00 |
| 2.00 × 1.00 m × 1.00 m deep | nr | 3350.00 |
| | | |
| Excavate and construct water feature of circular shape; lay 100 mm hardcore base and 150 mm concrete 1:2:4 site mixed; line base and vertical face with butyl liner 0.75 micron and construct vertical sides of reinforced blockwork; rendering two coats; anchor the liner behind blockwork; install pumps, balancing tanks and all connections to mains supply | | |
| Surround water body with ornamental fountain surround | | |
|     Architectural Heritage; Great Westwood; 6.00 m diameter 350 mm high | nr | 13000.00 |
|     Haddonstone; circular pool surround; internal diameter 1780 mm; kerb features continuous moulding enriched with ovolvo and palmette designs; inclusive of plinth and integral conch shell vases flanked by dolphins | nr | 6700.00 |
| | | |
| **8.5.3 ORNAMENTAL BUILDINGS** | | |
| | | |
| **Excavate and dispose excavated material; lay concrete base 150 thick; erect garden building** | | |
| Pavillion; Cast stone; Haddonstone; floor of Yorkstone diamond sawn 6 sides | | |
|     Venetian Folly L9400; Tuscan columns, pedimented arch, quoins and optional balustrading; 4.70 × 3.10 | nr | 18000.00 |
| Cast stone; Haddonstone temples with domed roof and stepped floor | | |
|     Small classical; 2.54 m diameter | nr | 13000.00 |
|     Large classical 3.19 m diameter | nr | 18500.00 |
| Temple; Soild limestone; Architectural Heritage | | |
|     Estate temple 2.70 m diameter; floor of Yorkstone; diamond sawn 6 sides | nr | 21500.00 |
| Pergola; Solid limestone columns with oak beams; Architectural Heritage | | |
|     Estate temple 2.70 m diameter; floor of Yorkstone; diamond sawn 6 sides | nr | 15000.00 |

## 8.6 EXTERNAL DRAINAGE

| Item<br>Excluding site overheads and profit | Unit | Range £ |
|---|---|---|
| **8.6.1 SURFACE WATER AND FOUL WATER DRAINAGE** | | |
| **PIPE LAYING** | | |
| **PVC-u pipe laying; excavate trench 300 mm wide by excavator; lay bedding; lay surface water drain pipe to depth shown; backfill to 150 mm above pipe with gravel or sand rejects; backfill with excavated material to ground level** | | |
| Trench 750 mm deep; pipe 640–590 mm deep | | |
|   110 mm | m | 15.00 |
|   110 mm; short lengths | m | 16.70 |
|   160 mm | m | 18.30 |
|   160 mm; short lengths | m | 26.00 |
| Trench 900 mm deep; pipe 640–590 mm deep | | |
|   110 mm | m | 25.00 |
|   110 mm; short lengths | m | 27.00 |
|   160 mm | m | 28.50 |
|   160 mm; short lengths | m | 36.00 |
| Trench 1.20 m deep; pipe 940–890 mm deep | | |
|   110 mm | m | 28.00 |
|   110 mm; short lengths | m | 30.00 |
|   160 mm | m | 31.00 |
|   160 mm; short lengths | m | 39.00 |
| Trench 1.50 m deep; pipe 940–890 mm deep | | |
|   110 mm | m | 32.00 |
|   110 mm; short lengths | m | 39.00 |
|   160 mm | m | 35.00 |
|   160 mm; short lengths | m | 42.50 |
| **Vitrified clay pipe laying; excavate trench by excavator; backfill with excavated material to ground level; lay surface water drain pipe to depth shown; note that the use of clay pipes precludes the use of bedding materials** | | |
| Trench 750 mm deep; pipe 640–590 mm deep | | |
|   100 mm | m | 16.60 |
|   100 mm; short lengths | m | 18.10 |
|   160 mm | m | 25.00 |
|   160 mm; short lengths | m | 32.50 |
| Trench 1.20 m deep; pipe 940–890 mm deep; bedding on reject sand | | |
|   110 mm | m | 30.00 |
|   110 mm; short lengths | m | 31.00 |
|   160 mm | m | 36.00 |
|   160 mm; short lengths | m | 45.50 |
| **ACCESS CHAMBERS** | | |
| **Excavate inspection chamber by machine; lay base of concrete 150 thick; allow for half section pipework and benchings** | | |
| Precast concrete inspection chamber | | |
|   600 × 400 × 600 mm deep | nr | 460.00 |
|   600 × 400 × 900 mm deep | nr | 490.00 |

## 8.6 EXTERNAL DRAINAGE

| Item<br>Excluding site overheads and profit | Unit | Range £ |
|---|---|---|
| **8.6.1 SURFACE WATER AND FOUL WATER DRAINAGE – cont** | | |
| **Excavate inspection chamber by machine – cont** | | |
| Polypropylene inspection chamber | | |
| Mini access chamber 600 mm deep | nr | 370.00 |
| 475 mm diameter × 900 mm deep; polymer cover | nr | 560.00 |
| 475 mm diameter × 900 mm deep; ductile iron cover with screw down lid | nr | 610.00 |
| **GULLIES** | | |
| **Gullies; vitrified clay; excavate by hand; supply and set gully in concrete (C10P); connect to drainage system with flexible joints; backfilling with excavated material to 250 mm below finished level. Lay 150 mm type 1 to receive surface treatments (not included)** | | |
| Trapped mud (dirt) gully; complete with galvanized bucket and cast iron hinged locking grate and frame | | |
| loading to 1 tonne; 100 or 150 mm outlet | each | 400.00 |
| loading to 5 tonne; 100 mm outlet | each | 460.00 |
| loading to 5 tonne; 150 mm outlet | each | 475.00 |
| Trapped mud (dirt) gully with rodding eye; complete with galvanized bucket and cast iron hinged locking grate and frame | | |
| 100 outlet; 300 mm internal diameter; 300 mm internal depth | each | 395.00 |
| 150 outlet; 400 mm internal diameter; 750 mm internal depth | each | 425.00 |
| **Concrete road gully** | | |
| Excavate and lay 100 mm concrete base (1:3:6) 150 x150 mm to suit given invert level of drain; supply and connect trapped precast concrete road gully set in concrete surround; connect to vitrified clay or uPVC drainage system with flexible joints; supply and fix straight bar dished top cast iron grating and frame; bedded in cement:sand mortar (1:3) | | |
| 375 mm diameter × 750 mm deep | each | 400.00 |
| 450 mm diameter × 750 mm deep | each | 405.00 |
| 450 mm diameter × 1.05 m deep | each | 435.00 |
| **Gullies uPVC** | | |
| Excavate and lay 100 mm concrete (C20P) base 150 × 150 mm to suit given invert level of drain; connect to drainage system; backfill with DoT Type 1 granular fill; install gully; complete with cast iron grate and frame | | |
| yard gully 300 mm diameter × 600 mm deep | each | 375.00 |
| trapped PVC-u gully; 110 mm diameter × 215 mm deep; ductile iron frame | each | 175.00 |
| bottle gully 228 × 228 × 642 mm deep | each | 205.00 |
| bottle gully 228 × 228 × 317 mm deep | each | 190.00 |
| **LINEAR DRAINAGE** | | |
| **Linear drainage to design sensitive areas** | | |
| Excavate trench by machine; lay Aco Brickslot channel drain on concrete base and surround to falls; all to manufacturers specifications | | |
| paving surround to both sides of channel | m | 210.00 |

## 8.6 EXTERNAL DRAINAGE

| Item<br>Excluding site overheads and profit | Unit | Range £ |
|---|---|---|
| **Linear drainage to pedestrian area** | | |
| Excavate trench by machine; lay Aco Multidrain MD polymer concrete channel drain on concrete base and surround to falls; all to manufacturers specifications; paving surround to channel with brick paving PC £300.00/1000 | | |
|    Brickslot galvanized grating; paving to both sides | m | 210.00 |
|    Brickslot stainless steel grating; paving to both sides | m | 390.00 |
|    slotted galvanized steel grating; paving surround to one side of channel | m | 150.00 |
| Excavate trench by machine; lay Aco Multidrain PPD recycled polypropylene channel drain on concrete base and surround to falls; all to manufacturers specifications; paving surround to channel with brick paving PC £300.00/1000 | | |
|    Heelguard composite black; paving surround to both sides of channel | m | 150.00 |
| **Linear drainage to light vehicular area** | | |
| Excavate trench by machine; lay Aco Multidrain PPD recycled polypropylene channel drain on concrete base and surround to falls; all to manufacturers specifications; paving surround to channel with brick paving PC £300.00/1000 | | |
|    Heelguard composite black; paving surround to both sides of channel | m | 180.00 |
|    ductile iron; paving surround to both sides of channel | m | 250.00 |
| **Accessories for channel drain** | | |
| Sump unit with sediment bucket | each | 190.00 |
| End cap; inlet/outlet | each | 7.30 |
| **MANHOLES** | | |
| Brick manhole; excavate pit including earthwork support and working space disposal of surplus spoil to dump on site not exceeding 100 m; lay concrete (1:2:4) base 1500 mm diameter × 200 mm thick; reinforced with mesh reinforcement; 110 mm vitrified clay channels; benching in concrete (1:3:6) allowing one outlet and two inlets for 110 mm diameter pipe; construct inspection chamber 1 brick thick walls of engineering brick Class B; backfill with excavated material; complete with 2 nr cast iron step irons | | |
| 1200 × 1200 × 1200 mm | | |
|    cover slab of precast concrete | each | 1450.00 |
|    access cover; Group 2; 600 × 450 mm | each | 1475.00 |
| 1200 × 1200 × 1500 mm | | |
|    access cover; Group 2; 600 × 450 mm | each | 1850.00 |
|    recessed cover 5 tonne load; 600 × 450 mm; filled with block paviors | each | 1975.00 |
| 1200 × 1200 × 2000 mm; Walls 327.5 thick | | |
|    access cover; Group 2; 600 × 450 mm | each | 2800.00 |
|    recessed cover 5 tonne load; 600 × 450 mm; filled with block paviors | each | 2900.00 |
| 1500 × 1500 × 2500 mm; Walls 327.5 thick | | |
|    access cover; Group 2; 600 × 450 mm | each | 4100.00 |
|    recessed cover 5 tonne load; 600 × 450 mm; filled with block paviors | each | 4250.00 |

## 8.6 EXTERNAL DRAINAGE

| Item<br>Excluding site overheads and profit | Unit | Range £ |
|---|---|---|
| **8.6.1 SURFACE WATER AND FOUL WATER DRAINAGE – cont** | | |
| Concrete manholes; excavate pit including earthwork support and working space disposal of surplus spoil to dump on site not exceeding 100 m; lay concrete (1:2:4) base 1500 mm diameter × 200 mm thick; reinforced with mesh reinforcement; 110 mm vitrified clay channels; benching in concrete (1:3:6) allowing one outlet and two inlets for 110 mm diameter pipe; construct inspection chamber precast concrete rings; surround in concrete 150 mm thick backfill with excavated material; complete with step irons | | |
| Manhole diameter 900 mm | | |
|   1.00 m deep | each | 1075.00 |
| Manhole diameter 1.05 m | | |
|   1.00 m deep | each | 1250.00 |
|   2.00 m deep | each | 1650.00 |
|   3.00 m deep | each | 2025.00 |
| Manhole diameter 1.20 m | | |
|   1.00 m deep | each | 1475.00 |
|   2.00 m deep | each | 1950.00 |
|   3.00 m deep | each | 2400.00 |
| Manhole diameter 1.50 m | | |
|   1.00 m deep | each | 1925.00 |
|   2.00 m deep | each | 2500.00 |
|   3.00 m deep | each | 3050.00 |
| **8.6.4 LAND DRAINAGE** | | |
| **SUBSOIL DRAINAGE – BY MACHINE** | | |
| Main drain; remove 150 mm topsoil and deposit alongside trench, excavate drain trench by machine and lay flexible perforated drain; lay bed of gravel rejects 100 mm; backfill with gravel rejects or similar to within 150 mm of finished ground level; complete fill with topsoil; remove surplus spoil to approved dump on site | | |
| Main drain 160 mm supplied in 35 m lengths | | |
|   450 mm deep | 100 m | 820.00 |
|   600 mm deep | 100 m | 950.00 |
|   900 mm deep | 100 m | 1575.00 |
| Extra for couplings | each | 2.80 |
| 100 mm main drain supplied in 100 m lengths | | |
|   450 mm deep | 100 m | 1175.00 |
|   600 mm deep | 100 m | 1550.00 |
|   900 mm deep | 100 m | 2500.00 |
| Extra for couplings | each | 2.25 |
| Laterals to mains above; herringbone pattern; excavation and backfilling as above; inclusive of connecting lateral to main drain | | |
| 160 mm pipe to 450 mm deep trench | | |
|   laterals at 1.0 m centres | 100 m² | 820.00 |
|   laterals at 2.0 m centres | 100 m² | 410.00 |
|   laterals at 3.0 m centres | 100 m² | 270.00 |
|   laterals at 5.0 m centres | 100 m² | 165.00 |
|   laterals at 10.0 m centres | 100 m² | 82.00 |

## 8.6 EXTERNAL DRAINAGE

| Item<br>Excluding site overheads and profit | Unit | Range £ |
|---|---|---|
| 160 mm pipe to 600 mm deep trench | | |
|    laterals at 1.0 m centres | 100 m² | 950.00 |
|    laterals at 2.0 m centres | 100 m² | 475.00 |
|    laterals at 3.0 m centres | 100 m² | 315.00 |
|    laterals at 5.0 m centres | 100 m² | 190.00 |
|    laterals at 10.0 m centres | 100 m² | 95.00 |
| 160 mm pipe to 900 mm deep trench | | |
|    laterals at 1.0 m centres | 100 m² | 1575.00 |
|    laterals at 2.0 m centres | 100 m² | 790.00 |
|    laterals at 3.0 m centres | 100 m² | 520.00 |
|    laterals at 5.0 m centres | 100 m² | 315.00 |
|    laterals at 10.0 m centres | 100 m² | 160.00 |
| Extra for 160/160 mm couplings connecting laterals to main drain | | |
|    laterals at 1.0 m centres | 10 m | 110.00 |
|    laterals at 2.0 m centres | 10 m | 56.00 |
|    laterals at 3.0 m centres | 10 m | 37.00 |
|    laterals at 5.0 m centres | 10 m | 22.00 |
|    laterals at 10.0 m centres | 10 m | 11.20 |
| 100 mm pipe to 450 mm deep trench | | |
|    laterals at 1.0 m centres | 100 m² | 1175.00 |
|    laterals at 2.0 m centres | 100 m² | 590.00 |
|    laterals at 3.0 m centres | 100 m² | 390.00 |
|    laterals at 5.0 m centres | 100 m² | 240.00 |
|    laterals at 10.0 m centres | 100 m² | 120.00 |
| 100 mm pipe to 600 mm deep trench | | |
|    laterals at 1.0 m centres | 100 m² | 1550.00 |
|    laterals at 2.0 m centres | 100 m² | 780.00 |
|    laterals at 3.0 m centres | 100 m² | 510.00 |
|    laterals at 5.0 m centres | 100 m² | 310.00 |
|    laterals at 10.0 m centres | 100 m² | 155.00 |
| 100 mm pipe to 900 mm deep trench | | |
|    laterals at 1.0 m centres | 100 m² | 2500.00 |
|    laterals at 2.0 m centres | 100 m² | 1275.00 |
|    laterals at 3.0 m centres | 100 m² | 840.00 |
|    laterals at 5.0 m centres | 100 m² | 510.00 |
|    laterals at 10.0 m centres | 100 m² | 250.00 |
| 80 mm pipe to 450 mm deep trench | | |
|    laterals at 1.0 m centres | 100 m² | 1125.00 |
|    laterals at 2.0 m centres | 100 m² | 570.00 |
|    laterals at 3.0 m centres | 100 m² | 375.00 |
|    laterals at 5.0 m centres | 100 m² | 225.00 |
|    laterals at 10.0 m centres | 100 m² | 110.00 |
| 80 mm pipe to 600 mm deep trench | | |
|    laterals at 1.0 m centres | 100 m² | 1500.00 |
|    laterals at 2.0 m centres | 100 m² | 750.00 |
|    laterals at 3.0 m centres | 100 m² | 500.00 |
|    laterals at 5.0 m centres | 100 m² | 300.00 |
|    laterals at 10.0 m centres | 100 m² | 150.00 |

## 8.6 EXTERNAL DRAINAGE

| Item<br>Excluding site overheads and profit | Unit | Range £ |
|---|---|---|
| **8.6.4 LAND DRAINAGE – cont** | | |
| **Laterals to mains above – cont** | | |
| 80 mm pipe to 900 mm deep trench | | |
|    laterals at 1.0 m centres | 100 m² | 2475.00 |
|    laterals at 2.0 m centres | 100 m² | 1250.00 |
|    laterals at 3.0 m centres | 100 m² | 820.00 |
|    laterals at 5.0 m centres | 100 m² | 495.00 |
|    laterals at 10.0 m centres | 100 m² | 250.00 |
| Extra for 100/80 mm junctions connecting laterals to main drain | | |
|    laterals at 1.0 m centres | 10 m | 52.00 |
|    laterals at 2.0 m centres | 10 m | 26.00 |
|    laterals at 3.0 m centres | 10 m | 17.30 |
|    laterals at 5.0 m centres | 10m | 10.40 |
|    laterals at 10.0 m centres | 10m | 5.20 |
| **SUBSOIL DRAINAGE – BY HAND** | | |
| **Main drain; remove 150 mm topsoil and deposit alongside trench; excavate drain trench by machine and lay flexible perforated drain; lay bed of gravel rejects 100 mm; backfill with gravel rejects or similar to within 150 mm of finished ground level; complete fill with topsoil; remove surplus spoil to approved dump on site** | | |
| Main drain 160 mm supplied in 35 m lengths | | |
|    450 mm deep | m | 31.50 |
|    600 mm deep | m | 40.00 |
|    900 mm deep | m | 55.00 |
| Extra for couplings | each | 2.80 |
| As above but with 100 mm main drain supplied in 100 m lengths | | |
|    450 mm deep | m | 20.50 |
|    600 mm deep | m | 27.00 |
|    900 mm deep | m | 38.00 |
| Extra for couplings | each | 2.25 |
| **Laterals to mains above; herringbone pattern; excavation and backfilling as above; inclusive of connecting lateral to main drain** | | |
| 160 mm pipe to 450 mm deep trench | | |
|    laterals at 1.0 m centres | m² | 31.50 |
|    laterals at 2.0 m centres | m² | 24.00 |
| 160 mm pipe to 600 mm deep trench | | |
|    laterals at 1.0 m centres | m² | 40.00 |
|    laterals at 2.0 m centres | m² | 30.00 |
| 100 mm pipe to 450 mm deep trench | | |
|    laterals at 1.0 m centres | m² | 14.30 |
|    laterals at 2.0 m centres | m² | 9.05 |
| 100 mm pipe to 600 mm deep trench | | |
|    laterals at 1.0 m centres | m² | 27.00 |
|    laterals at 2.0 m centres | m² | 19.80 |

## 8.6 EXTERNAL DRAINAGE

| Item<br>Excluding site overheads and profit | Unit | Range £ |
|---|---|---|
| 80 mm pipe to 450 mm deep trench | | |
|    laterals at 1.0 m centres | m$^2$ | 13.80 |
|    laterals at 2.0 m centres | m$^2$ | 8.50 |
| 80 mm pipe to 600 mm deep trench | | |
|    laterals at 1.0 m centres | m$^2$ | 26.00 |
|    laterals at 2.0 m centres | m$^2$ | 19.20 |
| Extra for 100/80 mm couplings connecting laterals to main drain | | |
|    laterals at 1.0 m centres | 10 m | 52.00 |
|    laterals at 2.0 m centres | 10 m | 26.00 |
| **DRAINAGE DITCHES** | | |
| **Excavate and form ditch and bank with 45° sides in light to medium soils; all widths taken at bottom of ditch** | | |
| 300 mm wide × 600 mm deep | 100 m | 120.00 |
| 600 mm wide × 900 mm deep | 100 m | 150.00 |
| 1.20 m wide × 900 mm deep | 100 m | 1125.00 |
| 1.50 m wide × 1.20 m deep | 100 m | 1875.00 |
| **Clear and bottom existing ditch average 1.50 m deep, trim back vegetation and remove debris to licensed tip, lay jointed concrete pipes; including bedding, haunching and topping with 150 mm concrete; 11.50 N/mm$^2$ – 40 mm aggregate; backfill with approved spoil from site** | | |
| Pipes 300 mm diameter | 100 m | 7900.00 |
| Pipes 450 mm diameter | 100 m | 10000.00 |
| Pipes 600 mm diameter | 100 m | 13000.00 |
| **LAND DRAINS** | | |
| Excavate trench by excavator; lay Type 2 bedding; backfill to 150 mm above pipe with gravel rejects; lay non-woven geofabric and fill with topsoil to ground level | | |
| Trench 600 deep | | |
|    160 mm PVC-u drainpipe | 100 m | 2175.00 |
|    110 mm PVC-u drainpipe | 100 m | 1875.00 |
|    150 mm vitrified clay | 100 m | 2900.00 |
|    100 mm vitrified clay | 100 m | 2025.00 |
| **SOAKAWAYS** | | |
| Construct soakaway from perforated concrete rings; excavation, casting in situ concrete ring beam base; filling with gravel 250 mm deep; placing perforated concrete rings; surrounding with geofabric and backfilling to external surround of soakaway with 250 mm wide granular surround and excavated material; step irons and cover slab; inclusive of all earthwork retention and disposal off site of surplus material | | |
| 900 mm diameter | | |
|    1.00 m deep | nr | 770.00 |
|    2.00 m deep | nr | 1450.00 |
| 1200 mm diameter | | |
|    1.00 m deep | nr | 1025.00 |
|    2.00 m deep | nr | 1850.00 |

## 8.6 EXTERNAL DRAINAGE

| Item<br>Excluding site overheads and profit | Unit | Range £ |
|---|---|---|
| **8.6.4 LAND DRAINAGE – cont** | | |
| **Construct soakaway from perforated concrete rings – cont**<br>2400 mm diameter | | |
| 1.00 m deep | nr | 3100.00 |
| 2.00 m deep | nr | 5200.00 |

# PART 4

# Prices for Measured Works

**INTRODUCTION**

Typical Project Profile

| | |
|---|---|
| Contract value | £100,000.00–£900,000.00 |
| Labour rate (see page 3) | £23.75 per hour |
| Bricklayer | £30.00 per hour |
| Labour rate for maintenance contracts | £19.00 per hour |
| Number of site staff | 35 |
| Project area | 6000 m² |
| Project location | Outer London |
| Project components | 50% hard landscape<br>50% soft landscape and planting |
| Access to works areas | Very good |
| Contract | Main contract |
| Delivery of materials | Full loads |
| Profit and site overheads | Excluded |

## NEW ITEMS 2017

| Item<br>Excluding site overheads and profit | PC<br>£ | Labour<br>hours | Labour<br>£ | Plant<br>£ | Material<br>£ | Unit | Total<br>rate £ |
|---|---|---|---|---|---|---|---|
| **5 EXCAVATION AND FILLING – EARTHWORK RETENTION** | | | | | | | |
| **Manhole boxes** | | | | | | | |
| Manhole boxes to excavations (not included); Timber plywood with timber braces to shore excavations to 1.00 m deep; opposing faces distance less than 1.00 m | | | | | | | |
| 1.00 m deep | – | 0.50 | 23.75 | – | 9.35 | m² | 33.10 |
| 2.00 m deep | | | | | | | |
| **Trench boxes; A-Plant** | | | | | | | |
| Manhole box soil retention | – | 0.50 | 35.63 | 166.30 | – | week | 201.93 |
| Placement of manhole protection system to 3.0 m deep to enable earthwork retention; maximum distance between opposing faces 1.8 m | – | 0.50 | 35.63 | 225.30 | – | week | 260.93 |
| **6 GROUND REMEDIATION – RETAINING WALL SYSTEMS** | | | | | | | |
| **Retaining Walls; FP McCann Ltd; Easi-Bloc; (Lego Block); precast self-locking concrete gravity wall system; storage bays, partitions, retaining walls, revetments; to firm and levelled base; concrete or granular; min. 800 mm wide (not included)** | | | | | | | |
| Blocks 450 mm high laid stretcher bond | | | | | | | |
| 1.20 m × 600 mm (750 kg per block) | 148.00 | 0.46 | 32.99 | 21.57 | 148.00 | m² | 202.56 |
| 600 mm × 600 mm (375 kg per block) | 177.60 | 0.50 | 35.63 | 23.30 | 177.60 | m² | 236.53 |
| **28 ROOF FINISHINGS – NOTES** | | | | | | | |
| **Notes** | | | | | | | |
| The following systems which are compliant with European Technical Approval ETA 13/0668 (Kits for Green roofs) are indicated as such. | | | | | | | |
| The rates below assume that all elements are in position in the works location; no allowances for access, lifting or materials movement to the roof areas. | | | | | | | |

## NEW ITEMS 2017

| Item<br>Excluding site overheads and profit | PC £ | Labour hours | Labour £ | Plant £ | Material £ | Unit | Total rate £ |
|---|---|---|---|---|---|---|---|
| **Root barriers; Zinco Green Roof Systems Ltd** | | | | | | | |
| Where waterproofing does not contain a root barrier | | | | | | | |
| WSB 100-PO Weldable reinforced flexible polyolefin. 1.1 mm thick. Root resistant tested according to FLL | 13.47 | – | – | – | – | m² | – |
| WSF 40 High pressure polyethylene 0.34 mm thick | 2.31 | 0.02 | 1.43 | – | 2.31 | m² | 3.74 |
| **28 ROOF FINISHINGS – EXTENSIVE GREEN ROOF ELEMENTS** | | | | | | | |
| **Extensive Sedum type Green Roof System; Zinco Green Roof Systems Ltd.** System for planting with Sedum plugs or overlaying with pre-grown cultivated Sedum Carpet; (ETA 13/0668) | | | | | | | |
| Flat roofs with no standing water up to 10° or for Inverted roofs and as an alternative to single layer systems. Drainage, water retention and protection mat with attached filter sheet. 20 mm thick. | | | | | | | |
| Zinco Drainage Element Fixodrain XD20; 3 l/m² reservoir | 4.80 | 0.04 | 2.85 | – | 4.80 | m² | 7.65 |
| Zinco System Substrate Growing medium Sedum Carpet – laid to Fixodrain XD20 at min. 70 mm thickness | 100.00 | 0.67 | 47.50 | – | 100.00 | m² | 147.50 |
| Zinco Premium Sedum Carpet Mat – Laid over System substrate | 20.00 | 0.07 | 4.75 | – | 20.00 | m² | 24.75 |
| **Extensive Rockery type Green Roof System; Zinco Green Roof Systems Ltd.** For planting with Sedum plugs or rockery type plants; flat roofs with no standing water up to 10°. (Not suitable for inverted roofs); Suitable system depth 90 mm. System saturated weight 110 kg/m²; system water storage capacity: 36 l/m²; (ETA 13/0668) | | | | | | | |
| Non-rotting synthetic fibre mat for protection of the waterproofing layer and provision of additional water reservoir of 5 l/m² | | | | | | | |
| Zinco Moisture mat/protection fleece SSM45 | 1.95 | 0.04 | 2.85 | – | 1.95 | m² | 4.80 |

## NEW ITEMS 2017

| Item<br>Excluding site overheads and profit | PC<br>£ | Labour<br>hours | Labour<br>£ | Plant<br>£ | Material<br>£ | Unit | Total<br>rate £ |
|---|---|---|---|---|---|---|---|
| **28 ROOF FINISHINGS – EXTENSIVE GREEN ROOF ELEMENTS – cont** | | | | | | | |
| **Extensive Rockery type Green Roof System – cont** | | | | | | | |
| Pressure resistant drainage and water storage element; recycled polyolefin; laid to SSM45; 25 mm thick with diffusion openings; water storage capacity 3 l/m². | | | | | | | |
| Zinco Drainage Element Fixodrain FD25-E | 6.00 | 0.04 | 2.85 | – | 6.00 | m² | **8.85** |
| Zinco Filter Sheet SF – Rolled to drainage layer FD25-E | 0.80 | 0.04 | 2.85 | – | 0.80 | m² | **3.65** |
| Zinco System Substrate Growing medium 'Rockery Type Plants laid to Filter Fleece SF min. 70 mm thick | 100.00 | 0.67 | 47.50 | – | 100.00 | m³ | **147.50** |
| Planting to substrates | | | | | | | |
| Zinco Premium Sedum Carpet Mat | 22.00 | 0.04 | 2.85 | – | 22.00 | m² | **24.85** |
| Zinco Seed Mix – Meadow Scents; 15 g/m² | 0.66 | 0.01 | 0.71 | – | 0.66 | m² | **1.37** |
| **Extensive Rockery type Green Roof System for ZERO degree roofs; Zinco Green Roof Systems Ltd; for Zero degree roofs where isolated standing water may be present; planting with Sedum plugs or rockery type plants; system depth 150 mm; saturated weight: 105 kg/m²; water storage capacity; 33 l/m²** | | | | | | | |
| Non rotting synthetic fibre mat for protection of the waterproofing layer and provision of additional water reservoir of 3 l/m² | | | | | | | |
| Zinco Moisture mat/protection fleece TSM32 | 1.76 | 0.04 | 2.85 | – | 1.76 | m² | **4.61** |
| Drainage Element; laid to TSM32; ensures that a gap is maintained between any standing water and the soil substrate and the necessary aeration of the root zone; expanded polystyrene; 50 mm thick; lightweight (0.6 kg/m²); water storage capacity of 3 l/m² | | | | | | | |
| Zinco Drainage Element; Floraset FS50 | 7.59 | 0.04 | 2.85 | – | 7.59 | m² | **10.44** |

## NEW ITEMS 2017

| Item Excluding site overheads and profit | PC £ | Labour hours | Labour £ | Plant £ | Material £ | Unit | Total rate £ |
|---|---|---|---|---|---|---|---|
| Zinco Filter Sheet SF – Rolled to drainage layer FS50 | 0.8 | 0.01 | 0.47 | – | 0.8 | m² | 1.27 |
| Zinco System Substrate Growing medium Rockery Type Plants – laid to filter fleece SF; min 70 mm thick Planting to substrates | 100.00 | 0.67 | 47.50 | – | 100.00 | m³ | 147.50 |
| Zinco Premium Sedum Carpet Mat | 22.00 | 0.04 | 2.85 | – | 22.00 | m² | 24.85 |
| Zinco Seed Mix – Meadow Scents; 15 g/m² | 0.66 | 0.01 | 0.71 | – | 0.66 | m² | 1.37 |

**28 ROOF FINISHINGS – INTENSIVE GREEN ROOF ELEMENTS**

**Flat roofs up to 10°; Intensive Simple and Wildflower Green Roof System; Zinco Green Roof Systems Ltd; (not suitable for Inverted roofs); for planting with small shrubs, blooming perennials and wildflowers; can be combined with patios and walkways if required. System depth – 145 mm–200 mm; saturated weight – 145 kg/m²–245 kg/m²; water storage capacity – 62 l/m²–80 l/m²; (ETA 13/0668)**

| | | | | | | | |
|---|---|---|---|---|---|---|---|
| Non-rotting synthetic fibre mat for protection of the waterproofing layer and provision of additional water reservoir of 5 l/m² | | | | | | | |
| Moisture mat/protection fleece SSM45 | 1.95 | 0.04 | 2.85 | – | 1.95 | m² | 4.80 |
| Pressure resistant drainage and water storage element made from recycled polyolefin laid to SSM45; 40 mm thick with diffusion openings and water storage capacity of 6 l/m². | | | | | | | |
| Fixodrain FD40-E | 7.26 | 0.04 | 2.85 | – | 7.26 | m² | 10.11 |
| Zinco Filter Sheet SF – Rolled to drainage layer FD40-E | 0.88 | 0.04 | 2.85 | – | 0.88 | m² | 3.73 |
| System Substrate; growing medium Heather with Lavender; laid to Filter Fleece SF at 100 mm thickness up to 200 mm | 125.00 | 0.20 | 14.25 | – | 125.00 | m³ | 139.25 |

## NEW ITEMS 2017

| Item<br>Excluding site overheads and profit | PC<br>£ | Labour<br>hours | Labour<br>£ | Plant<br>£ | Material<br>£ | Unit | Total<br>rate £ |
|---|---|---|---|---|---|---|---|
| **28 ROOF FINISHINGS – INTENSIVE GREEN ROOF ELEMENTS – cont** | | | | | | | |
| **Flat roofs up to 10°; Intensive Simple and Wildflower Green Roof System; Zinco Green Roof Systems Ltd; (suitable for Inverted roofs); for planting with small shrubs, blooming perennials and wildflowers; can be combined with patios and walkways if required. System depth – 145 mm–200 mm; saturated weight – 145 kg/m$^2$–245 kg/m$^2$; water storage capacity – 62 l/m$^2$–80 l/m$^2$.** | | | | | | | |
| Water repellent, air permeable membrane used as a diffusionable separation layer; laid above waterproofing and separation membrane; rolled over the insulation | | | | | | | |
| Separation Layer; TGV21 | 0.82 | 0.04 | 2.85 | – | 0.82 | m$^2$ | **3.67** |
| Pressure resistant drainage and water storage element made from recycled polyolefin laid to SSM45; 40 mm thick with diffusion openings and water storage capacity of 6 l/m$^2$. | | | | | | | |
| Fixodrain FD40-E; laid to TGV21 | 6.60 | 0.04 | 2.85 | – | 6.60 | m$^2$ | **9.45** |
| Zinco Filter Sheet SF – Rolled to drainage layer FD40-E | 0.88 | 0.04 | 2.85 | – | 0.88 | m$^2$ | **3.73** |
| System Substrate; growing medium 'Heather with Lavender'; laid to Filter Fleece SF at 100 mm thickness up to 200 mm | 125.00 | 0.67 | 47.50 | – | 125.00 | m$^3$ | **172.50** |
| **Flat roofs up to 10°; Intensive Simple and Wildflower Green Roof System; Zinco Green Roof Systems Ltd; (suitable for Inverted roofs); for planting with lawn, shrubs, small trees, and perennials; can be combined with patios and walkways if required. System depth – min 160 mm; saturated weight – 342 kg/m$^2$ ; water storage capacity – 110 l/m$^2$.** | | | | | | | |
| Water repellent, air permeable membrane used as a diffusionable separation layer; laid above waterproofing and separation membrane; rolled over the insulation | | | | | | | |
| Separation Layer; TGV21 | 0.82 | 0.04 | 2.85 | – | 0.82 | m$^2$ | **3.67** |

## NEW ITEMS 2017

| Item<br>Excluding site overheads and profit | PC<br>£ | Labour<br>hours | Labour<br>£ | Plant<br>£ | Material<br>£ | Unit | Total<br>rate £ |
|---|---|---|---|---|---|---|---|
| Pressure resistant drainage and water storage element made from thermoformed hard plastic; laid to TGV21; 60 mm thick filled with Zincolit Plus mineral aggregate | | | | | | | |
| Zincolit Plus; mineral aggregate infill; 27 l/m$^2$ | 2.97 | 0.03 | 1.78 | – | 2.97 | m$^2$ | **4.75** |
| Filter Sheet SF – Rolled to drainage layer FD 60 | 0.88 | 0.04 | 2.85 | – | 0.88 | m$^2$ | **3.73** |
| System Substrate Growing medium; Roof Garden laid to Filter Fleece SF min.200 mm thick dependent on planting required and depth capacities | 62.50 | 0.67 | 47.50 | – | 62.50 | m$^3$ | **110.00** |
| **Flat roofs up to 10°; Intensive Simple and Wildflower Green Roof System; Zinco Green Roof Systems Ltd; (suitable for Inverted roofs); for planting with lawn, shrubs, small trees, and perennials; System depth – min 270 mm; saturated weight – 365 kg/m$^2$ ; water storage capacity – 135 l/m$^2$; (ETA 13/0668)**<br>Moisture and protection mat; Synthetic fibre resistant to mechanical stress resistant for protection of waterproofing layer and provision of additional extra water reservoir of 4 l/m$^2$ | | | | | | | |
| Zinco Moisture mat/protection fleece ISM50. | 5.61 | 0.04 | 2.85 | – | 5.61 | m$^2$ | **8.46** |
| Zinco Drainage Element FD 60 – Laid to ISM50 | 13.64 | 0.04 | 2.85 | – | 13.64 | m$^2$ | **16.49** |
| Zincolit Plus mineral aggregate infill 27 l/m$^2$ | 2.97 | 0.03 | 1.78 | – | 2.97 | m$^2$ | **4.75** |
| Filter Sheet SF – Rolled to drainage layer FD 60 | 0.88 | 0.04 | 2.85 | – | 0.88 | m$^2$ | **3.73** |
| System Substrate Growing medium; Roof Garden laid to Filter Fleece SF min 200 mm thick dependent on planting required and depth capacities | 62.50 | 0.67 | 47.50 | – | 62.50 | m$^3$ | **110.00** |

## NEW ITEMS 2017

| Item<br>Excluding site overheads and profit | PC<br>£ | Labour<br>hours | Labour<br>£ | Plant<br>£ | Material<br>£ | Unit | Total<br>rate £ |
|---|---|---|---|---|---|---|---|
| **28 ROOF FINISHINGS – INTENSIVE GREEN ROOF ELEMENTS – cont** | | | | | | | |
| **LAWN Green Roof System; Zinco Green Roof Systems Ltd.** Lawns on roofs where build up depth is restricted; Flat roofs – 0° up to 5°; Lawn surface treatment; System depth 150 mm system saturated weight min 165 kg/m². System water storage capacity; from 65 l/m² | | | | | | | |
| Protection Layer – polypropylene mat, mechanical stress resistant for protection of the waterproofing layer | | | | | | | |
|    Zinco Filter sheet PV | 2.69 | 0.04 | 2.85 | – | 2.69 | m² | **5.54** |
| Drainage and distribution element; Water storage, distribution and drainage element made from thermoformed recycled hard plastic; 45 mm thick and preformed with clipping points for Zinco Irrigation Dripperline 100-L1 | | | | | | | |
|    Zinco Aquatec 45 – laid to Filter Sheet PV | 18.30 | 0.07 | 4.75 | – | 18.30 | m² | **23.05** |
| Drippers; clipped into Aquatec 45 and linked to Zinco irrigation manager BM-4 | | | | | | | |
|    Zinco Dripperline 100-L1 | 2.69 | 0.07 | 4.75 | – | 2.69 | m² | **7.44** |
| Programmable Irrigation management unit in lockable steel box for automated irrigation of lawn and green roofs; linked to dripperlines 100-L1 | | | | | | | |
|    Zinco Irrigation Manager BM-4 | 1527.90 | 2.00 | 142.50 | – | 1667.90 | nr | **1810.40** |
| Water distributing polyester fleece with capillary effective fibres; laid over Aquatec 45 containing dripperlines 100-L1 | | | | | | | |
|    Zinco Wicking Mat DV 40 | 9.85 | 0.03 | 2.04 | – | 9.85 | m² | **11.89** |
| System Substrate Growing medium; Minimum thickness 100 mm up to 150 mm laid over Wicking mat DV 40 | | | | | | | |
|    Zinco LAWN | 18.75 | 0.07 | 4.75 | – | 18.75 | m² | **23.50** |
| Top dressing; 15 mm layer laid over LAWN substrate | | | | | | | |
|    ZINCOHUM | 1.43 | 0.01 | 0.95 | – | 1.43 | m² | **2.38** |

## NEW ITEMS 2017

| Item  Excluding site overheads and profit | PC £ | Labour hours | Labour £ | Plant £ | Material £ | Unit | Total rate £ |
|---|---|---|---|---|---|---|---|
| **28 ROOF FINISHINGS – PITCHED GREEN ROOFS** | | | | | | | |
| Extensive Green Roof System for Pitched roofs; for pitched roofs with slopes from 10° up to 25°; Zinco Green Roof Systems Ltd; (degrees suitable for planting with Sedum plugs or overlaying with pre-grown cultivated Sedum Carpet. System depth – 130 mm; system saturated weight – 115 kg/m²; system water storage capacity – 38 l/m²; (Waterproofing layer must contain a root barrier as a separate root barrier cannot be used on pitched roofs) | | | | | | | |
| Non-rotting synthetic fibre mat for protection of the waterproofing layer and provision of additional water reservoir of 7 l/m²; laid to water proofing layer | | | | | | | |
| Protection Mat BSM64 | 2.55 | 0.10 | 7.13 | – | 2.55 | m² | **9.68** |
| Lightweight (1 kg/m²) drainage element of expanded polystyrene; 75 mm thick laid with large studs upwards; laid to BSM64 | | | | | | | |
| Zinco Drainage Element Floraset FS75 | 10.25 | 0.10 | 7.13 | – | 10.25 | m² | **17.38** |
| Zinco System Substrate Growing medium infilled to drainage board and to a level of 50 mm above the large studs | | | | | | | |
| Rockery Type Plants – laid to FS75 | 8.25 | 0.10 | 7.13 | – | 8.25 | m² | **15.38** |
| Erosion net; laid over substrate to aid prevention of wind erosion | | | | | | | |
| Zinco Jute Anti Erosion Net JEG | 0.90 | 0.07 | 4.75 | – | 0.90 | m² | **5.65** |
| Vegetation layer; laid over System substrate | | | | | | | |
| Zinco Premium Sedum Carpet Mat | 20.00 | 0.07 | 4.75 | – | 20.00 | m² | **24.75** |

## NEW ITEMS 2017

| Item<br>Excluding site overheads and profit | PC<br>£ | Labour<br>hours | Labour<br>£ | Plant<br>£ | Material<br>£ | Unit | Total<br>rate £ |
|---|---|---|---|---|---|---|---|
| **28 ROOF FINISHINGS – PITCHED GREEN ROOFS – cont** | | | | | | | |
| **Extensive Green Roof System for Steep Pitched roofs with slopes from 25° up to 35°; Zinco Green Roof Systems Ltd.** Pitched roofs with slopes from 25° up to 35°; Suitable for overlay with pre-grown cultivated Sedum Carpet. System height – 120 mm; system saturated weight – 155 kg/m$^2$; water storage capacity – 64 l/m$^2$ (depending on slope); (Waterproofing layer must contain a root barrier as a separate root barrier cannot be used on pitched roofs) | | | | | | | |
| Protection mat; non-rotting needle fleeced synthetic mat for protection of the waterproofing layer and provision of additional water reservoir of 12 l/m$^2$ | | | | | | | |
|   WSM150 – laid to waterproofing layer | 5.95 | 0.10 | 7.13 | – | 5.95 | m$^2$ | **13.08** |
| Drainage element; Stable Grid element 100 mm high made from recycled polyethylene. Grid dimension 625 mm with integrated connecting plug | | | | | | | |
|   Zinco Drainage Element Georaster – Laid to WSM150 | 14.65 | 0.13 | 9.50 | – | 14.65 | m$^2$ | **24.15** |
| Zinco System Substrate Growing medium Heather with Lavender filled to Georaster | | | | | | | |
|   Zinco Premium Sedum Carpet Mat – Laid over Georaster | 20.00 | 0.20 | 14.25 | – | 20.00 | m$^2$ | **34.25** |
| **Shear barriers and brackets for pitched roofs; Zinco Green Roof Systems Ltd.** Elements in pitched roof constructions transfer shear forces into the roof construction. Structural calculations required to establish the exact location and fixing of the shear barriers and brackets | | | | | | | |
| Zinco Eaves Profile and support brackets | | | | | | | |
|   Eaves Profile TRP140 E; Stainless steel angle 140 mm high × 3.00 m long | 15.47 | 0.08 | 5.94 | – | 15.47 | m | **21.41** |
|   Support bracket TSH100 | 22.20 | 0.50 | 35.63 | – | 22.20 | nr | **57.83** |
|   Shear fix bracket LF300 | 83.00 | 0.50 | 35.63 | – | 83.00 | nr | **118.63** |

## NEW ITEMS 2017

| Item<br>Excluding site overheads and profit | PC<br>£ | Labour<br>hours | Labour<br>£ | Plant<br>£ | Material<br>£ | Unit | Total<br>rate £ |
|---|---|---|---|---|---|---|---|
| **33 DRAINAGE ABOVE GROUND – CULVERTS** | | | | | | | |
| **Culverts; precast box sections FP McCann Ltd; For directing watercourses and underpasses; to trenches or prepared bases (not included) on 200 mm prepared granular or concrete base blinded with 50 mm sand** | | | | | | | |
| Plain box culvert; 800 mm high; width | | | | | | | |
| 1.20 m | 575.00 | 3.00 | 213.75 | 65.83 | 583.51 | m | 863.09 |
| 1.50 m | 605.00 | 3.50 | 249.38 | 82.28 | 613.51 | m | 945.17 |
| Plain box culvert; 1.00 m high; width | | | | | | | |
| 1.20 m | 600.00 | 4.00 | 285.00 | 98.74 | 608.51 | m | 992.25 |
| 1.50 m | 700.00 | 4.50 | 320.63 | 115.19 | 708.51 | m | 1144.33 |
| 1.80 m | 775.00 | 6.00 | 427.50 | 164.56 | 783.51 | m | 1375.57 |
| Plain box culvert; 1.20 m high; width | | | | | | | |
| 1.50 m | 740.00 | 6.00 | 427.50 | 164.56 | 748.51 | m | 1340.57 |
| 1.80 m | – | 8.00 | 570.00 | 197.47 | 808.51 | m | 1575.98 |
| Extra over for cantilevered animal ledges | 200.00 | – | – | – | 200.00 | m | 200.00 |
| **34 DRAINAGE BELOW GROUND – HEADWALLS** | | | | | | | |
| **Headwalls; FP McCann Ltd; Precast concrete headwalls to outflows; Placing to prepared graded and blinded 300 mm thick (not included); connection of outfall pipe; 150 mm thick wall with non-standard opening; excludes standard grating** | | | | | | | |
| Size on plan 1.618 m tapering to 889 mm × 1.32 m x 663 mm high | | | | | | | |
| Small Headwall; pipe size up to 300 mm | 638.60 | 4.00 | 285.00 | 46.59 | 701.00 | ea | 1032.59 |
| Size on plan 1.878 m tapering to 1.026 m x 1.37 m × 998 mm high | | | | | | | |
| Medium Headwall; pipe size up to 450 mm | 1040.30 | 7.00 | 498.75 | 139.78 | 1137.95 | ea | 1776.48 |
| Size on plan 2.351 m tapering to 1.656 m × 2.150 m x 1.258 m high | | | | | | | |
| Large Headwall; pipe size up to 900 mm | 1534.70 | 8.00 | 570.00 | 279.56 | 1534.70 | ea | 2384.26 |

## NEW ITEMS 2017

| Item<br>Excluding site overheads and profit | PC<br>£ | Labour<br>hours | Labour<br>£ | Plant<br>£ | Material<br>£ | Unit | Total<br>rate £ |
|---|---|---|---|---|---|---|---|
| **34 DRAINAGE BELOW GROUND – CONCRETE DRAINAGE** | | | | | | | |
| **Inspection chambers (House – HIC); precast concrete units; to excavations, earthwork reinforcement, in situ bases; benchings all priced separately; includes cover slab** | | | | | | | |
| Precast concrete inspection chamber units; FP McCann Ltd; bedding, jointing and pointing in cement:mortar (1:3); 600 × 450 mm internally | | | | | | | |
|   600 mm deep | 50.45 | 6.50 | 154.38 | – | 57.17 | nr | **211.55** |
|   900 mm deep | 67.20 | 7.00 | 166.25 | – | 72.31 | nr | **238.56** |
| Drainage chambers; FP McCann Ltd; 1200 × 750 mm; no base unit; depth of invert | | | | | | | |
|   1.00 m deep | 185.32 | 3.00 | 213.75 | – | 193.83 | nr | **407.58** |
|   1.50 m deep | 237.04 | 3.67 | 261.25 | – | 250.66 | nr | **511.91** |
|   2.00 m deep | 288.76 | 4.17 | 296.88 | – | 305.79 | nr | **602.67** |
| **Bases to chamber ring manholes; in situ concrete; to excavated ground (not included); on 50 mm ST2 blinding on 225 mm mass concrete to receive bottom manhole ring blinded with 50 mm ST2 blinding** | | | | | | | |
| Diameter of manhole | | | | | | | |
|   900 mm | – | 4.00 | 190.00 | – | 33.96 | nr | **223.96** |
|   1.05 m | – | 4.00 | 190.00 | – | 41.09 | nr | **231.09** |
|   1.20 m | – | 4.50 | 213.75 | – | 57.39 | nr | **271.14** |
|   1.50 m | – | 5.50 | 261.25 | – | 86.93 | nr | **348.18** |
| Bottom manhole chamber ring; FP McCann Ltd; including pipe entrances to bases (not included); forming pipe entrances to manhole ring; seating and sealing to base; connecting pipes (not included); depth 1.00 m; diameter | | | | | | | |
|   900 mm | 78.45 | 3.00 | 213.75 | 93.19 | 86.96 | nr | **393.90** |
|   1.05 m | 82.71 | 3.25 | 231.56 | 93.19 | 91.22 | nr | **415.97** |
|   1.20 m | 100.98 | 4.00 | 285.00 | 116.48 | 109.49 | nr | **510.97** |
|   1.50 m | 176.30 | 4.50 | 320.63 | 139.78 | 184.81 | nr | **645.22** |
| Benchings to manholes; 400 mm concrete; diameter of manhole | | | | | | | |
|   900 mm | – | 0.75 | 35.63 | – | 46.74 | nr | **82.37** |
|   1.05 m | – | 0.75 | 35.63 | – | 65.43 | nr | **101.06** |
|   1.20 m | – | 0.90 | 42.75 | – | 84.13 | nr | **126.88** |
|   1.50 m | – | 1.10 | 52.25 | – | 132.76 | nr | **185.01** |

## NEW ITEMS 2017

| Item<br>Excluding site overheads and profit | PC<br>£ | Labour<br>hours | Labour<br>£ | Plant<br>£ | Material<br>£ | Unit | Total<br>rate £ |
|---|---|---|---|---|---|---|---|
| **Easi-Base Polypropylene liner; FP McCann Ltd;** complete with prefabricated benching and channels as one unit; Easi-Base diameter | | | | | | | |
| 1200 mm | 427.45 | 1.00 | 47.50 | 46.59 | 427.45 | nr | **521.54** |
| 1500 mm | 854.90 | 1.00 | 47.50 | 46.59 | 854.90 | nr | **948.99** |
| **Concrete manhole chamber rings; FP McCann Ltd;** standard ring units to excavated ground (not included); sealed with bitumen tape; tongue and groove units installed to 1.00 m deep base units (not included); surrounding in concrete min 150 thick and backfilling; Note: The final depths shown below show operations onto the 1.00 m deep base rings above | | | | | | | |
| Internal diameter 900 mm; final manhole depth | | | | | | | |
| 1.00 m deep | – | 1.00 | 71.25 | 46.59 | 86.31 | nr | **204.15** |
| Internal diameter 1.05 mm; final manhole depth | | | | | | | |
| 1.00 m deep | – | 1.00 | 71.25 | 93.19 | 86.31 | nr | **250.75** |
| 2.00 m deep | 82.71 | 2.50 | 178.13 | 163.08 | 246.81 | nr | **588.02** |
| 3.00 m deep | 165.42 | 4.00 | 285.00 | 186.38 | 406.07 | nr | **877.45** |
| Internal diameter 1200 mm; final manhole depth | | | | | | | |
| 1.00 m deep | – | 1.00 | 71.25 | 139.78 | 86.31 | nr | **297.34** |
| 2.00 m deep | 100.98 | 3.00 | 213.75 | 186.38 | 265.08 | nr | **665.21** |
| 3.00 m deep | 201.96 | 4.00 | 285.00 | 279.56 | 442.61 | nr | **1007.17** |
| Internal diameter 1500 mm | | | | | | | |
| 1.00 m deep | – | 2.00 | 142.50 | 139.78 | 86.31 | nr | **368.59** |
| 2.00 m deep | 176.30 | 4.00 | 285.00 | 186.38 | 340.40 | nr | **811.78** |
| 3.00 m deep | 352.60 | 5.00 | 356.25 | 279.56 | 593.25 | nr | **1229.06** |
| **Concrete manhole chamber rings; Wide wall; FP McCann Ltd;** 130 mm thick precast manholes; (concrete surround on backfill not required) to excavated ground (not included); Water-tight, complete with sealant; to prepared base (or Easi-Base priced elsewhere) | | | | | | | |
| Internal diameter 1200 mm | | | | | | | |
| 1.00 m deep | 169.95 | 3.00 | 213.75 | 139.78 | 169.95 | nr | **523.48** |
| 2.00 m deep | 339.90 | 4.00 | 285.00 | 186.38 | 339.90 | nr | **811.28** |
| 3.00 m deep | 509.85 | 6.00 | 427.50 | 232.97 | 509.85 | nr | **1170.32** |

## NEW ITEMS 2017

| Item<br>Excluding site overheads and profit | PC<br>£ | Labour<br>hours | Labour<br>£ | Plant<br>£ | Material<br>£ | Unit | Total<br>rate £ |
|---|---|---|---|---|---|---|---|
| **34 DRAINAGE BELOW GROUND – CONCRETE DRAINAGE – cont** | | | | | | | |
| **Concrete manhole chamber rings – cont** | | | | | | | |
| Internal diameter 1500 mm | | | | | | | |
| 1.00 m deep | 267.80 | 3.00 | 213.75 | 186.38 | 276.31 | nr | **676.44** |
| 2.00 m deep | 535.60 | 5.00 | 356.25 | 279.56 | 544.11 | nr | **1179.92** |
| 3.00 m deep | 803.40 | 7.00 | 498.75 | 326.16 | 811.91 | nr | **1636.82** |
| **Cover slabs for chambers or shaft sections; FP McCann Ltd; heavy duty** | | | | | | | |
| 900 mm diameter internally | 79.70 | 0.67 | 15.82 | – | 79.70 | nr | **95.52** |
| 1050 mm diameter internally | 85.53 | 2.00 | 47.50 | 17.08 | 85.53 | nr | **150.11** |
| 1200 mm diameter internally | 103.74 | 1.00 | 23.75 | 17.08 | 103.74 | nr | **144.57** |
| 1500 mm diameter internally | 188.81 | 1.00 | 23.75 | 17.08 | 188.81 | nr | **229.64** |
| 1800 mm diameter internally | 310.01 | 2.00 | 47.50 | 38.44 | 310.01 | nr | **395.95** |
| **35 SITE WORKS – PAVINGS – PRECAST CONCRETE STEPS** | | | | | | | |
| **Precast concrete steps; FP McCann Ltd; Precast units laid to graded, blinded and levelled bases (not included); to receive surface treatments priced separately; waist 200 mm thick; finished sizes with paving surface treatments of 40 mm thick including bedding** | | | | | | | |
| 5 Risers; 280 mm deep x 175 mm high | | | | | | | |
| 1.20 m wide | 303.98 | 3.00 | 213.75 | 32.91 | 303.98 | nr | **550.64** |
| 2.00 m wide | 607.95 | 4.00 | 285.00 | 65.83 | 607.95 | nr | **958.78** |
| 2.50 m wide | 759.98 | 5.00 | 356.25 | 164.56 | 759.98 | nr | **1280.79** |
| 3.00 m wide | 911.92 | 7.00 | 498.75 | 164.56 | 911.92 | nr | **1575.23** |
| 4.00 m wide | 1215.90 | 8.00 | 570.00 | 230.39 | 1215.90 | nr | **2016.29** |
| 10 Risers; 280 mm deep x 175 mm high | | | | | | | |
| 1.20 m wide | 729.60 | 4.00 | 285.00 | 65.83 | 729.60 | nr | **1080.43** |
| 2.00 m wide | 1216.05 | 6.00 | 427.50 | 131.65 | 1216.05 | nr | **1775.20** |
| 2.50 m wide | 1520.10 | 8.00 | 570.00 | 263.30 | 1520.10 | nr | **2353.40** |
| 3.00 m wide | 1824.08 | 8.00 | 570.00 | 263.30 | 1824.08 | nr | **2657.38** |
| 4.00 m wide | 2432.10 | 8.00 | 570.00 | 263.30 | 2432.10 | nr | **3265.40** |
| 5 Risers; 450 mm deep x 175 mm high | | | | | | | |
| 1.20 m wide | 559.65 | 3.00 | 213.75 | 32.91 | 559.65 | nr | **806.31** |
| 2.00 m wide | 932.70 | 4.00 | 285.00 | 65.83 | 932.70 | nr | **1283.53** |
| 2.50 m wide | 1165.95 | 5.00 | 356.25 | 164.56 | 1165.95 | nr | **1686.76** |
| 3.00 m wide | 1057.05 | 7.00 | 498.75 | 164.56 | 1057.05 | nr | **1720.36** |
| 4.00 m wide | 1865.47 | 8.00 | 570.00 | 230.39 | 1865.47 | nr | **2665.86** |

## NEW ITEMS 2017

| Item<br>Excluding site overheads and profit | PC<br>£ | Labour<br>hours | Labour<br>£ | Plant<br>£ | Material<br>£ | Unit | Total<br>rate £ |
|---|---|---|---|---|---|---|---|
| 10 Risers; 450 mm deep × 175 mm high | | | | | | | |
|   1.20 m wide | 1119.60 | 3.00 | 213.75 | 32.91 | 1119.60 | nr | **1366.26** |
|   2.00 m wide | 1866.00 | 4.00 | 285.00 | 65.83 | 1866.00 | nr | **2216.83** |
|   2.50 m wide | 2332.50 | 5.00 | 356.25 | 164.56 | 2332.50 | nr | **2853.31** |
|   3.00 m wide | 2797.50 | 7.00 | 498.75 | 164.56 | 2797.50 | nr | **3460.81** |
|   4.00 m wide | 3727.50 | 8.00 | 570.00 | 230.39 | 3727.50 | nr | **4527.89** |
| Extra over for integrated landing in step flights; 1.00 m deep | | | | | | | |
|   1.20 m wide | – | – | – | – | 180.00 | nr | **180.00** |
|   2.00 m wide | – | – | – | – | 300.00 | nr | **300.00** |
|   2.50 m wide | – | – | – | – | 375.00 | nr | **375.00** |
|   3.00 m wide | – | – | – | – | 450.00 | nr | **450.00** |
|   4.00 m wide | – | – | – | – | 600.00 | nr | **600.00** |

### 35 SITE WORKS – PAVINGS – BLOCK AND BEAM SYSTEMS

Beam and Block suspended base for external use; FP McCann Ltd; Pre-stressed precast concrete beam; profile 115 mm wide at base, 80 mm wide at top 152 mm deep; laid to perimeter or spanned support retaining walls (not included); block infill of 440 × 225 × 100 mm solid blocks; screed of 75 mm concrete reinforced with mesh to receive surface treatments

| Item | PC £ | Labour hours | Labour £ | Plant £ | Material £ | Unit | Total rate £ |
|---|---|---|---|---|---|---|---|
| Block centres between beams | | | | | | | |
|   520 mm; blocks laid 440 mm wide | 10.50 | 0.30 | 14.36 | 5.89 | 26.53 | m² | **46.78** |
|   408 mm; alternate blocks 440 mm wide and 215 mm wide between beams | 13.34 | 0.36 | 17.28 | 7.78 | 27.49 | m² | **52.55** |
|   295 mm; blocks laid 215 mm wide between beams | 21.00 | 0.36 | 17.28 | 11.78 | 34.01 | m² | **63.07** |

### 35 SITE WORKS – PLAYING SURFACES – PLAYING FIELDS

Grassform Ltd – Sports field construction

Note: The playing fields in this section are constructed to a lower standard than the specialist sports fields elsewhere in this book and would compy with standards for general amenity in parks, recreation grounds or school playing fields.

**NEW ITEMS 2017**

| Item<br>Excluding site overheads and profit | PC<br>£ | Labour<br>hours | Labour<br>£ | Plant<br>£ | Material<br>£ | Unit | Total<br>rate £ |
|---|---|---|---|---|---|---|---|
| **35 SITE WORKS – PLAYING SURFACES – PLAYING FIELDS – cont** | | | | | | | |
| **Surface preparations to existing playing area** | | | | | | | |
| Apply herbicide and strip existing turf 25 mm thick | – | – | – | 270.00 | 25.60 | 1000 m² | **295.60** |
| Disposal green waste off site | – | – | – | 48.16 | 750.00 | 1000 m² | **798.16** |
| **Cultivation and soil preparation** | | | | | | | |
| Light to medium soils | | | | | | | |
| Decompaction operations by tractor; ripping and breaking up ground | – | – | – | 232.50 | – | 1000 m² | **232.50** |
| Cutivation and stone burying harrow and power raking in single operation | – | – | – | 540.00 | – | 1000 m² | **540.00** |
| Clay soils | | | | | | | |
| Decompaction operations by tractor; ripping and breaking up ground | – | – | – | 270.00 | – | 1000 m² | **270.00** |
| Strip and remove 100 mm thick surface | – | – | – | 192.63 | – | 1000 m² | **192.63** |
| Disposal on site | – | – | – | 259.18 | – | 1000 m² | **259.18** |
| Disposal off site 100 mm thick arisings | – | 2.78 | 66.03 | – | 3081.09 | 1000 m² | **3147.12** |
| Cutivation, stone burying, harrow and power raking in single operation | – | – | – | 600.00 | – | 1000 m² | **600.00** |
| Imported new topsoil rootzone 150 mm deep | – | – | – | 1348.49 | 5940.00 | 1000 m² | **7288.49** |
| Drainage operations | | | | | | | |
| Sand grooving or sand slitting to enable drainage | – | – | – | – | – | 1000 m² | **–** |
| Surface operations to prepared ground | | | | | | | |
| fertilizing 35 g/m² | 42.00 | 3.00 | 71.25 | – | 42.00 | 1000 m² | **113.25** |
| seeding operations by mechanical seeder; sportsfield seed 35g/m² by mechanical seeder | 230.77 | – | – | 40.00 | 230.77 | 1000 m² | **270.77** |
| **35 SITE WORKS – STREET FURNITURE – ARCHITECTURAL METALWORK** | | | | | | | |
| **Bespoke Planters; John Booth Engineering**; planters to required size and shape; with seamless polished corners; preformed drainage holes and lip detail; inclusive of placement; drainage layer 50 mm thick and potting soil | | | | | | | |

## NEW ITEMS 2017

| Item<br>Excluding site overheads and profit | PC<br>£ | Labour<br>hours | Labour<br>£ | Plant<br>£ | Material<br>£ | Unit | Total<br>rate £ |
|---|---|---|---|---|---|---|---|
| Galvanized 1.5 mm steel; powder coated | | | | | | | |
| 500 mm × 500 mm × 500 mm | 325.00 | 0.50 | 23.75 | – | 331.31 | nr | 355.06 |
| 1500 mm × 500 mm × 500 mm | 403.00 | 0.80 | 38.00 | – | 420.60 | nr | 458.60 |
| 1800 mm × 750 mm × 600 mm | 513.00 | 1.60 | 76.00 | 16.46 | 551.71 | nr | 644.17 |
| 2500 mm × 1000 mm × 600 mm | 629.00 | 1.50 | 71.25 | 16.46 | 706.74 | nr | 794.45 |
| Corten 2.0 mm | | | | | | | |
| 500 mm × 500 mm × 500 mm | 336.00 | 0.50 | 23.75 | – | 342.31 | nr | 366.06 |
| 1500 mm × 500 mm × 500 mm | 427.00 | 0.80 | 38.00 | – | 444.60 | nr | 482.60 |
| 1800 mm × 750 mm × 600 mm | 551.00 | 1.60 | 76.00 | 16.46 | 589.71 | nr | 682.17 |
| 2500 mm × 1000 mm × 600 mm | 687.00 | 1.50 | 71.25 | 16.46 | 764.74 | nr | 852.45 |
| Stainless steel 1.5 mm; Grained finish | | | | | | | |
| 500 mm × 500 mm × 500 mm | 416.00 | 0.50 | 23.75 | – | 422.31 | nr | 446.06 |
| 1500 mm × 500 mm × 500 mm | 602.00 | 0.80 | 38.00 | – | 619.60 | nr | 657.60 |
| 1800 mm × 750 mm × 600 mm | 828.00 | 1.60 | 76.00 | 16.46 | 866.71 | nr | 959.17 |
| 2500 mm × 1000 mm × 600 mm | 1106.00 | 1.50 | 71.25 | 16.46 | 1183.74 | nr | 1271.45 |
| **37 SOFT LANDSCAPING – MATURE MULTI-STEM TREES** | | | | | | | |
| **Tree planting; Multi-stem trees; Hillier Nureries; Airpot container grown trees; semi-mature** | | | | | | | |
| Betula Pendula; including backfilling with excavated material (other operations not included) | | | | | | | |
| Betula Pendula Multistem; 300–400 cm | 135.00 | 3.00 | 213.75 | 83.79 | 135.00 | nr | 432.54 |
| Betula Pendula Multistem; 400–500 cm | 250.00 | 3.25 | 231.56 | 83.79 | 250.00 | nr | 565.35 |
| Betula Pendula Multistem; 500–600 cm | 375.00 | 3.50 | 249.38 | 88.58 | 375.00 | nr | 712.96 |
| Betula Pendula Multistem; 600–700 cm | 850.00 | 3.75 | 267.19 | 91.87 | 850.00 | nr | 1209.06 |
| Betula Pendula Multistem; 700–800 cm | 1500.00 | 4.00 | 285.00 | 93.51 | 1500.00 | nr | 1878.51 |
| Betula Pendula Multistem; 800–900 cm | 1900.00 | 4.50 | 320.63 | 102.90 | 1900.00 | nr | 2323.53 |
| Betula jacquemontii; including backfilling with excavated material (other operations not included) | | | | | | | |
| Betula Jacquemontii Multistem; 300–400 cm | 300.00 | 3.00 | 213.75 | 83.79 | 300.00 | nr | 597.54 |
| Betula Jacquemontii Multistem; 400–500 cm | 450.00 | 3.25 | 231.56 | 83.79 | 450.00 | nr | 765.35 |
| Betula Jacquemontii Multistem; 500–600 cm | 750.00 | 3.50 | 249.38 | 88.58 | 750.00 | nr | 1087.96 |
| Betula Jacquemontii Multistem; 600–700 cm | 1500.00 | 4.00 | 285.00 | 93.51 | 1500.00 | nr | 1878.51 |
| Betula Jacquemontii Multistem; 700–800 cm | 1950.00 | 4.50 | 320.63 | 102.90 | 1950.00 | nr | 2373.53 |

**NEW ITEMS 2017**

| Item<br>Excluding site overheads and profit | PC<br>£ | Labour<br>hours | Labour<br>£ | Plant<br>£ | Material<br>£ | Unit | Total<br>rate £ |
|---|---|---|---|---|---|---|---|
| **37 SOFT LANDSCAPING –**<br>**SPECIMEN ORNAMENTAL**<br>**SHRUBS**<br>Market prices of high impact specimen shrubs; Griffin Nurseries; Prices are varied and the prices below reflect the price ranges for plants as described; container sizes | | | | | | | |
| 25 litre container | 40.00 | – | – | – | 40.00 | nr | **40.00** |
| 25 lite container; specialist feature plant specimen | 54.00 | – | – | – | 54.00 | nr | **54.00** |
| 40 litre container | 68.00 | – | – | – | 68.00 | nr | **68.00** |
| 40 litre container; specialist feature plant specimen | 105.00 | – | – | – | 105.00 | nr | **105.00** |
| 80 litre container | 95.00 | – | – | – | 95.00 | nr | **95.00** |
| 80 litre container; specialist feature plant specimen | 195.00 | – | – | – | 195.00 | nr | **195.00** |
| 130 litre container; specialist feature plant specimen | 295.00 | – | – | – | 295.00 | nr | **295.00** |

*Prices for Measured Works – Major Works*

## 1 PRELIMINARIES

| Item<br>Excluding site overheads and profit | PC<br>£ | Labour<br>hours | Labour<br>£ | Plant<br>£ | Material<br>£ | Unit | Total<br>rate £ |
|---|---|---|---|---|---|---|---|
| **1.2 EMPLOYER'S REQUIREMENTS: TENDERING/SUBLETTING/SUPPLY** | | | | | | | |
| **Tendering costs for an employed estimator for the acquisition of a landscape main or subcontract; inclusive of measurement, sourcing of materials and suppliers, cost and area calculations, pre- and post-tender meetings, bid compliance method statements and submission documents** | | | | | | | |
| Remeasureable contract from prepared bills; contract value | | | | | | | |
| £25,000 | – | – | – | – | – | nr | 216.00 |
| £50,000 | – | – | – | – | – | nr | 360.00 |
| £100,000 | – | – | – | – | – | nr | 864.00 |
| £200,000 | – | – | – | – | – | nr | 1440.00 |
| £500,000 | – | – | – | – | – | nr | 1872.00 |
| £1,000,000 | – | – | – | – | – | nr | 2880.00 |
| Lump sum contract from specifications and drawings only; contract value | | | | | | | |
| £25,000 | – | – | – | – | – | nr | 576.00 |
| £50,000 | – | – | – | – | – | nr | 720.00 |
| £100,000 | – | – | – | – | – | nr | 1440.00 |
| £200,000 | – | – | – | – | – | nr | 2592.00 |
| £500,000 | – | – | – | – | – | nr | 3600.00 |
| £1,000,000 | – | – | – | – | – | nr | 5760.00 |
| **Material costs for tendering** | | | | | | | |
| Drawing and plan printing and distribution costs; plans issued by employer on CD; project value | | | | | | | |
| £30,000 | – | – | – | – | 19.00 | nr | 19.00 |
| £30,000 to £80,000 | – | – | – | – | 47.50 | nr | 47.50 |
| £80,000 to £150,000 | – | – | – | – | 95.00 | nr | 95.00 |
| £150,000 to £300,000 | – | – | – | – | 133.00 | nr | 133.00 |
| £300,000 to £1,000,000 | – | – | – | – | 228.00 | nr | 228.00 |
| **Programmes; tender stage programmes** | | | | | | | |
| Allow for production of an outline works programmes for submission with the tenders; project value | | | | | | | |
| £30,000 | – | 1.50 | 52.50 | – | – | nr | 52.50 |
| £50,000 | – | 2.00 | 70.00 | – | – | nr | 70.00 |
| £75,000 | – | 2.50 | 87.50 | – | – | nr | 87.50 |
| £100,000 | – | 3.00 | 105.00 | – | – | nr | 105.00 |
| £200,000 | – | 4.00 | 140.00 | – | – | nr | 140.00 |
| £500,000 | – | 4.50 | 157.50 | – | – | nr | 157.50 |
| £1,000,000 | – | 6.00 | 210.00 | – | – | nr | 210.00 |

## 1 PRELIMINARIES

| Item<br>Excluding site overheads and profit | PC<br>£ | Labour<br>hours | Labour<br>£ | Plant<br>£ | Material<br>£ | Unit | **Total<br>rate £** |
|---|---|---|---|---|---|---|---|
| **1.2 EMPLOYER'S REQUIREMENTS: TENDERING/SUBLETTING/SUPPLY – cont** | | | | | | | |
| **Programmes – cont** | | | | | | | |
| Method statements for private or commercial projects where award is not points based; provide detailed method statements on all aspects of the works; project value | | | | | | | |
| £30,000 | – | 2.50 | 87.50 | – | – | nr | **87.50** |
| £50,000 | – | 3.50 | 122.50 | – | – | nr | **122.50** |
| £75,000 | – | 4.00 | 140.00 | – | – | nr | **140.00** |
| £100,000 | – | 5.00 | 175.00 | – | – | nr | **175.00** |
| £200,000 | – | 6.00 | 210.00 | – | – | nr | **210.00** |
| £500,000 | – | 7.00 | 245.00 | – | – | nr | **245.00** |
| £1,000,000 | – | 8.00 | 280.00 | – | – | nr | **280.00** |
| **Method statements for local authority type projects where award is points rated; provide detailed method statements on all aspects of the works; project value** | | | | | | | |
| £30,000 | – | 5.00 | 175.00 | – | – | nr | **175.00** |
| £50,000 | – | 5.00 | 175.00 | – | – | nr | **175.00** |
| £75,000 | – | 6.00 | 210.00 | – | – | nr | **210.00** |
| £100,000 | – | 7.00 | 245.00 | – | – | nr | **245.00** |
| £200,000 | – | 8.00 | 280.00 | – | – | nr | **280.00** |
| £500,000 | – | 9.00 | 315.00 | – | – | nr | **315.00** |
| £1,000,000 | – | 16.00 | 560.00 | – | – | nr | **560.00** |
| **Health and safety** | | | | | | | |
| Produce health and safety file including preliminary meeting and subsequent progress meetings with external planning officer in connection with health and safety; project value | | | | | | | |
| £35,000 | – | 8.00 | 280.00 | – | – | nr | **280.00** |
| £75,000 | – | 12.00 | 420.00 | – | – | nr | **420.00** |
| £100,000 | – | 16.00 | 560.00 | – | – | nr | **560.00** |
| £200,000 to £500,000 | – | 40.00 | 1400.00 | – | – | nr | **1400.00** |
| Maintain health and safety file for project duration; project value | | | | | | | |
| £35,000 | – | 4.00 | 140.00 | – | – | week | **140.00** |
| £75,000 | – | 4.00 | 140.00 | – | – | week | **140.00** |
| £100,000 | – | 8.00 | 280.00 | – | – | week | **280.00** |
| £200,000 to £500,000 | – | 8.00 | 280.00 | – | – | week | **280.00** |

## 1 PRELIMINARIES

| Item<br>Excluding site overheads and profit | PC<br>£ | Labour<br>hours | Labour<br>£ | Plant<br>£ | Material<br>£ | Unit | Total<br>rate £ |
|---|---|---|---|---|---|---|---|
| Produce written risk assessments on all areas of operations within the scope of works of the contract; project value | | | | | | | |
| £35,000 | – | 2.00 | 70.00 | – | – | nr | 70.00 |
| £75,000 | – | 3.00 | 105.00 | – | – | nr | 105.00 |
| £100,000 | – | 5.00 | 175.00 | – | – | nr | 175.00 |
| £200,000 to £500,000 | – | 8.00 | 280.00 | – | – | nr | 280.00 |
| Produce COSHH assessments on all substances to be used in connection with the contract; project value | | | | | | | |
| £35,000 | – | 1.50 | 52.50 | – | – | nr | 52.50 |
| £75,000 | – | 2.00 | 70.00 | – | – | nr | 70.00 |
| £100,000 | – | 3.00 | 105.00 | – | – | nr | 105.00 |
| £200,000 to £500,000 | – | 3.00 | 105.00 | – | – | nr | 105.00 |
| **1.2.4 EMPLOYER'S REQUIREMENTS: SECURITY/SAFETY/PROTECTION** | | | | | | | |
| **Temporary security fence; HSS Hire; mesh framed unclimbable fencing; including precast concrete supports and couplings** | | | | | | | |
| Weekly hire; 2.85 × 2.00 m high | | | | | | | |
| weekly hire rate | – | – | – | 2.26 | – | m | 2.26 |
| erection of fencing; labour only | – | 0.10 | 2.38 | – | – | m | 2.38 |
| removal of fencing loading to collection vehicle | – | 0.07 | 1.58 | – | – | m | 1.58 |
| delivery charge | – | – | – | 0.83 | – | m | 0.83 |
| return haulage charge | – | – | – | 0.63 | – | m | 0.63 |
| **1.2.6 SUBMISSION OF SAMPLES** | | | | | | | |
| **Set up sample panels for the works inclusive of arranging delivery of sample materials; paving samples on 100 mm base** | | | | | | | |
| Costs of labours only | | | | | | | |
| brick paving panel (pointed) | – | 6.00 | 142.50 | – | – | m² | 142.50 |
| brick paving panel (butt jointed) | – | 4.00 | 95.00 | – | – | m² | 95.00 |
| block paving panel | – | 4.00 | 95.00 | – | – | m² | 95.00 |
| stone slab paving panel | – | 6.00 | 142.50 | – | – | m² | 142.50 |
| cladding panel | – | 7.00 | 166.25 | – | – | m² | 166.25 |
| brick wall panel | – | 5.00 | 118.75 | – | – | m² | 118.75 |
| render panel | – | 3.50 | 83.13 | – | – | m² | 83.13 |
| paint panel | – | 3.00 | 71.25 | – | – | m² | 71.25 |

**1 PRELIMINARIES**

| Item<br>Excluding site overheads and profit | PC<br>£ | Labour<br>hours | Labour<br>£ | Plant<br>£ | Material<br>£ | Unit | **Total<br>rate £** |
|---|---|---|---|---|---|---|---|
| **1.2.7 CONTRACTOR'S GENERAL COST ITEMS: MECHANICAL PLANT** | | | | | | | |
| **Southern Conveyors; moving only of granular material by belt conveyor; conveyors fitted with troughed belts, receiving hopper, front and rear undercarriages and driven by electrical and air motors; support work for conveyor installation (scaffolding), delivery, collection and installation all excluded** | | | | | | | |
| Delivery collection haulage and setup charges | | | | | | | |
| Conveyors up to 10 m long | – | – | – | 430.00 | – | nr | **430.00** |
| Conveyors up to 20 m long | – | – | – | 780.00 | – | nr | **780.00** |
| Conveyors up to 50 m long | – | – | – | 2180.00 | – | nr | **2180.00** |
| Conveyor belt width 400 mm; mechanically loaded and removed at offload point; conveyor length | | | | | | | |
| up to 5 m | – | 1.50 | 35.63 | 15.97 | – | m$^3$ | **51.60** |
| 10 m | – | 1.50 | 35.63 | 16.44 | – | m$^3$ | **52.07** |
| 12.5 m | – | 1.50 | 35.63 | 16.99 | – | m$^3$ | **52.62** |
| 15 m | – | 1.50 | 35.63 | 17.38 | – | m$^3$ | **53.01** |
| 20 m | – | 1.50 | 35.63 | 18.31 | – | m$^3$ | **53.94** |
| 25 m | – | 1.50 | 35.63 | 19.49 | – | m$^3$ | **55.12** |
| 30 m | – | 1.50 | 35.63 | 20.66 | – | m$^3$ | **56.29** |
| Conveyor belt width 600 mm; mechanically loaded and removed at offload point; conveyor length | | | | | | | |
| up to 5 m | – | 1.00 | 23.75 | 10.91 | – | m$^3$ | **34.66** |
| 10 m | – | 1.00 | 23.75 | 11.43 | – | m$^3$ | **35.18** |
| 12.5 m | – | 1.00 | 23.75 | 11.80 | – | m$^3$ | **35.55** |
| 15 m | – | 1.00 | 23.75 | 12.21 | – | m$^3$ | **35.96** |
| 20 m | – | 1.00 | 23.75 | 12.99 | – | m$^3$ | **36.74** |
| 25 m | – | 1.00 | 23.75 | 14.30 | – | m$^3$ | **38.05** |
| 30 m | – | 1.00 | 23.75 | 15.34 | – | m$^3$ | **39.09** |
| Conveyor belt width 400 mm; mechanically loaded and removed by hand at offload point; conveyor length | | | | | | | |
| up to 5 m | – | 3.19 | 75.77 | 5.73 | – | m$^3$ | **81.50** |
| 10 m | – | 3.19 | 75.77 | 6.35 | – | m$^3$ | **82.12** |
| 12.5 m | – | 3.19 | 75.77 | 7.08 | – | m$^3$ | **82.85** |
| 15 m | – | 3.19 | 75.77 | 7.60 | – | m$^3$ | **83.37** |
| 20 m | – | 3.19 | 75.77 | 8.85 | – | m$^3$ | **84.62** |
| 25 m | – | 3.19 | 75.77 | 10.42 | – | m$^3$ | **86.19** |
| 30 m | – | 3.19 | 75.77 | 11.98 | – | m$^3$ | **87.75** |

## 1 PRELIMINARIES

| Item<br>Excluding site overheads and profit | PC<br>£ | Labour<br>hours | Labour<br>£ | Plant<br>£ | Material<br>£ | Unit | Total<br>rate £ |
|---|---|---|---|---|---|---|---|
| **Terranova Cranes Ltd; crane hire; materials handling and lifting; telescopic cranes supply and management; exclusive of roadway management or planning applications; prices below illustrate lifts of 1 tonne at maximum crane reach; lift cycle of 0.25 hours; mechanical filling of material skip if appropriate** | | | | | | | |
| 35 tonne mobile crane; lifting capacity of 1 tonne at 26 m; C.P.A hire (all management by hirer) | | | | | | | |
|   granular materials including concrete | – | 0.75 | 17.81 | 17.38 | – | tonne | **35.19** |
|   palletized or packed materials | – | 0.50 | 11.88 | 15.00 | – | tonne | **26.88** |
| 35 tonne mobile crane; lifting capacity of 1 tonne at 26 m; contract lift | | | | | | | |
|   granular materials or concrete | – | 0.75 | 17.81 | 44.88 | – | tonne | **62.69** |
|   palletized or packed materials | – | 0.50 | 11.88 | 42.50 | – | tonne | **54.38** |
| 50 tonne mobile crane; lifting capacity of 1 tonne at 34 m; C.P.A hire (all management by hirer) | | | | | | | |
|   granular materials including concrete | – | 0.75 | 17.81 | 22.38 | – | tonne | **40.19** |
|   palletized or packed materials | – | 1.00 | 23.75 | 20.00 | – | tonne | **43.75** |
| 50 tonne mobile crane; lifting capacity of 1 tonne at 34 m; contract lift | | | | | | | |
|   granular materials or concrete | – | 0.75 | 17.81 | 56.13 | – | tonne | **73.94** |
|   palletized or packed materials | – | 0.50 | 11.88 | 53.75 | – | tonne | **65.63** |
| 80 tonne mobile crane; lifting capacity of 1 tonne at 44 m; C.P.A. hire (all management by hirer) | | | | | | | |
|   granular materials including concrete | – | 0.75 | 17.81 | 29.88 | – | tonne | **47.69** |
|   palletized or packed materials | – | 0.50 | 11.88 | 27.50 | – | tonne | **39.38** |
| 80 tonne mobile crane; lifting capacity of 1 tonne at 44 m; contract lift | | | | | | | |
|   granular materials or concrete | – | 0.75 | 17.81 | 67.38 | – | tonne | **85.19** |
|   palletized or packed materials | – | 0.50 | 11.88 | 65.00 | – | tonne | **76.88** |

## 1 PRELIMINARIES

| Item<br>Excluding site overheads and profit | PC<br>£ | Labour<br>hours | Labour<br>£ | Plant<br>£ | Material<br>£ | Unit | Total<br>rate £ |
|---|---|---|---|---|---|---|---|
| **1.2.8 CONTRACTOR'S GENERAL COST ITEMS: TEMPORARY WORKS** | | | | | | | |
| **Eve Trakway; portable roadway systems** | | | | | | | |
| Temporary roadway system laid directly onto existing surface or onto PVC matting to protect existing surface; most systems are based on a weekly hire charge with transportation, installation and recovery charges included | | | | | | | |
| Heavy Duty Trakpanel; per panel (3.05 × 2.59 m) per week | – | – | – | – | – | m² | **5.44** |
| Outrigger Mats for use in conjunction with Heavy Duty Trakpanels; per set of 4 mats per week | – | – | – | – | – | set | **86.00** |
| Medium Duty Trakpanel; per panel (2.44 × 3.00 m) per week | – | – | – | – | – | m² | **5.87** |
| LD20 Eveolution; Light Duty Trakway; roll out system minimum delivery 50 m | – | – | – | – | – | m² | **5.73** |
| Terraplas Walkways; turf protection system; per section (1 × 1 m); per week | – | – | – | – | – | m² | **5.60** |
| **1.4 UNDERGROUND SERVICES SURVEY** | | | | | | | |
| **Ground penetrating radar surveys for artefacts, services and cable avoidance** | | | | | | | |
| Areas not exceeding 1000 m² | – | – | – | – | – | nr | **4900.00** |
| **1.4 BUILDING FABRIC SURVEY** | | | | | | | |
| **Surveys** | | | | | | | |
| Asbestos surveys | | | | | | | |
| type 3 asbestos survey (destructive testing) to single storey free standing buildings such as park buildings, storage sheds and the like; total floor area less than 100 m² | – | – | – | – | – | nr | **950.00** |
| extra over for the removal only of asbestos off site to tip (tipping charges only) | – | – | – | – | – | tonne | **120.00** |

## 1 PRELIMINARIES

| Item<br>Excluding site overheads and profit | PC<br>£ | Labour<br>hours | Labour<br>£ | Plant<br>£ | Material<br>£ | Unit | **Total<br>rate £** |
|---|---|---|---|---|---|---|---|
| **1.5 TENDER AND CONTRACT DOCUMENTS** | | | | | | | |
| Note: The preliminary requirements are different for all projects. Many are time related and many are set costs. Others vary depending on the project requirements. We have attempted to provide a guide to the cost related items in Section A of the National Building Specification. These costs are a guide only. Users of the book should evaluate each project separately. There are other preliminary cost items which may be added in to the preliminary costs of any project which may not be featured here. | | | | | | | |
| **Method statements** | | | | | | | |
| Provide detailed method statements on all aspects of the works; project value | | | | | | | |
| £30,000 | – | 2.50 | 87.50 | – | – | nr | **87.50** |
| £50,000 | – | 3.50 | 122.50 | – | – | nr | **122.50** |
| £75,000 | – | 4.00 | 140.00 | – | – | nr | **140.00** |
| £100,000 | – | 5.00 | 175.00 | – | – | nr | **175.00** |
| £200,000 | – | 6.00 | 210.00 | – | – | nr | **210.00** |
| **Tender and contract documents** | | | | | | | |
| Drawing and plan printing and distribution costs; plans issued by employer on CD; project value | | | | | | | |
| £30,000 | – | – | – | – | 19.00 | nr | **19.00** |
| £30,000 to £80,000 | – | – | – | – | 47.50 | nr | **47.50** |
| £80,000 to £150,000 | – | – | – | – | 95.00 | nr | **95.00** |
| £150,000 to £300,000 | – | – | – | – | 133.00 | nr | **133.00** |
| £300,000 to £1,000,000 | – | – | – | – | 228.00 | nr | **228.00** |
| Contract drawings for project management and distribution to suppliers, subcontractors and site staff | | | | | | | |
| £30,000 | – | – | – | – | 38.00 | nr | **38.00** |
| £30,000 to £80,000 | – | – | – | – | 95.00 | nr | **95.00** |
| £80,000 to £150,000 | – | – | – | – | 190.00 | nr | **190.00** |
| £150,000 to £300,000 | – | – | – | – | 266.00 | nr | **266.00** |
| £300,000 to £1,000,000 | – | – | – | – | 456.00 | nr | **456.00** |

## 1 PRELIMINARIES

| Item<br>Excluding site overheads and profit | PC<br>£ | Labour<br>hours | Labour<br>£ | Plant<br>£ | Material<br>£ | Unit | Total<br>rate £ |
|---|---|---|---|---|---|---|---|
| **1.5 THE CONTRACT/SUBCONTRACT** | | | | | | | |
| **Contract/subcontract evaluation** | | | | | | | |
| Due diligence on evaluation or examination of clauses in contract documents; evaluation and report by a suitably qualified quantity surveyor or legal advisor where necessary | | | | | | | |
|    minor works contract | – | 1.00 | 35.00 | – | – | nr | 35.00 |
|    JCLI | – | – | – | – | 70.00 | nr | 70.00 |
|    JCT intermediate contract or subcontract | – | 1.00 | 35.00 | – | 140.00 | nr | 175.00 |
| **1.6 DESIGN COSTS FOR LANDSCAPE WORKS** | | | | | | | |
| **Landscape design; Landscape architects; mixed hard and soft landscape; indicative prices for design project stages; brief; planning; detailed design and supervision during construction stage; the complexity ratings shown refer to the Landscape Institute tables published in this book** | | | | | | | |
| Commercial urban development; complexity rating 2; landscape construction value | | | | | | | |
|    £100,000.00 | – | – | – | – | – | nr | 10000.00 |
|    £250,000.00 | – | – | – | – | – | nr | 20000.00 |
|    £500,000.00 | – | – | – | – | – | nr | 35000.00 |
|    £1,000,000.00 | – | – | – | – | – | nr | 65000.00 |
| Hospitals and education; complexity rating 3; landscape construction value | | | | | | | |
|    £100,000.00 | – | – | – | – | – | nr | 11000.00 |
|    £250,000.00 | – | – | – | – | – | nr | 23000.00 |
|    £500,000.00 | – | – | – | – | – | nr | 40000.00 |
|    £1,000,000.00 | – | – | – | – | – | nr | 74000.00 |
| Public urban open space; complexity rating 4; landscape construction value | | | | | | | |
|    £100,000.00 | – | – | – | – | – | nr | 12000.00 |
|    £250,000.00 | – | – | – | – | – | nr | 26000.00 |
|    £500,000.00 | – | – | – | – | – | nr | 47500.00 |
|    £1,000,000.00 | – | – | – | – | – | nr | 85000.00 |
| Countryside or leisure open space; complexity rating 1 | | | | | | | |
|    £100,000.00 | – | – | – | – | – | nr | 9000.00 |
|    £250,000.00 | – | – | – | – | – | nr | 18000.00 |
|    £500,000.00 | – | – | – | – | – | nr | 32000.00 |
|    £1,000,000.00 | – | – | – | – | – | nr | 60000.00 |

## 1 PRELIMINARIES

| Item<br>Excluding site overheads and profit | PC<br>£ | Labour<br>hours | Labour<br>£ | Plant<br>£ | Material<br>£ | Unit | Total<br>rate £ |
|---|---|---|---|---|---|---|---|
| Reclamation project; complexity rating 4 | | | | | | | |
| £100,000.00 | – | – | – | – | – | nr | 12000.00 |
| £250,000.00 | – | – | – | – | – | nr | 26000.00 |
| £500,000.00 | – | – | – | – | – | nr | 45000.00 |
| £1,000,000.00 | – | – | – | – | – | nr | 80000.00 |
| Housing; complexity rating 3 | | | | | | | |
| £100,000.00 | – | – | – | – | – | nr | 11000.00 |
| £250,000.00 | – | – | – | – | – | nr | 23000.00 |
| £500,000.00 | – | – | – | – | – | nr | 40000.00 |
| £1,000,000.00 | – | – | – | – | – | nr | 74000.00 |
| **1.7 EMPLOYER'S REQUIREMENTS: PROVISION CONTENT AND USE OF DOCUMENTS** | | | | | | | |
| **Provision, content and use of documents** | | | | | | | |
| Supply as built drawings for elements of the project that may have carried from the original design drawings | | | | | | | |
| £35,000 | – | 2.00 | 76.00 | – | – | nr | 76.00 |
| £75,000 | – | 2.00 | 76.00 | – | – | nr | 76.00 |
| £100,000 | – | 4.00 | 152.00 | – | – | nr | 152.00 |
| £200,000 to £500,000 | – | 5.00 | 190.00 | – | – | nr | 190.00 |
| **Fees for Considerate Contractors Scheme (CCS)** | | | | | | | |
| Project value | | | | | | | |
| up to £100,000 | – | – | – | – | 100.00 | nr | 100.00 |
| £100,000–£500,000 | – | – | – | – | 200.00 | nr | 200.00 |
| £500,000–£5,000,000 | – | – | – | – | 400.00 | nr | 400.00 |
| over £5,000,000 | – | – | – | – | 600.00 | nr | 600.00 |
| Climatic conditions – keep records of temperature and rainfall; delays due to weather including descriptions of weather | | | | | | | |
| daily cost | – | 0.08 | 2.92 | – | – | day | 2.92 |
| weekly cost | – | 0.42 | 14.58 | – | – | day | 14.58 |
| **Programmes; master programme for the contract works** | | | | | | | |
| Allow for production of works programmes prior to the start of the works; project value | | | | | | | |
| £30,000 | – | 3.00 | 105.00 | – | – | nr | 105.00 |
| £50,000 | – | 6.00 | 210.00 | – | – | nr | 210.00 |
| £75,000 | – | 8.00 | 280.00 | – | – | nr | 280.00 |
| £100,000 | – | 10.00 | 350.00 | – | – | nr | 350.00 |
| £200,000 | – | 14.00 | 490.00 | – | – | nr | 490.00 |
| £500,000 | – | 15.00 | 525.00 | – | – | nr | 525.00 |
| £1,000,000 | – | 18.00 | 630.00 | – | – | nr | 630.00 |

## 1 PRELIMINARIES

| Item<br>Excluding site overheads and profit | PC<br>£ | Labour<br>hours | Labour<br>£ | Plant<br>£ | Material<br>£ | Unit | Total<br>rate £ |
|---|---|---|---|---|---|---|---|
| **1.7 EMPLOYER'S REQUIREMENTS: PROVISION CONTENT AND USE OF DOCUMENTS – cont** | | | | | | | |
| **Indicative staff resource chart** | | | | | | | |
| Project value | | | | | | | |
| up to £100,000 | – | 1.00 | 35.00 | – | – | nr | **35.00** |
| up to £500,000 | – | 1.50 | 52.50 | – | – | nr | **52.50** |
| up to £1,000,000 | – | 2.00 | 70.00 | – | – | nr | **70.00** |
| **Curriculum vitae of staff** | | | | | | | |
| Prepare and submit curriculum vitae of pre-construction and construction phase staff | | | | | | | |
| per staff member | – | 0.75 | 26.25 | – | – | nr | **26.25** |
| **1.7 EMPLOYER'S REQUIREMENTS: MANAGEMENT OF THE WORKS** | | | | | | | |
| Allow for updating the works programme during the course of the works; project value | | | | | | | |
| £30,000 | – | 1.00 | 35.00 | – | – | nr | **35.00** |
| £50,000 | – | 1.50 | 52.50 | – | – | nr | **52.50** |
| £75,000 | – | 2.00 | 70.00 | – | – | nr | **70.00** |
| £100,000 | – | 3.00 | 105.00 | – | – | nr | **105.00** |
| £200,000 | – | 5.00 | 175.00 | – | – | nr | **175.00** |
| **Setting out** | | | | | | | |
| Setting out for external works operations comprising hard and soft works elements; placing of pegs and string lines to Landscape Architect's drawings; surveying levels and placing level pegs; obtaining approval from the Landscape Architect to commence works; areas of entire site | | | | | | | |
| 1,000 m$^2$ | – | 5.00 | 118.75 | – | 8.00 | nr | **126.75** |
| 2,500 m$^2$ | – | 8.00 | 190.00 | – | 16.00 | nr | **206.00** |
| 5,000 m$^2$ | – | 8.00 | 190.00 | – | 16.00 | nr | **206.00** |
| 10,000 m$^2$ | – | 32.00 | 760.00 | – | 40.00 | nr | **800.00** |
| **1.11 CONTRACTOR'S GENERAL COST ITEMS: SITE ACCOMMODATION** | | | | | | | |
| **General** | | | | | | | |
| The following items are instances of the commonly found preliminary costs associated with external works contracts. The assumption is made that the external works contractor is subcontracted to a main contractor. | | | | | | | |

## 1 PRELIMINARIES

| Item<br>Excluding site overheads and profit | PC<br>£ | Labour<br>hours | Labour<br>£ | Plant<br>£ | Material<br>£ | Unit | Total<br>rate £ |
|---|---|---|---|---|---|---|---|
| **Elliot Hire; erect temporary accommodation and storage on concrete base measured separately** | | | | | | | |
| Prefabricated office hire; jackleg; open plan | | | | | | | |
|   3.6 x 2.4 m | – | – | – | 30.93 | – | week | **30.93** |
|   4.8 x 2.4 m | – | – | – | 34.09 | – | week | **34.09** |
| Armoured store; erect temporary secure storage container for tools and equipment | | | | | | | |
|   3.0 x 2.4 m | – | – | – | 14.00 | – | week | **14.00** |
|   3.6 x 2.4 m | – | – | – | 14.00 | – | week | **14.00** |
| Delivery and collection charges on site offices | | | | | | | |
|   delivery charge | – | – | – | 220.00 | – | load | **220.00** |
|   collection charge | – | – | – | 220.00 | – | load | **220.00** |
| **Toilet facilities** | | | | | | | |
| HSS Hire; serviced self-contained toilet delivered to and collected from site; maintained by toilet supply company | | | | | | | |
|   single chemical toilet including wash handbasin and water | – | – | – | 30.00 | – | week | **30.00** |
|   delivery and collection; each way | – | – | – | 35.00 | – | nr | **35.00** |
| **1.11 THE SITE/EXISTING BUILDINGS** | | | | | | | |
| **Existing mains and services; mark positions of existing mains and services; locate and mark; site area** | | | | | | | |
|   less than 500 m² | – | 1.50 | 35.63 | – | – | nr | **35.63** |
|   up to 1000 m² | – | 3.00 | 71.25 | – | – | nr | **71.25** |
|   up to 2000 m² | – | 8.00 | 190.00 | – | – | nr | **190.00** |
|   up to 4000 m² | – | 16.00 | 380.00 | – | – | nr | **380.00** |

**1 PRELIMINARIES**

| Item<br>Excluding site overheads and profit | PC<br>£ | Labour<br>hours | Labour<br>£ | Plant<br>£ | Material<br>£ | Unit | Total<br>rate £ |
|---|---|---|---|---|---|---|---|
| **1.11 THE SITE/EXISTING BUILDINGS – cont** | | | | | | | |
| **Access to the site** | | | | | | | |
| For pedestrians | | | | | | | |
| security kiosk | – | – | – | 170.00 | – | week | **170.00** |
| security guard | – | 40.00 | 420.00 | – | – | week | **420.00** |
| protected walkways; Heras fencing on 2 sides | – | – | – | 4.53 | – | m | **4.53** |
| **Parking** | | | | | | | |
| Parking expenses where vehicles do not park on the site area; per vehicle | | | | | | | |
| metropolitan area; city centre | – | – | – | – | – | week | **200.00** |
| metropolitan area; outer areas | – | – | – | – | – | week | **160.00** |
| suburban restricted parking areas | – | – | – | – | – | week | **40.00** |
| **Congestion charging** | | | | | | | |
| London only | – | – | – | – | – | week | **57.50** |
| **Site visit; pre-tender for purposes of understanding site and tender requirements; prices below are based on a single senior manager attending site** | | | | | | | |
| City location; average distance of travel 20 miles inclusive of travel and parking costs | | | | | | | |
| less than 500 m$^2$ | – | 4.00 | 140.00 | – | 39.50 | nr | **179.50** |
| up to 1000 m$^2$ | – | 5.00 | 175.00 | – | 41.50 | nr | **216.50** |
| up to 2000 m$^2$ | – | 6.00 | 210.00 | – | 47.50 | nr | **257.50** |
| up to 4000 m$^2$ | – | 8.00 | 280.00 | – | 55.50 | nr | **335.50** |
| Town or rural location; average distance of travel 20 miles inclusive of travel and parking costs | | | | | | | |
| less than 500 m$^2$ | – | 3.00 | 105.00 | – | 22.00 | nr | **127.00** |
| up to 1000 m$^2$ | – | 4.00 | 140.00 | – | 23.00 | nr | **163.00** |
| up to 2000 m$^2$ | – | 5.00 | 175.00 | – | 24.00 | nr | **199.00** |
| up to 4000 m$^2$ | – | 6.00 | 210.00 | – | 24.00 | nr | **234.00** |

## 2 OFF SITE MANUFACTURED MATERIALS, COMPONENTS AND BUILDINGS

| Item<br>Excluding site overheads and profit | PC<br>£ | Labour<br>hours | Labour<br>£ | Plant<br>£ | Material<br>£ | Unit | Total<br>rate £ |
|---|---|---|---|---|---|---|---|
| **PREFABRICATED BUILDINGS/ STRUCTURES/UNITS** | | | | | | | |
| Cast stone buildings; Haddonstone Ltd; ornamental garden buildings in Portland Bath or Terracotta finished cast stone; prices for stonework and facades only; excavations, foundations, reinforcement, concrete infill, roofing and floors all priced separately | | | | | | | |
| Pavilion Venetian Folly L9400; Tuscan columns, pedimented arch, quoins and optional balustrading | | | | | | | |
|    4184 mm high × 4728 mm wide × 3147 mm deep | 12060.00 | 175.00 | 4156.25 | 217.87 | 12242.42 | nr | **16616.54** |
| Pavilion L9300; Tuscan columns | | | | | | | |
|    3496 mm high × 3634 mm wide | 7320.00 | 144.00 | 3420.00 | 186.74 | 7502.42 | nr | **11109.16** |
| Small Classical Temple L9250; six column with fibreglass lead effect finish dome roof | | | | | | | |
|    overall height 3610 mm; diameter 2540 mm | 6330.00 | 130.00 | 3087.50 | 155.62 | 6512.42 | nr | **9755.54** |
| Large Classical Temple L9100; eight column with fibreglass lead effect finish dome roof | | | | | | | |
|    overall height 4664 mm; diameter 3190 mm | 11820.00 | 165.00 | 3918.75 | 155.62 | 12002.42 | nr | **16076.79** |
| Stepped floors to temples | | | | | | | |
|    single step; Small Classical Temple | 1160.00 | 24.00 | 570.00 | – | 1235.43 | nr | **1805.43** |
|    single step; Large Classical Temple | 1638.00 | 26.00 | 617.50 | – | 1728.51 | nr | **2346.01** |
| Stone structures; Architectural Heritage Ltd; hand carved from solid natural limestone with wrought iron domed roof, decorated frieze and base and integral seats; supply and erect only; excavations and concrete bases priced separately | | | | | | | |
| The Park Temple; five columns; 3500 mm high × 1650 mm diameter | 10000.00 | 96.00 | 2280.00 | 91.70 | 10000.00 | nr | **12371.70** |
| The Estate Temple; six columns; 4000 mm high × 2700 mm diameter | 17600.00 | 120.00 | 2850.00 | 183.40 | 17600.00 | nr | **20633.40** |

## 2 OFF SITE MANUFACTURED MATERIALS, COMPONENTS AND BUILDINGS

| Item<br>Excluding site overheads and profit | PC<br>£ | Labour<br>hours | Labour<br>£ | Plant<br>£ | Material<br>£ | Unit | Total<br>rate £ |
|---|---|---|---|---|---|---|---|
| **PREFABRICATED BUILDINGS/ STRUCTURES/UNITS – cont** | | | | | | | |
| **Stone structures; Architectural Heritage Ltd; hand carved from solid natural limestone with solid oak trelliage; prices for stonework and facades only; supply and erect only; excavations and concrete bases priced separately** | | | | | | | |
| The Pergola; 2240 mm high × 2640 mm wide × 6990 mm long | 10000.00 | 96.00 | 2280.00 | 366.80 | 10038.06 | nr | **12684.86** |
| **Ornamental stone structures; Architectural Heritage Ltd; setting to bases or plinths (not included)** | | | | | | | |
| The Obelisk; classic natural stone obelisk; tapering square form on panelled square base; 1860 mm high × 360 mm square | 1600.00 | 2.00 | 47.50 | – | 1600.00 | nr | **1647.50** |
| The Inverted Sundial Pedestal with Griffin Armillary Sphere | 2000.00 | 2.00 | 47.50 | – | 2000.00 | nr | **2047.50** |
| Large obelisk; 2.40 m high × 460 mm square; hand carved natural limestone obelisk of tapering square form raised upon a panelled square base supported by four spheres. Reproduction | 2200.00 | 1.00 | 23.75 | – | 2200.00 | nr | **2223.75** |

## 3 DEMOLITIONS

| Item<br>Excluding site overheads and profit | PC<br>£ | Labour<br>hours | Labour<br>£ | Plant<br>£ | Material<br>£ | Unit | Total<br>rate £ |
|---|---|---|---|---|---|---|---|
| **DEMOLITIONS/CLEARANCE** | | | | | | | |
| **Demolish existing structures; disposal off site; mechanical demolition; with 3 tonne excavator and dumper** | | | | | | | |
| Brick wall | | | | | | | |
| 112.5 mm thick | – | 0.07 | 1.58 | 2.57 | 6.74 | m² | **10.89** |
| 225 mm thick | – | 0.08 | 1.98 | 3.21 | 13.48 | m² | **18.67** |
| 337.5 mm thick | – | – | – | 9.00 | 22.62 | m² | **31.62** |
| 450 mm thick | – | – | – | 12.00 | 26.97 | m² | **38.97** |
| **Demolish existing structures; disposal off site mechanically loaded; all other works by hand** | | | | | | | |
| Brick wall | | | | | | | |
| 112.5 mm thick | – | 0.33 | 7.92 | – | 6.74 | m² | **14.66** |
| 225 mm thick | – | 0.50 | 11.88 | – | 13.48 | m² | **25.36** |
| 337.5 mm thick | – | 0.67 | 15.83 | – | 22.62 | m² | **38.45** |
| 450 mm thick | – | 1.00 | 23.75 | – | 26.97 | m² | **50.72** |
| **Demolish existing structures; disposal off site mechanically loaded; by diesel or electric breaker; all other works by hand** | | | | | | | |
| Brick wall | | | | | | | |
| 112.5 mm thick | – | 0.17 | 3.96 | 0.50 | 6.74 | m² | **11.20** |
| 225 mm thick | – | 0.20 | 4.75 | 0.60 | 13.48 | m² | **18.83** |
| 337.5 mm thick | – | 0.25 | 5.94 | 0.75 | 20.00 | m² | **26.69** |
| 450 mm thick | – | 0.33 | 7.92 | 0.99 | 26.97 | m² | **35.88** |
| **Break out concrete footings associated with freestanding walls; inclusive of all excavation and backfilling with excavated material; disposal off site mechanically loaded** | | | | | | | |
| By mechanical breaker; diesel or electric | | | | | | | |
| plain concrete | – | 1.50 | 35.62 | 9.87 | 30.81 | m³ | **76.30** |
| reinforced concrete | – | 2.50 | 59.38 | 15.00 | 59.93 | m³ | **134.31** |
| **Remove existing freestanding buildings; demolition by hand** | | | | | | | |
| Timber building with suspended timber floor; hardstanding or concrete base not included; disposal off site | | | | | | | |
| shed 6.0 m² | – | 2.00 | 47.50 | – | 62.42 | nr | **109.92** |
| shed 10.0 m² | – | 3.00 | 71.26 | – | 112.36 | nr | **183.62** |
| shed 15.0 m² | – | 3.50 | 83.13 | – | 174.79 | nr | **257.92** |

## 3 DEMOLITIONS

| Item<br>Excluding site overheads and profit | PC<br>£ | Labour<br>hours | Labour<br>£ | Plant<br>£ | Material<br>£ | Unit | Total<br>rate £ |
|---|---|---|---|---|---|---|---|
| **DEMOLITIONS/CLEARANCE – cont** | | | | | | | |
| **Remove existing freestanding buildings – cont** | | | | | | | |
| Timber building; insulated; with timber or concrete posts set in concrete, felt covered timber or tiled roof; internal walls cladding with timber or plasterboard; load arisings to skip | | | | | | | |
| timber structure 6.0 m² | – | 2.50 | 59.38 | – | 269.67 | nr | **329.05** |
| timber structure 12.0 m² | – | 3.50 | 83.13 | – | 377.04 | nr | **460.17** |
| timber structure 20.0 m² | – | 8.00 | 190.00 | 466.77 | 399.51 | nr | **1056.28** |
| **Demolition of freestanding brick buildings with tiled or sheet roof; concrete foundations measured separately; mechanical demolition; maximum distance to stockpile 25 m; inclusive for all access scaffolding and the like; maximum height of roof 4.0 m; inclusive of all doors, windows, guttering and down pipes; including disposal by grab** | | | | | | | |
| Half brick thick | | | | | | | |
| 10 m² | – | 8.00 | 190.00 | 160.91 | 167.23 | nr | **518.14** |
| 20 m² | – | 16.00 | 380.00 | 290.64 | 215.08 | nr | **885.72** |
| 1 brick thick | | | | | | | |
| 10 m² | – | 10.00 | 237.50 | 259.51 | 334.47 | nr | **831.48** |
| 20 m² | – | 16.00 | 380.00 | 389.36 | 430.15 | nr | **1199.51** |
| Cavity wall with blockwork inner skin and brick outer skin; insulated | | | | | | | |
| 10 m² | – | 12.00 | 285.11 | 358.23 | 464.27 | nr | **1107.61** |
| 20 m² | – | 20.00 | 475.00 | 488.14 | 598.14 | nr | **1561.28** |
| **Extra over to the above for disconnection of services** | | | | | | | |
| Electrical | | | | | | | |
| disconnection | – | 2.00 | 60.00 | – | – | nr | **60.00** |
| grub out cables and dispose; backfilling; by machine | – | – | – | 0.31 | 1.00 | m | **1.31** |
| grub out cables and dispose; backfilling; by hand | – | 0.50 | 11.88 | – | 1.00 | m | **12.88** |
| Water supply, foul or surface water drainage | | | | | | | |
| disconnection; capping off | – | 1.00 | 23.75 | – | 60.00 | nr | **83.75** |
| grub out pipes and dispose; backfilling; by machine | – | – | – | 0.31 | 1.00 | m | **1.31** |
| grub out pipes and dispose; backfilling; by hand | – | 0.50 | 11.88 | – | 1.00 | m | **12.88** |

## 3 DEMOLITIONS

| Item<br>Excluding site overheads and profit | PC<br>£ | Labour<br>hours | Labour<br>£ | Plant<br>£ | Material<br>£ | Unit | Total<br>rate £ |
|---|---|---|---|---|---|---|---|
| **Demolish existing fence; remove stakes or grub out posts as appropriate; remove to stock pile for removal off site; (not included)** | | | | | | | |
| Mechanical demolition; clear fence line | | | | | | | |
| chain link fence 1.20–1.50 m high | – | 0.01 | 0.47 | 0.55 | – | m | **1.02** |
| closeboard or timber panel fence 1.80 m high | – | 0.03 | 1.19 | 1.24 | – | m | **2.43** |
| Mechanical demolition; light shrubs or creepers in chainlinks | | | | | | | |
| chain link fence 1.20–1.50 m high | – | 0.01 | 0.59 | 0.71 | – | m | **1.30** |
| closeboard or timber panel fence 1.80 m high | – | 0.03 | 1.58 | 1.66 | – | m | **3.24** |
| Mechanical demolition; heavy shrubs or creepers, bramble and the like requiring clearance to enable removal | | | | | | | |
| fence 1.20–1.50 m high | – | 0.05 | 2.38 | 1.78 | – | m | **4.16** |
| closeboard or timber panel fence 1.80 m high | – | 0.05 | 2.38 | 2.63 | – | m | **5.01** |
| Demolition and site transport by hand; clear fence lines | | | | | | | |
| fence 1.20–1.50 m high | – | 0.03 | 1.58 | – | – | m | **1.58** |
| closeboard or timber panel fence 1.80 m high | – | 0.07 | 3.17 | – | – | m | **3.17** |
| Demolition and site transport by hand; light vegetation and creepers | | | | | | | |
| fence 1.20–1.50 m high | – | 0.05 | 2.38 | – | – | m | **2.38** |
| closeboard or timber panel fence 1.80 m high | – | 0.13 | 5.94 | – | – | m | **5.94** |
| Demolition and site transport by hand; heavy shrubs or creepers, bramble and the like requiring clearance to enable removal | | | | | | | |
| fence 1.20–1.50 m high | – | 0.13 | 5.94 | – | – | m | **5.94** |
| closeboard or timber panel fence 1.80 m high | – | 0.17 | 7.92 | – | – | m | **7.92** |
| **Demolition of posts and straining posts** | | | | | | | |
| Break out straining post; grub out concrete footings; remove to stockpile on site | | | | | | | |
| single straining post by machine | – | 0.25 | 5.94 | 13.04 | – | nr | **18.98** |
| double straining post by machine | – | 0.30 | 7.13 | 14.99 | – | nr | **22.12** |
| single straining post by hand | – | 1.00 | 47.50 | – | – | nr | **47.50** |
| double straining post by hand | – | 1.25 | 59.38 | – | – | nr | **59.38** |
| Break out gatepost; concrete steel or timber; girth not exceeding 150 mm; grub out concrete footings | | | | | | | |
| by machine | – | 0.25 | 5.94 | 13.04 | – | nr | **18.98** |
| by hand | – | 1.00 | 23.75 | – | – | nr | **23.75** |

## 3 DEMOLITIONS

| Item<br>Excluding site overheads and profit | PC<br>£ | Labour<br>hours | Labour<br>£ | Plant<br>£ | Material<br>£ | Unit | Total<br>rate £ |
|---|---|---|---|---|---|---|---|
| **DEMOLITIONS/CLEARANCE –<br>cont** | | | | | | | |
| **Site preparation** | | | | | | | |
| Felling and removing trees off site | | | | | | | |
| girth 600 mm–1.50 m (95–240 mm trunk diameter) | – | 2.50 | 178.13 | 18.20 | – | nr | **196.33** |
| girth 1.50–3.00 m (240–475 mm trunk diameter) | – | 10.00 | 712.50 | 72.80 | – | nr | **785.30** |
| girth 3.00–4.00 m (475–630 mm trunk diameter) | – | 16.00 | 1140.00 | 100.96 | – | nr | **1240.96** |
| Removing tree stumps | | | | | | | |
| girth 600 mm–1.50 m | – | 2.00 | 47.50 | 100.87 | – | nr | **148.37** |
| girth 1.50–3.00 m | – | 7.00 | 166.25 | 114.30 | – | nr | **280.55** |
| girth over 3.00 m | – | 12.00 | 285.00 | 195.94 | – | nr | **480.94** |
| Stump grinding; disposing to spoil heaps | | | | | | | |
| girth 600 mm–1.50 m | – | 2.00 | 47.50 | 32.86 | – | nr | **80.36** |
| girth 1.50–3.00 m | – | 2.50 | 59.38 | 65.72 | – | nr | **125.10** |
| girth over 3.00 m | – | 4.00 | 95.00 | 57.51 | – | nr | **152.51** |
| Clearing site vegetation | | | | | | | |
| mechanical clearance | – | 0.25 | 5.94 | 13.05 | – | 100 m² | **18.99** |
| hand clearance | – | 2.00 | 47.50 | – | – | 100 m² | **47.50** |
| Lifting turf for preservation | | | | | | | |
| sod cutter machine lift and stack | – | 0.75 | 17.81 | 7.08 | – | 100 m² | **24.89** |
| hand lift and stack | – | 8.33 | 197.91 | – | – | 100 m² | **197.91** |
| Lifting turf for disposal (disposal not included) | | | | | | | |
| by excavator; 5 tonne | – | – | – | 34.79 | – | 100 m² | **34.79** |
| by excavator; 8 tonne | – | – | – | 25.76 | – | 100 m² | **25.76** |
| by excavator; working with two 3 tonne dumpers | – | – | – | 28.23 | – | 100 m² | **28.23** |
| by excavator; working with two 5 tonne dumpers | – | – | – | 22.44 | – | 100 m² | **22.44** |
| hand lift and stack | – | 8.33 | 197.91 | – | – | 100 m² | **197.91** |
| **Site clearance; by machine; clear site of mature shrubs from existing cultivated beds; dig out roots by machine** | | | | | | | |
| Mixed shrubs in beds; planting centres 500 mm average | | | | | | | |
| height less than 1 m | – | 0.03 | 0.59 | 0.87 | – | m² | **1.46** |
| 1.00–1.50 m | – | 0.04 | 0.95 | 1.39 | – | m² | **2.34** |
| 1.50–2.00 m; pruning to ground level by hand | – | 0.10 | 2.38 | 3.48 | – | m² | **5.86** |
| 2.00–3.00 m; pruning to ground level by hand | – | 0.10 | 2.38 | 6.95 | – | m² | **9.33** |
| 3.00–4.00 m; pruning to ground level by hand | – | 0.20 | 4.75 | 11.59 | – | m² | **16.34** |

## 3 DEMOLITIONS

| Item<br>Excluding site overheads and profit | PC<br>£ | Labour<br>hours | Labour<br>£ | Plant<br>£ | Material<br>£ | Unit | Total<br>rate £ |
|---|---|---|---|---|---|---|---|
| **Site clearance; by hand; clear site of mature shrubs from existing cultivated beds; dig out roots** | | | | | | | |
| Mixed shrubs in beds; planting centres 500 mm average | | | | | | | |
| height less than 1 m | – | 0.33 | 7.92 | – | – | m² | **7.92** |
| 1.00–1.50 m | – | 0.50 | 11.88 | – | – | m² | **11.88** |
| 1.50–2.00 m | – | 1.00 | 23.75 | – | – | m² | **23.75** |
| 2.00–3.00 m | – | 2.00 | 47.50 | – | – | m² | **47.50** |
| 3.00–4.00 m | – | 3.00 | 71.26 | – | – | m² | **71.26** |
| **Mechanical ground clearance to large areas; Cleartrack (Evl) Limited; vegetation clearance of high capacity mobile mulching machines; 500 HP; 2.50 m head width; tracked; wheeled; skid steer and 360° operating** | | | | | | | |
| Site clearance and in situ mulching and incorporating to soil surface to existing area of scrub and small trees up to 150 mm diameter; cleared mulched material left on surface for clearance to stockpile (not included) | | | | | | | |
| Areas of existing grass, bramble and weed; average height not exceeding 1.00 m | – | – | – | – | – | 100 m² | **10.25** |
| Areas of existing small shrubs and weed; average height not exceeding 1.80 m | – | – | – | – | – | 100 m² | **15.38** |
| Areas of existing grass, bramble, weed and regular shrubs; varying heights to 3.00 m; max stem diameter 100 mm | – | – | – | – | – | 100 m² | **21.52** |
| Areas of existing grass, bramble, weed, regular shrubs and small trees; varying heights up and over 3.00 m; max stem diameter 150 mm | – | – | – | – | – | 100 m² | **55.35** |
| Spraying of vegetation; Glyphosate at 5 l/ha by tractor drawn spray equipment | – | – | – | – | – | 100 m² | **2.56** |
| **Excavated material; on site** | | | | | | | |
| In spoil heaps | | | | | | | |
| average 25 m distance | – | – | – | 2.44 | – | m³ | **2.44** |
| average 50 m distance | – | – | – | 2.82 | – | m³ | **2.82** |
| average 100 m distance (1 dumper) | – | – | – | 3.53 | – | m³ | **3.53** |
| average 100 m distance (2 dumpers) | – | – | – | 4.22 | – | m³ | **4.22** |
| average 200 m distance | – | – | – | 4.26 | – | m³ | **4.26** |
| average 200 m distance (2 dumpers) | – | – | – | 5.08 | – | m³ | **5.08** |

## 4 ALTERATIONS, REPAIRS AND CONSERVATION

| Item<br>Excluding site overheads and profit | PC<br>£ | Labour<br>hours | Labour<br>£ | Plant<br>£ | Material<br>£ | Unit | Total<br>rate £ |
|---|---|---|---|---|---|---|---|
| **CONSERVATION OF METAL RAILINGS** | | | | | | | |
| **Eura Conservation Ltd; conservation of metal railings; works to heritage conservation standards** | | | | | | | |
| Taking down and transporting existing metalwork to an off site workshop for conservation | | | | | | | |
|    pair of gates; maximum overall width 4.00 m | – | – | – | – | – | pair | 620.00 |
|    side screens to gates | – | – | – | – | – | pair | 505.00 |
|    railings; 1.20 m high; plain | – | – | – | – | – | m | 55.00 |
|    railings; 1.80 m high; with finials | – | – | – | – | – | m | 100.00 |
|    ornate cast or wrought iron railings | – | – | – | – | – | m | 115.00 |
| Conserving metalwork off site; inclusive of repairs to metalwork; rust removal, rubbing down and preparing for re-erection on site | | | | | | | |
|    gates; 2.50 m high; width not exceeding 4.00 m | – | – | – | – | – | m | 1150.00 |
|    railings; 1.20 m high; plain | – | – | – | – | – | m | 150.00 |
|    railings; 1.80 m high; with finials | – | – | – | – | – | m | 228.00 |
|    ornate cast or wrought iron railings | – | – | – | – | – | m | 452.00 |
| Transporting from store and re-erection of existing gates into locations recorded with position orientation and features of the original installation | | | | | | | |
|    gates; 2.50 m high; width not exceeding 4.00 m | – | – | – | – | – | nr | 1050.00 |
| Refixing existing railings previously taken down and repaired on site | | | | | | | |
|    railings; 1.20 m high; plain | – | – | – | – | – | m | 108.00 |
|    railings; 1.80 m high; with finials | – | – | – | – | – | m | 208.00 |
|    ornate cast or wrought iron railings | – | – | – | – | – | m | 260.00 |
| **Eura Conservation Ltd; supply of conservation grade railings to match existing materials on site; inclusive of all surveys, measurements and analysis of materials, installation methods and the like** | | | | | | | |
| Wavy bar railings | | | | | | | |
|    1.20 m high; plain | – | – | – | – | – | m | 101.00 |
|    1.70 m high; straight | – | – | – | – | – | m | 322.00 |
|    1.70 m high; curved | – | – | – | – | – | m | 390.00 |
|    1.80 m high; with finials | – | – | – | – | – | m | 320.00 |
|    ornate cast or wrought iron railings | – | – | – | – | – | m | 850.00 |
| Painting of railings: see section 29 | | | | | | | |

Prices for Measured Works – Major Works

## 5 EXCAVATING AND FILLING

| Item<br>Excluding site overheads and profit | PC<br>£ | Labour<br>hours | Labour<br>£ | Plant<br>£ | Material<br>£ | Unit | Total<br>rate £ |
|---|---|---|---|---|---|---|---|
| **Clarification notes on labour costs in this section** | | | | | | | |
| **General groundworks team**<br>Generally a three man team is used in this section; The column Labour hours reports team hours. The column Labour £ reports the total cost of the team for the unit of work shown | | | | | | | |
| 3 man team | – | 1.00 | 71.25 | – | – | hr | 71.25 |
| banksman | – | 1.00 | 23.75 | – | – | hr | 23.75 |
| **MACHINE SELECTION** | | | | | | | |
| Road Equipment Ltd; machine volumes for excavating/filling only and placing excavated material alongside or to a dumper; no bulkages are allowed for in the material volumes; these rates should be increased by user-preferred percentages to suit prevailing site conditions; the figures in the next section for 'Excavation mechanical' and filling allow for the use of banksmen within the rates shown below | | | | | | | |
| 1.5 tonne excavators: digging volume | | | | | | | |
| 1 cycle/minute; 0.04 m$^3$ | – | 0.42 | 9.90 | 3.10 | – | m$^3$ | 13.00 |
| 2 cycles/minute; 0.08 m$^3$ | – | 0.21 | 4.95 | 1.89 | – | m$^3$ | 6.84 |
| 3 cycles/minute; 0.12 m$^3$ | – | 0.14 | 3.30 | 1.51 | – | m$^3$ | 4.81 |
| 3 tonne excavators; digging volume | | | | | | | |
| 1 cycle/minute; 0.13 m$^3$ | – | 0.13 | 3.04 | 1.49 | – | m$^3$ | 4.53 |
| 2 cycles/minute; 0.26 m$^3$ | – | 0.06 | 1.52 | 2.34 | – | m$^3$ | 3.86 |
| 3 cycles/minute; 0.39 m$^3$ | – | 0.04 | 1.01 | 1.02 | – | m$^3$ | 2.03 |
| 5 tonne excavators; digging volume | | | | | | | |
| 1 cycle/minute; 0.28 m$^3$ | – | 0.06 | 1.41 | 1.92 | – | m$^3$ | 3.33 |
| 2 cycles/minute; 0.56 m$^3$ | – | 0.03 | 0.71 | 1.72 | – | m$^3$ | 2.43 |
| 3 cycles/minute; 0.84 m$^3$ | – | 0.02 | 0.47 | 1.80 | – | m$^3$ | 2.27 |
| 7 tonne excavators; supplied with operator; digging volume | | | | | | | |
| 1 cycle/minute; 0.28 m$^3$ | – | 0.06 | 1.41 | 3.28 | – | m$^3$ | 4.69 |
| 2 cycles/minute; 0.56 m$^3$ | – | 0.03 | 0.71 | 1.61 | – | m$^3$ | 2.32 |
| 3 cycles/minute; 0.84 m$^3$ | – | 0.02 | 0.47 | 1.84 | – | m$^3$ | 2.31 |
| 21 tonne excavators; supplied with operator; digging volume | | | | | | | |
| 1 cycle/minute; 1.21 m$^3$ | – | – | – | 0.96 | – | m$^3$ | 0.96 |
| 2 cycles/minute; 2.42 m$^3$ | – | – | – | 0.40 | – | m$^3$ | 0.40 |
| 3 cycles/minute; 3.63 m$^3$ | – | – | – | 0.27 | – | m$^3$ | 0.27 |

## 5 EXCAVATING AND FILLING

| Item<br>Excluding site overheads and profit | PC<br>£ | Labour<br>hours | Labour<br>£ | Plant<br>£ | Material<br>£ | Unit | Total<br>rate £ |
|---|---|---|---|---|---|---|---|
| **MACHINE SELECTION – cont** | | | | | | | |
| **Road Equipment Ltd – cont** | | | | | | | |
| Backhoe loader; excavating; JCB 3 CX rear bucket capacity 0.28 m$^3$ | | | | | | | |
|   1 cycle/minute; 0.28 m$^3$ | – | – | – | 2.73 | – | m$^3$ | 2.73 |
|   2 cycles/minute; 0.56 m$^3$ | – | – | – | 1.37 | – | m$^3$ | 1.37 |
|   3 cycles/minute; 0.84 m$^3$ | – | – | – | 0.91 | – | m$^3$ | 0.91 |
| Backhoe loader; loading from stockpile; JCB 3 CX front bucket capacity 1.00 m$^3$ | | | | | | | |
|   1 cycle/minute; 1.00 m$^3$ | – | – | – | 0.77 | – | m$^3$ | 0.77 |
|   2 cycles/minute; 2.00 m$^3$ | – | – | – | 0.38 | – | m$^3$ | 0.38 |
| **Note:** All volumes below are based on excavated 'earth' moist at 1,997 kg/m$^3$ solid or 1,598 kg/m$^3$ loose; a 25% bulkage factor has been used; the weight capacities below exceed the volume capacities of the machine in most cases; see the memorandum section at the back of this book for further weights of materials. | | | | | | | |
| **Dumpers; Road Equipment Ltd** | | | | | | | |
| 1 tonne high tip skip loader; volume 0.485 m$^3$ (775 kg) | | | | | | | |
|   5 loads per hour | – | 0.41 | 9.79 | 2.06 | – | m$^3$ | 11.85 |
|   7 loads per hour | – | 0.29 | 7.00 | 1.53 | – | m$^3$ | 8.53 |
|   10 loads per hour | – | 0.21 | 4.90 | 1.13 | – | m$^3$ | 6.03 |
| 2.5 tonne dumper; excavated material volume 1.3 m$^3$ | | | | | | | |
|   4 loads per hour | – | – | – | 5.90 | – | m$^3$ | 5.90 |
|   5 loads per hour | – | – | – | 4.72 | – | m$^3$ | 4.72 |
|   7 loads per hour | – | – | – | 4.20 | – | m$^3$ | 4.20 |
|   10 loads per hour | – | – | – | 2.30 | – | m$^3$ | 2.30 |
| 6 tonne dumper; maximum volume 3.40 m$^3$ (5.4 tonnes); available volume 3.77 m$^3$ | | | | | | | |
|   4 loads per hour | – | 0.07 | 1.57 | 0.40 | – | m$^3$ | 1.97 |
|   5 loads per hour | – | 0.05 | 1.26 | 0.33 | – | m$^3$ | 1.59 |
|   7 loads per hour | – | 0.04 | 0.90 | 0.25 | – | m$^3$ | 1.15 |
|   10 loads per hour | – | 0.03 | 0.63 | 0.20 | – | m$^3$ | 0.83 |

## 5 EXCAVATING AND FILLING

| Item<br>Excluding site overheads and profit | PC<br>£ | Labour<br>hours | Labour<br>£ | Plant<br>£ | Material<br>£ | Unit | Total<br>rate £ |
|---|---|---|---|---|---|---|---|
| **EXCAVATING** | | | | | | | |
| **Market prices of topsoil; prices shown include for 20% settlement** | | | | | | | |
| Multiple source screened topsoil | – | – | – | – | 39.60 | m³ | **39.60** |
| Single source topsoil; British Sugar PLC | – | – | – | – | 39.60 | m³ | **39.60** |
| High grade topsoil for planting | – | – | – | – | 78.00 | m³ | **78.00** |
| Note: The figures in this section relate to the machine capacities shown earlier in this section. The figures below however allow for dig efficiency based on depth. A banksman is allowed for in all excavation build and disposal costs shown. Bulkages of 25% allowed; adjustments should be made for different soil types. | | | | | | | |
| **Excavating; mechanical; topsoil for preservation** | | | | | | | |
| 3 tonne tracked excavator (bucket volume 0.13 m³) | | | | | | | |
| average depth 100 mm | – | – | – | 0.99 | – | m² | **0.99** |
| average depth 150 mm | – | – | – | 1.12 | – | m² | **1.12** |
| average depth 200 mm | – | – | – | 1.48 | – | m² | **1.48** |
| average depth 250 mm | – | – | – | 1.65 | – | m² | **1.65** |
| average depth 300 mm | – | – | – | 1.98 | – | m² | **1.98** |
| JCB Sitemaster 3CX (bucket volume 0.28 m³) | | | | | | | |
| average depth 100 mm | – | 1.00 | 23.75 | 50.44 | – | 100 m² | **74.19** |
| average depth 150 mm | – | 1.25 | 29.69 | 63.04 | – | 100 m² | **92.73** |
| average depth 200 mm | – | 1.78 | 42.27 | 89.77 | – | 100 m² | **132.04** |
| average depth 250 mm | – | 1.90 | 45.13 | 95.83 | – | 100 m² | **140.96** |
| average depth 300 mm | – | 2.00 | 47.50 | 100.87 | – | 100 m² | **148.37** |
| **Excavating; mechanical; to reduce levels** | | | | | | | |
| 5 tonne excavator (bucket volume 0.28 m³) | | | | | | | |
| maximum depth not exceeding 0.25 m | – | 0.07 | 1.66 | 2.72 | – | m³ | **4.38** |
| maximum depth not exceeding 1.00 m | – | 0.05 | 1.13 | 1.85 | – | m³ | **2.98** |
| maximum depth not exceeding 2.00 m | – | 0.06 | 1.41 | 2.31 | – | m³ | **3.72** |
| JCB Sitemaster 3CX (bucket volume 0.28 m³) | | | | | | | |
| maximum depth not exceeding 0.25 m | – | 0.07 | 1.66 | 3.53 | – | m³ | **5.19** |
| maximum depth not exceeding 1.00 m | – | 0.06 | 1.41 | 3.00 | – | m³ | **4.41** |

## 5 EXCAVATING AND FILLING

| Item<br>Excluding site overheads and profit | PC<br>£ | Labour<br>hours | Labour<br>£ | Plant<br>£ | Material<br>£ | Unit | Total<br>rate £ |
|---|---|---|---|---|---|---|---|
| **EXCAVATING – cont** | | | | | | | |
| **Excavating – cont** | | | | | | | |
| 7 tonne tracked excavator (bucket volume 0.28 m³) | | | | | | | |
| maximum depth not exceeding 1.00 m | – | 0.06 | 1.48 | 2.91 | – | m³ | **4.39** |
| maximum depth not exceeding 2.00 m | – | 0.07 | 1.70 | 3.33 | – | m³ | **5.03** |
| maximum depth not exceeding 3.00 m | – | 0.09 | 2.16 | 4.24 | – | m³ | **6.40** |
| 21 tonne 360 tracked excavator (bucket volume 1.21 m³) | | | | | | | |
| maximum depth not exceeding 1.00 m | – | 0.01 | 0.26 | 0.63 | – | m³ | **0.89** |
| maximum depth not exceeding 2.00 m | – | 0.02 | 0.39 | 0.95 | – | m³ | **1.34** |
| maximum depth not exceeding 3.00 m | – | 0.03 | 0.79 | 1.92 | – | m³ | **2.71** |
| Pits; 3 tonne tracked excavator | | | | | | | |
| maximum depth not exceeding 0.25 m | – | 0.33 | 7.92 | 11.85 | – | m³ | **19.77** |
| maximum depth not exceeding 1.00 m | – | 0.25 | 5.94 | 8.89 | – | m³ | **14.83** |
| maximum depth not exceeding 2.00 m | – | 0.40 | 9.50 | 14.22 | – | m³ | **23.72** |
| Pits; 8 tonne tracked excavator | | | | | | | |
| maximum depth not exceeding 2.00 m | – | 0.40 | 9.50 | 18.64 | – | m³ | **28.14** |
| maximum depth not exceeding 3.00 m | – | 0.50 | 11.88 | 23.30 | – | m³ | **35.18** |
| Trenches; width not exceeding 0.30 m; 3 tonne excavator | | | | | | | |
| maximum depth not exceeding 0.25 m | – | 1.00 | 23.75 | 4.69 | – | m³ | **28.44** |
| maximum depth not exceeding 1.00 m | – | 0.69 | 16.27 | 3.21 | – | m³ | **19.48** |
| maximum depth not exceeding 2.00 m | – | 0.60 | 14.25 | 2.81 | – | m³ | **17.06** |
| Trenches; width exceeding 0.30 m; 3 tonne excavator | | | | | | | |
| maximum depth not exceeding 0.25 m | – | 0.60 | 14.25 | 2.81 | – | m³ | **17.06** |
| maximum depth not exceeding 1.00 m | – | 0.50 | 11.88 | 2.34 | – | m³ | **14.22** |
| maximum depth not exceeding 2.00 m | – | 0.38 | 9.13 | 1.80 | – | m³ | **10.93** |

## 5 EXCAVATING AND FILLING

| Item<br>Excluding site overheads and profit | PC £ | Labour hours | Labour £ | Plant £ | Material £ | Unit | Total rate £ |
|---|---|---|---|---|---|---|---|
| Extra over any types of excavating irrespective of depth for breaking out existing materials; JCB with breaker attachment | | | | | | | |
|   hard rock | – | 0.50 | 11.88 | 97.95 | – | m³ | 109.83 |
|   concrete | – | 0.50 | 11.88 | 36.73 | – | m³ | 48.61 |
|   reinforced concrete | – | 1.00 | 23.75 | 59.24 | – | m³ | 82.99 |
|   brickwork, blockwork or stonework | – | 0.25 | 5.94 | 36.73 | – | m³ | 42.67 |
| Extra over any types of excavating irrespective of depth for breaking out existing hard pavings; JCB with breaker attachment | | | | | | | |
|   concrete; 100 mm thick | – | – | – | 2.69 | – | m² | 2.69 |
|   concrete; 150 mm thick | – | – | – | 4.49 | – | m² | 4.49 |
|   concrete; 200 mm thick | – | – | – | 5.39 | – | m² | 5.39 |
|   concrete; 300 mm thick | – | – | – | 8.08 | – | m² | 8.08 |
|   reinforced concrete; 100 mm thick | – | 0.08 | 1.98 | 3.09 | – | m² | 5.07 |
|   reinforced concrete; 150 mm thick | – | 0.08 | 1.78 | 4.25 | – | m² | 6.03 |
|   reinforced concrete; 200 mm thick | – | 0.10 | 2.38 | 5.67 | – | m² | 8.05 |
|   reinforced concrete; 300 mm thick | – | 0.15 | 3.56 | 8.50 | – | m² | 12.06 |
|   tarmacadam; 75 mm thick | – | – | – | 2.69 | – | m² | 2.69 |
|   tarmacadam and hardcore; 150 mm thick | – | – | – | 4.31 | – | m² | 4.31 |
| Extra over any types of excavating irrespective of depth for taking up | | | | | | | |
|   precast concrete paving slabs | – | 0.07 | 1.58 | 0.61 | – | m² | 2.19 |
|   natural stone paving | – | 0.10 | 2.38 | 0.91 | – | m² | 3.29 |
|   cobbles | – | 0.13 | 2.97 | 1.13 | – | m² | 4.10 |
|   brick paviors | – | 0.13 | 2.97 | 1.13 | – | m² | 4.10 |
| **Excavating; hand** | | | | | | | |
| Topsoil for preservation; loading to barrows | | | | | | | |
|   average depth 100 mm | – | 0.08 | 5.64 | – | – | m² | 5.64 |
|   average depth 150 mm | – | 0.12 | 8.46 | – | – | m² | 8.46 |
|   average depth 200 mm | – | 0.19 | 13.54 | – | – | m² | 13.54 |
|   average depth 250 mm | – | 0.24 | 16.93 | – | – | m² | 16.93 |
|   average depth 300 mm | – | 0.29 | 20.31 | – | – | m² | 20.31 |
| **Excavating; hand** | | | | | | | |
| Topsoil to reduce levels | | | | | | | |
|   maximum depth not exceeding 0.25 m | – | 0.79 | 56.43 | – | – | m³ | 56.43 |
|   maximum depth not exceeding 1.00 m | – | 1.03 | 73.36 | – | – | m³ | 73.36 |
| Pits | | | | | | | |
|   maximum depth not exceeding 0.25 m | – | 0.88 | 62.70 | – | – | m³ | 62.70 |
|   maximum depth not exceeding 1.00 m | – | 1.14 | 81.51 | – | – | m³ | 81.51 |
|   maximum depth not exceeding 2.00 m (includes earthwork support) | – | 2.29 | 163.02 | – | – | m³ | 228.48 |

## 5 EXCAVATING AND FILLING

| Item<br>Excluding site overheads and profit | PC<br>£ | Labour<br>hours | Labour<br>£ | Plant<br>£ | Material<br>£ | Unit | Total<br>rate £ |
|---|---|---|---|---|---|---|---|
| **EXCAVATING – cont** | | | | | | | |
| **Excavating – cont** | | | | | | | |
| Trenches; width not exceeding 0.30 m | | | | | | | |
| maximum depth not exceeding 0.25 m | – | 0.94 | 67.17 | – | – | m³ | 67.17 |
| maximum depth not exceeding 1.00 m | – | 1.23 | 87.49 | – | – | m³ | 87.49 |
| maximum depth not exceeding 2.00 m (includes earthwork support) | – | 1.23 | 87.49 | – | – | m³ | 120.22 |
| Trenches; width exceeding 0.30 m wide | | | | | | | |
| maximum depth not exceeding 0.25 m | – | 0.94 | 67.17 | – | – | m³ | 67.17 |
| maximum depth not exceeding 1.00 m | – | 1.32 | 94.05 | – | – | m³ | 94.05 |
| maximum depth not exceeding 2.00 m (includes earthwork support) | – | 1.98 | 141.07 | – | – | m³ | 206.53 |
| Extra over any types of excavating irrespective of depth for breaking out existing materials; hand held pneumatic breaker | | | | | | | |
| rock | – | 1.65 | 117.56 | 30.22 | – | m³ | 147.78 |
| concrete | – | 0.82 | 58.78 | 15.11 | – | m³ | 73.89 |
| reinforced concrete | – | 1.32 | 94.05 | 29.31 | – | m³ | 123.36 |
| brickwork, blockwork or stonework | – | 0.50 | 35.27 | 9.07 | – | m³ | 44.34 |
| **Earthwork Retention** | | | | | | | |
| Manhole boxes to excavations (not included); Timber plywood with timber braces to shore excavations to 1.00 m deep; opposing faces distance less than 1.00 m | | | | | | | |
| 1.00 m deep | – | 0.50 | 23.75 | – | 9.35 | m² | 33.10 |
| 2.00 m deep | – | 1.00 | 47.50 | – | 9.35 | m² | 56.85 |
| **Trench boxes; A-Plant** | | | | | | | |
| Trench box soil retention; mechanical placement of hydraulic trenchboxes | – | 0.50 | 35.63 | 166.30 | – | week | 201.93 |
| Manhole protection system to 3.0 m deep to enable earthwork retention; maximum distance between opposing faces 1.8 m | – | 0.50 | 35.63 | 225.30 | – | week | 260.93 |
| **FILLING** | | | | | | | |
| **Filling to make up levels; mechanical** | | | | | | | |
| Arising from the excavations | | | | | | | |
| average thickness not exceeding 0.25 m | – | 0.07 | 1.75 | 3.42 | – | m³ | 5.17 |
| average thickness less than 500 mm | – | 0.06 | 1.48 | 2.91 | – | m³ | 4.39 |
| average thickness 1.00 m | – | 0.05 | 1.10 | 2.16 | – | m³ | 3.26 |

## 5 EXCAVATING AND FILLING

| Item<br>Excluding site overheads and profit | PC<br>£ | Labour<br>hours | Labour<br>£ | Plant<br>£ | Material<br>£ | Unit | Total<br>rate £ |
|---|---|---|---|---|---|---|---|
| Obtained from on site spoil heaps; average 25 m distance; multiple handling | | | | | | | |
| average thickness less than 250 mm | – | – | – | 8.99 | – | m³ | 8.99 |
| average thickness less than 500 mm | – | – | – | 7.40 | – | m³ | 7.40 |
| average thickness 1.00 m | – | – | – | 6.30 | – | m³ | 6.30 |
| Obtained off site; planting quality topsoil PC £51.00/m³ | | | | | | | |
| average thickness less than 250 mm | – | – | – | 8.99 | 39.60 | m³ | 48.59 |
| average thickness less than 500 mm | – | – | – | 7.40 | 39.60 | m³ | 47.00 |
| average thickness 1.00 m | – | – | – | 6.30 | 39.60 | m³ | 45.90 |
| Obtained off site; hardcore; PC £20.00/m³ | | | | | | | |
| average thickness less than 250 mm | – | – | – | 10.49 | 27.00 | m³ | 37.49 |
| average thickness less than 500 mm | – | – | – | 7.87 | 27.00 | m³ | 34.87 |
| average thickness 1.00 m | – | – | – | 6.62 | 27.00 | m³ | 33.62 |
| **Filling to make up levels; hand** | | | | | | | |
| Arising from the excavations | | | | | | | |
| average thickness exceeding 0.25 m; depositing in layers 150 mm maximum thickness | – | 0.60 | 14.25 | ' – | – | m³ | 14.25 |
| Obtained from on site spoil heaps; average 25 m distance; multiple handling | | | | | | | |
| average thickness exceeding 0.25 m; depositing in layers 150 mm maximum thickness | – | 1.00 | 23.75 | – | – | m³ | 23.75 |

## 5 EXCAVATING AND FILLING

| Item<br>Excluding site overheads and profit | PC<br>£ | Labour<br>hours | Labour<br>£ | Plant<br>£ | Material<br>£ | Unit | Total<br>rate £ |
|---|---|---|---|---|---|---|---|
| **DISPOSAL** | | | | | | | |
| Note: Most commercial site disposal is carried out by 20 tonne, 8 wheeled vehicles. It has been customary to calculate disposal from construction sites in terms of full 15 m³ loads. Spon's research has found that based on weights of common materials such as clean hardcore and topsoil, vehicles could not load more than 12 m³ at a time. Most hauliers do not make it apparent that their loads are calculated by weight and not by volume. The rates below reflect these lesser volumes which are limited by the 20 tonne limit. | | | | | | | |
| The volumes shown below are based on volumes 'in the solid'. Weight and bulking factors have been applied. For further information please see the weights of typical materials in the Earthworks section of the Memoranda at the back of this book. | | | | | | | |
| Disposal; mechanical (all rates include banksman) | | | | | | | |
| Excavated material; off site; to tip; mechanically loaded (360 Excavator) | | | | | | | |
| inert (clean landfill, rubble concrete); loose | – | 0.03 | 0.66 | – | 30.81 | m³ | **31.47** |
| inert (clean excavated material); compacted | – | 0.03 | 0.66 | – | 38.51 | m³ | **39.17** |
| slightly contaminated – non-hazardous | – | 0.03 | 0.66 | – | 50.48 | m³ | **51.14** |
| hardcore | – | 0.02 | 0.49 | 0.95 | 15.30 | m³ | **16.74** |
| macadam | – | 0.02 | 0.49 | 0.95 | 26.55 | m³ | **27.99** |
| clean concrete | – | 0.03 | 0.66 | 1.27 | 18.90 | m³ | **20.83** |
| broken out compacted materials such as road bases and the like | – | 0.03 | 0.66 | 1.27 | 46.22 | m³ | **48.15** |
| Light soils and loams (bulking factor of 1.25); 40 tonnes (2 loads per hour) | | | | | | | |
| soil (sandy and loam); dry | – | 0.03 | 0.66 | – | 36.97 | m³ | **37.63** |
| soil (sandy and loam); wet | – | 0.08 | 1.85 | 0.90 | 37.97 | m³ | **40.72** |
| soil (clay); dry | – | 0.03 | 0.66 | – | 43.14 | m³ | **43.80** |
| soil (clay); wet | – | 0.03 | 0.66 | 1.27 | 46.22 | m³ | **48.15** |

## 5 EXCAVATING AND FILLING

| Item<br>Excluding site overheads and profit | PC<br>£ | Labour<br>hours | Labour<br>£ | Plant<br>£ | Material<br>£ | Unit | Total<br>rate £ |
|---|---|---|---|---|---|---|---|
| **Other materials** | | | | | | | |
| rubbish (mixed loads) | – | 0.03 | 0.79 | 1.53 | 48.33 | m³ | 50.65 |
| green waste | – | 0.03 | 0.79 | 1.53 | 37.50 | m³ | 39.82 |
| As above but allowing for 3 loads (36 m³, 60 tonne) per hour removed | | | | | | | |
| inert material | – | 0.03 | 0.66 | 1.27 | 61.01 | m³ | 62.94 |
| Excavated material; off site to tip; mechanically loaded by grab; capacity of load 13 m³ (18 tonne) | | | | | | | |
| inert material | – | – | – | – | – | m³ | 32.45 |
| hardcore | – | – | – | – | – | m³ | 43.76 |
| macadam | – | – | – | – | – | m³ | 24.00 |
| clean concrete | – | – | – | – | – | m³ | 29.50 |
| soil (sandy and loam); dry | – | – | – | – | – | m³ | 32.45 |
| soil (sandy and loam); wet | – | – | – | – | – | m³ | 34.42 |
| broken out compacted materials such as road bases and the like | – | – | – | – | – | m³ | 36.06 |
| soil (clay); dry | – | – | – | – | – | m³ | 42.12 |
| soil (clay); wet | – | – | – | – | – | m³ | 43.76 |
| rubbish (mixed loads) | – | – | – | – | – | m³ | 40.38 |
| green waste | – | – | – | – | – | tonne | 37.50 |
| **Disposal by skip; 6 yd³ (4.6 m³)** | | | | | | | |
| Excavated material loaded to skip | | | | | | | |
| by machine | – | – | – | 59.68 | – | m³ | 59.68 |
| by hand | – | 3.00 | 71.25 | 57.06 | – | m³ | 128.31 |
| **Excavated material; on site** | | | | | | | |
| In spoil heaps | | | | | | | |
| average 25 m distance | – | – | – | 2.44 | – | m³ | 2.44 |
| average 50 m distance | – | – | – | 2.82 | – | m³ | 2.82 |
| average 100 m distance (1 dumper) | – | – | – | 3.53 | – | m³ | 3.53 |
| average 100 m distance (2 dumpers) | – | – | – | 4.22 | – | m³ | 4.22 |
| average 200 m distance | – | – | – | 4.26 | – | m³ | 4.26 |
| average 200 m distance (2 dumpers) | – | – | – | 5.08 | – | m³ | 5.08 |
| Spreading on site | | | | | | | |
| average 25 m distance | – | 0.02 | 0.53 | 6.15 | – | m³ | 6.68 |
| average 50 m distance | – | 0.04 | 0.93 | 6.90 | – | m³ | 7.83 |
| average 100 m distance | – | 0.06 | 1.47 | 8.42 | – | m³ | 9.89 |
| average 200 m distance | – | 0.12 | 2.85 | 7.87 | – | m³ | 10.72 |
| **Disposal; hand** | | | | | | | |
| Excavated material; on site; in spoil heaps | | | | | | | |
| average 25 m distance | – | 2.40 | 57.00 | – | – | m³ | 57.00 |
| average 50 m distance | – | 2.64 | 62.70 | – | – | m³ | 62.70 |
| average 100 m distance | – | 3.00 | 71.25 | – | – | m³ | 71.25 |
| average 200 m distance | – | 3.60 | 85.50 | – | – | m³ | 85.50 |
| Excavated material; spreading on site | | | | | | | |
| average 25 m distance | – | 2.64 | 62.70 | – | – | m³ | 62.70 |
| average 50 m distance | – | 3.00 | 71.25 | – | – | m³ | 71.25 |
| average 100 m distance | – | 3.60 | 85.50 | – | – | m³ | 85.50 |
| average 200 m distance | – | 4.20 | 99.75 | – | – | m³ | 99.75 |

## 5 EXCAVATING AND FILLING

| Item<br>Excluding site overheads and profit | PC<br>£ | Labour<br>hours | Labour<br>£ | Plant<br>£ | Material<br>£ | Unit | Total<br>rate £ |
|---|---|---|---|---|---|---|---|
| **DISPOSAL – cont** | | | | | | | |
| **Disposal of material from site clearance operations** | | | | | | | |
| Shrubs and groundcovers less than 1.00 m height; disposal by 15 m$^3$ self loaded truck | | | | | | | |
| deciduous shrubs; not chipped; winter | – | 0.05 | 1.19 | 0.77 | 12.50 | m$^2$ | **14.46** |
| deciduous shrubs; chipped; winter | – | 0.05 | 1.19 | 1.79 | 1.88 | m$^2$ | **4.86** |
| evergreen or deciduous shrubs; chipped; summer | – | 0.08 | 1.78 | 1.03 | 7.50 | m$^2$ | **10.31** |
| Shrubs 1.00–2.00 m height; disposal by 15 m$^3$ self loaded truck | | | | | | | |
| deciduous shrubs; not chipped; winter | – | 0.17 | 3.96 | 0.77 | 37.50 | m$^2$ | **42.23** |
| deciduous shrubs; chipped; winter | – | 0.25 | 5.94 | 1.03 | 7.50 | m$^2$ | **14.47** |
| evergreen or deciduous shrubs; chipped; summer | – | 0.30 | 7.13 | 1.12 | 15.00 | m$^2$ | **23.25** |
| Shrubs or hedges 2.00–3.00 m height; disposal by 15 m$^3$ self loaded truck | | | | | | | |
| deciduous plants non-woody growth; not chipped; winter | – | 0.25 | 5.94 | 0.77 | 37.50 | m$^2$ | **44.21** |
| deciduous shrubs; chipped; winter | – | 0.50 | 11.88 | 1.03 | 18.75 | m$^2$ | **31.66** |
| evergreen or deciduous shrubs non-woody growth; chipped; summer | – | 0.25 | 5.94 | 1.03 | 7.50 | m$^2$ | **14.47** |
| evergreen or deciduous shrubs woody growth; chipped; summer | – | 0.67 | 15.83 | 1.30 | 22.50 | m$^2$ | **39.63** |
| Shrubs and groundcovers less than 1.00 m height; disposal to spoil heap | | | | | | | |
| deciduous shrubs; not chipped; winter | – | 0.05 | 1.19 | – | – | m$^2$ | **1.19** |
| deciduous shrubs; chipped; winter | – | 0.05 | 1.19 | 0.27 | – | m$^2$ | **1.46** |
| evergreen or deciduous shrubs; chipped; summer | – | 0.08 | 1.78 | 0.27 | – | m$^2$ | **2.05** |
| Shrubs 1.00–2.00 m height; disposal to spoil heap | | | | | | | |
| deciduous shrubs; not chipped; winter | – | 0.17 | 3.96 | – | – | m$^2$ | **3.96** |
| deciduous shrubs; chipped; winter | – | 0.25 | 5.94 | 0.27 | – | m$^2$ | **6.21** |
| evergreen or deciduous shrubs; chipped; summer | – | 0.30 | 7.13 | 0.36 | – | m$^2$ | **7.49** |
| Shrubs or hedges 2.00–3.00 m height; disposal to spoil heaps | | | | | | | |
| deciduous plants non-woody growth; not chipped; winter | – | 0.25 | 5.94 | – | – | m$^2$ | **5.94** |
| deciduous shrubs; chipped; winter | – | 0.50 | 11.88 | 0.27 | – | m$^2$ | **12.15** |
| evergreen or deciduous shrubs non-woody growth; chipped; summer | – | 0.67 | 15.83 | 0.53 | – | m$^2$ | **16.36** |
| evergreen or deciduous shrubs woody growth; chipped; summer | – | 1.50 | 35.62 | 1.07 | – | m$^2$ | **36.69** |

## 6 GROUND REMEDIATION AND SOIL STABILIZATION

| Item<br>Excluding site overheads and profit | PC<br>£ | Labour<br>hours | Labour<br>£ | Plant<br>£ | Material<br>£ | Unit | Total<br>rate £ |
|---|---|---|---|---|---|---|---|
| **CLARIFICATION NOTES ON LABOUR COSTS IN THIS SECTION** | | | | | | | |
| **General groundworks team** Generally a three man team is used in this section; The column Labour hours reports team hours. The column Labour £ reports the total cost of the team for the unit of work shown | | | | | | | |
| 3 man team | – | 1.00 | 71.25 | – | – | hr | **71.25** |
| **RETAINING WALLS AND REVETMENTS** | | | | | | | |
| **Preamble** Earth retaining and stabilization can take numerous forms depending on the height, soil type and loadings. The main types include gravity walls, reinforced soil and soil nailing. Prices for these items depend on quantity, access for the installation and the strength of the soils. Estimates should be obtained from the manufacturer when site conditions and layout have been determined. | | | | | | | |
| **Retaining walls; FP McCann Concrete** Retaining walls of units with plain concrete finish; prices based on 24 tonne loads but other quantities available (excavation, temporary shoring, foundations and backfilling not included) | | | | | | | |
| 1000 mm wide × 1000 mm high | 130.00 | 0.50 | 23.75 | 8.16 | 168.61 | m | **200.52** |
| 1000 mm wide × 1500 mm high | 165.00 | 0.42 | 20.09 | 8.16 | 212.36 | m | **240.61** |
| 1000 mm wide × 2000 mm high | 275.00 | 0.42 | 20.12 | 8.16 | 340.86 | m | **369.14** |
| 1000 mm wide × 2500 mm high | 355.00 | 0.42 | 20.11 | 8.16 | 446.18 | m | **474.45** |
| 1000 mm wide × 3000 mm high | 385.00 | 0.42 | 20.09 | 48.16 | 487.70 | m | **555.95** |

## 6 GROUND REMEDIATION AND SOIL STABILIZATION

| Item<br>Excluding site overheads and profit | PC<br>£ | Labour<br>hours | Labour<br>£ | Plant<br>£ | Material<br>£ | Unit | Total<br>rate £ |
|---|---|---|---|---|---|---|---|
| **RETAINING WALLS AND REVETMENTS – cont** | | | | | | | |
| **Retaining Walls; FP McCann Ltd; Easi-Bloc; (Lego Block); precast self-locking concrete gravity wall system; storage bays, partitions retaining walls, revetments; to firm and levelled base; concrete or granular; min. 800 mm wide (not included)** | | | | | | | |
| Blocks 450 mm high laid stretcher bond | | | | | | | |
| 1.20 m × 600 mm (750 kg per block) | 148.00 | 0.46 | 32.99 | 21.57 | 148.00 | m² | **202.56** |
| 600 mm × 600 mm (375 kg per block) | 177.60 | 0.50 | 35.63 | 23.30 | 177.60 | m² | **236.53** |
| **Retaining walls; Tensar International**<br>Tensartech TW1 retaining wall system; modular dry laid concrete blocks; 220 × 400 mm long × 150 mm high connected to Tensar RE geogrid with proprietary connectors; geogrid laid horizontally within the fill at 300 mm centres; on 150 × 450 mm concrete foundation; filling with imported granular material | | | | | | | |
| 1.00 m high | 92.25 | 1.00 | 47.50 | 12.81 | 100.47 | m² | **160.78** |
| 2.00 m high | 92.25 | 1.33 | 63.33 | 12.81 | 100.47 | m² | **176.61** |
| 3.00 m high | 92.25 | 1.50 | 71.25 | 12.81 | 100.47 | m² | **184.53** |
| **Retaining walls; Grass Concrete Ltd**<br>Betoflor precast concrete landscape retaining walls including soil filling to pockets (excavation, concrete foundations, backfilling stones to rear of walls and planting not included) | | | | | | | |
| Betoflor interlocking units; 250 mm long × 250 mm wide × 200 mm modular deep; in walls 250 mm wide | – | – | – | – | – | m² | **101.32** |
| extra over Betoflor interlocking units for colours | – | – | – | – | – | m² | **8.50** |
| Betoatlas earth retaining walls; 250 mm long × 500 mm wide × 200 mm modular deep; in walls 500 mm wide | – | – | – | – | – | m² | **175.39** |
| extra over Betoatlas interlocking units for colours | – | – | – | – | – | m² | **17.00** |

## 6 GROUND REMEDIATION AND SOIL STABILIZATION

| Item<br>Excluding site overheads and profit | PC<br>£ | Labour<br>hours | Labour<br>£ | Plant<br>£ | Material<br>£ | Unit | Total<br>rate £ |
|---|---|---|---|---|---|---|---|
| **Retaining walls; Forticrete Ltd**<br>Keystone precast concrete block retaining wall; geogrid included for walls over 1.00 m high; excavation, concrete foundation, stone backfill to rear of wall all measured separately; includes design | | | | | | | |
| 1.0 m high | 113.74 | 0.80 | 57.00 | – | 113.74 | m² | **170.74** |
| 2.0 m high | 113.74 | 0.80 | 57.00 | 7.78 | 126.87 | m² | **191.65** |
| 3.0 m high | 113.74 | 0.80 | 57.00 | 10.37 | 135.98 | m² | **203.35** |
| 4.0 m high | 113.74 | 0.80 | 57.00 | 12.45 | 143.05 | m² | **212.50** |
| **Retaining walls; Forticrete Ltd**<br>Stepoc Blocks; interlocking blocks; 10 mm reinforcing laid loose horizontally to preformed notches and vertical reinforcing nominal size 10 mm fixed to starter bars; infilling with concrete; foundations and starter bars measured separately | | | | | | | |
| type 325; 400 × 225 × 325 mm | 58.38 | 0.47 | 33.25 | – | 80.66 | m² | **113.91** |
| type 256; 400 × 225 × 256 mm | 54.99 | 0.40 | 28.50 | – | 75.26 | m² | **103.76** |
| type 200; 400 × 225 × 200 mm | 45.96 | 0.33 | 23.75 | – | 62.21 | m² | **85.96** |
| **Timber log retaining walls; AVS Fencing Supplies Ltd; timber posts are kiln dried redwood with 15 year guarantee**<br>Machine rounded softwood logs to trenches priced separately; disposal of excavated material priced separately; inclusive of 75 mm hardcore blinding to trench and backfilling trench with site mixed concrete 1:3:6; geofabric pinned to rear of logs; heights of logs above ground | | | | | | | |
| 500 mm (constructed from 1.80 m lengths) | 32.35 | 0.50 | 35.63 | – | 43.65 | m | **79.28** |
| 1.20 m (constructed from 1.80 m lengths) | 64.70 | 0.43 | 30.87 | – | 90.83 | m | **121.70** |
| 1.60 m (constructed from 2.40 m lengths) | 81.90 | 0.83 | 59.35 | – | 114.56 | m | **173.91** |
| 2.00 m (constructed from 3.00 m lengths) | 102.50 | 1.17 | 83.12 | – | 141.70 | m | **224.82** |
| As above but with 150 mm machine rounded timbers | | | | | | | |
| 500 mm | 44.30 | 0.83 | 59.35 | – | 55.60 | m | **114.95** |
| 1.20 m | 147.52 | 0.58 | 41.56 | – | 173.65 | m | **215.21** |
| 1.60 m | 147.67 | 1.00 | 71.25 | – | 180.33 | m | **251.58** |

## 6 GROUND REMEDIATION AND SOIL STABILIZATION

| Item<br>Excluding site overheads and profit | PC<br>£ | Labour<br>hours | Labour<br>£ | Plant<br>£ | Material<br>£ | Unit | Total<br>rate £ |
|---|---|---|---|---|---|---|---|
| **RETAINING WALLS AND REVETMENTS – cont** | | | | | | | |
| **Timber log retaining walls – cont** | | | | | | | |
| As above but with 200 mm machine rounded timbers | | | | | | | |
| 1.80 m (constructed from 2.40 m lengths) | 150.00 | 1.33 | 95.00 | – | 189.20 | m | **284.20** |
| **Railway sleeper walls; AVS Fencing Supplies Ltd; retaining wall from railway sleepers; fixed with steel galvanized pins 12 mm driven into the ground; sleepers laid flat** | | | | | | | |
| Grade 1 hardwood; | | | | | | | |
| 2590 × 250 × 150 mm | | | | | | | |
| 150 mm; 1 sleeper high | 14.01 | 0.17 | 11.87 | – | 14.41 | m | **26.28** |
| 300 mm; 2 sleepers high | 28.02 | 0.33 | 23.75 | – | 28.82 | m | **52.57** |
| 450 mm; 3 sleepers high | 41.69 | 0.50 | 35.63 | – | 42.74 | m | **78.37** |
| 600 mm; 4 sleepers high | 56.16 | 0.67 | 47.50 | – | 57.21 | m | **104.71** |
| Grade 1 hardwood as above but with 2 nr galvanized angle iron stakes set into concrete internally and screwed to the inside face of the sleepers | | | | | | | |
| 750 mm; 5 sleepers high | 69.48 | 0.83 | 59.37 | – | 94.47 | m² | **153.84** |
| 900 mm; 6 sleepers high | 83.16 | 0.92 | 65.31 | – | 110.15 | m² | **175.46** |
| New pine softwood; | | | | | | | |
| 2400 × 250 × 125 mm | | | | | | | |
| 120 mm; 1 sleeper high | 11.06 | 0.17 | 11.88 | – | 11.46 | m | **23.34** |
| 240 mm; 2 sleepers high | 22.13 | 0.33 | 23.75 | – | 22.93 | m | **46.68** |
| 360 mm; 3 sleepers high | 33.19 | 0.50 | 35.63 | – | 34.24 | m | **69.87** |
| 480 mm; 4 sleepers high | 44.26 | 0.67 | 47.50 | – | 45.31 | m | **92.81** |
| New pine softwood as above but with 2 nr galvanized angle iron stakes set into concrete internally and screwed to the inside face of the sleepers | | | | | | | |
| 600 mm; 5 sleepers high | 55.32 | 0.83 | 59.37 | – | 80.31 | m² | **139.68** |
| 720 mm; 6 sleepers high | 66.38 | 0.92 | 65.31 | – | 93.37 | m² | **158.68** |
| New oak hardwood | | | | | | | |
| 2600 × 220 × 130 mm | | | | | | | |
| 130 mm; 1 sleeper high | 13.70 | 0.17 | 11.87 | – | 14.10 | m | **25.97** |
| 260 mm; 2 sleepers high | 27.39 | 0.33 | 23.75 | – | 28.19 | m | **51.94** |
| 390 mm; 3 sleepers high | 41.09 | 5.00 | 356.25 | – | 42.14 | m | **398.39** |
| 520 mm; 4 sleepers high | 54.78 | 0.67 | 47.50 | – | 55.83 | m | **103.33** |
| New oak hardwood as above but with 2 nr galvanized angle iron stakes set into concrete internally and screwed to the inside face of the sleepers | | | | | | | |
| 640 mm; 5 sleepers high | 68.48 | 0.83 | 59.37 | – | 93.47 | m² | **152.84** |
| 760 mm; 6 sleepers high | 82.18 | 9.00 | 641.25 | – | 109.16 | m² | **750.41** |

## 6 GROUND REMEDIATION AND SOIL STABILIZATION

| Item<br>Excluding site overheads and profit | PC<br>£ | Labour<br>hours | Labour<br>£ | Plant<br>£ | Material<br>£ | Unit | Total<br>rate £ |
|---|---|---|---|---|---|---|---|
| **Excavate foundation trench; set railway sleepers vertically on end in concrete 1:3:6 continuous foundation to 33.3% of their length to form retaining wall** | | | | | | | |
| Grade 1 hardwood; finished height above ground level | | | | | | | |
| 300 mm | 37.26 | 1.00 | 71.25 | 1.95 | 43.61 | m | **116.81** |
| 500 mm | 54.00 | 1.00 | 71.25 | 1.95 | 60.35 | m | **133.55** |
| 600 mm | 75.06 | 1.00 | 71.25 | 1.95 | 86.37 | m | **159.57** |
| 750 mm | 81.00 | 1.17 | 83.13 | 1.95 | 92.31 | m | **177.39** |
| 1.00 m | 92.25 | 1.25 | 89.06 | 1.95 | 104.55 | m | **195.56** |
| **Excavate and place vertical steel universal beams 165 mm wide in concrete base at 2.59 m centres; fix railway sleepers set horizontally between beams to form horizontal fence or retaining wall** | | | | | | | |
| Grade 1 hardwood; bay length 2.59 m | | | | | | | |
| 500 mm high; 2 sleepers | 17.82 | 0.83 | 59.37 | – | 93.43 | bay | **152.80** |
| 750 mm high; 3 sleepers | 26.19 | 0.87 | 61.75 | – | 139.20 | bay | **200.95** |
| 1.00 m high; 4 sleepers | 35.10 | 1.00 | 71.25 | – | 181.80 | bay | **253.05** |
| 1.25 m high; 5 sleepers | 44.82 | 1.17 | 83.13 | – | 231.83 | bay | **314.96** |
| 1.50 m high; 6 sleepers | 52.65 | 1.00 | 71.25 | – | 276.66 | bay | **347.91** |
| 1.75 m high; 7 sleepers | 61.29 | 1.00 | 71.25 | – | 285.30 | bay | **356.55** |
| **Waterside revetments; Willowbank Services Ltd; natural engineering solutions; ecological sustainable waterside stabilization systems** | | | | | | | |
| Live willow spiling; 1 m high uprights at 0.50 m centres; woven geotextile backed; including filling with site based topsoil behind spiling to top of bank | | | | | | | |
| less than 50 m | – | – | – | – | – | m | **161.83** |
| 50–100 m | – | – | – | – | – | m | **126.42** |
| Brushwood faggot and pre-planted coir revetments; 900 mm high revetment; 300 mm diameter faggot; 300 mm diameter pre-planted coir roll; staked in position with 1.8 m tanalized posts; 2 posts per m; faggots and coir fixed to posts with wire | | | | | | | |
| less than 50 m | – | – | – | – | – | m | **145.01** |
| 50–100 m | – | – | – | – | – | m | **124.84** |

## 6 GROUND REMEDIATION AND SOIL STABILIZATION

| Item<br>Excluding site overheads and profit | PC<br>£ | Labour<br>hours | Labour<br>£ | Plant<br>£ | Material<br>£ | Unit | Total<br>rate £ |
|---|---|---|---|---|---|---|---|
| **RETAINING WALLS AND REVETMENTS – cont** | | | | | | | |
| **Waterside revetments – cont**<br>Brushwood faggot revetments;<br>900 mm high revetment; 2 nr 300 mm<br>diameter faggot staked in position<br>with 1.8 m tanalized posts; 2 posts<br>per m; faggots and coir fixed to posts<br>with wire | | | | | | | |
| less than 50 m | – | – | – | – | – | m | **124.48** |
| 50–100 m | – | – | – | – | – | m | **86.24** |
| **GENERALLY** | | | | | | | |
| **Soil stabilization – General**<br>Preamble: Earth-retaining and<br>stabilizing materials are often specified<br>as part of the earth-forming work in<br>landscape contracts, therefore this<br>section lists a number of products<br>specially designed for large-scale earth<br>control. There are two types: rigid units<br>for structural retention of earth on steep<br>slopes; and flexible meshes and sheets<br>for control of soil erosion where<br>structural strength is not required.<br>Prices for these items depend on<br>quantity, difficulty of access to the site<br>and availability of suitable filling<br>material; estimates should be obtained<br>from the manufacturer when the site<br>conditions have been determined. | | | | | | | |
| **GRADING AND SURFACE PREPARATION** | | | | | | | |
| **Grading operations; surface previously excavated to reduce levels to prepare to receive subsequent treatments; grading to accurate levels and falls 20 mm tolerances**<br>Clay or heavy soils or hardcore | | | | | | | |
| 5 tonne excavator | – | 0.04 | 0.95 | 0.19 | – | m² | **1.14** |
| JCB Sitemaster 3CX | – | 0.02 | 0.47 | 1.01 | – | m² | **1.48** |
| 21 tonne 360 tracked excavator | – | 0.01 | 0.24 | 0.58 | – | m² | **0.82** |
| by hand | – | 0.10 | 2.38 | – | – | m² | **2.38** |

## 6 GROUND REMEDIATION AND SOIL STABILIZATION

| Item<br>Excluding site overheads and profit | PC<br>£ | Labour<br>hours | Labour<br>£ | Plant<br>£ | Material<br>£ | Unit | Total<br>rate £ |
|---|---|---|---|---|---|---|---|
| Loamy topsoils | | | | | | | |
| 5 tonne excavator | – | 0.03 | 0.63 | 0.13 | – | $m^2$ | **0.76** |
| JCB Sitemaster 3CX | – | 0.01 | 0.32 | 0.67 | – | $m^2$ | **0.99** |
| 21 tonne 360 tracked excavator | – | 0.01 | 0.18 | 0.43 | – | $m^2$ | **0.61** |
| by hand | – | 0.05 | 1.19 | – | – | $m^2$ | **1.19** |
| Sand or graded granular materials | | | | | | | |
| 5 tonne excavator | – | 0.02 | 0.47 | 0.10 | – | $m^2$ | **0.57** |
| JCB Sitemaster 3CX | – | 0.01 | 0.24 | 0.50 | – | $m^2$ | **0.74** |
| 21 tonne 360 tracked excavator | – | 0.01 | 0.14 | 0.33 | – | $m^2$ | **0.47** |
| by hand | – | 0.03 | 0.79 | – | – | $m^2$ | **0.79** |
| **Grading operations to bottoms of excavations or trenches to receive foundations or bases; surface previously excavated to reduce levels to prepare to receive subsequent treatments; grading to accurate levels and falls 20 mm tolerances** | | | | | | | |
| Trenches | | | | | | | |
| 5 tonne excavator | – | 0.32 | 13.46 | 0.64 | – | $m^2$ | **14.10** |
| by hand | – | 0.33 | 15.83 | – | – | $m^2$ | **15.83** |
| Excavated areas to receive bases or formwork | | | | | | | |
| 5 tonne excavator | – | 0.13 | 5.54 | 0.32 | – | $m^2$ | **5.86** |
| by hand | – | 0.17 | 7.92 | – | – | $m^2$ | **7.92** |
| **Grading operations; surface recently filled to raise levels to prepare to receive subsequent treatments; grading to accurate levels and falls 20 mm tolerances** | | | | | | | |
| Clay or heavy soils or hardcore | | | | | | | |
| 5 tonne excavator | – | 0.03 | 0.71 | 0.14 | – | $m^2$ | **0.85** |
| JCB Sitemaster 3CX | – | 0.02 | 0.36 | 0.76 | – | $m^2$ | **1.12** |
| 21 tonne 360 tracked excavator | – | 0.01 | 0.16 | 0.39 | – | $m^2$ | **0.55** |
| by hand | – | 0.08 | 1.98 | – | – | $m^2$ | **1.98** |
| Loamy topsoils | | | | | | | |
| 5 tonne excavator | – | 0.02 | 0.49 | 0.10 | – | $m^2$ | **0.59** |
| JCB Sitemaster 3CX | – | 0.01 | 0.24 | 0.52 | – | $m^2$ | **0.76** |
| 21 tonne 360 tracked excavator | – | 0.01 | 0.16 | 0.39 | – | $m^2$ | **0.55** |
| by hand | – | 0.04 | 0.95 | – | – | $m^2$ | **0.95** |
| Sand or graded granular materials | | | | | | | |
| 5 tonne excavator | – | 0.02 | 0.36 | 0.07 | – | $m^2$ | **0.43** |
| JCB Sitemaster 3CX | – | 0.01 | 0.18 | 0.38 | – | $m^2$ | **0.56** |
| 21 tonne 360 tracked excavator | – | 0.01 | 0.12 | 0.29 | – | $m^2$ | **0.41** |
| by hand | – | 0.03 | 0.68 | – | – | $m^2$ | **0.68** |

## 6 GROUND REMEDIATION AND SOIL STABILIZATION

| Item<br>Excluding site overheads and profit | PC<br>£ | Labour<br>hours | Labour<br>£ | Plant<br>£ | Material<br>£ | Unit | Total<br>rate £ |
|---|---|---|---|---|---|---|---|
| **GRADING AND SURFACE PREPARATION – cont** | | | | | | | |
| **Surface grading; Agripower Ltd; using laser controlled equipment** | | | | | | | |
| Maximum variation of existing ground level | | | | | | | |
| 75 mm; laser box grader and tractor | – | – | – | – | – | ha | **1431.00** |
| 100 mm; laser controlled low ground pressure bulldozer | – | – | – | – | – | ha | **2279.00** |
| **Fraise mowing; scarification of existing turf layer to rootzone level; Agripower Ltd** | | | | | | | |
| Cart arisings to tip on site | | | | | | | |
| average thickness 50–75 mm | – | – | – | – | – | ha | **3140.78** |
| **Cultivating** | | | | | | | |
| Ripping up subsoil; using approved subsoiling machine; minimum depth 250 mm below topsoil; at 1.20 m centres; in | | | | | | | |
| gravel or sandy clay | – | – | – | 2.20 | – | 100 m$^2$ | **2.20** |
| soil compacted by machines | – | – | – | 2.44 | – | 100 m$^2$ | **2.44** |
| clay | – | – | – | 3.67 | – | 100 m$^2$ | **3.67** |
| chalk or other soft rock | – | – | – | 3.67 | – | 100 m$^2$ | **3.67** |
| Extra for subsoiling at 1 m centres | – | – | – | 1.10 | – | 100 m$^2$ | **1.10** |
| Breaking up existing ground; using pedestrian operated tine cultivator or rotavator; loam or sandy soil | | | | | | | |
| 100 mm deep | – | 0.22 | 5.22 | 0.79 | – | 100 m$^2$ | **6.01** |
| 150 mm deep | – | 0.28 | 6.53 | 0.99 | – | 100 m$^2$ | **7.52** |
| 200 mm deep | – | 0.37 | 8.71 | 1.32 | – | 100 m$^2$ | **10.03** |
| As above but in heavy clay or wet soils | | | | | | | |
| 100 mm deep | – | 0.44 | 10.45 | 1.59 | – | 100 m$^2$ | **12.04** |
| 150 mm deep | – | 0.66 | 15.68 | 2.38 | – | 100 m$^2$ | **18.06** |
| 200 mm deep | – | 0.82 | 19.59 | 2.98 | – | 100 m$^2$ | **22.57** |
| Breaking up existing ground; using tractor drawn tine cultivator or rotavator | | | | | | | |
| 100 mm deep | – | – | – | 1.56 | – | 100 m$^2$ | **1.56** |
| 150 mm deep | – | – | – | 1.95 | – | 100 m$^2$ | **1.95** |
| 200 mm deep | – | – | – | 2.59 | – | 100 m$^2$ | **2.59** |
| 400 mm deep | – | – | – | 7.80 | – | 100 m$^2$ | **7.80** |
| Final levelling to ploughed and rotavated ploughed ground; using disc, drag, or chain harrow | | | | | | | |
| 4 passes | – | – | – | 1.42 | – | 100 m$^2$ | **1.42** |
| Rolling cultivated ground lightly; using self-propelled agricultural roller | – | 0.06 | 1.32 | 0.61 | – | 100 m$^2$ | **1.93** |

## 6 GROUND REMEDIATION AND SOIL STABILIZATION

| Item<br>Excluding site overheads and profit | PC<br>£ | Labour<br>hours | Labour<br>£ | Plant<br>£ | Material<br>£ | Unit | Total<br>rate £ |
|---|---|---|---|---|---|---|---|
| Importing and storing selected and approved topsoil; 20 tonne load = 11.88 m$^3$ average | | | | | | | |
| 20 tonne loads | 39.60 | – | – | – | 39.60 | m$^3$ | **39.60** |
| **Surface treatments** | | | | | | | |
| Compacting | | | | | | | |
| bottoms of excavations | – | 0.01 | 0.12 | 0.02 | – | m$^2$ | **0.14** |
| Surface preparation | | | | | | | |
| Trimming surfaces of cultivated ground to final levels; removing roots, stones and debris exceeding 50 mm in any direction to tip off site; slopes less than 15° | | | | | | | |
| clean ground with minimal stone content | – | 0.25 | 5.94 | – | – | 100 m$^2$ | **5.94** |
| slightly stony; 0.5 kg stones per m$^2$ | – | 0.33 | 7.91 | – | 0.01 | 100 m$^2$ | **7.92** |
| very stony; 1.0–3.0 kg stones per m$^2$ | – | 0.50 | 11.88 | – | 0.03 | 100 m$^2$ | **11.91** |
| clearing mixed, slightly contaminated rubble; inclusive of roots and vegetation | – | 0.50 | 11.88 | – | 0.09 | 100 m$^2$ | **11.97** |
| clearing brick-bats, stones and clean rubble | – | 0.60 | 14.25 | – | 0.08 | 100 m$^2$ | **14.33** |

## 10 CRIB WALLS, GABIONS AND REINFORCED EARTH

| Item<br>Excluding site overheads and profit | PC<br>£ | Labour<br>hours | Labour<br>£ | Plant<br>£ | Material<br>£ | Unit | Total<br>rate £ |
|---|---|---|---|---|---|---|---|
| **CRIB WALLS, GABIONS AND REINFORCED EARTH** | | | | | | | |
| **Crib walls; Phi Group** | | | | | | | |
| Permacrib (timber) or Andacrib (concrete) crib walling system; machine filled infill with crushed rock; inclusive of concrete footing and rear wall drain; excluding excavation and backfill material | | | | | | | |
| retaining walls up to 2.0 m | – | – | – | – | – | m² | 210.00 |
| retaining walls up to 4.0 m | – | – | – | – | – | m² | 235.00 |
| retaining walls up to 6.0 m | – | – | – | – | – | m² | 255.00 |
| **Retaining walls** | | | | | | | |
| Wire mesh gabions; galvanized mesh 80 × 100 mm; filling with broken stones 125–200 mm size; wire down securely to manufacturer's instructions; filling front face by hand | | | | | | | |
| 2 × 1 × 0.50 m | 25.00 | 0.67 | 47.50 | 5.44 | 80.65 | nr | 133.59 |
| 2 × 1 × 1.0 m | 34.00 | 1.33 | 95.00 | 10.89 | 145.30 | nr | 251.19 |
| PVC coated gabions | | | | | | | |
| 2 × 1 × 0.5 m | 27.44 | 0.67 | 47.50 | 5.44 | 83.09 | nr | 136.03 |
| 2 × 1 × 1.0 m | 38.50 | 1.33 | 95.00 | 10.89 | 149.80 | nr | 255.69 |
| Mattress gabions | | | | | | | |
| 6 × 2 × 0.17 m | 94.90 | 1.00 | 71.25 | 8.16 | 208.43 | nr | 287.84 |
| 6 × 2 × 0.23 m | 103.10 | 1.50 | 106.88 | 10.89 | 256.69 | nr | 374.46 |
| 6 × 2 × 0.30 m | 121.08 | 2.00 | 142.50 | 12.25 | 321.42 | nr | 476.17 |
| **Embankments; Tensar International** | | | | | | | |
| Embankments; reinforced with Tensar Geogrid RE520; geogrid laid horizontally within fill to 100% of vertical height of slope at 1.00 m centres; filling with excavated material | | | | | | | |
| slopes less than 45°; Tensartech Natural Green System | 2.36 | 0.61 | 43.65 | 1.88 | 2.36 | m³ | 47.89 |
| slopes exceeding 45°; Tensartech Greenslope System | 5.08 | 0.06 | 4.40 | 2.92 | 5.08 | m³ | 12.40 |
| extra over to slopes exceeding 45° for temporary shuttering; shuttering measured per m² of vertical elevation | – | – | – | – | 12.00 | m² | 12.00 |

## 10 CRIB WALLS, GABIONS AND REINFORCED EARTH

| Item<br>Excluding site overheads and profit | PC<br>£ | Labour<br>hours | Labour<br>£ | Plant<br>£ | Material<br>£ | Unit | Total<br>rate £ |
|---|---|---|---|---|---|---|---|
| Embankments; reinforced with Tensar Geogrid RE540; geogrid laid horizontally within fill to 100% of vertical height of slope at 1.00 m centres; filling with excavated material | | | | | | | |
|    slopes less than 45°; Tensartech Natural Green System | 2.92 | 0.61 | 43.65 | 1.88 | 2.92 | m³ | **48.45** |
|    slopes exceeding 45°; Tensartech Greenslope System | 6.28 | 0.06 | 4.41 | 2.92 | 6.28 | m³ | **13.61** |
|    extra over to slopes exceeding 45° for temporary shuttering; shuttering measured per m² of vertical elevation | – | – | – | – | 12.00 | m² | **12.00** |
| Extra for Tensar Mat 400; erosion control mats; to faces of slopes of 45° or less; filling with 20 mm fine topsoil; seeding | 3.95 | 0.01 | 0.47 | 0.25 | 5.38 | m² | **6.10** |
| Embankments; reinforced with Tensar Geogrid RE540; geogrid laid horizontally within fill to 50% of horizontal length of slopes at specified centres; filling with imported fill PC £12.00/m³ | | | | | | | |
|    300 mm centres; slopes up to 45°; Tensartech Natural Green System | 4.85 | 0.08 | 5.54 | 2.18 | 19.25 | m³ | **26.97** |
|    600 mm centres; slopes up to 45°; Tensartech Natural Green System | 2.42 | 0.06 | 4.41 | 2.92 | 16.82 | m³ | **24.15** |
|    extra over to slopes exceeding 45° for temporary shuttering; shuttering measured per m² of vertical elevation | – | – | – | – | 12.00 | m² | **12.00** |
| Extra for Tensar Mat 400; erosion control mats; to faces of slopes of 45° or less; filling with 20 mm fine topsoil; seeding | 3.95 | 0.01 | 0.47 | 0.25 | 5.38 | m² | **6.10** |
| Embankments; reinforced with Tensar Geogrid RE540; geogrid laid horizontally within fill; filling with excavated material; wrapping around at faces | | | | | | | |
|    300 mm centres; slopes exceeding 45°; Tensartech Natural Green System | 9.72 | 0.06 | 4.40 | 1.88 | 9.72 | m³ | **16.00** |
|    600 mm centres; slopes exceeding 45°; Tensartech Greenslope System | 4.85 | 0.06 | 4.40 | 1.88 | 4.85 | m³ | **11.13** |
|    extra over to slopes exceeding 45° for temporary shuttering; shuttering measured per m² of vertical elevation | – | – | – | – | 12.00 | m² | **12.00** |
| Extra over embankments for bagwork face supports for slopes exceeding 45° | – | 0.17 | 11.87 | – | 14.00 | m² | **25.87** |
| Extra over embankments for seeding of bags | – | 0.03 | 2.37 | – | 0.33 | m² | **2.70** |
| Extra over embankments for Bodkin joints | – | – | – | – | 1.10 | m | **1.10** |

## 10 CRIB WALLS, GABIONS AND REINFORCED EARTH

| Item<br>Excluding site overheads and profit | PC<br>£ | Labour<br>hours | Labour<br>£ | Plant<br>£ | Material<br>£ | Unit | Total<br>rate £ |
|---|---|---|---|---|---|---|---|
| **CRIB WALLS, GABIONS AND REINFORCED EARTH – cont** | | | | | | | |
| **Anchoring systems; surface stabilization; Platipus Anchors Ltd; centres and depths shown should be verified with a design engineer; they may vary either way dependent on circumstances** | | | | | | | |
| Concrete revetments; hard anodized stealth anchors with stainless steel accessories installed to a depth of 1 m | | | | | | | |
| SO4 anchors to a depth of 1.00–1.50 m at 1.00 m centres | 26.63 | 0.17 | 7.92 | 1.51 | 26.63 | m² | **36.06** |
| Loadlocking and crimping tool; for stressing and crimping the anchors | | | | | | | |
| hire rate per week | 72.93 | – | – | – | 72.93 | nr | **72.93** |
| Surface erosion; geotextile anchoring on slopes; aluminium alloy Stealth anchors | | | | | | | |
| SO4/SO6 to a depth of 1.0–2.0 m at 2.00 m centres | 9.10 | 1.00 | 47.50 | 1.51 | 9.10 | m² | **58.11** |
| Brick block or in situ concrete distressed retaining walls; anchors installed in two rows at varying tensions between 18–50 kN along the length of the retaining wall; core drilling of wall not included | | | | | | | |
| SO8/BO6/BO8 bronze Stealth and Bat anchors installed in combination with stainless steel accessories to a depth of 6 m; average cost based on 1.50 m centres; long-term solution | 382.02 | 0.25 | 11.88 | 1.51 | 382.02 | m² | **395.41** |
| SO8/BO6/BO8 cast SG iron Stealth and Bat anchors installed in combination with galvanized steel accessories to a depth of 6 m; average cost based on 1.50 m centres; short-term solution | 163.22 | 0.25 | 11.88 | 1.51 | 163.22 | m² | **176.61** |
| Timber retaining wall support; anchors installed 19 kN along the length of the retaining wall | | | | | | | |
| SO6/SO8 bronze Stealth anchors including load plate and wege grip; installed in combination with stainless steel accessories to a depth of 3.0 m; 1.00 m centres; average cost | 184.80 | – | – | – | – | m² | **–** |

## 10 CRIB WALLS, GABIONS AND REINFORCED EARTH

| Item<br>Excluding site overheads and profit | PC<br>£ | Labour<br>hours | Labour<br>£ | Plant<br>£ | Material<br>£ | Unit | Total<br>rate £ |
|---|---|---|---|---|---|---|---|
| **Gabions; anchors for support and stability to moving, overturning or rotating gabion retaining walls** | | | | | | | |
| SO8/BO6 anchors including base plate at 1.00 m centres to a depth of 4.00 m | 289.41 | 0.17 | 11.88 | 1.51 | 289.41 | m² | 302.80 |
| Soil nailing of geofabrics; anchors fixed through surface of fabric or erosion control surface treatment (not included) | | | | | | | |
| SO8/BO6 anchors including base plate at 1.50 m centres to a depth of 3.00 m | 115.76 | 0.08 | 5.94 | 1.01 | 115.76 | m² | 122.71 |
| **Green faced reinforced soil; Phi Group**<br>Textomur reinforced soil system embankments; to reinforced soil slopes 55–70° to the horizontal; including facing and geogrid; excluding backfill material | | | | | | | |
| retaining walls up to 3.0 m | – | – | – | – | – | m² | 145.00 |
| retaining walls up to 6.0 m | – | – | – | – | – | m² | 150.00 |
| **Split face concrete blocks reinforced soil; Phi Group**<br>Modular blocks reinforced soil system embankments; to reinforced soil slopes near vertical; including geogrid; excluding backfill material | | | | | | | |
| retaining walls up to 3.0 m | – | – | – | – | – | m² | 210.00 |
| retaining walls up to 6.0 m | – | – | – | – | – | m² | 255.00 |
| **Embankments; Grass Concrete Ltd**<br>Grasscrete; in situ reinforced concrete surfacing; to 20 mm thick sand blinding layer (not included); including soiling and seeding | | | | | | | |
| GC1; 100 mm thick | – | – | – | – | 40.56 | m² | 40.56 |
| GC2; 150 mm thick | – | – | – | – | 52.07 | m² | 52.07 |
| Grassblock 103; solid matrix precast concrete blocks; to 20 mm thick sand blinding layer; excluding edge restraint; including soiling and seeding | | | | | | | |
| 406 × 406 × 103 mm; fully interlocking | – | – | – | – | 51.12 | m² | 51.12 |

## 10 CRIB WALLS, GABIONS AND REINFORCED EARTH

| Item<br>Excluding site overheads and profit | PC<br>£ | Labour<br>hours | Labour<br>£ | Plant<br>£ | Material<br>£ | Unit | Total<br>rate £ |
|---|---|---|---|---|---|---|---|
| **CRIB WALLS, GABIONS AND REINFORCED EARTH – cont** | | | | | | | |
| **Flexible sheet materials; Tensar International** | | | | | | | |
| Tensar Mat 400 erosion mats; 3.0–4.5 m wide; securing with Tensar pegs; lap rolls 100 mm; anchors at top and bottom of slopes; in trenches | 4.40 | 0.01 | 0.39 | – | 4.40 | $m^2$ | **4.79** |
| Topsoil filling to Tensar Mat; including brushing and raking | 1.03 | 0.01 | 0.39 | 0.09 | 1.03 | $m^2$ | **1.51** |
| Tensar Bi-axial Geogrid; to graded compacted base; filling with 200 mm granular fill; compacting (turf or paving to surfaces not included); 400 mm laps | | | | | | | |
| TriAx 150; 39 × 39 mm mesh | 1.40 | – | 0.32 | 0.16 | 7.16 | $m^2$ | **7.64** |
| TriAx 160; 39 × 39 mm mesh | 1.82 | – | 0.32 | 0.16 | 7.58 | $m^2$ | **8.06** |
| TriAx 170; 39 × 39 mm mesh | 2.15 | – | 0.32 | 0.16 | 7.91 | $m^2$ | **8.39** |
| **Flexible sheet materials; Terram Ltd** | | | | | | | |
| Terram synthetic fibre filter fabric; to graded base (not included) | | | | | | | |
| Terram 1000; 0.70 mm thick; mean water flow 50 l/$m^2$/s | 0.41 | 0.01 | 0.39 | – | 0.41 | $m^2$ | **0.80** |
| Terram 2000; 1.00 m thick; mean water flow 33 l/$m^2$/s | 0.62 | 0.01 | 0.39 | – | 0.62 | $m^2$ | **1.01** |
| Terram Minipack | 0.71 | – | 0.08 | – | 0.71 | $m^2$ | **0.79** |
| **Flexible sheet materials; Greenfix Ltd** | | | | | | | |
| Greenfix; erosion control mats; 10–15 mm thick; fixing with 4 nr crimped pins in accordance with manufacturer's instructions; to graded surface (not included) | | | | | | | |
| unseeded Eromat 1; 2.40 m wide | 103.47 | 0.67 | 47.50 | – | 167.47 | 100 $m^2$ | **214.97** |
| unseeded Eromat 2; 2.40 m wide | 130.69 | 0.67 | 47.50 | – | 194.69 | 100 $m^2$ | **242.19** |
| unseeded Eromat 3; 2.40 m wide | 147.03 | 0.67 | 47.50 | – | 211.03 | 100 $m^2$ | **258.53** |
| seeded Covamat 1; 2.40 m wide | 166.25 | 0.67 | 47.50 | – | 230.25 | 100 $m^2$ | **277.75** |
| seeded Covamat 2; 2.40 m wide | 210.00 | 0.67 | 47.50 | – | 274.00 | 100 $m^2$ | **321.50** |
| seeded Covamat 3; 2.40 m wide | 227.50 | 0.67 | 47.50 | – | 291.50 | 100 $m^2$ | **339.00** |
| Bioroll; 300 mm diameter; to river banks and revetments | 19.00 | 0.04 | 2.97 | – | 21.47 | m | **24.44** |
| Extra over Greenfix erosion control mats for fertilizer applied at 70 g/$m^2$ | 7.81 | 0.07 | 4.75 | – | 7.81 | 100 $m^2$ | **12.56** |
| Extra over Greenfix erosion control mats for Geojute; fixing with steel pins | 1.32 | 0.01 | 0.53 | – | 1.32 | $m^2$ | **1.85** |
| Extra over Greenfix erosion control mats for laying to slopes exceeding 30° | – | – | – | – | – | 25% | – |

## 10 CRIB WALLS, GABIONS AND REINFORCED EARTH

| Item<br>Excluding site overheads and profit | PC<br>£ | Labour<br>hours | Labour<br>£ | Plant<br>£ | Material<br>£ | Unit | Total<br>rate £ |
|---|---|---|---|---|---|---|---|
| **Extra for the following operations** | | | | | | | |
| Spreading 25 mm approved topsoil | | | | | | | |
|   by machine | 0.99 | – | 0.15 | 0.16 | 0.99 | m$^2$ | **1.30** |
|   by hand | 0.99 | 0.01 | 0.83 | – | 0.99 | m$^2$ | **1.82** |
| Grass seed; PC £4.50/kg; spreading in | | | | | | | |
| two operations; by hand | | | | | | | |
|   35 g/m$^2$ | 15.75 | 0.06 | 3.96 | – | 15.75 | 100 m$^2$ | **19.71** |
|   50 g/m$^2$ | 22.50 | 0.06 | 3.96 | – | 22.50 | 100 m$^2$ | **26.46** |
|   70 g/m$^2$ | 31.50 | 0.06 | 3.96 | – | 31.50 | 100 m$^2$ | **35.46** |
|   100 g/m$^2$ | 45.00 | 0.07 | 4.75 | – | 45.00 | 100 m$^2$ | **49.75** |
|   125 g/m$^2$ | 56.25 | 0.67 | 47.50 | – | 56.25 | 100 m$^2$ | **103.75** |
| Extra over seeding by hand for slopes | | | | | | | |
| over 30° (allowing for the actual area | | | | | | | |
| but measured in plan) | | | | | | | |
|   35 g/m$^2$ | 2.34 | – | 0.04 | – | 2.34 | 100 m$^2$ | **2.38** |
|   50 g/m$^2$ | 3.38 | – | 0.04 | – | 3.38 | 100 m$^2$ | **3.42** |
|   70 g/m$^2$ | 4.72 | – | 0.04 | – | 4.72 | 100 m$^2$ | **4.76** |
|   100 g/m$^2$ | 6.75 | – | 0.04 | – | 6.75 | 100 m$^2$ | **6.79** |
|   125 g/m$^2$ | 8.41 | – | 0.04 | – | 8.41 | 100 m$^2$ | **8.45** |
| Grass seed; PC £4.50/kg; spreading in | | | | | | | |
| two operations; by machine | | | | | | | |
|   35 g/m$^2$ | 15.75 | – | – | 0.52 | 15.75 | 100 m$^2$ | **16.27** |
|   50 g/m$^2$ | 22.50 | – | – | 0.52 | 22.50 | 100 m$^2$ | **23.02** |
|   70 g/m$^2$ | 31.50 | – | – | 0.52 | 31.50 | 100 m$^2$ | **32.02** |
|   100 g/m$^2$ | 45.00 | – | – | 0.52 | 45.00 | 100 m$^2$ | **45.52** |
|   125 kg/ha | 562.50 | – | – | 51.99 | 562.50 | ha | **614.49** |
|   150 kg/ha | 675.00 | – | – | 51.99 | 675.00 | ha | **726.99** |
|   200 kg/ha | 900.00 | – | – | 51.99 | 900.00 | ha | **951.99** |
|   250 kg/ha | 1125.00 | – | – | 51.99 | 1125.00 | ha | **1176.99** |
|   300 kg/ha | 1350.00 | – | – | 51.99 | 1350.00 | ha | **1401.99** |
|   350 kg/ha | 1575.00 | – | – | 51.99 | 1575.00 | ha | **1626.99** |
|   400 kg/ha | 1800.00 | – | – | 51.99 | 1800.00 | ha | **1851.99** |
|   500 kg/ha | 2250.00 | – | – | 51.99 | 2250.00 | ha | **2301.99** |
|   700 kg/ha | 3150.00 | – | – | 51.99 | 3150.00 | ha | **3201.99** |
|   1400 kg/ha | 6300.00 | – | – | 51.99 | 6300.00 | ha | **6351.99** |
| Extra over seeding by machine for | | | | | | | |
| slopes over 30° (allowing for the actual | | | | | | | |
| area but measured in plan) | | | | | | | |
|   35 g/m$^2$ | 2.36 | – | – | 0.08 | 2.36 | 100 m$^2$ | **2.44** |
|   50 g/m$^2$ | 3.38 | – | – | 0.08 | 3.38 | 100 m$^2$ | **3.46** |
|   70 g/m$^2$ | 4.72 | – | – | 0.08 | 4.72 | 100 m$^2$ | **4.80** |
|   100 g/m$^2$ | 6.75 | – | – | 0.08 | 6.75 | 100 m$^2$ | **6.83** |
|   125 kg/ha | 84.38 | – | – | 7.80 | 84.38 | ha | **92.18** |
|   150 kg/ha | 101.25 | – | – | 7.80 | 101.25 | ha | **109.05** |
|   200 kg/ha | 135.00 | – | – | 7.80 | 135.00 | ha | **142.80** |
|   250 kg/ha | 168.75 | – | – | 7.80 | 168.75 | ha | **176.55** |
|   300 kg/ha | 202.50 | – | – | 7.80 | 202.50 | ha | **210.30** |
|   350 kg/ha | 236.25 | – | – | 7.80 | 236.25 | ha | **244.05** |
|   400 kg/ha | 270.00 | – | – | 7.80 | 270.00 | ha | **277.80** |
|   500 kg/ha | 337.50 | – | – | 7.80 | 337.50 | ha | **345.30** |
|   700 kg/ha | 472.50 | – | – | 7.80 | 472.50 | ha | **480.30** |
|   1400 kg/ha | 945.00 | – | – | 7.80 | 945.00 | ha | **952.80** |

## 11 IN SITU CONCRETE WORKS

| Item<br>Excluding site overheads and profit | PC<br>£ | Labour<br>hours | Labour<br>£ | Plant<br>£ | Material<br>£ | Unit | **Total<br>rate £** |
|---|---|---|---|---|---|---|---|
| **CLARIFICATION NOTES ON LABOUR COSTS IN THIS SECTION** | | | | | | | |
| **In situ concrete team**<br>Generally a three man team is used in this section; The column Labour hours reports team hours. The column Labour £ reports the total cost of the team for the unit of work shown | | | | | | | |
| 3 man team | – | 1.00 | 71.25 | – | – | hr | **71.25** |
| 2 man team | – | 1.00 | 47.50 | – | – | hr | **47.50** |
| **CONCRETE MIX INFORMATION** | | | | | | | |
| **General**<br>The concrete mixes used here are referred to as Designed, Standard and Designated mixes. | | | | | | | |
| **Designed mix**<br>User specifies the performance of the concrete. Producer is responsible for selecting the appropriate mix. Strength testing is essential. | | | | | | | |
| **Prescribed mix**<br>User specifies the mix constituents and is responsible for ensuring that the concrete meets performance requirements. Accurate mix proportions are essential. | | | | | | | |
| **Standard mix**<br>Made with a restricted range of materials. Specification to include the proposed use of the material as well as the standard mix reference, type of cement, type and size of aggregate, and slump (workability). Quality assurance is required. | | | | | | | |
| **Designated mix**<br>Producer to hold current product conformity certification and quality approval. Quality assurance is essential. The mix must not be modified. | | | | | | | |

## 11 IN SITU CONCRETE WORKS

| Item<br>Excluding site overheads and profit | PC<br>£ | Labour<br>hours | Labour<br>£ | Plant<br>£ | Material<br>£ | Unit | Total<br>rate £ |
|---|---|---|---|---|---|---|---|
| **CONCRETE MIXES** | | | | | | | |
| Note: The mixes below are all mixed in 113 litre mixers. | | | | | | | |
| Concrete mixes; mixed on site; costs for producing concrete; prices for commonly used mixes for various types of work; based on bulk load 20 tonne rates for aggregates | | | | | | | |
| Roughest type mass concrete such as footings and road haunchings; 300 mm thick | | | | | | | |
| 1:3:6 | 100.17 | – | – | – | 100.17 | m³ | **100.17** |
| As above but aggregates delivered in 10 tonne loads | | | | | | | |
| 1:3:6 | 113.38 | – | – | – | 113.38 | m³ | **113.38** |
| As above but aggregates delivered in 850 kg bulk bags | | | | | | | |
| 1:3:6 | 142.64 | – | – | – | 142.64 | m³ | **142.64** |
| Most ordinary use of concrete such as mass walls above ground, road slabs, etc. and general reinforced concrete work | | | | | | | |
| 1:2:4 | – | – | – | – | 115.88 | m³ | **115.88** |
| As above but aggregates delivered in 10 tonne loads | | | | | | | |
| 1:2:4 | 123.48 | – | – | – | 123.48 | m³ | **123.48** |
| As above but aggregates delivered in 850 kg bulk bags | | | | | | | |
| 1:2:4 | – | – | – | – | 152.74 | m³ | **152.74** |
| Watertight floors, pavements, walls, tanks, pits, steps, paths, surface of two course roads; reinforced concrete where extra strength is required | | | | | | | |
| 1:1.5:3 | 109.46 | 1.20 | 28.43 | – | 109.46 | m³ | **137.89** |
| As above but aggregates delivered in 10 tonne loads | | | | | | | |
| 1:1.5:3 | 117.06 | 1.20 | 28.43 | – | 117.06 | m³ | **145.49** |
| As above but aggregates delivered in 850 kg bulk bags | | | | | | | |
| 1:1.5:3 | 146.32 | 1.20 | 28.43 | – | 146.32 | m³ | **174.75** |

## 11 IN SITU CONCRETE WORKS

| Item<br>Excluding site overheads and profit | PC<br>£ | Labour<br>hours | Labour<br>£ | Plant<br>£ | Material<br>£ | Unit | Total<br>rate £ |
|---|---|---|---|---|---|---|---|
| **IN SITU CONCRETE** | | | | | | | |
| **Plain in situ concrete; site mixed;**<br>**10 N/mm² – 40 mm aggregate (1:3:6)**<br>**(aggregate delivery indicated)** | | | | | | | |
| Foundations | | | | | | | |
| ordinary Portland cement; 20 tonne ballast loads | 102.67 | 0.33 | 23.51 | – | 102.67 | m³ | **126.18** |
| ordinary Portland cement; 10 tonne ballast loads | 116.22 | 0.33 | 23.51 | – | 116.22 | m³ | **139.73** |
| ordinary Portland cement; 850 kg bulk bags | 146.21 | 0.33 | 23.51 | – | 146.21 | m³ | **169.72** |
| Foundations; poured on or against earth or unblinded hardcore | | | | | | | |
| ordinary Portland cement; 20 tonne ballast loads | 105.17 | 0.33 | 23.51 | – | 105.17 | m³ | **128.68** |
| ordinary Portland cement; 10 tonne ballast loads | 119.05 | 0.33 | 23.51 | – | 119.05 | m³ | **142.56** |
| ordinary Portland cement; 850 kg bulk bags | 149.77 | 0.33 | 23.51 | – | 149.77 | m³ | **173.28** |
| Isolated foundations | | | | | | | |
| ordinary Portland cement; 20 tonne ballast loads | 102.67 | 0.33 | 23.51 | – | 102.67 | m³ | **126.18** |
| ordinary Portland cement; 850 kg bulk bags | 146.21 | 0.33 | 23.51 | – | 146.21 | m³ | **169.72** |
| **Plain in situ concrete; site mixed;**<br>**21 N/mm² – 20 mm aggregate (1:2:4)** | | | | | | | |
| Foundations | | | | | | | |
| ordinary Portland cement; 20 tonne ballast loads | 118.78 | 0.33 | 23.51 | – | 118.78 | m³ | **142.29** |
| ordinary Portland cement; 850 kg bulk bags | 156.56 | 0.33 | 23.51 | – | 156.56 | m³ | **180.07** |
| Foundations; poured on or against earth or unblinded hardcore | | | | | | | |
| ordinary Portland cement; 20 tonne ballast loads | 121.67 | 0.33 | 23.51 | – | 121.67 | m³ | **145.18** |
| ordinary Portland cement; 850 kg bulk bags | 160.37 | 0.33 | 23.51 | – | 160.37 | m³ | **183.88** |
| Isolated foundations | | | | | | | |
| ordinary Portland cement; 20 tonne ballast loads | 118.78 | 0.33 | 23.51 | – | 118.78 | m³ | **142.29** |
| ordinary Portland cement; 850 kg bulk bags | 156.56 | 0.33 | 23.51 | – | 156.56 | m³ | **180.07** |

## 11 IN SITU CONCRETE WORKS

| Item<br>Excluding site overheads and profit | PC<br>£ | Labour<br>hours | Labour<br>£ | Plant<br>£ | Material<br>£ | Unit | Total<br>rate £ |
|---|---|---|---|---|---|---|---|
| **Ready mix concrete; concrete mixed on site; Euromix** | | | | | | | |
| Ready mix concrete mixed on site; placed by barrow not more than 25 m distance (16 barrows/m³) | | | | | | | |
|   concrete 1:3:6 | 96.50 | 0.25 | 17.81 | – | 96.50 | m³ | **114.31** |
|   concrete 1:2:4 | 98.00 | 0.25 | 17.81 | – | 98.00 | m³ | **115.81** |
|   concrete C30 | 99.50 | 0.25 | 17.81 | – | 99.50 | m³ | **117.31** |
| **Reinforced in situ concrete; site mixed; 21 N/mm² – 20 mm aggregate (1:2:4); aggregates delivered in 10 tonne loads** | | | | | | | |
| Foundations | | | | | | | |
|   ordinary Portland cement | 118.78 | 0.67 | 47.45 | – | 118.78 | m³ | **166.23** |
| Foundations; poured on or against earth or unblinded hardcore | | | | | | | |
|   ordinary Portland cement | 118.78 | 0.67 | 47.45 | – | 118.78 | m³ | **166.23** |
| Isolated foundations | | | | | | | |
|   ordinary Portland cement | 118.78 | 0.83 | 59.35 | – | 118.78 | m³ | **178.13** |
| **Plain in situ concrete; ready mixed; Tarmac Southern; 10 N/mm mixes; suitable for mass concrete fill and blinding** | | | | | | | |
| Foundations | | | | | | | |
|   C10; 10 N/mm² | 87.50 | 0.20 | 14.25 | – | 87.50 | m³ | **101.75** |
|   ST 2; 10 N/mm² | 88.20 | 0.20 | 14.25 | – | 88.20 | m³ | **102.45** |
| Foundations; poured on or against earth or unblinded hardcore | | | | | | | |
|   C10; 10 N/mm² | 87.50 | 0.20 | 14.25 | – | 87.50 | m³ | **101.75** |
|   ST 2; 10 N/mm² | 90.30 | 0.20 | 14.25 | – | 90.30 | m³ | **104.55** |
| Isolated foundations | | | | | | | |
|   C10; 10 N/mm² | 87.50 | 0.67 | 47.45 | – | 87.50 | m³ | **134.95** |
|   ST 2; 10 N/mm² | 88.20 | 0.67 | 47.45 | – | 88.20 | m³ | **135.65** |
| **Plain in situ concrete; ready mixed; Hanson Concrete; 15 N/mm mixes; suitable for oversite below suspended slabs and strip footings in non aggressive soils** | | | | | | | |
| Foundations | | | | | | | |
|   C15; 15 N/mm² | 95.55 | 0.20 | 14.25 | – | 95.55 | m³ | **109.80** |
|   ST 3; 15 N/mm² | 94.50 | 0.20 | 14.25 | – | 94.50 | m³ | **108.75** |
| Foundations; poured on or against earth or unblinded hardcore | | | | | | | |
|   C15; 15 N/mm² | 91.00 | 0.20 | 14.25 | – | 91.00 | m³ | **105.25** |
|   ST 3; 15 N/mm² | 90.00 | 0.20 | 14.25 | – | 90.00 | m³ | **104.25** |

## 11 IN SITU CONCRETE WORKS

| Item<br>Excluding site overheads and profit | PC<br>£ | Labour<br>hours | Labour<br>£ | Plant<br>£ | Material<br>£ | Unit | Total<br>rate £ |
|---|---|---|---|---|---|---|---|
| **IN SITU CONCRETE – cont** | | | | | | | |
| **Plain in situ concrete – cont** | | | | | | | |
| Isolated foundations | | | | | | | |
| C15; 15 N/mm$^2$ | 91.00 | 0.67 | 47.45 | – | 91.00 | m$^3$ | **138.45** |
| ST 3; 15 N/mm$^2$ | 90.00 | 0.67 | 47.45 | – | 90.00 | m$^3$ | **137.45** |
| **Plain in situ concrete; ready mixed; Tarmac Southern; air entrained mixes suitable for paving** | | | | | | | |
| Beds or slabs; house drives, parking and external paving | | | | | | | |
| PAV 1; 35 N/mm$^2$; Designated mix | 106.58 | 0.25 | 17.81 | – | 106.58 | m$^3$ | **124.39** |
| Beds or slabs; heavy duty external paving | | | | | | | |
| PAV 2; 40 N/mm$^2$; Designated mix | 110.25 | 0.25 | 17.81 | – | 110.25 | m$^3$ | **128.06** |
| **Reinforced in situ concrete; ready mixed; Tarmac Southern; 35 N/mm$^2$ mix; suitable for foundations in class 2 sulphate conditions** | | | | | | | |
| Foundations | | | | | | | |
| RC 35; Designated mix | 100.48 | 0.20 | 14.25 | – | 100.48 | m$^3$ | **114.73** |
| Foundations; poured on or against earth or unblinded hardcore | | | | | | | |
| RC 35; Designated mix | 102.88 | 0.20 | 14.25 | – | 102.88 | m$^3$ | **117.13** |
| Isolated foundations | | | | | | | |
| RC 35; Designated mix | 95.70 | 0.67 | 47.45 | – | 95.70 | m$^3$ | **143.15** |
| **Reinforced in situ concrete; 21 N/mm$^2$ – 20 mm aggregate (1:2:4); aggregates delivered in 20 tonne loads** | | | | | | | |
| Site mixed concrete; aggregates in 20 tonne loads; Walls; thickness not exceeding 150 mm | | | | | | | |
| 1:2:4; 21N/mm$^2$; 20 mm aggregate | 118.78 | 0.67 | 47.45 | 2.64 | 118.78 | m$^3$ | **168.87** |
| Ready mixed concrete | | | | | | | |
| C10 Concrete | 87.50 | 0.33 | 23.75 | 2.64 | 87.50 | m$^3$ | **113.89** |
| C15 Concrete | 95.55 | 0.33 | 23.75 | 2.64 | 95.55 | m$^3$ | **121.94** |
| C20 Concrete | 96.08 | 0.33 | 23.75 | 2.64 | 96.08 | m$^3$ | **122.47** |
| C25 Concrete | 100.48 | 0.33 | 23.75 | 2.64 | 100.48 | m$^3$ | **126.87** |
| ST 2; 10 N/mm$^2$ | 88.20 | 0.33 | 23.75 | 2.64 | 88.20 | m$^3$ | **114.59** |
| ST 3; 15 N/mm$^2$ | 94.50 | 0.33 | 23.75 | 2.64 | 94.50 | m$^3$ | **120.89** |
| RC 35; Designated mix | 100.48 | 0.33 | 23.75 | 2.64 | 100.48 | m$^3$ | **126.87** |
| Walls; thickness not exceeding 150–450 mm | | | | | | | |
| Site mixed concrete; aggregates in 20 tonne loads 1:2:4; 21 N/mm$^2$; 20 mm aggregate | 118.78 | 0.50 | 35.63 | 2.64 | 118.78 | m$^3$ | **157.05** |

## 11 IN SITU CONCRETE WORKS

| Item<br>Excluding site overheads and profit | PC<br>£ | Labour<br>hours | Labour<br>£ | Plant<br>£ | Material<br>£ | Unit | Total<br>rate £ |
|---|---|---|---|---|---|---|---|
| Ready mixed concrete | | | | | | | |
|   C10 Concrete | 87.50 | 0.20 | 14.25 | 2.64 | 87.50 | m$^3$ | **104.39** |
|   C15 Concrete | 95.55 | 0.20 | 14.25 | 2.64 | 95.55 | m$^3$ | **112.44** |
|   ST 2; 10 N/mm$^2$ | 88.20 | 0.20 | 14.25 | 2.64 | 88.20 | m$^3$ | **105.09** |
|   ST 3; 15 N/mm$^2$ | 94.50 | 0.20 | 14.25 | 2.64 | 94.50 | m$^3$ | **111.39** |
|   RC 35; Designated mix | 100.48 | 0.20 | 14.25 | 2.64 | 100.48 | m$^3$ | **117.37** |
| Adjustments for site mixed concrete; add the following amounts for aggregates delivered in different load sizes | | | | | | | |
|   850 kg bulk bags | – | – | – | – | 36.86 | m$^3$ | **36.86** |
|   10 tonne loads | – | – | – | – | 7.60 | m$^3$ | **7.60** |

### PLAIN VERTICAL FORMWORK

**Plain vertical formwork; basic finish**
Sides of foundations

| Item | PC | Labour hours | Labour £ | Plant £ | Material £ | Unit | Total rate £ |
|---|---|---|---|---|---|---|---|
|   height not exceeding 250 mm | – | 0.25 | 11.88 | – | 1.31 | m | **13.19** |
|   height 250–500 mm | – | 0.25 | 11.88 | – | 2.20 | m | **14.08** |
|   height 500 mm–1.00 m | – | 0.33 | 15.83 | – | 7.35 | m | **23.18** |
|   height exceeding 1.00 m | – | 0.50 | 23.75 | – | 10.30 | m$^2$ | **34.05** |
| Sides of foundations; left in | | | | | | | |
|   height not exceeding 250 mm | – | 0.20 | 9.50 | – | 4.80 | m | **14.30** |
|   height 250–500 mm | – | 0.20 | 9.50 | – | 9.19 | m | **18.69** |
|   height 500 mm–1.00 m | – | 0.25 | 11.88 | – | 18.37 | m | **30.25** |
|   height over 1.00 m | – | 0.40 | 19.00 | – | 8.32 | m$^2$ | **27.32** |
| Walls | | | | | | | |
|   height not exceeding 250 mm | – | 0.31 | 14.84 | – | 3.31 | m | **18.15** |
|   height 250–500 mm | – | 0.33 | 15.83 | – | 6.20 | m | **22.03** |
|   height 500 mm–1.00 m | – | 0.42 | 19.79 | – | 15.35 | m | **35.14** |
|   height exceeding 1.00 m | – | 0.75 | 35.63 | – | 22.30 | m$^2$ | **57.93** |
| Walls curved to 6.00 radius | | | | | | | |
|   height not exceeding 250 mm | – | 0.38 | 17.81 | – | 3.34 | m | **21.15** |
|   height 250–500 mm | – | 0.40 | 19.00 | – | 6.26 | m | **25.26** |
|   height 500 mm–1.00 m | – | 0.50 | 23.75 | – | 15.47 | m | **39.22** |
|   height exceeding 1.00 m | – | 0.90 | 42.75 | – | 22.42 | m$^2$ | **65.17** |
| Walls curved to 3.00 radius | | | | | | | |
|   height not exceeding 250 mm | – | 0.47 | 22.27 | – | 3.31 | m | **25.58** |
|   height 250–500 mm | – | 0.50 | 23.75 | – | 6.20 | m | **29.95** |
|   height 500 mm–1.00 m | – | 0.62 | 29.68 | – | 15.35 | m | **45.03** |
|   height exceeding 1.00 m | – | 1.13 | 53.44 | – | 22.30 | m$^2$ | **75.74** |

## 11 IN SITU CONCRETE WORKS

| Item<br>Excluding site overheads and profit | PC<br>£ | Labour<br>hours | Labour<br>£ | Plant<br>£ | Material<br>£ | Unit | Total<br>rate £ |
|---|---|---|---|---|---|---|---|
| **BAR REINFORCEMENT** | | | | | | | |
| The rates for reinforcement shown for steel bars below are based on prices which would be supplied on a typical landscape contract. The steel prices shown have been priced on a selection of steel delivered to site where the total order quantity is in the region of 2 tonnes. The assumption is that should larger quantities be required, the work would fall outside the scope of the typical landscape contract defined in the front of this book. Keener rates can be obtained for larger orders. | | | | | | | |
| **Reinforcement bars; hot rolled plain round mild steel; straight or bent** | | | | | | | |
| Bars | | | | | | | |
| 8 mm nominal size | 500.00 | 13.50 | 641.25 | – | 500.00 | tonne | **1141.25** |
| 10 mm nominal size | 500.00 | 13.00 | 617.50 | – | 500.00 | tonne | **1117.50** |
| 12 mm nominal size | 500.00 | 12.50 | 593.75 | – | 500.00 | tonne | **1093.75** |
| 16 mm nominal size | 500.00 | 12.00 | 570.00 | – | 500.00 | tonne | **1070.00** |
| 20 mm nominal size | 500.00 | 11.50 | 546.25 | – | 500.00 | tonne | **1046.25** |
| **Reinforcement bar to concrete formwork** | | | | | | | |
| 8 mm bar | | | | | | | |
| 100 ccs | 1.95 | 0.17 | 7.84 | – | 1.95 | $m^2$ | **9.79** |
| 200 ccs | 0.95 | 0.13 | 5.94 | – | 0.95 | $m^2$ | **6.89** |
| 300 ccs | 0.65 | 0.08 | 3.56 | – | 0.65 | $m^2$ | **4.21** |
| 10 mm bar | | | | | | | |
| 100 ccs | 3.05 | 0.17 | 7.84 | – | 3.05 | $m^2$ | **10.89** |
| 200 ccs | 1.50 | 0.13 | 5.94 | – | 1.50 | $m^2$ | **7.44** |
| 300 ccs | 1.00 | 0.08 | 3.56 | – | 1.00 | $m^2$ | **4.56** |
| 12 mm bar | | | | | | | |
| 100 ccs | 4.40 | 0.17 | 7.84 | – | 4.40 | $m^2$ | **12.24** |
| 200 ccs | 2.20 | 0.13 | 5.94 | – | 2.20 | $m^2$ | **8.14** |
| 300 ccs | 1.45 | 0.08 | 3.56 | – | 1.45 | $m^2$ | **5.01** |
| 16 mm bar | | | | | | | |
| 100 ccs | 7.85 | 0.20 | 9.50 | – | 7.85 | $m^2$ | **17.35** |
| 200 ccs | 3.95 | 0.17 | 7.84 | – | 3.95 | $m^2$ | **11.79** |
| 300 ccs | 2.65 | 0.13 | 5.94 | – | 2.65 | $m^2$ | **8.59** |
| 25 mm bar | | | | | | | |
| 100 ccs | 19.25 | 0.20 | 9.50 | – | 19.25 | $m^2$ | **28.75** |
| 200 ccs | 9.60 | 0.13 | 5.94 | – | 9.60 | $m^2$ | **15.54** |
| 300 ccs | 6.40 | 0.13 | 5.94 | – | 6.40 | $m^2$ | **12.34** |
| 32 mm bar | | | | | | | |
| 100 ccs | 31.55 | 0.20 | 9.50 | – | 31.55 | $m^2$ | **41.05** |
| 200 ccs | 15.75 | 0.17 | 7.84 | – | 15.75 | $m^2$ | **23.59** |
| 300 ccs | 10.50 | 0.13 | 5.94 | – | 10.50 | $m^2$ | **16.44** |

## 11 IN SITU CONCRETE WORKS

| Item<br>Excluding site overheads and profit | PC<br>£ | Labour<br>hours | Labour<br>£ | Plant<br>£ | Material<br>£ | Unit | **Total<br>rate £** |
|---|---|---|---|---|---|---|---|
| **Reinforcement fabric; lapped; in beds or suspended slabs** | | | | | | | |
| Fabric | | | | | | | |
| A98 (1.54 kg/m$^2$) | 0.94 | 0.11 | 5.22 | – | 0.94 | m$^2$ | **6.16** |
| A142 (2.22 kg/m$^2$) | 1.54 | 0.11 | 5.22 | – | 1.54 | m$^2$ | **6.76** |
| A193 (3.02 kg/m$^2$) | 1.87 | 0.11 | 5.22 | – | 1.87 | m$^2$ | **7.09** |
| A252 (3.95 kg/m$^2$) | 2.44 | 0.12 | 5.70 | – | 2.44 | m$^2$ | **8.14** |
| A393 (6.16 kg/m$^2$) | 3.79 | 0.14 | 6.65 | – | 3.79 | m$^2$ | **10.44** |

## 13 PRECAST CONCRETE

| Item<br>Excluding site overheads and profit | PC<br>£ | Labour<br>hours | Labour<br>£ | Plant<br>£ | Material<br>£ | Unit | Total<br>rate £ |
|---|---|---|---|---|---|---|---|
| **SUSPENDED WALKWAY** | | | | | | | |
| **Precast concrete units white cement to form suspended walkway 1.5 m wide; walkway supported on precast walls 650 mm high** | | | | | | | |
| Standard units | | | | | | | |
|   2250 × 1242 mm; 150 mm thick | 414.10 | 0.50 | 35.63 | 19.44 | 414.10 | m | **469.17** |
|   Junction units 4432 × 1242 with splayed edge | 923.14 | 0.50 | 35.63 | 38.88 | 923.14 | m | **997.65** |
|   Splayed units to meet direction change | 337.34 | 0.50 | 35.63 | 19.44 | 337.34 | $m^2$ | **392.41** |
| Precast walkway units hoisted by tower crane to upper podium levels | | | | | | | |
| Standard units | | | | | | | |
|   2250 × 1242 mm; 150 mm thick | 414.10 | 1.25 | 89.06 | – | 414.10 | $m^2$ | **503.16** |
|   Junction units 4432 × 1242 mm with splayed edge | 923.14 | 2.00 | 142.50 | – | 923.14 | $m^2$ | **1065.64** |
|   Splayed units to meet direction change average price | 337.34 | 2.00 | 142.50 | – | 337.34 | nr | **479.84** |

## 14 MASONRY

| Item<br>Excluding site overheads and profit | PC<br>£ | Labour<br>hours | Labour<br>£ | Plant<br>£ | Material<br>£ | Unit | Total<br>rate £ |
|---|---|---|---|---|---|---|---|
| **MARKET PRICES OF MATERIALS** | | | | | | | |
| **Cement; Builder Centre** | | | | | | | |
| Portland cement | – | – | – | – | 4.40 | 25 kg | 4.40 |
| Sulphate resistant cement | – | – | – | – | 5.70 | 25 kg | 5.70 |
| White cement | – | – | – | – | 8.77 | 25 kg | 8.77 |
| **Sand; Builder Centre** | | | | | | | |
| Building sand | | | | | | | |
| loose | 39.60 | – | – | – | 39.60 | m³ | 39.60 |
| 850 kg bulk bags | – | – | – | – | 36.90 | nr | 36.90 |
| 850 kg bulk bags | – | – | – | – | 71.95 | m³ | 71.95 |
| Sharp sand | | | | | | | |
| loose | – | – | – | – | 40.95 | m³ | 40.95 |
| 850 kg bulk bags | – | – | – | – | 36.90 | nr | 36.90 |
| 850 kg bulk bags | – | – | – | – | 71.95 | m³ | 71.95 |
| **Sand; Yeoman Aggregates Ltd** | | | | | | | |
| Sharp sand | – | – | – | – | 22.75 | tonne | 22.75 |
| Sharp sand | – | – | – | – | 36.40 | m³ | 36.40 |
| Soft sand | – | – | – | – | 22.00 | tonne | 22.00 |
| Soft sand | – | – | – | – | 34.40 | m³ | 34.40 |
| **CLARIFICATION NOTES ON LABOUR COSTS IN THIS SECTION** | | | | | | | |
| **Brick/Block walling team** | | | | | | | |
| Generally a three man team is used in this section; The column Labour hours reports team hours. The column Labour £ reports the total cost of the team for the unit of work shown | | | | | | | |
| 3 man team | – | 1.00 | 71.25 | – | – | hr | 71.25 |
| 2 man team | – | 1.00 | 47.50 | – | – | hr | 47.50 |
| **BRICK/BLOCK WALLING DATA** | | | | | | | |
| Note: Batching quantities for these mortar mixes may be found in the Tables and Memoranda section of this book. | | | | | | | |

## 14 MASONRY

| Item<br>Excluding site overheads and profit | PC<br>£ | Labour<br>hours | Labour<br>£ | Plant<br>£ | Material<br>£ | Unit | Total<br>rate £ |
|---|---|---|---|---|---|---|---|
| **BRICK/BLOCK WALLING DATA – cont** | | | | | | | |
| **Mortar mixes for brickwork and pavings; all mortars mixed on site and priced at delivery to waiting wheelbarrows at the cement mixer; cement and lime in 25 kg bags; aggregates as described** | | | | | | | |
| Mortar grade 1:3; aggregates delivered in 20 tonne load; cement:sand volumes shown below (bags:litres) | | | | | | | |
|    85 litre mixer (1.84:93.5) | – | 0.75 | 17.81 | – | 132.88 | m³ | **150.69** |
|    150 litre mixer (3.24:165) | – | 0.50 | 11.88 | – | 132.88 | m³ | **144.76** |
| Mortar grade 1:3; aggregates delivered in 10 tonne load; cement:sand volumes shown below (bags:litres) | | | | | | | |
|    85 litre mixer (1.84:93.5) | – | 0.75 | 17.81 | – | 152.46 | m³ | **170.27** |
|    150 litre mixer (3.24:165) | – | 0.50 | 11.88 | – | 152.46 | m³ | **164.34** |
| Mortar grade 1:3; aggregates delivered in 850 kg bulk bags; cement:sand volumes shown below (bags:litres) | | | | | | | |
|    85 litre mixer (1.84:93.5) | – | 0.75 | 17.81 | – | 171.42 | m³ | **189.23** |
|    150 litre mixer (3.24:165) | – | 0.50 | 11.88 | – | 171.42 | m³ | **183.30** |
| Mortar grade 1:4; aggregates delivered in 20 tonne load; cement:sand volumes shown below (bags:litres) | | | | | | | |
|    85 litre mixer (1.36:102) | – | 0.75 | 17.81 | – | 111.68 | m³ | **129.49** |
|    150 litre mixer (2.4:180) | – | 0.50 | 11.88 | – | 111.68 | m³ | **123.56** |
| Mortar grade 1:4; aggregates delivered in 10 tonne load; cement:sand volumes shown below (bags:litres) | | | | | | | |
|    85 litre mixer (1.36:102) | – | 0.75 | 17.81 | – | 133.04 | m³ | **150.85** |
|    150 litre mixer (2.4:180) | – | 0.50 | 11.88 | – | 133.04 | m³ | **144.92** |
| Mortar grade 1:4; aggregates delivered in 850 kg bulk bags; cement:sand volumes shown below (bags:litres) | | | | | | | |
|    85 litre mixer (1.36:102) | – | 0.75 | 17.81 | – | 153.79 | m³ | **171.60** |
|    150 litre mixer (2.4:180) | – | 0.50 | 11.88 | – | 153.79 | m³ | **165.67** |
| Mortar grade 1:1:6; aggregates delivered in 20 tonne load; cement:lime:sand volumes shown below (bags:bags:litres) | | | | | | | |
|    85 litre mixer;(0.9:0.44:93.5) | – | 0.75 | 17.81 | – | 139.80 | m³ | **157.61** |
|    150 litre mixer; (1.6: 0.8: 165) | – | 0.50 | 11.88 | – | 139.80 | m³ | **151.68** |
| Mortar grade 1:1:6; aggregates delivered in 10 tonne load; cement:lime:sand volumes shown below (bags:bags:litres) | | | | | | | |
|    85 litre mixer (0.9:0.44:93.5) | – | 0.75 | 17.81 | – | 164.60 | m³ | **182.41** |
|    150 litre mixer (1.6:0.8:165) | – | 0.50 | 11.88 | – | 164.60 | m³ | **176.48** |

## 14 MASONRY

| Item<br>Excluding site overheads and profit | PC<br>£ | Labour<br>hours | Labour<br>£ | Plant<br>£ | Material<br>£ | Unit | Total<br>rate £ |
|---|---|---|---|---|---|---|---|
| Mortar grade 1:1:6; aggregates delivered in 850 kg bulk bags; cement: lime:sand volumes shown below (bags: bags:litres) | | | | | | | |
|   85 litre mixer (0.9:0.44:93.5) | – | 0.75 | 17.81 | – | 178.35 | m$^3$ | 196.16 |
|   150 litre mixer (1.6:0.8:165) | – | 0.50 | 11.88 | – | 178.35 | m$^3$ | 190.23 |
| Mortar grade 1:2:9; aggregates delivered in 20 tonne load; cement:lime: sand volumes shown below (bags:bags: litres) | | | | | | | |
|   85 litre mixer (0.7:0.5:102) | – | 0.75 | 17.81 | – | 139.30 | m$^3$ | 157.11 |
|   150 litre mixer (1.2:0.9:180) | – | 0.50 | 11.88 | – | 135.86 | m$^3$ | 147.74 |
| Mortar grade 1:2:9; aggregates delivered in 10 tonne load; cement:lime: sand volumes shown below (bags:bags: litres) | | | | | | | |
|   85 litre mixer (0.7:0.5:102) | – | 0.75 | 17.81 | – | 160.66 | m$^3$ | 178.47 |
|   150 litre mixer (1.2:0.9:180) | – | 0.50 | 11.88 | – | 160.66 | m$^3$ | 172.54 |
| Mortar grade 1:2:9; aggregates delivered in 850 kg bulk bags; cement: lime:sand volumes shown below (bags: bags:litres) | | | | | | | |
|   85 litre mixer (0.7:0.5:102) | – | 0.75 | 17.81 | – | 181.41 | m$^3$ | 199.22 |
|   150 litre mixer (1.2:0.9:180) | – | 0.50 | 11.88 | – | 181.05 | m$^3$ | 192.93 |
| **Variation in brick prices (area × price difference × value shown below)** | | | | | | | |
| Add or subtract the following amounts for every £1.00/1000 difference in the PC price of the measured items below | | | | | | | |
|   half brick thick | – | – | – | – | 6.30 | m$^2$ | 6.30 |
|   one brick thick | – | – | – | – | 12.60 | m$^2$ | 12.60 |
|   one and a half brick thick | – | – | – | – | 18.90 | m$^2$ | 18.90 |
|   two brick thick | – | – | – | – | 25.20 | m$^2$ | 25.20 |
| **Mortar (1:3) required per m$^2$ of brickwork; brick size 215 × 102.5 × 65 mm** | | | | | | | |
| Half brick wall (103 mm); 58 bricks/m$^2$ | | | | | | | |
|   no frog | – | – | – | – | 3.41 | m$^2$ | 3.41 |
|   single frog | – | – | – | – | 3.93 | m$^2$ | 3.93 |
|   double frog | – | – | – | – | 4.65 | m$^2$ | 4.65 |
| 2 × half brick cavity wall (270 mm); 116 bricks/ m$^2$ | | | | | | | |
|   no frog | – | – | – | – | 6.79 | m$^2$ | 6.79 |
|   single frog | – | – | – | – | 8.05 | m$^2$ | 8.05 |
|   double frog | – | – | – | – | 9.84 | m$^2$ | 9.84 |
| One brick wall (215 mm); 116 bricks/m$^2$ | | | | | | | |
|   no frog | – | – | – | – | 8.22 | m$^2$ | 8.22 |
|   single frog | – | – | – | – | 9.84 | m$^2$ | 9.84 |
|   double frog | – | – | – | – | 11.44 | m$^2$ | 11.44 |

## 14 MASONRY

| Item<br>Excluding site overheads and profit | PC<br>£ | Labour<br>hours | Labour<br>£ | Plant<br>£ | Material<br>£ | Unit | Total<br>rate £ |
|---|---|---|---|---|---|---|---|
| **BRICK/BLOCK WALLING DATA – cont** | | | | | | | |
| **Mortar (1:3) required per m² of brickwork – cont** | | | | | | | |
| One and a half brick wall (328 mm) 174 bricks/m² | | | | | | | |
| no frog | – | – | – | – | 11.27 | m² | **11.27** |
| single frog | – | – | – | – | 13.23 | m² | **13.23** |
| double frog | – | – | – | – | 15.73 | m² | **15.73** |
| **Mortar (1:3) required per m² of blockwork; blocks 440 × 215 mm** | | | | | | | |
| Block thickness | | | | | | | |
| 100 mm | – | – | – | – | 1.26 | m² | **1.26** |
| 140 mm | – | – | – | – | 1.62 | m² | **1.62** |
| hollow blocks 440 × 215 mm | – | – | – | – | 0.41 | m² | **0.41** |
| **Movement of materials** | | | | | | | |
| Loading to wheelbarrows and transporting to location; per 215 mm thick walls; maximum distance 25 m | – | 0.42 | 9.90 | – | – | m² | **9.90** |
| **Adjustments for mortars in the brick prices below** | | | | | | | |
| Difference for cement:mortar (1:3) in lieu of gauged mortar | | | | | | | |
| half brick thick | – | – | – | – | -0.27 | m² | **-0.27** |
| one brick thick | – | – | – | – | -0.68 | m² | **-0.68** |
| one and a half brick thick | – | – | – | – | -0.92 | m² | **-0.92** |
| two brick thick | – | – | – | – | -1.37 | m² | **-1.37** |
| **BRICK WALLING AND BRICK PIERS** | | | | | | | |
| **Note: frog types and mortar quantities used in the calculations below are based on bricks with single frogs. Please make adjustments for mortar volumes used with varying brick frog formats. See data above for mortar quantities for various frog types.** | | | | | | | |
| Delivery volumes of aggregates for mortars are based on 10 tonne deliveries. The figures below show the approximate areas of walling which would be constructed mixed with building sand from a 10 tonne (6.25 m³) load of building sand | | | | | | | |
| half brick wall = 256 m² | | | | | | | |
| one brick wall = 124 m² | | | | | | | |

## 14 MASONRY

| Item<br>Excluding site overheads and profit | PC<br>£ | Labour<br>hours | Labour<br>£ | Plant<br>£ | Material<br>£ | Unit | Total<br>rate £ |
|---|---|---|---|---|---|---|---|
| **Class B engineering bricks; PC £360.00/1000; double Flemish bond in cement:mortar (1:3)** Offloading mechanically; loading to wheelbarrows; transporting to location maximum 25 m distance per 215 mm thick walls | – | 0.42 | 9.90 | – | – | m² | **9.90** |
| Construct walls | | | | | | | |
|   half brick thick | – | 0.60 | 50.25 | – | 32.10 | m² | **82.35** |
|   one brick thick | – | 1.20 | 100.50 | – | 68.69 | m² | **169.19** |
|   one and a half brick thick | – | 1.80 | 150.75 | – | 98.91 | m² | **249.66** |
|   two brick thick | – | 2.40 | 200.92 | – | 134.00 | m² | **334.92** |
| Construct walls; curved; mean radius 6 m | | | | | | | |
|   half brick thick | – | 0.90 | 75.38 | – | 32.04 | m² | **107.42** |
|   one brick thick | – | 1.80 | 150.75 | – | 66.06 | m² | **216.81** |
| Construct walls; curved; mean radius 1.50 m | | | | | | | |
|   half brick thick | – | 1.20 | 100.49 | – | 32.10 | m² | **132.59** |
|   one brick thick | – | 3.60 | 193.50 | – | 66.06 | m² | **259.56** |
| Construct walls; tapering; one face battering; average | | | | | | | |
|   one and a half brick thick | – | 2.22 | 185.62 | – | 97.65 | m² | **283.27** |
|   two brick thick | – | 2.96 | 247.48 | – | 132.13 | m² | **379.61** |
| Construct walls; battering (retaining) | | | | | | | |
|   one and a half brick thick | – | 2.22 | 185.62 | – | 97.65 | m² | **283.27** |
|   two brick thick | – | 2.96 | 247.48 | – | 132.13 | m² | **379.61** |
| Projections; vertical | | | | | | | |
|   one brick × half brick | – | 2.33 | 195.39 | – | 6.89 | m | **202.28** |
|   one brick × one brick | – | 0.47 | 39.03 | – | 13.79 | m | **52.82** |
|   one and a half brick × one brick | – | 0.70 | 58.63 | – | 20.69 | m | **79.32** |
|   two brick by one brick | – | 0.77 | 64.15 | – | 27.58 | m | **91.73** |
| Walls; half brick thick | | | | | | | |
|   in honeycomb bond | – | 0.60 | 50.25 | – | 21.86 | m² | **72.11** |
|   in quarter bond | – | 0.56 | 46.48 | – | 30.23 | m² | **76.71** |
| **Facing bricks; PC £400.00/1000; English garden wall bond; in gauged mortar (1:1:6); facework one side** Mechanically offloading; maximum 25 m distance; loading to wheelbarrows and transporting to location; walls | – | 0.42 | 9.90 | – | – | m² | **9.90** |
| Construct walls | | | | | | | |
|   half brick thick | – | 0.60 | 50.17 | – | 31.02 | m² | **81.19** |
|   half brick thick (using site cut snap headers to form bond) | – | 0.80 | 67.33 | – | 31.02 | m² | **98.35** |
|   one brick thick | – | 1.20 | 100.50 | – | 62.04 | m² | **162.54** |
|   one and a half brick thick | – | 1.80 | 150.75 | – | 87.30 | m² | **238.05** |
|   two brick thick | – | 2.40 | 201.00 | – | 118.47 | m² | **319.47** |

## 14 MASONRY

| Item<br>Excluding site overheads and profit | PC<br>£ | Labour<br>hours | Labour<br>£ | Plant<br>£ | Material<br>£ | Unit | Total<br>rate £ |
|---|---|---|---|---|---|---|---|
| **BRICK WALLING AND BRICK PIERS** – cont | | | | | | | |
| **Facing bricks** – cont | | | | | | | |
| Walls; curved; mean radius 6 m | | | | | | | |
|   half brick thick | – | 0.90 | 75.38 | – | 29.21 | m² | **104.59** |
|   one brick thick | – | 1.80 | 150.75 | – | 150.73 | m² | **301.48** |
| Walls; curved; mean radius 1.50 m | | | | | | | |
|   half brick thick | – | 1.20 | 100.50 | – | 29.81 | m² | **130.31** |
|   one brick thick | – | 2.40 | 201.00 | – | 151.93 | m² | **352.93** |
| Walls; tapering; one face battering; average | | | | | | | |
|   one and a half brick thick | – | 2.20 | 184.25 | – | 92.70 | m² | **276.95** |
|   two brick thick | – | 3.00 | 251.25 | – | 125.67 | m² | **376.92** |
| Walls; battering (retaining) | | | | | | | |
|   one and a half brick thick | – | 2.00 | 167.50 | – | 92.70 | m² | **260.20** |
|   two brick thick | – | 2.70 | 226.13 | – | 116.54 | m² | **342.67** |
| Projections; vertical | | | | | | | |
|   one brick × half brick | – | 0.23 | 19.26 | – | 6.24 | m | **25.50** |
|   one brick × one brick | – | 0.50 | 41.88 | – | 12.47 | m | **54.35** |
|   one and a half brick × one brick | – | 0.70 | 58.63 | – | 18.72 | m | **77.35** |
|   two brick × one brick | – | 0.80 | 67.00 | – | 26.05 | m | **93.05** |
| Brickwork fair faced both sides; facing bricks in gauged mortar (1:1:6) | | | | | | | |
|   extra for fair face both sides; flush, struck, weathered, or bucket-handle pointing | – | 0.67 | 15.83 | – | – | m² | **15.83** |
| **Bricks; PC £800.00/1000; English garden wall bond; in gauged mortar (1:1:6)** | | | | | | | |
| Walls | | | | | | | |
|   half brick thick (stretcher bond) | – | 0.60 | 50.25 | – | 53.21 | m² | **103.46** |
|   half brick thick (using site cut snap headers to form bond) | – | 0.80 | 67.00 | – | 53.21 | m² | **120.21** |
|   one brick thick | – | 1.20 | 100.50 | – | 110.92 | m² | **211.42** |
|   one and a half brick thick | – | 1.80 | 150.75 | – | 161.82 | m² | **312.57** |
|   two brick thick | – | 2.40 | 201.00 | – | 219.65 | m² | **420.65** |
| Walls; curved; mean radius 6 m | | | | | | | |
|   half brick thick | – | 0.90 | 75.38 | – | 54.41 | m² | **129.79** |
|   one brick thick | – | 2.40 | 201.00 | – | 111.75 | m² | **312.75** |
| Walls; curved; mean radius 1.50 m | | | | | | | |
|   half brick thick | – | 0.53 | 44.39 | – | 59.21 | m² | **103.60** |
|   one brick thick | – | 1.07 | 89.36 | – | 121.34 | m² | **210.70** |
| Walls; stretcher bond; wall ties at 450 mm centres vertically and horizontally | | | | | | | |
|   one brick thick | 0.98 | 1.09 | 91.29 | – | 110.81 | m² | **202.10** |
|   two brick thick | 2.94 | 1.64 | 137.35 | – | 222.59 | m² | **359.94** |

## 14 MASONRY

| Item<br>Excluding site overheads and profit | PC<br>£ | Labour<br>hours | Labour<br>£ | Plant<br>£ | Material<br>£ | Unit | Total<br>rate £ |
|---|---|---|---|---|---|---|---|
| **Brickwork fair faced both sides; facing bricks in gauged mortar (1:1:6)** | | | | | | | |
| extra for fair face both sides; flush, struck, weathered or bucket-handle pointing | – | 0.67 | 15.83 | – | – | m² | 15.83 |
| **Brick copings** | | | | | | | |
| Copings; all brick headers-on-edge; two angles rounded 53 mm radius; flush pointing top and both sides as work proceeds; one brick wide; horizontal | | | | | | | |
| machine-made specials | 48.05 | 0.16 | 13.73 | – | 49.15 | m | 62.88 |
| handmade specials | 61.80 | 0.16 | 13.73 | – | 62.89 | m | 76.62 |
| Extra over copings for two courses machine-made tile creasings; projecting 25 mm each side; 260 mm wide copings; horizontal | 6.24 | 0.17 | 14.24 | – | 6.78 | m | 21.02 |
| Copings; all brick headers-on-edge; flush pointing top and both sides as work proceeds; one brick wide; horizontal | | | | | | | |
| engineering bricks PC £360.00/1000 | 6.30 | 0.16 | 13.73 | – | 6.30 | m | 20.03 |
| facing bricks PC £400.00/1000 | 5.73 | 0.16 | 13.73 | – | 5.73 | m | 19.46 |
| facing bricks PC £800.00/1000 | 10.66 | 0.16 | 13.73 | – | 10.66 | m | 24.39 |
| **Isolated brick piers in English bond; in gauged mortar 1:1:6** | | | | | | | |
| Engineering brick PC £360.00/1000 | | | | | | | |
| one brick thick (225 mm) | – | 2.33 | 195.39 | – | 67.65 | m² | 263.04 |
| one and a half brick thick (337.5 mm) | – | 3.00 | 251.25 | – | 101.47 | m² | 352.72 |
| two brick thick (450 mm) | – | 3.33 | 279.14 | – | 137.11 | m² | 416.25 |
| three brick thick (675 mm) | – | 4.13 | 346.14 | – | 202.94 | m² | 549.08 |
| Engineering brick PC £360.00/1000 (not SMM) | | | | | | | |
| one brick (225 mm) | – | 0.54 | 45.61 | – | 15.57 | m | 61.18 |
| one and a half brick thick (337.5 mm) | – | 1.00 | 83.75 | – | 43.75 | m | 127.50 |
| two brick thick (450 mm) | – | 1.50 | 125.63 | – | 71.10 | m | 196.73 |
| three brick thick (675 mm) | – | 2.79 | 233.66 | – | 147.65 | m | 381.31 |
| Brick PC £400.00/1000 | | | | | | | |
| one brick thick (225 mm) | – | 2.33 | 195.39 | – | 61.34 | m² | 256.73 |
| one and a half brick thick (337.5 mm) | – | 3.00 | 251.25 | – | 92.02 | m² | 343.27 |
| two brick thick (450 mm) | – | 3.33 | 279.14 | – | 124.51 | m² | 403.65 |
| three brick thick (675 mm) | – | 4.13 | 346.14 | – | 184.03 | m² | 530.17 |
| Brick PC £400.00/1000 (not SMM) | | | | | | | |
| one brick (225 mm) | – | 0.54 | 45.56 | – | 14.10 | m | 59.66 |
| one and a half brick thick (337.5 mm) | – | 1.00 | 83.75 | – | 40.56 | m | 124.31 |
| two brick thick (450 mm) | – | 1.50 | 125.63 | – | 65.43 | m | 191.06 |
| three brick thick (675 mm) | – | 2.79 | 233.66 | – | 134.90 | m | 368.56 |

## 14 MASONRY

| Item<br>Excluding site overheads and profit | PC<br>£ | Labour<br>hours | Labour<br>£ | Plant<br>£ | Material<br>£ | Unit | Total<br>rate £ |
|---|---|---|---|---|---|---|---|
| **BRICK WALLING AND BRICK PIERS** – cont | | | | | | | |
| **Isolated brick piers in English bond** – cont | | | | | | | |
| Brick PC £800.00/1000 | | | | | | | |
| one brick thick (225 mm) | – | 2.33 | 195.39 | – | 111.75 | m² | **307.14** |
| one and a half brick thick (337.5 mm) | – | 3.00 | 251.25 | – | 167.62 | m² | **418.87** |
| two brick thick (450 mm) | – | 3.33 | 279.14 | – | 225.31 | m² | **504.45** |
| three brick thick (675 mm) | – | 4.13 | 346.14 | – | 335.23 | m² | **681.37** |
| Brick PC £800.00/1000 (not SMM) | | | | | | | |
| one brick (225 mm) | – | 0.54 | 45.56 | – | 25.86 | m | **71.42** |
| one and a half brick thick (337.5 mm) | – | 1.00 | 83.75 | – | 66.08 | m | **149.83** |
| two brick thick (450 mm) | – | 1.50 | 125.63 | – | 110.79 | m | **236.42** |
| three brick thick (675 mm) | – | 2.79 | 233.66 | – | 236.96 | m | **470.62** |
| **RESTORATION BRICKWORK** | | | | | | | |
| **Carefully dismantle existing coping to walls; Salvage copings (50%); Stack for re-use and dispose of balance off site** | | | | | | | |
| Walls over 1.5 m high to 2.40 m high | | | | | | | |
| one brick thick | – | 0.50 | 26.88 | 2.57 | 0.33 | m | **29.78** |
| one and a half brick thick | – | 0.75 | 40.31 | 2.57 | 0.50 | m | **43.38** |
| two brick thick | – | 1.00 | 53.75 | 2.57 | 0.67 | m | **56.99** |
| Walls under 1.5 m high | | | | | | | |
| one brick thick | – | 0.50 | 26.88 | – | 0.33 | m | **27.21** |
| one and a half brick thick | – | 0.75 | 40.31 | – | 0.50 | m | **40.81** |
| two brick thick | – | 1.00 | 53.75 | – | 0.67 | m | **54.42** |
| **Dismantle existing wall; salvage bricks and dispose damaged material off site (25%)** | | | | | | | |
| half brick thick | – | – | – | – | – | m | – |
| one brick thick | – | 0.70 | 37.63 | 0.41 | 2.30 | m | **40.34** |
| one and a half brick thick | – | 1.30 | 69.88 | 0.41 | 3.45 | m | **73.74** |
| two brick thick | – | 1.80 | 96.75 | 0.41 | 4.60 | m | **101.76** |
| **Cut out damaged brickwork in isolated areas** | | | | | | | |
| half brick thick | – | 0.17 | 3.96 | – | – | brick | **3.96** |
| one brick thick to 2 bricks thick | – | 0.14 | 3.39 | – | – | brick | **3.39** |
| Isolated areas under 0.5 m² | | | | | | | |
| half brick thick | – | 0.50 | 11.88 | – | – | m² | **11.88** |
| one brick thick to 2 bricks thick per side | – | 0.40 | 9.50 | – | – | m² | **9.50** |

## 14 MASONRY

| Item<br>Excluding site overheads and profit | PC<br>£ | Labour<br>hours | Labour<br>£ | Plant<br>£ | Material<br>£ | Unit | Total<br>rate £ |
|---|---|---|---|---|---|---|---|
| **Isolated areas over 1.0 m²** | | | | | | | |
| half brick thick | – | 1.00 | 23.75 | – | – | m² | 23.75 |
| one brick thick to 2 bricks thick | – | 0.80 | 19.00 | – | – | m² | 19.00 |
| **Cut out existing damaged pointing and replace with new mortar pointing** | | | | | | | |
| Half brick thick walls in strecher bond | | | | | | | |
| in standard mortar | 0.29 | 0.42 | 9.90 | – | 0.29 | m² | 10.19 |
| in lime mortar | 1.51 | 0.45 | 10.69 | – | 1.51 | m² | 12.20 |
| One brick thick walls in English garden wall bond | | | | | | | |
| in standard mortar | 0.63 | 0.58 | 13.85 | – | 0.63 | m² | 14.48 |
| in lime mortar | 3.02 | 0.58 | 13.85 | – | 3.02 | m² | 16.87 |
| Flemish bond | | | | | | | |
| in standard mortar | 0.66 | 0.69 | 16.29 | – | 0.66 | m² | 16.95 |
| in lime mortar | 3.02 | 0.69 | 16.29 | – | 3.02 | m² | 19.31 |
| **Replace damaged brickwork in existing wall; Cut out damaged bricks and replace with new bricks to match; reclaimed bricks PC £600.00/1000** | | | | | | | |
| One brick thick; one side fair faced | | | | | | | |
| per brick | 0.60 | 0.27 | 6.33 | – | 0.69 | nr | 7.02 |
| larger areas | 72.00 | 2.50 | 104.38 | – | 86.69 | m² | 191.07 |
| One brick thick both sides fair faced | | | | | | | |
| per brick | 0.60 | 0.28 | 6.73 | – | 0.69 | nr | 7.42 |
| larger areas | 72.00 | 3.50 | 158.13 | – | 86.69 | m² | 244.82 |
| One and a half brick thick; one side fair faced | | | | | | | |
| per brick | 1.20 | 0.57 | 13.44 | – | 1.37 | nr | 14.81 |
| larger areas | 108.00 | 5.50 | 235.63 | – | 130.04 | m² | 365.67 |
| **Construct new wall sections to match existing walls in lime mortar; Facing bricks; PC £1000.00/1000; English garden wall bond; in lime mortar** | | | | | | | |
| Facing bricks; PC £1000.00/1000; English garden wall bond; | | | | | | | |
| half brick thick | 28.22 | 0.64 | 53.73 | – | 89.72 | m² | 143.45 |
| half brick thick (using site cut snap headers to form bond) | 28.22 | 0.84 | 70.70 | – | 89.72 | m² | 160.42 |
| one brick thick | 53.42 | 1.26 | 105.53 | – | 176.42 | m² | 281.95 |
| one and a half brick thick | 78.62 | 1.89 | 158.29 | – | 263.12 | m² | 421.41 |
| two brick thick | 103.82 | 2.52 | 211.05 | – | 349.82 | m² | 560.87 |
| Walls; curved; mean radius 6 m | | | | | | | |
| half brick thick | 28.22 | 0.94 | 79.14 | – | 91.22 | m² | 170.36 |
| one brick thick | 53.42 | 1.89 | 158.29 | – | 179.42 | m² | 337.71 |

## 14 MASONRY

| Item<br>Excluding site overheads and profit | PC<br>£ | Labour<br>hours | Labour<br>£ | Plant<br>£ | Material<br>£ | Unit | Total<br>rate £ |
|---|---|---|---|---|---|---|---|
| **RESTORATION BRICKWORK –<br>cont** | | | | | | | |
| **Construct new wall sections to match existing walls in lime mortar –<br>cont** | | | | | | | |
| Walls; curved; mean radius 1.50 m | | | | | | | |
|   half brick thick | 28.22 | 1.26 | 105.53 | – | 92.72 | m² | **198.25** |
|   one brick thick | 53.42 | 2.52 | 211.05 | – | 182.42 | m² | **393.47** |
| Walls; tapering; one face battering; average | | | | | | | |
|   one and a half brick thick | 78.62 | 2.31 | 193.46 | – | 276.62 | m² | **470.08** |
|   two brick thick | 103.82 | 3.15 | 263.81 | – | 367.82 | m² | **631.63** |
| Walls; battering (retaining) | | | | | | | |
|   one and a half brick thick | 78.62 | 2.10 | 175.88 | – | 276.62 | m² | **452.50** |
|   two brick thick | 78.62 | 2.83 | 237.43 | – | 342.62 | m² | **580.05** |
| Projections; vertical | | | | | | | |
|   one brick × half brick | 8.62 | 0.24 | 20.23 | – | 22.62 | m | **42.85** |
|   one brick × one brick | 14.22 | 0.53 | 43.97 | – | 42.21 | m | **86.18** |
|   one and a half brick × one brick | 19.82 | 0.73 | 61.56 | – | 61.82 | m | **123.38** |
|   two brick × one brick | 25.42 | 0.84 | 70.35 | – | 81.42 | m | **151.77** |
| Brickwork fair faced both sides; facing bricks in gauged mortar (1:1:6) | | | | | | | |
|   extra for fair face both sides; flush, struck, weathered, or bucket-handle pointing | 3.02 | 0.70 | 16.62 | – | 3.02 | m² | **19.64** |
| **Brick copings; lime mortar bedding and jointing** | | | | | | | |
| Copings; all brick headers-on-edge; two angles rounded 53 mm radius; flush pointing top and both sides as work proceeds; one brick wide; horizontal | | | | | | | |
|   machine-made specials PC £3.50 each | 53.99 | 0.17 | 14.42 | – | 53.99 | m | **68.41** |
|   handmade specials PC £4.50 each | 70.42 | 0.17 | 14.42 | – | 70.42 | m | **84.84** |
| Copings; all brick headers-on-edge; flush pointing top and both sides as work proceeds; one brick wide; horizontal | | | | | | | |
|   reclaimed bricks PC £1000.00 /1000 | 5.83 | 0.17 | 14.42 | – | 19.49 | m | **33.91** |
| **Isolated brick piers in English bond; in gauged mortar 1:1:6** | | | | | | | |
| Brick PC £1000.00/1000 | | | | | | | |
|   one brick thick (225 mm) | 53.42 | 2.45 | 205.16 | – | 179.42 | m² | **384.58** |
|   one and a half brick thick (337.5 mm) | 14.22 | 3.15 | 263.81 | – | 203.22 | m² | **467.03** |
|   two brick thick (450 mm) | 103.82 | 3.50 | 293.10 | – | 355.82 | m² | **648.92** |
|   three brick thick (675 mm) | 154.22 | 4.34 | 363.44 | – | 532.22 | m² | **895.66** |

## 14 MASONRY

| Item<br>Excluding site overheads and profit | PC<br>£ | Labour<br>hours | Labour<br>£ | Plant<br>£ | Material<br>£ | Unit | Total<br>rate £ |
|---|---|---|---|---|---|---|---|
| Brick PC £1000.00/1000 (not SMM) | | | | | | | |
| one brick (225 mm) | 13.81 | 0.57 | 47.84 | – | 43.22 | m | **91.06** |
| one and a half brick thick (337.5 mm) | 92.36 | 1.05 | 87.94 | – | 92.36 | m | **180.30** |
| two brick thick (450 mm) | 49.44 | 1.57 | 131.91 | – | 162.84 | m | **294.75** |
| three brick thick (675 mm) | 108.18 | 2.93 | 245.35 | – | 363.33 | m | **608.68** |
| **BLOCK WALLING AND BLOCK PIERS** | | | | | | | |
| **Dense aggregate concrete blocks; Tarmac Topblock or other equal and approved; in gauged mortar (1:2:9); one course underground** | | | | | | | |
| Walls | | | | | | | |
| Solid blocks 7 N/mm² | | | | | | | |
| 440 × 215 × 100 mm thick | 5.90 | 0.40 | 33.50 | – | 7.26 | m² | **40.76** |
| 440 × 215 × 140 mm thick | 11.23 | 0.44 | 37.10 | – | 12.76 | m² | **49.86** |
| Solid blocks 7 N/mm² laid flat | | | | | | | |
| 440 × 100 × 215 mm thick | 12.03 | 1.07 | 89.61 | – | 13.56 | m² | **103.17** |
| Hollow concrete blocks | | | | | | | |
| 440 × 215 × 215 mm thick | 18.19 | 0.43 | 36.26 | – | 19.72 | m² | **55.98** |
| Filling of hollow concrete blocks with concrete as work proceeds; tamping and compacting | | | | | | | |
| 440 × 215 × 215 mm thick | 21.18 | 0.07 | 5.53 | – | 21.18 | m² | **26.71** |
| **Isolated blockwork piers; to receive facing treatments (not included)** | | | | | | | |
| Solid blocks; 440 × 100 × 215 mm thick; laid on-flat; 7 N/mm² | | | | | | | |
| 450 × 450 mm (SMM measurement) | 23.73 | 2.00 | 107.50 | – | 24.92 | m² | **132.42** |
| 450 × 450 mm (not SMM) | 10.69 | 0.90 | 48.38 | – | 11.23 | m | **59.61** |
| **NATURAL STONE WALLING** | | | | | | | |
| **Expert stonework team** | | | | | | | |
| Generally a three man team is used in this section; The column Labour hours reports team hours. The column Labour £ reports the total cost of the team for the unit of work shown | | | | | | | |
| 2 man team | – | 1.00 | 47.50 | – | – | hr | **47.50** |

## 14 MASONRY

| Item<br>Excluding site overheads and profit | PC<br>£ | Labour<br>hours | Labour<br>£ | Plant<br>£ | Material<br>£ | Unit | Total<br>rate £ |
|---|---|---|---|---|---|---|---|
| **NATURAL STONE WALLING – cont** | | | | | | | |
| **Dry stone walling – General** | | | | | | | |
| Preamble: In rural areas where natural stone is a traditional material, it may be possible to use dry stone walling or dyking as an alternative to fences or brick walls. Many local authorities are willing to meet the extra cost of stone walling in areas of high landscape value, and they may hold lists of available craftsmen. DSWA Office, Westmorland County Showground, Lane Farm, Crooklands, Milnthorpe, Cumbria, LA7 7NH; Tel: 01539 567953; E-mail: information@dswa.org.uk | | | | | | | |
| Note: Traditional walls are not built on concrete foundations. | | | | | | | |
| Dry stone wall; wall on concrete foundation (not included); dry stone coursed wall inclusive of locking stones and filling to wall with broken stone or rubble; walls up to 1.20 m high; battered; 2 sides fair faced | | | | | | | |
|   Yorkstone | – | 3.25 | 154.38 | – | 71.74 | m² | **226.12** |
|   Cotswold stone | – | 3.25 | 154.38 | – | 87.69 | m² | **242.07** |
|   Purbeck | – | 3.25 | 154.38 | – | 95.65 | m² | **250.03** |
| **Granite walls** | | | | | | | |
| Granite random rubble walls; laid dry | | | | | | | |
|   200 mm thick; single faced | 52.99 | 3.33 | 158.18 | – | 52.99 | m² | **211.17** |
| Granite walls; one face battering to 50°; pointing faces | | | | | | | |
|   450 mm (average) thick | 128.17 | 2.50 | 118.75 | – | 129.24 | m² | **247.99** |
| **RECONSTITUTED STONE WALLING** | | | | | | | |
| **Reconstituted stone reproduction balustrades; Arundel Stone Sussex Ltd; moulds to match existing or new balustrades** | | | | | | | |
| Pier plinths; 400 mm high × 524 bases to receive shafts to builder's work (not included) | | | | | | | |
|   525 × 525 × 440 mm high | 116.66 | 0.25 | 11.88 | – | 116.66 | nr | **128.54** |
|   525 × 525 × 540 mm high | 300.98 | 0.25 | 11.88 | – | 300.98 | nr | **312.86** |

## 14 MASONRY

| Item<br>Excluding site overheads and profit | PC<br>£ | Labour<br>hours | Labour<br>£ | Plant<br>£ | Material<br>£ | Unit | Total<br>rate £ |
|---|---|---|---|---|---|---|---|
| Pier shafts; at specified centres along the balustrade fixed to upstand (not included); subsequently filled with concrete | | | | | | | |
|    Piers 400 × 400 × 475 mm high | 124.23 | 1.00 | 47.50 | – | 128.16 | nr | **175.66** |
|    Piers 400 × 400 × 650 mm high | 696.90 | 1.25 | 59.38 | – | 701.09 | nr | **760.47** |
| Pier caps to Pier shafts; | | | | | | | |
|    575 × 575 × 150 mm | 168.67 | 0.25 | 11.88 | – | 171.19 | nr | **183.07** |
| **Bottle balustrades reconstituted stone; Arundel Stone Sussex Ltd; to prepared brickwork walls or bases (not included)** | | | | | | | |
| Plinths to receive balustrade components; fixed to existing base or to low walls to receive balustrade components | | | | | | | |
|    360 wide × 440 mm high | 99.31 | 0.75 | 35.63 | – | 99.93 | m | **135.56** |
|    350 wide × 150 mm high | 75.75 | 0.75 | 35.63 | – | 76.36 | m | **111.99** |
| Bottle balustrades comprising reconstituted bottle balusters; fixed to plinths with steel rods and mortar, top rail 1200 × 335 × 150 mm to plinths high from standard moulds; balusters | | | | | | | |
|    175 × 175 × 475 mm high; at 210 mm centres | 288.40 | 0.80 | 38.00 | – | 288.40 | m | **326.40** |
|    175 × 175 × 475 mm high; at 250 mm centres | 253.46 | 0.75 | 35.63 | – | 253.46 | m | **289.09** |
|    square balusters 180 × 600 mm high at 210 mm centres | 78.74 | 0.80 | 38.00 | – | 321.14 | m | **359.14** |
|    square balusters 180 × 600 mm high at 250 mm centres | 78.74 | 0.75 | 35.63 | – | 280.74 | m | **316.37** |
|    half bottles abutting walls or piers at ends 775 mm high above plinth | 28.53 | 0.17 | 7.92 | – | 28.53 | nr | **36.45** |
| Raked balustrades to steps; Bottle balustrades comprising top rail and balusters to plinths and with reconstituted stone bottles at 214 centres from non-standard moulds | | | | | | | |
|    Overall height varies at landings to raked sections between 1.30 m to 1.0 m high | 883.95 | 2.40 | 114.00 | – | 883.95 | m | **997.95** |

## 14 MASONRY

| Item<br>Excluding site overheads and profit | PC<br>£ | Labour<br>hours | Labour<br>£ | Plant<br>£ | Material<br>£ | Unit | Total<br>rate £ |
|---|---|---|---|---|---|---|---|
| **RECONSTITUTED STONE WALLING** – cont | | | | | | | |
| **Haddonstone Ltd; cast stone piers; ornamental masonry in Portland Bath or Terracotta finished cast stone** | | | | | | | |
| Gate Pier S120; to foundations and underground work measured separately; concrete infill | | | | | | | |
| S120G base unit to pier; | | | | | | | |
| 699 × 699 × 172 mm | 177.01 | 0.25 | 17.81 | – | 182.67 | nr | **200.48** |
| S120F/F shaft base unit; | | | | | | | |
| 533 × 533 × 280 mm | 140.18 | 0.33 | 23.73 | – | 146.06 | nr | **169.79** |
| S120E/E main shaft unit; | | | | | | | |
| 533 × 533 × 280 mm; nr of units required dependent on height of pier | 140.18 | 0.33 | 23.73 | – | 146.06 | nr | **169.79** |
| S120D/D top shaft unit; | | | | | | | |
| 33 × 533 × 280 mm | 140.18 | 0.33 | 23.73 | – | 146.06 | nr | **169.79** |
| S120C pier cap unit; | | | | | | | |
| 737 × 737 × 114 mm | 166.32 | 0.17 | 11.83 | – | 172.19 | nr | **184.02** |
| S120B pier block unit; base for finial; | | | | | | | |
| 533 × 533 × 64 mm | 71.28 | 0.11 | 7.84 | – | 71.50 | nr | **79.34** |
| Pier blocks; flat to receive gate finial | | | | | | | |
| S100B; 440 × 440 × 63 mm | 46.33 | 0.11 | 7.84 | – | 46.84 | nr | **54.68** |
| S120B; 546 × 546 × 64 mm | 71.28 | 0.11 | 7.84 | – | 71.79 | nr | **79.63** |
| S150B; 330 × 330 × 51 mm | 24.95 | 0.11 | 7.84 | – | 25.46 | nr | **33.30** |
| Pier caps; part weathered | | | | | | | |
| S100C; 915 × 915 × 150 mm | 395.60 | 0.25 | 17.81 | – | 396.63 | nr | **414.44** |
| S120C; 737 × 737 × 114 mm | 166.32 | 0.17 | 11.83 | – | 166.83 | nr | **178.66** |
| S150C; 584 × 584 × 120 mm | 121.18 | 0.17 | 11.83 | – | 121.69 | nr | **133.52** |
| Pier caps; weathered | | | | | | | |
| S230C; 1029 × 1029 × 175 mm | 519.16 | 0.17 | 11.83 | – | 519.67 | nr | **531.50** |
| S215C; 687 × 687 × 175 mm | 228.10 | 0.17 | 11.83 | – | 228.61 | nr | **240.44** |
| S210C; 584 × 584 × 175 mm | 142.56 | 0.17 | 11.83 | – | 143.07 | nr | **154.90** |
| Pier strings | | | | | | | |
| S100S; 800 × 800 × 55 mm | 135.43 | 1.66 | 118.28 | – | 135.74 | nr | **254.02** |
| S120S; 555 × 555 × 44 mm | 61.78 | 0.17 | 11.83 | – | 62.09 | nr | **73.92** |
| S150S; 457 × 457 × 48 mm | 38.02 | 0.17 | 11.83 | – | 38.33 | nr | **50.16** |
| Balls and bases | | | | | | | |
| E150A Ball 535 mm and E150C collared base | 400.00 | 0.17 | 11.83 | – | 400.31 | nr | **412.14** |
| E120A Ball 330 mm and E120C collared base | 154.00 | 0.17 | 11.83 | – | 154.31 | nr | **166.14** |
| E110A Ball 230 mm and E110C collared base | 104.00 | 0.17 | 11.83 | – | 104.31 | nr | **116.14** |
| E100A Ball 170 mm and E100B plain base | 70.00 | 0.17 | 11.83 | – | 70.31 | nr | **82.14** |

## 14 MASONRY

| Item<br>Excluding site overheads and profit | PC<br>£ | Labour<br>hours | Labour<br>£ | Plant<br>£ | Material<br>£ | Unit | Total<br>rate £ |
|---|---|---|---|---|---|---|---|
| **Haddonstone Ltd; cast stone copings; ornamental masonry in Portland Bath or Terracotta finished cast stone** | | | | | | | |
| Copings for walls; bedded, jointed and pointed in approved coloured cement:lime mortar 1:2:9 | | | | | | | |
| T100 weathered coping; 102 mm high × 178 mm wide × 914 mm | 41.86 | 0.11 | 7.91 | – | 42.29 | m | **50.20** |
| T140 weathered coping; 102 mm high × 337 mm wide × 914 mm | 69.32 | 0.11 | 7.91 | – | 69.76 | m | **77.67** |
| T200 weathered coping; 127 mm high × 508 mm wide × 750 mm | 104.64 | 0.11 | 7.91 | – | 105.08 | m | **112.99** |
| T170 weathered coping; 108 mm high × 483 mm wide × 914 mm | 112.49 | 0.11 | 7.91 | – | 112.93 | m | **120.84** |
| T340 raked coping; 75–100 mm high × 290 mm wide × 900 mm | 64.09 | 0.11 | 7.91 | – | 64.53 | m | **72.44** |
| T310 raked coping; 76–89 mm high × 381 mm wide × 914 mm | 74.56 | 0.11 | 7.91 | – | 74.99 | m | **82.90** |
| **Slope-Loc – Anderton Concrete Ltd** | | | | | | | |
| Dry stacked random sized units 150–400 mm long × 150 mm high cast stone wall mechanically interlocked with fibreglass pins; constructed to levelling pad of coarse compacted granular material (not included) back filled behind the elevation with 300 mm wide granular drainage material; walls to 5.00 m high (retaining walls over heights shown below require individual design) | | | | | | | |
| Slope-Loc 300 × 245 × 125 mm; gravity wall up to 850 mm; | 52.55 | 1.00 | 47.50 | 38.88 | 56.87 | m² | **143.25** |
| Slope-Loc 300 × 245 × 125 mm; reinforced wall; inclusive of design and reinforcement grids; retaining up to 1.70 m | 109.15 | 1.00 | 47.50 | 38.88 | 113.47 | m² | **199.85** |
| **Retaining wall; as above but reinforced with Tensar Geogrid RE520 laid between every two courses horizontally into the face of the excavation (excavation not included)** | | | | | | | |
| Near vertical; 250 mm thick; 1.50 m of geogrid length | | | | | | | |
| up to 1.20 m high; 2 layers of geogrid | 52.55 | 0.50 | 35.63 | – | 59.30 | m² | **94.93** |
| 1.20–1.50 m high; 3 layers of geogrid | 52.55 | 0.67 | 47.45 | – | 62.67 | m² | **110.12** |
| 1.50–1.80 m high; 4 layers of geogrid | 52.55 | 0.75 | 53.44 | – | 66.05 | m² | **119.49** |

## 14 MASONRY

| Item<br>Excluding site overheads and profit | PC<br>£ | Labour<br>hours | Labour<br>£ | Plant<br>£ | Material<br>£ | Unit | Total<br>rate £ |
|---|---|---|---|---|---|---|---|
| **RECONSTITUTED STONE WALLING** – cont | | | | | | | |
| **Retaining wall** – cont | | | | | | | |
| Battered; maximum 1:3 slope | | | | | | | |
| up to 1.20 m high; 3 layers of geogrid | 52.55 | 0.83 | 59.35 | – | 62.67 | m$^2$ | **122.02** |
| 1.20–1.50 m high; 4 layers of geogrid | 52.55 | 0.83 | 59.35 | – | 66.05 | m$^2$ | **125.40** |
| 1.50–1.80 m high; 5 layers of geogrid | 52.55 | 1.00 | 71.25 | – | 68.86 | m$^2$ | **140.11** |
| **DAMP PROOF COURSES** | | | | | | | |
| **Damp proof courses; pitch polymer; 150 mm laps** | | | | | | | |
| Horizontal | | | | | | | |
| width not exceeding 225 mm | 1.12 | 1.14 | 27.07 | – | 2.99 | m | **30.06** |
| width exceeding 225 mm | 5.53 | 0.58 | 13.78 | – | 7.41 | m$^2$ | **21.19** |
| Vertical | | | | | | | |
| width not exceeding 225 mm | 1.12 | 1.72 | 40.85 | – | 2.99 | m | **43.84** |
| **Two courses slates in cement:mortar (1:3)** | | | | | | | |
| Horizontal | | | | | | | |
| width exceeding 225 mm | 9.49 | 3.46 | 82.17 | – | 14.60 | m$^2$ | **96.77** |
| Vertical | | | | | | | |
| width exceeding 225 mm | 9.49 | 5.18 | 123.03 | – | 14.60 | m$^2$ | **137.63** |
| **COPINGS AND PIER CAPS** | | | | | | | |
| **Precast concrete coping** | | | | | | | |
| Copings; once weathered; twice grooved | | | | | | | |
| 152 × 75 mm | 5.79 | 0.40 | 9.50 | – | 6.10 | m | **15.60** |
| 178 × 65 mm | 8.00 | 0.40 | 9.50 | – | 8.43 | m | **17.93** |
| 305 × 75 mm | 11.36 | 0.50 | 11.88 | – | 12.04 | m | **23.92** |
| Pier caps; four sides weathered | | | | | | | |
| 305 × 305 mm | 7.25 | 1.00 | 23.75 | – | 7.45 | nr | **31.20** |
| 381 × 381 mm | 7.68 | 1.00 | 23.75 | – | 7.88 | nr | **31.63** |
| 533 × 533 mm | 7.47 | 1.20 | 28.50 | – | 7.67 | nr | **36.17** |

Prices for Measured Works – Major Works

## 19 WATERPROOFING

| Item<br>Excluding site overheads and profit | PC<br>£ | Labour<br>hours | Labour<br>£ | Plant<br>£ | Material<br>£ | Unit | Total<br>rate £ |
|---|---|---|---|---|---|---|---|
| **SPECIALIST WATERPROOF RENDERING** | | | | | | | |
| **Sika 1 waterproof rendering; steel trowelled** | | | | | | | |
| Walls; 18 mm thick; three coats; to concrete base | | | | | | | |
| width exceeding 300 mm | – | – | – | – | – | m² | 46.60 |
| width not exceeding 300 mm | – | – | – | – | – | m² | 49.10 |
| Walls; 18 mm thick; three coats; to concrete base | | | | | | | |
| width exceeding 300 mm | – | – | – | – | – | m² | 43.50 |
| width not exceeding 300 mm | – | – | – | – | – | m² | 46.00 |
| **TANKING AND DAMP-PROOFING** | | | | | | | |
| **Tanking and damp-proofing; mastic asphalt; type T1097; Bituchem** | | | | | | | |
| 13 mm thick; one coat covering; to concrete base; flat; work subsequently covered | | | | | | | |
| width exceeding 300 mm | – | – | – | – | – | m² | 14.40 |
| 20 mm thick; two coat coverings; to concrete base; flat; work subsequently covered | | | | | | | |
| width exceeding 300 mm | – | – | – | – | – | m² | 18.00 |
| 30 mm thick; three coat coverings; to concrete base; flat; work subsequently covered | | | | | | | |
| width exceeding 300 mm | – | – | – | – | – | m² | 42.00 |
| 13 mm thick; two coat coverings; to brickwork base; vertical; work subsequently covered | | | | | | | |
| width exceeding 300 mm | – | – | – | – | – | m² | 27.60 |
| 20 mm thick; three coat coverings; to brickwork base; vertical; work subsequently covered | | | | | | | |
| width exceeding 300 mm | – | – | – | – | – | m² | 29.40 |
| Internal angle fillets; work subsequently covered | – | – | – | – | – | m | 5.50 |
| Turning asphalt nibs into grooves; 20 mm deep | – | – | – | – | – | m | 3.60 |

## 19 WATERPROOFING

| Item<br>Excluding site overheads and profit | PC<br>£ | Labour<br>hours | Labour<br>£ | Plant<br>£ | Material<br>£ | Unit | Total<br>rate £ |
|---|---|---|---|---|---|---|---|
| **COLD APPLIED BITUMINOUS EMULSION** | | | | | | | |
| **Tanking and damp-proofing; Ruberoid Building Products; Synthaprufe cold applied bituminous emulsion waterproof coating** | | | | | | | |
| Synthaprufe; to smooth finished concrete or screeded slabs; flat; blinding with sand | | | | | | | |
| two coats | 6.34 | 0.22 | 5.28 | – | 6.52 | m$^2$ | **11.80** |
| three coats | 9.50 | 0.31 | 7.36 | – | 9.69 | m$^2$ | **17.05** |
| Synthaprufe; to fair faced brickwork with flush joints, rendered brickwork or smooth finished concrete walls; vertical | | | | | | | |
| two coats | 7.13 | 0.29 | 6.79 | – | 7.31 | m$^2$ | **14.10** |
| three coats | 10.45 | 0.40 | 9.50 | – | 10.64 | m$^2$ | **20.14** |
| **LIQUID ASPHALT** | | | | | | | |
| **Tanking and damp-proofing; RIW Ltd** | | | | | | | |
| Liquid asphaltic composition; to smooth finished concrete screeded slabs or screeded slabs; flat | | | | | | | |
| two coats | 5.18 | 0.33 | 7.92 | – | 5.18 | m$^2$ | **13.10** |
| Liquid asphaltic composition; fair faced brickwork with flush joints, rendered brickwork or smooth finished concrete walls; vertical | | | | | | | |
| two coats | 5.18 | 0.50 | 11.88 | – | 5.18 | m$^2$ | **17.06** |
| Heviseal; to smooth finished concrete or screeded slabs; to surfaces of ponds, tanks or planters; flat | | | | | | | |
| two coats | 15.31 | 0.33 | 7.92 | – | 15.31 | m$^2$ | **23.23** |
| Heviseal; to fair faced brickwork with flush joints, rendered brickwork or smooth finished concrete walls; to surfaces of retaining walls, ponds, tanks or planters; vertical | | | | | | | |
| two coats | 15.31 | 0.50 | 11.88 | – | 15.31 | m$^2$ | **27.19** |
| **SELF ADHESIVE** | | | | | | | |
| **Tanking and damp-proofing; Grace Construction Products** | | | | | | | |
| Bitu-thene 3000; 1.50 mm thick; overlapping and bonding; including sealing all edges | | | | | | | |
| to concrete slabs; flat | 7.04 | 0.25 | 5.94 | – | 7.78 | m$^2$ | **13.72** |
| to brick/concrete walls; vertical | 7.04 | 0.40 | 9.50 | – | 7.78 | m$^2$ | **17.28** |

## 21 CLADDING AND COVERING

| Item<br>Excluding site overheads and profit | PC<br>£ | Labour<br>hours | Labour<br>£ | Plant<br>£ | Material<br>£ | Unit | Total<br>rate £ |
|---|---|---|---|---|---|---|---|
| **NATURAL STONE SLAB CLADDING/FEATURES** | | | | | | | |
| **Sawn Yorkstone cladding; Johnsons Wellfield Quarries Ltd**<br>Six sides sawn stone; rubbed face; sawn and jointed edges; fixed to blockwork (not included) with stainless steel fixings; Ancon Ltd; grade 304 stainless steel frame cramp and dowel 7 mm; cladding units drilled 4 × to receive dowels | | | | | | | |
| 440 × 200 × 50 mm thick | 95.40 | 1.87 | 44.41 | – | 98.80 | m² | **143.21** |
| **CAST STONE SLAB CLADDING/ FEATURES** | | | | | | | |
| **Cast stone cladding; Haddonstone Ltd**<br>Reconstituted stone in Portland Bath or Terracotta; fixed to blockwork or concrete (not included) with stainless steel fixings; stainless steel frame cramp and dowel M 6 mm; cladding units drilled 4 × to receive dowels | | | | | | | |
| 1000 × 300 × 50 mm thick | 45.00 | 1.87 | 44.41 | – | 48.40 | m² | **92.81** |

## 28 FLOOR, WALL, CEILING AND ROOF FINISHINGS

| Item<br>Excluding site overheads and profit | PC<br>£ | Labour<br>hours | Labour<br>£ | Plant<br>£ | Material<br>£ | Unit | Total<br>rate £ |
|---|---|---|---|---|---|---|---|
| **PLASTERED/RENDERED/ ROUGHCAST COATINGS** | | | | | | | |
| **Cement:lime:sand (1:1:6); 19 mm thick; two coats; wood floated finish** | | | | | | | |
| Walls | | | | | | | |
| width exceeding 300 mm; to brickwork or blockwork base | – | – | – | – | – | m² | 14.25 |
| Extra over cement:sand:lime (1:1:6) coatings for decorative texture finish with water repellent cement | | | | | | | |
| combed or floated finish | – | – | – | – | – | m² | 3.80 |
| **STONE/CONCRETE/QUARRY/ CERAMIC TILING/MOSAIC** | | | | | | | |
| **Ceramic tiles; unglazed slip resistant; various colours and textures; jointing** | | | | | | | |
| Floors | | | | | | | |
| level or to falls only not exceeding 15° from horizontal; 150 × 150 × 8 mm thick | 21.00 | 1.00 | 23.75 | – | 24.41 | m² | 48.16 |
| level or to falls only not exceeding 15° from horizontal; 150 × 150 × 12 mm thick | 26.25 | 1.00 | 23.75 | – | 29.66 | m² | 53.41 |
| **Clay tiles – General**<br>Preamble: Typical specification – Clay tiles shall be reasonably true to shape, flat, free from flaws, frost resistant and true to sample approved by the Landscape Architect prior to laying. Quarry tiles (or semi-vitrified tiles) shall be of external quality, either heather brown or blue, to size specified, laid on 1:2:4 concrete, with 20 mm maximum aggregate 100 mm thick, on 100 mm hardcore. The hardened concrete should be well wetted and the surplus water taken off. Clay tiles shall be thoroughly wetted immediately before laying and then drained and shall be bedded to 19 mm thick cement:sand (1:3) screed. Joints should be approximately 4 mm (or 3 mm for vitrified tiles) grouted in cement:sand (1:2) and cleaned off immediately. | | | | | | | |

## 28 FLOOR, WALL, CEILING AND ROOF FINISHINGS

| Item<br>Excluding site overheads and profit | PC<br>£ | Labour<br>hours | Labour<br>£ | Plant<br>£ | Material<br>£ | Unit | Total<br>rate £ |
|---|---|---|---|---|---|---|---|
| **Quarry tiles; external quality; including bedding; jointing**<br>Floors | | | | | | | |
| level or to falls only not exceeding 15° from horizontal; 150 × 150 × 12.5 mm thick; heather brown; PC £15.00/m$^2$ | – | 0.80 | 19.00 | – | 19.16 | m$^2$ | **38.16** |
| level or to falls only not exceeding 15° from horizontal; 225 × 225 × 29 mm thick; heather brown PC £25.00/m$^2$ | – | 0.67 | 15.83 | – | 29.66 | m$^2$ | **45.49** |
| level or to falls only not exceeding 15° from horizontal; 150 × 150 × 12.5 mm thick; blue/black PC £16.00/m$^2$ | – | 1.00 | 23.75 | – | 20.21 | m$^2$ | **43.96** |
| level or to falls only not exceeding 15° from horizontal; 194 × 194 × 12.5 mm thick; heather brown PC £15.00/m$^2$ | – | 0.80 | 19.00 | – | 19.16 | m$^2$ | **38.16** |
| **EXTENSIVE GREEN ROOFS** | | | | | | | |
| **Extensive Sedum type Green Roof System; Zinco Green Roof Systems Ltd. System for planting with Sedum plugs or overlaying with pre-grown cultivated Sedum Carpet; (ETA 13/ 0668)**<br>Flat roofs with no standing water up to 10° or for Inverted roofs and as an alternative to single layer systems. Drainage, water retention and protection mat with attached filter sheet; 20 mm thick. | | | | | | | |
| Zinco Drainage Element Fixodrain XD20; 3 l/m$^2$ reservoir | 4.80 | 0.04 | 2.85 | – | 4.80 | m$^2$ | **7.65** |
| Zinco System Substrate Growing medium Sedum Carpet – laid to Fixodrain XD20 at min. 70 mm thickness | 100.00 | 0.67 | 47.50 | – | 100.00 | m$^2$ | **147.50** |
| Zinco Premium Sedum Carpet Mat – laid over System substrate | 20.00 | 0.07 | 4.75 | – | 20.00 | m$^2$ | **24.75** |

## 28 FLOOR, WALL, CEILING AND ROOF FINISHINGS

| Item<br>Excluding site overheads and profit | PC<br>£ | Labour<br>hours | Labour<br>£ | Plant<br>£ | Material<br>£ | Unit | Total<br>rate £ |
|---|---|---|---|---|---|---|---|
| **EXTENSIVE GREEN ROOFS – cont** | | | | | | | |
| **Extensive Rockery type Green Roof System; Zinco Green Roof Systems Ltd. For planting with Sedum plugs or rockery type plants; flat roofs with no standing water up to 10°. (Not suitable for inverted roofs); Suitable system depth 90 mm. System saturated weight 110 kg/m$^2$; system water storage capacity: 36 l/m$^2$; (ETA 13/0668)** | | | | | | | |
| Non-rotting synthetic fibre mat for protection of the waterproofing layer and provision of additional water reservoir of 5 l/m$^2$ | | | | | | | |
|    Zinco Moisture mat/protection fleece SSM45 | 1.95 | 0.04 | 2.85 | – | 1.95 | m$^2$ | **4.80** |
| Pressure resistant drainage and water storage element; recycled polyolefin; laid to SSM45; 25 mm thick with diffusion openings; water storage capacity 3 l/m$^2$ | | | | | | | |
|    Zinco Drainage Element Fixodrain FD25-E | 6.00 | 0.04 | 2.85 | – | 6.00 | m$^2$ | **8.85** |
|    Zinco Filter Sheet SF – Rolled to drainage layer FD25-E | 0.80 | 0.04 | 2.85 | – | 0.80 | m$^2$ | **3.65** |
|    Zinco System Substrate Growing medium Rockery Type Plants laid to Filter Fleece SF min. 70 mm thick | 100.00 | 0.67 | 47.50 | – | 100.00 | m$^3$ | **147.50** |
| Planting to substrates | | | | | | | |
|    Zinco Premium Sedum Carpet Mat | 22.00 | 0.04 | 2.85 | – | 22.00 | m$^2$ | **24.85** |
|    Zinco Seed Mix – Meadow Scents; 15 g/m$^2$ | 0.66 | 0.01 | 0.71 | – | 0.66 | m$^2$ | **1.37** |
| **Extensive Rockery type Green Roof System for ZERO degree roofs; Zinco Green Roof Systems Ltd; for 0° roofs where isolated standing water may be present; planting with Sedum plugs or rockery type plants; system depth 150 mm; saturated weight: 105 kg/m$^2$; water storage capacity; 33 l/m$^2$** | | | | | | | |
| Non-rotting synthetic fibre mat for protection of the waterproofing layer and provision of additional water reservoir of 3 l/m$^2$ | | | | | | | |
|    Zinco Moisture mat/protection fleece TSM32 | 1.76 | 0.04 | 2.85 | – | 1.76 | m$^2$ | **4.61** |

## 28 FLOOR, WALL, CEILING AND ROOF FINISHINGS

| Item<br>Excluding site overheads and profit | PC<br>£ | Labour<br>hours | Labour<br>£ | Plant<br>£ | Material<br>£ | Unit | Total<br>rate £ |
|---|---|---|---|---|---|---|---|
| Drainage Element; laid to TSM32; ensures that a gap is maintained between any standing water and the soil substrate and the necessary aeration of the root zone; expanded polystyrene; 50 mm thick; lightweight (0.6 kg/m$^2$); water storage capacity of 3 l/m$^2$ | | | | | | | |
|   Zinco Drainage Element; Floraset FS50 | 7.59 | 0.04 | 2.85 | – | 7.59 | m$^2$ | 10.44 |
|   Zinco Filter Sheet SF – Rolled to drainage layer FS50 | | | | | | | |
|   Zinco System Substrate Growing medium Rockery Type Plants – laid to filter fleece SF; min 70 mm thick | 100.00 | 0.67 | 47.50 | – | 100.00 | m$^3$ | 147.50 |
| Planting to substrates | | | | | | | |
|   Zinco Premium Sedum Carpet Mat | 22.00 | 0.04 | 2.85 | – | 22.00 | m$^2$ | 24.85 |
|   Zinco Seed Mix – Meadow Scents; 15 g/m$^2$ | 0.66 | 0.01 | 0.71 | – | 0.66 | m$^2$ | 1.37 |
| **Extensive Green Roof System; BBS Green Roofing Ltd; low maintenance Sedum system for flat roof 1:60 min falls up to 15°** | | | | | | | |
| Moisture/protection layer; underlay moisture layer membrane protection | | | | | | | |
|   BBSU | – | – | – | – | – | m$^2$ | 2.60 |
| Drainage layer; BBS TDC recycled polypropylene; providing water reservoir, multidirectional drainage | | | | | | | |
|   BBS TCD20 20 mm deep | – | – | – | – | – | m$^2$ | 9.56 |
| Filter sheet rolled to drainage layer | | | | | | | |
|   BBSF filter sheet | – | – | – | – | – | m$^2$ | 2.80 |
| Growing media | | | | | | | |
|   BBS EXT; extensive substrate; lightweight growing medium laid to filter sheet; 80 mm thick | – | – | – | – | – | m$^2$ | 14.44 |
|   BBS SED; pre-grown Sedum turf | – | – | – | – | – | m$^2$ | 23.09 |
| **Extensive Lightweight Green Roof System; BBS Green Roofing Ltd; low maintenance Sedum nutrimat system for flat roof 1:60 min falls up to 15°; 60 kg/m$^2$ saturated** | | | | | | | |
| Moisture/Protection layer – Underlay moisture layer which protects the membrane | | | | | | | |
|   BBSU | – | – | – | – | – | m$^2$ | 2.60 |

## 28 FLOOR, WALL, CEILING AND ROOF FINISHINGS

| Item<br>Excluding site overheads and profit | PC<br>£ | Labour<br>hours | Labour<br>£ | Plant<br>£ | Material<br>£ | Unit | Total<br>rate £ |
|---|---|---|---|---|---|---|---|
| **EXTENSIVE GREEN ROOFS – cont** | | | | | | | |
| **Extensive Lightweight Green Roof System – cont** | | | | | | | |
| Drainage/reservoir/root-zone/filter – recycled PUR foam hydroponic mat | | | | | | | |
|   BBS Nutrimat | – | – | – | – | – | $m^2$ | **19.88** |
|   BBS EXT – Extensive substrate; lightweight growing medium laid to filter sheet 30 mm | – | – | – | – | – | $m^2$ | **8.20** |
|   BBS EXT – Extensive substrate; lightweight growing medium laid to filter sheet 80 mm | – | – | – | – | – | $m^2$ | **14.44** |
|   BBS SED – Pre-grown Sedum turf | – | – | – | – | – | $m^2$ | **23.09** |
| **EXTENSIVE PITCH ROOF SYSTEMS** | | | | | | | |
| **Extensive Green Roof System for Pitched roofs;** for pitched roofs with slopes from 10° up to 25°; Zinco Green Roof Systems Ltd; (degrees suitable for planting with Sedum plugs or overlaying with pre-grown cultivated Sedum Carpet). System depth – 130 mm; system saturated weight – 115 kg/$m^2$; system water storage capacity – 38 l/$m^2$; (Waterproofing layer must contain a root barrier as a separate root barrier cannot be used on pitched roofs) Non-rotting synthetic fibre mat for protection of the waterproofing layer and provision of additional water reservoir of 7 l/$m^2$; laid to water proofing layer | | | | | | | |
|   Protection Mat BSM64 | 2.55 | 0.10 | 7.13 | – | 2.55 | $m^2$ | **9.68** |
| Lightweight (1 kg/$m^2$) drainage element of expanded polystyrene. 75 mm thick laid with large studs upwards; laid to BSM64 | | | | | | | |
|   Zinco Drainage Element Floraset FS75 | 10.25 | 0.10 | 7.13 | – | 10.25 | $m^2$ | **17.38** |
| Zinco System Substrate Growing medium infilled to drainage board and to a level of 50 mm above the large studs. | | | | | | | |
|   Rockery Type Plants – laid to FS75 | 8.25 | 0.10 | 7.13 | – | 8.25 | $m^2$ | **15.38** |

## 28 FLOOR, WALL, CEILING AND ROOF FINISHINGS

| Item<br>Excluding site overheads and profit | PC<br>£ | Labour<br>hours | Labour<br>£ | Plant<br>£ | Material<br>£ | Unit | Total<br>rate £ |
|---|---|---|---|---|---|---|---|
| Erosion net; laid over substrate to aid prevention of wind erosion | | | | | | | |
|    Zinco Jute Anti Erosion Net JEG | 0.90 | 0.07 | 4.75 | – | 0.90 | m² | **5.65** |
| Vegetation layer; laid over System substrate | | | | | | | |
|    Zinco Premium Sedum Carpet Mat | 20.00 | 0.07 | 4.75 | – | 20.00 | m² | **24.75** |
| **Extensive Green Roof System for Steep Pitched roofs with slopes from 25° up to 35°; Zinco Green Roof Systems Ltd. Pitched roofs with slopes from 25° up to 35°; Suitable for overlay with pre-grown cultivated Sedum Carpet. System height – 120 mm; system saturated weight – 155 kg/m²; water storage capacity – 64 l/m² (depending on slope); (Waterproofing layer must contain a root barrier as a separate root barrier cannot be used on pitched roofs)** | | | | | | | |
| Protection mat; non-rotting needle fleeced synthetic mat for protection of the waterproofing layer and provision of additional water reservoir of 12 l/m² | | | | | | | |
|    WSM150 – laid to waterproofing layer | 5.95 | 0.10 | 7.13 | – | 5.95 | m² | **13.08** |
| Drainage element; Stable Grid element 100 mm high made from recycled polyethylene. Grid dimension 625 mm with integrated connecting plug | | | | | | | |
|    Zinco Drainage Element Georaster – Laid to WSM150 | 14.65 | 0.13 | 9.50 | – | 14.65 | m² | **24.15** |
| Zinco System Substrate Growing medium Heather with Lavender filled to Georaster. | | | | | | | |
|    Zinco Premium Sedum Carpet Mat – Laid over Georaster | 20.00 | 0.20 | 14.25 | – | 20.00 | m² | **34.25** |

## 28 FLOOR, WALL, CEILING AND ROOF FINISHINGS

| Item<br>Excluding site overheads and profit | PC<br>£ | Labour<br>hours | Labour<br>£ | Plant<br>£ | Material<br>£ | Unit | Total<br>rate £ |
|---|---|---|---|---|---|---|---|
| **EXTENSIVE PITCH ROOF SYSTEMS – cont** | | | | | | | |
| **Shear barriers and brackets for pitched roofs; Zinco Green Roof Systems Ltd.** Elements in pitched roof constructions transfer shear forces into the roof construction. Structural calculations required to establish the exact location and fixing of the shear barriers and brackets | | | | | | | |
| Zinco Eaves Profile and support brackets | | | | | | | |
| Eaves Profile TRP140 E; Stainless steel angle 140 mm high × 3.00 m long | 15.47 | 0.08 | 5.94 | – | 15.47 | m | **21.41** |
| Support bracket TSH100 | 22.20 | 0.50 | 35.63 | – | 22.20 | nr | **57.83** |
| Shear fix bracket LF300 | 83.00 | 0.50 | 35.63 | – | 83.00 | nr | **118.63** |
| **Extensive Lightweight Green Roof System; BBS Green Roofing Ltd**; low maintenance Sedum system for Pitch roof 3° to 20°. 60 kg/m$^2$ saturated NB// 20°–50° requires extra restraints | | | | | | | |
| BBSFG Nutrimat – Fully Grown Sedum Tile – modular assembly to form monolithic green roof layer low maintenance | | | | | | | |
| Sedum system; 60 kg/m$^2$ saturated | – | – | – | – | – | m$^2$ | **68.51** |
| **INTENSIVE GREEN ROOFS** | | | | | | | |
| **Flat roofs up to 10°; Intensive Simple and Wildflower Green Roof System; Zinco Green Roof Systems Ltd**; (not suitable for Inverted roofs); for planting with small shrubs, blooming perennials and wildflowers; can be combined with patios and walkways if required. System depth – 145 mm–200 mm; saturated weight – 145 kg/m$^2$–245 Kg/m$^2$; water storage capacity – 62 l/m$^2$–80 l/m$^2$; (ETA 13/0668) | | | | | | | |
| Non-rotting synthetic fibre mat for protection of the waterproofing layer and provision of additional water reservoir of 5 l/m$^2$ | | | | | | | |
| Moisture mat/protection fleece SSM45 | 1.95 | 0.04 | 2.85 | – | 1.95 | m$^2$ | **4.80** |

## 28 FLOOR, WALL, CEILING AND ROOF FINISHINGS

| Item<br>Excluding site overheads and profit | PC<br>£ | Labour<br>hours | Labour<br>£ | Plant<br>£ | Material<br>£ | Unit | Total<br>rate £ |
|---|---|---|---|---|---|---|---|
| Pressure resistant drainage and water storage element made from recycled polyolefin laid to SSM45; 40 mm thick with diffusion openings and water storage capacity of 6 l/m². | | | | | | | |
|    Fixodrain FD40-E | 7.26 | 0.04 | 2.85 | – | 7.26 | m² | **10.11** |
| Zinco Filter Sheet SF – Rolled to drainage layer FD40-E | 0.88 | 0.04 | 2.85 | – | 0.88 | m² | **3.73** |
| System Substrate; growing medium Heather with Lavender; laid to Filter Fleece SF at 100 mm thickness up to 200 mm | 125.00 | 0.20 | 14.25 | – | 125.00 | m³ | **139.25** |
| **Flat roofs up to 10°; Intensive Simple and Wildflower Green Roof System; Zinco Green Roof Systems Ltd; (suitable for Inverted roofs); for planting with small shrubs, blooming perennials and wildflowers; can be combined with patios and walkways if required. System depth – 145 mm–200 mm; saturated weight – 145 kg/m²–245 Kg/m²; water storage capacity – 62 l/m²–80 l/m².** | | | | | | | |
| Water repellent, air permeable membrane used as a diffusionable separation layer; laid above waterproofing and separation membrane; rolled over the insulation | | | | | | | |
|    Separation Layer; TGV21 | 0.82 | 0.04 | 2.85 | – | 0.82 | m² | **3.67** |
| Pressure resistant drainage and water storage element made from recycled polyolefin laid to SSM45; 40 mm thick with diffusion openings and water storage capacity of 6 l/m² | | | | | | | |
|    Fixodrain FD40-E; laid to TGV21 | 6.60 | 0.04 | 2.85 | – | 6.60 | m² | **9.45** |
| Zinco Filter Sheet SF – Rolled to drainage layer FD40-E | 0.88 | 0.04 | 2.85 | – | 0.88 | m² | **3.73** |
| System Substrate; growing medium Heather with Lavender; laid to Filter Fleece SF at 100 mm thickness up to 200 mm | 125.00 | 0.67 | 47.50 | – | 125.00 | m³ | **172.50** |

## 28 FLOOR, WALL, CEILING AND ROOF FINISHINGS

| Item<br>Excluding site overheads and profit | PC<br>£ | Labour<br>hours | Labour<br>£ | Plant<br>£ | Material<br>£ | Unit | Total<br>rate £ |
|---|---|---|---|---|---|---|---|
| **INTENSIVE GREEN ROOFS – cont** | | | | | | | |
| Flat roofs up to 10°; Intensive Simple and Wildflower Green Roof System; Zinco Green Roof Systems Ltd; (suitable for Inverted roofs); for planting with lawn, shrubs, small trees, and perennials; can be combined with patios and walkways if required. System depth – min 160 mm; saturated weight – 342 kg/m$^2$ ; water storage capacity – 110 l/m$^2$. | | | | | | | |
| Water repellent, air permeable membrane used as a diffusionable separation layer; laid above waterproofing and separation membrane; rolled over the insulation | | | | | | | |
| Separation Layer; TGV21 | 0.82 | 0.04 | 2.85 | – | 0.82 | m$^2$ | **3.67** |
| Pressure resistant drainage and water storage element made from thermoformed hard plastic; laid to TGV21; 60 mm thick filled with Zincolit Plus mineral aggregate | | | | | | | |
| Zincolit Plus; mineral aggregate infill; 27 l/m$^2$ | 2.97 | 0.03 | 1.78 | – | 2.97 | m$^2$ | **4.75** |
| Filter Sheet SF – Rolled to drainage layer FD 60 | 0.88 | 0.04 | 2.85 | – | 0.88 | m$^2$ | **3.73** |
| System Substrate Growing medium; Roof Garden laid to Filter Fleece SF min.200 mm thick dependent on planting required and depth capacities | 62.50 | 0.67 | 47.50 | – | 62.50 | m$^3$ | **110.00** |

## 28 FLOOR, WALL, CEILING AND ROOF FINISHINGS

| Item<br>Excluding site overheads and profit | PC<br>£ | Labour<br>hours | Labour<br>£ | Plant<br>£ | Material<br>£ | Unit | Total<br>rate £ |
|---|---|---|---|---|---|---|---|
| **Flat roofs up to 10°; Intensive Simple and Wildflower Green Roof System; Zinco Green Roof Systems Ltd; (suitable for Inverted roofs); for planting with lawn, shrubs, small trees, and perennials; System depth – min 270 mm; saturated weight – 365 kg/m$^2$ ; water storage capacity – 135 l/m$^2$. (ETA 13/0668)**<br>Moisture and protection mat; Synthetic fibre resistant to mechanical stress resistant for protection of waterproofing layer and provision of additional extra water reservoir of 4 l/m$^2$ | | | | | | | |
|    Zinco Moisture mat/protection fleece ISM50 | 5.61 | 0.04 | 2.85 | – | 5.61 | m$^2$ | 8.46 |
|    Zinco Drainage Element FD 60 – Laid to ISM50 | 13.64 | 0.04 | 2.85 | – | 13.64 | m$^2$ | 16.49 |
|    Zincolit Plus mineral aggregate infill 27 Litres per m$^2$ | 2.97 | 0.03 | 1.78 | – | 2.97 | m$^2$ | 4.75 |
|    Filter Sheet SF – Rolled to drainage layer FD 60 | 0.88 | 0.04 | 2.85 | – | 0.88 | m$^2$ | 3.73 |
|    System Substrate Growing medium; Roof Garden laid to Filter Fleece SF min 200 mm thick dependent on planting required and depth capacities | 62.50 | 0.67 | 47.50 | – | 62.50 | m$^3$ | 110.00 |
| **LAWN Green Roof System; Zinco Green Roof Systems Ltd. Lawns on roofs where build up depth is restricted; Flat roofs – 0° up to 5°; Lawn surface treatment; System depth 150 mm; system saturated weight min 165 kg/m$^2$; System water storage capacity; from 65 l/m$^2$**<br>Protection Layer – polypropylene mat, mechanical stress resistant for protection of the waterproofing layer | | | | | | | |
|    Zinco Filter sheet PV | 2.69 | 0.04 | 2.85 | – | 2.69 | m$^2$ | 5.54 |
| Drainage and distribution element; Water storage, distribution and drainage element made from thermoformed recycled hard plastic; 45 mm thick and preformed with clipping points for Zinco Irrigation Dripperline 100-L1 | | | | | | | |
|    Zinco Aquatec 45 – laid to Filter Sheet PV | 18.30 | 0.07 | 4.75 | – | 18.30 | m$^2$ | 23.05 |

## 28 FLOOR, WALL, CEILING AND ROOF FINISHINGS

| Item<br>Excluding site overheads and profit | PC<br>£ | Labour<br>hours | Labour<br>£ | Plant<br>£ | Material<br>£ | Unit | Total<br>rate £ |
|---|---|---|---|---|---|---|---|
| **INTENSIVE GREEN ROOFS – cont** | | | | | | | |
| **LAWN Green Roof System – cont** | | | | | | | |
| Drippers; clipped into Aquatec 45 and linked to Zinco irrigation manager BM-4 | | | | | | | |
|   Zinco Dripperline 100-L1 | 2.69 | 0.07 | 4.75 | – | 2.69 | m² | **7.44** |
| Programmable Irrigation management unit in lockable steel box for automated irrigation of lawn and green roofs; linked to dripperlines 100-L1 | | | | | | | |
|   Zinco Irrigation Manager BM-4 | 1527.90 | 2.00 | 142.50 | – | 1667.90 | nr | **1810.40** |
| Water distributing polyester fleece with capillary effective fibres; laid over Aquatec 45 containing dripperlines 100-L1 | | | | | | | |
|   Zinco Wicking Mat DV 40 | 9.85 | 0.03 | 2.04 | – | 9.85 | m² | **11.89** |
| System Substrate Growing medium; Minimum thickness 100 mm up to 150 mm laid over Wicking mat DV 40 | | | | | | | |
|   Zinco LAWN | 18.75 | 0.07 | 4.75 | – | 18.75 | m² | **23.50** |
| Top dressing; 15 mm layer laid over LAWN substrate | | | | | | | |
|   ZINCOHUM | 1.43 | 0.01 | 0.95 | – | 1.43 | m² | **2.38** |
| **Intensive Green Roof; BBS Green Roof Systems; soil based systems able to provide a variety of hard and soft landscaping; laid to the surface of an unprepared roof deck** | | | | | | | |
| **Intensive Green Roof Systems; BBS Green roofing Ltd; components laid to the inverted build-up; insulation over waterproof rootproof membrane** | | | | | | | |
| Drainage layer; BBS TDC; recycled polypropylene; providing water reservoir, multidirectional drainage and mechanical damage protection | | | | | | | |
|   BBS TDC40; 40 mm deep | – | – | – | – | – | m² | **14.70** |
|   BBS TDC60; 60 mm deep | – | – | – | – | – | m² | **18.20** |
| Optional drainage unfill to BBSTDC layers | | | | | | | |
|   BBS TDC 40; 16 l/m² | – | – | – | – | – | m² | **5.80** |
|   BBS TDC60; 26 l/m² | – | – | – | – | – | m² | **7.60** |
| Filter sheet rolled out to drainage layer | | | | | | | |
|   BBSF filter sheet | – | – | – | – | – | m² | **2.80** |

## 28 FLOOR, WALL, CEILING AND ROOF FINISHINGS

| Item<br>Excluding site overheads and profit | PC<br>£ | Labour<br>hours | Labour<br>£ | Plant<br>£ | Material<br>£ | Unit | Total<br>rate £ |
|---|---|---|---|---|---|---|---|
| Growing medium and substrates | | | | | | | |
| BBS INT – Intensive substrate; lightweight growing medium laid to filter sheet 300 mm deep | – | – | – | – | – | m² | **38.20** |
| BBS SMi – Semi-intensive substrate; Lightweight growing meduim laid to filter sheet 200 mm deep | | | | | | m² | **26.76** |
| **LANDSCAPE OPTIONS FOR GREEN ROOF SYSTEMS** | | | | | | | |
| **Green Roof Drainage** | | | | | | | |
| Outlet inspection chambers | | | | | | | |
| BBS CHM; 150 mm deep | – | – | – | – | – | nr | **70.00** |
| BBS CHM; 250 mm deep | – | – | – | – | – | nr | **85.00** |
| BBS CHM; 350 mm deep | – | – | – | – | – | nr | **95.00** |

**Preamble**
Preamble: A variety of systems are available which address all the varied requirements for a successful Green Roof. For installation by approved contractors only. The prices shown are for budgeting purposes only, as each installation is site specific and may incorporate some or all of the resources shown. Specifiers should verify that the systems specified include for design liability and inspections by the suppliers. The systems below assume commercial insulation levels are required to the space below the proposed Green Roof. Extensive Green Roofs are those of generally lightweight construction with low maintenance planting and shallow soil designed for aesthetics only; Intensive Green Roofs are designed to allow use for recreation and trafficking. They require more maintenance and allow a greater variety of surfaces and plant types.
The following systems which are compliant with European Technical Approval ETA 13/0668 (Kits for Green roofs) are indicated as such.
The rates below assume that all elements are in position in the works location; no allowances for access, lifting or materials movement to the roof areas.

## 28 FLOOR, WALL, CEILING AND ROOF FINISHINGS

| Item<br>Excluding site overheads and profit | PC<br>£ | Labour<br>hours | Labour<br>£ | Plant<br>£ | Material<br>£ | Unit | Total<br>rate £ |
|---|---|---|---|---|---|---|---|
| **LANDSCAPE OPTIONS FOR GREEN ROOF SYSTEMS – cont** | | | | | | | |
| | | | | | | | |
| **Root barriers; Zinco Green Roof Systems Ltd** | | | | | | | |
| Where waterproofing does not contain a root barrier | | | | | | | |
| WSB 100-PO Weldable reinforced flexible polyolefin. 1.1 mm thick. Root resistant tested according to FLL | 13.47 | – | – | – | – | $m^2$ | – |
| WSF 40 High pressure polyethylene 0.34 mm thick | 2.31 | 0.02 | 1.43 | – | 2.31 | $m^2$ | 3.74 |
| | | | | | | | |
| **WATERPROOFING AND DRAINAGE OPTIONS** | | | | | | | |
| | | | | | | | |
| **Warm Roof waterproofing systems; to roof surfaces to receive Green Roof systems** | | | | | | | |
| Vapour barriers | | | | | | | |
| Aluminium lined vapour barrier; 2 mm thick; bonded in hot bitumen to the roof deck | – | – | – | – | – | $m^2$ | 11.16 |
| Korklite insulation bonded to the vapour barrier in hot bitumen; U value dependent; 80 mm thick | – | – | – | – | – | $m^2$ | 21.13 |
| Universal 2 mm thick underlayer; fully bonded to the insulation | – | – | – | – | – | $m^2$ | 9.09 |
| Anti-Root cap sheet impregnated with root resisting chemical bonded to the underlayer | – | – | – | – | – | $m^2$ | 18.70 |
| | | | | | | | |
| **Waterproofing to upstands** | | | | | | | |
| Vapour Barrier | | | | | | | |
| up to 200 mm high | – | – | – | – | – | m | 19.11 |
| up to 400 mm high | – | – | – | – | – | m | 26.34 |
| up to 600 mm high | – | – | – | – | – | m | 31.23 |
| | | | | | | | |
| **Inverted waterproofing systems; to roof surfaces to receive Green Roof systems** | | | | | | | |
| Monolithic hot melt rubberized bitumen; applied in two 3 mm layers incorporating a polyester reinforcing sheet with 4 mm thick protection sheet and chemically impregnated root barrier; fully bonded; applied to plywood or suitably prepared wood float finish and primed concrete deck or screeds | | | | | | | |
| 10 mm thick | – | – | – | – | – | $m^2$ | 35.62 |

## 28 FLOOR, WALL, CEILING AND ROOF FINISHINGS

| Item<br>Excluding site overheads and profit | PC<br>£ | Labour<br>hours | Labour<br>£ | Plant<br>£ | Material<br>£ | Unit | Total<br>rate £ |
|---|---|---|---|---|---|---|---|
| Roofmate; extruded polystyrene insulation; optional system; thickness to suit required U value; calculated at design stage; indicative thicknesses; laid to Hydrotech 6125 | | | | | | | |
| 0.25 U value; average requirement 120 mm | – | – | – | – | – | m² | **24.69** |

## 29 DECORATION

| Item<br>Excluding site overheads and profit | PC<br>£ | Labour<br>hours | Labour<br>£ | Plant<br>£ | Material<br>£ | Unit | Total<br>rate £ |
|---|---|---|---|---|---|---|---|
| **PAINTINGS/CLEAR FINISHING** | | | | | | | |
| **Eura Conservation Ltd; restoration of railings; works carried out off site; excludes removal of railings from site** | | | | | | | |
| Shotblast railings; remove all traces of paint and corrosion; apply temporary protective primer; measured overall two sides | | | | | | | |
| to plain railings | – | – | – | – | – | m² | **43.00** |
| to decorative railings | – | – | – | – | – | m² | **87.00** |
| Paint new galvanized railings in situ with three coats system | | | | | | | |
| vertical bar railings or gates | – | – | – | – | – | m² | **11.50** |
| vertical bar railings or gates with dog mesh panels | – | – | – | – | – | m² | **12.50** |
| gate posts not exceeding 300 mm girth | – | – | – | – | – | m² | **6.25** |
| Paint previously painted railings; shotblasted off site | | | | | | | |
| vertical bar railings or gates | – | – | – | – | – | m² | **14.50** |
| gate posts not exceeding 300 mm girth | – | – | – | – | – | m² | **6.25** |
| **Prepare; touch up primer; two undercoats and one finishing coat of gloss oil paint; on metal surfaces** | | | | | | | |
| General surfaces | | | | | | | |
| girth exceeding 300 mm | 3.39 | 0.67 | 15.83 | – | 3.39 | m² | **19.22** |
| isolated surfaces; girth not exceeding 300 mm | 10.27 | 0.40 | 9.50 | – | 10.27 | m | **19.77** |
| isolated areas not exceeding 0.50 m² irrespective of girth | 6.78 | 0.67 | 15.83 | – | 6.78 | nr | **22.61** |
| Ornamental railings; each side measured separately | | | | | | | |
| girth exceeding 300 mm | 3.39 | 0.57 | 13.57 | – | 3.39 | m² | **16.96** |
| **Prepare; one coat primer; two finishing coats of gloss paint; on wood surfaces** | | | | | | | |
| New wood surfaces | | | | | | | |
| girth exceeding 300 mm | 2.46 | 0.80 | 19.00 | – | 2.46 | m² | **21.46** |
| isolated surfaces; girth not exceeding 300 mm | 0.82 | 1.00 | 23.75 | – | 0.82 | m | **24.57** |
| isolated areas not exceeding 0.50 m² irrespective of girth | 3.12 | 0.50 | 11.88 | – | 3.12 | nr | **15.00** |

## 29 DECORATION

| Item<br>Excluding site overheads and profit | PC<br>£ | Labour<br>hours | Labour<br>£ | Plant<br>£ | Material<br>£ | Unit | Total<br>rate £ |
|---|---|---|---|---|---|---|---|
| Previously painted wood surfaces | | | | | | | |
| girth exceeding 300 mm | 1.20 | 0.50 | 11.88 | – | 1.20 | m² | **13.08** |
| isolated surfaces; girth not exceeding 300 mm | 0.40 | 0.67 | 15.83 | – | 0.40 | m | **16.23** |
| isolated areas not exceeding 0.50 m² irrespective of girth | 2.40 | 0.40 | 9.50 | – | 2.40 | nr | **11.90** |
| **Fences and sheds; prepare; two coats of Protek wood preserver on wood surfaces** | | | | | | | |
| Planed surfaces | | | | | | | |
| girth exceeding 300 mm | 0.79 | 0.07 | 1.58 | – | 0.79 | m² | **2.37** |
| isolated surfaces; girth not exceeding 300 mm | 0.26 | 0.17 | 3.96 | – | 0.26 | m | **4.22** |
| isolated areas not exceeding 0.50 m² irrespective of girth | 0.79 | 0.25 | 5.94 | – | 0.79 | nr | **6.73** |
| Sawn surfaces | | | | | | | |
| girth exceeding 300 mm | 0.30 | 0.10 | 2.38 | – | 0.30 | m² | **2.68** |
| isolated surfaces; girth not exceeding 300 mm | 3.97 | 0.17 | 3.96 | – | 3.97 | m | **7.93** |
| isolated areas not exceeding 0.50 m² irrespective of girth | 1.19 | 0.25 | 5.94 | – | 1.19 | nr | **7.13** |
| **Fences and sheds; prepare; two coats of Sadolin wood preserver on wood surfaces** | | | | | | | |
| Planed surfaces | | | | | | | |
| girth exceeding 300 mm | 1.54 | 0.07 | 1.58 | – | 1.54 | m² | **3.12** |
| isolated surfaces; girth not exceeding 300 mm | 0.51 | 0.17 | 3.96 | – | 0.51 | m | **4.47** |
| isolated areas not exceeding 0.50 m² irrespective of girth | 0.77 | 0.25 | 5.94 | – | 0.77 | nr | **6.71** |
| Sawn surfaces | | | | | | | |
| girth exceeding 300 mm | 1.90 | 0.10 | 2.38 | – | 1.90 | m² | **4.28** |
| isolated surfaces; girth not exceeding 300 mm | 0.82 | 0.17 | 3.96 | – | 0.82 | m | **4.78** |
| isolated areas not exceeding 0.50 m² irrespective of girth | 1.24 | 0.25 | 5.94 | – | 1.24 | nr | **7.18** |
| **Prepare; proprietary solution primer; two coats of dark stain; on wood surfaces** | | | | | | | |
| General surfaces | | | | | | | |
| girth exceeding 300 mm | 1.92 | 0.17 | 3.96 | – | 1.92 | m² | **5.88** |
| isolated surfaces; girth not exceeding 300 mm | 0.64 | 0.13 | 2.97 | – | 0.64 | m | **3.61** |

## 33 DRAINAGE ABOVE GROUND

| Item<br>Excluding site overheads and profit | PC<br>£ | Labour<br>hours | Labour<br>£ | Plant<br>£ | Material<br>£ | Unit | Total<br>rate £ |
|---|---|---|---|---|---|---|---|
| **PAINTINGS/CLEAR FINISHING – cont** | | | | | | | |
| **Two coats Weathershield; to clean, dry surfaces; in accordance with manufacturer's instructions** | | | | | | | |
| Brick or block walls | | | | | | | |
| girth exceeding 300 mm | 1.25 | 0.40 | 9.50 | – | 1.25 | m² | **10.75** |
| Cement render or concrete walls | | | | | | | |
| girth exceeding 300 mm | 1.25 | 0.33 | 7.92 | – | 1.25 | m² | **9.17** |
| **DRAINAGE TO ROOF DECKS AND PLANTERS** | | | | | | | |
| **Leca (light expanded clay aggregate); drainage aggregate to roofdecks and planters** | | | | | | | |
| Placed mechanically to planters; average 100 mm thick; by mechanical plant tipped into planters | | | | | | | |
| aggregate size 10–20 mm; delivered in 30 m³ loads | 49.00 | 0.40 | 9.50 | 5.15 | 49.00 | m³ | **63.65** |
| aggregate size 10–20 mm; delivered in 70 m³ loads | 38.00 | 0.20 | 4.75 | 6.80 | 38.00 | m³ | **49.55** |
| Placed by light aggregate blower (maximum 40 m) | | | | | | | |
| aggregate size 10–20 mm; delivered in 30 m³ loads | 49.00 | 0.14 | 3.39 | – | 53.59 | m³ | **56.98** |
| aggregate size 10–20 mm; delivered in 55 m³ loads on blower vehicle | 41.48 | 0.14 | 3.39 | – | 41.48 | m³ | **44.87** |
| aggregate size 10–20 mm; delivered in 70 m³ loads | 38.00 | 0.14 | 3.39 | – | 42.59 | m³ | **45.98** |
| By hand | | | | | | | |
| aggregate size 10–20 mm; delivered in 30 m³ loads | 49.00 | 1.33 | 31.67 | – | 49.00 | m³ | **80.67** |
| aggregate size 10–20 mm; delivered in 70 m³ loads | 38.00 | 1.33 | 31.67 | – | 38.00 | m³ | **69.67** |
| **Drainage boards laid to insulated slabs on roof decks; boards laid below growing medium and granulated drainage layer and geofabric (all not included) to collect and channel water to drainage outlets (not included)** | | | | | | | |
| Alumasc Floradrain; polyethylene irrigation/drainage layer; inclusive of geofabric laid over the surface of the drainage board | | | | | | | |
| Floradrain FD40; 0.96 × 2.08 m panels | 12.05 | 0.05 | 1.19 | – | 13.20 | m² | **14.39** |
| Floradrain FD60; 1.00 × 2.00 m panel | 21.42 | 0.07 | 1.58 | – | 22.57 | m² | **24.15** |

## 34 DRAINAGE BELOW GROUND

| Item<br>Excluding site overheads and profit | PC<br>£ | Labour<br>hours | Labour<br>£ | Plant<br>£ | Material<br>£ | Unit | Total<br>rate £ |
|---|---|---|---|---|---|---|---|
| **TRENCHES/PIPEWAYS/PITS FOR BURIED ENGINEERING** | | | | | | | |
| **Excavating trenches; using 3 tonne tracked excavator; to receive pipes; grading bottoms; earthwork support; filling with excavated material to within 150 mm of finished surfaces and compacting; completing fill with topsoil; disposal of surplus soil** | | | | | | | |
| Services not exceeding 200 mm nominal size | | | | | | | |
| average depth of run not exceeding 0.50 m | 1.78 | 0.12 | 2.85 | 1.04 | 1.78 | m | **5.67** |
| average depth of run not exceeding 0.75 m | 1.78 | 0.16 | 3.87 | 1.44 | 1.78 | m | **7.09** |
| average depth of run not exceeding 1.00 m | 1.78 | 0.28 | 6.73 | 2.51 | 1.78 | m | **11.02** |
| average depth of run not exceeding 1.25 m | 1.78 | 0.38 | 9.10 | 3.37 | 1.78 | m | **14.25** |
| **Excavating trenches; using 3 tonne tracked excavator; to receive pipes; grading bottoms; earthwork support; filling with imported granular material and compacting; disposal of surplus soil** | | | | | | | |
| Services not exceeding 200 mm nominal size | | | | | | | |
| average depth of run not exceeding 0.50 m | 2.59 | 0.09 | 2.06 | 0.72 | 7.21 | m | **9.99** |
| average depth of run not exceeding 0.75 m | 3.89 | 0.11 | 2.57 | 0.90 | 10.82 | m | **14.29** |
| average depth of run not exceeding 1.00 m | 5.18 | 0.14 | 3.32 | 1.18 | 14.42 | m | **18.92** |
| average depth of run not exceeding 1.25 m | 6.48 | 0.23 | 5.43 | 2.02 | 18.03 | m | **25.48** |
| **Excavating trenches; using 3 tonne tracked excavator; to receive pipes; grading bottoms; earthwork support; filling with lean mix concrete; disposal of surplus soil** | | | | | | | |
| Services not exceeding 200 mm nominal size | | | | | | | |
| average depth of run not exceeding 0.50 m | 12.60 | 0.11 | 2.53 | 0.38 | 18.15 | m | **21.06** |
| average depth of run not exceeding 0.75 m | 18.90 | 0.13 | 3.09 | 0.47 | 27.22 | m | **30.78** |
| average depth of run not exceeding 1.00 m | 25.20 | 0.17 | 3.96 | 0.63 | 36.29 | m | **40.88** |

## 34 DRAINAGE BELOW GROUND

| Item<br>Excluding site overheads and profit | PC<br>£ | Labour<br>hours | Labour<br>£ | Plant<br>£ | Material<br>£ | Unit | Total<br>rate £ |
|---|---|---|---|---|---|---|---|
| **TRENCHES/PIPEWAYS/PITS FOR BURIED ENGINEERING – cont** | | | | | | | |
| **Excavating trenches – cont** | | | | | | | |
| Services not exceeding 200 mm nominal size – cont | | | | | | | |
| average depth of run not exceeding 1.25 m | 31.50 | 0.23 | 5.34 | 0.94 | 45.36 | m | **51.64** |
| **Earthwork support; providing support to opposing faces of excavation; moving along as work proceeds** | | | | | | | |
| Maximum depth not exceeding 2.00 m | | | | | | | |
| trenchbox; distance between opposing faces not exceeding 2.00 m | – | 0.80 | 19.00 | 30.45 | – | m | **49.45** |
| timber; distance between opposing faces not exceeding 500 mm | – | 0.50 | 23.75 | – | 2.65 | m | **26.40** |
| **ACCESS COVERS AND FRAMES** | | | | | | | |
| **Load Classes for Access Covers** | | | | | | | |
| FACTA (Fabricated Access Cover Trade Association) class: | | | | | | | |
| A – 0.5 tonne maximum slow moving wheel load | | | | | | | |
| AA – 1.5 tonne maximum slow moving wheel load | | | | | | | |
| AAA – 2.5 tonne maximum slow moving wheel load | | | | | | | |
| B – 5 tonne maximum slow moving wheel load | | | | | | | |
| C – 6.5 tonne maximum slow moving wheel load | | | | | | | |
| D – 11 tonne maximum slow moving wheel load | | | | | | | |
| **Access covers and frames; solid top; galvanized; Steelway Brickhouse; Bristeel; bedding frame in cement: mortar (1:3); cover in grease and sand; clear opening sizes; base size shown in brackets (50 mm depth)** | | | | | | | |
| FACTA AA; single seal | | | | | | | |
| 450 × 450 mm (520 × 520 mm) | 75.15 | 1.50 | 35.63 | – | 87.28 | nr | **122.91** |
| 600 × 450 mm (670 × 520 mm) | 83.41 | 1.80 | 42.75 | – | 96.89 | nr | **139.64** |
| 600 × 600 mm (670 × 670 mm) | 89.18 | 2.00 | 47.50 | – | 105.35 | nr | **152.85** |

## 34 DRAINAGE BELOW GROUND

| Item<br>Excluding site overheads and profit | PC<br>£ | Labour<br>hours | Labour<br>£ | Plant<br>£ | Material<br>£ | Unit | Total<br>rate £ |
|---|---|---|---|---|---|---|---|
| FACTA AA; double seal | | | | | | | |
| 450 × 450 mm (560 × 560 mm) | 122.09 | 1.50 | 35.63 | – | 134.22 | nr | **169.85** |
| 600 × 450 mm (710 × 560 mm) | 134.15 | 1.80 | 42.75 | – | 147.63 | nr | **190.38** |
| 600 × 600 mm (710 × 710 mm) | 146.36 | 2.00 | 47.50 | – | 162.53 | nr | **210.03** |
| FACTA AAA; single seal | | | | | | | |
| 450 × 450 mm (520 × 520 mm) | 115.75 | 1.50 | 35.63 | – | 127.88 | nr | **163.51** |
| 600 × 450 mm (670 × 520 mm) | 124.92 | 1.80 | 42.75 | – | 138.40 | nr | **181.15** |
| 600 × 600 mm (670 × 670 mm) | 133.10 | 2.00 | 47.50 | – | 149.27 | nr | **196.77** |
| FACTA AAA; double seal | | | | | | | |
| 450 mm × 450 mm (560 × 560) | 147.36 | 1.50 | 35.63 | – | 159.49 | nr | **195.12** |
| 600 mm × 450 mm (710 × 560) | 161.00 | 1.80 | 42.75 | – | 174.48 | nr | **217.23** |
| 600 mm × 600 mm (710 × 710) | 179.30 | 2.00 | 47.50 | – | 195.47 | nr | **242.97** |
| FACTA B; single seal | | | | | | | |
| 450 × 450 mm (520 × 520 mm) | 97.15 | 1.50 | 35.63 | – | 109.28 | nr | **144.91** |
| 600 × 450 mm (670 × 520 mm) | 121.29 | 1.80 | 42.75 | – | 134.77 | nr | **177.52** |
| 600 × 600 mm (670 × 670 mm) | 141.00 | 2.00 | 47.50 | – | 157.17 | nr | **204.67** |
| FACTA B; double seal | | | | | | | |
| 450 × 450 mm (560 × 560 mm) | 140.20 | 1.50 | 35.63 | – | 152.33 | nr | **187.96** |
| 600 × 450 mm (710 × 560 mm) | 156.08 | 1.80 | 42.75 | – | 169.56 | nr | **212.31** |
| 600 × 600 mm (710 × 710 mm) | 187.43 | 2.00 | 47.50 | – | 203.60 | nr | **251.10** |
| Access covers and frames; recessed; galvanized; Steelway Brickhouse; Bripave; bedding frame in cement:mortar (1:3); cover in grease and sand; clear opening sizes; base size shown in brackets (for block depths 50 mm, 65 mm, 80 mm or 100 mm) | | | | | | | |
| FACTA AA | | | | | | | |
| 450 × 450 mm (562 × 562 mm) | 217.84 | 1.50 | 35.63 | – | 229.97 | nr | **265.60** |
| 600 × 450 mm (712 × 562 mm) | 238.97 | 1.80 | 42.75 | – | 252.45 | nr | **295.20** |
| 600 × 600 mm (712 × 712 mm) | 247.68 | 2.00 | 47.50 | – | 263.29 | nr | **310.79** |
| 750 × 600 mm (862 × 712 mm) | 276.37 | 2.40 | 57.00 | – | 293.89 | nr | **350.89** |
| FACTA B | | | | | | | |
| 450 × 450 mm (562 × 562 mm) | 218.22 | 1.50 | 35.63 | – | 230.35 | nr | **265.98** |
| 600 × 450 mm (712 × 562 mm) | 253.68 | 1.80 | 42.75 | – | 267.16 | nr | **309.91** |
| 600 × 600 mm (712 × 712 mm) | 256.84 | 2.00 | 47.50 | – | 273.01 | nr | **320.51** |
| 750 × 600 mm (862 × 712 mm) | 257.67 | 2.40 | 57.00 | – | 275.19 | nr | **332.19** |
| FACTA D | | | | | | | |
| 450 × 450 mm (562 × 562 mm) | 244.57 | 1.50 | 35.63 | – | 256.70 | nr | **292.33** |
| 600 × 450 mm (712 × 562 mm) | 259.85 | 1.80 | 42.75 | – | 273.33 | nr | **316.08** |
| 600 × 600 mm (712 × 712 mm) | 268.47 | 2.00 | 47.50 | – | 284.64 | nr | **332.14** |
| 750 × 600 mm (862 × 712 mm) | 326.73 | 2.40 | 57.00 | – | 344.25 | nr | **401.25** |

## 34 DRAINAGE BELOW GROUND

| Item<br>Excluding site overheads and profit | PC<br>£ | Labour<br>hours | Labour<br>£ | Plant<br>£ | Material<br>£ | Unit | Total<br>rate £ |
|---|---|---|---|---|---|---|---|
| **ACCESS COVERS AND FRAMES – cont** | | | | | | | |
| **Access covers and frames; Jones of Oswestry; bedding frame in cement:mortar (1:3); cover in grease and sand; clear opening sizes** | | | | | | | |
| Access covers and frames; Suprabloc; to paved areas; filling with blocks cut and fitted to match surrounding paving | | | | | | | |
| pedestrian weight; 300 × 300 mm | 199.21 | 2.50 | 59.38 | – | 207.38 | nr | **266.76** |
| pedestrian weight; 450 × 450 mm | 339.92 | 2.50 | 59.38 | – | 339.92 | nr | **399.30** |
| pedestrian weight; 450 × 600 mm | 378.58 | 2.50 | 59.38 | – | 387.09 | nr | **446.47** |
| light vehicular weight; 300 × 300 mm | 255.22 | 2.40 | 57.00 | – | 263.39 | nr | **320.39** |
| light vehicular weight; 450 × 450 mm | 196.77 | 3.00 | 71.25 | – | 206.14 | nr | **277.39** |
| light vehicular weight; 450 × 600 mm | 390.92 | 4.00 | 95.00 | – | 399.09 | nr | **494.09** |
| heavy vehicular weight; 300 × 300 mm | 255.79 | 3.00 | 71.25 | – | 264.30 | nr | **335.55** |
| heavy vehicular weight; 450 × 600 mm | 405.26 | 3.00 | 71.25 | – | 414.63 | nr | **485.88** |
| heavy vehicular weight; 600 × 600 mm | 475.60 | 4.00 | 95.00 | – | 489.22 | nr | **584.22** |
| Extra over manhole frames and covers for | | | | | | | |
| filling recessed manhole covers with brick paviors; PC £450.00/1000 | 17.60 | 1.00 | 23.75 | – | 18.21 | m² | **41.96** |
| filling recessed manhole covers with vehicular paving blocks; PC £10.51/m² | 10.51 | 0.75 | 17.81 | – | 11.12 | m² | **28.93** |
| filling recessed manhole covers with concrete paving flags; PC £16.20/m² | 16.20 | 0.35 | 8.31 | – | 16.56 | m² | **24.87** |
| **CHANNELS** | | | | | | | |
| **Best quality vitrified clay half section channels; Hepworth Plc; bedding and jointing in cement:mortar (1:2)** | | | | | | | |
| Channels; straight | | | | | | | |
| 100 mm | 8.17 | 0.80 | 19.00 | – | 11.46 | m | **30.46** |
| 150 mm | 13.60 | 1.00 | 23.75 | – | 16.89 | m | **40.64** |
| 225 mm | 30.54 | 1.35 | 32.06 | – | 33.82 | m | **65.88** |
| 300 mm | 62.69 | 1.80 | 42.75 | – | 65.97 | m | **108.72** |
| Bends; 15, 30, 45 or 90° | | | | | | | |
| 100 mm bends | 7.36 | 0.75 | 17.81 | – | 9.00 | nr | **26.81** |
| 150 mm bends | 12.72 | 0.90 | 21.38 | – | 15.19 | nr | **36.57** |
| 225 mm bends | 49.28 | 1.20 | 28.50 | – | 52.57 | nr | **81.07** |
| 300 mm bends | 100.46 | 1.10 | 26.13 | – | 104.58 | nr | **130.71** |

## 34 DRAINAGE BELOW GROUND

| Item<br>Excluding site overheads and profit | PC<br>£ | Labour<br>hours | Labour<br>£ | Plant<br>£ | Material<br>£ | Unit | Total<br>rate £ |
|---|---|---|---|---|---|---|---|
| **Best quality vitrified clay channels; Hepworth Plc; bedding and jointing in cement:mortar (1:2)** | | | | | | | |
| Branch bends; 15, 30, 45 or 90°; left or right hand | | | | | | | |
| 100 mm | 7.36 | 0.75 | 17.81 | – | 9.00 | nr | **26.81** |
| 150 mm | 12.72 | 0.90 | 21.38 | – | 15.19 | nr | **36.57** |
| **CLARIFICATION ON LABOUR COSTS IN THIS SECTION** | | | | | | | |
| **General groundworks team** | | | | | | | |
| Generally a three man team is used in this section; The column Labour hours reports team hours. The column Labour £ reports the total cost of the team for the unit of work shown | | | | | | | |
| 3 man team | – | 1.00 | 71.25 | – | – | hr | **71.25** |
| **EXCAVATION FOR DRAINAGE** | | | | | | | |
| **Machine excavation** | | | | | | | |
| Excavating pits; starting from ground level; works exclude for earthwork retention | | | | | | | |
| maximum depth not exceeding 1.00 m | – | 0.50 | 11.88 | 17.78 | – | m³ | **29.66** |
| maximum depth not exceeding 2.00 m | – | 0.50 | 11.88 | 29.16 | – | m³ | **41.04** |
| maximum depth not exceeding 4.00 m | – | 0.50 | 11.88 | 75.65 | – | m³ | **87.53** |
| Disposal of excavated material; depositing on site in permanent spoil heaps; average 50 m | – | 0.04 | 0.99 | 2.36 | – | m³ | **3.35** |
| Filling to excavations; obtained from on site spoil heaps; average thickness not exceeding 0.25 m | – | 0.13 | 3.17 | 6.72 | – | m³ | **9.89** |
| **Hand excavation** | | | | | | | |
| Excavating pits; starting from ground level; works exclude for earthwork retention | | | | | | | |
| maximum depth not exceeding 1.00 m | – | 1.20 | 57.00 | – | – | m³ | **57.00** |
| maximum depth not exceeding 2.00 m | – | 1.50 | 71.25 | – | – | m³ | **71.25** |
| **Works to pits or drainage excavations** | | | | | | | |
| Surface treatments; compacting; bottoms of excavations | – | 0.05 | 1.19 | – | – | m² | **1.19** |

## 34 DRAINAGE BELOW GROUND

| Item<br>Excluding site overheads and profit | PC<br>£ | Labour<br>hours | Labour<br>£ | Plant<br>£ | Material<br>£ | Unit | Total<br>rate £ |
|---|---|---|---|---|---|---|---|
| **EXCAVATION FOR DRAINAGE – cont** | | | | | | | |
| **Earthwork support** | | | | | | | |
| Manhole boxes to excavations (not included); Timber plywood with timber braces to shore excavations to 1.00 m deep; opposing faces distance less than 1.00 m | | | | | | | |
| 1.00 m deep | – | 0.50 | 23.75 | – | 9.35 | $m^2$ | 33.10 |
| 2.00 m deep | – | 1.00 | 47.50 | – | 9.35 | $m^2$ | 56.85 |
| **Trench boxes; A-Plant** | | | | | | | |
| Manhole box soil retention | – | 0.50 | 35.63 | 166.30 | – | week | 201.93 |
| Placement of manhole protection system to 3.0 m deep to enable earthwork retention; maximum distance between opposing faces 1.8 m | – | 0.50 | 35.63 | 225.30 | – | week | 260.93 |
| **GULLIES AND INTERCEPTION TRAPS** | | | | | | | |
| **Intercepting traps; Hepworth Plc** | | | | | | | |
| Vitrified clay; inspection arms; brass stoppers; iron levers; chains and staples; galvanized; staples cut and pinned to brickwork; cement:mortar (1:2) joints to vitrified clay pipes and channels; bedding and surrounding in concrete; 11.50 N/mm²–40 mm aggregate; cutting and fitting brickwork; making good facings | | | | | | | |
| 100 mm inlet; 100 mm outlet | 106.89 | 3.00 | 71.25 | – | 115.49 | nr | 186.74 |
| 150 mm inlet; 150 mm outlet | 154.12 | 2.00 | 47.50 | – | 167.66 | nr | 215.16 |
| **Gullies; concrete; FP McCann** | | | | | | | |
| **Concrete road gullies; trapped; cement:mortar (1:2) joints to concrete pipes; bedding and surrounding in concrete; 11.50 N/mm²–40 mm aggregate; rodding eye; stoppers** | | | | | | | |
| 375 mm diameter × 750 mm deep | 48.02 | 3.00 | 142.50 | – | 71.28 | nr | 213.78 |
| 375 mm diameter × 900 mm deep | 49.99 | 3.00 | 142.50 | – | 73.25 | nr | 215.75 |
| 450 mm diameter × 750 mm deep | 50.01 | 3.00 | 142.50 | – | 73.27 | nr | 215.77 |
| 450 mm diameter × 900 mm deep | 54.18 | 3.00 | 142.50 | – | 77.44 | nr | 219.94 |
| 450 mm diameter × 1.05 m deep | 58.40 | 3.50 | 166.25 | – | 81.66 | nr | 247.91 |
| 450 mm diameter × 1.20 m deep | 75.10 | 4.00 | 190.00 | – | 98.36 | nr | 288.36 |
| Accessories for clay gullies | | | | | | | |
| coverslab | 21.37 | 0.25 | 11.88 | – | 21.37 | nr | 33.25 |
| seating rings | 22.58 | – | – | – | 22.58 | nr | 22.58 |

## 34 DRAINAGE BELOW GROUND

| Item<br>Excluding site overheads and profit | PC<br>£ | Labour<br>hours | Labour<br>£ | Plant<br>£ | Material<br>£ | Unit | Total<br>rate £ |
|---|---|---|---|---|---|---|---|
| **Gullies; vitrified clay; Hepworth Plc; bedding in concrete; 11.50 N/mm²– 40 mm aggregate** | | | | | | | |
| Yard gullies (mud); trapped; domestic duty (up to 1 tonne) | | | | | | | |
|   100 mm outlet; 100 mm diameter; 225 mm internal width; 585 mm internal depth | 139.00 | 3.50 | 83.13 | – | 139.71 | nr | **222.84** |
|   150 mm outlet; 100 mm diameter; 225 mm internal width; 585 mm internal depth | 139.00 | 3.50 | 83.13 | – | 139.71 | nr | **222.84** |
| Yard gullies (mud); trapped; medium duty (up to 5 tonnes) | | | | | | | |
|   100 mm outlet; 100 mm diameter; 225 mm internal width; 585 mm internal depth | 196.51 | 3.50 | 83.13 | – | 197.21 | nr | **280.34** |
|   150 mm outlet; 100 mm diameter; 225 mm internal width; 585 mm internal depth | 215.31 | 3.50 | 83.13 | – | 216.01 | nr | **299.14** |
| Combined filter and silt bucket for yard gullies | | | | | | | |
|   225 mm diameter | 50.26 | – | – | – | 50.26 | nr | **50.26** |
| Road gullies; trapped with rodding eye | | | | | | | |
|   100 mm outlet; 300 mm internal diameter; 600 mm internal depth | – | 3.50 | 83.13 | – | 134.39 | nr | **217.52** |
|   150 mm outlet; 300 mm internal diameter; 600 mm internal depth | 136.91 | 3.50 | 83.13 | – | 137.61 | nr | **220.74** |
|   150 mm outlet; 400 mm internal diameter; 750 mm internal depth | 158.78 | 3.50 | 83.13 | – | 159.49 | nr | **242.62** |
|   150 mm outlet; 450 mm internal diameter; 900 mm internal depth | 214.82 | 3.50 | 83.13 | – | 215.53 | nr | **298.66** |
| Hinged gratings and frames for gullies; alloy | | | | | | | |
|   193 mm for 150 mm diameter gully | – | – | – | – | 33.32 | nr | **33.32** |
|   120 × 120 mm | – | – | – | – | 11.85 | nr | **11.85** |
|   150 × 150 mm | – | – | – | – | 21.43 | nr | **21.43** |
|   230 × 230 mm | – | – | – | – | 39.18 | nr | **39.18** |
|   316 × 316 mm | – | – | – | – | 103.87 | nr | **103.87** |
| Hinged gratings and frames for gullies; cast iron | | | | | | | |
|   265 mm for 225 mm diameter gully | – | – | – | – | 66.56 | nr | **66.56** |
|   150 × 150 mm | – | – | – | – | 21.43 | nr | **21.43** |
|   230 × 230 mm | – | – | – | – | 39.18 | nr | **39.18** |
|   316 × 316 mm | – | – | – | – | 103.87 | nr | **103.87** |

## 34 DRAINAGE BELOW GROUND

| Item<br>Excluding site overheads and profit | PC<br>£ | Labour<br>hours | Labour<br>£ | Plant<br>£ | Material<br>£ | Unit | Total<br>rate £ |
|---|---|---|---|---|---|---|---|
| **GULLIES AND INTERCEPTION TRAPS – cont** | | | | | | | |
| **Universal gully trap PVC-u; Wavin Plastics Ltd; OsmaDrain system; bedding in concrete; 11.50 N/mm²–40 mm aggregate** | | | | | | | |
| Universal gully fitting; comprising gully trap only | | | | | | | |
| 110 mm outlet; 110 mm diameter; 205 mm internal depth | 14.87 | 3.50 | 83.13 | – | 15.57 | nr | **98.70** |
| Vertical inlet hopper; c\w plastic grate | | | | | | | |
| 272 × 183 mm | 19.98 | 0.25 | 5.94 | – | 19.98 | nr | **25.92** |
| Sealed access hopper | | | | | | | |
| 110 × 110 mm | 46.99 | 0.25 | 5.94 | – | 46.99 | nr | **52.93** |
| **Universal gully PVC-u; Wavin Plastics Ltd; OsmaDrain system; accessories to universal gully trap** | | | | | | | |
| Hoppers; backfilling with clean granular material; tamping; surrounding in lean mix concrete | | | | | | | |
| plain hopper; with 110 mm spigot; 150 mm long | 16.34 | 0.40 | 9.50 | – | 16.76 | nr | **26.26** |
| vertical inlet hopper; with 110 mm spigot; 150 mm long | 19.98 | 0.40 | 9.50 | – | 19.98 | nr | **29.48** |
| sealed access hopper; with 110 mm spigot; 150 mm long | 46.99 | 0.40 | 9.50 | – | 46.99 | nr | **56.49** |
| plain hopper; solvent weld to trap | 11.16 | 0.40 | 9.50 | – | 11.16 | nr | **20.66** |
| vertical inlet hopper; solvent wed to trap | 19.17 | 0.40 | 9.50 | – | 19.17 | nr | **28.67** |
| sealed access cover; PVC-u | 23.93 | 0.10 | 2.38 | – | 23.93 | nr | **26.31** |
| **Gullies PVC-u; Wavin Plastics Ltd; OsmaDrain system; bedding in concrete; 11.50 N/mm²–40 mm aggregate** | | | | | | | |
| Bottle gully; providing access to the drainage system for cleaning | | | | | | | |
| bottle gully; 228 × 228 × 317 mm deep | 99.18 | 0.50 | 11.88 | – | 99.78 | nr | **111.66** |
| sealed access cover; PVC-u; 217 × 217 mm | 30.84 | 0.10 | 2.38 | – | 30.84 | nr | **33.22** |
| grating; ductile iron; 215 × 215 mm | 23.78 | 0.10 | 2.38 | – | 23.78 | nr | **26.16** |
| bottle gully riser; 325 mm | 5.65 | 0.50 | 11.88 | – | 6.95 | nr | **18.83** |
| Yard gully; trapped; 300 mm diameter × 600 mm deep; including catchment bucket and ductile iron cover and frame; medium duty loading | | | | | | | |
| 305 mm diameter × 600 mm deep | 246.18 | 2.50 | 59.38 | – | 250.05 | nr | **309.43** |

## 34 DRAINAGE BELOW GROUND

| Item<br>Excluding site overheads and profit | PC<br>£ | Labour<br>hours | Labour<br>£ | Plant<br>£ | Material<br>£ | Unit | Total<br>rate £ |
|---|---|---|---|---|---|---|---|
| **Kerbs to gullies** | | | | | | | |
| One course Class B engineering bricks to four sides; rendering in cement: mortar (1:3); dished to gully gratings | | | | | | | |
| 150 × 150 mm | 2.25 | 0.33 | 7.92 | – | 2.68 | nr | **10.60** |
| **INSPECTION CHAMBERS AND MANHOLES** | | | | | | | |
| **Inspection chambers; in situ concrete; excavations; earthwork reinforcement all not included** | | | | | | | |
| Beds; plain in situ concrete; 11.50 N/mm²–40 mm aggregate | | | | | | | |
| thickness not exceeding 150 mm | – | 2.00 | 47.50 | – | 115.88 | m³ | **163.38** |
| thickness 150–450 mm | – | 1.75 | 41.56 | – | 115.88 | m³ | **157.44** |
| Benchings in bottoms; plain in situ concrete; 25.50 N/mm²–20 mm aggregate | | | | | | | |
| thickness 150–450 mm | – | 2.00 | 47.50 | – | 115.88 | m³ | **163.38** |
| Isolated cover slabs; reinforced In situ concrete; 21.00 N/mm²–20 mm aggregate | | | | | | | |
| thickness not exceeding 150 mm | 115.88 | 4.00 | 95.00 | – | 115.88 | m³ | **210.88** |
| Fabric reinforcement; A193 (3.02 kg/m²) in cover slabs | 1.87 | 0.06 | 1.49 | – | 1.87 | m² | **3.36** |
| Formwork to reinforced in situ concrete; isolated cover slabs | | | | | | | |
| soffits; horizontal | – | 3.28 | 77.90 | – | 4.25 | m² | **82.15** |
| height not exceeding 250 mm | – | 0.97 | 23.04 | – | 1.71 | m | **24.75** |
| **Inspection chambers (House – HIC); precast concrete units; to excavations, earthwork reinforcement, in situ bases benchings all priced separately; includes cover slab** | | | | | | | |
| Precast concrete inspection chamber units; FP McCann Ltd; bedding, jointing and pointing in cement:mortar (1:3); 600 × 450 mm internally | | | | | | | |
| 600 mm deep | 50.45 | 6.50 | 154.38 | – | 57.17 | nr | **211.55** |
| 900 mm deep | 67.20 | 7.00 | 166.25 | – | 72.31 | nr | **238.56** |
| Drainage chambers; FP McCann Ltd; 1200 × 750 mm; no base unit; depth of invert | | | | | | | |
| 1.00 m deep | 185.32 | 3.00 | 213.75 | – | 193.83 | nr | **407.58** |
| 1.50 m deep | 237.04 | 3.67 | 261.25 | – | 250.66 | nr | **511.91** |
| 2.00 m deep | 288.76 | 4.17 | 296.88 | – | 305.79 | nr | **602.67** |

## 34 DRAINAGE BELOW GROUND

| Item<br>Excluding site overheads and profit | PC<br>£ | Labour<br>hours | Labour<br>£ | Plant<br>£ | Material<br>£ | Unit | Total<br>rate £ |
|---|---|---|---|---|---|---|---|
| **INSPECTION CHAMBERS AND MANHOLES – cont** | | | | | | | |
| | | | | | | | |
| **Bases to chamber ring manholes; in situ concrete; to excavated ground (not included); on 50 mm ST2 blinding on 225 mm mass concrete to receive bottom manhole ring blinded with 50 mm ST2 blinding** | | | | | | | |
| Diameter of manhole | | | | | | | |
| 900 mm | – | 4.00 | 190.00 | – | 33.96 | nr | **223.96** |
| 1.05 m | – | 4.00 | 190.00 | – | 41.09 | nr | **231.09** |
| 1.20 m | – | 4.50 | 213.75 | – | 57.39 | nr | **271.14** |
| 1.50 m | – | 5.50 | 261.25 | – | 86.93 | nr | **348.18** |
| Bottom manhole chamber ring; FP McCann Ltd; including pipe entrances to bases (not included); forming pipe entrances to manhole ring; seating and sealing to base; connecting pipes (not included); depth 1.00 m; diameter | | | | | | | |
| 900 mm | 78.45 | 3.00 | 213.75 | 93.19 | 86.96 | nr | **393.90** |
| 1.05 m | 82.71 | 3.25 | 231.56 | 93.19 | 91.22 | nr | **415.97** |
| 1.20 m | 100.98 | 4.00 | 285.00 | 116.48 | 109.49 | nr | **510.97** |
| 1.50 m | 176.30 | 4.50 | 320.63 | 139.78 | 184.81 | nr | **645.22** |
| Benchings to manholes; 400 mm concrete; diameter of manhole | | | | | | | |
| 900 mm | – | 0.75 | 35.63 | – | 46.74 | nr | **82.37** |
| 1.05 m | – | 0.75 | 35.63 | – | 65.43 | nr | **101.06** |
| 1.20 m | – | 0.90 | 42.75 | – | 84.13 | nr | **126.88** |
| 1.50 m | – | 1.10 | 52.25 | – | 132.76 | nr | **185.01** |
| Easi-Base Polypropylene liner; FP McCann Ltd; complete with prefabricated benching and channels as one unit; Easi-Base diameter | | | | | | | |
| 1200 mm | 427.45 | 1.00 | 47.50 | 46.59 | 427.45 | nr | **521.54** |
| 1500 mm | 854.90 | 1.00 | 47.50 | 46.59 | 854.90 | nr | **948.99** |

## 34 DRAINAGE BELOW GROUND

| Item<br>Excluding site overheads and profit | PC<br>£ | Labour<br>hours | Labour<br>£ | Plant<br>£ | Material<br>£ | Unit | Total<br>rate £ |
|---|---|---|---|---|---|---|---|
| **Concrete manhole chamber rings; FP McCann Ltd; standard ring units to excavated ground (not included); sealed with bitumen tape; tongue and groove units installed to 1.00 m deep base units (not included); surrounding in concrete min 150 thick and backfilling; Note: The final depths shown below show operations onto the 1.00 m deep base rings above** | | | | | | | |
| Internal diameter 900 mm; final manhole depth | | | | | | | |
| 1.00 m deep | – | 1.00 | 71.25 | 46.59 | 86.31 | nr | 204.15 |
| Internal diameter 1.05 mm; final manhole depth | | | | | | | |
| 1.00 m deep | – | 1.00 | 71.25 | 93.19 | 86.31 | nr | 250.75 |
| 2.00 m deep | 82.71 | 2.50 | 178.13 | 163.08 | 246.81 | nr | 588.02 |
| 3.00 m deep | 165.42 | 4.00 | 285.00 | 186.38 | 406.07 | nr | 877.45 |
| Internal diameter 1200 mm; final manhole depth | | | | | | | |
| 1.00 m deep | – | 1.00 | 71.25 | 139.78 | 86.31 | nr | 297.34 |
| 2.00 m deep | 100.98 | 3.00 | 213.75 | 186.38 | 265.08 | nr | 665.21 |
| 3.00 m deep | 201.96 | 4.00 | 285.00 | 279.56 | 442.61 | nr | 1007.17 |
| Internal diameter 1500 mm | | | | | | | |
| 1.00 m deep | – | 2.00 | 142.50 | 139.78 | 86.31 | nr | 368.59 |
| 2.00 m deep | 176.30 | 4.00 | 285.00 | 186.38 | 340.40 | nr | 811.78 |
| 3.00 m deep | 352.60 | 5.00 | 356.25 | 279.56 | 593.25 | nr | 1229.06 |
| **Concrete manhole chamber rings; Wide wall; FP McCann Ltd; 130 mm thick precast manholes; (concrete surround on backfill not required) to excavated ground (not included); Water-tight, complete with sealant; to prepared base (or Easi-Base priced elsewhere)** | | | | | | | |
| Internal diameter 1200 mm | | | | | | | |
| 1.00 m deep | 169.95 | 3.00 | 213.75 | 139.78 | 169.95 | nr | 523.48 |
| 2.00 m deep | 339.90 | 4.00 | 285.00 | 186.38 | 339.90 | nr | 811.28 |
| 3.00 m deep | 509.85 | 6.00 | 427.50 | 232.97 | 509.85 | nr | 1170.32 |
| Internal diameter 1500 mm | | | | | | | |
| 1.00 m deep | 267.80 | 3.00 | 213.75 | 186.38 | 276.31 | nr | 676.44 |
| 2.00 m deep | 535.60 | – | – | – | – | nr | – |
| 3.00 m deep | 803.40 | 7.00 | 498.75 | 326.16 | 811.91 | nr | 1636.82 |
| **Cover slabs for chambers or shaft sections; FP McCann Ltd; heavy duty** | | | | | | | |
| 900 mm diameter internally | 79.70 | 0.67 | 15.82 | – | 79.70 | nr | 95.52 |
| 1050 mm diameter internally | 85.53 | 2.00 | 47.50 | 17.08 | 85.53 | nr | 150.11 |
| 1200 mm diameter internally | 103.74 | 1.00 | 23.75 | 17.08 | 103.74 | nr | 144.57 |

## 34 DRAINAGE BELOW GROUND

| Item<br>Excluding site overheads and profit | PC<br>£ | Labour<br>hours | Labour<br>£ | Plant<br>£ | Material<br>£ | Unit | Total<br>rate £ |
|---|---|---|---|---|---|---|---|
| **INSPECTION CHAMBERS AND MANHOLES – cont** | | | | | | | |
| **Cover slabs for chambers or shaft sections – cont** | | | | | | | |
| 1500 mm diameter internally | 188.81 | 1.00 | 23.75 | 17.08 | 188.81 | nr | **229.64** |
| 1800 mm diameter internally | 310.01 | 2.00 | 47.50 | 38.44 | 310.01 | nr | **395.95** |
| **Brickwork** | | | | | | | |
| Walls to manholes; bricks; PC £400.00/1000; in cement:mortar (1:3) | | | | | | | |
|    one brick thick | 50.40 | 3.00 | 71.25 | – | 60.62 | m$^2$ | **131.87** |
|    one and a half brick thick | 75.60 | 4.00 | 95.00 | – | 90.92 | m$^2$ | **185.92** |
|    two brick thick projection of footing or the like | 100.80 | 4.80 | 114.00 | – | 121.23 | m$^2$ | **235.23** |
| Walls to manholes; engineering bricks; in cement:mortar (1:3) | | | | | | | |
|    one brick thick | 56.70 | 3.00 | 71.25 | – | 66.92 | m$^2$ | **138.17** |
|    one and a half brick thick | 85.05 | 4.00 | 95.00 | – | 100.37 | m$^2$ | **195.37** |
|    two brick thick projection of footing or the like | 113.40 | 4.80 | 114.00 | – | 133.83 | m$^2$ | **247.83** |
| Extra over common or engineering bricks in any mortar for fair face; flush pointing as work proceeds; English bond walls or the like | – | 0.13 | 3.17 | – | – | m$^2$ | **3.17** |
| In situ finishings; cement:sand mortar (1:3); steel trowelled; 13 mm one coat work to manhole walls; to brickwork or blockwork base; over 300 mm wide | – | 0.80 | 19.00 | – | 3.41 | m$^2$ | **22.41** |
| Building into brickwork; ends of pipes; making good facings or renderings | | | | | | | |
|    small | – | 0.20 | 4.75 | – | – | nr | **4.75** |
|    large | – | 0.30 | 7.13 | – | – | nr | **7.13** |
|    extra large | – | 0.40 | 9.50 | – | – | nr | **9.50** |
|    extra large; including forming ring arch cover | – | 0.50 | 11.88 | – | – | nr | **11.88** |
| **Inspection chambers; polypropylene; Hepworth Plc** | | | | | | | |
| Mini access chamber; up to 600 mm deep; including cover and frame | | | | | | | |
|    300 mm diameter × 600 mm deep; three 100/110 mm inlets | 177.66 | 3.00 | 71.25 | – | 180.66 | nr | **251.91** |
| Up to 1200 mm deep; including polymer cover and frame with screw down lid | | | | | | | |
|    475 mm diameter × 940 mm deep; five 100/110 mm inlets; supplied with four stoppers in inlets | 262.65 | 4.00 | 95.00 | – | 350.57 | nr | **445.57** |
| Extra for square ductile iron cover and frame to 1 tonne load; screw down lid | 129.45 | – | – | – | 44.53 | nr | **44.53** |

## 34 DRAINAGE BELOW GROUND

| Item<br>Excluding site overheads and profit | PC<br>£ | Labour<br>hours | Labour<br>£ | Plant<br>£ | Material<br>£ | Unit | Total<br>rate £ |
|---|---|---|---|---|---|---|---|
| **Step irons; drainage systems; malleable cast iron; galvanized; building into joints** | | | | | | | |
| General purpose pattern; for one brick walls | 4.03 | 0.17 | 4.04 | – | 4.03 | nr | **8.07** |
| Accessories in PVC-u | | | | | | | |
| 110 mm screwed access cover | 2.52 | – | – | – | 2.52 | nr | **2.52** |
| 110 mm rodding eye | 17.99 | 0.50 | 11.88 | – | 23.00 | nr | **34.88** |
| Kerbs; to gullies; in one course Class B engineering bricks; to four sides; rendering in cement:mortar (1:3); dished to gully gratings | 3.78 | 1.00 | 23.75 | – | 4.80 | nr | **28.55** |
| **CAST IRON DRAINAGE** | | | | | | | |
| **Cast iron pipes and fittings; PAM; Timesaver** | | | | | | | |
| 100 mm cast iron pipes double spiggot; in trenches (trenches not included) | | | | | | | |
| laid straight | 52.56 | 0.11 | 5.28 | – | 52.56 | m | **57.84** |
| short runs under 3.00 m | 71.06 | 0.17 | 7.92 | – | 71.06 | m | **78.98** |
| Extra over 100 mm cast iron pipes for | | | | | | | |
| bends; medium radius – 87.5°–60° | 107.41 | 0.33 | 7.84 | – | 107.41 | nr | **115.25** |
| bends; medium radius – 45°–10° | 99.61 | 0.33 | 7.84 | – | 99.61 | nr | **107.45** |
| bends; long radius | 141.29 | 0.25 | 5.94 | – | 141.29 | nr | **147.23** |
| plain branch | 124.35 | 0.25 | 5.94 | – | 124.35 | nr | **130.29** |
| 150 mm cast iron pipes; in trenches (trenches not included) | | | | | | | |
| laid straight | 91.38 | 0.13 | 5.94 | – | 91.38 | m | **97.32** |
| short runs under 3.00 m | 113.80 | 0.25 | 11.88 | – | 113.80 | m | **125.68** |
| Extra over 150 mm cast iron pipes for | | | | | | | |
| bends; medium radius – 87.5°–60° | 136.26 | 0.33 | 7.84 | – | 136.26 | nr | **144.10** |
| bends; medium radius – 45°–10° | 162.83 | 0.33 | 7.84 | – | 162.83 | nr | **170.67** |
| 225 mm cast iron pipes; in trenches (trenches not included) | | | | | | | |
| laid straight | 317.05 | 0.20 | 9.50 | – | 317.05 | m | **326.55** |
| short runs under 3.00 m | 317.05 | 0.50 | 23.75 | – | 317.05 | m | **340.80** |
| Extra over 225 mm cast iron pipes for | | | | | | | |
| bends; medium radius – 87.5°–10° | – | 0.50 | 11.88 | – | 400.95 | nr | **412.83** |
| **Cast iron inspection chambers; PAM; Timesaver; drainage systems bolted flat covers; bedding in cement: mortar (1:3); mechanical coupling joints** | | | | | | | |
| 100 × 100 mm | | | | | | | |
| one branch each side | 242.74 | 1.50 | 71.25 | – | 299.94 | nr | **371.19** |
| two branches each side | 477.86 | 2.50 | 118.75 | – | 646.03 | nr | **764.78** |

## 34 DRAINAGE BELOW GROUND

| Item<br>Excluding site overheads and profit | PC<br>£ | Labour<br>hours | Labour<br>£ | Plant<br>£ | Material<br>£ | Unit | Total<br>rate £ |
|---|---|---|---|---|---|---|---|
| **CAST IRON DRAINAGE – cont** | | | | | | | |
| **Cast iron inspection chambers – cont** | | | | | | | |
| 100 × 150 mm | | | | | | | |
|   one branch each side | 437.24 | 2.00 | 95.00 | – | 437.58 | nr | **532.58** |
|   two branch each side | 828.69 | 2.50 | 118.75 | – | 829.04 | nr | **947.79** |
| 150 × 150 mm | | | | | | | |
|   one branch each side | 399.10 | 2.50 | 118.75 | – | 533.91 | nr | **652.66** |
|   two branches each side | 796.92 | 3.00 | 142.50 | – | 998.97 | nr | **1141.47** |
| **Accessories in Cast iron** | | | | | | | |
| Gullies; including concrete bed and surround | | | | | | | |
|   Gully trap with surface access 100 mm × 87.5° | 253.70 | 1.00 | 47.50 | – | 284.57 | nr | **332.07** |
|   Gully trap with 225 mm inlet | 222.77 | 1.00 | 47.50 | – | 253.64 | nr | **301.14** |
|   Trapless gully | 249.15 | 1.00 | 47.50 | – | 280.02 | nr | **327.52** |
|   Gully trap square 225 × 225 mm with sediment trap and grate | 502.54 | 1.00 | 47.50 | – | 533.41 | nr | **580.91** |
|   Deans trap gully 600 mm deep inclusive of sediment pan | 587.58 | 1.00 | 47.50 | – | 618.45 | nr | **665.95** |
|   Road Gully including hinged grate and frame | 677.14 | 3.00 | 142.50 | – | 708.01 | nr | **850.51** |
| **LINEAR DRAINAGE** | | | | | | | |
| **Marshalls Plc; Mini Beany combined kerb and channel drainage system; to trenches (not included)** | | | | | | | |
| Precast concrete drainage channel base; 185–385 mm deep; bedding, jointing and pointing in cement:mortar (1:3); on 150 mm deep concrete (ready mixed) foundation; including haunching with in situ concrete; 11.50 N/mm$^2$ – 40 mm aggregate one side; channels 250 mm wide × 1.00 m long | | | | | | | |
|   straight; 1.00 m long | 33.97 | 1.00 | 23.75 | – | 38.10 | m | **61.85** |
|   straight; 500 mm long | 38.08 | 1.05 | 24.94 | – | 42.21 | m | **67.15** |
|   radial; 30/10 or 9/6 internal or external | 35.19 | 1.33 | 31.67 | – | 39.32 | m | **70.99** |
|   angles 45 or 90° | 101.59 | 1.00 | 23.75 | – | 105.72 | nr | **129.47** |
| Mini Beany Top Block; perforated kerb unit to drainage channel above; natural grey | | | | | | | |
|   straight | 40.37 | 0.33 | 7.92 | – | 41.87 | m | **49.79** |
|   radial; 30/10 or 9/6 internal or external | 25.07 | 0.50 | 11.88 | – | 26.57 | m | **38.45** |
|   angles 45 or 90° | 101.14 | 0.50 | 11.88 | – | 102.64 | nr | **114.52** |

## 34 DRAINAGE BELOW GROUND

| Item<br>Excluding site overheads and profit | PC<br>£ | Labour<br>hours | Labour<br>£ | Plant<br>£ | Material<br>£ | Unit | Total<br>rate £ |
|---|---|---|---|---|---|---|---|
| Mini Beany; outfalls; two section concrete trapped outfall with Mini Beany cast iron access cover and frame; to concrete foundation | | | | | | | |
|    high capacity outfalls; silt box 150/225 mm outlet; two section trapped outfall silt box and cast iron access cover | 330.70 | 1.00 | 23.75 | – | 702.81 | nr | 726.56 |
|    inline side or end outlet; outfall 150 mm; 2 section concrete trapped outfall; cast iron Mini Beany access cover and frame | 548.20 | 1.00 | 23.75 | – | 549.05 | nr | 572.80 |
| Ancillaries to Mini Beany | | | | | | | |
|    end cap | 20.06 | 0.25 | 5.94 | – | 20.06 | nr | 26.00 |
|    end cap outlets | 52.84 | 0.25 | 5.94 | – | 52.84 | nr | 58.78 |
| **Precast concrete channels; Charcon Hard Landscaping; on 150 mm deep concrete foundations; including haunching with in situ concrete; 21.00 N/mm² – 20 mm aggregate; both sides** | | | | | | | |
| Charcon Highway Precast drainage channel | | | | | | | |
|    Safetycurb; 250 × 250 mm | 44.92 | 0.67 | 15.83 | – | 64.62 | m | 80.45 |
| Charcon Safeticurb; slotted safety channels; for heavy vehicles | | | | | | | |
|    DBM; 248 × 248 mm | 80.12 | 0.80 | 19.00 | – | 99.82 | m | 118.82 |
|    Clearway; 324 × 257 mm | 91.43 | 0.80 | 19.00 | – | 111.13 | m | 130.13 |
| Silt box tops; concrete frame; cast iron grid lids; type 1; set over gully | | | | | | | |
|    457 × 610 mm | 409.46 | 2.00 | 47.50 | – | 410.62 | nr | 458.12 |
| **Slot and Channel drains** | | | | | | | |
| Loadings for slot drains | | | | | | | |
| A15; 1.5 tonne; pedestrian | | | | | | | |
| B125; 12.5 tonne; domestic use | | | | | | | |
| C250; 25 tonne; car parks, supermarkets, industrial units | | | | | | | |
| D400; 40 tonne; highways | | | | | | | |
| E600; 60 tonne; forklifts | | | | | | | |
| **Fully galvanized hexagonal format slot drains; Gatic Ltd; for paving or asphalt use; 269 mm depth** | | | | | | | |
| 'Facade drain' fixed to building facade to collect water falling from the building facade | | | | | | | |
|    115 mm maximum flow rate 19.6 litres/second | 37.00 | 0.33 | 15.83 | – | 37.00 | m | 52.83 |

## 34 DRAINAGE BELOW GROUND

| Item<br>Excluding site overheads and profit | PC<br>£ | Labour<br>hours | Labour<br>£ | Plant<br>£ | Material<br>£ | Unit | Total<br>rate £ |
|---|---|---|---|---|---|---|---|
| **LINEAR DRAINAGE – cont** | | | | | | | |
| **Fully galvanized hexagonal format slot drains – cont** | | | | | | | |
| Slotdrain channels; Hexagonally profiled fully galvanized slotdrain system; In trenched (not included) surrounded by concrete | | | | | | | |
|   Pavslot 10 mm throat 100 mm deep | 45.00 | 0.33 | 15.83 | – | 47.75 | m | **63.58** |
|   Pavslot 10 mm throat 150 mm deep | 50.00 | 0.33 | 15.83 | – | 52.88 | m | **68.71** |
|   Gatic Access points at 1nr per 30 m per 30 m | 5.33 | 0.25 | 11.88 | – | 8.07 | m | **19.95** |
|   Gatic outlets | 170.00 | 1.00 | 47.50 | – | 172.75 | nr | **220.25** |
| **Slot drains; ACO Technologies;** laid to concrete bed C25 on compacted granular base on 200 mm deep concrete bed; haunched with 200 mm concrete surround; all in 750 × 430 mm wide trench with compacted 200 mm granular base surround (excavation and subbase not included) | | | | | | | |
| ACO MultiDrain M100PPD; recycled polypropylene drainage channel; range of gratings to complement installations which require discreet slot drainage | | | | | | | |
|   142 mm wide × 150 mm deep | 74.70 | 0.83 | 19.79 | – | 87.31 | m | **107.10** |
| Accessories for M100PPD | | | | | | | |
|   connectors; vertical outlet 110 mm | 17.25 | – | – | – | 17.25 | nr | **17.25** |
|   connectors; vertical outlet 160 mm | 17.25 | – | – | – | 17.25 | nr | **17.25** |
|   sump units; 110 mm | 132.85 | 1.50 | 35.63 | – | 156.03 | nr | **191.66** |
|   universal gully and bucket; 440 × 440 × 1315 mm deep | 793.05 | 3.00 | 71.25 | – | 836.50 | nr | **907.75** |
| **Channel drains; ACO Technologies;** ACO MultiDrain MD polymer concrete channel drainage system; traditional channel and grate drainage solution | | | | | | | |
| ACO MultiDrain MD Brickslot; offset galvanized slot drain grating for M100PPD; load class C250 | | | | | | | |
|   brickslot; galvanized steel; 1.00 m | 85.00 | – | – | – | 85.00 | m | **85.00** |
| ACO MultiDrain MD Brickslot; offset slot drain grating; load class C250 – 400 | | | | | | | |
|   brickslot; galvanized steel; 1000 mm | 158.80 | 1.20 | 28.50 | – | 158.80 | m | **187.30** |
|   brickslot; stainless steel; 1000 mm | 173.80 | 1.20 | 28.50 | – | 347.60 | m | **376.10** |
| Accessories for MultiDrain MD | | | | | | | |
|   sump unit; complete with sediment bucket and access unit | 132.85 | 1.00 | 23.75 | – | 270.20 | nr | **293.95** |
|   end cap; closing piece | 7.30 | – | – | – | 7.30 | nr | **7.30** |
|   end cap; inlet/outlet | 7.30 | – | – | – | 7.30 | nr | **7.30** |

## 34 DRAINAGE BELOW GROUND

| Item<br>Excluding site overheads and profit | PC<br>£ | Labour<br>hours | Labour<br>£ | Plant<br>£ | Material<br>£ | Unit | Total<br>rate £ |
|---|---|---|---|---|---|---|---|
| **ACO Drainlock Gratings for M100PPD and M100D system** | | | | | | | |
| A15 loading | | | | | | | |
|   slotted galvanized steel | 23.05 | 0.15 | 3.56 | – | 23.05 | m | **26.61** |
|   perforated galvanized steel | 42.10 | 0.15 | 3.56 | – | 42.10 | m | **45.66** |
| C250 loading | | | | | | | |
|   Heelguard composite black; 500 mm long with security locking | 55.10 | – | – | – | 55.10 | m | **55.10** |
|   Intercept; ductile iron; 500 mm long | 126.50 | – | – | – | 126.50 | m | **126.50** |
|   slotted galvanized steel; 1.00 m long | 59.60 | – | – | – | 59.60 | m | **59.60** |
|   perforated galvanized steel; 1.00 m long | 65.10 | – | – | – | 65.10 | m | **65.10** |
|   mesh galvanized steel; 1.00 m long | 44.60 | – | – | – | 44.60 | m | **44.60** |
| **Wade Ltd; stainless steel channel drains; specialized applications; bespoke manufacture** | | | | | | | |
| Drain in stainless steel; c/w tie in lugs and inbuilt falls to 100 mm spigot outlet; stainless steel gratings | | | | | | | |
|   NE channel; Ref 12430; secured gratings | 420.00 | 1.20 | 28.50 | – | 432.61 | m | **461.11** |
| **PIPE LAYING** | | | | | | | |
| **Excavating trenches; to receive pipes; grading bottoms to falls; backfilling with excavated material and compacting; disposal of surplus material off site; volumes allow for bedding materials which are priced separately below** | | | | | | | |
| Trenches 300 mm wide | | | | | | | |
|   depth of pipe 750 mm | – | – | – | 9.89 | 5.25 | m³ | **15.14** |
|   depth of pipe 900 mm | – | – | – | 11.30 | 6.64 | m³ | **17.94** |
|   depth of pipe 1.20 m | – | – | – | 11.86 | 9.41 | m³ | **21.27** |
|   depth of pipe 1.50 m | – | – | – | 13.18 | 12.19 | m³ | **25.37** |
|   depth of pipe 2.00 m | – | – | – | 8.64 | 16.81 | m³ | **25.45** |
| Trenches 300 mm wide | | | | | | | |
|   depth of pipe 750 mm | – | – | – | 3.30 | 3.47 | m | **6.77** |
|   depth of pipe 900 mm | – | – | – | 4.00 | 4.16 | m | **8.16** |
|   depth of pipe 1.20 m | – | – | – | 5.34 | 5.55 | m | **10.89** |
|   depth of pipe 1.50 m | – | – | – | 6.68 | 6.93 | m | **13.61** |
|   depth of pipe 2.00 m | – | – | – | 8.64 | 16.81 | m | **25.45** |

## 34 DRAINAGE BELOW GROUND

| Item<br>Excluding site overheads and profit | PC<br>£ | Labour<br>hours | Labour<br>£ | Plant<br>£ | Material<br>£ | Unit | Total<br>rate £ |
|---|---|---|---|---|---|---|---|
| **PIPE LAYING – cont** | | | | | | | |
| **Earthwork support; providing support to opposing faces of excavation; moving along as work proceeds** | | | | | | | |
| Maximum depth not exceeding 2.00 m | | | | | | | |
| distance between opposing faces not exceeding 2.00 m | – | 0.80 | 19.00 | 27.32 | – | m | **46.32** |
| **Excavating trenches; using 3 tonne tracked excavator; to receive pipes; grading bottoms; earthwork support; filling with excavated material to within 150 mm of finished surfaces and compacting; completing fill with topsoil; disposal of surplus soil** | | | | | | | |
| Services not exceeding 200 mm nominal size | | | | | | | |
| average depth of run not exceeding 0.50 m | 1.78 | 0.12 | 2.85 | 1.04 | 1.78 | m | **5.67** |
| average depth of run not exceeding 0.75 m | 1.78 | 0.16 | 3.87 | 1.44 | 1.78 | m | **7.09** |
| average depth of run not exceeding 1.00 m | 1.78 | 0.28 | 6.73 | 2.51 | 1.78 | m | **11.02** |
| average depth of run not exceeding 1.25 m | 1.49 | 0.38 | 9.10 | 3.37 | 1.49 | m | **13.96** |
| **Granular beds to trenches; lay granular material to trenches excavated separately; to receive pipes (not included)** | | | | | | | |
| 300 mm wide × 100 mm thick | | | | | | | |
| reject sand | – | 0.05 | 1.19 | 0.23 | 3.79 | m | **5.21** |
| reject gravel | – | 0.05 | 1.19 | 0.23 | 3.44 | m | **4.86** |
| shingle 40 mm aggregate | – | 0.05 | 1.19 | 0.23 | 4.30 | m | **5.72** |
| sharp sand | 4.50 | 0.05 | 1.19 | 0.23 | 4.50 | m | **5.92** |
| 300 mm wide × 150 mm thick | | | | | | | |
| reject sand | – | 0.08 | 1.78 | 0.35 | 5.69 | m | **7.82** |
| reject gravel | – | 0.08 | 1.78 | 0.35 | 5.16 | m | **7.29** |
| shingle 40 mm aggregate | – | 0.08 | 1.78 | 0.35 | 6.45 | m | **8.58** |
| sharp sand | 6.76 | 0.08 | 1.78 | 0.35 | 6.76 | m | **8.89** |

## 34 DRAINAGE BELOW GROUND

| Item<br>Excluding site overheads and profit | PC<br>£ | Labour<br>hours | Labour<br>£ | Plant<br>£ | Material<br>£ | Unit | Total<br>rate £ |
|---|---|---|---|---|---|---|---|
| **Excavating trenches; using 3 tonne tracked excavator; to receive pipes; grading bottoms; earthwork support; filling with imported granular material type 2 and compacting; disposal of surplus soil** | | | | | | | |
| Services not exceeding 200 mm nominal size | | | | | | | |
| average depth of run not exceeding 0.50 m | 2.59 | 0.09 | 2.06 | 0.72 | 8.14 | m | **10.92** |
| average depth of run not exceeding 0.75 m | 3.89 | 0.11 | 2.57 | 0.90 | 12.21 | m | **15.68** |
| average depth of run not exceeding 1.00 m | 5.18 | 0.14 | 3.32 | 1.18 | 16.27 | m | **20.77** |
| average depth of run not exceeding 1.25 m | 6.48 | 0.23 | 5.43 | 2.02 | 20.34 | m | **27.79** |
| **Excavating trenches; using 3 tonne tracked excavator; to receive pipes; grading bottoms; earthwork support; filling with concrete, ready mixed ST2; disposal of surplus soil** | | | | | | | |
| Services not exceeding 200 mm nominal size | | | | | | | |
| average depth of run not exceeding 0.50 m | 12.92 | 0.11 | 2.53 | 0.38 | 18.47 | m | **21.38** |
| average depth of run not exceeding 0.75 m | 19.37 | 0.13 | 3.09 | 0.47 | 27.69 | m | **31.25** |
| average depth of run not exceeding 1.00 m | 25.83 | 0.17 | 3.96 | 0.63 | 36.92 | m | **41.51** |
| average depth of run not exceeding 1.25 m | 32.29 | 0.23 | 5.34 | 0.94 | 46.15 | m | **52.43** |
| **Earthwork support; providing support to opposing faces of excavation; moving along as work proceeds** | | | | | | | |
| Maximum depth not exceeding 2.00 m | | | | | | | |
| trenchbox; distance between opposing faces not exceeding 2.00 m | – | 0.80 | 19.00 | 27.32 | – | m | **46.32** |
| timber; distance between opposing faces not exceeding 500 mm | – | 0.20 | 9.50 | – | 0.02 | m | **9.52** |
| **Clay pipes and fittings; Hepworth Plc; Supersleve** | | | | | | | |
| 100 mm clay pipes; polypropylene slip coupling; in trenches (trenches not included) | | | | | | | |
| laid straight | 4.52 | 0.25 | 5.94 | – | 6.58 | m | **12.52** |
| short runs under 3.00 m | 4.52 | 0.31 | 7.42 | – | 6.58 | m | **14.00** |

## 34 DRAINAGE BELOW GROUND

| Item<br>Excluding site overheads and profit | PC<br>£ | Labour<br>hours | Labour<br>£ | Plant<br>£ | Material<br>£ | Unit | Total<br>rate £ |
|---|---|---|---|---|---|---|---|
| **PIPE LAYING – cont** | | | | | | | |
| **Clay pipes and fittings – cont** | | | | | | | |
| Extra over 100 mm clay pipes for | | | | | | | |
| bends; 15–90°; single socket | 9.51 | 0.25 | 5.94 | – | 9.51 | nr | **15.45** |
| junction; 45 or 90°; double socket | 20.05 | 0.25 | 5.94 | – | 20.05 | nr | **25.99** |
| slip couplings; polypropylene | 3.30 | 0.08 | 1.98 | – | 3.30 | nr | **5.28** |
| gully with P trap; 100 mm; | | | | | | | |
| 154 × 154 mm plastic grating | 23.15 | 1.00 | 23.75 | – | 38.17 | nr | **61.92** |
| 150 mm clay pipes; polypropylene slip coupling; in trenches (trenches not included) | | | | | | | |
| laid straight | 14.14 | 0.30 | 7.13 | – | 14.14 | m | **21.27** |
| short runs under 3.00 m | 14.14 | 0.60 | 14.25 | – | 14.14 | m | **28.39** |
| Extra over 150 mm clay pipes for | | | | | | | |
| bends; 15–90° | 11.97 | 0.28 | 6.65 | – | 17.74 | nr | **24.39** |
| junction; 45 or 90°; 100 × 150 mm | 16.02 | 0.40 | 9.50 | – | 27.55 | nr | **37.05** |
| junction; 45 or 90°; 150 × 150 mm | 17.58 | 0.40 | 9.50 | – | 29.12 | nr | **38.62** |
| slip couplings; polypropylene | 5.77 | 0.05 | 1.19 | – | 5.77 | nr | **6.96** |
| tapered pipe; 100–150 mm | 18.11 | 0.50 | 11.88 | – | 18.11 | nr | **29.99** |
| tapered pipe; 150–225 mm | 46.21 | 0.50 | 11.88 | – | 46.21 | nr | **58.09** |
| socket adaptor; connection to traditional pipes and fittings | 12.14 | 0.33 | 7.84 | – | 17.91 | nr | **25.75** |
| Accessories in clay | | | | | | | |
| access pipe; 150 mm | 45.12 | – | – | – | 56.66 | nr | **56.66** |
| rodding eye; 150 mm | 43.16 | 0.50 | 11.88 | – | 48.17 | nr | **60.05** |
| gully with P traps; 150 mm; | | | | | | | |
| 154 × 154 mm plastic grating | 50.92 | 0.80 | 19.00 | – | 58.69 | nr | **77.69** |
| **PVC-u pipes and fittings; Wavin Plastics Ltd; OsmaDrain; laid to and covered over with imported granular material** | | | | | | | |
| 110 mm PVC-u pipes; in trenches (trenches not included) | | | | | | | |
| laid straight | 5.08 | 0.08 | 1.90 | 1.78 | 7.22 | m | **10.90** |
| short runs under 3.00 m | 5.83 | 0.12 | 2.85 | 1.78 | 7.97 | m | **12.60** |
| Extra over 110 mm PVC-u pipes for | | | | | | | |
| bends; short radius | 6.75 | 0.25 | 5.94 | – | 6.75 | nr | **12.69** |
| bends; long radius | 7.22 | 0.25 | 5.94 | – | 7.22 | nr | **13.16** |
| junctions; equal; double socket | 5.34 | 0.25 | 5.94 | – | 5.34 | nr | **11.28** |
| slip couplings | 3.95 | 0.25 | 5.94 | – | 3.95 | nr | **9.89** |
| adaptors to clay | 7.38 | 0.50 | 11.88 | – | 7.38 | nr | **19.26** |
| 160 mm PVC-u pipes; in trenches (trenches not included) | | | | | | | |
| laid straight | 8.35 | 0.08 | 1.90 | 1.78 | 10.50 | m | **14.18** |
| short runs under 3.00 m | 14.86 | 0.12 | 2.85 | 1.78 | 17.01 | m | **21.64** |
| Extra over 160 mm PVC-u pipes for | | | | | | | |
| socket bend; double; 90 or 45° | 19.91 | 0.20 | 4.75 | – | 19.91 | nr | **24.66** |
| socket bend; double; 15 or 30° | 15.56 | 0.20 | 4.75 | – | 15.56 | nr | **20.31** |
| socket bend; single; 87.5 or 45° | 6.12 | 0.20 | 4.75 | – | 6.12 | nr | **10.87** |

## 34 DRAINAGE BELOW GROUND

| Item<br>Excluding site overheads and profit | PC<br>£ | Labour<br>hours | Labour<br>£ | Plant<br>£ | Material<br>£ | Unit | Total<br>rate £ |
|---|---|---|---|---|---|---|---|
| socket bend; single; 15 or 30° | 6.12 | 0.20 | 4.75 | – | 6.12 | nr | **10.87** |
| bends; short radius | 16.13 | 0.25 | 5.94 | – | 22.88 | nr | **28.82** |
| bends; long radius | 32.93 | 0.25 | 5.94 | – | 32.93 | nr | **38.87** |
| junctions; single | 19.97 | 0.33 | 7.92 | – | 19.97 | nr | **27.89** |
| pipe coupler | 8.06 | 0.05 | 1.19 | – | 8.06 | nr | **9.25** |
| slip couplings PVC-u | 1.87 | 0.05 | 1.19 | – | 1.87 | nr | **3.06** |
| adaptors to clay | 7.75 | 0.50 | 11.88 | – | 7.75 | nr | **19.63** |
| level invert reducer | 4.72 | 0.50 | 11.88 | – | 4.72 | nr | **16.60** |
| spiggot | 11.78 | 0.20 | 4.75 | – | 11.78 | nr | **16.53** |
| **CATCHWATER OR FRENCH DRAINS** | | | | | | | |
| Catchwater or French drains; laying pipes to drain excavated separately; 100 mm diameter non-coilable perforated plastic pipes; including straight jointing; pipes laid with perforations uppermost; lining trench; wrapping pipes with filter fabric; backfilling with shingle | | | | | | | |
| Width 300 mm | | | | | | | |
| depth 450 mm | 197.44 | 9.10 | 216.13 | 38.88 | 711.36 | 100 m | **966.37** |
| depth 600 mm | 201.56 | 9.84 | 233.70 | 60.73 | 899.75 | 100 m | **1194.18** |
| depth 750 mm | 205.86 | 10.60 | 251.75 | 76.98 | 1088.33 | 100 m | **1417.06** |
| depth 900 mm | 210.07 | 11.34 | 269.32 | 92.81 | 1276.82 | 100 m | **1638.95** |
| depth 1000 mm | 212.88 | 11.84 | 281.20 | 103.50 | 1402.48 | 100 m | **1787.18** |
| depth 1200 mm | 218.49 | 12.84 | 304.95 | 124.89 | 1653.79 | 100 m | **2083.63** |
| Width 450 mm | | | | | | | |
| depth 450 mm | 203.76 | 10.44 | 247.95 | 112.06 | 994.09 | 100 m | **1354.10** |
| depth 600 mm | 210.07 | 11.34 | 269.32 | 92.81 | 1276.82 | 100 m | **1638.95** |
| depth 750 mm | 216.38 | 12.46 | 295.93 | 116.76 | 1559.54 | 100 m | **1972.23** |
| depth 900 mm | 222.70 | 13.60 | 323.00 | 141.14 | 1842.27 | 100 m | **2306.41** |
| depth 1000 mm | 226.91 | 14.34 | 340.57 | 156.96 | 2030.75 | 100 m | **2528.28** |
| depth 1200 mm | 235.33 | 15.84 | 376.20 | 189.04 | 2407.72 | 100 m | **2972.96** |
| depth 1500 mm | 247.95 | 18.10 | 429.88 | 237.37 | 2973.18 | 100 m | **3640.43** |
| depth 2000 mm | 269.00 | 21.84 | 518.70 | 317.35 | 3915.60 | 100 m | **4751.65** |
| Width 600 mm | | | | | | | |
| depth 450 mm | 210.07 | 11.24 | 266.95 | 92.81 | 1276.82 | 100 m | **1636.58** |
| depth 600 mm | 218.49 | 12.84 | 304.95 | 124.89 | 1653.79 | 100 m | **2083.63** |
| depth 750 mm | 226.91 | 14.34 | 340.57 | 156.96 | 2030.75 | 100 m | **2528.28** |
| depth 900 mm | 235.33 | 15.84 | 376.20 | 189.04 | 2407.72 | 100 m | **2972.96** |
| depth 1000 mm | 232.54 | 16.84 | 399.95 | 210.43 | 2650.64 | 100 m | **3261.02** |
| depth 1200 mm | 252.16 | 18.84 | 447.45 | 253.19 | 3161.66 | 100 m | **3862.30** |
| depth 1500 mm | 269.00 | 21.84 | 518.70 | 746.35 | 3915.60 | 100 m | **5180.65** |
| depth 2000 mm | 297.06 | 19.84 | 471.20 | 573.97 | 5172.16 | 100 m | **6217.33** |
| Width 900 mm | | | | | | | |
| depth 450 mm | 200.72 | 13.60 | 323.00 | 141.14 | 1820.29 | 100 m | **2284.43** |
| depth 600 mm | 235.33 | 7.92 | 188.10 | 189.04 | 2407.72 | 100 m | **2784.86** |
| depth 750 mm | 247.95 | 9.05 | 214.94 | 237.37 | 2973.18 | 100 m | **3425.49** |

## 34 DRAINAGE BELOW GROUND

| Item<br>Excluding site overheads and profit | PC<br>£ | Labour<br>hours | Labour<br>£ | Plant<br>£ | Material<br>£ | Unit | Total<br>rate £ |
|---|---|---|---|---|---|---|---|
| **CATCHWATER OR FRENCH DRAINS** – cont | | | | | | | |
| **Catchwater or French drains** – cont | | | | | | | |
| Width 900 mm – cont | | | | | | | |
| depth 900 mm | 260.58 | 10.17 | 241.54 | 285.27 | 3538.63 | 100 m | **4065.44** |
| depth 1000 mm | 269.00 | 10.92 | 259.35 | 317.35 | 3915.60 | 100 m | **4492.30** |
| depth 1200 mm | 285.84 | 12.42 | 294.98 | 381.50 | 4669.54 | 100 m | **5346.02** |
| depth 1500 mm | 311.09 | 14.67 | 348.41 | 477.73 | 5800.44 | 100 m | **6626.58** |
| depth 2000 mm | 353.19 | 18.92 | 449.35 | 781.12 | 7685.28 | 100 m | **8915.75** |
| depth 2500 mm | 395.28 | 22.17 | 526.54 | 941.51 | 9570.13 | 100 m | **11038.18** |
| depth 3000 mm | 437.37 | 25.92 | 615.60 | 958.89 | 11454.97 | 100 m | **13029.46** |
| **Catchwater or French drains; laying pipes to drains excavated separately; 160 mm diameter non-coilable perforated plastic pipes; including straight jointing; pipes laid with perforations uppermost; lining trench; wrapping pipes with filter fabric; backfilling with shingle** | | | | | | | |
| Width 600 mm | | | | | | | |
| depth 450 mm | 393.30 | 11.20 | 266.00 | 89.82 | 1426.47 | 100 m | **1782.29** |
| depth 600 mm | 401.72 | 12.70 | 301.63 | 121.89 | 1803.43 | 100 m | **2226.95** |
| depth 750 mm | 410.13 | 14.20 | 337.25 | 153.97 | 2180.40 | 100 m | **2671.62** |
| depth 900 mm | 418.55 | 15.70 | 372.88 | 186.05 | 2557.37 | 100 m | **3116.30** |
| depth 1000 mm | 424.17 | 16.70 | 396.63 | 207.43 | 2808.68 | 100 m | **3412.74** |
| depth 1200 mm | 435.39 | 18.70 | 444.13 | 250.33 | 3311.31 | 100 m | **4005.77** |
| depth 1500 mm | 452.23 | 21.70 | 515.38 | 314.36 | 4065.25 | 100 m | **4894.99** |
| depth 2000 mm | 480.29 | 26.70 | 634.13 | 421.28 | 5321.81 | 100 m | **6377.22** |
| depth 2500 mm | 508.35 | 31.70 | 752.88 | 528.20 | 6578.37 | 100 m | **7859.45** |
| depth 3000 mm | 536.42 | 36.70 | 871.63 | 635.13 | 7834.93 | 100 m | **9341.69** |
| Width 900 mm | | | | | | | |
| depth 450 mm | 200.72 | 13.60 | 323.00 | 141.14 | 1820.29 | 100 m | **2284.43** |
| depth 600 mm | 213.35 | 15.84 | 376.20 | 189.04 | 2385.74 | 100 m | **2950.98** |
| depth 750 mm | 225.98 | 18.10 | 429.88 | 237.37 | 2951.20 | 100 m | **3618.45** |
| depth 900 mm | 238.60 | 20.34 | 483.07 | 285.27 | 3516.65 | 100 m | **4284.99** |
| depth 1000 mm | 247.02 | 21.84 | 518.70 | 317.35 | 3893.62 | 100 m | **4729.67** |
| depth 1200 mm | 263.86 | 24.84 | 589.95 | 381.50 | 4647.56 | 100 m | **5619.01** |
| depth 1500 mm | 289.12 | 29.34 | 696.83 | 477.73 | 5778.46 | 100 m | **6953.02** |
| depth 2000 mm | 331.21 | 36.84 | 874.95 | 638.12 | 7663.31 | 100 m | **9176.38** |
| depth 2500 mm | 373.30 | 44.34 | 1053.08 | 798.51 | 9548.15 | 100 m | **11399.74** |
| depth 3000 mm | 415.40 | 51.84 | 1231.20 | 958.89 | 11432.99 | 100 m | **13623.08** |

## 34 DRAINAGE BELOW GROUND

| Item<br>Excluding site overheads and profit | PC<br>£ | Labour<br>hours | Labour<br>£ | Plant<br>£ | Material<br>£ | Unit | Total<br>rate £ |
|---|---|---|---|---|---|---|---|
| **DITCHING** | | | | | | | |
| **Ditching; clear silt and bottom ditch not exceeding 1.50 m deep; strim back vegetation; disposing to spoil heaps; by machine** | | | | | | | |
| Up to 1.50 m wide at top | – | 6.00 | 142.50 | 28.13 | – | 100 m | **170.63** |
| 1.50–2.50 m wide at top | – | 8.00 | 190.00 | 37.50 | – | 100 m | **227.50** |
| 2.50–4.00 m wide at top | – | 9.00 | 213.75 | 42.19 | – | 100 m | **255.94** |
| **Ditching; clear only vegetation from ditch not exceeding 1.50 m deep; disposing to spoil heaps; by strimmer** | | | | | | | |
| Up to 1.50 m wide at top | – | 5.00 | 118.75 | 7.15 | – | 100 m | **125.90** |
| 1.50–2.50 m wide at top | – | 6.00 | 142.50 | 12.87 | – | 100 m | **155.37** |
| 2.50–4.00 m wide at top | – | 7.00 | 166.25 | 20.02 | – | 100 m | **186.27** |
| **Ditching; clear silt from ditch not exceeding 1.50 m deep; trimming back vegetation; disposing to spoil heaps; by hand** | | | | | | | |
| Up to 1.50 m wide at top | – | 15.00 | 356.25 | 7.15 | – | 100 m | **363.40** |
| 1.50–2.50 m wide at top | – | 27.00 | 641.25 | 12.87 | – | 100 m | **654.12** |
| 2.50–4.00 m wide at top | – | 42.00 | 997.50 | 20.02 | – | 100 m | **1017.52** |
| **Ditching; excavating and forming ditch and bank to given profile (normally 45°); in loam or sandy loam; by machine** | | | | | | | |
| Width 300 mm | | | | | | | |
| depth 600 mm | – | 3.70 | 87.88 | 34.69 | – | 100 m | **122.57** |
| depth 900 mm | – | 5.20 | 123.50 | 48.75 | – | 100 m | **172.25** |
| depth 1200 mm | – | 7.20 | 171.00 | 67.50 | – | 100 m | **238.50** |
| depth 1500 mm | – | 9.40 | 223.25 | 77.11 | – | 100 m | **300.36** |
| Width 600 mm | | | | | | | |
| depth 600 mm | – | – | – | 101.23 | – | 100 m | **101.23** |
| depth 900 mm | – | – | – | 150.22 | – | 100 m | **150.22** |
| depth 1200 mm | – | – | – | 204.10 | – | 100 m | **204.10** |
| depth 1500 mm | – | – | – | 293.91 | – | 100 m | **293.91** |
| Width 900 mm | | | | | | | |
| depth 600 mm | – | – | – | 481.43 | – | 100 m | **481.43** |
| depth 900 mm | – | – | – | 848.23 | – | 100 m | **848.23** |
| depth 1200 mm | – | – | – | 1139.83 | – | 100 m | **1139.83** |
| depth 1500 mm | – | – | – | 1412.18 | – | 100 m | **1412.18** |
| Width 1200 mm | | | | | | | |
| depth 600 mm | – | – | – | 756.52 | – | 100 m | **756.52** |
| depth 900 mm | – | – | – | 1123.33 | – | 100 m | **1123.33** |
| depth 1200 mm | – | – | – | 1487.83 | – | 100 m | **1487.83** |
| depth 1500 mm | – | – | – | 1906.44 | – | 100 m | **1906.44** |

## 34 DRAINAGE BELOW GROUND

| Item<br>Excluding site overheads and profit | PC<br>£ | Labour<br>hours | Labour<br>£ | Plant<br>£ | Material<br>£ | Unit | Total<br>rate £ |
|---|---|---|---|---|---|---|---|
| **DITCHING – cont** | | | | | | | |
| **Ditching – cont** | | | | | | | |
| Width 1500 mm | | | | | | | |
|   depth 600 mm | – | – | – | 933.05 | – | 100 m | 933.05 |
|   depth 900 mm | – | – | – | 1422.27 | – | 100 m | 1422.27 |
|   depth 1200 mm | – | – | – | 1891.31 | – | 100 m | 1891.31 |
|   depth 1500 mm | – | – | – | 2370.45 | – | 100 m | 2370.45 |
| Extra for ditching in clay | – | – | – | – | – | 20% | – |
| **Ditching; excavating and forming ditch and bank to given profile (normal 45°); in loam or sandy loam; by hand** | | | | | | | |
| Width 300 mm | | | | | | | |
|   depth 600 mm | – | 36.00 | 855.00 | – | – | 100 m | 855.00 |
|   depth 900 mm | – | 42.00 | 997.50 | – | – | 100 m | 997.50 |
|   depth 1200 mm | – | 56.00 | 1330.00 | – | – | 100 m | 1330.00 |
|   depth 1500 mm | – | 70.00 | 1662.50 | – | – | 100 m | 1662.50 |
| Width 600 mm | | | | | | | |
|   depth 600 mm | – | 56.00 | 1330.00 | – | – | 100 m | 1330.00 |
|   depth 900 mm | – | 84.00 | 1995.00 | – | – | 100 m | 1995.00 |
|   depth 1200 mm | – | 112.00 | 2660.00 | – | – | 100 m | 2660.00 |
|   depth 1500 mm | – | 140.00 | 3325.00 | – | – | 100 m | 3325.00 |
| Width 900 mm | | | | | | | |
|   depth 600 mm | – | 84.00 | 1995.00 | – | – | 100 m | 1995.00 |
|   depth 900 mm | – | 126.00 | 2992.50 | – | – | 100 m | 2992.50 |
|   depth 1200 mm | – | 168.00 | 3990.00 | – | – | 100 m | 3990.00 |
|   depth 1500 mm | – | 210.00 | 4987.50 | – | – | 100 m | 4987.50 |
| Width 1200 mm | | | | | | | |
|   depth 600 mm | – | 112.00 | 2660.00 | – | – | 100 m | 2660.00 |
|   depth 900 mm | – | 168.00 | 3990.00 | – | – | 100 m | 3990.00 |
|   depth 1200 mm | – | 224.00 | 5320.00 | – | – | 100 m | 5320.00 |
|   depth 1500 mm | – | 280.00 | 6650.00 | – | – | 100 m | 6650.00 |
| Extra for ditching in clay | – | – | – | – | – | 50% | – |
| Extra for ditching in compacted soil | – | – | – | – | – | 90% | – |
| **Piped ditching** | | | | | | | |
| Jointed concrete pipes; FP McCann Ltd; including bedding, haunching and topping with 150 mm concrete; 11.50 N/mm$^2$ – 40 mm aggregate; to existing ditch | | | | | | | |
|   300 mm diameter | 21.84 | 0.67 | 15.83 | 10.09 | 42.82 | m | 68.74 |
|   450 mm diameter | 32.55 | 0.67 | 15.83 | 10.09 | 64.03 | m | 89.95 |
|   600 mm diameter | 52.29 | 0.67 | 15.83 | 10.09 | 96.98 | m | 122.90 |
|   900 mm diameter | 138.60 | 1.00 | 23.75 | 15.13 | 162.29 | m | 201.17 |

## 34 DRAINAGE BELOW GROUND

| Item<br>Excluding site overheads and profit | PC<br>£ | Labour<br>hours | Labour<br>£ | Plant<br>£ | Material<br>£ | Unit | Total<br>rate £ |
|---|---|---|---|---|---|---|---|
| Jointed concrete pipes; FP McCann Ltd; including bedding, haunching and topping with 150 mm concrete; 11.50 N/mm$^2$ – 40 mm aggregate; to existing ditch | | | | | | | |
|   1200 mm diameter | 237.87 | 1.00 | 23.75 | 15.13 | 268.87 | m | 307.75 |
|   extra over jointed concrete pipes for bends to 45° | 218.40 | 0.67 | 15.83 | 12.61 | 227.85 | nr | 256.29 |
|   extra over jointed concrete pipes for single junctions 300 mm diameter | 149.63 | 0.67 | 15.83 | 12.61 | 159.08 | nr | 187.52 |
|   extra over jointed concrete pipes for single junctions 450 mm diameter | 221.81 | 0.67 | 15.83 | 11.46 | 231.26 | nr | 258.55 |
|   extra over jointed concrete pipes for single junctions 600 mm diameter | 359.52 | 0.67 | 15.83 | 11.46 | 368.97 | nr | 396.26 |
|   extra over jointed concrete pipes for single junctions 900 mm diameter | 607.74 | 0.67 | 15.83 | 11.46 | 617.19 | nr | 644.48 |
|   extra over jointed concrete pipes for single junctions 1200 mm diameter | 928.41 | 0.67 | 15.83 | 11.46 | 937.86 | nr | 965.15 |
| **Outfalls** | | | | | | | |
| Reinforced concrete outfalls to water course; flank walls; for 150 mm drain outlets; overall dimensions | | | | | | | |
|   Small headwall FP McCann Ltd | 638.60 | 4.00 | 285.00 | 46.59 | 701.00 | m | 1032.59 |
| **LAND DRAINS** | | | | | | | |
| **Market prices of backfilling materials** | | | | | | | |
| Sand | 41.40 | – | – | – | 41.40 | m$^3$ | 41.40 |
| Gravel rejects | 36.02 | – | – | – | 36.02 | m$^3$ | 36.02 |
| Imported topsoil (allowing for 20% settlement) | – | – | – | – | 39.60 | m$^3$ | 39.60 |
| **Land drainage; calculation table** | | | | | | | |
| For calculation of drainage per hectare the following table can be used; rates show the lengths of drains per unit and not the value | | | | | | | |
| Lateral drains | | | | | | | |
|   10.00 m centres | – | – | – | – | – | m/ha | 1000.00 |
|   15.00 m centres | – | – | – | – | – | m/ha | 650.00 |
|   25.00 m centres | – | – | – | – | – | m/ha | 400.00 |
|   30.00 m centres | – | – | – | – | – | m/ha | 330.00 |
| Main drains | | | | | | | |
|   1 nr (at 100 m centres) | – | – | – | – | – | m/ha | 100.00 |
|   2 nr (at 50 m centres) | – | – | – | – | – | m/ha | 200.00 |
|   3 nr (at 33.3 m centres) | – | – | – | – | – | m/ha | 300.00 |
|   4 nr (at 25 m centres) | – | – | – | – | – | m/ha | 400.00 |

## 34 DRAINAGE BELOW GROUND

| Item<br>Excluding site overheads and profit | PC<br>£ | Labour<br>hours | Labour<br>£ | Plant<br>£ | Material<br>£ | Unit | Total<br>rate £ |
|---|---|---|---|---|---|---|---|
| **LAND DRAINS – cont** | | | | | | | |
| **Land drainage; excavating** | | | | | | | |
| Removing 150 mm depth of topsoil; 300 mm wide; depositing beside trench; by machine | – | 1.50 | 35.63 | 14.06 | – | 100 m | **49.69** |
| Removing 150 mm depth of topsoil; 300 mm wide; depositing beside trench; by hand | – | 8.00 | 190.00 | – | – | 100 m | **190.00** |
| **Excavated material; disposal on site** | | | | | | | |
| In spoil heaps | | | | | | | |
| average 25 m distance | – | – | – | 2.44 | – | m$^3$ | **2.44** |
| average 50 m distance | – | – | – | 2.82 | – | m$^3$ | **2.82** |
| average 100 m distance (1 dumper) | – | – | – | 3.53 | – | m$^3$ | **3.53** |
| average 100 m distance (2 dumpers) | – | – | – | 4.22 | – | m$^3$ | **4.22** |
| average 200 m distance | – | – | – | 4.26 | – | m$^3$ | **4.26** |
| average 200 m distance (2 dumpers) | – | – | – | 5.08 | – | m$^3$ | **5.08** |
| Removing excavated material from site to tip; mechanically loaded | | | | | | | |
| excavated material and clean hardcore rubble | – | – | – | 1.91 | 29.20 | m$^3$ | **31.11** |
| **Land drainage; Agripower Ltd; excavating trenches (with minimum project size of 1000 m) by trenching machine; spreading arisings on site; laying perforated pipe; as shown below; backfilling with shingle to 350 mm from surface** | | | | | | | |
| Width 150 mm; 80 mm perforated pipe depth 750 mm | – | – | – | – | – | m | **6.64** |
| Width 175 mm; 100 mm perforated pipe depth 800 mm | – | – | – | – | – | m | **7.05** |
| Width 250 mm; 150 mm perforated pipe depth 900 mm | – | – | – | – | – | m | **10.28** |
| **Land drainage; excavating for drains; by 5 tonne mini-excavator; including disposing spoil to spoil heaps not exceeding 100 m; boning to levels by laser** | | | | | | | |
| Width 150–225 mm | | | | | | | |
| depth 450 mm | – | 9.67 | 229.58 | 330.13 | – | 100 m | **559.71** |
| depth 600 mm | – | 13.00 | 308.75 | 495.19 | – | 100 m | **803.94** |
| depth 700 mm | – | 15.50 | 368.13 | 618.99 | – | 100 m | **987.12** |
| depth 900 mm | – | 23.00 | 546.25 | 990.39 | – | 100 m | **1536.64** |
| Width 300 mm | | | | | | | |
| depth 450 mm | – | 9.67 | 229.58 | 330.13 | – | 100 m | **559.71** |
| depth 700 mm | – | 14.00 | 332.50 | 544.71 | – | 100 m | **877.21** |
| depth 900 mm | – | 23.00 | 546.25 | 990.39 | – | 100 m | **1536.64** |

## 34 DRAINAGE BELOW GROUND

| Item<br>Excluding site overheads and profit | PC<br>£ | Labour<br>hours | Labour<br>£ | Plant<br>£ | Material<br>£ | Unit | Total<br>rate £ |
|---|---|---|---|---|---|---|---|
| depth 1000 mm | – | 25.00 | 593.75 | 1089.43 | – | 100 m | **1683.18** |
| depth 1200 mm | – | 25.00 | 593.75 | 1237.99 | – | 100 m | **1831.74** |
| depth 1500 mm | – | 31.00 | 736.25 | 1386.54 | – | 100 m | **2122.79** |
| **Land drainage; excavating for drains; by 7 tonne tracked excavator; including disposing spoil to spoil heaps not exceeding 100 m** | | | | | | | |
| Width 225 mm | | | | | | | |
| depth 450 mm | – | 7.00 | 166.25 | 92.31 | – | 100 m | **258.56** |
| depth 600 mm | – | 7.17 | 170.21 | 96.16 | – | 100 m | **266.37** |
| depth 700 mm | – | 7.35 | 174.51 | 100.34 | – | 100 m | **274.85** |
| depth 900 mm | – | 7.76 | 184.35 | 109.90 | – | 100 m | **294.25** |
| depth 1000 mm | – | 8.00 | 190.00 | 115.39 | – | 100 m | **305.39** |
| depth 1200 mm | – | 8.26 | 196.25 | 121.46 | – | 100 m | **317.71** |
| Width 300 mm | | | | | | | |
| depth 450 mm | – | 7.35 | 174.51 | 100.34 | – | 100 m | **274.85** |
| depth 600 mm | – | 7.76 | 184.35 | 109.90 | – | 100 m | **294.25** |
| depth 700 mm | – | 8.26 | 196.25 | 121.46 | – | 100 m | **317.71** |
| depth 900 mm | – | 9.25 | 219.69 | 144.24 | – | 100 m | **363.93** |
| depth 1000 mm | – | 10.14 | 240.89 | 164.85 | – | 100 m | **405.74** |
| depth 1200 mm | – | 11.00 | 261.25 | 184.63 | – | 100 m | **445.88** |
| Width 600 mm | | | | | | | |
| depth 450 mm | – | 13.00 | 308.75 | 230.78 | – | 100 m | **539.53** |
| depth 600 mm | – | 14.11 | 335.14 | 256.42 | – | 100 m | **591.56** |
| depth 700 mm | – | 16.33 | 387.92 | 307.71 | – | 100 m | **695.63** |
| depth 900 mm | – | 17.29 | 410.54 | 329.69 | – | 100 m | **740.23** |
| depth 1000 mm | – | 18.38 | 436.63 | 355.05 | – | 100 m | **791.68** |
| depth 1200 mm | – | 21.18 | 503.07 | 419.60 | – | 100 m | **922.67** |
| **Land drainage; excavating for drains; by hand, including disposing spoil to spoil heaps not exceeding 100 m** | | | | | | | |
| Width 150 mm | | | | | | | |
| depth 450 mm | – | 0.44 | 10.46 | – | – | m | **10.46** |
| depth 600 mm | – | 0.59 | 13.96 | – | – | m | **13.96** |
| depth 700 mm | – | 0.69 | 16.27 | – | – | m | **16.27** |
| depth 900 mm | – | 0.88 | 20.93 | – | – | m | **20.93** |
| Width 225 mm | | | | | | | |
| depth 450 mm | – | 0.66 | 15.70 | – | – | m | **15.70** |
| depth 600 mm | – | 0.88 | 20.93 | – | – | m | **20.93** |
| depth 700 mm | – | 1.03 | 24.42 | – | – | m | **24.42** |
| depth 900 mm | – | 1.32 | 31.40 | – | – | m | **31.40** |
| Width 300 mm | | | | | | | |
| depth 450 mm | – | 0.88 | 20.93 | – | – | m | **20.93** |
| depth 600 mm | – | 1.18 | 27.91 | – | – | m | **27.91** |
| depth 700 mm | – | 1.37 | 32.56 | – | – | m | **32.56** |
| Width 375 mm | | | | | | | |
| depth 450 mm | – | 1.10 | 26.16 | – | – | m | **26.16** |
| depth 600 mm | – | 1.47 | 34.88 | – | – | m | **34.88** |
| depth 700 mm | – | 1.71 | 40.70 | – | – | m | **40.70** |

## 34 DRAINAGE BELOW GROUND

| Item<br>Excluding site overheads and profit | PC<br>£ | Labour<br>hours | Labour<br>£ | Plant<br>£ | Material<br>£ | Unit | Total<br>rate £ |
|---|---|---|---|---|---|---|---|
| **LAND DRAINS – cont** | | | | | | | |
| **Land drainage; pipe laying** | | | | | | | |
| **Wavin Plastics Ltd; flexible plastic perforated pipes in trenches (not included); to a minimum depth of 450 mm (couplings not included)** | | | | | | | |
| OsmaDrain; flexible plastic perforated pipes in trenches (not included); to a minimum depth of 450 mm (couplings not included) | | | | | | | |
| 80 mm diameter; available in 100 m coil | 86.10 | 2.00 | 47.50 | – | 86.10 | 100 m | **133.60** |
| 100 mm diameter; available in 100 m coil | 138.58 | 2.00 | 47.50 | – | 138.58 | 100 m | **186.08** |
| 160 mm diameter; available in 35 m coil | 334.97 | 2.00 | 47.50 | – | 334.97 | 100 m | **382.47** |
| WavinCoil; plastic pipe junctions | | | | | | | |
| 100 × 80 mm | 4.01 | 0.05 | 1.19 | – | 4.01 | nr | **5.20** |
| 160 × 160 mm | 10.00 | 0.05 | 1.19 | – | 10.00 | nr | **11.19** |
| WavinCoil; couplings for flexible pipes | | | | | | | |
| 100 mm diameter | 1.47 | 0.03 | 0.79 | – | 1.47 | nr | **2.26** |
| 160 mm diameter | 1.98 | 0.03 | 0.79 | – | 1.98 | nr | **2.77** |
| **Land drainage; backfilling trench after laying pipes with gravel rejects or similar; blind filling with ash or sand; topping with 150 mm imported topsoil from dumps not exceeding 100 m; by machine** | | | | | | | |
| Width 150 mm | | | | | | | |
| depth 450 mm | – | 3.30 | 78.38 | 81.37 | 231.05 | 100 m | **390.80** |
| depth 600 mm | – | 4.30 | 102.13 | 106.59 | 310.95 | 100 m | **519.67** |
| depth 900 mm | – | 6.30 | 149.63 | 157.02 | 452.11 | 100 m | **758.76** |
| Width 225 mm | | | | | | | |
| depth 600 mm | – | 6.45 | 153.19 | 159.88 | 466.60 | 100 m | **779.67** |
| depth 900 mm | – | 9.45 | 224.44 | 235.54 | 687.66 | 100 m | **1147.64** |
| **Land drainage; backfilling trench after laying pipes with gravel rejects or similar, blind filling with ash or sand; topping with 150 mm topsoil from dumps not exceeding 100 m; by hand** | | | | | | | |
| Width 150 mm | | | | | | | |
| depth 450 mm | – | 0.18 | 4.28 | – | 2.15 | m | **6.43** |
| depth 600 mm | – | 24.84 | 589.95 | – | 310.95 | 100 m | **900.90** |
| depth 750 mm | – | 31.05 | 737.44 | – | 279.62 | 100 m | **1017.06** |
| depth 900 mm | – | 37.26 | 884.92 | – | 452.11 | 100 m | **1337.03** |

## 34 DRAINAGE BELOW GROUND

| Item<br>Excluding site overheads and profit | PC<br>£ | Labour<br>hours | Labour<br>£ | Plant<br>£ | Material<br>£ | Unit | Total<br>rate £ |
|---|---|---|---|---|---|---|---|
| Width 225 mm | | | | | | | |
| depth 450 mm | – | 27.94 | 663.58 | – | 356.23 | 100 m | **1019.81** |
| depth 600 mm | – | 37.26 | 884.92 | – | 466.60 | 100 m | **1351.52** |
| depth 750 mm | – | 46.57 | 1106.04 | – | 577.29 | 100 m | **1683.33** |
| depth 900 mm | – | 55.89 | 1327.39 | – | 687.66 | 100 m | **2015.05** |
| Width 375 mm | | | | | | | |
| depth 450 mm | – | 46.57 | 1106.04 | – | 593.49 | 100 m | **1699.53** |
| depth 600 mm | – | 61.10 | 1451.13 | – | 777.55 | 100 m | **2228.68** |
| depth 750 mm | – | 77.63 | 1843.71 | – | 961.93 | 100 m | **2805.64** |
| depth 900 mm | – | 93.15 | 2212.31 | – | 1145.99 | 100 m | **3358.30** |
| **MARKET PRICES OF BACKFILLING MATERIALS** | | | | | | | |
| Market prices of commonly used materials used for backfilling | | | | | | | |
| The prices below show the market price with standard settlement factors allowed for | | | | | | | |
| Sharp sand | – | – | – | – | 43.00 | m³ | **43.00** |
| Gravel (washed river or pit) | – | – | – | – | 39.77 | m³ | **39.77** |
| Shingle | – | – | – | – | 45.66 | m³ | **45.66** |
| Topsoil | – | – | – | – | 39.60 | m³ | **39.60** |
| Gravel rejects | – | – | – | – | 33.57 | m³ | **33.57** |
| Selected granular material 6F2 | – | – | – | – | 14.40 | m³ | **14.40** |
| **SAND SLITTING/GROOVING** | | | | | | | |
| **Sand slitting; Agripower Ltd**<br>Drainage slits; at 1.00 m centres; using spinning disc trenching machine; backfilling to 100 mm of surface with pea gravel; finished surface with medium grade sand 100; arisings to be loaded, hauled and tipped on site; minimum pitch size 6000 m² | | | | | | | |
| 250 mm depth | – | – | – | – | – | m | **2.05** |
| 300 mm depth | – | – | – | – | – | m | **2.25** |
| 400 mm depth | – | – | – | – | – | m | **3.38** |
| **Sand grooving or banding; Agripower Ltd**<br>Drainage bands at 260 mm m centres; using Blec Sandmaster; to existing grass at 260 mm centres; minimum pitch size 4000 m² | | | | | | | |
| sand banding or sand grooving | – | – | – | – | – | m² | **1.14** |

## 34 DRAINAGE BELOW GROUND

| Item<br>Excluding site overheads and profit | PC<br>£ | Labour<br>hours | Labour<br>£ | Plant<br>£ | Material<br>£ | Unit | Total<br>rate £ |
|---|---|---|---|---|---|---|---|
| **SOAKAWAYS** | | | | | | | |
| Preamble: Flat rate hourly rainfall = 50 mm/hr and assumes 100 impermeability of the run-off area. A storage capacity of the soakaway should be ⅓ of the hourly rainfall. Formulae for calculating soakaway depths are provided in the publications mentioned below and in the Tables and Memoranda section of this publication. The design of soakaways is dependent on, amongst other factors, soil conditions, permeability, groundwater level and runoff. The definitive documents for design of soakaways are CIRIA 156 and BRE Digest 365 dated September 1991. The suppliers of the systems below will assist through their technical divisions. Excavation earthwork support of pits and disposal not included. | | | | | | | |
| **Soakaway excavation** | | | | | | | |
| Pits; 3 tonne tracked excavator | | | | | | | |
| maximum depth not exceeding 1.00 m | – | 0.25 | 5.94 | 8.89 | – | m$^3$ | **14.83** |
| maximum depth not exceeding 2.00 m | – | 0.40 | 9.50 | 14.22 | – | m$^3$ | **23.72** |
| Pits; 8 tonne tracked excavator | | | | | | | |
| maximum depth not exceeding 2.00 m | – | 0.40 | 9.50 | 18.64 | – | m$^3$ | **28.14** |
| maximum depth not exceeding 3.00 m | – | 0.50 | 11.88 | 23.30 | – | m$^3$ | **35.18** |
| **Disposal** | | | | | | | |
| Excavated material; off site; to tip; mechanically loaded (8 tonne excavator | | | | | | | |
| inert | – | – | – | – | – | m$^3$ | **30.81** |
| **In situ concrete ring beam foundations to base of soakaway; 300 mm wide × 250 mm deep; poured on or against earth or unblinded hardcore** | | | | | | | |
| Internal diameters of rings | | | | | | | |
| 900 mm | 23.37 | 2.00 | 95.00 | – | 23.37 | nr | **118.37** |
| 1200 mm | 31.15 | 2.25 | 106.88 | – | 31.15 | nr | **138.03** |
| 1500 mm | 38.93 | 2.50 | 118.75 | – | 38.93 | nr | **157.68** |
| 2400 mm | 62.30 | 2.75 | 130.63 | – | 62.30 | nr | **192.93** |

## 34 DRAINAGE BELOW GROUND

| Item<br>Excluding site overheads and profit | PC<br>£ | Labour<br>hours | Labour<br>£ | Plant<br>£ | Material<br>£ | Unit | Total<br>rate £ |
|---|---|---|---|---|---|---|---|
| **Concrete manhole rings; Milton Pipes Ltd; perforations and step irons to concrete rings at manufacturers recommended centres; placing of concrete ring manholes to in situ concrete ring beams (1:3:6) (not included); filling and surrounding base with gravel 225 mm deep (not included)** | | | | | | | |
| Ring diameter 900 mm | | | | | | | |
| 1.00 m deep; volume 636 litres | 95.29 | 0.63 | 29.69 | 20.38 | 226.79 | nr | **276.86** |
| 1.50 m deep; volume 954 litres | 142.94 | 0.82 | 39.19 | 53.82 | 362.57 | nr | **455.58** |
| 2.00 m deep; volume 1272 litres | 190.58 | 0.90 | 42.75 | 53.82 | 481.26 | nr | **577.83** |
| Ring diameter 1200 mm | | | | | | | |
| 1.00 m deep; volume 1131 litres | 124.48 | 1.50 | 71.25 | 48.92 | 295.23 | nr | **415.40** |
| 1.50 m deep; volume 1696 litres | 186.72 | 2.25 | 106.88 | 73.39 | 437.06 | nr | **617.33** |
| 2.00 m deep; volume 2261 litres | 248.96 | 2.50 | 118.75 | 73.39 | 578.89 | nr | **771.03** |
| 2.50 m deep; volume 2827 litres | 311.20 | 3.00 | 142.50 | 97.85 | 660.87 | nr | **901.22** |
| Ring diameter 1500 mm | | | | | | | |
| 1.00 m deep; volume 1767 litres | 200.76 | 2.25 | 106.88 | 48.92 | 412.23 | nr | **568.03** |
| 1.50 m deep; volume 2651 litres | 301.14 | 3.00 | 142.50 | 68.49 | 609.30 | nr | **820.29** |
| 2.00 m deep; volume 3534 litres | 401.52 | 3.50 | 166.25 | 68.49 | 806.37 | nr | **1041.11** |
| 2.50 m deep; volume 4418 litres | 311.20 | 3.75 | 178.13 | 122.31 | 812.74 | nr | **1113.18** |
| Ring diameter 2400 mm | | | | | | | |
| 1.00 m deep; volume 4524 litres | 771.12 | 3.00 | 142.50 | 48.92 | 1059.54 | nr | **1250.96** |
| 1.50 m deep; volume 6786 litres | 1156.68 | 3.75 | 178.13 | 68.49 | 1575.99 | nr | **1822.61** |
| 2.00 m deep; volume 9048 litres | 1542.24 | 4.00 | 190.00 | 68.49 | 2100.99 | nr | **2359.48** |
| 2.50 m deep; volume 11310 litres | 1927.80 | 4.50 | 213.75 | 122.31 | 2617.44 | nr | **2953.50** |
| Extra over for | | | | | | | |
| 250 mm depth chamber ring | – | – | – | – | – | 100% | – |
| 500 mm depth chamber ring | – | – | – | – | – | 50% | – |
| **Cover slabs to soakaways** | | | | | | | |
| Heavy duty precast concrete | | | | | | | |
| 900 mm diameter | 157.56 | 0.50 | 35.63 | 24.46 | 157.56 | nr | **217.65** |
| 1200 mm diameter | 176.66 | 0.50 | 35.63 | 24.46 | 176.66 | nr | **236.75** |
| 1500 mm diameter | 286.23 | 0.50 | 35.63 | 24.46 | 286.23 | nr | **346.32** |
| 2400 mm diameter | 1026.75 | 0.50 | 35.63 | 24.46 | 1026.75 | nr | **1086.84** |
| Step irons to concrete chamber rings | 39.48 | – | – | – | 39.48 | m | **39.48** |
| Extra over soakaways for filter wrapping with a proprietary filter membrane | | | | | | | |
| 900 mm diameter × 1.00 m deep | 1.44 | 0.50 | 35.63 | – | 1.44 | nr | **37.07** |
| 900 mm diameter × 2.00 m deep | 2.88 | 0.75 | 53.44 | – | 2.88 | nr | **56.32** |
| 1050 mm diameter × 1.00 m deep | 1.68 | 0.75 | 53.44 | – | 1.68 | nr | **55.12** |
| 1050 mm diameter × 2.00 m deep | 3.36 | 1.00 | 71.25 | – | 3.36 | nr | **74.61** |
| 1200 mm diameter × 1.00 m deep | 1.92 | 1.00 | 71.25 | – | 1.92 | nr | **73.17** |
| 1200 mm diameter × 2.00 m deep | 3.83 | 1.25 | 89.06 | – | 3.83 | nr | **92.89** |
| 1500 mm diameter × 1.00 m deep | 2.39 | 1.25 | 89.06 | – | 2.39 | nr | **91.45** |
| 1500 mm diameter × 2.00 m deep | 4.78 | 1.25 | 89.06 | – | 4.78 | nr | **93.84** |
| 1800 mm diameter × 1.00 m deep | 2.87 | 1.50 | 106.88 | – | 2.87 | nr | **109.75** |
| 1800 mm diameter × 2.00 m deep | 5.74 | 1.63 | 115.78 | – | 5.74 | nr | **121.52** |

## 34 DRAINAGE BELOW GROUND

| Item<br>Excluding site overheads and profit | PC<br>£ | Labour<br>hours | Labour<br>£ | Plant<br>£ | Material<br>£ | Unit | Total<br>rate £ |
|---|---|---|---|---|---|---|---|
| **SOAKAWAYS – cont** | | | | | | | |
| **Gravel surrounding to concrete ring soakaway** | | | | | | | |
| 40 mm aggregate backfilled to vertical face of soakaway wrapped with geofabric (not included) | | | | | | | |
| 250 mm thick | 40.95 | 0.10 | 2.38 | 14.45 | 40.95 | m³ | **57.78** |
| **Backfilling to face of soakaway; carefully compacting as work proceeds** | | | | | | | |
| Arising from the excavations | | | | | | | |
| average thickness exceeding 0.25 m; depositing in layers 150 mm maximum thickness | – | 0.03 | 0.79 | 4.82 | – | m³ | **5.61** |
| **Aquacell soakaway; Wavin Plastics Ltd; preformed polypropylene soakaway infiltration crate units; to trenches; surrounded by geotextile and 40 mm aggregate laid 100 mm thick in trenches (excavation, disposal and backfilling not included)** | | | | | | | |
| 1.00 m × 500 × 400 mm; internal volume 190 litres | | | | | | | |
| 4 crates; 2.00 m × 1.00 m × 400 mm; 760 litres | 212.75 | 0.80 | 38.00 | 21.53 | 268.54 | nr | **328.07** |
| 8 crates; 2.00 m × 1.00 m × 800 mm; 1520 litres | 425.50 | 1.60 | 76.00 | 27.68 | 494.25 | nr | **597.93** |
| 12 crates; 6.00 m × 500 mm × 800 mm; 2280 litres | 638.26 | 2.40 | 114.00 | 57.40 | 769.67 | nr | **941.07** |
| 16 crates; 4.00 m × 1.00 m × 800 mm; 3040 litres | 851.01 | 3.20 | 152.00 | 51.25 | 969.46 | nr | **1172.71** |
| 20 crates; 5.00 m × 1.00 m × 800 mm; 3800 litres | 1063.76 | 4.00 | 190.00 | 50.74 | 1191.54 | nr | **1432.28** |
| 30 crates; 15.00 m × 1.00 m × 400 mm; 5700 litres | 1595.64 | 6.00 | 285.00 | 138.38 | 1908.17 | nr | **2331.55** |
| 60 crates; 15.00 m × 1.00 m × 800 mm; 11400 litres | 3191.28 | 10.00 | 475.00 | 180.92 | 3614.30 | nr | **4270.22** |
| **Geofabric surround to Aquacell units; Terram Ltd** | | | | | | | |
| Terram synthetic fibre filter fabric; to face of concrete rings (not included); anchoring whilst backfilling (not included) | | | | | | | |
| Terram 1000; 0.70 mm thick; mean water flow 50 l/m²/s | 0.49 | 0.03 | 1.19 | – | 0.49 | m² | **1.68** |

## 34 DRAINAGE BELOW GROUND

| Item<br>Excluding site overheads and profit | PC<br>£ | Labour<br>hours | Labour<br>£ | Plant<br>£ | Material<br>£ | Unit | Total<br>rate £ |
|---|---|---|---|---|---|---|---|
| **CULVERTS** | | | | | | | |
| **Culverts; precast box sections FP McCann Ltd; For directing watercourses and underpasses; to trenches or prepared bases (not included) on 200 mm prepared granular or concrete base blinded with 50 mm sand** | | | | | | | |
| Plain box culvert; 800 mm high; width | | | | | | | |
| 1.20 m | 575.00 | 3.00 | 213.75 | 65.83 | 583.51 | m | 863.09 |
| 1.50 m | 605.00 | 3.50 | 249.38 | 82.28 | 613.51 | m | 945.17 |
| Plain box culvert; 1.00 m high; width | | | | | | | |
| 1.20 m | 600.00 | 4.00 | 285.00 | 98.74 | 608.51 | m | 992.25 |
| 1.50 m | 700.00 | 4.50 | 320.63 | 115.19 | 708.51 | m | 1144.33 |
| 1.80 m | 775.00 | 6.00 | 427.50 | 164.56 | 783.51 | m | 1375.57 |
| Plain box culvert; 1.20 m high; width | | | | | | | |
| 1.50 m | 740.00 | 6.00 | 427.50 | 164.56 | 748.51 | m | 1340.57 |
| 1.80 m | 800.00 | 8.00 | 570.00 | 197.47 | 808.51 | m | 1575.98 |
| Extra over for cantilevered animal ledges | 200.00 | – | – | – | 200.00 | m | 200.00 |
| **HEADWALLS** | | | | | | | |
| **Headwalls; FP McCann Ltd; Precast concrete headwalls to outflows; Placing to prepared graded and blinded 300 mm thick (not included); connection of outfall pipe; 150 mm thick wall with non-standard opening; excludes standard grating** | | | | | | | |
| Size on plan 1.618 m tapering to 889 mm × 1.32 m × 663 mm high | | | | | | | |
| Small Headwall; pipe size up to 300 mm | 638.60 | 4.00 | 285.00 | 46.59 | 701.00 | nr | 1032.59 |
| Size on plan 1.878 m tapering to 1.026 m × 1.37 m × 998 mm high | | | | | | | |
| Medium Headwall; pipe size up to 450 mm | 1040.30 | 7.00 | 498.75 | 139.78 | 1137.95 | nr | 1776.48 |
| Size on plan 2.351 m tapering to 1.656 m × 2.150 m × 1.258 m high | | | | | | | |
| Large Headwall; pipe size up to 900 mm | 1534.70 | 8.00 | 570.00 | 279.56 | 1534.70 | nr | 2384.26 |

## 34 DRAINAGE BELOW GROUND

| Item<br>Excluding site overheads and profit | PC<br>£ | Labour<br>hours | Labour<br>£ | Plant<br>£ | Material<br>£ | Unit | Total<br>rate £ |
|---|---|---|---|---|---|---|---|
| **SPORTS AND AMENITY DRAINAGE – cont** | | | | | | | |
| **SPORTS AND AMENITY DRAINAGE** | | | | | | | |
| Sports or amenity drainage; Agripower Ltd; excavating trenches (with minimum project size of 1000 m) by trenching machine; arisings to spoil heap on site maximum 100 m; backfill with shingle to within 125 mm from surface and with rootzone to ground level | | | | | | | |
| Width 145 mm; 80 mm perforated pipe depth 550 mm | – | – | – | – | – | m | **8.42** |
| Width 145 mm; 100 mm perforated pipe depth 650 mm | – | – | – | – | – | m | **9.67** |
| Width 145 mm; 150 mm perforated pipe depth 700 mm | – | – | – | – | – | 100 m | **12.46** |

## 35 SITE WORKS – BASES

| Item<br>Excluding site overheads and profit | PC<br>£ | Labour<br>hours | Labour<br>£ | Plant<br>£ | Material<br>£ | Unit | Total<br>rate £ |
|---|---|---|---|---|---|---|---|
| **HARDCORE BASES** | | | | | | | |
| **Hardcore bases; obtained off site; PC £20.00/m³** | | | | | | | |
| By machine | | | | | | | |
| 100 mm thick | 2.00 | 0.05 | 1.19 | 1.19 | 2.00 | m² | 4.38 |
| 150 mm thick | 3.00 | 0.07 | 1.58 | 1.55 | 3.00 | m² | 6.13 |
| 200 mm thick | 4.00 | 0.08 | 1.90 | 1.62 | 4.00 | m² | 7.52 |
| 300 mm thick | 6.00 | 0.08 | 1.98 | 1.73 | 6.00 | m² | 9.71 |
| exceeding 300 mm thick | 24.00 | 0.17 | 3.96 | 3.97 | 24.00 | m³ | 31.93 |
| By hand | | | | | | | |
| 100 mm thick | 2.00 | 0.20 | 4.75 | 0.12 | 2.00 | m² | 6.87 |
| 150 mm thick | 3.60 | 0.30 | 7.13 | 0.12 | 3.60 | m² | 10.85 |
| 200 mm thick | 4.80 | 0.40 | 9.50 | 0.20 | 4.80 | m² | 14.50 |
| 300 mm thick | 7.20 | 0.60 | 14.25 | 0.12 | 7.20 | m² | 21.57 |
| exceeding 300 mm thick | 24.00 | 2.00 | 47.50 | 0.41 | 24.00 | m³ | 71.91 |
| Hardcore; difference for each £1.00 increase/decrease in PC price per m³; price will vary with type and source of hardcore | | | | | | | |
| average 75 mm thick | – | – | – | – | 0.07 | m² | 0.07 |
| average 100 mm thick | – | – | – | – | 0.10 | m² | 0.10 |
| average 150 mm thick | – | – | – | – | 0.14 | m² | 0.14 |
| average 200 mm thick | – | – | – | – | 0.19 | m² | 0.19 |
| average 250 mm thick | – | – | – | – | 0.24 | m² | 0.24 |
| average 300 mm thick | – | – | – | – | 0.28 | m² | 0.28 |
| exceeding 300 mm thick | – | – | – | – | 1.04 | m³ | 1.04 |
| **TYPE 1 BASES** | | | | | | | |
| **Type 1 granular fill base; PC £20.00 /tonne (£42.90/m³ compacted)** | | | | | | | |
| By machine | | | | | | | |
| 100 mm thick | 4.95 | 0.02 | 1.48 | 0.60 | 4.95 | m² | 7.03 |
| 150 mm thick | 7.42 | 0.03 | 1.78 | 0.60 | 7.42 | m² | 9.80 |
| 250 mm thick | 12.38 | 0.03 | 2.04 | 1.19 | 12.38 | m² | 15.61 |
| over 250 mm thick | 49.50 | 0.20 | 14.25 | 3.66 | 49.50 | m³ | 67.41 |
| By hand (mechanical compaction) | | | | | | | |
| 100 mm thick | 3.30 | 0.07 | 4.75 | 0.04 | 3.30 | m² | 8.09 |
| 150 mm thick | 7.42 | 0.08 | 5.94 | 0.06 | 7.42 | m² | 13.42 |
| 250 mm thick | 12.38 | 0.13 | 8.91 | 0.10 | 12.38 | m² | 21.39 |
| over 250 mm thick | 24.75 | 0.20 | 14.25 | 0.10 | 24.75 | m³ | 39.10 |
| **Type 1 (crushed concrete) granular fill base; PC £14.00/tonne (£30.80/m³ compacted)** | | | | | | | |
| By machine | | | | | | | |
| 100 mm thick | 3.08 | 0.02 | 1.48 | 0.39 | 3.08 | m² | 4.95 |
| 150 mm thick | 4.62 | 0.03 | 1.78 | 0.60 | 4.62 | m² | 7.00 |
| 250 mm thick | 7.70 | 0.03 | 2.04 | 0.99 | 7.70 | m² | 10.73 |
| over 250 mm thick | 14.00 | 0.20 | 14.25 | 3.66 | 14.00 | m³ | 31.91 |

## 35 SITE WORKS – BASES

| Item<br>Excluding site overheads and profit | PC<br>£ | Labour<br>hours | Labour<br>£ | Plant<br>£ | Material<br>£ | Unit | Total<br>rate £ |
|---|---|---|---|---|---|---|---|
| **TYPE 1 BASES – cont** | | | | | | | |
| **Type 1 (crushed concrete) granular fill base – cont** | | | | | | | |
| By hand (mechanical compaction) | | | | | | | |
| 100 mm thick | 3.08 | 0.07 | 4.75 | 0.04 | 3.08 | m² | **7.87** |
| 150 mm thick | 4.62 | 0.08 | 5.94 | 0.06 | 4.62 | m² | **10.62** |
| 250 mm thick | 7.70 | 0.13 | 8.91 | 0.10 | 7.70 | m² | **16.71** |
| over 250 mm thick | 14.00 | 0.20 | 14.25 | 0.10 | 14.00 | m³ | **28.35** |
| **CELLULAR CONFINED BASES** | | | | | | | |
| **Geoweb; Greenfix; cellular confinement system for load support or permeable bases below pavings** | | | | | | | |
| To graded and compacted substrate (not included); on 50 mm sharp sand base filled with angular drainage aggregate 20–5 mm | | | | | | | |
| Geoweb 75 mm depth | 8.42 | 0.05 | 3.24 | 1.54 | 15.02 | m² | **19.80** |
| Geoweb 100 mm depth | 9.68 | 0.05 | 3.56 | 1.74 | 17.71 | m² | **23.01** |
| Geoweb 150 mm depth | 12.28 | 0.06 | 3.96 | 1.83 | 22.89 | m² | **28.68** |
| Geoweb 200 mm depth | 16.02 | 0.08 | 5.94 | 1.99 | 29.22 | m² | **37.15** |
| Geoweb 300 mm depth | 19.62 | 0.10 | 7.13 | 2.32 | 37.98 | m² | **47.43** |
| **SURFACE TREATMENTS TO GRANULAR BASES** | | | | | | | |
| **Surface treatments** | | | | | | | |
| Sand blinding; to hardcore base (not included); 25 mm thick | 1.02 | 0.03 | 0.79 | – | 1.02 | m² | **1.81** |
| Sand blinding; to hardcore base (not included); 50 mm thick | 2.05 | 0.05 | 1.19 | – | 2.05 | m² | **3.24** |
| Filter fabrics; to hardcore base (not included) | 0.43 | 0.01 | 0.24 | – | 0.43 | m² | **0.67** |

## 35 SITE WORKS – KERBS AND EDGINGS

| Item<br>Excluding site overheads and profit | PC<br>£ | Labour<br>hours | Labour<br>£ | Plant<br>£ | Material<br>£ | Unit | Total<br>rate £ |
|---|---|---|---|---|---|---|---|
| **FOUNDATIONS TO KERBS** | | | | | | | |
| **Foundations to kerbs** | | | | | | | |
| Excavating trenches; width 300 mm; 3 tonne excavator; disposal off site | | | | | | | |
|   depth 200 mm | – | – | – | 2.59 | 1.85 | m | **4.44** |
|   depth 300 mm | – | – | – | 3.73 | 2.77 | m | **6.50** |
|   depth 400 mm | – | – | – | 4.07 | 3.70 | m | **7.77** |
| By hand; barrowing to spoil heap and disposal off site | | | | | | | |
|   depth 300 | – | 0.06 | 4.28 | – | 30.81 | m | **35.09** |
|   depth 400 | – | 0.08 | 5.70 | – | 30.81 | m | **36.51** |
| Excavating trenches; width 450 mm; 3 tonne excavator; disposal off site | | | | | | | |
|   depth 300 mm | – | – | – | 4.66 | 4.16 | m | **8.82** |
|   depth 400 mm | – | – | – | 5.33 | 5.55 | m | **10.88** |
| **Foundations; to kerbs, edgings or channels; in situ concrete; 21 N/mm² – 20 mm aggregate ((1:2:4) site mixed); one side against earth face, other against formwork (not included); site mixed concrete** | | | | | | | |
| Site mixed concrete | | | | | | | |
|   150 mm wide × 100 mm deep | – | 0.04 | 3.17 | – | 1.74 | m | **4.91** |
|   150 mm wide × 150 mm deep | – | 0.06 | 3.95 | – | 2.61 | m | **6.56** |
|   200 mm wide × 150 mm deep | – | 0.07 | 4.75 | – | 3.48 | m | **8.23** |
|   300 mm wide × 150 mm deep | – | 0.06 | 4.16 | – | 5.21 | m | **9.37** |
|   600 mm wide × 200 mm deep | – | 0.06 | 4.15 | – | 13.91 | m | **18.06** |
| Ready mixed concrete | | | | | | | |
|   150 mm wide × 100 mm deep | – | 0.04 | 3.17 | – | 1.37 | m | **4.54** |
|   150 mm wide × 150 mm deep | – | 0.06 | 3.95 | – | 2.06 | m | **6.01** |
|   200 mm wide × 150 mm deep | – | 0.07 | 4.75 | – | 2.75 | m | **7.50** |
|   300 mm wide × 150 mm deep | – | 0.08 | 5.54 | – | 4.12 | m | **9.66** |
|   600 mm wide × 200 mm deep | – | 0.10 | 6.78 | – | 10.98 | m | **17.76** |
| Formwork; sides of foundations (this will usually be required to one side of each kerb foundation adjacent to road subbases) | | | | | | | |
|   100 mm deep | – | 0.01 | 0.99 | – | 0.12 | m | **1.11** |
|   150 mm deep | – | 0.01 | 0.99 | – | 0.18 | m | **1.17** |

## 35 SITE WORKS – KERBS AND EDGINGS

| Item<br>Excluding site overheads and profit | PC<br>£ | Labour<br>hours | Labour<br>£ | Plant<br>£ | Material<br>£ | Unit | Total<br>rate £ |
|---|---|---|---|---|---|---|---|
| **PRECAST CONCRETE KERBS** | | | | | | | |
| **Precast concrete kerbs, channels, edgings etc.; Marshalls Mono; bedding, jointing and pointing in cement:mortar (1:3); including haunching with in situ concrete; 11.50 N/mm$^2$ – 40 mm aggregate one side** | | | | | | | |
| Kerbs; straight | | | | | | | |
| 150 × 305 mm; HB1 | 11.68 | 0.17 | 11.88 | – | 14.69 | m | **26.57** |
| 125 × 255 mm; HB2; SP | 5.55 | 0.15 | 10.55 | – | 8.56 | m | **19.11** |
| 125 × 150 mm; BN | 3.61 | 0.13 | 9.50 | – | 6.62 | m | **16.12** |
| Dropper kerbs; left and right handed | | | | | | | |
| 125 × 255–150 mm; DL1 or DR1 | 8.04 | 0.17 | 11.88 | – | 11.05 | m | **22.93** |
| 125 × 255–150 mm; DL2 or DR2 | 8.04 | 0.17 | 11.88 | – | 11.05 | m | **22.93** |
| Quadrant kerbs | | | | | | | |
| 305 mm radius | 14.08 | 0.17 | 11.88 | – | 16.08 | nr | **27.96** |
| 455 mm radius | 14.95 | 0.17 | 11.88 | – | 17.95 | nr | **29.83** |
| Straight kerbs or channels; to radius; 125 × 255 mm | | | | | | | |
| 3.00 m radius (5 units per quarter circle) | 10.69 | 0.22 | 15.83 | – | 13.80 | m | **29.63** |
| 4.50 m radius (2 units per quarter circle) | 10.69 | 0.20 | 14.26 | – | 13.80 | m | **28.06** |
| 6.10 m radius (11 units per quarter circle) | 10.69 | 0.19 | 13.77 | – | 13.80 | m | **27.57** |
| 7.60 m radius (14 units per quarter circle) | 10.69 | 0.19 | 13.57 | – | 13.80 | m | **27.37** |
| 9.15 m radius (17 units per quarter circle) | 10.69 | 0.19 | 13.20 | – | 13.80 | m | **27.00** |
| 10.70 m radius (20 units per quarter circle) | 10.69 | 0.19 | 13.20 | – | 13.80 | m | **27.00** |
| 12.20 m radius (22 units per quarter circle) | 10.69 | 0.18 | 12.50 | – | 13.80 | m | **26.30** |
| Kerbs; Conservation Kerb units; to simulate natural granite kerbs | | | | | | | |
| 255 × 150 × 914 mm; laid flat | 26.28 | 0.19 | 13.57 | – | 31.67 | m | **45.24** |
| 150 × 255 × 914 mm; laid vertical | 26.28 | 0.19 | 13.57 | – | 30.79 | m | **44.36** |
| 145 × 255 mm; radius internal 3.25 m | 33.91 | 0.28 | 19.79 | – | 38.42 | m | **58.21** |
| 150 × 255 mm; radius external 3.40 m | 32.39 | 0.28 | 19.79 | – | 36.89 | m | **56.68** |
| 145 × 255 mm; radius internal 6.50 m | 30.30 | 0.22 | 15.83 | – | 34.80 | m | **50.63** |
| 145 × 255 mm; radius external 6.70 m | 30.13 | 0.22 | 15.83 | – | 34.63 | m | **50.46** |
| 150 × 255 mm; radius internal 9.80 m | 29.96 | 0.22 | 15.83 | – | 34.47 | m | **50.30** |
| 150 × 255 mm; radius external 10.00 m | 28.40 | 0.22 | 15.83 | – | 32.91 | m | **48.74** |
| 305 × 305 × 255 mm; solid quadrants | 36.49 | 0.19 | 13.57 | – | 41.00 | nr | **54.57** |

## 35 SITE WORKS – KERBS AND EDGINGS

| Item<br>Excluding site overheads and profit | PC<br>£ | Labour<br>hours | Labour<br>£ | Plant<br>£ | Material<br>£ | Unit | Total<br>rate £ |
|---|---|---|---|---|---|---|---|
| **Marshalls Mono; small element precast concrete kerb system; Keykerb Large (KL) upstand of 100–125 mm; on 150 mm concrete foundation including haunching with in situ concrete (1:3:6) one side** | | | | | | | |
| Bullnosed or half battered; | | | | | | | |
| 100 × 127 × 200 mm | | | | | | | |
| laid straight | 14.50 | 0.27 | 19.00 | – | 15.55 | m | **34.55** |
| radial blocks laid to curve; 8 blocks/ | | | | | | | |
| ¼ circle – 500 mm radius | 11.12 | 0.33 | 23.75 | – | 12.17 | m | **35.92** |
| radial blocks laid to curve; 8 radial blocks, alternating 8 standard blocks/ | | | | | | | |
| ¼ circle – 1000 mm radius | 18.37 | 0.42 | 29.69 | – | 19.42 | m | **49.11** |
| radial blocks, alternating 16 standard blocks/¼ circle – 1500 mm radius | 16.91 | 0.50 | 35.63 | – | 17.96 | m | **53.59** |
| internal angle 90° | 6.13 | 0.07 | 4.75 | – | 6.13 | nr | **10.88** |
| external angle | 6.13 | 0.07 | 4.75 | – | 6.13 | nr | **10.88** |
| drop crossing kerbs; KL half battered to KL Splay; LH and RH | 25.62 | 0.33 | 23.75 | – | 26.67 | pair | **50.42** |
| drop crossing kerbs; KL half battered to KS bullnosed | 24.92 | 0.33 | 23.75 | – | 25.97 | pair | **49.72** |
| **Marshalls Mono; small element precast concrete kerb system; Keykerb Small (KS) upstand of 25–50 mm; on 150 mm concrete foundation including haunching with in situ concrete (1:3:6) one side** | | | | | | | |
| Half battered | | | | | | | |
| laid straight | 10.19 | 0.27 | 19.00 | – | 11.24 | m | **30.24** |
| radial blocks laid to curve; 8 blocks/ | | | | | | | |
| ¼ circle; 500 mm radius | 8.36 | 0.33 | 23.75 | – | 9.41 | m | **33.16** |
| radial blocks laid to curve; 8 radial blocks, alternating 8 standard blocks/ | | | | | | | |
| ¼ circle; 1000 mm radius | 13.45 | 0.42 | 29.69 | – | 14.51 | m | **44.20** |
| radial blocks, alternating 16 standard blocks/¼ circle; 1500 mm radius | 14.40 | 0.50 | 35.63 | – | 15.45 | m | **51.08** |
| internal angle 90° | 6.13 | 0.07 | 4.75 | – | 6.13 | nr | **10.88** |
| external angle | 6.13 | 0.07 | 4.75 | – | 6.13 | nr | **10.88** |

## 35 SITE WORKS – KERBS AND EDGINGS

| Item<br>Excluding site overheads and profit | PC<br>£ | Labour<br>hours | Labour<br>£ | Plant<br>£ | Material<br>£ | Unit | **Total<br>rate £** |
|---|---|---|---|---|---|---|---|
| **PRECAST CONCRETE EDGINGS** | | | | | | | |
| **Precast concrete edging units; including haunching with in situ concrete; 11.50 N/mm² – 40 mm aggregate both sides** | | | | | | | |
| Edgings; rectangular, bullnosed or chamfered | | | | | | | |
|   50 × 150 mm | 2.44 | 0.11 | 7.92 | – | 7.45 | m | **15.37** |
|   125 × 150 mm bullnosed | 3.93 | 0.11 | 7.92 | – | 8.94 | m | **16.86** |
|   50 × 200 mm | 3.49 | 0.11 | 7.92 | – | 8.50 | m | **16.42** |
|   50 × 250 mm | 3.86 | 0.11 | 7.92 | – | 8.87 | m | **16.79** |
|   50 × 250 mm flat top | 4.28 | 0.11 | 7.92 | – | 9.29 | m | **17.21** |
| Conservation edgings | | | | | | | |
|   150 × 915 × 63 mm | 13.55 | 0.14 | 10.18 | – | 16.66 | ea | **26.84** |
| **Precast concrete block edgings; PC £10.51/m²; 200 × 100 × 60 mm; on prepared base (not included); haunching one side** | | | | | | | |
| Edgings; butt joints | | | | | | | |
|   stretcher course | 1.05 | 0.06 | 3.95 | – | 1.26 | m | **5.21** |
|   header course | 2.10 | 0.09 | 6.33 | – | 2.52 | m | **8.85** |
| **STONE KERBS** | | | | | | | |
| **Dressed natural stone kerbs – General** | | | | | | | |
| Preamble: The kerbs are to be good, sound and uniform in texture and free from defects; worked straight or to radius, square and out of wind, with the top front and back edges parallel or concentric to the dimensions specified. All drill and pick holes shall be removed from dressed faces. Standard dressings shall be designated as either fine picked, single axed or nidged or rough punched. | | | | | | | |
| **Dressed natural stone kerbs; CED Ltd; on concrete foundations (not included); including haunching with in situ concrete; 11.50 N/mm² – 40 mm aggregate one side** | | | | | | | |
| Granite kerbs; 125 × 250 mm | | | | | | | |
|     special quality; straight; random lengths | 21.00 | 0.27 | 19.00 | – | 24.63 | m | **43.63** |

## 35 SITE WORKS – KERBS AND EDGINGS

| Item<br>Excluding site overheads and profit | PC<br>£ | Labour<br>hours | Labour<br>£ | Plant<br>£ | Material<br>£ | Unit | Total<br>rate £ |
|---|---|---|---|---|---|---|---|
| Granite kerbs; 125 × 250 mm; curved to mean radius 3 m | | | | | | | |
|   special quality; random lengths | 29.03 | 0.30 | 21.61 | – | 32.66 | m | 54.27 |
| **STONE EDGINGS** | | | | | | | |
| **New granite sett edgings; CED Ltd; 100 × 100 × 100 mm; bedding in cement:mortar (1:4); including haunching with in situ concrete; 11.50 N/mm$^2$ – 40 mm aggregate one side; concrete foundations (not included)** | | | | | | | |
| Edgings 100 × 100 × 100 mm; ref 1R 'better quality' with 20 mm pointing gaps | | | | | | | |
|   100 mm wide | 4.34 | 0.44 | 31.59 | – | 7.84 | m | **39.43** |
|   2 rows; 220 mm wide | 8.68 | 0.74 | 52.73 | – | 13.21 | m | **65.94** |
|   3 rows; 340 mm wide | 13.02 | 1.00 | 71.25 | – | 16.58 | m | **87.83** |
| Edgings 100 × 100 × 200 mm; ref 1R 'better quality' with 20 mm pointing gaps | | | | | | | |
|   100 mm wide | 4.42 | 0.25 | 17.81 | – | 7.92 | m | **25.73** |
|   2 rows; 220 mm wide | 8.83 | 0.50 | 35.63 | – | 13.36 | m | **48.99** |
|   3 rows; 340 mm wide | 13.25 | 0.75 | 53.44 | – | 16.81 | m | **70.25** |
| Edgings 100 × 100 × 100 mm; ref 8R 'standard quality' with 20 mm pointing gaps | | | | | | | |
|   100 mm wide | 3.37 | 0.44 | 31.59 | – | 6.87 | m | **38.46** |
|   2 rows; 220 mm wide | 6.73 | 0.74 | 52.73 | – | 11.26 | m | **63.99** |
|   3 rows; 340 mm wide | 10.10 | 1.00 | 71.25 | – | 13.66 | m | **84.91** |
| Edgings 100 × 100 × 200 mm; ref 8R 'standard quality' with 20 mm pointing gaps | | | | | | | |
|   100 mm wide | 3.59 | 0.25 | 17.81 | – | 7.09 | m | **24.90** |
|   2 rows; 220 mm wide | 7.18 | 0.50 | 35.63 | – | 11.71 | m | **47.34** |
|   3 rows; 340 mm wide | 10.77 | 0.75 | 53.44 | – | 14.33 | m | **67.77** |
| **Reclaimed granite setts edgings; CED Ltd; 100 × 100 mm; bedding in cement:mortar (1:4); including haunching with in situ concrete; 11.50 N/mm$^2$ – 40 mm aggregate one side; on concrete foundations (not included)** | | | | | | | |
| Edgings; with 20 mm pointing gaps | | | | | | | |
|   100 mm wide | 3.58 | 0.44 | 31.59 | – | 7.08 | m | **38.67** |
|   2 rows; 220 mm wide | 7.17 | 0.74 | 52.73 | – | 11.69 | m | **64.42** |
|   3 rows; 340 mm wide | 10.75 | 1.00 | 71.25 | – | 14.31 | m | **85.56** |

## 35 SITE WORKS – KERBS AND EDGINGS

| Item<br>Excluding site overheads and profit | PC<br>£ | Labour<br>hours | Labour<br>£ | Plant<br>£ | Material<br>£ | Unit | Total<br>rate £ |
|---|---|---|---|---|---|---|---|
| **STONE EDGINGS – cont** | | | | | | | |
| **Natural Yorkstone edgings; Johnsons Wellfield Quarries; sawn six sides; 50 mm thick; on prepared base measured separately; bedding on 25 mm cement:sand (1:3); cement:sand (1:3) joints** | | | | | | | |
| Edgings | | | | | | | |
| 100 mm wide × random lengths | 8.90 | 0.17 | 11.88 | – | 9.33 | m | **21.21** |
| 100 × 100 mm | 8.90 | 0.17 | 11.88 | – | 13.16 | m | **25.04** |
| 100 × 200 mm | 17.81 | 0.17 | 11.88 | – | 22.06 | m | **33.94** |
| 250 mm wide × random lengths | 18.09 | 0.13 | 9.50 | – | 22.35 | m | **31.85** |
| 500 mm wide × random lengths | 36.17 | 0.11 | 7.92 | – | 40.43 | m | **48.35** |
| Yorkstone edgings; 600 mm long × 250 mm wide; cut to radius | | | | | | | |
| 1.00 m to 3.00 | 50.09 | 0.17 | 11.88 | – | 50.51 | m | **62.39** |
| 3.00 m to 5.00 m | 50.09 | 0.15 | 10.55 | – | 50.51 | m | **61.06** |
| exceeding 5.00 m | 50.09 | 0.13 | 9.50 | – | 50.51 | m | **60.01** |
| **Natural Yorkstone edgings; Johnsons Wellfield Quarries; sawn six sides; 75 mm thick; on prepared base measured separately; bedding on 25 mm cement:sand (1:3); cement:sand (1:3) joints** | | | | | | | |
| Edgings | | | | | | | |
| 100 mm wide × random lengths | 10.57 | 0.20 | 14.25 | – | 11.00 | m | **25.25** |
| 100 × 100 mm | 1.06 | 0.20 | 14.25 | – | 5.31 | m | **19.56** |
| 100 × 200 mm | 2.11 | 0.17 | 11.88 | – | 6.37 | m | **18.25** |
| 250 mm wide × random lengths | 20.87 | 0.17 | 11.88 | – | 25.13 | m | **37.01** |
| 500 mm wide × random lengths | 41.74 | 0.13 | 9.50 | – | 45.99 | m | **55.49** |
| Edgings; 600 mm long × 250 mm wide; cut to radius | | | | | | | |
| 1.00 m to 3.00 | 64.00 | 0.20 | 14.25 | – | 64.42 | m | **78.67** |
| 3.00 m to 5.00 m | 64.00 | 0.17 | 11.88 | – | 64.42 | m | **76.30** |
| exceeding 5.00 m | 64.00 | 0.15 | 10.55 | – | 64.42 | m | **74.97** |
| **Natural stone, slate or granite flag edgings; CED Ltd; on prepared base (not included); bedding on 25 mm cement:sand (1:3); cement:sand (1:3) joints** | | | | | | | |
| Edgings; silver grey | | | | | | | |
| 100 mm wide × random lengths | 2.02 | 0.33 | 23.75 | – | 2.45 | m | **26.20** |
| 100 × 100 mm | 2.31 | 0.50 | 35.63 | – | 2.74 | m | **38.37** |
| 100 mm long × 200 mm wide | 3.76 | 0.20 | 14.25 | – | 4.19 | m | **18.44** |
| 250 mm wide × random lengths | 3.98 | 0.25 | 17.81 | – | 8.23 | m | **26.04** |
| 300 mm wide × random lengths | 4.34 | 0.25 | 17.81 | – | 8.59 | m | **26.40** |

## 35 SITE WORKS – KERBS AND EDGINGS

| Item<br>Excluding site overheads and profit | PC<br>£ | Labour<br>hours | Labour<br>£ | Plant<br>£ | Material<br>£ | Unit | Total<br>rate £ |
|---|---|---|---|---|---|---|---|
| **BRICK EDGINGS** | | | | | | | |
| **Brick or block stretchers; bedding in cement:mortar (1:4); on 150 mm deep concrete foundations (not included);** | | | | | | | |
| Single course | | | | | | | |
|   concrete paving blocks; PC £11.29/m²; 200 × 100 × 60 mm | 2.26 | 0.10 | 7.30 | – | 3.10 | m | **10.40** |
|   engineering bricks; PC £360.00/1000; 215 × 102.5 × 65 mm | 2.10 | 0.13 | 9.50 | – | 2.94 | m | **12.44** |
|   paving bricks; PC £450.00/1000; 215 × 102.5 × 65 mm | 2.00 | 0.13 | 9.50 | – | 2.84 | m | **12.34** |
| Two courses | | | | | | | |
|   concrete paving blocks; PC £11.29 /m²; 200 × 100 × 60 mm | 4.52 | 0.13 | 9.50 | – | 6.04 | m | **15.54** |
|   engineering bricks; PC £360.00/1000; 215 × 102.5 × 65 mm | 4.20 | 0.19 | 13.57 | – | 5.72 | m | **19.29** |
|   paving bricks; PC £450.00/1000; 215 × 102.5 × 65 mm | 4.00 | 0.19 | 13.57 | – | 5.52 | m | **19.09** |
| Three courses | | | | | | | |
|   concrete paving blocks; PC £11.29/m²; 200 × 100 × 60 mm | 6.77 | 0.15 | 10.55 | – | 8.30 | m | **18.85** |
|   engineering bricks; PC £360.00/1000; 215 × 102.5 × 65 mm | 6.30 | 0.22 | 15.83 | – | 7.82 | m | **23.65** |
|   paving bricks; PC £450.00/1000; 215 × 102.5 × 65 mm | 6.00 | 0.22 | 15.83 | – | 7.52 | m | **23.35** |
| **Bricks on edge; bedding in cement: mortar (1:4); on 150 mm deep concrete foundations** | | | | | | | |
| One brick wide | | | | | | | |
|   engineering bricks; 215 × 102.5 × 65 mm | 6.30 | 0.19 | 13.57 | – | 7.99 | m | **21.56** |
|   paving bricks; 215 × 102.5 × 65 mm | 6.00 | 0.19 | 13.57 | – | 7.69 | m | **21.26** |
| Two courses; stretchers laid on edge; 225 mm wide | | | | | | | |
|   engineering bricks; 215 × 102.5 × 65 mm | 12.60 | 0.40 | 28.50 | – | 15.80 | m | **44.30** |
|   paving bricks; 215 × 102.5 × 65 mm | 12.00 | 0.40 | 28.50 | – | 15.20 | m | **43.70** |
| **Edge restraints; to brick paving; on prepared base (not included); 65 mm thick bricks; PC £300.00/1000** | | | | | | | |
| Header course | | | | | | | |
|   200 × 100 mm; butt joints | 3.00 | 0.09 | 6.33 | – | 3.41 | m | **9.74** |
|   200x 100 mm; mortar joints | 2.80 | 0.17 | 11.88 | – | 5.40 | m | **17.28** |
|   200 × 100 mm × 50; on edge; mortar joints | 4.19 | 0.23 | 16.63 | – | 5.01 | m | **21.64** |
|   215 × 102.5 mm mortar joints | 2.67 | 0.17 | 11.88 | – | 3.48 | m | **15.36** |
|   215 × 102.5 mm × 65; on edge; mortar joints | 4.00 | 0.25 | 17.82 | – | 4.81 | m | **22.63** |

## 35 SITE WORKS – KERBS AND EDGINGS

| Item<br>Excluding site overheads and profit | PC<br>£ | Labour<br>hours | Labour<br>£ | Plant<br>£ | Material<br>£ | Unit | Total<br>rate £ |
|---|---|---|---|---|---|---|---|
| **BRICK EDGINGS – cont** | | | | | | | |
| **Edge restraints – cont** | | | | | | | |
| Stretcher course; on flat or on edge | | | | | | | |
| 200 × 100 mm; butt joints | 1.50 | 0.06 | 3.95 | – | 1.70 | m | **5.65** |
| 210 × 105 mm; mortar joints | 1.36 | 0.11 | 7.92 | – | 1.75 | m | **9.67** |
| **Haunching** | | | | | | | |
| Haunching with in situ concrete; 11.50 N/mm$^2$ – 40 mm aggregate one side | | | | | | | |
| one side | – | 0.13 | 5.94 | – | 3.43 | m | **9.37** |
| **Variation in brick prices; add or subtract the following amounts for every £100/1000 difference in the PC price** | | | | | | | |
| Edgings; mortar jointed | | | | | | | |
| 100 mm wide | 0.50 | – | – | – | 0.50 | m | **0.50** |
| 200 mm wide | 0.95 | – | – | – | 0.95 | m | **0.95** |
| 102.5 mm wide | 0.47 | – | – | – | 0.47 | m | **0.47** |
| 215 mm wide | 0.93 | – | – | – | 0.93 | m | **0.93** |
| **CHANNELS** | | | | | | | |
| **Channels; bedding in cement:mortar (1:3); joints pointed flush; on concrete foundations (not included)** | | | | | | | |
| Three courses stretchers; 350 mm wide; quarter bond to form dished channels | | | | | | | |
| engineering bricks; PC £360.00/1000; 215 × 102.5 × 65 mm | 6.30 | 0.33 | 23.75 | – | 7.82 | m | **31.57** |
| paving bricks; PC £450.00/1000; 215 × 102.5 × 65 mm | 6.00 | 0.33 | 23.75 | – | 7.52 | m | **31.27** |
| Three courses granite setts; 340 mm wide; to form dished channels | | | | | | | |
| 340 mm wide | 12.03 | 0.67 | 47.50 | – | 15.53 | m | **63.03** |
| **Precast concrete channels etc.; Marshalls Mono; bedding, jointing and pointing in cement:mortar (1:3); including haunching with in situ concrete; 11.50 N/mm$^2$ – 40 mm aggregate one side** | | | | | | | |
| Channels; square | | | | | | | |
| 125 × 225 × 915 mm long; CS1 | 4.81 | 0.13 | 9.50 | – | 15.65 | m | **25.15** |
| 125 × 150 × 915 mm long; CS2 | 4.17 | 0.13 | 9.50 | – | 13.18 | m | **22.68** |

## 35 SITE WORKS – KERBS AND EDGINGS

| Item<br>Excluding site overheads and profit | PC<br>£ | Labour<br>hours | Labour<br>£ | Plant<br>£ | Material<br>£ | Unit | Total<br>rate £ |
|---|---|---|---|---|---|---|---|
| Channels; dished | | | | | | | |
| 305 × 150 × 915 mm long; CD | 14.35 | 0.13 | 9.50 | – | 26.06 | m | **35.56** |
| 150 × 100 × 915 mm | 8.42 | 0.13 | 9.50 | – | 15.08 | m | **24.58** |
| **TIMBER EDGINGS** | | | | | | | |
| **Timber edging boards; fixed with 50 × 50 × 750 mm timber pegs at 1000 mm centres (excavations and hardcore under edgings not included)** | | | | | | | |
| Straight | | | | | | | |
| 22 × 150 mm treated softwood edge boards | 1.96 | 0.07 | 3.17 | – | 1.96 | m | **5.13** |
| 50 × 150 mm treated softwood edge boards | 2.89 | 0.10 | 4.75 | – | 2.89 | m | **7.64** |
| Curved | | | | | | | |
| 22 × 150 mm treated softwood edge boards | 1.96 | 0.13 | 5.94 | – | 1.96 | m | **7.90** |
| 50 × 150 mm treated softwood edge boards | 2.89 | 0.14 | 6.79 | – | 2.89 | m | **9.68** |
| **METAL EDGINGS** | | | | | | | |
| **AluExcel Asphalt edge L-shaped aluminium edging; Kinley Systems; extruded aluminium alloy L shaped edging with 5.33 mm exposed upper lip; edging fixed to roadway base and edge profile with 250 mm steel fixing spike; laid to straight or curvilinear road edge; subsequently filled with macadam (not included); Rigid or Flexible profiles** | | | | | | | |
| Depth of macadam | | | | | | | |
| 40 mm | 9.28 | 0.07 | 3.17 | – | 9.28 | m | **12.45** |
| 50 mm | 10.11 | 0.07 | 3.17 | – | 10.11 | m | **13.28** |
| 65 mm | 10.60 | 0.07 | 3.39 | – | 10.60 | m | **13.99** |
| 75 mm | 11.88 | 0.08 | 3.65 | – | 11.88 | m | **15.53** |
| 100 mm | 14.78 | 0.08 | 3.96 | – | 14.78 | m | **18.74** |

## 35 SITE WORKS – KERBS AND EDGINGS

| Item<br>Excluding site overheads and profit | PC<br>£ | Labour<br>hours | Labour<br>£ | Plant<br>£ | Material<br>£ | Unit | Total<br>rate £ |
|---|---|---|---|---|---|---|---|
| **METAL EDGINGS – cont** | | | | | | | |
| **AluExcel L-shaped aluminium edging; Kinley Systems; heavy duty straight profile edging; for edgings to soft landscape beds or turf areas; 3.2 mm × 100 mm high; 3.2 mm thick with 4.75 mm exposed upper lip; fixed to form straight or curvilinear edge with 305 mm fixing spike; Rigid or Flexible profiles** | | | | | | | |
| Milled aluminium | | | | | | | |
| 100 mm deep | – | 0.10 | 4.75 | – | 14.42 | m | **19.17** |
| **Permaloc Permastrip; Kinley Systems; heavy duty L-shaped profile maintenance strip; 3.2 mm × 89 mm high with 5.2 mm exposed top lip; for straight or gentle curves on paths or bed turf interfaces; fixed to form straight or curvilinear edge with standard 305 mm stake; other stake lengths available** | | | | | | | |
| Milled aluminium | | | | | | | |
| 75 mm deep | 11.88 | 0.10 | 4.75 | – | 11.88 | m | **16.63** |
| **Permaloc Proline; Kinley Systems; medium duty straight profiled maintenance strip; 3.2 mm × 102 mm high with 3.18 mm exposed top lip; for straight or gentle curves on paths or bed turf interfaces; fixed to form straight or curvilinear edge with standard 305 mm stake; other stake lengths available** | | | | | | | |
| Milled aluminium | | | | | | | |
| 50 mm deep | – | 0.10 | 4.75 | – | 9.86 | m | **14.61** |
| **Mild steel edgings; 6 mm galvanized steel** | | | | | | | |
| Bespoke edgings; complete with bolts for adjoining sections and mild steel pins at 1.50 m centres | | | | | | | |
| 100 mm straight | 18.00 | 0.07 | 3.17 | – | 18.00 | m | **21.17** |
| 150 mm straight | 26.00 | 0.10 | 4.75 | – | 26.00 | m | **30.75** |
| 300 mm straight | 33.00 | 0.13 | 5.94 | – | 33.00 | m | **38.94** |

## 35 SITE WORKS – PAVINGS

| Item<br>Excluding site overheads and profit | PC<br>£ | Labour<br>hours | Labour<br>£ | Plant<br>£ | Material<br>£ | Unit | Total<br>rate £ |
|---|---|---|---|---|---|---|---|
| **SPECIALIZED MORTARS FOR PAVINGS** | | | | | | | |
| **SteinTec Ltd** | | | | | | | |
| Primer for specialized mortar; SteinTec Tuffbond priming mortar immediately prior to paving material being placed on bedding mortar | | | | | | | |
| maximum thickness 1.5 mm (1.5 kg/m$^2$) | 1.55 | 0.02 | 0.47 | – | 1.55 | m$^2$ | **2.02** |
| Tuffbed; 2 pack hydraulic mortar for paving applications | | | | | | | |
| 30 mm thick | 8.48 | 0.17 | 3.96 | – | 8.48 | m$^2$ | **12.44** |
| 40 mm thick | 11.31 | 0.20 | 4.75 | – | 11.31 | m$^2$ | **16.06** |
| 50 mm thick | 14.14 | 0.25 | 5.94 | – | 14.14 | m$^2$ | **20.08** |
| SteinTec jointing mortar; joint size | | | | | | | |
| 100 × 100 × 100 mm; 10 mm joints; 31.24 kg/m$^2$ | 29.03 | 0.50 | 11.88 | – | 29.03 | m$^2$ | **40.91** |
| 100 × 100 × 100 mm; 20 mm joints; 31.24 kg/m$^2$ | 51.11 | 1.00 | 23.75 | – | 51.11 | m$^2$ | **74.86** |
| 200 × 100 × 65 mm; 10 mm joints; 15.7 kg/m$^2$ | 14.59 | 0.10 | 2.38 | – | 14.59 | m$^2$ | **16.97** |
| 215 × 112.5 × 65 mm; 10 mm joints; 14.3 kg/m$^2$ | 13.29 | 0.08 | 1.98 | – | 13.29 | m$^2$ | **15.27** |
| 300 × 450 × 65 mm; 10 mm joints; 6.24 kg/m$^2$ | 5.80 | 0.04 | 0.95 | – | 5.80 | m$^2$ | **6.75** |
| 300 × 450 × 65 mm; 20 mm joints; 11.98 kg/m$^2$ | 11.13 | 0.10 | 2.38 | – | 11.13 | m$^2$ | **13.51** |
| 450 × 450 × 65 mm; 20 mm joints; 9.75 kg/m$^2$ | 9.06 | 0.11 | 2.64 | – | 9.06 | m$^2$ | **11.70** |
| 450 × 600 × 65 mm; 20 mm joints; 8.6 kg/m$^2$ | 7.99 | 0.09 | 2.16 | – | 7.99 | m$^2$ | **10.15** |
| 600 × 600 × 65 mm; 20 mm joints; 7.43 kg/m$^2$ | 6.90 | 0.07 | 1.70 | – | 6.90 | m$^2$ | **8.60** |
| **EXCAVATION OF PAVINGS** | | | | | | | |
| **Excavation and path preparation** | | | | | | | |
| Excavating; 300 mm deep; to width of path; depositing excavated material at sides of excavation | | | | | | | |
| width 1.00 m | – | – | – | 3.37 | – | m$^2$ | **3.37** |
| width 1.50 m | – | – | – | 2.81 | – | m$^2$ | **2.81** |
| width 2.00 m | – | – | – | 2.40 | – | m$^2$ | **2.40** |
| width 3.00 m | – | – | – | 2.02 | – | m$^2$ | **2.02** |
| Excavating trenches; in centre of pathways; 100 mm flexible drain pipes; filling with clean broken stone or gravel rejects | | | | | | | |
| 300 × 450 mm deep | 5.62 | 0.03 | 2.37 | 1.68 | 5.62 | m | **9.67** |

## 35 SITE WORKS – PAVINGS

| Item<br>Excluding site overheads and profit | PC<br>£ | Labour<br>hours | Labour<br>£ | Plant<br>£ | Material<br>£ | Unit | Total<br>rate £ |
|---|---|---|---|---|---|---|---|
| **EXCAVATION OF PAVINGS – cont** | | | | | | | |
| **Excavation and path preparation – cont** | | | | | | | |
| Hand trimming and compacting reduced surface of pathway; by machine | | | | | | | |
| width 1.00 m | – | 0.02 | 1.19 | 0.12 | – | m | **1.31** |
| width 1.50 m | – | 0.01 | 1.05 | 0.11 | – | m | **1.16** |
| width 2.00 m | – | 0.01 | 0.95 | 0.10 | – | m | **1.05** |
| width 3.00 m | – | 0.01 | 0.95 | 0.10 | – | m | **1.05** |
| Permeable membranes; to trimmed and compacted surface of pathway | | | | | | | |
| Terram 1000 | 0.43 | 0.01 | 0.48 | – | 0.43 | m$^2$ | **0.91** |
| **FILLING TO MAKE UP LEVELS FOR PAVINGS AND SURFACES** | | | | | | | |
| **Filling to make up levels** | | | | | | | |
| Obtained off site; hardcore; PC £20.00/m$^3$ | | | | | | | |
| 150 mm thick | 3.20 | 0.01 | 0.95 | 0.60 | 3.20 | m$^2$ | **4.75** |
| Type 1 granular fill base; PC £20.00/ tonne (£42.90/m$^3$ compacted) | | | | | | | |
| 100 mm thick | 4.95 | 0.01 | 0.66 | 0.46 | 4.95 | m$^2$ | **6.07** |
| 150 mm thick | 7.42 | 0.01 | 0.59 | 0.69 | 7.42 | m$^2$ | **8.70** |
| **Surface treatments** | | | | | | | |
| Sand blinding; to hardcore (not included) | | | | | | | |
| 50 mm thick | 2.25 | 0.01 | 0.95 | – | 2.25 | m$^2$ | **3.20** |
| Filter fabric; to hardcore (not included) | 0.43 | – | 0.24 | – | 0.43 | m$^2$ | **0.67** |
| **BRICK PAVINGS** | | | | | | | |
| **Bricks – General** | | | | | | | |
| Preamble: Bricks shall be hard, well burnt, non-dusting, resistant to frost and sulphate attack and true to shape, size and sample. | | | | | | | |
| **Movement of materials** | | | | | | | |
| Mechanically offloading bricks; loading wheelbarrows; transporting maximum 25 m distance | – | 0.07 | 4.75 | – | – | m$^2$ | **4.75** |

## 35 SITE WORKS – PAVINGS

| Item<br>Excluding site overheads and profit | PC<br>£ | Labour<br>hours | Labour<br>£ | Plant<br>£ | Material<br>£ | Unit | Total<br>rate £ |
|---|---|---|---|---|---|---|---|
| **Edge restraints; to brick paving; on prepared base (not included); 65 mm thick bricks; PC £300.00/ 1000; haunching one side** | | | | | | | |
| Header course | | | | | | | |
| 200 × 100 mm; butt joints | 3.00 | 0.09 | 6.33 | – | 6.02 | m | **12.35** |
| 210 × 105 mm; mortar joints | 2.67 | 0.17 | 11.88 | – | 6.09 | m | **17.97** |
| Stretcher course | | | | | | | |
| 200 × 100 mm; butt joints | 1.50 | 0.06 | 3.95 | – | 4.31 | m | **8.26** |
| 210 × 105 mm; mortar joints | 1.36 | 0.11 | 7.92 | – | 4.36 | m | **12.28** |
| **Variation in brick prices; add or subtract the following amounts for every £100/1000 difference in the PC price** | | | | | | | |
| Edgings; mortar jointed | | | | | | | |
| 100 mm wide | 0.50 | – | – | – | 0.50 | m | **0.50** |
| 200 mm wide | 0.95 | – | – | – | 0.95 | m | **0.95** |
| 102.5 mm wide | 0.47 | – | – | – | 0.47 | m | **0.47** |
| 215 mm wide | 0.93 | – | – | – | 0.93 | m | **0.93** |
| **Clay brick pavings; on prepared base (not included); bedding on 50 mm sharp sand; kiln dried sand joints** | | | | | | | |
| Pavings; 200 × 100 × 65 mm wirecut chamfered paviors | | | | | | | |
| bricks; PC £450.00/1000 | 23.06 | 0.46 | 32.90 | 0.23 | 25.66 | $m^2$ | **58.79** |
| **Clay brick pavings; 200 × 100 × 50 mm; laid to running stretcher or stack bond only; on prepared base (not included); bedding on cement:sand (1:4) pointing mortar as work proceeds** | | | | | | | |
| PC £600.00/1000 | | | | | | | |
| laid on edge | 48.81 | 1.59 | 113.10 | – | 57.95 | $m^2$ | **171.05** |
| laid on edge but pavior 65 mm thick | 41.00 | 1.27 | 90.47 | – | 50.14 | $m^2$ | **140.61** |
| laid flat | 26.62 | 0.73 | 52.25 | – | 32.07 | $m^2$ | **84.32** |
| PC £500.00/1000 | | | | | | | |
| laid on edge | 40.67 | 1.59 | 113.10 | – | 49.82 | $m^2$ | **162.92** |
| laid on edge but pavior 65 mm thick | 34.16 | 1.27 | 90.47 | – | 43.31 | $m^2$ | **133.78** |
| laid flat | 22.19 | 0.73 | 52.25 | – | 27.63 | $m^2$ | **79.88** |
| PC £400.00/1000 | | | | | | | |
| laid flat | 17.75 | 0.73 | 52.25 | – | 23.20 | $m^2$ | **75.45** |
| laid on edge | 32.54 | 1.59 | 113.10 | – | 41.68 | $m^2$ | **154.78** |
| laid on edge but pavior 65 mm thick | 27.33 | 1.27 | 90.47 | – | 36.47 | $m^2$ | **126.94** |
| PC £300.00/1000 | | | | | | | |
| laid on edge | 24.40 | 1.59 | 113.10 | – | 33.55 | $m^2$ | **146.65** |
| laid on edge but pavior 65 mm thick | 20.50 | 1.27 | 90.47 | – | 29.64 | $m^2$ | **120.11** |
| laid flat | 13.31 | 0.73 | 52.25 | – | 18.76 | $m^2$ | **71.01** |

## 35 SITE WORKS – PAVINGS

| Item<br>Excluding site overheads and profit | PC<br>£ | Labour<br>hours | Labour<br>£ | Plant<br>£ | Material<br>£ | Unit | Total<br>rate £ |
|---|---|---|---|---|---|---|---|
| **BRICK PAVINGS – cont** | | | | | | | |
| **Clay brick pavings;**<br>**200 × 100 × 50 mm; butt jointed laid**<br>**herringbone or basketweave pattern**<br>**only; on prepared base (not**<br>**included); bedding on 50 mm sharp**<br>**sand** | | | | | | | |
| PC £600.00/1000 | | | | | | | |
|   laid flat | 30.75 | 0.46 | 32.90 | 0.30 | 33.35 | m² | **66.55** |
| PC £500.00/1000 | | | | | | | |
|   laid flat | 25.63 | 0.46 | 32.90 | 0.30 | 28.22 | m² | **61.42** |
| PC £400.00/1000 | | | | | | | |
|   laid flat | 20.50 | 0.46 | 32.90 | 0.30 | 23.10 | m² | **56.30** |
| PC £300.00/1000 | | | | | | | |
|   laid flat | 15.38 | 0.46 | 32.90 | 0.30 | 17.97 | m² | **51.17** |
| **Clay brick pavings;**<br>**215 × 102.5 × 65 mm; on prepared**<br>**base (not included); bedding on**<br>**cement:sand (1:4) pointing mortar as**<br>**work proceeds** | | | | | | | |
| Paving bricks; PC £600.00/1000;<br>herringbone bond | | | | | | | |
|   laid on edge | 36.44 | 1.19 | 84.43 | – | 44.48 | m² | **128.91** |
|   laid flat | 23.70 | 0.79 | 56.33 | – | 31.74 | m² | **88.07** |
| Paving bricks; PC £600.00/1000;<br>basketweave bond | | | | | | | |
|   laid on edge | 36.44 | 0.79 | 56.29 | – | 44.48 | m² | **100.77** |
|   laid flat | 23.70 | 0.53 | 37.53 | – | 31.74 | m² | **69.27** |
| Paving bricks; PC £600.00/1000;<br>running or stack bond | | | | | | | |
|   laid on edge | 36.44 | 0.63 | 45.04 | – | 44.48 | m² | **89.52** |
|   laid flat | 23.70 | 0.42 | 30.02 | – | 31.74 | m² | **61.76** |
| Paving bricks; PC £500.00/1000;<br>herringbone bond | | | | | | | |
|   laid on edge | 30.37 | 1.19 | 84.43 | – | 34.98 | m² | **119.41** |
|   laid flat | 19.75 | 0.79 | 56.29 | – | 27.79 | m² | **84.08** |
| Paving bricks; PC £500.00/1000;<br>basketweave bond | | | | | | | |
|   laid on edge | 30.37 | 0.79 | 56.29 | – | 34.98 | m² | **91.27** |
|   laid flat | 19.75 | 0.53 | 37.53 | – | 27.79 | m² | **65.32** |
| Paving bricks; PC £500.00/1000;<br>running or stack bond | | | | | | | |
|   laid on edge | 30.37 | 0.63 | 45.04 | – | 34.98 | m² | **80.02** |
|   laid flat | 19.75 | 0.42 | 30.02 | – | 27.79 | m² | **57.81** |
| Paving bricks; PC £400.00/1000;<br>herringbone bond | | | | | | | |
|   laid on edge | 24.29 | 1.19 | 84.43 | – | 28.91 | m² | **113.34** |
|   laid flat | 15.80 | 0.79 | 56.29 | – | 23.69 | m² | **79.98** |

## 35 SITE WORKS – PAVINGS

| Item<br>Excluding site overheads and profit | PC<br>£ | Labour<br>hours | Labour<br>£ | Plant<br>£ | Material<br>£ | Unit | Total<br>rate £ |
|---|---|---|---|---|---|---|---|
| Paving bricks; PC £400.00/1000; basketweave bond | | | | | | | |
| laid on edge | 24.29 | 0.79 | 56.29 | – | 28.91 | m² | **85.20** |
| laid flat | 15.80 | 0.53 | 37.53 | – | 23.69 | m² | **61.22** |
| Paving bricks; PC £400.00/1000; running or stack bond | | | | | | | |
| laid on edge | 24.29 | 0.63 | 45.04 | – | 28.91 | m² | **73.95** |
| laid flat | 15.80 | 0.42 | 30.02 | – | 23.69 | m² | **53.71** |
| Paving bricks; PC £300.00/1000; herringbone bond | | | | | | | |
| laid on edge | 17.77 | 1.19 | 84.43 | – | 22.39 | m² | **106.82** |
| laid flat | 11.85 | 0.79 | 56.29 | – | 19.89 | m² | **76.18** |
| Paving bricks; PC £300.00/1000; basketweave bond | | | | | | | |
| laid on edge | 17.77 | 0.79 | 56.29 | – | 22.39 | m² | **78.68** |
| laid flat | 11.85 | 0.53 | 37.53 | – | 19.89 | m² | **57.42** |
| Paving bricks; PC £300.00/1000; running or stack bond | | | | | | | |
| laid on edge | 17.77 | 0.63 | 45.03 | – | 22.39 | m² | **67.42** |
| laid flat | 11.85 | 0.42 | 30.02 | – | 19.89 | m² | **49.91** |
| Cutting | | | | | | | |
| curved cutting | – | 0.15 | 10.55 | 5.25 | – | m | **15.80** |
| raking cutting | – | 0.11 | 7.92 | 4.48 | – | m | **12.40** |
| **Add or subtract the following amounts for every £10.00/1000 difference in the prime cost of bricks** | | | | | | | |
| Butt joints | | | | | | | |
| 200 × 100 mm | – | – | – | – | 0.50 | m² | **0.50** |
| 215 × 102.5 mm | – | – | – | – | 0.45 | m² | **0.45** |
| 10 mm mortar joints | | | | | | | |
| 200 × 100 mm | – | – | – | – | 0.43 | m² | **0.43** |
| 215 × 102.5 mm | – | – | – | – | 0.40 | m² | **0.40** |

## 35 SITE WORKS – PAVINGS

| Item<br>Excluding site overheads and profit | PC<br>£ | Labour<br>hours | Labour<br>£ | Plant<br>£ | Material<br>£ | Unit | Total<br>rate £ |
|---|---|---|---|---|---|---|---|
| **CELLULAR CONFINED GRANULAR PAVINGS** | | | | | | | |
| **StablePATH; Cedar Nursery;** honeycomb shaped cellular polypropylene containment grids for retaining filled decorative aggregate surfaces (not included) for driveways, cycle paths and footpaths; to graded compacted substrate (not included) and 50 mm sharp sand base; integrated geofabric base | | | | | | | |
| Foot traffic, wheelchairs, bikes and pushchairs | | | | | | | |
| StablePATH; 1200 × 800 × 25 mm deep | 9.25 | 0.08 | 2.97 | – | 9.25 | m² | **12.22** |
| Driveways: vehicles, wheelchairs, bikes and pushchairs (requires compacted granular base to support vehicle loadings) | | | | | | | |
| StablePAVE; 2400 × 1200 × 40 mm deep | 2.05 | 0.08 | 3.23 | – | 10.04 | m² | **13.27** |
| Aggregate filling to StablePATH | | | | | | | |
| shingle | 1.11 | 0.03 | 2.37 | 1.95 | 1.11 | m² | **5.43** |
| decorative aggregate; CED Ltd; PC £61.00/tonne | 2.75 | 0.03 | 2.37 | 1.95 | 2.75 | m² | **7.07** |
| Aggregate filling to StablePAVE | | | | | | | |
| shingle | 1.78 | 0.03 | 2.00 | 2.36 | 1.78 | m² | **6.14** |
| decorative aggregate; CED Ltd; PC £61.00/tonne | 4.39 | 0.03 | 2.00 | 2.36 | 4.39 | m² | **8.75** |
| **StableDRIVE; Cedar Nursery;** honeycomb shaped polypropylene cellular containment system for retaining filled decorative aggregate surfaces for driveways, cyclepaths and footpaths; on compacted subgrade and subbase (both not included) | | | | | | | |
| Driveways: vehicles, wheelchairs, bikes and pushchairs (requires compacted granular base to support vehicle loadings) | | | | | | | |
| StableDRIVE; 600 × 500 × 32 mm deep | 2.05 | 0.08 | 3.42 | – | 12.44 | m² | **15.86** |
| Aggregate filling to StableDRIVE | | | | | | | |
| shingle | 1.43 | 0.03 | 2.23 | 2.85 | 1.43 | m² | **6.51** |
| decorative aggregate; CED Ltd; PC £61.00/tonne | 3.51 | 0.03 | 2.23 | 2.85 | 3.51 | m² | **8.59** |

## 35 SITE WORKS – PAVINGS

| Item<br>Excluding site overheads and profit | PC<br>£ | Labour<br>hours | Labour<br>£ | Plant<br>£ | Material<br>£ | Unit | Total<br>rate £ |
|---|---|---|---|---|---|---|---|
| **StableGRASS 38 for Grass reinforcement**<br>Polyethylene soil stabilizing panels; to soil surfaces brought to grade (not included); filling with excavated material | | | | | | | |
| panels; 50 cm × 50 cm × 38 mm deep | 8.30 | 0.03 | 2.37 | 2.62 | 8.30 | m² | **13.29** |
| Extra for filling StableGRASS with topsoil; seeding with rye grass at 50 g/m² | 2.00 | 0.07 | 4.77 | 0.19 | 2.00 | m² | **6.96** |
| **Cellular pavings**<br>Honeycomb cellular polyproylene interconnecting paviors with integral downstead anti-shear cleats including topsoil but excluding edge restraints; to granular subbase (not included) | | | | | | | |
| 635 × 330 × 42 mm overall laid to a module of 622 × 311 × 32 mm | – | – | – | – | – | m² | **32.20** |
| **GRASS CONCRETE** | | | | | | | |
| **Grass concrete – General**<br>Preamble: Grass seed should be a perennial ryegrass mixture, with the proportion depending on expected traffic. Hardwearing winter sportsground mixtures are suitable for public areas. Loose gravel, shingle or sand is liable to be kicked out of the blocks; rammed hoggin or other stabilized material should be specified. | | | | | | | |
| **Grass concrete; Grass Concrete Ltd; on blinded granular Type 1 subbase (not included)**<br>Grasscrete in situ concrete continuously reinforced surfacing; including expansion joints at 10 m centres; soiling; seeding | | | | | | | |
| GC2; 150 mm thick; traffic up to 40.00 tonnes | – | – | – | – | – | m² | **44.73** |
| GC1; 100 mm thick; traffic up to 13.30 tonnes | – | – | – | – | – | m² | **35.61** |
| GC3; 76 mm thick; traffic up to 4.30 tonnes | – | – | – | – | – | m² | **30.58** |

## 35 SITE WORKS – PAVINGS

| Item<br>Excluding site overheads and profit | PC<br>£ | Labour<br>hours | Labour<br>£ | Plant<br>£ | Material<br>£ | Unit | **Total<br>rate £** |
|---|---|---|---|---|---|---|---|
| **GRASS CONCRETE – cont** | | | | | | | |
| Grass concrete; Grass Concrete Ltd; 406 × 406 mm blocks; on 20 mm sharp sand; on blinded MOT type 1 subbase (not included); level and to falls only; including filling with topsoil; seeding with dwarf rye grass at PC £4.50/kg | | | | | | | |
| Pavings | | | | | | | |
|   GB103; 103 mm thick | 24.48 | 0.40 | 9.50 | 0.19 | 27.58 | m$^2$ | **37.27** |
|   GB83; 83 mm thick | 23.43 | 0.36 | 8.64 | 0.19 | 26.53 | m$^2$ | **35.36** |
| Grass concrete; Charcon Hard Landscaping; on 25 mm sharp sand; including filling with topsoil; seeding with dwarf rye grass at PC £4.50/kg | | | | | | | |
| Grassgrid grass/concrete paving blocks | | | | | | | |
|   366 × 274 × 100 mm thick | 20.15 | 0.38 | 8.91 | 0.19 | 23.25 | m$^2$ | **32.35** |
| Grass concrete; Marshalls Plc; on 25 mm sharp sand; including filling with topsoil; seeding with dwarf rye grass at PC £4.50/kg | | | | | | | |
| Concrete grass pavings | | | | | | | |
|   Grassguard 130; for light duty applications (80 mm prepared base not included) | 23.24 | 0.38 | 8.91 | 0.19 | 26.56 | m$^2$ | **35.66** |
|   Grassguard 160; for medium duty applications (80–150 mm prepared base not included) | 23.26 | 0.46 | 10.96 | 0.19 | 26.58 | m$^2$ | **37.73** |
|   Grassguard 180; for heavy duty applications (150 mm prepared base not included) | 29.30 | 0.60 | 14.25 | 0.19 | 32.63 | m$^2$ | **47.07** |
| **Full mortar bedding** | | | | | | | |
| Extra over pavings for bedding on 25 mm cement:sand (1:4); in lieu of spot bedding on sharp sand | – | 0.03 | 0.59 | – | 1.87 | m$^2$ | **2.46** |

## 35 SITE WORKS – PAVINGS

| Item<br>Excluding site overheads and profit | PC<br>£ | Labour<br>hours | Labour<br>£ | Plant<br>£ | Material<br>£ | Unit | Total<br>rate £ |
|---|---|---|---|---|---|---|---|
| **COBBLE PAVINGS** | | | | | | | |
| **Cobble pavings – General** | | | | | | | |
| Cobbles should be embedded by hand, tight-butted, endwise to a depth of 60% of their length. A dry grout of rapid-hardening cement:sand (1:2) shall be brushed over the cobbles until the interstices are filled to the level of the adjoining paving. Surplus grout shall then be brushed off and a light, fine spray of water applied over the area. | | | | | | | |
| **Cobble pavings** | | | | | | | |
| Cobbles; to present a uniform colour in panels; or varied in colour as required | | | | | | | |
|    Scottish Beach Cobbles; 50–75 mm | 27.64 | 1.11 | 79.17 | – | 35.14 | m² | **114.31** |
|    Scottish Beach Cobbles; 75–100 mm | 40.37 | 0.83 | 59.37 | – | 47.86 | m² | **107.23** |
|    Scottish Beach Cobbles; 100–200 mm | 37.38 | 2.67 | 95.00 | – | 44.88 | m² | **139.88** |
| **GRANULAR PAVINGS** | | | | | | | |
| **Footpath gravels; porous self binding gravel** | | | | | | | |
| CED Ltd; Cedec gravel; self-binding; laid to inert (non-limestone) base measured separately; compacting | | | | | | | |
|    red, silver or gold; 50 mm thick | 12.37 | 0.01 | 0.59 | 0.36 | 12.37 | m² | **13.32** |
| Grundon Ltd; Coxwell self-binding path gravels laid and compacted to excavation or base measured separately | | | | | | | |
|    50 mm thick | 4.31 | 0.01 | 0.59 | 0.36 | 4.31 | m² | **5.26** |
| Breedon Special Aggregates; Golden Gravel or equivalent; rolling wet; on hardcore base (not included); for pavements; to falls and crossfalls and to slopes not exceeding 15° from horizontal; over 300 mm wide | | | | | | | |
|    50 mm thick | 8.53 | 0.01 | 0.59 | 0.29 | 8.53 | m² | **9.41** |
|    75 mm thick | 12.80 | 0.03 | 2.37 | 0.37 | 12.80 | m² | **15.54** |
| Breedon Special Aggregates; Wayfarer specially formulated fine gravel for use on golf course pathways | | | | | | | |
|    50 mm thick | 7.76 | 0.01 | 0.59 | 0.29 | 7.76 | m² | **8.64** |
|    75 mm thick | 11.64 | 0.02 | 1.19 | 0.49 | 11.64 | m² | **13.32** |

## 35 SITE WORKS – PAVINGS

| Item<br>Excluding site overheads and profit | PC<br>£ | Labour<br>hours | Labour<br>£ | Plant<br>£ | Material<br>£ | Unit | Total<br>rate £ |
|---|---|---|---|---|---|---|---|
| **GRANULAR PAVINGS – cont** | | | | | | | |
| **Footpath gravels – cont** | | | | | | | |
| Hoggin (stabilized); PC £31.65/m³ on hardcore base (not included); to falls and crossfalls and to slopes not exceeding 15° from horizontal; over 300 mm wide | | | | | | | |
| 100 mm thick | 4.75 | 0.01 | 0.79 | 0.48 | 4.75 | m² | **6.02** |
| 150 mm thick | 7.12 | 0.02 | 1.19 | 0.71 | 7.12 | m² | **9.02** |
| Ballast; as dug; watering; rolling; on hardcore base (not included) | | | | | | | |
| 100 mm thick | 5.13 | 0.01 | 0.79 | 0.48 | 5.13 | m² | **6.40** |
| 150 mm thick | 7.69 | 0.02 | 1.19 | 0.71 | 7.69 | m² | **9.59** |
| **Footpath gravels; porous loose gravels** | | | | | | | |
| Breedon Special Aggregates; Breedon Buff decorative limestone chippings in Warwick Gold, Coastal cream | | | | | | | |
| 50 mm thick | 5.76 | – | 0.24 | 0.09 | 5.76 | m² | **6.09** |
| 75 mm thick | 11.21 | – | 0.30 | 0.12 | 11.21 | m² | **11.63** |
| Breedon Special Aggregates; Brindle, Moorland Black, Shierglas and Galloway chippings | | | | | | | |
| 50 mm thick | 6.99 | – | 0.24 | 0.09 | 6.99 | m² | **7.32** |
| 75 mm thick | 10.50 | – | 0.30 | 0.12 | 10.50 | m² | **10.92** |
| Breedon Special Aggregates; Criggion Green decorative chippings 20 mm | | | | | | | |
| 50 mm thick | 5.76 | – | 0.24 | 0.09 | 5.76 | m² | **6.09** |
| 75 mm thick | – | – | 0.30 | 0.12 | 8.65 | m² | **9.07** |
| Breedon Special Aggregates; 40 mm slate chippings; plum/blue | | | | | | | |
| 50 mm thick | 6.52 | – | 0.24 | 0.09 | 6.52 | m² | **6.85** |
| 75 mm thick | 9.79 | – | 0.30 | 0.12 | 9.79 | m² | **10.21** |
| Washed shingle; on prepared base (not included) | | | | | | | |
| 25–50 mm size; 25 mm thick | 1.17 | 0.01 | 0.59 | 0.08 | 1.17 | m² | **1.84** |
| 25–50 mm size; 75 mm thick | 3.52 | 0.03 | 1.78 | 0.25 | 3.52 | m² | **5.55** |
| 50–75 mm size; 25 mm thick | 1.17 | 0.01 | 0.48 | 0.09 | 1.17 | m² | **1.74** |
| 50–75 mm size; 75 mm thick | 3.52 | 0.02 | 1.58 | 0.31 | 3.52 | m² | **5.41** |
| Pea shingle; on prepared base (not included) | | | | | | | |
| 10–15 mm size; 25 mm thick | 1.08 | 0.01 | 0.42 | 0.08 | 1.08 | m² | **1.58** |
| 5–10 mm size; 75 mm thick | 3.23 | 0.02 | 1.25 | 0.25 | 3.23 | m² | **4.73** |

## 35 SITE WORKS – PAVINGS

| Item<br>Excluding site overheads and profit | PC<br>£ | Labour<br>hours | Labour<br>£ | Plant<br>£ | Material<br>£ | Unit | Total<br>rate £ |
|---|---|---|---|---|---|---|---|
| **PRECAST CONCRETE BLOCK PAVINGS** | | | | | | | |
| **Precast concrete block edgings; PC £10.51/m$^2$; 200 × 100 × 60 mm; on prepared base (not included); haunching one side** | | | | | | | |
| Edgings; butt joints | | | | | | | |
| stretcher course | 1.05 | 0.06 | 3.95 | – | 3.86 | m | 7.81 |
| header course | 2.10 | 0.09 | 6.33 | – | 5.12 | m | 11.45 |
| **Precast concrete vehicular paving blocks; Marshalls Plc; on prepared base (not included); on 50 mm compacted sharp sand bed; blocks laid in 7 mm loose sand and vibrated; joints filled with sharp sand and vibrated; level and to falls only** | | | | | | | |
| Trafica paving blocks; | | | | | | | |
| 450 × 450 × 70 mm | | | | | | | |
| Perfecta finish; colour natural | 43.51 | 0.17 | 11.88 | 0.10 | 45.76 | m$^2$ | 57.74 |
| Perfecta finish; colour buff | 50.43 | 0.17 | 11.88 | 0.10 | 52.68 | m$^2$ | 64.66 |
| Saxon finish; colour natural | 35.98 | 0.17 | 11.88 | 0.10 | 38.23 | m$^2$ | 50.21 |
| Saxon finish; colour buff | 39.16 | 0.17 | 11.88 | 0.10 | 41.41 | m$^2$ | 53.39 |
| **Precast concrete paving blocks; Keyblok Marshalls Plc; on prepared base (not included); on 50 mm compacted sharp sand bed; blocks laid in 7 mm loose sand and vibrated; joints filled with sharp sand and vibrated; level and to falls only** | | | | | | | |
| Herringbone bond | | | | | | | |
| 200 × 100 × 60 mm; natural grey | 11.04 | 0.28 | 19.96 | 0.10 | 13.40 | m$^2$ | 33.46 |
| 200 × 100 × 60 mm; colours | 11.85 | 0.28 | 19.96 | 0.10 | 14.22 | m$^2$ | 34.28 |
| 200 × 100 × 80 mm; natural grey | 12.55 | 0.28 | 19.96 | 0.10 | 14.92 | m$^2$ | 34.98 |
| 200 × 100 × 80 mm; colours | 13.04 | 0.28 | 19.96 | 0.10 | 15.41 | m$^2$ | 35.47 |
| Basketweave bond | | | | | | | |
| 200 × 100 × 60 mm; natural grey | 11.04 | 0.25 | 17.81 | 0.10 | 13.40 | m$^2$ | 31.31 |
| 200 × 100 × 60 mm; colours | 11.85 | 0.25 | 17.81 | 0.10 | 14.22 | m$^2$ | 32.13 |
| 200 × 100 × 80 mm; natural grey | 12.55 | 0.25 | 17.81 | 0.10 | 14.92 | m$^2$ | 32.83 |
| 200 × 100 × 80 mm; colours | 13.04 | 0.25 | 17.81 | 0.10 | 15.41 | m$^2$ | 33.32 |

## 35 SITE WORKS – PAVINGS

| Item<br>Excluding site overheads and profit | PC<br>£ | Labour<br>hours | Labour<br>£ | Plant<br>£ | Material<br>£ | Unit | Total<br>rate £ |
|---|---|---|---|---|---|---|---|
| **PRECAST CONCRETE BLOCK PAVINGS – cont** | | | | | | | |
| **Precast concrete paving blocks; Charcon Hard Landscaping; on prepared base (not included); on 50 mm compacted sharp sand bed; blocks laid in 7 mm loose sand and vibrated; joints filled with sharp sand and vibrated; level and to falls only** | | | | | | | |
| Europa concrete blocks | | | | | | | |
|   200 × 100 × 60 mm; natural grey | 10.72 | 0.25 | 17.81 | 0.10 | 13.09 | m² | **31.00** |
|   200 × 100 × 60 mm; colours | 11.18 | 0.25 | 17.81 | 0.10 | 13.55 | m² | **31.46** |
| Andover concrete blocks | | | | | | | |
|   200 × 100 × 50 mm; textured grey | 29.24 | 0.25 | 17.81 | 0.10 | 31.61 | m² | **49.52** |
|   200 × 100 × 80 mm; textured grey | 29.24 | 0.25 | 17.81 | 0.10 | 31.61 | m² | **49.52** |
| **Concrete cobble paviors; Charcon Hard Landscaping; Concrete Products; on prepared base (not included); bedding on 50 mm sand; kiln dried sand joints swept in** | | | | | | | |
| Paviors | | | | | | | |
|   Woburn blocks; 100–201 × 134 × 80 mm; random sizes | 22.33 | 0.29 | 20.36 | 0.10 | 24.70 | m² | **45.16** |
|   Woburn blocks; 100–201 × 134 × 80 mm; single size | 22.33 | 0.25 | 17.81 | 0.10 | 24.70 | m² | **42.61** |
|   Woburn blocks; 100–201 × 134 × 60 mm; random sizes | 17.85 | 0.22 | 15.83 | 0.10 | 20.21 | m² | **36.14** |
|   Woburn blocks; 100–201 × 134 × 60 mm; single size | 17.83 | 0.21 | 14.84 | 0.10 | 20.20 | m² | **35.14** |
| **Concrete setts; on 25 mm sand; compacted; vibrated; joints filled with sand; natural or coloured; well rammed hardcore base (not included)** | | | | | | | |
| Marshalls Plc; Tegula cobble paving | | | | | | | |
|   60 mm thick; random sizes | 24.82 | 0.19 | 13.57 | 0.10 | 27.40 | m² | **41.07** |
|   60 mm thick; single size | 24.82 | 0.15 | 10.79 | 0.10 | 27.40 | m² | **38.29** |
|   80 mm thick; random sizes | 28.34 | 0.19 | 13.57 | 0.10 | 30.91 | m² | **44.58** |
|   80 mm thick; single size | 28.34 | 0.15 | 10.79 | 0.10 | 30.91 | m² | **41.80** |
|   cobbles; 80 × 80 × 60 mm thick; traditional | 42.52 | 0.19 | 13.20 | 0.10 | 45.09 | m² | **58.39** |
| **Cobbles** | | | | | | | |
| Charcon Hard Landscaping; Country setts | | | | | | | |
|   100 mm thick; random sizes | 42.02 | 0.33 | 23.75 | 0.10 | 44.59 | m² | **68.44** |
|   100 mm thick; single size | 43.07 | 0.22 | 15.83 | 0.10 | 45.64 | m² | **61.57** |

## 35 SITE WORKS – PAVINGS

| Item<br>Excluding site overheads and profit | PC<br>£ | Labour<br>hours | Labour<br>£ | Plant<br>£ | Material<br>£ | Unit | Total<br>rate £ |
|---|---|---|---|---|---|---|---|
| **SUDS PERMEABLE BLOCK PAVING** | | | | | | | |
| **SUDS paving; Concrete Block Permeable Paving (CBPP)**<br>Note: Sustainable Drainage Systems (SUDS) aim to reduce flood risk by controlling the rate and volume of surface water run-off from developments. Permeable Paving is a SUDS technique. In addition to managing the quantity of surface water run-off more effectively, one of the primary benefits of CBPP is the potential enhancement of water quality. Permeable pavements improve water quality by mirroring nature in providing filtration and allowing for natural biodegradation of hydrocarbons and the dilution of other contaminants as water passes through the system. | | | | | | | |
| **Bases for Sustainable Urban Drainage System (SUDS) subbase for use below impervious paving; Aggregate Industries Ltd; Bardon DrainAgg 20**<br>4/20 mm open graded material laid loose on prepared sub-grade; within edge restraints to paving area (not included); graded to levels and falls as per paving manufacturer's instructions | | | | | | | |
| 150 mm thick | 6.37 | 0.01 | 0.59 | 0.69 | 6.37 | m² | **7.65** |
| 200 mm thick | 8.49 | 0.02 | 1.19 | 0.93 | 8.49 | m² | **10.61** |
| 250 mm thick | 10.61 | 0.02 | 1.48 | 1.15 | 10.61 | m² | **13.24** |
| 350 mm thick | 14.86 | 0.03 | 2.04 | 1.58 | 14.86 | m² | **18.48** |
| 450 mm thick | 19.10 | 0.04 | 2.85 | 2.21 | 19.10 | m² | **24.16** |

## 35 SITE WORKS – PAVINGS

| Item<br>Excluding site overheads and profit | PC<br>£ | Labour<br>hours | Labour<br>£ | Plant<br>£ | Material<br>£ | Unit | Total<br>rate £ |
|---|---|---|---|---|---|---|---|
| **SUDS PERMEABLE BLOCK PAVING – cont** | | | | | | | |
| **Subbase replacement system for SUDS paving Charcon Permavoid; Aggregate Industries Ltd; geocellular, interlocking, high strength; high void capacity (95%); typically a 4:1 ratio when compared with alternative granular systems; allows for water storage to be confined at shallow depth within the subbase layer; inclusive of connection, ties, pins; does not include for encapsulation in geotextile/geomembranes** | | | | | | | |
| 708 × 354 × 150 mm | | | | | | | |
| 150 mm thick | 65.51 | 0.06 | 3.97 | – | 65.51 | m² | **69.48** |
| 300 mm thick | 135.94 | 0.11 | 7.92 | – | 135.94 | m² | **143.86** |
| Encapsulation of Permavoid; 150 mm thick systems; geomembrane geofabric or combination of both for filtration or attenuation of water; please see the manufacturer's design guide | | | | | | | |
| Charcon geomembrane, heat welded | 16.09 | 0.02 | 1.19 | – | 16.09 | m² | **17.28** |
| Charcon Permatex 300 geotextile | 10.16 | 0.11 | 7.92 | – | 10.16 | m² | **18.08** |
| Charcon Permafilter Pollution Control geotextile | 5.35 | 0.01 | 0.79 | – | 5.35 | m² | **6.14** |
| Encapsulation of Permavoid; 300 mm thick systems; geomembrane geofabric or combination of both for filtration or attenuation of water; please see the manufacturer's design guide | | | | | | | |
| Charcon geomembrane, heat welded | 18.19 | 0.02 | 1.32 | – | 18.19 | m² | **19.51** |
| Charcon Permatex 300 geotextile | 11.48 | 0.01 | 0.79 | – | 11.48 | m² | **12.27** |
| Charcon Permafilter Pollution Control geotextile | 5.04 | 0.01 | 0.85 | – | 5.04 | m² | **5.89** |

## 35 SITE WORKS – PAVINGS

| Item<br>Excluding site overheads and profit | PC<br>£ | Labour<br>hours | Labour<br>£ | Plant<br>£ | Material<br>£ | Unit | Total<br>rate £ |
|---|---|---|---|---|---|---|---|
| **Pervious surfacing materials; Aggregate Industries Ltd; Charcon Infilta CBPP**<br>Rectangular permeable block paving system; incorporating a 5 mm spacer design that provides a 5 mm void allowing ingress of water through to the subbase storage system (priced separately); various colours; bedded on 50 mm thick 2–6.3 mm clean, angular, free draining uncompacted aggregate; joints infilled with 3 mm clean grit | | | | | | | |
| Infilta; 200 × 100 × 80 mm thick; natural grey | 15.17 | 0.15 | 10.55 | 0.10 | 18.14 | m² | **28.79** |
| Infilta; 200 × 100 × 80 mm thick; coloured | 15.17 | 0.15 | 10.55 | 0.10 | 18.14 | m² | **28.79** |
| **PRECAST CONCRETE SLAB PAVINGS** | | | | | | | |
| **Slab paving; precast concrete pavings; Charcon Hard Landscaping; on prepared subbase (not included); bedding on 25 mm thick cement:sand mortar (1:4); butt joints; straight both ways; on 50 mm thick sharp sand base** | | | | | | | |
| Pavings; natural grey | | | | | | | |
| 450 × 450 × 50 mm chamfered | 16.20 | 0.15 | 10.55 | – | 22.02 | m² | **32.57** |
| 600 × 300 × 50 mm | 13.83 | 0.15 | 10.55 | – | 19.65 | m² | **30.20** |
| 400 × 400 × 65 mm chamfered | 19.81 | 0.13 | 9.50 | – | 25.63 | m² | **35.13** |
| 450 × 600 × 50 mm | 12.56 | 0.15 | 10.55 | – | 18.37 | m² | **28.92** |
| 600 × 600 × 50 mm | 9.56 | 0.13 | 9.50 | – | 15.37 | m² | **24.87** |
| 750 × 600 × 50 mm | 9.04 | 0.13 | 9.50 | – | 14.86 | m² | **24.36** |
| 900 × 600 × 50 mm | 7.68 | 0.13 | 9.50 | – | 13.50 | m² | **23.00** |
| Pavings; coloured | | | | | | | |
| 450 × 600 × 50 mm | 18.07 | 0.15 | 10.55 | – | 23.89 | m² | **34.44** |
| 400 × 400 × 65 mm chamfered | 19.81 | 0.13 | 9.50 | – | 25.63 | m² | **35.13** |
| 600 × 600 × 50 mm | 12.67 | 0.13 | 9.50 | – | 18.49 | m² | **27.99** |
| 750 × 600 × 50 mm | 12.76 | 0.13 | 9.50 | – | 18.57 | m² | **28.07** |
| 900 × 600 × 50 mm | 10.62 | 0.13 | 9.50 | – | 16.44 | m² | **25.94** |

## 35 SITE WORKS – PAVINGS

| Item<br>Excluding site overheads and profit | PC<br>£ | Labour<br>hours | Labour<br>£ | Plant<br>£ | Material<br>£ | Unit | Total<br>rate £ |
|---|---|---|---|---|---|---|---|
| **PRECAST CONCRETE SLAB PAVINGS – cont**<br>**Precast concrete pavings; Charcon Hard Landscaping; on prepared subbase (not included); bedding on 25 mm thick cement: sand mortar (1:4); butt joints; straight both ways; jointing in cement:sand (1:3) brushed in; on 50 mm thick sharp sand base**<br>Appalacian rough textured exposed aggregate pebble paving | | | | | | | |
| 600 × 600 × 65 mm | 29.01 | 0.17 | 11.88 | – | 32.78 | m² | **44.66** |
| **Pavings; Marshalls Plc; spot bedding on 5 nr pads of cement:sand mortar (1:4); on sharp sand**<br>Blister Tactile pavings; specially textured slabs for guidance of blind pedestrians; red or buff | | | | | | | |
| 400 × 400 × 50 mm | 39.33 | 0.17 | 11.88 | – | 41.46 | m² | **53.34** |
| 450 × 450 × 50 mm | 32.70 | 0.17 | 11.88 | – | 34.82 | m² | **46.70** |
| Extra labours for laying a selection of the above sizes to random rectangular pattern | – | 0.33 | 7.92 | – | – | m² | **7.92** |
| **Marshalls Plc; La Linia Paving; fine textured exposed aggregate 80 mm thick pavings in various sizes to designed laying patterns; laid to 50 mm sharp sand bed on Type 1 base all priced separately**<br>Bonded laying patterns; 300 × 300 mm | | | | | | | |
| light granite/anthracite basalt | 40.64 | 0.25 | 17.81 | – | 40.64 | m² | **58.45** |
| Random scatter pattern incorporating 100 × 200 mm, 200 × 200 mm and 300 × 200 mm units | | | | | | | |
| light granite/anthracite basalt | 39.65 | 0.33 | 23.75 | – | 39.65 | m² | **63.40** |
| **Precast concrete slab pavings; Marshalls Plc; Conservation paving; Natural granite aggregate appearance on prepared subbase measured separately; bedding on 25 mm thick cement:sand mortar (1:4); pointed straight both ways Steintec jointing mixture**<br>Smooth; Silver grey; (Other colours available) | | | | | | | |
| 450 × 300 × 50 mm | 53.62 | 0.33 | 23.75 | – | 62.05 | m² | **85.80** |
| 450 × 600 × 50 mm | 56.81 | 0.30 | 21.59 | – | 64.19 | m² | **85.78** |
| 450 × 450 × 50 mm | 35.75 | 0.32 | 22.99 | – | 43.48 | m² | **66.47** |

## 35 SITE WORKS – PAVINGS

| Item<br>Excluding site overheads and profit | PC<br>£ | Labour<br>hours | Labour<br>£ | Plant<br>£ | Material<br>£ | Unit | Total<br>rate £ |
|---|---|---|---|---|---|---|---|
| 600 × 600 × 63 mm | 59.27 | 0.27 | 19.00 | – | 66.29 | m² | 85.29 |
| Conservation Smooth Paving | | | | | | | |
| 450 × 450 × 50 mm – Square | 55.14 | 0.32 | 22.99 | – | 62.88 | m² | 85.87 |
| 450 × 450 × 50 mm – Chamfered | 53.80 | 0.32 | 22.99 | – | 61.53 | m² | 84.52 |
| 600 × 600 × 50 mm – Square | 41.94 | 0.27 | 19.00 | – | 48.96 | m² | 67.96 |
| 600 × 600 × 50 mm – Chamfered | 40.92 | 0.27 | 19.00 | – | 47.94 | m² | 66.94 |
| 900 × 300 × 50 mm – Chamfered | 40.74 | 0.27 | 19.00 | – | 47.76 | m² | 66.76 |
| Conservation smooth -Trafica | | | | | | | |
| Trafica 300 × 300 × 60 mm – Chamfered | 64.57 | 0.36 | 25.44 | – | 73.69 | m² | 99.13 |
| Conservation Textured Paving | | | | | | | |
| 450 × 450 × 50 mm | 45.55 | 0.32 | 22.99 | – | 53.28 | m² | 76.27 |
| 400 × 400 × 50 mm | 45.94 | 0.33 | 23.75 | – | 54.37 | m² | 78.12 |
| 600 × 600 × 50 mm | 35.78 | 0.27 | 19.00 | – | 42.80 | m² | 61.80 |
| 900 × 300 × 50 mm | 34.45 | 0.27 | 19.00 | – | 41.47 | m² | 60.47 |
| 900 × 600 × 50 mm | 32.91 | 0.25 | 17.81 | – | 39.58 | m² | 57.39 |
| Conservation textured – Trafica | | | | | | | |
| Trafica 400 × 400 × 65 mm – Chamfered | 54.81 | 0.33 | 23.75 | – | 63.87 | m² | 87.62 |
| Trafica 300 × 300 × 60 mm | 51.36 | 0.36 | 25.44 | – | 60.48 | m² | 85.92 |
| Trafica 400 × 400 × 65 mm | 50.48 | 0.33 | 23.75 | – | 59.54 | m² | 83.29 |
| Trafica 450 × 450 × 70 mm | 52.15 | 0.32 | 22.99 | – | 61.53 | m² | 84.52 |
| **Precast concrete slab pavings; Marshalls Plc; Fibre reinforced pavings to match other slab paving families; on prepared subbase measured separately; bedding on 25 mm thick cement:sand mortar (1:4); pointed straight both ways Steintec jointing mixture** | | | | | | | |
| Fibre Reinforced Paving | | | | | | | |
| Conservation 600 × 600 × 63 mm Smooth | 57.62 | 0.27 | 19.00 | – | 64.64 | m² | 83.64 |
| Conservation 600 × 600 × 63 mm Textured | 53.90 | 0.27 | 19.00 | – | 60.92 | m² | 79.92 |
| Perfecta 600 × 600 × 63 mm Smooth | 42.74 | 0.27 | 19.00 | – | 49.76 | m² | 68.76 |
| Saxon 600 × 600 × 63 mm Textured | 32.99 | 0.27 | 19.00 | – | 40.01 | m² | 59.01 |
| Standard 600 × 600 × 63 mm Pimple | 18.39 | 0.27 | 19.00 | – | 25.41 | m² | 44.41 |

## 35 SITE WORKS – PAVINGS

| Item<br>Excluding site overheads and profit | PC<br>£ | Labour<br>hours | Labour<br>£ | Plant<br>£ | Material<br>£ | Unit | Total<br>rate £ |
|---|---|---|---|---|---|---|---|
| **PRECAST CONCRETE SLAB PAVINGS – cont** | | | | | | | |
| **Precast concrete slab pavings; Marshalls Plc; Charnwood paving; Natural larger granite aggregate appearance on prepared subbase measured separately; bedding on 25 mm thick cement:sand mortar (1:4); pointed straight both ways Steintec jointing mixture** | | | | | | | |
| Charnwood Textured Paving | | | | | | | |
| Trafica 300 × 300 mm | 49.88 | 0.36 | 25.44 | – | 59.00 | m² | **84.44** |
| Trafica 450 × 450 × 70 mm | 37.37 | 0.32 | 22.99 | – | 46.26 | m² | **69.25** |
| Trafica 400 × 400 × 65 mm | 37.75 | 0.33 | 23.75 | – | 46.80 | m² | **70.55** |
| 450 × 450 × 50 mm | 30.91 | 0.32 | 22.99 | – | 38.65 | m² | **61.64** |
| 600 × 600 × 60 mm | 26.02 | 0.27 | 19.00 | – | 33.04 | m² | **52.04** |
| **Precast concrete slab pavings; Marshalls Plc; Contemporary linear paving range for the urban environment; Incorporating modern granite aggregates; Suitable for use in pedestrian and occasional traffic applications on prepared subbase measured separately; bedding on 25 mm thick cement:sand mortar (1:4); pointed straight both ways Steintec jointing mixture** | | | | | | | |
| Celestia Smooth Paving | | | | | | | |
| 600 × 200 × 50 mm | 51.65 | 0.25 | 17.81 | – | 61.87 | ea | **79.68** |
| 600 × 600 × 50 mm | 53.00 | 0.27 | 19.00 | – | 60.02 | ea | **79.02** |
| 900 × 300 × 50 mm | 51.62 | 0.27 | 19.00 | – | 60.09 | ea | **79.09** |
| **Sustainable underground drainage paving; Marshalls Plc; Concrete block paving to 50 mm thick 6 mm aggregate compacted bedding layer; base layer of open graded crushed rock (OGCR); not included** | | | | | | | |
| Priora Permeable Paving | | | | | | | |
| Priora ML45 200 × 100 × 80 mm – Chamfered | 18.40 | 0.13 | 9.26 | 2.05 | 20.65 | m² | **31.96** |
| Pallas 600 × 300 × 80 mm – Chamfered | 65.83 | 0.25 | 17.81 | – | 73.56 | m² | **91.37** |

## 35 SITE WORKS – PAVINGS

| Item<br>Excluding site overheads and profit | PC<br>£ | Labour<br>hours | Labour<br>£ | Plant<br>£ | Material<br>£ | Unit | Total<br>rate £ |
|---|---|---|---|---|---|---|---|
| **DETERRENT PAVINGS** | | | | | | | |
| **Pedestrian deterrent pavings; Marshalls Plc; on prepared base (not included); bedding on 25 mm cement:sand (1:3); cement:sand (1:3) joints** | | | | | | | |
| Blister paving | | | | | | | |
| 450 × 450 × 50 mm | 32.30 | 0.08 | 5.94 | – | 37.14 | $m^2$ | **43.08** |
| Contour deterrent; granite sett appearance | | | | | | | |
| 498 × 498 × 75 mm | 134.76 | 0.11 | 7.92 | – | 139.60 | $m^2$ | **147.52** |
| **Pedestrian deterrent pavings; Townscape Products Ltd; on prepared base (not included); bedding on 25 mm cement:sand (1:3); cement:sand (1:3) joints** | | | | | | | |
| Milford paving Slab finished in a smooth prestige mix | | | | | | | |
| 500 × 500 × 50 mm | 82.72 | 0.17 | 11.88 | – | 87.56 | $m^2$ | **99.44** |
| Natural Aggregate Vibro-cast paving | | | | | | | |
| 400 × 400 × 50 mm | 102.12 | 0.17 | 11.88 | – | 106.96 | $m^2$ | **118.84** |
| Prestige Vibro-cast paving | | | | | | | |
| 400 × 400 × 50 mm | 92.86 | 0.27 | 19.00 | – | 97.70 | $m^2$ | **116.70** |
| **Edge restraints; to block paving; on prepared base (not included); 200 × 100 × 80 mm; PC £11.95/$m^2$; haunching one side** | | | | | | | |
| Header course | | | | | | | |
| 200 × 100 mm; butt joints | 2.39 | 0.09 | 6.33 | – | 5.41 | m | **11.74** |
| Stretcher course | | | | | | | |
| 200 × 100 mm; butt joints | 1.20 | 0.06 | 3.95 | – | 4.00 | m | **7.95** |

## 35 SITE WORKS – PAVINGS

| Item<br>Excluding site overheads and profit | PC<br>£ | Labour<br>hours | Labour<br>£ | Plant<br>£ | Material<br>£ | Unit | Total<br>rate £ |
|---|---|---|---|---|---|---|---|
| **STONE PAVINGS** | | | | | | | |
| **Natural stone paving – General**<br>Preamble: Provide paving slabs of the specified thickness in random sizes but not less than 25 slabs per 10 m² of surface area, to be laid in parallel courses with joints alternately broken and laid to falls. | | | | | | | |
| **Natural stone, slate or granite flag pavings; CED Ltd; on prepared base (not included); bedding on 25 mm cement:sand (1:3); cement:sand (1:3) joints**<br>Yorkstone; riven laid random rectangular | | | | | | | |
| new slabs; 40–60 mm thick | 104.36 | 0.57 | 40.72 | – | 108.84 | m² | **149.56** |
| reclaimed slabs; Cathedral grade; 50–75 mm thick | 110.88 | 0.93 | 66.50 | – | 115.36 | m² | **181.86** |
| Donegal quartzite slabs; standard tiles | | | | | | | |
| 200 mm × random lengths × 15–25 mm | 64.26 | 1.17 | 83.13 | – | 68.74 | m² | **151.87** |
| 250 mm × random lengths × 15–25 mm | 64.26 | 1.10 | 78.38 | – | 68.74 | m² | **147.12** |
| 300 mm × random lengths × 15–25 mm | 64.26 | 1.03 | 73.62 | – | 68.74 | m² | **142.36** |
| 350 mm × random lengths × 15–25 mm | 64.26 | 0.93 | 66.50 | – | 68.74 | m² | **135.24** |
| 400 mm × random lengths × 15–25 mm | 64.26 | 0.83 | 59.37 | – | 68.74 | m² | **128.11** |
| 450 mm × random lengths × 15–25 mm | 64.26 | 0.78 | 55.42 | – | 68.74 | m² | **124.16** |
| **Natural Yorkstone pavings; Johnsons Wellfield Quarries; sawn six sides; 50 mm thick; on prepared base measured separately; bedding on 25 mm cement:sand (1:3); cement:sand (1:3) joints**<br>Paving | | | | | | | |
| laid to random rectangular pattern | 73.03 | 0.57 | 40.71 | – | 77.29 | m² | **118.00** |
| laid to coursed laying pattern; 3 sizes | 72.34 | 0.57 | 40.95 | – | 76.60 | m² | **117.55** |
| Paving; single size | | | | | | | |
| 600 × 600 mm | 77.91 | 0.28 | 20.19 | – | 82.17 | m² | **102.36** |
| 600 × 400 mm | 77.91 | 0.33 | 23.75 | – | 82.17 | m² | **105.92** |
| 300 × 200 mm | 83.47 | 0.67 | 47.50 | – | 87.73 | m² | **135.23** |
| 215 × 102.5 mm | 89.04 | 0.83 | 59.37 | – | 93.30 | m² | **152.67** |

## 35 SITE WORKS – PAVINGS

| Item<br>Excluding site overheads and profit | PC<br>£ | Labour<br>hours | Labour<br>£ | Plant<br>£ | Material<br>£ | Unit | Total<br>rate £ |
|---|---|---|---|---|---|---|---|
| Paving; cut to template off site;<br>600 × 600 mm; radius | | | | | | | |
| 1.00 m | 185.50 | 1.11 | 79.17 | – | 189.76 | m² | 268.93 |
| 2.50 m | 185.50 | 0.67 | 47.50 | – | 189.76 | m² | 237.26 |
| 5.00 m | 185.50 | 0.67 | 47.50 | – | 189.76 | m² | 237.26 |
| **Natural Yorkstone pavings;**<br>**Johnsons Wellfield Quarries; sawn**<br>**six sides; 75 mm thick; on prepared**<br>**base measured separately; bedding**<br>**on 25 mm cement:sand (1:3); cement:**<br>**sand (1:3) joints** | | | | | | | |
| Paving | | | | | | | |
| laid to random rectangular pattern | 83.47 | 0.32 | 22.56 | – | 87.73 | m² | 110.29 |
| laid to coursed laying pattern; 3 sizes | 83.47 | 0.32 | 22.56 | – | 87.73 | m² | 110.29 |
| Paving; single size | | | | | | | |
| 600 × 600 mm | 89.04 | 0.32 | 22.56 | – | 93.30 | m² | 115.86 |
| 600 × 400 mm | 89.04 | 0.32 | 22.56 | – | 93.30 | m² | 115.86 |
| 300 × 200 mm | 94.61 | 0.25 | 17.81 | – | 98.86 | m² | 116.67 |
| 215 × 102.5 mm | 105.73 | 0.83 | 59.37 | – | 109.99 | m² | 169.36 |
| Paving; cut to template off site;<br>600 × 600 mm; radius | | | | | | | |
| 1.00 m | 243.80 | 1.33 | 95.00 | – | 248.06 | m² | 343.06 |
| 2.50 m | 243.80 | 0.83 | 59.37 | – | 248.06 | m² | 307.43 |
| 5.00 m | 243.80 | 0.83 | 59.37 | – | 248.06 | m² | 307.43 |
| **CED Ltd; Indian sandstone, riven**<br>**pavings or edgings; calibrated**<br>**+/-26 mm thick; on prepared base**<br>**measured separately; bedding on**<br>**25 mm cement:sand (1:3); cement:**<br>**sand (1:3) joints** | | | | | | | |
| Paving | | | | | | | |
| laid to random rectangular pattern | 18.38 | 0.80 | 57.01 | – | 22.63 | m² | 79.64 |
| laid to coursed laying pattern; 3 sizes | 18.38 | 0.67 | 47.50 | – | 22.63 | m² | 70.13 |
| Paving; single size | | | | | | | |
| 600 × 600 mm | 18.38 | 0.33 | 23.75 | – | 22.63 | m² | 46.38 |
| 600 × 400 mm | 18.38 | 0.42 | 29.69 | – | 22.63 | m² | 52.32 |
| 400 × 400 mm | 18.38 | 0.56 | 39.59 | – | 22.63 | m² | 62.22 |
| **Natural stone, slate or granite flag**<br>**pavings; CED Ltd; on prepared base**<br>**(not included); bedding on 25 mm**<br>**cement:sand (1:3); cement:sand (1:3)**<br>**joints** | | | | | | | |
| Granite paving; sawn six sides; textured<br>top | | | | | | | |
| new slabs; silver grey; 50 mm thick | 42.93 | 0.57 | 40.71 | – | 47.41 | m² | 88.12 |
| new slabs; blue grey; 50 mm thick | 42.93 | 0.57 | 40.71 | – | 47.41 | m² | 88.12 |
| new slabs; yellow grey; 50 mm thick | 47.80 | 0.57 | 40.71 | – | 52.27 | m² | 92.98 |
| new slabs; black; 50 mm thick | 74.80 | 0.57 | 40.71 | – | 79.28 | m² | 119.99 |

## 35 SITE WORKS – PAVINGS

| Item<br>Excluding site overheads and profit | PC<br>£ | Labour<br>hours | Labour<br>£ | Plant<br>£ | Material<br>£ | Unit | Total<br>rate £ |
|---|---|---|---|---|---|---|---|
| **STONE PAVINGS** – cont | | | | | | | |
| **Natural stone, slate or granite flag pavings** – cont | | | | | | | |
| Edgings; silver grey | | | | | | | |
| 100 mm wide × random lengths | 2.02 | 0.33 | 23.75 | – | 2.45 | m | **26.20** |
| 100 × 100 mm | 2.31 | 0.50 | 35.63 | – | 2.74 | m | **38.37** |
| 100 mm long × 200 mm wide | 3.76 | 0.20 | 14.25 | – | 4.19 | m | **18.44** |
| 250 mm wide × random lengths | 3.98 | 0.25 | 17.81 | – | 8.23 | m | **26.04** |
| 300 mm wide × random lengths | 4.34 | 0.25 | 17.81 | – | 8.59 | m | **26.40** |
| **STONE SETTS** | | | | | | | |
| **Granite setts; bedding on 25 mm cement:sand (1:3)** | | | | | | | |
| Natural granite setts; 100 × 100 mm to 125 × 150 mm × 150–250 mm length; riven surface; silver grey | | | | | | | |
| new; standard grade | 35.61 | 0.67 | 47.50 | – | 45.83 | m$^2$ | **93.33** |
| new; high grade | 43.81 | 0.67 | 47.50 | – | 54.03 | m$^2$ | **101.53** |
| reclaimed; cleaned | 36.26 | 0.67 | 47.50 | – | 46.47 | m$^2$ | **93.97** |
| **Granite setts; bedding on 30 mm SteinTec mortar; surfaces primed with SteinTec primer** | | | | | | | |
| Natural granite setts; 100 × 100 mm to 125 × 150 mm × 150–250 mm length; riven surface; silver grey | | | | | | | |
| new; standard grade | 45.64 | 0.85 | 51.94 | – | 45.64 | m$^2$ | **97.58** |
| new; high grade | 53.84 | 0.85 | 51.94 | – | 53.84 | m$^2$ | **105.78** |
| reclaimed; cleaned | 46.29 | 0.85 | 51.94 | – | 46.29 | m$^2$ | **98.23** |
| **RECONSTITUTED STONE PAVINGS** | | | | | | | |
| **Reconstituted Yorkstone aggregate pavings; Marshalls Plc; Saxon on prepared subbase measured separately; bedding on 25 mm thick cement:sand mortar (1:4); on 50 mm thick sharp sand base** | | | | | | | |
| Square and rectangular paving in buff; butt joints straight both ways | | | | | | | |
| 300 × 300 × 35 mm | 42.96 | 0.27 | 19.00 | – | 49.56 | m$^2$ | **68.56** |
| 600 × 300 × 35 mm | 28.45 | 0.22 | 15.44 | – | 35.05 | m$^2$ | **50.49** |
| 450 × 450 × 50 mm | 34.02 | 0.25 | 17.81 | – | 40.63 | m$^2$ | **58.44** |
| 600 × 600 × 35 mm | 21.28 | 0.17 | 11.88 | – | 27.89 | m$^2$ | **39.77** |
| 600 × 600 × 50 mm | 26.80 | 0.18 | 13.06 | – | 33.40 | m$^2$ | **46.46** |

## 35 SITE WORKS – PAVINGS

| Item<br>Excluding site overheads and profit | PC<br>£ | Labour<br>hours | Labour<br>£ | Plant<br>£ | Material<br>£ | Unit | Total<br>rate £ |
|---|---|---|---|---|---|---|---|
| Square and rectangular paving in natural; butt joints straight both ways | | | | | | | |
| 300 × 300 × 35 mm | 35.19 | 0.27 | 19.00 | – | 41.79 | m² | 60.79 |
| 450 × 450 × 50 mm | 26.51 | 0.23 | 16.62 | – | 33.11 | m² | 49.73 |
| 600 × 300 × 35 mm | 24.98 | 0.25 | 17.81 | – | 31.58 | m² | 49.39 |
| 600 × 600 × 35 mm | 18.37 | 0.17 | 11.88 | – | 24.97 | m² | 36.85 |
| 600 × 600 × 50 mm | 22.54 | 0.20 | 14.25 | – | 29.15 | m² | 43.40 |
| **RESIN BOUND SURFACING** | | | | | | | |
| Resin bound paving; Sureset Ltd; fully mixed permeable decorative aggregate paving; laid to macadam binder and base course (not included); thickness dependent on loading application and aggregate size | | | | | | | |
| 18 mm thick – 6 mm aggregate | | | | | | | |
| Natural gravel | | | | | | | |
| areas 100–300 m² | – | – | – | – | – | m² | 50.00 |
| areas 300–500 m² | – | – | – | – | – | m² | 46.00 |
| areas over 500 m² | – | – | – | – | – | m² | 42.00 |
| Crushed rock | | | | | | | |
| areas 100–300 m² | – | – | – | – | – | m² | 52.00 |
| areas 300–500 m² | – | – | – | – | – | m² | 48.00 |
| areas over 500 m² | – | – | – | – | – | m² | 45.00 |
| Marble | | | | | | | |
| areas 100–300 m² | – | – | – | – | – | m² | 58.00 |
| areas 300–500 m² | – | – | – | – | – | m² | 52.00 |
| areas over 500 m² | – | – | – | – | – | m² | 50.00 |
| Recycled glass | | | | | | | |
| areas 100–300 m² | – | – | – | – | – | m² | 64.00 |
| areas 300–500 m² | – | – | – | – | – | m² | 66.00 |
| areas over 500 m² | – | – | – | – | – | m² | 60.00 |
| Spectrum | | | | | | | |
| areas 100–300 m² | – | – | – | – | – | m² | 76.00 |
| areas 300–500 m² | – | – | – | – | – | m² | 72.00 |
| areas over 500 m² | – | – | – | – | – | m² | 68.00 |
| Resin bound macadam pavings; Bituchem; to pedestrian or vehicular hard landscape areas; laid to base course (not included) | | | | | | | |
| Natratex wearing course; clear resin bound macadam | | | | | | | |
| 25 mm thick to pedestrian areas | 20.00 | 0.04 | 2.97 | – | 20.00 | m² | 22.97 |
| 30 mm thick to vehicular areas | 25.00 | 0.01 | 0.79 | – | 25.00 | m² | 25.79 |

## 35 SITE WORKS – PAVINGS

| Item<br>Excluding site overheads and profit | PC<br>£ | Labour<br>hours | Labour<br>£ | Plant<br>£ | Material<br>£ | Unit | Total<br>rate £ |
|---|---|---|---|---|---|---|---|
| **RESIN BOUND SURFACING – cont** | | | | | | | |
| **Resin bound macadam pavings – cont** | | | | | | | |
| Colourtex coloured resin bound macadam | | | | | | | |
| 25 mm thick | 20.00 | 0.04 | 2.97 | – | 20.00 | m² | **22.97** |
| **Resin bound paving to individual tree pits; SureSet Ltd; fully mixed permeable decorative aggregate paving; laid to well compacted stone Type 3 (not included)** | | | | | | | |
| 30–40 mm thick – average tree pit size 1.50 × 1.50 tree pit aggregate | | | | | | | |
| Natural gravel | | | | | | | |
| areas 100–300 m² | – | – | – | – | – | nr | **95.00** |
| **RESIN BONDED SURFACING** | | | | | | | |
| **Bonded aggregates; Addagrip Terraco Ltd; natural decorative resin bonded surface dressing laid to concrete, macadam or to plywood panels priced separately** | | | | | | | |
| Primer coat to macadam or concrete base | – | 0.04 | 2.97 | – | 5.00 | m² | **7.97** |
| Brittany Bronze; 1–3 mm | | | | | | | |
| buff adhesive | – | 0.04 | 2.97 | – | 22.00 | m² | **24.97** |
| red adhesive | – | – | – | – | – | m² | **22.00** |
| green adhesive | – | – | – | – | – | m² | **22.00** |
| Brittany Bronze; 2–5 mm | | | | | | | |
| buff adhesive | – | – | – | – | – | m² | **25.00** |
| Chinese bauxite; 1–3 mm | | | | | | | |
| buff adhesive | – | – | – | – | – | m² | **25.00** |
| **Bound aggregates; Addagrip Terraco Ltd; natural decorative resin bound surface dressing laid to concrete, macadam or to plywood panels priced separately** | | | | | | | |
| Chocolate Terrabound; 6 mm | | | | | | | |
| 16 mm depth | – | – | – | – | – | m² | **39.00** |
| Terracotta 6 mm | | | | | | | |
| 16 mm or 18 mm depth | – | – | – | – | – | m² | **45.00** |
| Midnight Grey; 6 mm | | | | | | | |
| 15 mm depth | – | – | – | – | – | m² | **45.00** |

## 35 SITE WORKS – PAVINGS

| Item<br>Excluding site overheads and profit | PC<br>£ | Labour<br>hours | Labour<br>£ | Plant<br>£ | Material<br>£ | Unit | Total<br>rate £ |
|---|---|---|---|---|---|---|---|
| Chocolate; 6 mm | | | | | | | |
|   15 mm depth | – | – | – | – | – | m² | 45.00 |
|   18 mm depth | – | – | – | – | – | m² | 55.00 |
| Autumn gold 3 mm | | | | | | | |
|   12 or 16 mm | – | – | – | – | – | m² | 45.00 |
| Terrabase; rustic bound porous paving | | | | | | | |
|   16 mm deep | – | – | – | – | – | m² | 70.00 |

**MACADAM SURFACING**

**Coated macadam/asphalt roads/ pavings – General**
Preamble: The prices for all in situ finishings to roads and footpaths include for work to falls, crossfalls or slopes not exceeding 15° from horizontal; for laying on prepared bases (not included) and for rolling with an appropriate roller.
Users should note the new terminology for the surfaces described below which is to European standard descriptions. The now redundant descriptions for each course are shown in brackets.

**Macadam surfacing; Spadeoak Construction Co Ltd; surface (wearing) course; 20 mm of 6 mm dense bitumen macadam**

| | | | | | | | |
|---|---|---|---|---|---|---|---|
| Machine lay; areas 1000 m² and over | | | | | | | |
|   limestone aggregate | – | – | – | – | – | m² | 7.47 |
|   granite aggregate | – | – | – | – | – | m² | 7.74 |
|   red | – | – | – | – | – | m² | 13.85 |
| Hand lay; areas 400 m² and over | | | | | | | |
|   limestone aggregate | – | – | – | – | – | m² | 6.87 |
|   granite aggregate | – | – | – | – | – | m² | 10.15 |
|   red | – | – | – | – | – | m² | 16.31 |

**Macadam surfacing; Spadeoak Construction Co Ltd; surface (wearing) course; 30 mm of 10 mm dense bitumen macadam**

| | | | | | | | |
|---|---|---|---|---|---|---|---|
| Machine lay; areas 1000 m² and over | | | | | | | |
|   limestone aggregate | – | – | – | – | – | m² | 9.30 |
|   granite aggregate | – | – | – | – | – | m² | 9.43 |
|   red | – | – | – | – | – | m² | 18.02 |
| Hand lay; areas 400 m² and over | | | | | | | |
|   limestone aggregate | – | – | – | – | – | m² | 11.49 |
|   granite aggregate | – | – | – | – | – | m² | 14.11 |
|   red | – | – | – | – | – | m² | 22.34 |

## 35 SITE WORKS – PAVINGS

| Item<br>Excluding site overheads and profit | PC<br>£ | Labour<br>hours | Labour<br>£ | Plant<br>£ | Material<br>£ | Unit | Total<br>rate £ |
|---|---|---|---|---|---|---|---|
| **MACADAM SURFACING – cont** | | | | | | | |
| **Macadam surfacing; Spadeoak Construction Co Ltd; surface (wearing) course; 40 mm of 10 mm dense bitumen macadam** | | | | | | | |
| Machine lay; areas 1000 m² and over | | | | | | | |
|   limestone aggregate | – | – | – | – | – | m² | **11.83** |
|   granite aggregate | – | – | – | – | – | m² | **12.01** |
|   red | – | – | – | – | – | m² | **25.19** |
| Hand lay; areas 400 m² and over | | | | | | | |
|   limestone aggregate | – | – | – | – | – | m² | **15.40** |
|   granite aggregate | – | – | – | – | – | m² | **15.50** |
|   red | – | – | – | – | – | m² | **27.60** |
| **Macadam surfacing; Spadeoak Construction Co Ltd; binder (base) course; 50 mm of 20 mm dense bitumen macadam** | | | | | | | |
| Machine lay; areas 1000 m² and over | | | | | | | |
|   limestone aggregate | – | – | – | – | – | m² | **11.88** |
|   granite aggregate | – | – | – | – | – | m² | **12.06** |
| Hand lay; areas 400 m² and over | | | | | | | |
|   limestone aggregate | – | – | – | – | – | m² | **14.23** |
|   granite aggregate | – | – | – | – | – | m² | **14.45** |
| **Macadam surfacing; Spadeoak Construction Co Ltd; binder (base) course; 60 mm of 20 mm dense bitumen macadam** | | | | | | | |
| Machine lay; areas 1000 m² and over | | | | | | | |
|   limestone aggregate | – | – | – | – | – | m² | **13.05** |
|   granite aggregate | – | – | – | – | – | m² | **13.28** |
| Hand lay; areas 400 m² and over | | | | | | | |
|   limestone aggregate | – | – | – | – | – | m² | **15.50** |
|   granite aggregate | – | – | – | – | – | m² | **15.72** |
| **Macadam surfacing; Spadeoak Construction Co Ltd; base (roadbase) course; 75 mm of 28 mm dense bitumen macadam** | | | | | | | |
| Machine lay; areas 1000 m² and over | | | | | | | |
|   limestone aggregate | – | – | – | – | – | m² | **16.02** |
|   granite aggregate | – | – | – | – | – | m² | **16.31** |
| Hand lay; areas 400 m² and over | | | | | | | |
|   limestone aggregate | – | – | – | – | – | m² | **18.65** |
|   granite aggregate | – | – | – | – | – | m² | **18.97** |

## 35 SITE WORKS – PAVINGS

| Item<br>Excluding site overheads and profit | PC<br>£ | Labour<br>hours | Labour<br>£ | Plant<br>£ | Material<br>£ | Unit | Total<br>rate £ |
|---|---|---|---|---|---|---|---|
| **Macadam surfacing; Spadeoak Construction Co Ltd; base (roadbase) course; 100 mm of 28 mm dense bitumen macadam** | | | | | | | |
| Machine lay; areas 1000 m² and over | | | | | | | |
|   limestone aggregate | – | – | – | – | – | m² | 19.85 |
|   granite aggregate | – | – | – | – | – | m² | 20.25 |
| Hand lay; areas 400 m² and over | | | | | | | |
|   limestone aggregate | – | – | – | – | – | m² | 22.74 |
|   granite aggregate | – | – | – | – | – | m² | 23.14 |
| **Macadam surfacing; Spadeoak Construction Co Ltd; base (roadbase) course; 150 mm of 28 mm dense bitumen macadam in two layers** | | | | | | | |
| Machine lay; areas 1000 m² and over | | | | | | | |
|   limestone aggregate | – | – | – | – | – | m² | 31.29 |
|   granite aggregate | – | – | – | – | – | m² | 31.88 |
| Hand lay; areas 400 m² and over | | | | | | | |
|   limestone aggregate | – | – | – | – | – | m² | 36.54 |
|   granite aggregate | – | – | – | – | – | m² | 37.14 |
| **Base (roadbase) course; 200 mm of 28 mm dense bitumen macadam in two layers** | | | | | | | |
| Machine lay; areas 1000 m² and over | | | | | | | |
|   limestone aggregate | – | – | – | – | – | m² | 39.77 |
|   granite aggregate | – | – | – | – | – | m² | 40.55 |
| Hand lay; areas 400 m² and over | | | | | | | |
|   limestone aggregate | – | – | – | – | – | m² | 40.07 |
|   granite aggregate | – | – | – | – | – | m² | 41.38 |
| **Resin bound macadam pavings; Bituchem; to pedestrian or vehicular hard landscape areas; laid to base course (not included)** | | | | | | | |
| Natratex wearing course; clear resin bound macadam | | | | | | | |
|   25 mm thick to pedestrian areas | 20.00 | 0.04 | 2.97 | – | 20.00 | m² | 22.97 |
|   30 mm thick to vehicular areas | 25.00 | 0.01 | 0.79 | – | 25.00 | m² | 25.79 |
| Colourtex coloured resin bound macadam | | | | | | | |
|   25 mm thick | 20.00 | 0.04 | 2.97 | – | 20.00 | m² | 22.97 |

## 35 SITE WORKS – PAVINGS

| Item<br>Excluding site overheads and profit | PC<br>£ | Labour<br>hours | Labour<br>£ | Plant<br>£ | Material<br>£ | Unit | Total<br>rate £ |
|---|---|---|---|---|---|---|---|
| **MARKING OF CAR PARKS** | | | | | | | |
| Car parking space division strips; laid hot at 115° C; on bitumen macadam surfacing | | | | | | | |
| minimum daily rate | – | – | – | – | – | item | **500.00** |
| Stainless metal road studs | | | | | | | |
| 100 × 100 mm | 5.90 | 0.08 | 5.94 | – | 5.90 | nr | **11.84** |
| **SUSPENDED PAVING** | | | | | | | |
| **Beam and Block suspended base for external use; FP McCann Ltd; Prestressed precast concrete beam; profile 115 mm wide at base, 80 mm wide at top 152 mm deep; laid to perimeter or spanned support retaining walls (not included); block infill of 440 × 225 × 100 mm solid blocks; screed of 75 mm concrete reinforced with mesh to receive surface treatments** | | | | | | | |
| Block centres between beams | | | | | | | |
| 520 mm; blocks laid 440 mm wide | 10.50 | 0.30 | 14.36 | 5.89 | 26.53 | m² | **46.78** |
| 408 mm; alternate blocks 440 mm wide and 215 mm wide between beams | 13.34 | 0.36 | 17.28 | 7.78 | 27.49 | m² | **52.55** |
| 295 mm; blocks laid 215 mm wide between beams | 21.00 | 0.36 | 17.28 | 11.78 | 34.01 | m² | **63.07** |
| **Suspended paving slabs over slab podiums; uPVC suspension units; Placement of pedestals and suspended slabs (slabs not included)** DHP 6; 170–290 mm above slab level | | | | | | | |
| Slab size: 300 × 300 mm | 171.70 | 0.25 | 11.88 | – | 171.70 | m² | **183.58** |
| Slab size: 450 × 450 mm | 75.75 | 0.17 | 7.92 | – | 75.75 | m² | **83.67** |
| Slab size: 600 × 600 mm | 70.70 | 0.14 | 6.79 | – | 70.70 | m² | **77.49** |
| **FORMWORK FOR CONCRETE PAVINGS** | | | | | | | |
| **Formwork for in situ concrete** | | | | | | | |
| Sides of foundations | | | | | | | |
| height not exceeding 250 mm | 0.29 | 0.01 | 0.79 | – | 0.42 | m | **1.21** |
| height 250–500 mm | 0.39 | 0.01 | 0.95 | – | 0.68 | m | **1.63** |
| height 500 mm–1.00 m | 0.39 | 0.02 | 1.19 | – | 0.78 | m | **1.97** |
| height exceeding 1.00 m | 1.17 | 0.67 | 47.50 | – | 4.40 | m² | **51.90** |
| Extra over formwork for curved work | | | | | | | |
| 6 m radius | – | 0.25 | 11.88 | – | – | m | **11.88** |

## 35 SITE WORKS – PAVINGS

| Item<br>Excluding site overheads and profit | PC<br>£ | Labour<br>hours | Labour<br>£ | Plant<br>£ | Material<br>£ | Unit | Total<br>rate £ |
|---|---|---|---|---|---|---|---|
| **REINFORCEMENT FOR CONCRETE PAVINGS** | | | | | | | |
| **Reinforcement; fabric; side laps 150 mm; head laps 300 mm; mesh 200 × 200 mm; in roads, footpaths or pavings** | | | | | | | |
| Fabric | | | | | | | |
| A142 (2.22 kg/m$^2$) | 1.54 | 0.03 | 1.97 | – | 1.54 | m$^2$ | 3.51 |
| A193 (3.02 kg/m$^2$) | 1.87 | 0.02 | 1.26 | – | 1.87 | m$^2$ | 3.13 |
| **IN SITU CONCRETE PAVINGS** | | | | | | | |
| **Unreinforced concrete; on prepared subbase (not included)** | | | | | | | |
| Roads; 21.00 N/mm$^2$ – 20 mm aggregate (1:2:4) mechanically mixed on site | | | | | | | |
| 100 mm thick | 12.17 | 0.04 | 2.97 | – | 12.17 | m$^2$ | 15.14 |
| 150 mm thick | 18.25 | – | 0.14 | – | 18.25 | m$^2$ | 18.39 |
| **Reinforced in situ concrete; mechanically mixed on site; normal Portland cement; on hardcore base (not included); reinforcement (not included)** | | | | | | | |
| Roads; 11.50 N/mm$^2$ – 40 mm aggregate (1:3:6) | | | | | | | |
| 100 mm thick | 10.27 | 0.03 | 2.37 | – | 10.27 | m$^2$ | 12.64 |
| 150 mm thick | 15.41 | 0.20 | 14.25 | – | 15.41 | m$^2$ | 29.66 |
| 200 mm thick | 21.03 | 0.27 | 19.00 | – | 21.03 | m$^2$ | 40.03 |
| 250 mm thick | 25.66 | 0.33 | 23.75 | 0.30 | 25.66 | m$^2$ | 49.71 |
| 300 mm thick | 30.80 | 0.40 | 28.50 | 0.30 | 30.80 | m$^2$ | 59.60 |
| Roads; 21.00 N/mm$^2$ – 20 mm aggregate (1:2:4) | | | | | | | |
| 100 mm thick | 11.88 | 0.13 | 9.50 | – | 11.88 | m$^2$ | 21.38 |
| 150 mm thick | 17.82 | 0.20 | 14.25 | – | 17.82 | m$^2$ | 32.07 |
| 200 mm thick | 23.76 | 0.27 | 19.00 | – | 23.76 | m$^2$ | 42.76 |
| 250 mm thick | 29.69 | 0.33 | 23.75 | 0.30 | 29.69 | m$^2$ | 53.74 |
| 300 mm thick | 35.63 | 0.40 | 28.50 | 0.30 | 35.63 | m$^2$ | 64.43 |
| Roads; 25.00 N/mm$^2$ – 20 mm aggregate C 25/30 ready mixed | | | | | | | |
| 100 mm thick | 9.79 | 0.13 | 9.50 | – | 9.79 | m$^2$ | 19.29 |
| 150 mm thick | 14.69 | 0.20 | 14.25 | – | 14.69 | m$^2$ | 28.94 |
| 200 mm thick | 19.58 | 0.27 | 19.00 | – | 19.58 | m$^2$ | 38.58 |
| 250 mm thick | 24.47 | 0.33 | 23.75 | 0.30 | 24.47 | m$^2$ | 48.52 |
| 300 mm thick | 29.37 | 0.40 | 28.50 | 0.30 | 29.37 | m$^2$ | 58.17 |

## 35 SITE WORKS – PAVINGS

| Item<br>Excluding site overheads and profit | PC<br>£ | Labour<br>hours | Labour<br>£ | Plant<br>£ | Material<br>£ | Unit | Total<br>rate £ |
|---|---|---|---|---|---|---|---|
| **IN SITU CONCRETE PAVINGS –** **cont** | | | | | | | |
| **Reinforced in situ concrete; ready mixed; discharged directly into location from supply vehicle; normal Portland cement; on hardcore base (not included); reinforcement (not included)** | | | | | | | |
| Roads; 11.50 N/mm$^2$ – 40 mm aggregate (1:3:6) | | | | | | | |
| 100 mm thick | 8.40 | 0.05 | 3.80 | – | 8.40 | m$^2$ | **12.20** |
| 150 mm thick | 12.60 | 0.08 | 5.70 | – | 12.60 | m$^2$ | **18.30** |
| 200 mm thick | 16.80 | 0.12 | 8.55 | – | 16.80 | m$^2$ | **25.35** |
| 250 mm thick | 21.00 | 0.18 | 12.82 | 0.30 | 21.00 | m$^2$ | **34.12** |
| 300 mm thick | 25.20 | 0.22 | 15.68 | 0.30 | 25.20 | m$^2$ | **41.18** |
| Roads; 21.00 N/mm$^2$ – 20 mm aggregate (1:2:4) | | | | | | | |
| 100 mm thick | 9.10 | 2.08 | 148.44 | – | 9.10 | m$^2$ | **157.54** |
| 150 mm thick | 13.65 | 0.08 | 5.70 | – | 13.65 | m$^2$ | **19.35** |
| 200 mm thick | 18.20 | 0.12 | 8.55 | – | 18.20 | m$^2$ | **26.75** |
| 250 mm thick | 22.75 | 0.18 | 12.82 | 0.30 | 22.75 | m$^2$ | **35.87** |
| 300 mm thick | 27.30 | 0.22 | 15.68 | 0.30 | 27.30 | m$^2$ | **43.28** |
| Roads; 26.00 N/mm$^2$ – 20 mm aggregate (1:1.5:3) | | | | | | | |
| 100 mm thick | 9.57 | 0.05 | 3.80 | – | 9.57 | m$^2$ | **13.37** |
| 150 mm thick | 14.36 | 0.08 | 5.70 | – | 14.36 | m$^2$ | **20.06** |
| 200 mm thick | 18.20 | 0.12 | 8.55 | – | 18.20 | m$^2$ | **26.75** |
| 250 mm thick | 23.93 | 0.18 | 12.82 | 0.30 | 23.93 | m$^2$ | **37.05** |
| 300 mm thick | 28.71 | 0.22 | 15.68 | 0.30 | 28.71 | m$^2$ | **44.69** |
| Roads; PAV1 concrete – 35 N/mm$^2$; Designated mix | | | | | | | |
| 100 mm thick | 10.40 | 0.05 | 3.80 | – | 10.40 | m$^2$ | **14.20** |
| 150 mm thick | 15.61 | 0.08 | 5.70 | – | 15.61 | m$^2$ | **21.31** |
| 200 mm thick | 20.81 | 0.12 | 8.55 | – | 20.81 | m$^2$ | **29.36** |
| 250 mm thick | 26.00 | 0.18 | 12.82 | 0.30 | 26.00 | m$^2$ | **39.12** |
| 300 mm thick | 31.21 | 0.22 | 15.68 | 0.30 | 31.21 | m$^2$ | **47.19** |
| **Concrete sundries** | | | | | | | |
| Treating surfaces of unset concrete; grading to cambers; tamping with 75 mm thick steel shod tamper or similar | – | 0.13 | 6.33 | – | – | m$^2$ | **6.33** |

## 35 SITE WORKS – PAVINGS

| Item<br>Excluding site overheads and profit | PC<br>£ | Labour<br>hours | Labour<br>£ | Plant<br>£ | Material<br>£ | Unit | Total<br>rate £ |
|---|---|---|---|---|---|---|---|
| **EXPANSION JOINTS IN PAVINGS** | | | | | | | |
| 13 mm thick joint filler; formwork | | | | | | | |
| width or depth not exceeding | | | | | | | |
| 150 mm | 1.08 | 0.04 | 2.12 | – | 1.97 | m | **4.09** |
| width or depth 150–300 mm | 2.17 | 0.06 | 2.65 | – | 3.94 | m | **6.59** |
| width or depth 300–450 mm | 3.25 | 0.07 | 3.18 | – | 5.91 | m | **9.09** |
| 25 mm thick joint filler; formwork | | | | | | | |
| width or depth not exceeding | | | | | | | |
| 150 mm | 3.00 | 0.04 | 2.12 | – | 3.89 | m | **6.01** |
| width or depth 150–300 mm | 6.00 | 0.06 | 2.65 | – | 7.77 | m | **10.42** |
| width or depth 300–450 mm | 9.00 | 0.07 | 3.18 | – | 11.66 | m | **14.84** |
| Sealants; sealing top 25 mm of joint with rubberized bituminous compound | 1.25 | 0.06 | 2.65 | – | 1.25 | m | **3.90** |
| **WOODCHIP SURFACES** | | | | | | | |
| **Wood chip surfaces; Melcourt Industries Ltd (items labelled FSC are Forest Stewardship Council certified)** | | | | | | | |
| Wood chips; to surface of pathways by machine; material delivered in 80 m³ loads; levelling and spreading by hand (excavation and preparation not included) | | | | | | | |
| Walk Chips; 100 mm thick; FSC (25 m³ loads) | 4.12 | 0.01 | 0.79 | 0.42 | 4.12 | m² | **5.33** |
| Walk Chips; 100 mm thick; FSC (80 m³ loads) | 2.75 | 0.01 | 0.79 | 0.42 | 2.75 | m² | **3.96** |
| Woodfibre; 100 mm thick; FSC (80 m³ loads) | 2.57 | 0.01 | 0.79 | 0.42 | 2.57 | m² | **3.78** |
| **TIMBER DECKING** | | | | | | | |
| **Market prices of timber decking materials; Exterior Decking Ltd** | | | | | | | |
| Hardwood decking; stainless steel screw fixings included in rates shown | | | | | | | |
| Ipe; pre-drilled and countersunk | 131.27 | – | – | – | – | m² | – |
| Hardwood decking; Exterpark invisible fixings included in rates shown | | | | | | | |
| Ipe; 21 mm thick | 145.25 | – | – | – | – | m² | – |
| Ipe; 28 mm thick | 209.51 | – | – | – | – | m² | – |
| Ipe; 35 mm thick | 277.23 | – | – | – | – | m² | – |
| Teak Rustic FSC 21 mm thick | 161.12 | – | – | – | – | m² | – |
| Merbau 21 mm | 123.13 | – | – | – | – | m² | – |
| Merbau 28 mm | 155.94 | – | – | – | – | m² | – |
| Merbau 35 mm | 180.62 | – | – | – | – | m² | – |

## 35 SITE WORKS – PAVINGS

| Item<br>Excluding site overheads and profit | PC<br>£ | Labour<br>hours | Labour<br>£ | Plant<br>£ | Material<br>£ | Unit | Total<br>rate £ |
|---|---|---|---|---|---|---|---|
| **TIMBER DECKING – cont** | | | | | | | |
| **Market prices of timber decking materials – cont** | | | | | | | |
| Composite decking; stainless steel screw fixings included in rates shown | | | | | | | |
| Millboard 176 mm wide | 90.59 | – | – | – | – | m$^2$ | – |
| Millboard 196 mm wide | 92.81 | – | – | – | 92.81 | m$^2$ | 92.81 |
| Softwood decking | | | | | | | |
| Pine smooth; 145 × 38 mm | 45.96 | – | – | – | – | m$^2$ | – |
| **Timber decking; Exterior Decking Ltd; timber decking substructure/ carcassing costs** | | | | | | | |
| Roof decks exclusive of beams; timber craned (not included); joists at 400 mm centres | | | | | | | |
| 47 × 47 mm | – | – | – | – | – | m$^2$ | 37.80 |
| 47 × 100 mm | – | – | – | – | – | m$^2$ | 40.92 |
| 47 × 150 mm | – | – | – | – | – | m$^2$ | 44.09 |
| pedestals up to 150 mm | – | – | – | – | – | m$^2$ | 29.86 |
| Commercial applications; decks with double beams at 300 mm centres | | | | | | | |
| 47 × 100 mm | – | – | – | – | – | m$^2$ | 64.68 |
| 47 × 150 mm | – | – | – | – | – | m$^2$ | 80.91 |
| Domestic garden applications; decks with double beams at 300 mm centres | | | | | | | |
| 47 × 100 mm | – | – | – | – | – | m$^2$ | 76.23 |
| 47 × 150 mm | – | – | – | – | – | m$^2$ | 80.91 |
| Structural posts to decks; tanalized treated softwood; rebated including excavation and setting in concrete | | | | | | | |
| 100 × 100 mm; single beam; single rebate | – | – | – | – | – | m | 11.17 |
| 150 × 150 mm; double beam; double rebate | – | – | – | – | – | m | 15.26 |
| **Timber decking; Exterior Decking Ltd; timber decking surfaces; fixed to substructure (not included)** | | | | | | | |
| Softwood deck timbers; screw fixed | | | | | | | |
| pine; smooth; 145 × 38 mm fixed with 65 × 5 mm coated deck screws | – | – | – | – | – | m$^2$ | 43.20 |
| Hardwood deck timbers; screw fixed | | | | | | | |
| Ipe; S4S E4E; 145 × 21 mm; pre-drilled | – | – | – | – | – | m$^2$ | 127.07 |
| Hardwood deck timbers; invisibly fixed | | | | | | | |
| Ipe; 21 mm thick | – | – | – | – | – | m$^2$ | 145.52 |
| Ipe; 28 mm thick | – | – | – | – | – | m$^2$ | 209.78 |
| Ipe; 35 mm thick | – | – | – | – | – | m$^2$ | 277.50 |

## 35 SITE WORKS – PAVINGS

| Item<br>Excluding site overheads and profit | PC<br>£ | Labour<br>hours | Labour<br>£ | Plant<br>£ | Material<br>£ | Unit | Total<br>rate £ |
|---|---|---|---|---|---|---|---|
| Teak Rustic FSC 21 mm | – | – | – | – | – | m² | **161.39** |
| Merbau 21 mm | – | – | – | – | – | m² | **123.32** |
| Merbau 28 mm | – | – | – | – | – | m² | **156.21** |
| Merbau 35 mm | – | – | – | – | – | m² | **180.89** |
| Composite deck surface; Millboard | | | | | | | |
| 146 mm wide; fixed with stainless screws 4.5 × 70 mm | – | – | – | – | – | m² | **90.59** |
| 196 mm wide; fixed with stainless screws 4.5 × 70 mm | – | – | – | – | – | m² | **90.59** |
| Surface treatments to timber decks; Owatrol decking oil | | | | | | | |
| Seasonite | – | – | – | – | – | m² | **3.21** |
| Textrol | – | – | – | – | – | m² | **5.13** |
| **Timber decking** | | | | | | | |
| Supports for timber decking; softwood joists to receive decking boards; joists at 400 mm centres; Southern Yellow pine | | | | | | | |
| 38 × 88 mm | 15.95 | 1.00 | 23.75 | – | 17.69 | m² | **41.44** |
| 47 × 100 mm | 9.63 | 1.00 | 23.75 | – | 11.37 | m² | **35.12** |
| 47 × 150 mm | 13.75 | 1.00 | 23.75 | – | 15.49 | m² | **39.24** |
| Hardwood decking; Yellow Balau; grooved or smooth; 6 mm joints | | | | | | | |
| deck boards; 90 mm wide × 19 mm thick | 43.31 | 1.00 | 23.75 | – | 45.29 | m² | **69.04** |
| deck boards; 145 mm wide × 21 mm thick | 82.43 | 1.00 | 23.75 | – | 84.41 | m² | **108.16** |
| deck boards; 145 mm wide × 28 mm thick | 53.81 | 1.00 | 23.75 | – | 55.79 | m² | **79.54** |
| Hardwood decking; Ipe; smooth; 6 mm joints | | | | | | | |
| deck boards; 90 mm wide × 19 mm thick | 66.66 | 1.00 | 23.75 | – | 68.64 | m² | **92.39** |
| deck boards; 145 mm wide × 19 mm thick | 83.05 | 1.00 | 23.75 | – | 85.03 | m² | **108.78** |
| AVS; Profile; Western Red Cedar; 6 mm joints smooth or reeded | | | | | | | |
| ex 125 mm wide × 38 mm thick | 32.68 | 1.00 | 23.75 | – | 34.66 | m² | **58.41** |
| ex 150 mm wide × 32 mm thick | 56.26 | 1.00 | 23.75 | – | 58.24 | m² | **81.99** |
| Handrails and base rail; fixed to posts at 2.00 m centres | | | | | | | |
| posts; 100 × 100 × 1370 mm high | 26.44 | 1.00 | 23.75 | – | 26.89 | m | **50.64** |
| posts; turned; 1220 mm high | 27.83 | 1.00 | 23.75 | – | 28.28 | m | **52.03** |
| Handrails; balusters | | | | | | | |
| square balusters at 100 mm centres | 24.00 | 0.50 | 11.88 | – | 24.45 | m | **36.33** |
| square balusters at 300 mm centres | 7.99 | 0.33 | 7.91 | – | 8.36 | m | **16.27** |
| turned balusters at 100 mm centres | 55.00 | 0.50 | 11.88 | – | 55.45 | m | **67.33** |
| turned balusters at 300 mm centres | 18.32 | 0.33 | 7.84 | – | 18.68 | m | **26.52** |

## 35 SITE WORKS – PAVINGS

| Item<br>Excluding site overheads and profit | PC<br>£ | Labour<br>hours | Labour<br>£ | Plant<br>£ | Material<br>£ | Unit | **Total<br>rate £** |
|---|---|---|---|---|---|---|---|
| **STEPS** | | | | | | | |
| **In situ concrete steps; ready mixed concrete; Gen 1; on 100 mm hardcore base; base concrete thickness 100 mm treads and risers cast simultaneously; inclusive of A142 mesh within the concrete; step riser 170 mm high; inclusive of all shuttering and concrete labours; floated finish; to graded ground (not included)** | | | | | | | |
| Treads 250 mm wide | | | | | | | |
| one riser | 5.91 | 1.32 | 94.05 | – | 7.92 | m | **101.97** |
| two risers | 9.46 | 1.98 | 141.07 | – | 13.65 | m | **154.72** |
| three risers | 13.69 | 2.64 | 188.10 | – | 21.72 | m | **209.82** |
| four risers | 18.39 | 3.96 | 282.15 | – | 27.43 | m | **309.58** |
| six risers | 23.29 | 5.28 | 376.20 | – | 31.64 | m | **407.84** |
| Lay surface to concrete in situ steps above | | | | | | | |
| granite treads; pre-cut off site; 305 mm wide × 50 mm thick bullnose finish inclusive of 20 mm overhang; granite riser 20 mm thick all laid on 15 mm mortar bed | 62.43 | 0.66 | 47.02 | – | 63.71 | m | **110.73** |
| riven Yorkstone tread and riser 50 mm thick; cut on site | 43.64 | 0.82 | 58.78 | – | 44.92 | m | **103.70** |
| riven Yorkstone treads 225 mm wide × 50 mm thick cut on site; two courses brick risers laid on flat stretcher bond; bricks PC £500.00/1000 | 21.35 | 0.99 | 70.54 | – | 27.39 | m | **97.93** |
| Marshalls Saxon paving slab 450 × 450 mm cut to 225 mm wide × 38 mm thick; cut on site; two courses brick risers laid on flat stretcher bond; bricks PC £500.00/1000 | 6.01 | 0.99 | 70.54 | – | 12.05 | m | **82.59** |
| Treads 450 mm wide | | | | | | | |
| one riser | 10.63 | 1.65 | 117.56 | – | 13.49 | m | **131.05** |
| two risers | 21.72 | 2.31 | 164.59 | – | 27.61 | m | **192.20** |
| three risers | 30.70 | 2.97 | 211.61 | – | 38.93 | m | **250.54** |
| four risers | 53.09 | 4.29 | 305.66 | – | 64.85 | m | **370.51** |
| six risers | 67.48 | 5.94 | 423.23 | – | 77.59 | m | **500.82** |

## 35 SITE WORKS – PAVINGS

| Item Excluding site overheads and profit | PC £ | Labour hours | Labour £ | Plant £ | Material £ | Unit | Total rate £ |
|---|---|---|---|---|---|---|---|
| Lay surface to concrete in situ steps | | | | | | | |
| granite treads; pre-cut off site; 470 mm wide × 50 mm thick bullnose finish inclusive of 20 mm overhang; granite riser 20 mm thick all laid on 15 mm mortar bed | 0.94 | 1.32 | 94.05 | – | 2.54 | m | 96.59 |
| riven Yorkstone tread and riser 50 mm thick; cut on site | 59.29 | 1.49 | 105.81 | – | 60.89 | m | 166.70 |
| riven Yorkstone treads 460 mm wide 50 mm thick cut on site; two courses brick risers laid on flat stretcher bond; bricks PC £500.00/1000 | 43.64 | 1.65 | 117.56 | – | 49.94 | m | 167.50 |
| Marshalls Saxon paving slab 450 × 450 × 38 mm thick cut on site; two courses brick risers laid on flat stretcher bond; bricks PC £500.00/1000 | 12.01 | 0.91 | 64.66 | – | 18.05 | m | 82.71 |
| **Precast concrete steps; FP McCann Ltd; Precast units laid to graded blinded and levelled bases (not included); to receive surface treatments priced separately; waist 200 mm thick; finished sizes with paving surface treatments of 40 mm thick including bedding** | | | | | | | |
| 5 Risers; 280 mm deep × 175 mm high | | | | | | | |
| 1.20 m wide | 303.98 | 3.00 | 213.75 | 32.91 | 303.98 | nr | 550.64 |
| 2.00 m wide | 607.95 | 4.00 | 285.00 | 65.83 | 607.95 | nr | 958.78 |
| 2.50 m wide | 759.98 | 5.00 | 356.25 | 164.56 | 759.98 | nr | 1280.79 |
| 3.00 m wide | 911.92 | 7.00 | 498.75 | 164.56 | 911.92 | nr | 1575.23 |
| 4.00 m wide | 1215.90 | 8.00 | 570.00 | 230.39 | 1215.90 | nr | 2016.29 |
| 10 Risers; 280 mm deep × 175 mm high | | | | | | | |
| 1.20 m wide | 729.60 | 4.00 | 285.00 | 65.83 | 729.60 | nr | 1080.43 |
| 2.00 m wide | 1216.05 | 6.00 | 427.50 | 131.65 | 1216.05 | nr | 1775.20 |
| 2.50 m wide | 1520.10 | 8.00 | 570.00 | 263.30 | 1520.10 | nr | 2353.40 |
| 3.00 m wide | 1824.08 | 8.00 | 570.00 | 263.30 | 1824.08 | nr | 2657.38 |
| 4.00 m wide | 2432.10 | 8.00 | 570.00 | 263.30 | 2432.10 | nr | 3265.40 |
| 5 Risers; 450 mm deep × 175 mm high | | | | | | | |
| 1.20 m wide | 559.65 | 3.00 | 213.75 | 32.91 | 559.65 | nr | 806.31 |
| 2.00 m wide | 932.70 | 4.00 | 285.00 | 65.83 | 932.70 | nr | 1283.53 |
| 2.50 m wide | 1165.95 | 5.00 | 356.25 | 164.56 | 1165.95 | nr | 1686.76 |
| 3.00 m wide | 1057.05 | 7.00 | 498.75 | 164.56 | 1057.05 | nr | 1720.36 |
| 4.00 m wide | 1865.47 | 8.00 | 570.00 | 230.39 | 1865.47 | nr | 2665.86 |
| 10 Risers; 450 mm deep × 175 mm high | | | | | | | |
| 1.20 m wide | 1119.60 | 3.00 | 213.75 | 32.91 | 1119.60 | nr | 1366.26 |
| 2.00 m wide | 1866.00 | 4.00 | 285.00 | 65.83 | 1866.00 | nr | 2216.83 |
| 2.50 m wide | 2332.50 | 5.00 | 356.25 | 164.56 | 2332.50 | nr | 2853.31 |
| 3.00 m wide | 2797.50 | 7.00 | 498.75 | 164.56 | 2797.50 | nr | 3460.81 |
| 4.00 m wide | 3727.50 | 8.00 | 570.00 | 230.39 | 3727.50 | nr | 4527.89 |

## 35 SITE WORKS – PAVINGS

| Item<br>Excluding site overheads and profit | PC<br>£ | Labour<br>hours | Labour<br>£ | Plant<br>£ | Material<br>£ | Unit | Total<br>rate £ |
|---|---|---|---|---|---|---|---|
| **STEPS – cont** | | | | | | | |
| **Precast concrete steps – cont** | | | | | | | |
| Extra over for integrated landing in step flights; 1.00 m deep | | | | | | | |
| 1.20 m wide | – | – | – | – | 180.00 | nr | **180.00** |
| 2.00 m wide | – | – | – | – | 300.00 | nr | **300.00** |
| 2.50 m wide | – | – | – | – | 375.00 | nr | **375.00** |
| 3.00 m wide | – | – | – | – | 450.00 | nr | **450.00** |
| 4.00 m wide | – | – | – | – | 600.00 | nr | **600.00** |
| **Step treads; Marshalls Plc; Precast step treads to match paving types** | | | | | | | |
| Conservation | | | | | | | |
| Step Tread 450 × 450 × 50 mm – Chamfered | 138.86 | 1.00 | 71.25 | – | 140.79 | m | **212.04** |
| Step Tread 450 × 450 × 50 mm | 132.73 | 1.00 | 71.25 | – | 134.65 | m | **205.90** |
| Step Tread 400 × 400 × 65 mm | 148.13 | 1.05 | 74.81 | – | 150.05 | m | **224.86** |
| Step Unit 350 × 1000 × 150 mm | 224.90 | 1.00 | 71.25 | – | 226.82 | m | **298.07** |
| Step Unit 400 × 1000 × 150 mm | 240.50 | 1.00 | 71.25 | – | 242.42 | m | **313.67** |
| Saxon Step Tread 400 × 400 × 50 mm | 123.45 | 1.05 | 74.81 | – | 125.37 | m | **200.18** |
| Saxon Step Tread 450 × 450 × 50 mm | 115.85 | 1.00 | 71.25 | – | 117.78 | m | **189.03** |

## 35 SITE WORKS – STREET FURNITURE

| Item<br>Excluding site overheads and profit | PC<br>£ | Labour<br>hours | Labour<br>£ | Plant<br>£ | Material<br>£ | Unit | Total<br>rate £ |
|---|---|---|---|---|---|---|---|
| **BARRIERS** | | | | | | | |
| **Barriers – General** | | | | | | | |
| Preamble: The provision of car and truck control barriers may form part of the landscape contract. Barriers range from simple manual counterweighted poles to fully automated remote-control security gates, and the exact degree of control required must be specified. Complex barriers may need special maintenance and repair. | | | | | | | |
| **Barriers; Barriers Direct** | | | | | | | |
| Manually operated pole barriers; counterbalance; to tubular steel supports; bolting to concrete foundation (foundation not included); aluminium boom; various finishes, including arm rest | | | | | | | |
| clear opening up to 3.00 m | 680.61 | 6.00 | 142.50 | – | 705.64 | nr | **848.14** |
| clear opening up to 4.00 m | 711.48 | 6.00 | 142.50 | – | 736.51 | nr | **879.01** |
| clear opening 5.00 m | 743.82 | 6.00 | 142.50 | – | 768.85 | nr | **911.35** |
| clear opening 6.00 m | 771.75 | 6.00 | 142.50 | – | 796.78 | nr | **939.28** |
| clear opening 7.00 m | 845.25 | 6.00 | 142.50 | – | 870.28 | nr | **1012.78** |
| Electrically operated pole barriers; enclosed fan-cooled motor; double worm reduction gear; overload clutch coupling with remote controls; aluminium boom; various finishes, including catch pole (exclusive of electrical connections by electrician) | | | | | | | |
| Barriers Direct 2.65 m boom with catchpost | 1582.00 | 6.00 | 142.50 | – | 1607.03 | nr | **1749.53** |
| Barriers Direct 4.5 m boom with catchpost | 1582.00 | 6.00 | 142.50 | – | 1607.03 | nr | **1749.53** |
| Barriers Direct 6.0 m boom with catchpost | 1582.00 | 6.00 | 142.50 | – | 1607.03 | nr | **1749.53** |
| **BASES FOR STREET FURNITURE** | | | | | | | |
| **Bases for street furniture** | | | | | | | |
| Excavating; filling with concrete 1:3:6; bases for street furniture | | | | | | | |
| 300 × 450 × 500 mm deep | – | 0.75 | 17.81 | – | 9.26 | nr | **27.07** |
| 300 × 600 × 500 mm deep | – | 1.11 | 26.36 | – | 12.34 | nr | **38.70** |
| 300 × 900 × 500 mm deep | – | 1.67 | 39.66 | – | 18.51 | nr | **58.17** |
| 1750 × 900 × 300 mm deep | – | 2.33 | 55.34 | – | 64.80 | nr | **120.14** |
| 2000 × 900 × 300 mm deep | – | 2.67 | 63.41 | – | 74.06 | nr | **137.47** |
| 2400 × 900 × 300 mm deep | – | 3.20 | 76.00 | – | 88.87 | nr | **164.87** |
| 2400 × 1000 × 300 mm deep | – | 3.56 | 84.55 | – | 98.74 | nr | **183.29** |

## 35 SITE WORKS – STREET FURNITURE

| Item<br>Excluding site overheads and profit | PC<br>£ | Labour<br>hours | Labour<br>£ | Plant<br>£ | Material<br>£ | Unit | Total<br>rate £ |
|---|---|---|---|---|---|---|---|
| **BASES FOR STREET FURNITURE – cont** | | | | | | | |
| **Bases for street furniture – cont** | | | | | | | |
| Precast concrete flags; to concrete bases (not included); bedding and jointing in cement:mortar (1:4) | | | | | | | |
| 450 × 600 × 50 mm | 12.56 | 1.17 | 27.71 | – | 20.10 | $m^2$ | **47.81** |
| Precast concrete paving blocks; to concrete bases (not included); bedding in sharp sand; butt joints | | | | | | | |
| 200 × 100 × 65 mm | 10.77 | 0.50 | 11.88 | – | 13.13 | $m^2$ | **25.01** |
| 200 × 100 × 80 mm | 11.95 | 0.50 | 11.88 | – | 14.40 | $m^2$ | **26.28** |
| Engineering paving bricks; to concrete bases (not included); bedding and jointing in sulphate-resisting cement: lime:sand mortar (1:1:6) | | | | | | | |
| over 300 mm wide | 18.63 | 0.56 | 13.34 | – | 34.23 | $m^2$ | **47.57** |
| Edge restraints to pavings; haunching in concrete (1:3:6) | | | | | | | |
| 200 × 300 mm | 6.01 | 0.10 | 2.38 | – | 6.01 | m | **8.39** |
| Bases; Earth Anchors Ltd | | | | | | | |
| Rootfast ancillary anchors; A1; 500 mm long × 25 mm diameter and top strap F2; including bolting to site furniture (site furniture not included) | 9.50 | 0.50 | 11.88 | – | 9.50 | set | **21.38** |
| installation tool for above | 23.50 | – | – | – | 23.50 | nr | **23.50** |
| Rootfast ancillary anchors; A4; heavy duty 40 mm square fixed head anchors; including bolting to site furniture (site furniture not included) | 29.00 | 0.33 | 7.92 | – | 29.00 | set | **36.92** |
| Rootfast ancillary anchors; F4; vertical socket; including bolting to site furniture (site furniture not included) | 13.07 | 0.05 | 1.19 | – | 13.07 | set | **14.26** |
| Rootfast ancillary anchors; F3; horizontal socket; including bolting to site furniture (site furniture not included) | 14.35 | 0.05 | 1.19 | – | 14.35 | set | **15.54** |
| installation tools for the above | 68.00 | – | – | – | 68.00 | nr | **68.00** |
| Rootfast ancillary anchors; A3; anchored bases; including bolting to site furniture (site furniture not included) | 48.17 | 0.33 | 7.92 | – | 48.17 | set | **56.09** |
| installation tools for the above | 68.00 | – | – | – | 68.00 | nr | **68.00** |

## 35 SITE WORKS – STREET FURNITURE

| Item<br>Excluding site overheads and profit | PC<br>£ | Labour<br>hours | Labour<br>£ | Plant<br>£ | Material<br>£ | Unit | Total<br>rate £ |
|---|---|---|---|---|---|---|---|
| **BINS: LITTER/WASTE/DOG WASTE/SALT AND GRIT** | | | | | | | |
| **Dog waste bins** | | | | | | | |
| Earth Anchors Ltd | | | | | | | |
|   HG45A; steel; 45 l; earth anchored; post mounted | 107.00 | 0.33 | 7.92 | – | 107.00 | nr | **114.92** |
|   HG45A; steel; 45 l; as above with pedal operation | 137.00 | 0.33 | 7.92 | – | 137.00 | nr | **144.92** |
| Furnitubes International Ltd | | | | | | | |
|   PED 701; Pedigree; post mounted cast iron dog waste bins; 1250 mm total height above ground; 400 mm square bin | 546.00 | 0.75 | 17.81 | – | 553.42 | nr | **571.23** |
|   LUK745 P; Lucky; steel dog bin; post mounted; 47 l | 315.00 | 1.50 | 35.63 | – | 322.42 | nr | **358.05** |
|   LUK745 W; Lucky; steel dog bin; wall mounted; 47 l | 340.00 | 0.75 | 17.81 | – | 347.42 | nr | **365.23** |
|   TER801 P; Terrier; polythene dog bin; post mounted; 40 l | 99.00 | 1.50 | 35.63 | – | 106.42 | nr | **142.05** |
|   TER801 W; Terrier; polythene dog bin; wall mounted; 40 l | 90.00 | 0.75 | 17.81 | – | 97.42 | nr | **115.23** |
| **Litter bins; precast concrete in textured white or exposed aggregate finish; with wire baskets and drainage holes** | | | | | | | |
| Bins; Marshalls Plc | | | | | | | |
|   Boulevard 700; concrete circular litter bin | 505.81 | 0.50 | 11.88 | – | 505.81 | nr | **517.69** |
| Bins; Neptune Outdoor Furniture Ltd | | | | | | | |
|   AR16; 42 l | 450.00 | 0.50 | 11.88 | – | 450.00 | nr | **461.88** |
|   AR14; 100 l | 500.00 | 0.50 | 11.88 | – | 500.00 | nr | **511.88** |
| Bins; Townscape Products Ltd | | | | | | | |
|   Braunton; with basket and canopy; finished in natural round exposed aggregate | 160.00 | 1.50 | 35.63 | – | 160.00 | nr | **195.63** |
| **Litter bins; metal; stove-enamelled perforated metal for holder and container** | | | | | | | |
| Bins; Townscape Products Ltd | | | | | | | |
|   Sherringham Grade 316 stainless steel | 885.00 | 0.33 | 7.92 | – | 885.00 | nr | **892.92** |
|   Voltan Large round steel powder coated finish | 703.00 | 0.33 | 7.92 | – | 703.00 | nr | **710.92** |

## 35 SITE WORKS – STREET FURNITURE

| Item<br>Excluding site overheads and profit | PC<br>£ | Labour<br>hours | Labour<br>£ | Plant<br>£ | Material<br>£ | Unit | Total<br>rate £ |
|---|---|---|---|---|---|---|---|
| **BINS: LITTER/WASTE/DOG WASTE/SALT AND GRIT – cont** | | | | | | | |
| **Litter bins; all-steel** | | | | | | | |
| Bins; Marshall Plc | | | | | | | |
|   MSF Central; steel litter bin | 363.55 | 1.00 | 23.75 | – | 363.55 | nr | **387.30** |
|   MSF Central; stainless steel litter bin | 694.35 | 0.67 | 15.83 | – | 694.35 | nr | **710.18** |
| Bins; Furnitubes International Ltd | | | | | | | |
|   Wave Bin; WVB 440; free standing 55 l liners; 440 mm diameter × 850 mm high | 460.00 | 0.50 | 11.88 | – | 461.64 | nr | **473.52** |
|   Wave Bin WVB 520; free standing 85 l liners; 520 mm diameter × 850 mm high; cast iron plinth | 470.00 | 0.50 | 11.88 | – | 471.64 | nr | **483.52** |
|   Wave Bin; WVB 440 S304; free standing; steel with satin finish; cast bronze plinth; open top; 55 l | 770.00 | 0.50 | 11.88 | – | 770.00 | nr | **781.88** |
|   Wave Bin; WVB 520 S304; free standing; steel with satin finish; cast bronze plinth; open top; 80 l | 830.00 | 0.50 | 11.88 | – | 830.00 | nr | **841.88** |
|   Liverpool Bin; LVR520 S304; free standing; stainless steel; side opening; 965 mm high; 125 l | 1200.00 | 0.50 | 11.88 | – | 1200.00 | nr | **1211.88** |
| Bins; Earth Anchors Ltd | | | | | | | |
|   Ranger; 100 l; pedestal mounted | 445.00 | 0.50 | 11.88 | – | 445.00 | nr | **456.88** |
|   Big Ben; 82 l; steel frame and liner; colour coated finish; earth anchored | 268.00 | 1.00 | 23.75 | – | 268.00 | nr | **291.75** |
|   Beau; 42 l; steel frame and liner; colour coated finish; earth anchored | 249.00 | 1.00 | 23.75 | – | 249.00 | nr | **272.75** |
| Bins; Townscape Products Ltd | | | | | | | |
|   Sandringham steel and timber with weather cover | 1412.00 | 1.00 | 23.75 | – | 1412.00 | nr | **1435.75** |
| **Litter bins; cast iron** | | | | | | | |
| Bins; Marshalls Plc | | | | | | | |
|   MSF5501 Heritage; cast iron litter bin | 722.58 | 1.00 | 23.75 | 2.27 | 722.58 | nr | **748.60** |
| Bins; Furnitubes International Ltd | | | | | | | |
|   Covent Garden COV 702; side opening, 500 mm square × 1050 mm high; 105 l capacity | 295.00 | 0.50 | 11.88 | – | 299.77 | nr | **311.65** |
|   Covent Garden COV 803; side opening, 500 mm diameter × 1025 mm high; 85 l capacity | 249.50 | 0.50 | 11.88 | – | 254.27 | nr | **266.15** |
|   Covent Garden COV 912; open top; 500 mm A/F octagonal × 820 mm high; 85 l capacity | 412.00 | 0.50 | 11.88 | – | 416.77 | nr | **428.65** |
|   Albert ALB 800; open top; 400 mm diameter × 845 mm high; 55 l capacity | 328.00 | 0.50 | 11.88 | – | 332.77 | nr | **344.65** |

## 35 SITE WORKS – STREET FURNITURE

| Item<br>Excluding site overheads and profit | PC<br>£ | Labour<br>hours | Labour<br>£ | Plant<br>£ | Material<br>£ | Unit | Total<br>rate £ |
|---|---|---|---|---|---|---|---|
| **Bins; Broxap Street Furniture** | | | | | | | |
| Derby Hercules; post mounted; steel; 40 l | 126.25 | 1.00 | 23.75 | – | 131.02 | nr | **154.77** |
| **Litter bins; timber faced; hardwood slatted casings with removable metal litter containers; ground or wall fixing** | | | | | | | |
| Bins; Lister Lutyens Co Ltd | | | | | | | |
| Monmouth; 675 mm high × 450 mm wide; freestanding | 401.00 | 0.33 | 7.92 | – | 401.00 | nr | **408.92** |
| Bins; Woodscape Ltd | | | | | | | |
| square; 580 × 580 × 950 mm high; with lockable lid | 660.00 | 0.50 | 11.88 | – | 660.00 | nr | **671.88** |
| round; 580 mm diameter × 950 mm high; with lockable lid | 792.00 | 0.50 | 11.88 | – | 792.00 | nr | **803.88** |
| **Plastic litter and grit bins; glassfibre reinforced polyester grit bins; yellow body; hinged lids** | | | | | | | |
| Bins; Wybone Ltd; Victoriana glass fibre; cast iron effect litter bins; including lockable liner | | | | | | | |
| LBV/2; 521 × 521 × 673 mm high; open top; square shape; 0.078 m³ capacity | 156.99 | 1.00 | 23.75 | – | 161.76 | nr | **185.51** |
| LVC/3; 457 mm diameter × 648 mm high; open top; drum shape; 0.084 m³ capacity; with lockable liner | 360.99 | 1.00 | 23.75 | – | 365.76 | nr | **389.51** |
| **Grit bins; Furnitubes International Ltd** | | | | | | | |
| Grit and salt bin; yellow glass fibre; hinged lid; 170 l | 95.00 | – | – | – | – | nr | – |
| **Cigarette bins** | | | | | | | |
| Furnitubes International Ltd | | | | | | | |
| ZEN 275 Zenith cigarette bin; stainless steel; wall or post mounted; 1.7 l | 40.00 | 0.75 | 17.81 | – | 47.42 | nr | **65.23** |
| LVR250 PC Liverpool cigarette bin; steel; wall mounted; 1.9 l | 33.00 | 0.50 | 11.88 | – | 33.00 | nr | **44.88** |
| SMK500F Smoke King cigarette bin; cast aluminium; bolt down; 3.5 l | 220.00 | 0.75 | 17.81 | – | 227.42 | nr | **245.23** |

## 35 SITE WORKS – STREET FURNITURE

| Item<br>Excluding site overheads and profit | PC<br>£ | Labour<br>hours | Labour<br>£ | Plant<br>£ | Material<br>£ | Unit | Total<br>rate £ |
|---|---|---|---|---|---|---|---|
| **BOLLARDS** | | | | | | | |
| **Excavating; for bollards and barriers; by hand** | | | | | | | |
| Holes for bollards | | | | | | | |
| 400 × 400 × 400 mm; disposing off site | – | 0.42 | 9.90 | – | 2.37 | nr | **12.27** |
| 600 × 600 × 600 mm; disposing off site | – | 0.97 | 23.04 | – | 6.66 | nr | **29.70** |
| **Concrete bollards – General** | | | | | | | |
| Preamble: Precast concrete bollards are available in a very wide range of shapes and sizes. The bollards listed here are the most commonly used sizes and shapes; manufacturer's catalogues should be consulted for the full range. Most manufacturers produce bollards to match their suites of street furniture, which may include planters, benches, litter bins and cycle stands. Most parallel sided bollards can be supplied in removable form, with a reduced shank, precast concrete socket and lifting hole to permit removal with a bar. | | | | | | | |
| **Concrete bollards** | | | | | | | |
| Marshalls Plc; cylinder; straight or tapered; 200–400 mm diameter; plain grey concrete; setting into firm ground (excavating and backfilling not included) | | | | | | | |
| Bridgford; 915 mm high above ground | 100.48 | 2.00 | 47.50 | – | 113.00 | nr | **160.50** |
| Marshalls Plc; cylinder; straight or tapered; 200–400 mm diameter; Beadalite reflective finish; setting into firm ground (excavating and backfilling not included) | | | | | | | |
| Wexham concrete bollard; exposed silver grey | 203.23 | 2.00 | 47.50 | – | 209.96 | nr | **257.46** |
| Wexham Major concrete bollard; exposed silver grey | 316.13 | 2.00 | 47.50 | – | 328.65 | nr | **376.15** |
| Townscape Products Ltd; Precast concrete verge markers; various shapes; 400 mm high | | | | | | | |
| plain grey concrete | 53.55 | 1.00 | 23.75 | – | 58.36 | nr | **82.11** |
| white concrete | 56.15 | 1.00 | 23.75 | – | 60.96 | nr | **84.71** |
| exposed aggregate | 59.44 | 1.00 | 23.75 | – | 64.25 | nr | **88.00** |

## 35 SITE WORKS – STREET FURNITURE

| Item<br>Excluding site overheads and profit | PC<br>£ | Labour<br>hours | Labour<br>£ | Plant<br>£ | Material<br>£ | Unit | Total<br>rate £ |
|---|---|---|---|---|---|---|---|
| Townscape Products Ltd; Linby; natural grey concrete | | | | | | | |
| bollards; 220 mm diameter × 450 mm high | 70.90 | 1.00 | 23.75 | – | 74.79 | nr | **98.54** |
| seats; 1800 × 600 × 736 mm high | 472.16 | 2.00 | 47.50 | – | 472.16 | nr | **519.66** |
| **Bollards** | | | | | | | |
| Recycled plastic bollards; Furnitubes International Ltd | | | | | | | |
| Aberdeen Circular ABR150; 1000 × 150 mm diameter | 48.00 | 2.00 | 47.50 | – | 54.73 | nr | **102.23** |
| Aberdeen Square ABR140; 1000 × 140 × 140 mm | 30.00 | 2.00 | 47.50 | – | 36.73 | nr | **84.23** |
| Service bollards; Furnitubes International Ltd; stainless steel; for housing services connection points to electricity, water etc. (services connections not included) | | | | | | | |
| Kenton KEN717; 900 × 250 mm diameter | 513.00 | 2.00 | 47.50 | – | 519.73 | nr | **567.23** |
| Zenith ZEN707; 900 × 250 mm diameter | 632.00 | 2.00 | 47.50 | – | 638.73 | nr | **686.23** |
| **Other bollards** | | | | | | | |
| Removable parking posts; Marshalls Plc | | | | | | | |
| RT\RD4; domestic telescopic bollard | 186.29 | 2.00 | 47.50 | – | 193.02 | nr | **240.52** |
| RT\R8; heavy duty telescopic bollard | 265.32 | 2.00 | 47.50 | – | 272.05 | nr | **319.55** |
| Plastic bollards; Marshalls Plc | | | | | | | |
| Lismore; three ring recycled plastic bollard | 88.06 | 2.00 | 47.50 | – | 94.79 | nr | **142.29** |
| **Cast iron bollards – General** | | | | | | | |
| Preamble: The following bollards are particularly suitable for conservation areas. Logos for civic crests can be incorporated to order. | | | | | | | |
| **Cast iron bollards** | | | | | | | |
| Bollards; Furnitubes International Ltd (excavating and backfilling not included) | | | | | | | |
| Doric Round; 920 mm high × 170 mm diameter | 140.00 | 2.00 | 47.50 | – | 146.41 | nr | **193.91** |
| Manchester Round; 975 mm high; 225 mm square base | 160.00 | 2.00 | 47.50 | – | 166.41 | nr | **213.91** |
| Kenton; heavy duty galvanized steel; 900 mm high; 350 mm diameter | 240.00 | 2.00 | 47.50 | – | 246.41 | nr | **293.91** |

## 35 SITE WORKS – STREET FURNITURE

| Item<br>Excluding site overheads and profit | PC<br>£ | Labour<br>hours | Labour<br>£ | Plant<br>£ | Material<br>£ | Unit | Total<br>rate £ |
|---|---|---|---|---|---|---|---|
| **BOLLARDS – cont** | | | | | | | |
| **Cast iron bollards – cont** | | | | | | | |
| Bollards; Marshalls Plc (excavating and backfilling not included) | | | | | | | |
|   MSF103; cast iron bollard; Small Manchester | 163.71 | 2.00 | 47.50 | – | 170.44 | nr | **217.94** |
|   MSF102; cast iron bollard; Manchester | 163.71 | 2.00 | 47.50 | – | 170.44 | nr | **217.94** |
| **Cast iron bollards with rails – General** | | | | | | | |
| Preamble: The following cast iron bollards are suitable for conservation areas. | | | | | | | |
| **Cast iron bollards with rails** | | | | | | | |
| Cast iron posts with steel tubular rails; Broxap Street Furniture; setting into firm ground (excavating not included) | | | | | | | |
|   Weaver Polyurethane single rail post | 52.00 | 1.50 | 35.63 | – | 55.13 | nr | **90.76** |
|   Type C mild steel tubular rail; including connector | 8.33 | 0.03 | 0.79 | – | 9.72 | m | **10.51** |
| **Steel bollards** | | | | | | | |
| Steel bollards; Marshalls Plc (excavating and backfilling not included) | | | | | | | |
|   SSB01; stainless steel bollard; 101 × 1250 mm | 129.84 | 2.00 | 47.50 | – | 136.57 | nr | **184.07** |
|   RS001; stainless steel bollard; 114 × 1500 mm | 135.48 | 2.00 | 47.50 | – | 142.21 | nr | **189.71** |
|   RB119 Brunel; steel bollard; 168 × 1500 mm | 134.35 | 2.00 | 47.50 | – | 146.87 | nr | **194.37** |
| **Timber bollards** | | | | | | | |
| Woodscape Ltd; durable hardwood | | | | | | | |
|   RP 250/1500; 250 mm diameter × 1500 mm long | 319.22 | 1.00 | 23.75 | – | 319.50 | nr | **343.25** |
|   SP 250/1500; 250 mm square × 1500 mm long | 319.22 | 1.00 | 23.75 | – | 319.50 | nr | **343.25** |
|   SP 150/1200; 150 mm square × 1200 mm long | 107.42 | 1.00 | 23.75 | – | 107.70 | nr | **131.45** |
|   SP 125/750; 125 mm square × 750 mm long | 67.32 | 1.00 | 23.75 | – | 67.60 | nr | **91.35** |
|   RP 125/750; 125 mm diameter × 750 mm long | 67.32 | 1.00 | 23.75 | – | 67.60 | nr | **91.35** |

## 35 SITE WORKS – STREET FURNITURE

| Item<br>Excluding site overheads and profit | PC<br>£ | Labour<br>hours | Labour<br>£ | Plant<br>£ | Material<br>£ | Unit | Total<br>rate £ |
|---|---|---|---|---|---|---|---|
| **Deterrent bollards** | | | | | | | |
| Semi-mountable vehicle deterrent and kerb protection bollards; Furnitubes International Ltd (excavating and backfilling not included) | | | | | | | |
|    bell decorative | 575.00 | 2.00 | 47.50 | – | 584.13 | nr | **631.63** |
|    half bell | 630.00 | 2.00 | 47.50 | – | 639.13 | nr | **686.63** |
|    full bell ® | 750.00 | 2.00 | 47.50 | – | 759.13 | nr | **806.63** |
|    three quarter bell | 730.00 | 2.00 | 47.50 | – | 739.13 | nr | **786.63** |
| **Security bollards** | | | | | | | |
| Security bollards; Furnitubes International Ltd (excavating and backfilling not included) | | | | | | | |
|    Burr Bloc Type 6; removable steel security bollard; 750 mm high above ground; 410 × 285 mm | 538.00 | 2.00 | 47.50 | – | 559.64 | nr | **607.14** |
| **CYLCE HOLDERS/CYLCE SHELTERS** | | | | | | | |
| **Cycle holders** | | | | | | | |
| Cycle stands; Marshalls Plc | | | | | | | |
|    Sheffield; steel cycle stand; RCS1 | 50.81 | 0.50 | 11.88 | – | 50.81 | nr | **62.69** |
|    Sheffield; stainless steel cycle stand; RSCS1 | 136.61 | 0.50 | 11.88 | – | 136.61 | nr | **148.49** |
| Cycle holders; Autopa; VELOPA; galvanized steel | | | | | | | |
|    R; fixing to wall or post; making good | 28.00 | 1.00 | 23.75 | – | 37.54 | nr | **61.29** |
|    SR(V); fixing in ground; making good | 43.00 | 1.00 | 23.75 | – | 52.54 | nr | **76.29** |
|    Sheffield cycle stands; ragged steel | 55.00 | 1.00 | 23.75 | – | 64.54 | nr | **88.29** |
| Cycle holders; Townscape Products Ltd | | | | | | | |
|    Guardian cycle holders; tubular steel frame; setting in concrete; 1250 × 550 × 775 mm high; making good | 374.40 | 1.00 | 23.75 | – | 383.94 | nr | **407.69** |
|    Penny cycle stands; 600 mm diameter; exposed aggregate bollards with 8 nr cycle holders; in galvanized steel; setting in concrete; making good | 841.92 | 1.00 | 23.75 | – | 851.46 | nr | **875.21** |

## 35 SITE WORKS – STREET FURNITURE

| Item<br>Excluding site overheads and profit | PC<br>£ | Labour<br>hours | Labour<br>£ | Plant<br>£ | Material<br>£ | Unit | Total<br>rate £ |
|---|---|---|---|---|---|---|---|
| **CYLCE HOLDERS/CYLCE SHELTERS – cont** | | | | | | | |
| **Cycle holders – cont** | | | | | | | |
| Cycle holders; Broxap Street Furniture | | | | | | | |
| Neath cycle rack; BX/MW/AG; for 6 nr cycles; semi-vertical; galvanized and polyester powder coated; 1.32 m wide × 2.542 m long × 1.80 m high | 370.00 | 10.00 | 237.50 | – | 389.07 | nr | **626.57** |
| Wardale cycle shelter for 10 nr cycles; galvanized; 2.13 m wide × 3.05 m long × 2.15 m high | 1060.00 | 10.00 | 237.50 | – | 1079.07 | nr | **1316.57** |
| Toast Rack double sided free standing cycle rack; BX/MW/GH; for 10 nr cycles; galvanized and polyester coated; 3.25 m long | 235.00 | 4.00 | 95.00 | – | 247.51 | nr | **342.51** |
| **Cycle shelters** | | | | | | | |
| Furnitubes International Ltd; Academy freestanding bolt down shelter, galvanized steel frame and roof; 3450 mm overall width; 2150 mm maximum height; 2670 mm depth | | | | | | | |
| corrugated roof | 1880.00 | 6.00 | 142.50 | – | 1894.83 | nr | **2037.33** |
| clear polycarbonate roof | 1960.00 | 6.00 | 142.50 | – | 1974.83 | nr | **2117.33** |
| **Concrete cycle blocks; Townscape Products Ltd; on 100 mm concrete (1:2:4); on 150 mm hardcore; bedding in cement:sand (1:4)** | | | | | | | |
| Cycle blocks | | | | | | | |
| Cycle Bloc left hand; in natural round exposed aggregate | 55.79 | 0.25 | 5.94 | – | 56.74 | nr | **62.68** |
| Cycle Bloc right hand; in exposed aggregate concrete | 55.79 | 0.25 | 5.94 | – | 56.74 | nr | **62.68** |
| **FLAGPOLES** | | | | | | | |
| Flagpoles; ground mounted; in glass fibre; smooth white finish; including hinged baseplate, nylon halyard system and revolving gold onion finial; setting in concrete; to manufacturer's recommendations (excavating not included) | | | | | | | |
| 6 m high; external halyard system | 200.07 | 4.00 | 95.00 | – | 225.10 | nr | **320.10** |
| 6 m high; internal halyard system | 268.83 | 4.00 | 95.00 | – | 293.86 | nr | **388.86** |
| 8 m high; external halyard system | 228.38 | 6.00 | 142.50 | – | 253.41 | nr | **395.91** |
| 8 m high; internal halyard system | 296.83 | 6.00 | 142.50 | – | 321.86 | nr | **464.36** |
| 10 m high; external halyard system | 349.65 | 5.00 | 118.75 | – | 374.68 | nr | **493.43** |
| 10 m high; internal halyard system | 418.22 | 5.00 | 118.75 | – | 443.25 | nr | **562.00** |

## 35 SITE WORKS – STREET FURNITURE

| Item<br>Excluding site overheads and profit | PC<br>£ | Labour<br>hours | Labour<br>£ | Plant<br>£ | Material<br>£ | Unit | Total<br>rate £ |
|---|---|---|---|---|---|---|---|
| Flagpoles; wall mounted; vertical or angled poles; in glass fibre; smooth white finish; including base, top bracket, external nylon halyard rope, cleat and revolving gold onion finial | | | | | | | |
|    3 m pole | 138.99 | 2.00 | 47.50 | – | 138.99 | nr | **186.49** |
| Banner flagpoles; 90 mm diameter aluminium pole; suitable for 1 × 2 m banner | | | | | | | |
|    6 m high | 332.38 | 4.00 | 95.00 | – | 357.41 | nr | **452.41** |

**GENERALLY**

**Note:** The costs of delivery from the supplier are generally not included in these items. Readers should check with the supplier to verify the delivery costs.

**Furniture/equipment – General Preamble:** The following items include fixing to manufacturer's instructions; holding down bolts or other fittings and making good (excavating, backfilling and tarmac, concrete or paving bases not included).

**PERGOLAS**

| Pergolas; AVS Fencing Supplies Ltd; construct timber pergola; posts 150 × 150 mm × 2.40 m finished height in 600 mm deep minimum concrete pits; beams of 200 × 50 mm × 2.00 m wide; fixed to posts with dowels; rafters 200 × 38 mm notched to beams; inclusive of all mechanical excavation and disposal off site | | | | | | | |
|---|---|---|---|---|---|---|---|
| Pergola 2.00 m wide in green oak; posts at 2.00 m centres; dowel fixed | | | | | | | |
|    rafters at 600 mm centres | 125.31 | 8.40 | 217.00 | 3.91 | 201.97 | m | **422.88** |
|    rafters at 400 mm centres | 136.17 | 10.80 | 279.00 | 3.91 | 220.17 | m | **503.08** |
| Pergola 3.00 m wide in green oak; posts at 1.50 m centres | | | | | | | |
|    rafters at 600 mm centres | 165.71 | 7.20 | 193.50 | 4.69 | 255.48 | m | **453.67** |
|    rafters at 400 mm centres | 181.96 | 8.00 | 215.00 | 4.69 | 282.73 | m | **502.42** |
| Pergola 2.00 m wide in prepared softwood; beams fixed with bolts; posts at 2.00 m centres | | | | | | | |
|    rafters at 600 mm centres | 99.79 | 5.00 | 134.38 | 3.91 | 177.06 | m | **315.35** |
|    rafters at 400 mm centres | 118.93 | 6.00 | 161.25 | 3.91 | 203.54 | m | **368.70** |

## 35 SITE WORKS – STREET FURNITURE

| Item<br>Excluding site overheads and profit | PC<br>£ | Labour<br>hours | Labour<br>£ | Plant<br>£ | Material<br>£ | Unit | Total<br>rate £ |
|---|---|---|---|---|---|---|---|
| **PERGOLAS – cont** | | | | | | | |
| **Pergolas – cont** | | | | | | | |
| Pergola 3.00 m wide in prepared softwood; posts at 1.50 m centres | | | | | | | |
|    rafters at 600 mm centres | 87.78 | 7.20 | 193.50 | 4.69 | 177.55 | m | 375.74 |
|    rafters at 400 mm centres | 101.30 | 8.00 | 215.00 | 4.69 | 202.07 | m | 421.76 |
| **PICNIC TABLES/SUNDRY FURNITURE** | | | | | | | |
| **Picnic benches – General** | | | | | | | |
| Preamble: The following items include for fixing to ground to manufacturer's instructions or concreting in. | | | | | | | |
| **Picnic tables and benches** | | | | | | | |
| Picnic tables and benches; Broxap Street Furniture | | | | | | | |
|    Springfield timber picnic unit 1800 mm long softwood | 215.00 | 2.00 | 47.50 | – | 224.54 | nr | 272.04 |
| Picnic tables and benches; Woodscape Ltd | | | | | | | |
|    table and benches built in; 2 m long | 2431.00 | 2.00 | 47.50 | – | 2440.54 | nr | 2488.04 |
| **Picnic table; recycled plastic** | | | | | | | |
| Furnitubes International Ltd | | | | | | | |
|    Dundee picnic table; brown recycled plastic; 1.8 m long | 687.00 | 2.00 | 47.50 | – | 705.11 | nr | 752.61 |
| **Street furniture ranges** | | | | | | | |
| **Lifebuoy stations** | | | | | | | |
| Lifebuoy stations; Earth Anchors Ltd | | | | | | | |
|    Rootfast lifebuoy station complete with post AP44; SOLAS approved lifebuoys 590 mm and lifeline | 165.00 | 2.00 | 47.50 | – | 176.59 | nr | 224.09 |
|    installation tool for above | – | – | – | – | 82.00 | nr | 82.00 |
| **PLANT CONTAINERS** | | | | | | | |
| **Market prices of containers** | | | | | | | |
| **Plant containers; terracotta** | | | | | | | |
| Capital Garden Products Ltd | | | | | | | |
|    Large Pot LP63; weathered terracotta; 1170 × 1600 mm diameter | – | – | – | – | 1265.00 | nr | 1265.00 |
|    Large Pot LP38; weathered terracotta; 610 × 970 mm diameter | – | – | – | – | 495.00 | nr | 495.00 |
|    Large Pot LP23; weathered terracotta; 480 × 580 mm diameter | – | – | – | – | 265.00 | nr | 265.00 |

## 35 SITE WORKS – STREET FURNITURE

| Item<br>Excluding site overheads and profit | PC<br>£ | Labour<br>hours | Labour<br>£ | Plant<br>£ | Material<br>£ | Unit | Total<br>rate £ |
|---|---|---|---|---|---|---|---|
| Indian style Shimmer Pot 2322;<br>585 × 560 mm diameter | – | – | – | – | 215.00 | nr | **215.00** |
| Indian style Shimmer Pot 1717;<br>430 × 430 mm diameter | – | – | – | – | 132.00 | nr | **132.00** |
| Indian style Shimmer Pot 1314;<br>330 × 355 mm diameter | – | – | – | – | 115.00 | nr | **115.00** |
| **Plant containers; faux lead** | | | | | | | |
| Capital Garden Products Ltd | | | | | | | |
| Trough 2508 Tudor Rose;<br>620 × 220 × 230 mm high | – | – | – | – | 64.95 | nr | **64.95** |
| Tub 2004 Elizabethan; 510 mm square | – | – | – | – | 102.00 | nr | **102.00** |
| Tub 1513 Elizabethan; 380 mm square | – | – | – | – | 69.95 | nr | **69.95** |
| Tub 1601 Tudor Rose; 420 × 400 mm diameter | – | – | – | – | 63.00 | nr | **63.00** |
| **Plant containers; window boxes** | | | | | | | |
| Capital Garden Products Ltd | | | | | | | |
| Adam 5401; faux lead;<br>1370 × 270 × 210 mm high | – | – | – | – | 129.95 | nr | **129.95** |
| Oakleaf OAK24; terracotta;<br>610 × 230 × 240 mm high | – | – | – | – | 104.95 | nr | **104.95** |
| Swag 2402; faux lead;<br>610 × 200 × 210 mm high | – | – | – | – | 54.95 | nr | **54.95** |
| **Plant containers; timber** | | | | | | | |
| Plant containers; hardwood; Neptune Outdoor Furniture Ltd | | | | | | | |
| Woodland T38-4D;<br>1500 × 1500 × 900 mm high | – | – | – | – | 1270.00 | nr | **1270.00** |
| Woodland T38-3C;<br>1000 × 1500 × 700 mm high | – | – | – | – | 930.00 | nr | **930.00** |
| Woodland T38-2A;<br>1000 × 500 × 500 mm high | – | – | – | – | 490.00 | nr | **490.00** |
| Hudson T42-4D;<br>1500 × 1500 × 900 mm high | – | – | – | – | 1270.00 | nr | **1270.00** |
| Hudson T42-3C;<br>1000 × 1500 × 700 mm high | – | – | – | – | 930.00 | nr | **930.00** |
| Hudson T42-2A;<br>1000 × 500 × 500 mm high | – | – | – | – | 490.00 | nr | **490.00** |
| Plant containers; hardwood; Woodscape Ltd | | | | | | | |
| square; 900 × 900 × 420 mm high | – | – | – | – | 448.00 | nr | **448.00** |
| **Plant containers; precast concrete** | | | | | | | |
| Plant containers; Marshalls Plc | | | | | | | |
| Boulevard 700; circular base and ring | 704.52 | 2.00 | 47.50 | 25.22 | 704.52 | nr | **777.24** |
| Boulevard 1200; circular base and ring | 889.68 | 1.00 | 23.75 | 25.22 | 889.68 | nr | **938.65** |

## 35 SITE WORKS – STREET FURNITURE

| Item<br>Excluding site overheads and profit | PC<br>£ | Labour<br>hours | Labour<br>£ | Plant<br>£ | Material<br>£ | Unit | Total<br>rate £ |
|---|---|---|---|---|---|---|---|
| **PLANT CONTAINERS – cont** | | | | | | | |
| **Bespoke Planters; Outdoor Design; planters to required size and shape; with seamless polished corners; preformed drainage holes and lip detail; inclusive of placement; drainage layer 50 mm thick and potting soil** | | | | | | | |
| Galvanized 1.5 mm steel; powder coated | | | | | | | |
| 500 mm × 500 mm × 500 mm | 325.00 | 0.50 | 23.75 | – | 331.31 | nr | **355.06** |
| 1500 mm × 500 mm × 500 mm | 403.00 | 0.80 | 38.00 | – | 420.60 | nr | **458.60** |
| 1800 mm × 750 mm × 600 mm | 513.00 | 1.60 | 76.00 | 16.46 | 551.71 | nr | **644.17** |
| 2500 mm × 1000 mm × 600 mm | 629.00 | 1.50 | 71.25 | 16.46 | 706.74 | nr | **794.45** |
| Corten 2.0 mm | | | | | | | |
| 500 mm × 500 mm × 500 mm | 336.00 | 0.50 | 23.75 | – | 342.31 | nr | **366.06** |
| 1500 mm × 500 mm × 500 mm | 427.00 | 0.80 | 38.00 | – | 444.60 | nr | **482.60** |
| 1800 mm × 750 mm × 600 mm | 551.00 | 1.60 | 76.00 | 16.46 | 589.71 | nr | **682.17** |
| 2500 mm × 1000 mm × 600 mm | 687.00 | 1.50 | 71.25 | 16.46 | 764.74 | nr | **852.45** |
| Stainless steel 1.5 mm; Grained finish | | | | | | | |
| 500 mm × 500 mm × 500 mm | 416.00 | 0.50 | 23.75 | – | 422.31 | nr | **446.06** |
| 1500 mm × 500 mm × 500 mm | 602.00 | 0.80 | 38.00 | – | 619.60 | nr | **657.60** |
| 1800 mm × 750 mm × 600 mm | 828.00 | 1.60 | 76.00 | 16.46 | 866.71 | nr | **959.17** |
| 2500 mm × 1000 mm × 600 mm | 1106.00 | 1.50 | 71.25 | 16.46 | 1183.74 | nr | **1271.45** |
| **SEATING** | | | | | | | |
| **Outdoor seats** | | | | | | | |
| **CED Ltd; stone bench; Sinuous bench; stone type bench to organic S pattern; laid to concrete base; 500 mm high × 400 mm wide (not included)** | | | | | | | |
| 2.00 m long | 1585.98 | 4.00 | 95.00 | – | 1585.98 | nr | **1680.98** |
| 5.00 m long | 3893.08 | 8.00 | 190.00 | – | 3893.08 | nr | **4083.08** |
| 10.00 m long | 7783.79 | 11.00 | 261.25 | – | 7783.79 | nr | **8045.04** |
| **Outdoor seats; concrete framed – General**<br>Preamble: Prices for the following concrete framed seats and benches with hardwood slats include for fixing (where necessary) by bolting into existing paving or concrete bases (bases not included) or building into walls or concrete foundations (walls and foundations not included). Delivery generally not included. | | | | | | | |

## 35 SITE WORKS – STREET FURNITURE

| Item<br>Excluding site overheads and profit | PC<br>£ | Labour<br>hours | Labour<br>£ | Plant<br>£ | Material<br>£ | Unit | Total<br>rate £ |
|---|---|---|---|---|---|---|---|
| **Outdoor seats; concrete framed** | | | | | | | |
| Outdoor seats; Townscape Products Ltd | | | | | | | |
| Belgrave seat; exposed aggregate | | | | | | | |
| 1800 × 645 × 730 mm | 119.00 | 2.00 | 47.50 | – | 133.83 | nr | **181.33** |
| Maidstone seats; | | | | | | | |
| 1800 × 610 × 785 mm | 572.16 | 2.00 | 47.50 | – | 586.99 | nr | **634.49** |
| **Outdoor seats; concrete** | | | | | | | |
| Outdoor seats; Marshalls Ltd | | | | | | | |
| Boulevard 2000 concrete seat | 824.19 | 2.00 | 47.50 | – | 842.30 | nr | **889.80** |
| Outdoor seats; Furnitubes International Ltd | | | | | | | |
| Amesbury concrete seat, 2.2 m long | 890.00 | 2.00 | 47.50 | – | 908.11 | nr | **955.61** |
| Marlborough concrete seat, 1.8 m long | 820.00 | 2.00 | 47.50 | – | 838.11 | nr | **885.61** |
| **Outdoor seats; metal framed – General** | | | | | | | |
| Preamble: Metal framed seats with hardwood backs to various designs can be bolted to ground anchors. | | | | | | | |
| **Outdoor seats; metal framed** | | | | | | | |
| Outdoor seats; Furnitubes International Ltd | | | | | | | |
| NS6 Newstead; steel standards with iroko slats; 1.80 m long | 392.00 | 2.00 | 47.50 | – | 410.11 | nr | **457.61** |
| NEB6 New Forest Single Bench; cast iron standards with iroko slats; 1.83 m long | 345.00 | 2.00 | 47.50 | – | 363.11 | nr | **410.61** |
| EA6 Eastgate; cast iron standards with iroko slats; 1.86 m long | 450.00 | 2.00 | 47.50 | – | 468.11 | nr | **515.61** |
| NE6 New Forest Seat; cast iron standards with iroko slats; 1.83 m long | 430.00 | 2.00 | 47.50 | – | 448.11 | nr | **495.61** |
| Zenith long bench, satin stainless steel, iroko slats, 1.8 m long | 1100.00 | 2.00 | 47.50 | – | 1118.11 | nr | **1165.61** |
| Ashburton; steel with recycled plastic slats; 1.8 m long | 400.00 | – | – | – | – | nr | **–** |
| Outdoor seats; Neptune Outdoor Furniture Ltd; Kara; treated hardwood slats to heavy duty steel frame | | | | | | | |
| 1.50 m long | 570.00 | 2.00 | 47.50 | – | 579.54 | nr | **627.04** |
| 1.80 m long | 595.00 | 2.00 | 47.50 | – | 604.54 | nr | **652.04** |
| 2.00 m long | 620.00 | 2.00 | 47.50 | – | 629.54 | nr | **677.04** |
| Outdoor seats; Neptune Outdoor Furniture Ltd; Beaufort, powder coated steel frame with treated hardwood slats | | | | | | | |
| 1.50 m long | 535.00 | 2.00 | 47.50 | – | 544.54 | nr | **592.04** |
| 1.80 m long | 575.00 | 2.00 | 47.50 | – | 584.54 | nr | **632.04** |
| 2.00 m long | 595.00 | 2.00 | 47.50 | – | 604.54 | nr | **652.04** |

## 35 SITE WORKS – STREET FURNITURE

| Item<br>Excluding site overheads and profit | PC<br>£ | Labour<br>hours | Labour<br>£ | Plant<br>£ | Material<br>£ | Unit | Total<br>rate £ |
|---|---|---|---|---|---|---|---|
| **SEATING – cont** | | | | | | | |
| **Outdoor seats – cont** | | | | | | | |
| Outdoor seats; Neptune Outdoor Furniture Ltd; Avenue; treated hardwood slats to cast iron seat and bench ends | | | | | | | |
| 1.80 m long | 665.00 | 2.00 | 47.50 | – | 674.54 | nr | **722.04** |
| 2.10 m long | 720.00 | 2.00 | 47.50 | – | 729.54 | nr | **777.04** |
| Outdoor seats; Broxap Street Furniture; metal and timber | | | | | | | |
| Eastgate | 365.00 | 2.00 | 47.50 | – | 374.54 | nr | **422.04** |
| Outdoor seats; Earth Anchors Ltd; Forest-Saver; steel frame and recycled slats | | | | | | | |
| bench; 1.8 m | 205.00 | 1.00 | 23.75 | – | 205.00 | nr | **228.75** |
| seat; 1.8 m | 329.00 | 1.00 | 23.75 | – | 329.00 | nr | **352.75** |
| Outdoor seats; Earth Anchors Ltd; Evergreen; cast iron frame and recycled slats | | | | | | | |
| bench; 1.8 m | 448.00 | 1.00 | 23.75 | – | 448.00 | nr | **471.75** |
| seat; 1.8 m | 575.00 | 1.00 | 23.75 | – | 575.00 | nr | **598.75** |
| **Outdoor seats; all steel** | | | | | | | |
| Outdoor seats; Marshalls Plc | | | | | | | |
| MSF Central; stainless steel seat | 1371.77 | 0.50 | 11.88 | – | 1371.77 | nr | **1383.65** |
| MSF Central; steel seat | 700.00 | 2.00 | 47.50 | – | 700.00 | nr | **747.50** |
| MSF 502; cast iron Heritage seat | 604.03 | 2.00 | 47.50 | – | 604.03 | nr | **651.53** |
| Outdoor seats; Earth Anchors Ltd; Ranger | | | | | | | |
| bench; 1.8 m | 285.00 | 1.00 | 23.75 | – | 285.00 | nr | **308.75** |
| seat; 1.8 m | 431.00 | 1.00 | 23.75 | – | 431.00 | nr | **454.75** |
| **Outdoor seats; all timber** | | | | | | | |
| Outdoor seats; Woodcraft UK; Mendip; teak | | | | | | | |
| 1.524 m long | 604.00 | 1.00 | 23.75 | – | 626.75 | nr | **650.50** |
| 1.829 m long | 646.00 | 1.00 | 23.75 | – | 668.75 | nr | **692.50** |
| 2.438 m long (including centre leg) | 854.00 | 1.00 | 23.75 | – | 876.75 | nr | **900.50** |
| Outdoor seats; Woodcraft UK; Wykeham; hardwood | | | | | | | |
| 1.5 m | 408.00 | 2.00 | 47.50 | – | 430.75 | nr | **478.25** |
| Outdoor seats; Woodscape Ltd; solid hardwood | | | | | | | |
| seat type 3 with back; 2.00 m long; freestanding | 929.50 | 2.00 | 47.50 | – | 952.25 | nr | **999.75** |
| seat type 3 with back; 2.00 m long; building in | 1006.51 | 4.00 | 95.00 | – | 1029.26 | nr | **1124.26** |
| seat type 4; 2.00 m long; fixing to wall | 445.50 | 2.00 | 47.50 | – | 468.25 | nr | **515.75** |

## 35 SITE WORKS – STREET FURNITURE

| Item<br>Excluding site overheads and profit | PC<br>£ | Labour<br>hours | Labour<br>£ | Plant<br>£ | Material<br>£ | Unit | Total<br>rate £ |
|---|---|---|---|---|---|---|---|
| seat type 4 with back; 2.00 m long; fixing to wall | 544.51 | 3.00 | 71.25 | – | 549.43 | nr | 620.68 |
| seat type 4 with back; 2.50 m long; fixing to wall | 605.00 | 3.00 | 71.25 | – | 609.92 | nr | 681.17 |
| seat type 5 with back; 2.00 m long; freestanding | 957.00 | 2.00 | 47.50 | – | 979.75 | nr | 1027.25 |
| seat type 5 with back; 2.50 m long; freestanding | 1023.00 | 2.00 | 47.50 | – | 1045.75 | nr | 1093.25 |
| seat type 5 with back; 2.00 m long; building in | 1034.00 | 2.00 | 47.50 | – | 1056.75 | nr | 1104.25 |
| seat type 5 with back; 2.50 m long; building in | 1100.00 | 3.00 | 71.25 | – | 1122.75 | nr | 1194.00 |
| bench type 1; 2.00 m long; freestanding | 797.51 | 2.00 | 47.50 | – | 820.26 | nr | 867.76 |
| bench type 1; 2.00 m long; building in | 874.51 | 4.00 | 95.00 | – | 897.26 | nr | 992.26 |
| bench type 2; 2.00 m long; freestanding | 753.50 | 2.00 | 47.50 | – | 776.25 | nr | 823.75 |
| bench type 2; 2.50 m long; freestanding | 797.51 | 2.00 | 47.50 | – | 820.26 | nr | 867.76 |
| bench type 2; 2.00 m long; building in | 830.50 | 4.00 | 95.00 | – | 853.25 | nr | 948.25 |
| bench type 2; 2.50 m long; building in | 874.51 | 4.00 | 95.00 | – | 897.26 | nr | 992.26 |
| bench type 2; 2.00 m long overall; curved to 5 m radius; building in | 990.00 | 4.00 | 95.00 | – | 1012.75 | nr | 1107.75 |
| Outdoor seats; Neptune Outdoor Furniture Ltd; Southampton – hardwood slats to exposed aggregate concrete base | | | | | | | |
| 1.50 m long | 455.00 | 2.00 | 47.50 | – | 477.75 | nr | 525.25 |
| 1.80 m long | 470.00 | 2.00 | 47.50 | – | 492.75 | nr | 540.25 |
| 2.410 m long | 500.00 | 2.00 | 47.50 | – | 522.75 | nr | 570.25 |
| **Outdoor seats; tree benches/seats** | | | | | | | |
| Tree bench; Neptune Outdoor Furniture Ltd; Beaufort; hexagonal; timber | | | | | | | |
| SF34-15A; 1500 mm diameter | 1105.00 | 0.50 | 11.88 | – | 1105.00 | nr | 1116.88 |
| Tree seat; Neptune Outdoor Furniture Ltd; Beaufort; hexagonal; timber; with back | | | | | | | |
| SF32-10A; 720 mm diameter | 1270.00 | 0.50 | 11.88 | – | 1270.00 | nr | 1281.88 |
| SF32-20A; 1720 mm diameter | 1580.00 | 0.50 | 11.88 | – | 1580.00 | nr | 1591.88 |
| **Outdoor seats; recycled plastic** | | | | | | | |
| Furnitubes International Ltd | | | | | | | |
| Aberdeen seat, brown recycled plastic; 1 slat; 2 m long | 372.00 | 2.00 | 47.50 | – | 390.11 | nr | 437.61 |
| Aberdeen seat, brown recycled plastic; 2 slats; 2 m long | 529.00 | 2.00 | 47.50 | – | 547.11 | nr | 594.61 |

## 35 SITE WORKS – STREET FURNITURE

| Item<br>Excluding site overheads and profit | PC<br>£ | Labour<br>hours | Labour<br>£ | Plant<br>£ | Material<br>£ | Unit | Total<br>rate £ |
|---|---|---|---|---|---|---|---|
| **SEATING – cont** | | | | | | | |
| **Anti-skate board devices** | | | | | | | |
| Stainless steel; Furnitubes International Ltd | | | | | | | |
|   for fitting to seats | 20.00 | 0.20 | 4.75 | – | 20.00 | nr | **24.75** |
|   for fitting to benches | 20.00 | 0.20 | 4.75 | – | 20.00 | nr | **24.75** |
| **SIGNAGE** | | | | | | | |
| **Directional signage; cast aluminium** | | | | | | | |
| Signage; Furnitubes International Ltd | | | | | | | |
|   FFL1 Lancer; cast aluminium finials | 85.00 | 0.07 | 1.58 | – | 85.00 | nr | **86.58** |
|   FAAIS; arrow end type cast aluminium directional arms; single line; 90 mm wide | 134.00 | 2.00 | 47.50 | – | 134.00 | nr | **181.50** |
|   FAAID; arrow end type cast aluminium directional arms; double line; 145 mm wide | 170.00 | 0.13 | 3.17 | – | 170.00 | nr | **173.17** |
|   FAAIS; arrow end type cast aluminium directional arms; treble line; 200 mm wide | 190.00 | 0.20 | 4.75 | – | 190.00 | nr | **194.75** |
|   FCK1211G Kingston; composite standard root columns | 575.00 | 2.00 | 47.50 | – | 588.01 | nr | **635.51** |
| **Park signage** | | | | | | | |
| Entrance signs and map boards; vitreous enamel | | | | | | | |
|   entrance map board; 1250 × 1000 mm high with two support posts; all associated works to post bases | – | – | – | – | – | nr | **2200.00** |
|   information board with two locking cabinets; 1250 × 1000 mm high with two support posts; all associated works to post bases | – | – | – | – | – | nr | **2100.00** |
| **Miscellaneous park signage; vitreous enamel** | | | | | | | |
| 'No dogs' sign; 200 × 150 mm; fixing to fencing or gates | – | – | – | – | – | nr | **80.00** |
| 'Dog exercise area' sign; 300 × 400 mm; fixing to fencing or gates | – | – | – | – | – | nr | **110.00** |
| 'Nature conservation area' sign; 900 × 400 mm high with two support posts; all associated works to post bases | – | – | – | – | – | nr | **1140.00** |

## 35 SITE WORKS – STREET FURNITURE

| Item<br>Excluding site overheads and profit | PC<br>£ | Labour<br>hours | Labour<br>£ | Plant<br>£ | Material<br>£ | Unit | Total<br>rate £ |
|---|---|---|---|---|---|---|---|
| **SMOKING SHELTERS** | | | | | | | |
| **Smoking shelters**<br>Furnitubes International Ltd; Ashby freestanding bolt down; aluminium frame; PET glazing; 2050 mm length × 2330 mm high | | | | | | | |
| 3290 mm depth; maximum 3 people | 2047.32 | 4.00 | 95.00 | – | 2062.15 | nr | **2157.15** |
| 6050 mm depth; maximum 6 people | 2572.02 | 5.00 | 118.75 | – | 2586.85 | nr | **2705.60** |
| **STONES AND BOULDERS** | | | | | | | |
| **Standing stones; CED Ltd;** erect standing stones; vertical height above ground; in concrete base; including excavation setting in concrete to ⅓ depth and crane offload into position | | | | | | | |
| Purple schist | | | | | | | |
| 1.00 m high | 290.00 | 1.50 | 35.63 | 17.38 | 311.61 | nr | **364.62** |
| 1.25 m high | 290.00 | 1.50 | 35.63 | 17.38 | 307.15 | nr | **360.16** |
| 1.50 m high | 290.00 | 2.00 | 47.50 | 20.86 | 308.68 | nr | **377.04** |
| 2.00 m high | 290.00 | 3.00 | 71.25 | 34.76 | 344.24 | nr | **450.25** |
| 2.50 m high | 290.00 | 1.50 | 35.63 | 69.53 | 394.63 | nr | **499.79** |
| **Rockery stone – General**<br>Preamble: Rockery stone prices vary considerably with source, carriage, distance and load. Typical PC prices are in the range of £60–£80 per tonne collected. | | | | | | | |
| **Rockery stone; CED Ltd**<br>Boulders; maximum distance 25 m; by machine | | | | | | | |
| 750 mm diameter | 193.57 | 0.90 | 21.38 | 8.16 | 193.57 | nr | **223.11** |
| 1 m diameter | 193.57 | 2.00 | 47.50 | 16.33 | 193.57 | nr | **257.40** |
| 1.5 m diameter | 193.57 | 2.00 | 47.50 | 55.00 | 193.57 | nr | **296.07** |
| 2 m diameter | 193.57 | 2.00 | 47.50 | 55.00 | 193.57 | nr | **296.07** |
| Boulders; maximum distance 25 m; by hand | | | | | | | |
| 750 mm diameter | 193.57 | 0.75 | 17.81 | – | 193.57 | nr | **211.38** |
| 1 m diameter | 193.57 | 1.89 | 44.78 | – | 193.57 | nr | **238.35** |

## 35 SITE WORKS – STREET FURNITURE

| Item<br>Excluding site overheads and profit | PC<br>£ | Labour<br>hours | Labour<br>£ | Plant<br>£ | Material<br>£ | Unit | Total<br>rate £ |
|---|---|---|---|---|---|---|---|
| **TREE GRILLES/TREE PROTECTION** | | | | | | | |
| **Tree grilles; cast iron** | | | | | | | |
| Cast iron tree grilles; Furnitubes International Ltd | | | | | | | |
| GS 1070 Greenwich; two part; 1000 mm square × 700 mm diameter tree hole | 195.00 | 2.00 | 47.50 | – | 195.00 | nr | **242.50** |
| GC 1270 Greenwich; two part, 1200 mm diameter × 700 mm diameter tree hole | 153.00 | 2.00 | 47.50 | – | 153.00 | nr | **200.50** |
| Cast iron tree grilles; Marshalls Plc | | | | | | | |
| Heritage; cast iron grille plus frame; 1 × 1 m | 446.25 | 3.00 | 71.25 | – | 462.89 | nr | **534.14** |
| Cast iron tree grilles; Townscape Products Ltd | | | | | | | |
| Baltimore; 1200 mm square Ductile (SG) finished in Black RAL9004 | 530.64 | 2.00 | 47.50 | – | 530.64 | nr | **578.14** |
| Sunburst 1200 mm square with 600 mm central aperture; four part | 323.73 | 2.00 | 47.50 | – | 323.73 | nr | **371.23** |
| **Tree grilles; steel** | | | | | | | |
| Steel tree grilles; Furnitubes International Ltd | | | | | | | |
| GSF 102G Greenwich; steel tree grille frame for GS 1045; one part | 130.00 | 2.00 | 47.50 | – | 130.00 | nr | **177.50** |
| GSF 122G Greenwich; steel tree grille frame for GS 1270 and GC 1245; one part | 170.00 | 2.00 | 47.50 | – | 170.00 | nr | **217.50** |
| Note: Care must be taken to ensure that tree grids and guards are removed when trees grow beyond the specified diameter of guard. | | | | | | | |

## 35 SITE WORKS – SPORTS, PLAY EQUIPMENT, SURFACES

| Item<br>Excluding site overheads and profit | PC<br>£ | Labour<br>hours | Labour<br>£ | Plant<br>£ | Material<br>£ | Unit | Total<br>rate £ |
|---|---|---|---|---|---|---|---|
| **MARKET PRICES OF SPORTS AND PLAY SURFACING MATERIALS** | | | | | | | |
| **Market prices of surfacing materials** | | | | | | | |
| Surfacings; Melcourt Industries Ltd (items labelled FSC are Forest Stewardship Council certified) | | | | | | | |
| Playbark® 10/50; per 25 m³ load | – | – | – | – | 64.10 | m³ | 64.10 |
| Playbark® 10/50; per 80 m³ load | – | – | – | – | 51.10 | m³ | 51.10 |
| Playbark® 8/25; per 25 m³ load | – | – | – | – | 62.60 | m³ | 62.60 |
| Playbark® 8/25; per 80 m³ load | – | – | – | – | 49.60 | m³ | 49.60 |
| Playchips®; per 25 m³ load; FSC | – | – | – | – | 42.25 | m³ | 42.25 |
| Playchips®; per 80 m³ load; FSC | – | – | – | – | 29.25 | m³ | 29.25 |
| Kushyfall; per 25 m³ load; FSC | – | – | – | – | 40.20 | m³ | 40.20 |
| Kushyfall; per 80 m³ load; FSC | – | – | – | – | 27.20 | m³ | 27.20 |
| Softfall; per 25 m³ load | – | – | – | – | 31.90 | m³ | 31.90 |
| Softfall; per 80 m³ load | – | – | – | – | 18.90 | m³ | 18.90 |
| Playsand; per 10 t load | – | – | – | – | 102.06 | m³ | 102.06 |
| Playsand; per 20 t load | – | – | – | – | 89.82 | m³ | 89.82 |
| Walk Chips; per 25 m³ load; FSC | – | – | – | – | 39.20 | m³ | 39.20 |
| Walk Chips; per 80 m³ load; FSC | – | – | – | – | 26.20 | m³ | 26.20 |
| Woodfibre; per 25 m³ load; FSC | – | – | – | – | 37.45 | m³ | 37.45 |
| Woodfibre; per 80 m³ load; FSC | – | – | – | – | 24.45 | m³ | 24.45 |
| **SPORTS EQUIPMENT** | | | | | | | |
| **Sports equipment** | | | | | | | |
| Tennis posts; steel; suitable for hard or grass tennis courts; including winder, sockets and dust cap | | | | | | | |
| round | 227.65 | 1.00 | 23.75 | – | 227.65 | set | 251.40 |
| square | 232.15 | 1.00 | 23.75 | – | 232.15 | set | 255.90 |
| Tennis nets; not including posts or fixings | | | | | | | |
| Tournament | 137.95 | 3.00 | 71.25 | – | 146.29 | set | 217.54 |
| Match | 125.95 | 3.00 | 71.25 | – | 134.29 | set | 205.54 |
| Club | 102.95 | 3.00 | 71.25 | – | 111.29 | set | 182.54 |
| Football goals; full size; socketed; including international net supports and nets | | | | | | | |
| aluminium | 1415.00 | 4.00 | 95.00 | – | 1423.34 | set | 1518.34 |
| steel; heavyweight | 866.00 | 5.00 | 118.75 | – | 1097.76 | set | 1216.51 |
| Football goals; full size; freestanding; including nets | | | | | | | |
| aluminium | 2319.00 | 4.00 | 95.00 | – | 2319.00 | set | 2414.00 |
| steel | 1790.00 | 4.00 | 95.00 | – | 1790.00 | set | 1885.00 |

## 35 SITE WORKS – SPORTS, PLAY EQUIPMENT, SURFACES

| Item<br>Excluding site overheads and profit | PC<br>£ | Labour<br>hours | Labour<br>£ | Plant<br>£ | Material<br>£ | Unit | Total<br>rate £ |
|---|---|---|---|---|---|---|---|
| **SPORTS EQUIPMENT – cont** | | | | | | | |
| **Sports equipment – cont** | | | | | | | |
| Mini-soccer goals; freestanding; including nets | | | | | | | |
| aluminium | 864.55 | 4.00 | 95.00 | – | 864.55 | set | **959.55** |
| steel | 509.99 | 4.00 | 95.00 | – | 509.99 | set | **604.99** |
| Rugby posts; socketed | | | | | | | |
| aluminium; 10 m high | 1770.00 | 6.00 | 142.50 | – | 1792.25 | set | **1934.75** |
| aluminium; 12 m high | 1927.00 | 6.00 | 142.50 | – | 1949.25 | set | **2091.75** |
| steel; 12 m high | 1117.00 | 6.00 | 142.50 | – | 1139.25 | set | **1281.75** |
| Hockey goals; steel; including backboards and nets | | | | | | | |
| socketed | 978.30 | 1.00 | 23.75 | – | 978.30 | set | **1002.05** |
| freestanding | 968.40 | 1.00 | 23.75 | – | 968.40 | set | **992.15** |
| Cricket cages; steel; including netting | | | | | | | |
| freestanding | 1199.00 | 12.00 | 285.00 | – | 1199.00 | set | **1484.00** |
| wheelaway | 940.00 | 12.00 | 285.00 | – | 940.00 | set | **1225.00** |

**PLAYGROUND EQUIPMENT**

**Playground equipment – General**
Preamble: The range of equipment manufactured or available in the UK is so great that comprehensive coverage would be impossible, especially as designs, specifications and prices change fairly frequently. The following information should be sufficient to give guidance to anyone designing or equipping a playground. In comparing prices note that only outline specification details are given here and that other refinements which are not mentioned may be the reason for some difference in price between two apparently identical elements. The fact that a particular manufacturer does not appear under one item heading does not necessarily imply that they do not make it. Landscape designers are advised to check that equipment complies with ever more stringent safety standards before specifying.

**Playground equipment – Installation**
The rates below include for installation of the specified equipment by the manufacturers. Most manufacturers will offer an option to install the equipment they have supplied.

## 35 SITE WORKS – SPORTS, PLAY EQUIPMENT, SURFACES

| Item Excluding site overheads and profit | PC £ | Labour hours | Labour £ | Plant £ | Material £ | Unit | Total rate £ |
|---|---|---|---|---|---|---|---|
| **Play systems; Kompan Ltd; Galaxy; multiple play activity systems for non-prescribed play; for children 6-14 years; galvanized steel and high density polyethylene** | | | | | | | |
| GXY906 Adara; 14 different play activities | – | – | – | – | – | nr | 19022.04 |
| GXY8011 Sirius; 10 different play activities | – | – | – | – | – | nr | 11397.02 |
| **Sports and social areas; Freegame multi-use games areas; Kompan Ltd; enclosed sports areas; complete with surfacing boundary and goals and targets; galvanized steel framework with high density polyethylene panels, galvanized steel goals and equipment; surfacing priced separately** | | | | | | | |
| FRE1211 Classic Multigoal; 7 m Pitch; complete; suitable for multiple ball sports; suitable for use with natural artificial or hard landscape surfaces; fully enclosed including two end sports walls | – | – | – | – | – | nr | 4956.00 |
| FRE2110 Cosmos; 12 × 20 m | – | – | – | – | – | nr | 29162.06 |
| FRE2116 Cosmos; 19 × 36 m | – | – | – | – | – | nr | 42697.08 |
| FRE3000; Meeting Point; shelter or social area | – | – | – | – | – | nr | 5098.01 |
| **Swings – General** Preamble: Prices for the following vary considerably. Those given represent the middle of the range and include multiple swings and timber, rubber or tyre seats; ground fixing and priming only. | | | | | | | |
| **Swings** | | | | | | | |
| Swings; Wicksteed Ltd | | | | | | | |
| Flexi Swing; 1 bay 1 flat seat | – | – | – | – | – | nr | 2218.00 |
| Flexi Swing; 1 bay 2 flat seats | – | – | – | – | – | nr | 2723.00 |
| Flexi Swing; 1 bay 2 cradle seats | – | – | – | – | – | nr | 2888.00 |
| Flexi Swing; 1 bay 1 flat seat and 1 cradle seat | – | – | – | – | – | nr | 2805.00 |
| Hurricane Swing; 2 button seats | – | – | – | – | – | nr | 6623.02 |
| Pendulum Swing; 1 basket seat | – | – | – | – | – | nr | 8909.02 |
| 1.8 m Timber Swing; 2 cradle seats | – | – | – | – | – | nr | 1675.00 |
| 2.4 m Timber Swing; 2 cradle seats | – | – | – | – | – | nr | 1972.00 |
| 2.4 m Timber Swing; 2 flat seats | – | – | – | – | – | nr | 3140.00 |
| 2.4 m Birds Nest Timber Swing; 1 basket seat | – | – | – | – | – | nr | 1692.00 |

## 35 SITE WORKS – SPORTS, PLAY EQUIPMENT, SURFACES

| Item<br>Excluding site overheads and profit | PC<br>£ | Labour<br>hours | Labour<br>£ | Plant<br>£ | Material<br>£ | Unit | Total<br>rate £ |
|---|---|---|---|---|---|---|---|
| **PLAYGROUND EQUIPMENT –** cont | | | | | | | |
| **Swings – cont** | | | | | | | |
| Swings; Lappset UK Ltd | | | | | | | |
| 230050 Halo Swing frame and 2 rubber seats; chain 160 cm | – | – | – | – | – | nr | **3612.00** |
| Swings; Kompan Ltd | | | | | | | |
| M951P Sunflower swing; 1–3 years | – | – | – | – | – | nr | **1287.00** |
| M947P double swings; 1–6 years | – | – | – | – | – | nr | **1733.00** |
| M961P double swings; 6–12 years | – | – | – | – | – | nr | **2825.00** |
| **Slides** | | | | | | | |
| Slides; Wicksteed Ltd | | | | | | | |
| Pedestal slides; 3.40 m | – | – | – | – | – | nr | **3943.00** |
| Pedestal slides; 4.40 m | – | – | – | – | – | nr | **5134.02** |
| Pedestal slides; 5.80 m | – | – | – | – | – | nr | **5973.02** |
| Embankment slides; 3.40 m | – | – | – | – | – | nr | **3020.00** |
| Embankment slides; 4.40 m | – | – | – | – | – | nr | **3927.00** |
| Embankment slides; 5.80 m | – | – | – | – | – | nr | **4998.00** |
| Embankment slides; 7.30 m | – | – | – | – | – | nr | **6351.02** |
| Embankment slides; 9.10 m | – | – | – | – | – | nr | **7955.02** |
| Embankment slides; 11.00 m | – | – | – | – | – | nr | **9514.02** |
| Slides; Lappset UK Ltd | | | | | | | |
| 137338M Finno Slide | – | – | – | – | – | nr | **2604.00** |
| 175031 Clover slide | – | – | – | – | – | nr | **1744.00** |
| Slides; Kompan Ltd | | | | | | | |
| M351P slide | – | – | – | – | – | nr | **2341.00** |
| M326P Aladdin's Cave slide | – | – | – | – | – | nr | **2844.00** |
| **Moving equipment – General** | | | | | | | |
| Preamble: The following standard items of playground equipment vary considerably in quality and price; the following prices are middle of the range. | | | | | | | |
| **Moving equipment** | | | | | | | |
| Roundabouts; Wicksteed Ltd | | | | | | | |
| Dual axis roundabout; Dizzy | – | – | – | – | – | nr | **3838.00** |
| Dual axis roundabout; Whizzy | – | – | – | – | – | nr | **2614.00** |
| Dual axis roundabout; Solar Spinner | – | – | – | – | – | nr | **3243.00** |

## 35 SITE WORKS – SPORTS, PLAY EQUIPMENT, SURFACES

| Item<br>Excluding site overheads and profit | PC<br>£ | Labour<br>hours | Labour<br>£ | Plant<br>£ | Material<br>£ | Unit | Total<br>rate £ |
|---|---|---|---|---|---|---|---|
| **Roundabouts; Kompan Ltd**<br>Supernova GXY 916; multifunctional spinning and balancing disc; capacity approximately 15 children | – | – | – | – | – | nr | 4826.00 |
| **Seesaws**<br>Seesaws; Lappset UK Ltd | | | | | | | |
|   137004M Rumba seesaw | – | – | – | – | – | nr | 2551.00 |
|   137237M Finno seesaw | – | – | – | – | – | nr | 2049.00 |
| Seesaws; Wicksteed Ltd | | | | | | | |
|   Buddy Board | – | – | – | – | – | nr | 2373.00 |
|   Glow Worm | – | – | – | – | – | nr | 2029.00 |
|   Cobra | – | – | – | – | – | nr | 2367.00 |
|   Sidewinder | – | – | – | – | – | nr | 2171.00 |
| **Play sculptures – General**<br>Preamble: Many variants on the shapes of playground equipment are available, simulating spacecraft, trains, cars, houses etc., and these designs are frequently changed. The basic principles remain constant but manufacturers' catalogues should be checked for the latest styles. | | | | | | | |
| **Climbing equipment and play structures – General**<br>Preamble: Climbing equipment generally consists of individually designed modules. Play structures generally consist of interlinked and modular pieces of equipment and sculptures. These may consist of climbing, play and skill based modules, nets and various other activities. Both are set into either safety surfacing or defined sand pit areas. The equipment below outlines a range from various manufacturers. Individual catalogues should be consulted in each instance. Safety areas should be allowed round all equipment. | | | | | | | |
| **Climbing equipment and play structures**<br>Climbing equipment; Wicksteed Ltd | | | | | | | |
|   Funrun Fitness Trail; Under Starter's Orders; set of 12 units | – | – | – | – | – | nr | 20275.04 |

## 35 SITE WORKS – SPORTS, PLAY EQUIPMENT, SURFACES

| Item<br>Excluding site overheads and profit | PC<br>£ | Labour<br>hours | Labour<br>£ | Plant<br>£ | Material<br>£ | Unit | Total<br>rate £ |
|---|---|---|---|---|---|---|---|
| **PLAYGROUND EQUIPMENT –<br>cont** | | | | | | | |
| **Climbing equipment and play structures – cont** | | | | | | | |
| Climbing equipment; Lappset UK Ltd | | | | | | | |
|   137072M Finno Climbing Frame | – | – | – | – | – | nr | **1153.00** |
|   137740 Farmhouse | – | – | – | – | – | nr | **12031.00** |
|   230015 Halo Activity Tower | – | – | – | – | – | nr | **18215.00** |
|   175046 Clover Motoric Track | – | – | – | – | – | nr | **7344.00** |
| Hags-SMP Playgrounds Ltd | | | | | | | |
|   Nexus – The Core; multi-play structure | – | – | – | – | – | nr | **9730.00** |
| **Spring equipment** | | | | | | | |
| Spring based equipment for 1–8 year olds; Kompan Ltd | | | | | | | |
|   M101P Crazy Hen | – | – | – | – | – | nr | **611.00** |
|   M128P Crazy Daisy | – | – | – | – | – | nr | **905.00** |
|   M141P Spring Seesaw | – | – | – | – | – | nr | **2135.00** |
|   M155P Quartet Seesaw | – | – | – | – | – | nr | **1419.00** |
| Spring based equipment for under 12's; Lappset UK Ltd | | | | | | | |
|   010501 Horse Springer | – | – | – | – | – | nr | **716.00** |
| **Sandpits** | | | | | | | |
| Market prices | | | | | | | |
|   play pit sand; Boughton Loam Ltd | 122.40 | – | – | – | – | m$^3$ | – |
| Kompan Ltd | | | | | | | |
|   Basic550; 276 × 154 × 31 cm deep | – | – | – | – | – | nr | **1590.00** |
| **NATURAL SPORTS PITCHES** | | | | | | | |
| **Natural sports pitches; Agripower Ltd; sports pitches; gravel raft construction** | | | | | | | |
| Professional standard | | | | | | | |
|   football pitch; 6500 m$^2$ | – | – | – | – | – | nr | **159000.00** |
|   cricket wicket; county standard | – | – | – | – | – | nr | **31800.00** |
| Playing fields; Sport England Compliant | | | | | | | |
|   football pitch; 6500 m$^2$ | – | – | – | – | – | nr | **68900.00** |
|   cricket wicket; playing field standard | – | – | – | – | – | nr | **15900.00** |
|   track and infield | – | – | – | – | – | nr | **97520.00** |
| Existing area to 1:100 existing grade; 6500 m$^2$ | 1420.13 | – | – | 1204.00 | 1825.73 | m$^2$ | **3029.73** |

## 35 SITE WORKS – SPORTS, PLAY EQUIPMENT, SURFACES

| Item Excluding site overheads and profit | PC £ | Labour hours | Labour £ | Plant £ | Material £ | Unit | Total rate £ |
|---|---|---|---|---|---|---|---|
| **Sports field construction** | | | | | | | |
| *Note: The playing fields in this section are constructed to a lower standard than the specialist sports fields elsewhere in this book and would comply with standards for general amenity in parks, recreation grounds or school playing fields.* | | | | | | | |
| **Surface preparations to existing playing area; prices per 1000 m²** | | | | | | | |
| Apply herbicide and strip existing turf 25 mm thick | – | – | – | 270.00 | 25.60 | nr | 295.60 |
| Disposal green waste off site | – | – | – | 48.16 | 750.00 | nr | 798.16 |
| **Cultivation and soil preparation; prices per 1000 m²** | | | | | | | |
| Light to medium soils | | | | | | | |
| Decompaction operations by tractor; ripping and breaking up ground | – | – | – | 232.50 | – | nr | 232.50 |
| Cultivation and stone burying harrow and power raking in single operation | – | – | – | 540.00 | – | nr | 540.00 |
| Clay soils | | | | | | | |
| Decompaction operations by tractor; ripping and breaking up ground | – | – | – | 270.00 | – | nr | 270.00 |
| Strip and remove 100 mm thick surface | – | – | – | 192.63 | – | nr | 192.63 |
| Disposal on site | – | – | – | 259.18 | – | nr | 259.18 |
| Disposal off site 100 mm thick arisings | – | 2.78 | 66.03 | – | 3081.09 | nr | 3147.12 |
| Cultivation, stone burying, harrow and power raking in single operation | – | – | – | 600.00 | – | nr | 600.00 |
| Imported new topsoil rootzone 150 mm deep | – | – | – | 1348.49 | 5940.00 | nr | 7288.49 |
| Drainage operations | | | | | | | |
| Sand grooving or sand slitting to enable drainage | – | – | – | – | – | nr | – |
| Surface operations to prepared ground fertilizing 35 g/m² | 42.00 | 3.00 | 71.25 | – | 42.00 | nr | 113.25 |
| seeding operations by mechanical seeder; sportsfield seed 35 g/m² by mechanical seeder | 236.69 | – | – | 40.00 | 236.69 | nr | 276.69 |

## 35 SITE WORKS – SPORTS, PLAY EQUIPMENT, SURFACES

| Item<br>Excluding site overheads and profit | PC<br>£ | Labour<br>hours | Labour<br>£ | Plant<br>£ | Material<br>£ | Unit | Total<br>rate £ |
|---|---|---|---|---|---|---|---|
| **PLAY SURFACES** | | | | | | | |
| **Playgrounds; Rubaflex in situ playground surfacing, porous; on prepared stone/granular base Type 1 (not included) and macadam base course (not included)** | | | | | | | |
| Black | | | | | | | |
|   15 mm thick (0.50 m critical fall height) | – | – | – | – | – | m² | **32.00** |
|   35 mm thick (1.00 m critical fall height) | – | – | – | – | – | m² | **38.00** |
|   60 mm thick (1.50 m critical fall height) | – | – | – | – | – | m² | **42.00** |
| **Playgrounds; Wetpour safety surfacing in situ playground surfacing, porous; on prepared stone/granular base Type 1 (not included) and macadam base course (not included)** | | | | | | | |
| Coloured | | | | | | | |
|   15 mm thick (0.50 m critical fall height) | – | – | – | – | – | m² | **36.00** |
|   35 mm thick (1.00 m critical fall height) | – | – | – | – | – | m² | **40.00** |
|   60 mm thick (1.50 m critical fall height) | – | – | – | – | – | m² | **50.00** |
| **Playgrounds; Meon UK; porous safety surfaces; on prepared Type 1 base (not included)** | | | | | | | |
| Critical fall height 500 mm | | | | | | | |
|   natural colours | – | – | – | – | – | m² | **53.00** |
|   bright colours | – | – | – | – | – | m² | **50.00** |
| Critical fall height 1.70 m | | | | | | | |
|   natural colours | – | – | – | – | – | m² | **74.00** |
|   bright colours | – | – | – | – | – | m² | **68.00** |
| **Playgrounds Meon UK; Rubber crumb resin surfacing** | | | | | | | |
| Mulch resin surfacing; Black or colours | | | | | | | |
|   10–50 m² | – | – | – | – | – | m² | **70.00** |
|   over 50 m² | – | – | – | – | – | m² | **62.00** |

## 35 SITE WORKS – SPORTS, PLAY EQUIPMENT, SURFACES

| Item<br>Excluding site overheads and profit | PC<br>£ | Labour<br>hours | Labour<br>£ | Plant<br>£ | Material<br>£ | Unit | Total<br>rate £ |
|---|---|---|---|---|---|---|---|
| **Playgrounds; Melcourt Industries Ltd; specifiers and users should contact the supplier for performance specifications of the materials below** | | | | | | | |
| Play surfaces; on drainage layer (not included); minimum 300 mm settled depth; prices for 80 m$^3$ loads unless specified | | | | | | | |
| Playbark® 8/25; 8-25 mm particles; red/brown; 25 m$^3$ loads | 18.78 | 0.35 | 8.33 | – | 18.78 | m$^2$ | 27.11 |
| Playbark® 8/25; 8-25 mm particles; red/brown | 16.37 | 0.35 | 8.33 | – | 16.37 | m$^2$ | 24.70 |
| Playbark® 10/50; 10–50 mm particles; red/brown | 18.73 | 0.35 | 8.33 | – | 18.73 | m$^2$ | 27.06 |
| Playchips®; FSC graded woodchips | 9.75 | 0.35 | 8.33 | – | 9.75 | m$^2$ | 18.08 |
| Kushyfall; fibreised woodchips | 9.07 | 0.35 | 8.33 | – | 9.07 | m$^2$ | 17.40 |
| Softfall; conifer shavings | 6.30 | 0.35 | 8.33 | – | 6.30 | m$^2$ | 14.63 |
| **TENNIS COURTS** | | | | | | | |
| **Tennis courts; Spadeoak Ltd**<br>Hard playing surfaces to SAPCA Code of Practice minimum requirements; laid on 65 mm thick macadam base on 150 mm thick stone foundation; to include lines, nets, posts, brick edging, 2.75 m high fence and perimeter drainage (excluding excavation, levelling or additional foundation); based on court size 36.6 × 18.3 m (670 m$^2$) | | | | | | | |
| Durapore; all weather porous macadam and acrylic colour coating | – | – | – | – | – | m$^2$ | 63.47 |
| Tiger Turf Advantage Pro; sand filled artificial grass | – | – | – | – | – | m$^2$ | 87.04 |
| Tiger Turf Baseline; short pile sand filled artificial grass | – | – | – | – | – | m$^2$ | 86.55 |
| DecoColour; impervious acrylic hardcourt (200 mm foundation) | – | – | – | – | – | m$^2$ | 80.78 |
| DecoTurf; cushioned impervious acrylic tournament surface (200 mm foundation) | – | – | – | – | – | m$^2$ | 92.32 |
| Porous Kushion Kourt; porous cushioned acrylic surface | – | – | – | – | – | m$^2$ | 90.39 |
| EasiClay; synthetic clay system | – | – | – | – | – | m$^2$ | 94.23 |
| Canada Tenn; American green clay; fast dry surface (no macadam but including irrigation) | – | – | – | – | – | m$^2$ | 96.16 |

## 35 SITE WORKS – SPORTS, PLAY EQUIPMENT, SURFACES

| Item<br>Excluding site overheads and profit | PC<br>£ | Labour<br>hours | Labour<br>£ | Plant<br>£ | Material<br>£ | Unit | Total<br>rate £ |
|---|---|---|---|---|---|---|---|
| **SPORTS SURFACES** | | | | | | | |
| **Artificial surfaces and finishes – General**<br>Preamble: Advice should also be sought from the Technical Unit for Sport, the appropriate regional office of the Sports Council or the National Playing Fields Association. Some of the following prices include base work whereas others are for a specialist surface only on to a base prepared and costed separately. | | | | | | | |
| **Artificial sports pitches; Agripower Ltd; third generation rubber crumb**<br>Sports pitches to respective National governing body standards; inclusive of all excavation, drainage, lighting, fencing, etc. | | | | | | | |
| football pitch; 6500 m² | – | – | – | – | – | nr | 492900.00 |
| rugby union multi-use pitch; 120 × 75 m | – | – | – | – | – | nr | 625400.00 |
| hockey pitch; water based; 101.4 × 63 m | – | – | – | – | – | nr | 583000.00 |
| athletic track; 8 lane; International amateur athletic federation | – | – | – | – | – | nr | 583000.00 |
| Multi-sport pitches; inclusive of all excavation, drainage, lighting, fencing, etc. | | | | | | | |
| sand dressed; 101.4 × 63 m | – | – | – | – | – | nr | 413400.00 |
| sand filled; 101.4 × 63 m | – | – | – | – | – | nr | 360400.00 |
| polymeric (synthetic bound rubber) | – | – | – | – | – | m² | 137.80 |
| macadam | – | – | – | – | – | m² | 84.80 |
| **Sports areas; Agripower Ltd**<br>Sports tracks; polyurethane rubber surfacing; on bitumen-macadam (not included); prices for 5500 m² minimum | | | | | | | |
| International | – | – | – | – | – | m² | 42.40 |
| Club Grade | – | – | – | – | – | m² | 32.86 |
| Sports areas; multi-component polyurethane rubber surfacing; on bitumen-macadam; finished with polyurethane or acrylic coat; green or red | | | | | | | |
| Permaprene | – | – | – | – | – | m² | 43.46 |

## 35 SITE WORKS – WATER AND WATER FEATURES

| Item<br>Excluding site overheads and profit | PC<br>£ | Labour<br>hours | Labour<br>£ | Plant<br>£ | Material<br>£ | Unit | Total<br>rate £ |
|---|---|---|---|---|---|---|---|
| **CLARIFICATION NOTES ON LABOUR COSTS IN THIS SECTION** | | | | | | | |
| **General groundworks team**<br>Generally a three man team is used in this section; The column Labour hours reports team hours. The column Labour £ reports the total cost of the team for the unit of work shown | | | | | | | |
| 3 man team | – | 1.00 | 71.25 | – | – | hr | 71.25 |
| 2 man team | – | 1.00 | 47.50 | – | – | hr | 47.50 |
| **BOREHOLES** | | | | | | | |
| **Borehole drilling; Agripower Ltd**<br>Drilling operations to average 100 m depth; lining with 150 mm diameter sieve | | | | | | | |
| drilling to 100 m | – | – | – | – | – | nr | 12720.00 |
| rate per m over 100 m | – | – | – | – | – | m | 127.20 |
| pump, rising main, cable and kiosk | – | – | – | – | – | nr | 5300.00 |
| **COLD WATER** | | | | | | | |
| **Blue MDPE polythene pipes; type 50; for cold water services; with compression fittings; bedding on 100 mm DOT type 1 granular fill material**<br>Pipes | | | | | | | |
| 20 mm diameter | 0.48 | 0.08 | 1.90 | – | 4.93 | m | 6.83 |
| 25 mm diameter | 1.10 | 0.08 | 1.90 | – | 5.56 | m | 7.46 |
| 32 mm diameter | 1.35 | 0.08 | 1.90 | – | 5.80 | m | 7.70 |
| 50 mm diameter | 2.22 | 0.10 | 2.38 | – | 6.67 | m | 9.05 |
| 60 mm diameter | 3.55 | 0.10 | 2.38 | – | 8.01 | m | 10.39 |
| Hose union bib taps; including fixing to wall; making good surfaces | | | | | | | |
| 15 mm | 13.77 | 0.75 | 17.81 | – | 14.17 | nr | 31.98 |
| 22 mm | 18.44 | 0.75 | 17.81 | – | 19.04 | nr | 36.85 |
| Stopcocks; including fixing to wall; making good surfaces | | | | | | | |
| 15 mm | 8.06 | 0.75 | 17.81 | – | 8.46 | nr | 26.27 |
| 22 mm | 12.09 | 0.75 | 17.81 | – | 12.49 | nr | 30.30 |
| Standpipes; to existing 25 mm water mains | | | | | | | |
| 1.00 m high | – | – | – | – | – | nr | 225.00 |

## 35 SITE WORKS – WATER AND WATER FEATURES

| Item<br>Excluding site overheads and profit | PC<br>£ | Labour<br>hours | Labour<br>£ | Plant<br>£ | Material<br>£ | Unit | Total<br>rate £ |
|---|---|---|---|---|---|---|---|
| **COLD WATER – cont** | | | | | | | |
| **Blue MDPE polythene pipes – cont** | | | | | | | |
| Hose junction bib taps; to standpipes | | | | | | | |
| 19 mm | – | – | – | – | – | nr | **80.00** |
| **RAINWATER HARVESTING** | | | | | | | |
| **Rainwater harvesting tanks** | | | | | | | |
| Note: Rainwater harvesting tanks must be installed on granular or concrete bases. If installed below ground the tanks may require construction of drained underground chambers to support them. Please see the appropriate sections in this book for excavation, disposal, bases, retaining walls and drainage. | | | | | | | |
| **Excavation for underground tanks; excavation inclusive of earthwork retention for self supporting tanks** | | | | | | | |
| 7 tonne tracked excavator (bucket volume 0.28 m³) | | | | | | | |
| maximum depth not exceeding 1.00 m | – | 0.06 | 1.48 | 2.91 | – | m³ | **4.39** |
| maximum depth not exceeding 2.00 m | – | 0.07 | 1.70 | 3.33 | – | m³ | **5.03** |
| maximum depth not exceeding 3.00 m | – | 0.09 | 2.16 | 4.24 | – | m³ | **6.40** |
| **Disposal** | | | | | | | |
| Excavated material; off site; to tip; mechanically loaded (JCB) | | | | | | | |
| inert | – | – | – | – | – | m³ | **30.81** |
| **Type 1 granular fill base; PC £20.00/ tonne (£42.90/m³ compacted)** | | | | | | | |
| By machine | | | | | | | |
| 100 mm thick | 4.95 | 0.03 | 0.67 | 0.46 | 4.95 | m² | **6.08** |
| 150 mm thick | 7.42 | 0.03 | 0.59 | 0.69 | 7.42 | m² | **8.70** |
| **Backfilling to surround of rainwater tank; carefully compacting as work proceeds** | | | | | | | |
| Arising from the excavations | | | | | | | |
| average thickness exceeding 0.25 m; depositing in layers 150 mm maximum thickness | – | 0.03 | 0.79 | 6.05 | – | m³ | **6.84** |

## 35 SITE WORKS – WATER AND WATER FEATURES

| Item<br>Excluding site overheads and profit | PC<br>£ | Labour<br>hours | Labour<br>£ | Plant<br>£ | Material<br>£ | Unit | Total<br>rate £ |
|---|---|---|---|---|---|---|---|
| **Rainwater harvesting tanks;**<br>**Rainwater Harvesting Ltd; self**<br>**supporting underground or above**<br>**ground tanks**<br>Rainwater tank with max 1.00 m cover in pedestrian areas; complete with tank dome and pedestrian lid, submersible automatic pump system, supra filtration system, 25 m black and green rainwater pipe, rainwater labelling kit, 125 mm fine filter (excavation, backfilling base, trenching and inlet pipework all not included) | | | | | | | |
| 3700 litre; 2.44 × 1.65 × 1.95 m | 1813.85 | 3.00 | 142.50 | – | 1953.85 | nr | **2096.35** |
| 4500 litre; 2.44 × 1.84 × 1.84 m | 1940.27 | 3.50 | 166.25 | – | 2080.27 | nr | **2246.52** |
| 6500 litre; 2.68 × 2.02 × 2.29 m | 2553.29 | 4.00 | 190.00 | – | 2693.29 | nr | **2883.29** |
| 9000 litre; 2.44 × 1.84 × 1.88 m | 3462.58 | 4.50 | 213.75 | – | 3602.58 | nr | **3816.33** |
| 13000 litre – 2 × 6500 l tanks;<br>2.68 × 4.20 × 2.29 m | 4485.40 | 7.00 | 332.50 | – | 4625.40 | nr | **4957.90** |
| Lighter weight tank; complete with tank dome and pedestrian lid, submersible automatic pump system, supra filtration system, 25 m black and green rainwater pipe, rainwater labelling kit, 125 mm fine filter (excavation, backfilling base, trenching and inlet pipework all not included) | | | | | | | |
| 1650 litre; 2.10 × 1.05 × 1.22 m | 1159.44 | 5.00 | 237.50 | – | 1299.44 | nr | **1536.94** |
| 2650 litre; 2.10 × 1.30 × 1.50 m | 1559.72 | 6.00 | 285.00 | – | 1699.72 | nr | **1984.72** |
| Rainwater tank; above or below ground (pedestrian areas only) installation; extendable to multiple tanks | | | | | | | |
| 1600 litre; 1.35 m diameter × 1.60 m high | 1451.00 | 4.50 | 213.75 | – | 1591.00 | nr | **1804.75** |
| Low profile tanks for reduced excavation; complete with rainwater filters and pressure pump | | | | | | | |
| 1500 litre; 2.10 × 1.25 × 1.02 m | 1295.07 | 4.00 | 190.00 | – | 1295.07 | nr | **1629.07** |
| 3000 litre; 2.46 × 2.1 × 1.05 m | 1705.59 | 6.00 | 285.00 | – | 1845.59 | nr | **2130.59** |
| 5000 litre; 2.89 × 2.30 × 1.26 m | 2168.31 | 6.50 | 308.75 | – | 2308.31 | nr | **2617.06** |
| 10000 litre 2.89 × 4,60 × 1.26 m | 2523.42 | 7.00 | 332.50 | – | 2663.42 | nr | **2995.92** |
| **Rainwater Harvesting Ltd;**<br>**accessories for rainwater harvesting**<br>**tanks**<br>Downpipe filter for connection to roof downpipes | | | | | | | |
| for roof areas up to 50 m² | 15.00 | 0.25 | 5.94 | – | 15.00 | nr | **20.94** |
| for roof areas up to 80 m² | 27.56 | 0.25 | 5.94 | – | 27.56 | nr | **33.50** |
| for roof areas up to 100 m² | 88.19 | – | – | – | 88.19 | nr | **88.19** |

## 35 SITE WORKS – WATER AND WATER FEATURES

| Item<br>Excluding site overheads and profit | PC<br>£ | Labour<br>hours | Labour<br>£ | Plant<br>£ | Material<br>£ | Unit | Total<br>rate £ |
|---|---|---|---|---|---|---|---|
| **RAINWATER HARVESTING – cont** | | | | | | | |
| **Rainwater Harvesting Ltd – cont** | | | | | | | |
| Pumps; inclusive of electrical installation and connections | | | | | | | |
|    Rainwater direct pump package | 285.00 | 0.25 | 5.94 | – | 425.00 | nr | **430.94** |
|    HydroForce Series 3 pump | 274.53 | 0.75 | 20.94 | – | 274.53 | nr | **295.47** |
| Filters for rainwater tanks; complete with overflow siphons and rodent guard | 342.50 | – | – | – | 342.50 | nr | **342.50** |
| Extra over to cover rainwater tanks in geofabric prior to backfilling; 10 m² of terram per tank; average cost | – | 0.25 | 11.88 | – | 4.07 | nr | **15.95** |
| **IRRIGATION** | | | | | | | |
| **Irrigation Infrastructure and pipework; main or ring main supply** | | | | | | | |
| Excavate and lay mains supply pipe to supply irrigated area | | | | | | | |
|    PE 80 20 mm | – | – | – | – | – | m | **4.87** |
|    PE 80 25 mm | – | – | – | – | – | m | **5.06** |
|    Pro-Flow PEHD 63 mm PE | – | – | – | – | – | m | **8.68** |
|    Pro-Flow PEHD 50 mm PE | – | – | – | – | – | m | **7.09** |
|    Pro-Flow PEHD 40 mm PE | – | – | – | – | – | m | **6.11** |
|    Pro-Flow PEHD 32 mm PE | – | – | – | – | – | m | **5.49** |
| **Irrigation systems; head control to sprinkler stations** | | | | | | | |
| Valves installed at supply point on irrigation supply manifold; includes for pressure control and filtration | | | | | | | |
|    2 valve unit | – | – | – | – | – | nr | **132.03** |
|    4 valve unit | – | – | – | – | – | nr | **206.11** |
|    6 valve unit | – | – | – | – | – | nr | **291.49** |
|    12 valve unit | – | – | – | – | – | nr | **558.92** |
| Solenoid valve; 25 mm with chamber; extra over for each active station | – | – | – | – | – | nr | **106.16** |
| Cable to solenoid valves (alternative to valves on manifold as above) | | | | | | | |
|    4 core | – | – | – | – | – | m | **1.57** |
|    6 core | – | – | – | – | – | m | **7.63** |
|    12 core | – | – | – | – | – | m | **3.19** |
| **Irrigation; water supply and control** | | | | | | | |
| Header tank and submersible pump and pressure stat | | | | | | | |
|    1000 litre (500 gallon 25 mm/10 days to 1500 m²) 2 m³/hr pump | – | – | – | – | – | nr | **489.67** |
|    4540 litre (1000 gallon 25 mm/10 days to 3500 m²) | – | – | – | – | – | nr | **991.89** |
|    9080 litre (2000 gallon 25 mm/10 days to 7000 m²) | – | – | – | – | – | nr | **1642.27** |

## 35 SITE WORKS – WATER AND WATER FEATURES

| Item<br>Excluding site overheads and profit | PC<br>£ | Labour<br>hours | Labour<br>£ | Plant<br>£ | Material<br>£ | Unit | Total<br>rate £ |
|---|---|---|---|---|---|---|---|
| Electric multistation controllers 240 V | | | | | | | |
| WeatherMatic Smartline; 4 station; standard | – | – | – | – | – | nr | 149.21 |
| WeatherMatic Smartline; 6 station; standard | – | – | – | – | – | nr | 167.62 |
| WeatherMatic Smartline; 12 station; standard | – | – | – | – | – | nr | 284.18 |
| WeatherMatic Smartline; 12 station; radio controlled | – | – | – | – | – | nr | 583.80 |
| WeatherMatic Smartline; 12 station; with moisture sensor | – | – | – | – | – | nr | 981.43 |
| WeatherMatic Smartline; 12 station; radio controlled; with moisture sensor | – | – | – | – | – | nr | 1178.62 |
| **Irrigation sprinklers; station consisting of multiple sprinklers; inclusive of all trenching, wiring and connections to ring main or main supply** | | | | | | | |
| Sprayheads; WeatherMatic LX; including nozzle; 21 m² coverage | | | | | | | |
| 100 mm (4') | – | – | – | – | – | nr | 15.52 |
| 150 mm (6') | – | – | – | – | – | nr | 21.92 |
| 300 mm (12') | – | – | – | – | – | nr | 25.28 |
| Matched precipitation sprinklers; WeatherMatic LX | | | | | | | |
| MP 1000; 16 m² coverage | – | – | – | – | – | nr | 35.57 |
| MP 2000; 30 m² coverage | – | – | – | – | – | nr | 35.57 |
| MP 3000; 81 m² coverage | – | – | – | – | – | nr | 35.57 |
| Gear drive sprinklers; WeatherMatic; placed to provide head to head (100%) overlap; average | | | | | | | |
| T3; 121 m² coverage | – | – | – | – | – | nr | 46.72 |
| CT70; 255 m² coverage | – | – | – | – | – | nr | 82.36 |
| CT70SS; 361 m² coverage | – | – | – | – | – | nr | 91.46 |
| **Drip irrigation; driplines; inclusive of connections and draindown systems** | | | | | | | |
| Metzerplas TechLand 30 cm drip emitter | | | | | | | |
| fixed on soil surface | – | – | – | – | – | m | 1.07 |
| subsurface | – | – | – | – | – | m | 1.57 |
| Metzerplas TechLand 50 cm drip emitter | | | | | | | |
| fixed on soil surface | – | – | – | – | – | m | 0.90 |
| subsurface | – | – | – | – | – | m | 1.41 |
| **Commissioning and testing of irrigation system** | | | | | | | |
| per station | – | – | – | – | – | nr | 39.56 |

## 35 SITE WORKS – WATER AND WATER FEATURES

| Item<br>Excluding site overheads and profit | PC<br>£ | Labour<br>hours | Labour<br>£ | Plant<br>£ | Material<br>£ | Unit | Total<br>rate £ |
|---|---|---|---|---|---|---|---|
| **IRRIGATION – cont** | | | | | | | |
| **Annual maintenance costs of irrigation system** | | | | | | | |
| Call out charge per visit | – | – | – | – | – | nr | **329.59** |
| Extra over per station | – | – | – | – | – | nr | **13.19** |
| **Moisture Control moisture sensing control system** | | | | | | | |
| Large garden comprising 7000 m$^2$ and 24 stations with four moisture sensors | | | | | | | |
| turf only | – | – | – | – | – | nr | **13333.35** |
| turf/shrub beds; 70/30 | – | – | – | – | – | nr | **7777.80** |
| Medium garden comprising 3500 m$^2$ and 12 stations with two moisture sensors | | | | | | | |
| turf only | – | – | – | – | – | nr | **8888.91** |
| turf/shrub beds; 70/30 | – | – | – | – | – | nr | **11555.60** |
| Smaller gardens garden comprising 1000 m$^2$ and 6 stations with one moisture sensor | | | | | | | |
| turf only | – | – | – | – | – | nr | **5444.44** |
| turf/shrub beds; 70/30 | – | – | – | – | – | nr | **6000.03** |
| turf/shrub beds; 50/50 | – | – | – | – | – | nr | **6444.44** |
| **Leaky Pipe Systems Ltd; Leaky Pipe; moisture leaking pipe irrigation system** | | | | | | | |
| Main supply pipe inclusive of machine excavation; exclusive of connectors | | | | | | | |
| 20 mm LDPE polytubing | 0.64 | 0.05 | 1.19 | 0.63 | 0.64 | m | **2.46** |
| 16 mm LDPE polytubing | 0.54 | 0.05 | 1.19 | 0.63 | 0.54 | m | **2.36** |
| Water filters and cartridges | | | | | | | |
| No 10; 20 mm | – | – | – | – | 109.00 | nr | **109.00** |
| Big Blue and RR30 cartridge; 25 mm | – | – | – | – | 149.00 | nr | **149.00** |
| Water filters and pressure regulator sets; complete assemblies | | | | | | | |
| No 10; flow rate 3.1–82 litres per minute | – | – | – | – | 59.00 | nr | **59.00** |
| Leaky pipe hose; placed 150 mm subsurface for turf irrigation; distance between laterals 350 mm; excavation and backfilling priced separately | | | | | | | |
| LP12L low leak | 3.13 | 0.04 | 0.95 | – | 3.13 | m$^2$ | **4.08** |
| LP12H high leak | 2.74 | 0.04 | 0.95 | – | 2.74 | m$^2$ | **3.69** |
| Leaky pipe hose; laid to surface for landscape irrigation; distance between laterals 600 mm | | | | | | | |
| LP12L low leak | 1.87 | 0.03 | 0.59 | – | 1.87 | m$^2$ | **2.46** |
| LP12H high leak | 1.63 | 0.03 | 0.59 | – | 1.63 | m$^2$ | **2.22** |

## 35 SITE WORKS – WATER AND WATER FEATURES

| Item<br>Excluding site overheads and profit | PC<br>£ | Labour<br>hours | Labour<br>£ | Plant<br>£ | Material<br>£ | Unit | Total<br>rate £ |
|---|---|---|---|---|---|---|---|
| Leaky pipe hose; laid to surface for landscape irrigation; distance between laterals 900 mm | | | | | | | |
|   LP12L low leak | 1.25 | 0.02 | 0.40 | – | 1.25 | m² | 1.65 |
|   LP12H high leak | 1.09 | 0.02 | 0.40 | – | 1.09 | m² | 1.49 |
| Leaky pipe hose; laid to surface for tree irrigation laid around circumference of tree pit | | | | | | | |
|   LP12L low leak | 1.85 | 0.13 | 2.97 | – | 1.85 | nr | 4.82 |
|   LP12H high leak | 1.62 | 0.13 | 2.97 | – | 1.62 | nr | 4.59 |
| Accessories | | | | | | | |
|   automatic multi-station controller stations; inclusive of connections | 350.00 | 2.00 | 60.00 | – | 350.00 | nr | 410.00 |
|   solenoid valves; inclusive of wiring and connections to a multi-station controller; nominal distance from controller 25 m | 120.00 | 0.50 | 11.88 | – | 597.50 | nr | 609.38 |

### FOUNTAINS AND WATER FEATURES

**Waterfall construction; Fairwater Water Garden Design Consultants Ltd**
Stone placed on top of butyl liner; securing with concrete and dressing to form natural rock pools and edgings

| Item | PC £ | Labour hours | Labour £ | Plant £ | Material £ | Unit | Total rate £ |
|---|---|---|---|---|---|---|---|
|   Portland stone; m³ rate | – | – | – | – | – | m³ | 736.19 |
|   Portland stone; tonne rate | – | – | – | – | – | tonne | 407.64 |
| Balancing tank; blockwork construction; inclusive of recirculation pump and pond level control; 110 mm balancing pipe to pond; waterproofed with preformed polypropylene membrane; pipework mains water top-up and overflow | | | | | | | |
|   450 × 600 × 1000 mm | – | – | – | – | – | nr | 1537.50 |
| Extra for pump | | | | | | | |
|   2000 gallons per hour; submersible | – | – | – | – | – | nr | 220.38 |

**Water features; Fairwater Water Garden Design Consultants Ltd**
Natural stream; butyl lined level changes of 1.00 m; water pumped from lower pond (not included) via balancing tank; level changes via Purbeck stone water falls

| Item | PC £ | Labour hours | Labour £ | Plant £ | Material £ | Unit | Total rate £ |
|---|---|---|---|---|---|---|---|
|   20 m long stream | – | – | – | – | – | nr | 21525.02 |

## 35 SITE WORKS – WATER AND WATER FEATURES

| Item<br>Excluding site overheads and profit | PC<br>£ | Labour<br>hours | Labour<br>£ | Plant<br>£ | Material<br>£ | Unit | Total<br>rate £ |
|---|---|---|---|---|---|---|---|
| **FOUNTAINS AND WATER FEATURES – cont** | | | | | | | |
| **Water features – cont**<br>Bespoke freestanding water wall; steel or glass panel; self contained recirculation system and reservoir | | | | | | | |
| water wall; 2.00 m high × 850 mm wide | – | – | – | – | – | nr | 10250.01 |
| **INTERNAL WATER FEATURES** | | | | | | | |
| **Internal Water features; Fairwater Water Garden Design Consultants Ltd**<br>Stainless steel rill; powder coated black 3000 mm × 250 mm; 2 nr letterbox weirs to maintain water level in rill; WRAS compliant header tank; modified balancing tank with chemical dosing unit; circulation pump and pipework | | | | | | | |
| 3.00 m × 250 mm wide | – | – | – | – | – | nr | 4612.50 |
| **Brimming bowl with catchment tray; Fairwater Water Garden Design Consultants Ltd**<br>Marble brimming bowl. recessed Polypro catchment tray<br>1.50 × 1.50 × 120 mm; pipework; WRAS compliant header tank balancing tank with Chemical dosing unit; circulation pump | | | | | | | |
| 1.00 m diameter | – | – | – | – | – | nr | 3843.75 |
| Design mockup and commissioning on the above | – | – | – | – | – | nr | 3587.50 |
| Maintenance 12 months; monthly | – | – | – | – | – | nr | 1537.50 |
| **WATER AND RIVER BANK REMEDIATION/ECOLOGY** | | | | | | | |
| **River and lake bed dredging; Salix Limited; Specialist water reclamation engineers; Operations include for removal of excavated material away from the water body bank and spreading locally in layers not exceeding 300 mm deep** | | | | | | | |

## 35 SITE WORKS – WATER AND WATER FEATURES

| Item<br>Excluding site overheads and profit | PC<br>£ | Labour<br>hours | Labour<br>£ | Plant<br>£ | Material<br>£ | Unit | Total<br>rate £ |
|---|---|---|---|---|---|---|---|
| **By long reach excavator** | | | | | | | |
| Mobilization for desilting | – | – | – | – | – | nr | 3000.00 |
| Dewatering and mechanical excavation with long reach excavator and tracked excavator to stockpile on bank | – | – | – | – | – | m³ | 12.00 |
| Mobilization £12,000 craning in barges, pontoon and tug | – | – | – | – | – | nr | 12000.01 |
| Excavation by floating excavator, transport to bank by tug and barge | – | – | – | – | – | m³ | 24.80 |
| Pumping of silt to geotubes for containment and drying | – | – | – | – | – | m³ | 22.50 |
| **By Pump dredger** | | | | | | | |
| **By long reach excavator using dewatering techniques** | | | | | | | |
| Mobilization Truxor type pump dredger up to 1000 m³ | – | – | – | – | – | nr | 600.00 |
| Mobilization pontoon type pump dredger over 1000 m³ | – | – | – | – | – | nr | 5000.01 |
| Lake size 1.5 ha; 15000 m³ | – | – | – | – | – | item | 6000.01 |
| Lake size 3.0 ha; 30000 m³ | – | – | – | – | – | item | 14000.01 |
| **Bioengineering techniques for reclamation or establishment of waterways; Salix Limited** | | | | | | | |
| Erosion control | | | | | | | |
| Coir roll + Faggot | – | – | – | – | – | m² | 37.70 |
| Coir roll and rock roll | – | – | – | – | – | m² | 75.00 |
| Brushwood mattress revetments to encourage deposition of sediments | | | | | | | |
| 200 mm deep | 33.25 | 0.07 | 4.75 | – | 33.25 | m² | 38.00 |
| 500 mm deep | 56.00 | 0.10 | 7.13 | – | 56.00 | m² | 63.13 |
| **Habitat creation to waterways and water bodies; Salix Limited; Specialist works to SSSIs and SAC sites** | | | | | | | |
| Grading and enablement of river bank to allow access; temporary trackways or haul road and removal of same | – | 8.00 | 570.00 | 385.25 | 1117.56 | nr | 2072.81 |
| Creation of riffle bars and point bars using gravel beds | – | – | – | – | – | m² | 201.00 |
| **Wetland establishment; Salix limited** | | | | | | | |
| Restoration of riverbanks; to create habitats for wildlife | | | | | | | |
| greenfield sites | – | – | – | – | – | m² | 15.00 |
| brownfield sites | – | – | – | – | – | m² | 25.00 |

## 35 SITE WORKS – WATER AND WATER FEATURES

| Item<br>Excluding site overheads and profit | PC<br>£ | Labour<br>hours | Labour<br>£ | Plant<br>£ | Material<br>£ | Unit | Total<br>rate £ |
|---|---|---|---|---|---|---|---|
| **WATER AND RIVER BANK REMEDIATION/ECOLOGY – cont** | | | | | | | |
| **Wetland and revetment establishment and erosion control; Salix Limited** | | | | | | | |
| Preparation works to receive treatments below | | | | | | | |
| Grade out bank receive treatments | – | 0.05 | 2.38 | 1.74 | – | m$^2$ | **4.12** |
| Remove silt and accumulated deposits to maximum 300 mm deep within water line to reinstate freeflow of water along shoreline; Spread on river bank | – | 0.05 | 2.38 | 2.33 | – | m$^2$ | **4.71** |
| Placement of pre-grown coir rolls 200 mm diameter to base of revetments or within the water body to aid establishment of vegetation and habitat at the base of retained revetments (above) | | | | | | | |
| on tidal banks or wet river banks | 26.90 | 0.11 | 7.92 | – | 26.90 | m | **34.82** |
| on shorelines max 300 mm deep | 43.25 | 0.11 | 8.09 | – | 43.25 | m | **51.34** |
| on shorelines max 600 mm deep | 66.00 | 0.13 | 8.91 | – | 66.00 | m | **74.91** |
| on shorelines max 1.00 m deep | 91.50 | 0.13 | 9.50 | – | 91.50 | m | **101.00** |
| on shorelines max 1.50 m deep | 97.50 | 0.17 | 11.88 | – | 97.50 | m | **109.38** |
| Placement of pre-grown coir rolls 300 mm diameter to base of revetments or within the water body to aid establishment of vegetation and habitat at the base of retained revetments (above) | | | | | | | |
| on tidal banks or wet river banks | 29.40 | 0.11 | 7.92 | – | 29.40 | m | **37.32** |
| on shorelines max 300 mm deep | 45.75 | 0.11 | 8.09 | – | 45.75 | m | **53.84** |
| on shorelines max 600 mm deep | 66.00 | 0.13 | 8.91 | – | 66.00 | m | **74.91** |
| on shorelines max 1.00 m deep | 94.00 | 0.13 | 9.50 | – | 94.00 | m | **103.50** |
| on shorelines max 1.50 m deep | 100.00 | 0.17 | 11.88 | – | 100.00 | m | **111.88** |
| Stronger tidal demands – placement of pre-grown coir rolls 400 mm diameter to base of revetments or within the water body to aid establishment of vegetation and habitat at the base of retained revetments (above) | | | | | | | |
| on tidal banks or wet river banks | 57.00 | 0.13 | 8.91 | – | 57.00 | m | **65.91** |
| on shorelines max 300 mm deep | 61.00 | 0.13 | 8.91 | – | 61.00 | m | **69.91** |
| on shorelines max 600 mm deep | 84.50 | 0.17 | 11.88 | – | 103.50 | m | **115.38** |
| on shorelines max 1.00 m deep | 110.00 | 0.20 | 14.25 | – | 110.00 | m | **124.25** |
| on shorelines max 1.50 m deep | 116.00 | 0.20 | 14.25 | – | 116.00 | m | **130.25** |

## 35 SITE WORKS – WATER AND WATER FEATURES

| Item<br>Excluding site overheads and profit | PC<br>£ | Labour<br>hours | Labour<br>£ | Plant<br>£ | Material<br>£ | Unit | **Total<br>rate £** |
|---|---|---|---|---|---|---|---|
| Placement of pre-planted coir pallets to establish reedbeds to wetlands; habitat creation projects where mature vegetation needs to be established rapidly | | | | | | | |
| on tidal banks or wet river banks | 34.25 | 0.20 | 14.25 | – | 34.25 | m | **48.50** |
| on shorelines max 300 mm deep | 37.25 | 0.25 | 17.81 | – | 37.25 | m | **55.06** |
| unplanted | 23.00 | 0.25 | 17.81 | – | 65.00 | m² | **82.81** |
| Brushwood faggots fascine in Hazel or Willow; 300 mm diameter to form sediment traps or build up intertidal sediment | | | | | | | |
| 300 m diameter | – | 0.10 | 7.13 | – | 23.50 | m | **30.63** |
| 400 mm diameter | – | 0.10 | 7.13 | – | 25.50 | m | **32.63** |
| Rock rolls; permanent revetment for use around reservoirs, shorelines, lake edges, streams and river banks | | | | | | | |
| 300 mm diameter | – | 0.07 | 4.75 | 3.21 | 99.90 | m² | **107.86** |
| 400 mm diameter | – | 0.10 | 7.13 | 4.82 | 100.00 | m² | **111.95** |
| rock mattresses to retain and in conjuction with rock rolls on riverbank | – | 0.08 | 5.94 | 4.01 | 47.00 | m² | **56.95** |

## 35 SITE WORKS – WATER AND WATER FEATURES

| Item<br>Excluding site overheads and profit | PC<br>£ | Labour<br>hours | Labour<br>£ | Plant<br>£ | Material<br>£ | Unit | Total<br>rate £ |
|---|---|---|---|---|---|---|---|
| **LAKES AND PONDS** | | | | | | | |
| **Lakes and ponds – General** | | | | | | | |
| Preamble: The pressure of water against a retaining wall or dam is considerable, and where water retaining structures form part of the design of water features, the landscape architect is advised to consult a civil engineer. Artificially contained areas of water in raised reservoirs over 25,000 m$^3$ have to be registered with the local authority and their dams will have to be covered by a civil engineer's certificate of safety. | | | | | | | |
| **Typical linings – General** | | | | | | | |
| Preamble: In addition to the traditional methods of forming the linings of lakes and ponds in puddled clay or concrete, there are a number of lining materials available. They are mainly used for reservoirs but can also help to form comparatively economic water features especially in soil which is not naturally water retentive. Information on the construction of traditional clay puddle ponds can be obtained from the British Trust for Conservation Volunteers, 36 St. Mary's Street, Wallingford, Oxfordshire, OX10 0EU. Tel: (01491) 39766. The cost of puddled clay ponds depends on the availability of suitable clay, the type of hand or machine labour that can be used and the use to which the pond is to be put. | | | | | | | |
| **Lake liners; Fairwater Water Garden Design Consultants Ltd; to evenly graded surface of excavations (excavating not included); all stones over 75 mm; removing debris; including all welding and jointing of liner sheets** | | | | | | | |
| Geotextile underlay; inclusive of spot welding to prevent dragging | | | | | | | |
|    to water features | – | – | – | – | – | m$^2$ | 1.80 |
|    to lakes or large features; 1000 m$^2$ | – | – | – | – | – | nr | 1804.00 |

## 35 SITE WORKS – WATER AND WATER FEATURES

| Item<br>Excluding site overheads and profit | PC<br>£ | Labour<br>hours | Labour<br>£ | Plant<br>£ | Material<br>£ | Unit | Total<br>rate £ |
|---|---|---|---|---|---|---|---|
| **Butyl rubber liners; Varnamo; inclusive of site vulcanizing laid to geotextile above** | | | | | | | |
| 0.75 mm thick | – | – | – | – | – | m² | 8.07 |
| 0.75 mm thick; 1000 m² | – | – | – | – | – | nr | 8066.80 |
| 1.00 m thick | – | – | – | – | – | m² | 12.66 |
| 1.00 m thick; 1000 m² | – | – | – | – | – | nr | 12658.80 |
| **Lake liners; Butek Landline Ltd; Landflex geomembranes; to prepared surfaces (surfaces not included); all joints fully welded; installation by Butek Landline employees; prices per 1000 m²** | | | | | | | |
| Landflex HC polyethylene geomembranes | | | | | | | |
| 0.50 mm thick | – | – | – | – | – | nr | 3150.00 |
| 0.75 mm thick | – | – | – | – | – | nr | 3410.00 |
| 1.00 m thick | – | – | – | – | – | nr | 3600.00 |
| 1.50 m thick | – | – | – | – | – | nr | 5250.00 |
| 2.00 m thick | – | – | – | – | – | nr | 6300.00 |
| Landflex PVC geomembranes | | | | | | | |
| 0.75 mm thick | – | – | – | – | – | nr | 6000.00 |
| 1.00 m thick | – | – | – | – | – | nr | 8500.00 |
| Landflex G3000 Puncture resistant geotextile | – | – | – | – | – | nr | 1500.00 |
| **Lake liners; Monarflex** | | | | | | | |
| Low density polyethylene (LDPE) lake and reservoir lining system; welding on site by Monarflex technicians (surface preparation and backfilling not included) | | | | | | | |
| Blackline 500 | – | – | – | – | – | 100 m² | 931.70 |
| Blackline 750 | – | – | – | – | – | 100 m² | 1023.80 |
| Blackline 1000 | – | – | – | – | – | 100 m² | 1264.95 |
| **Operations over surfaces of lake liners** | | | | | | | |
| Dug ballast; evenly spread over excavation already brought to grade | | | | | | | |
| 150 mm thick | 676.50 | 2.00 | 47.50 | 32.66 | 676.50 | 100 m² | 756.66 |
| 200 mm thick | 902.00 | 3.00 | 71.25 | 48.98 | 902.00 | 100 m² | 1022.23 |
| 300 mm thick | 1353.00 | 3.50 | 83.13 | 57.15 | 1353.00 | 100 m² | 1493.28 |
| Imported topsoil; evenly spread over excavation | | | | | | | |
| 100 mm thick | 396.00 | 1.50 | 35.63 | 24.49 | 396.00 | 100 m² | 456.12 |
| 150 mm thick | 594.00 | 2.00 | 47.50 | 32.66 | 594.00 | 100 m² | 674.16 |
| 200 mm thick | 792.00 | 3.00 | 71.25 | 48.98 | 792.00 | 100 m² | 912.23 |
| Blinding existing subsoil with 50 mm sand | 225.22 | 1.00 | 23.75 | 32.66 | 225.22 | 100 m² | 281.63 |

## 35 SITE WORKS – WATER AND WATER FEATURES

| Item<br>Excluding site overheads and profit | PC<br>£ | Labour<br>hours | Labour<br>£ | Plant<br>£ | Material<br>£ | Unit | Total<br>rate £ |
|---|---|---|---|---|---|---|---|
| **LAKES AND PONDS – cont** | | | | | | | |
| **Operations over surfaces of lake liners – cont** | | | | | | | |
| Topsoil from excavation; evenly spread over excavation | | | | | | | |
| 100 mm thick | – | – | – | 24.49 | – | 100 m² | 24.49 |
| 200 mm thick | – | – | – | 32.66 | – | 100 m² | 32.66 |
| 300 mm thick | – | – | – | 48.98 | – | 100 m² | 48.98 |
| Extra over for screening topsoil using a Powergrid screener; removing debris | – | – | – | 5.04 | 1.54 | m³ | 6.58 |
| **Lake construction; Fairwater Water Garden Design Consultants Ltd; lake construction at ground level; excavation; forming of lake; commissioning; excavated material spread on site** | | | | | | | |
| Lined with existing site clay; natural edging with vegetation meeting the water | | | | | | | |
| 500 m² | – | – | – | – | – | nr | 10250.01 |
| 1000 m² | – | – | – | – | – | nr | 16912.52 |
| 1500 m² | – | – | – | – | – | nr | 23575.02 |
| 2000 m² | – | – | – | – | – | nr | 28700.03 |
| Lined with 0.75 mm butyl on geotextile underlay; natural edging with vegetation meeting the water | | | | | | | |
| 500 m² | – | – | – | – | – | nr | 15887.52 |
| 1000 m² | – | – | – | – | – | nr | 29725.03 |
| 1500 m² | – | – | – | – | – | nr | 38950.04 |
| 2000 m² | – | – | – | – | – | nr | 46637.55 |
| Extra over to the above for hard block edging to secure and protect liner; measured at lake perimeter | – | – | – | – | – | m | 48.65 |
| Lined with imported puddling clay; natural edging with vegetation meeting the water | | | | | | | |
| 500 m² | – | – | – | – | – | nr | 25112.52 |
| 1000 m² | – | – | – | – | – | nr | 43050.04 |
| 1500 m² | – | – | – | – | – | nr | 58937.56 |
| 2000 m² | – | – | – | – | – | nr | 73800.07 |

## 35 SITE WORKS – WATER AND WATER FEATURES

| Item<br>Excluding site overheads and profit | PC<br>£ | Labour<br>hours | Labour<br>£ | Plant<br>£ | Material<br>£ | Unit | Total<br>rate £ |
|---|---|---|---|---|---|---|---|
| **Ornamental pools – General** | | | | | | | |
| Preamble: Small pools may be lined with one of the materials mentioned under lakes and ponds, or may be in rendered brickwork, puddled clay or, for the smaller sizes, fibreglass. Most of these tend to be cheaper than waterproof concrete. Basic prices for various sizes of concrete pools are given in the Approximate Estimating Rates section. Prices for excavation, grading, mass concrete, and precast concrete retaining walls are given in the relevant sections. The manufacturers should be consulted before specifying the type and thickness of pool liner, as this depends on the size, shape and proposed use of the pool. The manufacturer's recommendation on foundations and construction should be followed. | | | | | | | |
| **Ornamental pools** | | | | | | | |
| Pool liners; to 50 mm sand blinding to excavation (excavating not included); all stones over 50 mm; removing debris from surfaces of excavation; including underlay, all welding and jointing of liner sheets | | | | | | | |
|    black polythene; 1000 gauge | – | – | – | – | – | m$^2$ | **6.17** |
|    blue polythene; 1000 gauge | – | – | – | – | – | m$^2$ | **8.16** |
|    coloured PVC; 1500 gauge | – | – | – | – | – | m$^2$ | **9.35** |
|    black PVC; 1500 gauge | – | – | – | – | – | m$^2$ | **7.06** |
|    black butyl; 0.75 mm thick | – | – | – | – | – | m$^2$ | **9.20** |
|    black butyl; 1.00 m thick | – | – | – | – | – | m$^2$ | **10.28** |
|    black butyl; 1.50 m thick | – | – | – | – | – | m$^2$ | **34.03** |
| Fine gravel; 100 mm; evenly spread over area of pool; by hand | 3.88 | 0.13 | 2.97 | – | 3.88 | m$^2$ | **6.85** |
| Selected topsoil from excavation; 100 mm; evenly spread over area of pool; by hand | – | 0.13 | 2.97 | – | – | m$^2$ | **2.97** |
| Extra over selected topsoil for spreading imported topsoil over area of pool; by hand | 39.60 | – | – | – | 39.60 | m$^3$ | **39.60** |

## 35 SITE WORKS – WATER AND WATER FEATURES

| Item<br>Excluding site overheads and profit | PC<br>£ | Labour<br>hours | Labour<br>£ | Plant<br>£ | Material<br>£ | Unit | Total<br>rate £ |
|---|---|---|---|---|---|---|---|
| **LAKES AND PONDS – cont** | | | | | | | |
| **Pool surrounds and ornament; Haddonstone Ltd; Portland Bath or terracotta cast stone** | | | | | | | |
| Pool surrounds; installed to pools or water feature construction priced separately; surrounds and copings to 112.5 mm internal brickwork | | | | | | | |
|   C4HSKVP half small pool surround; internal diameter 1780 mm; kerb features continuous moulding enriched with ovolvo and palmette designs; inclusive of plinth and integral conch shell vases flanked by dolphins | 1374.00 | 16.00 | 380.00 | – | 1388.30 | nr | **1768.30** |
|   C4SKVP small pool surround as above but with full circular construction; internal diameter 1780 mm | 2748.00 | 48.00 | 1140.00 | – | 2766.72 | nr | **3906.72** |
|   C4MKVP medium pool surround; internal diameter 2705 mm; inclusive of plinth and integral vases | 4122.00 | 48.00 | 1140.00 | – | 4161.47 | nr | **5301.47** |
|   C4XLKVP extra large pool surround; internal diameter 5450 mm | 5496.00 | 140.00 | 3325.00 | – | 5543.90 | nr | **8868.90** |
| Pool centre pieces and fountains; inclusive of plumbing and pumps | | | | | | | |
|   C251 Gothic Fountain and Gothic Upper Base A350; freestanding fountain | 1079.00 | 4.00 | 95.00 | – | 1082.77 | nr | **1177.77** |
|   HC521 Romanesque Fountain; freestanding bowl with self-circulating fountain; filled with cobbles; 815 mm diameter × 348 mm high | 475.00 | 2.00 | 47.50 | – | 575.65 | nr | **623.15** |
|   C300 Lion Fountain; 610 mm high on fountain base C305; 280 mm high | 241.00 | 2.00 | 47.50 | – | 244.77 | nr | **292.27** |
| **Wall fountains, watertanks and fountains; inclusive of installation drainage, automatic top up, pump and balancing tank** | | | | | | | |
| Capital Garden Products Ltd | | | | | | | |
|   Lion Wall Fountain F010; 970 × 940 × 520 mm | – | 4.00 | 95.00 | – | 1782.50 | nr | **1877.50** |
|   Dolphin Wall Fountain F001; 740 × 510 mm | – | 4.00 | 95.00 | – | 1632.50 | nr | **1727.50** |
|   Dutch Master Wall Fountain F012; 740 × 410 mm | – | 4.00 | 95.00 | – | 1623.50 | nr | **1718.50** |

## 35 SITE WORKS – WATER AND WATER FEATURES

| Item<br>Excluding site overheads and profit | PC<br>£ | Labour<br>hours | Labour<br>£ | Plant<br>£ | Material<br>£ | Unit | Total<br>rate £ |
|---|---|---|---|---|---|---|---|
| James II Watertank 2801;<br>730 × 730 × 760 mm high; 405 litres | – | 4.00 | 95.00 | – | 1903.50 | nr | **1998.50** |
| James II Fountain 4901 bp;<br>710 × 1300 × 1440 mm high; 655 litres | – | 4.00 | 95.00 | – | 1773.50 | nr | **1868.50** |
| **Fountain surrounds; Architectural Heritage Ltd; to pools constructed separately inclusive of installation, drainage, automatic top up, pump and balancing tank** | | | | | | | |
| Parterre Pool Surround; age-patinated artificial stone; reproduction; Overall diameter 3.65 m, overall height 280 mm | 3800.00 | 4.00 | 95.00 | – | 5337.50 | nr | **5432.50** |
| Great Westwood pool surround; age-patinated artificial stone pool; reproduction; Overall diameter 6 m; height 350 mm | – | 4.00 | 95.00 | – | 7537.50 | nr | **7632.50** |
| **Fountain kits; typical prices of submersible units comprising fountain pumps, fountain nozzles, underwater spotlights, nozzle extension armatures, underwater terminal boxes and electrical control panels** | | | | | | | |
| Single aerated white foamy water columns; ascending jet 70 mm diameter; descending water up to four times larger; jet height adjustable between 1.00–1.70 m | 4135.95 | – | – | – | 4734.95 | nr | **4734.95** |
| Single aerated white foamy water columns; ascending jet 110 mm diameter; descending water up to four times larger; jet height adjustable between 1.50–3.00 m | 6893.25 | – | – | – | 7818.05 | nr | **7818.05** |
| **SWIMMING POOLS** | | | | | | | |
| **Natural swimming pools; Fairwater Water Garden Design Consultants Ltd** | | | | | | | |
| Natural swimming pool of 50 m² area; shingle regeneration zone 50 m² planted with marginal and aquatic plants | – | – | – | – | – | nr | **66625.07** |

## 35 SITE WORKS – WATER AND WATER FEATURES

| Item<br>Excluding site overheads and profit | PC<br>£ | Labour<br>hours | Labour<br>£ | Plant<br>£ | Material<br>£ | Unit | Total<br>rate £ |
|---|---|---|---|---|---|---|---|
| **SWIMMING POOLS – cont** | | | | | | | |
| | | | | | | | |
| **Guncast Swimming Pools Ltd; excavation and blinding of swimming pools; works by machine inclusive of allowances for working space and wall thicknesses; excavation of all trenches for pool pipeworks and electrical services** | | | | | | | |
| Regular shaped pools | | | | | | | |
| 100 m² pool | – | – | – | – | – | nr | **9600.00** |
| 72 m² pool | – | – | – | – | – | nr | **7500.00** |
| 60 m² pool | – | – | – | – | – | nr | **7500.00** |
| Irregular shaped pools | | | | | | | |
| 100 m² pool | – | – | – | – | – | nr | **10080.00** |
| 72 m² pool | – | – | – | – | – | nr | **7875.00** |
| 60 m² pool | – | – | – | – | – | nr | **7875.00** |
| | | | | | | | |
| **Guncast Swimming Pools Ltd; construction of a rectangular swimming pool in situ concrete corner steps, graduated shell depths of 1.1–2.5 m, water depth of 1.0–2.4 m as standard skimmer pool, tiled corner steps; excludes pump house construction** | | | | | | | |
| Inclusive of drainage layer, formwork reinforcement and all Gunite construction, rendered internally and bespoke mosaic surface finish; bullnosed Indian sandstone copings | | | | | | | |
| pool size 17.0 × 6.0 m; 172,500 litre | – | – | – | – | – | nr | **149000.00** |
| pool size 12.0 × 6.0 m; 125,750 litre | – | – | – | – | – | nr | **112000.00** |
| pool size 12.0 × 5.0 m; 102,150 litre | – | – | – | – | – | nr | **99000.00** |
| | | | | | | | |
| **Guncast Swimming Pools Ltd; options and accessories for swimming pools** | | | | | | | |
| Underwater lighting | | | | | | | |
| 8 nr underwater floodlights 300 W 12 V; stainless steel | – | – | – | – | – | nr | **5500.00** |
| 2 nr underwater floodlights 300 W 12 V; stainless steel | – | – | – | – | – | nr | **1375.00** |
| Automatic pool cleaner | – | – | – | – | – | nr | **1550.00** |

## 35 SITE WORKS – WATER AND WATER FEATURES

| Item<br>Excluding site overheads and profit | PC<br>£ | Labour<br>hours | Labour<br>£ | Plant<br>£ | Material<br>£ | Unit | Total<br>rate £ |
|---|---|---|---|---|---|---|---|
| Automatic pool covers | | | | | | | |
| insulated automatic slatted cover; 12 × 6 m inclusive of additional surface skimmer and balance pipe in pit; housed in decking (not included) | – | – | – | – | – | nr | 20900.00 |
| solar cover; 17 × 6 m Ocea; housed in wall cave pit; logic motor; concealed with stainless steel frame and removable pvc panels; tiled or rendered to match pool | – | – | – | – | – | nr | 35800.00 |
| Driglide pool cover; 17 × 6 m selected colours on trackway | – | – | – | – | – | nr | 39540.00 |
| Ozone disinfection system | – | – | – | – | – | nr | 5625.00 |
| **Guncast Swimming Pools Ltd; refurbishment of an existing swimming pool; with Roman bay end steps; no leaks present in existing pool; existing pipework and existing filtration systems and pump** | | | | | | | |
| Empty pool, retain existing blockwork structure; retile pool; replace copings as Indian sandstone bullnose; prepare shell; fix mosaic; pressure test | | | | | | | |
| pool size 8.50 × 4.25 59,000 litre | – | – | – | – | – | nr | 46500.00 |
| **Guncast Swimming Pools Ltd; options and accessories for swimming pools** | | | | | | | |
| Pool heating options | | | | | | | |
| propane boiler 250,000 BTU | – | – | – | – | – | nr | 4130.00 |
| pool heating via propane boiler 400,000 BTU | – | – | – | – | – | nr | 5240.00 |
| heat pump 100 × 1000 mm concrete base | – | – | – | – | – | nr | 6400.00 |
| pool heating via heat pump | – | – | – | – | – | nr | 6400.00 |
| Diving board | – | – | – | – | – | nr | 2360.00 |
| Pool maintenance kit | | | | | | | |
| telescopic pole, wall/floor brush; skimmer net; flexible vacuum hose; underwater vacuum cleaner; pool water test kit; thermometer; pool treatment chemicals | – | – | – | – | – | nr | 920.00 |

## 36 FENCING

| Item<br>Excluding site overheads and profit | PC<br>£ | Labour<br>hours | Labour<br>£ | Plant<br>£ | Material<br>£ | Unit | Total<br>rate £ |
|---|---|---|---|---|---|---|---|
| **CLARIFICATION NOTES ON LABOUR COSTS IN THIS SECTION** | | | | | | | |
| **Fencing team**<br>Generally a two man fencing team is used in this section; The column Labour hours reports team hours. The column Labour £ reports the total cost of the team for the unit of fencing shown | | | | | | | |
| 2 man team | – | 1.00 | 47.50 | – | – | hr | **47.50** |
| **TEMPORARY FENCING** | | | | | | | |
| **Temporary security fence; HSS Hire; mesh framed unclimbable fencing; including precast concrete supports and couplings**<br>Weekly hire; 2.85 × 2.00 m high | | | | | | | |
| weekly hire rate | – | – | – | 2.26 | – | m | **2.26** |
| erection of fencing; labour only | – | 0.07 | 3.17 | – | – | m | **3.17** |
| removal of fencing loading to collection vehicle | – | 0.03 | 1.59 | – | – | m | **1.59** |
| delivery charge | – | – | – | 0.83 | – | m | **0.83** |
| return haulage charge | – | – | – | 0.63 | – | m | **0.63** |
| **Protective fencing; AVS Fencing Supplies Ltd**<br>Cleft chestnut rolled fencing; fixing to 100 mm diameter chestnut posts; driving into firm ground at 3 m centres | | | | | | | |
| 900 mm high | 5.16 | 0.05 | 2.53 | – | 11.69 | m | **14.22** |
| 1200 mm high | 6.62 | 0.08 | 3.80 | – | 13.15 | m | **16.95** |
| 1500 mm high; 3 strand | 8.29 | 0.11 | 5.07 | – | 14.82 | m | **19.89** |
| Enclosures; Earth Anchors | | | | | | | |
| Rootfast anchored galvanized steel enclosures post ADP 20-1000; 1000 mm high × 20 mm diameter with AA25-750 socket and padlocking ring | 35.87 | 0.05 | 2.38 | – | 38.57 | nr | **40.95** |
| steel cable, orange plastic coated | 1.25 | – | 0.05 | – | 1.25 | m | **1.30** |

## 36 FENCING

| Item<br>Excluding site overheads and profit | PC<br>£ | Labour<br>hours | Labour<br>£ | Plant<br>£ | Material<br>£ | Unit | Total<br>rate £ |
|---|---|---|---|---|---|---|---|
| **TIMBER FENCING** | | | | | | | |
| **Timber fencing; AVS Fencing Supplies Ltd; tanalized softwood fencing; all timber posts are kiln dried redwood with 15 year guarantee** | | | | | | | |
| Timber lap panels; pressure treated; fixed to timber posts 75 × 75 mm in 1:3:6 concrete; at 1.90 m centres | | | | | | | |
| 900 mm high | 10.54 | 0.33 | 15.83 | – | 19.41 | m | **35.24** |
| 1200 mm high | 10.96 | 0.38 | 17.81 | – | 19.36 | m | **37.17** |
| 1500 mm high | 10.96 | 0.40 | 19.00 | – | 19.84 | m | **38.84** |
| 1800 mm high | 11.06 | 0.45 | 21.38 | – | 21.37 | m | **42.75** |
| Timber lap panels; fixed to slotted concrete posts 100 × 100 mm 1:3:6 concrete at 1.88 m centres | | | | | | | |
| 900 mm high | 10.54 | 0.38 | 17.81 | – | 32.82 | m | **50.63** |
| 1200 mm high | 10.77 | 0.40 | 19.00 | – | 33.06 | m | **52.06** |
| 1500 mm high | 12.35 | 0.42 | 20.19 | – | 36.64 | m | **56.83** |
| 1800 mm high | 12.46 | 0.50 | 23.75 | – | 38.80 | m | **62.55** |
| extra for corner posts | 41.68 | 0.45 | 21.38 | – | 49.69 | nr | **71.07** |
| extra over panel fencing for 300 mm high trellis tops; slats at 100 mm centres; including additional length of posts | 8.09 | 0.50 | 23.75 | – | 8.09 | m | **31.84** |
| Close boarded fencing; to concrete posts 100 × 100 mm; 2 nr softwood arris rails; 100 × 22 mm softwood pales lapped 13 mm; including excavating and backfilling into firm ground at 3.00 m centres; setting in concrete 1:3:6; concrete gravel board 150 × 150 mm | | | | | | | |
| 1050 mm high | 16.28 | 0.55 | 26.13 | – | 19.28 | m | **45.41** |
| 1500 mm high | 17.42 | 0.57 | 27.31 | – | 20.42 | m | **47.73** |
| Close boarded fencing; to concrete posts 100 × 100 mm; 3 nr softwood arris rails; 100 × 22 mm softwood pales lapped 13 mm; including excavating and backfilling into firm ground at 3.00 m centres; setting in concrete 1:3:6 | | | | | | | |
| 1650 mm high | 20.39 | 0.63 | 29.69 | – | 23.40 | m | **53.09** |
| 1800 mm high | 20.81 | 0.65 | 30.88 | – | 23.82 | m | **54.70** |

## 36 FENCING

| Item<br>Excluding site overheads and profit | PC<br>£ | Labour<br>hours | Labour<br>£ | Plant<br>£ | Material<br>£ | Unit | Total<br>rate £ |
|---|---|---|---|---|---|---|---|
| **TIMBER FENCING – cont** | | | | | | | |
| Close boarded fencing; to timber redwood posts 100 × 100 mm; 2 nr softwood arris rails; 100 × 22 mm softwood pales lapped 13 mm; including excavating and backfilling into firm ground at 3.00 m centres; setting in concrete 1:3:6 | | | | | | | |
| 1350 mm high | 11.82 | 0.50 | 23.75 | – | 14.82 | m | **38.57** |
| 1650 mm high | 12.66 | 0.63 | 29.69 | – | 15.66 | m | **45.35** |
| 1800 mm high | 13.49 | 0.65 | 30.88 | – | 16.50 | m | **47.38** |
| Close boarded fencing; to timber posts 100 × 100 mm; 3 nr softwood arris rails; 100 × 22 mm softwood pales lapped 13 mm; including excavating and backfilling into firm ground at 3.00 m centres; setting in concrete 1:3:6 | | | | | | | |
| 1350 mm high | 13.10 | 0.48 | 22.62 | – | 16.10 | m | **38.72** |
| 1650 mm high | 13.94 | 0.65 | 30.88 | – | 16.94 | m | **47.82** |
| 1800 mm high | 14.77 | 0.68 | 32.06 | – | 17.78 | m | **49.84** |
| extra over for post 125 × 100 mm; for 1800 mm high fencing | 14.75 | – | – | – | 14.75 | m | **14.75** |
| extra over to the above for counter rail | 0.70 | – | – | – | 0.70 | m | **0.70** |
| extra over to the above for capping rail | 0.70 | 0.05 | 2.38 | – | 0.70 | m | **3.08** |
| Feather edge board fencing; to timber posts 100 × 100 mm; 2 nr softwood cant rails; 125 × 22 mm feather edge boards lapped 13 mm; including excavating and backfilling into firm ground at 3.00 m centres; setting in concrete 1:3:6 | | | | | | | |
| 1350 mm high | 10.57 | 0.28 | 13.30 | – | 11.93 | m | **25.23** |
| 1650 mm high | 11.49 | 0.28 | 13.30 | – | 12.84 | m | **26.14** |
| 1800 mm high | 12.33 | 0.28 | 13.30 | – | 13.68 | m | **26.98** |
| Feather edge board fencing; to timber posts 100 × 100 mm; 3 nr softwood cant rails; 125 × 22 mm feather edge boards lapped 13 mm; including excavating and backfilling into firm ground at 3.00 m centres; setting in concrete 1:3:6 | | | | | | | |
| 1350 mm high | 11.70 | 0.29 | 13.57 | – | 13.06 | m | **26.63** |
| 1650 mm high | 12.62 | 0.29 | 13.57 | – | 13.97 | m | **27.54** |
| 1800 mm high | 12.24 | 0.29 | 13.57 | – | 14.81 | m | **28.38** |
| extra over for post 125 × 100 mm; for 1800 mm high fencing | 14.75 | – | – | – | 14.75 | nr | **14.75** |
| extra over to the above for counter rail and capping | 1.93 | 0.05 | 2.38 | – | 1.93 | m | **4.31** |

*Prices for Measured Works – Major Works*

## 36 FENCING

| Item<br>Excluding site overheads and profit | PC<br>£ | Labour<br>hours | Labour<br>£ | Plant<br>£ | Material<br>£ | Unit | **Total<br>rate £** |
|---|---|---|---|---|---|---|---|
| Palisade fencing; 22 × 75 mm softwood vertical palings with pointed tops; nailing to two 50 × 100 mm horizontal softwood arris rails; morticed to 100 × 100 mm softwood posts with weathered tops at 3.00 m centres; setting in concrete | | | | | | | |
| 900 mm high | 12.03 | 0.45 | 21.38 | – | 15.04 | m | **36.42** |
| 1200 mm high | 10.57 | 0.47 | 22.56 | – | 13.57 | m | **36.13** |
| extra over for rounded tops | 0.79 | – | – | – | 0.79 | m | **0.79** |
| Post-and-rail fencing; 90 × 38 mm softwood horizontal rails; fixing with galvanized nails to 150 × 75 mm softwood posts; including excavating and backfilling into firm ground at 1.80 m centres; all treated timber | | | | | | | |
| 1200 mm high; 3 horizontal rails | 10.03 | 0.18 | 8.32 | – | 10.25 | m | **18.57** |
| 1200 mm high; 4 horizontal rails | 11.72 | 0.18 | 8.32 | – | 11.72 | m | **20.04** |
| Cleft rail fencing; chestnut adze tapered rails 2.80 m long; morticed into joints; to 125 × 100 mm softwood posts 1.95 m long; including excavating and backfilling into firm ground at 2.80 m centres | | | | | | | |
| two rails | 9.93 | 0.13 | 5.94 | – | 9.93 | m | **15.87** |
| three rails | 12.80 | 0.14 | 6.65 | – | 12.80 | m | **19.45** |
| four rails | 15.67 | 0.18 | 8.32 | – | 15.67 | m | **23.99** |
| Hit and Miss horizontal rail fencing; 87 × 38 mm top and bottom rails; 100 × 22 mm vertical boards arranged alternately on opposite side of rails; to 100 × 100 mm posts; including excavating and backfilling into firm ground; setting in concrete at 1.8 m centres | | | | | | | |
| treated softwood; 1600 mm high | 26.69 | 0.60 | 28.50 | – | 29.40 | m | **57.90** |
| treated softwood; 1800 mm high | 29.15 | 0.67 | 31.66 | – | 32.13 | m | **63.79** |
| treated softwood; 2000 mm high | 31.63 | 0.70 | 33.25 | – | 34.60 | m | **67.85** |
| **Trellis tops to fencing** | | | | | | | |
| Extra over screen fencing for 300 mm high trellis tops; slats at 100 mm centres; including additional length of posts | 8.09 | 0.05 | 2.38 | – | 8.09 | m | **10.47** |

## 36 FENCING

| Excluding site overheads and profit | PC £ | Labour hours | Labour £ | Plant £ | Material £ | Unit | Total rate £ |
|---|---|---|---|---|---|---|---|
| **TRELLIS** | | | | | | | |
| **Traditional trellis panels; The Garden Trellis Company; bespoke trellis panels for decorative, screening or security applications; timber planed all round; height of trellis 1800 mm** | | | | | | | |
| Freestanding panels; posts 70 × 70 mm set in concrete; timber frames mitred and grooved 45 × 34 mm; heights and widths to suit; slats 32 × 10 mm joinery quality tanalized timber at 100 mm ccs; elements fixed by galvanized staples; capping rail 70 × 34 mm | | | | | | | |
| HV68 horizontal and vertical slats; softwood | 129.38 | 1.00 | 23.75 | – | 136.64 | m | **160.39** |
| D68 diagonal slats; softwood | 164.52 | 1.00 | 23.75 | – | 171.78 | m | **195.53** |
| HV68 horizontal and vertical slats; hardwood iroko | 222.73 | 1.00 | 23.75 | – | 229.99 | m | **253.74** |
| D68 diagonal slats; hardwood iroko or Western Red Cedar | 228.60 | 1.00 | 23.75 | – | 235.86 | m | **259.61** |
| Trellis panels fixed to face of existing wall or railings; timber frames mitred and grooved 45 × 34 mm; heights and widths to suit; slats 32 × 10 mm joinery quality tanalized timber at 100 mm ccs; elements fixed by galvanized staples; capping rail 70 × 34 mm | | | | | | | |
| HV68 horizontal and vertical slats; softwood | 114.98 | 0.50 | 11.88 | – | 116.78 | m | **128.66** |
| D68 diagonal slats; softwood | 148.32 | 0.50 | 11.88 | – | 150.12 | m | **162.00** |
| HV68 horizontal and vertical slats; iroko or Western Red Cedar | 208.33 | 0.50 | 11.88 | – | 210.13 | m | **222.01** |
| D68 diagonal slats; iroko or Western Red Cedar | 212.40 | 0.50 | 11.88 | – | 214.20 | m | **226.08** |
| **Contemporary style trellis panels; The Garden Trellis Company; bespoke trellis panels for decorative, screening or security applications; timber planed all round** | | | | | | | |
| Freestanding panels 30/15; posts 70 × 70 mm set in concrete with 90 × 30 mm top capping; slats 30 × 14 mm with 15 mm gaps; vertical support at 450 mm centres | | | | | | | |
| joinery treated softwood | 155.70 | 1.00 | 23.75 | – | 162.96 | m | **186.71** |
| hardwood iroko or Western Red Cedar | 261.90 | 1.00 | 23.75 | – | 269.16 | m | **292.91** |

## 36 FENCING

| Item<br>Excluding site overheads and profit | PC<br>£ | Labour<br>hours | Labour<br>£ | Plant<br>£ | Material<br>£ | Unit | Total<br>rate £ |
|---|---|---|---|---|---|---|---|
| Panels 30/15 face fixed to existing wall or fence; posts 70 × 70 mm set in concrete with 90 × 30 mm top capping; slats 30 × 14 mm with 15 mm gaps; vertical support at 450 mm centres | | | | | | | |
| joinery treated softwood | 133.20 | 0.50 | 11.88 | – | 135.00 | m | **146.88** |
| hardwood iroko or Western Red Cedar | 239.40 | 0.50 | 11.88 | – | 241.20 | m | **253.08** |
| **Integral arches to ornamental trellis panels; The Garden Trellis Company** | | | | | | | |
| Arches to trellis panels in 45 × 34 mm grooved timbers to match framing; fixed to posts | | | | | | | |
| R450 ¼ circle; joinery treated softwood | 45.00 | 1.50 | 35.63 | – | 52.26 | nr | **87.89** |
| R450 ¼ circle; hardwood iroko or Western Red Cedar | 60.00 | 1.50 | 35.63 | – | 67.26 | nr | **102.89** |
| Arches to span 1800 mm wide | | | | | | | |
| joinery treated softwood | 165.83 | 1.50 | 35.63 | – | 173.09 | nr | **208.72** |
| hardwood iroko or Western Red Cedar | 141.66 | 1.50 | 35.63 | – | 148.92 | nr | **184.55** |
| Painting or staining of trellis panels; high quality coatings | | | | | | | |
| microporous opaque paint or spirit based stain | – | – | – | – | 30.00 | m$^2$ | **30.00** |
| **STRAINED WIRE FENCING/ FIELDS FENCING** | | | | | | | |
| **Posts for strained wire and wire mesh; AVS Fencing Supplies Ltd** | | | | | | | |
| Strained wire fencing; concrete inter posts only at 2750 mm centres; 610 mm below ground; excavating holes; filling with concrete; replacing topsoil; disposing surplus soil off site | | | | | | | |
| 900 mm high | 4.92 | 0.28 | 13.20 | 0.64 | 6.48 | m | **20.32** |
| 1200 mm high | 4.94 | 0.28 | 13.20 | 0.64 | 6.51 | m | **20.35** |
| 1800 mm high | 5.74 | 0.39 | 18.47 | 1.59 | 7.30 | m | **27.36** |
| Extra over strained wire fencing for concrete straining posts with one strut; posts and struts 610 mm below ground; struts, cleats, stretchers, winders, bolts and eye bolts; excavating holes; filling to within 150 mm of ground level with concrete (1:12) – 40 mm aggregate; replacing topsoil; disposing surplus soil off site | | | | | | | |
| 900 mm high | 32.90 | 0.34 | 15.91 | 1.43 | 121.33 | nr | **138.67** |
| 1200 mm high | 38.92 | 0.33 | 15.84 | 1.43 | 157.89 | nr | **175.16** |
| 1800 mm high | 66.37 | 0.33 | 15.84 | 1.43 | 187.27 | nr | **204.54** |

## 36 FENCING

| Item<br>Excluding site overheads and profit | PC<br>£ | Labour<br>hours | Labour<br>£ | Plant<br>£ | Material<br>£ | Unit | Total<br>rate £ |
|---|---|---|---|---|---|---|---|
| **STRAINED WIRE FENCING/ FIELDS FENCING – cont** | | | | | | | |
| **Posts for strained wire and wire mesh – cont** | | | | | | | |
| Extra over strained wire fencing for concrete straining posts with two struts; posts and struts 610 mm below ground; excavating holes; filling to within 150 mm of ground level with concrete (1:12) – 40 mm aggregate; replacing topsoil; disposing surplus soil off site | | | | | | | |
| 900 mm high | 51.50 | 0.46 | 21.61 | 1.83 | 213.60 | nr | **237.04** |
| 1200 mm high | 59.84 | 0.42 | 20.19 | 1.43 | 281.42 | nr | **303.04** |
| 1800 mm high | 96.14 | 0.42 | 20.19 | 1.43 | 325.52 | nr | **347.14** |
| Strained wire fencing; galvanized steel angle posts only at 2750 mm centres; 610 mm below ground; driving in | | | | | | | |
| 900 mm high; 40 × 40 × 5 mm | 2.64 | 0.02 | 0.72 | – | 2.64 | m | **3.36** |
| 1200 mm high; 40 × 40 × 5 mm | 3.84 | 0.02 | 0.79 | – | 3.84 | m | **4.63** |
| 1500 mm high; 40 × 40 × 5 mm | 5.11 | 0.03 | 1.44 | – | 5.11 | m | **6.55** |
| 1800 mm high; 40 × 40 × 5 mm | 5.50 | 0.04 | 1.73 | – | 5.50 | m | **7.23** |
| Galvanized steel straining posts with two struts for strained wire fencing; setting in concrete | | | | | | | |
| 900 mm high; 50 × 50 × 6 mm | 61.62 | 0.50 | 23.75 | – | 64.68 | nr | **88.43** |
| 1200 mm high; 50 × 50 × 6 mm | 74.41 | 0.50 | 23.75 | – | 77.47 | nr | **101.22** |
| 1500 mm high; 50 × 50 × 6 mm | 92.97 | 0.50 | 23.75 | – | 96.03 | nr | **119.78** |
| 1800 mm high; 50 × 50 × 6 mm | 96.34 | 0.50 | 23.75 | – | 99.40 | nr | **123.15** |
| **Fixing of fencing to posts** | | | | | | | |
| Strained wire; to posts (posts not included); 3 mm galvanized wire; fixing with galvanized stirrups | | | | | | | |
| 900 mm high; 2 wire | 0.65 | 0.02 | 0.79 | – | 0.75 | m | **1.54** |
| 1200 mm high; 3 wire | 0.97 | 0.02 | 1.11 | – | 1.07 | m | **2.18** |
| 1400 mm high; 3 wire | 0.97 | 0.02 | 1.11 | – | 1.07 | m | **2.18** |
| 1800 mm high; 3 wire | 0.97 | 0.02 | 1.11 | – | 1.07 | m | **2.18** |
| Barbed wire; to posts (posts not included); 3 mm galvanized wire; fixing with galvanized stirrups | | | | | | | |
| 900 mm high; 2 wire | 0.30 | 0.03 | 1.58 | – | 0.40 | m | **1.98** |
| 1200 mm high; 3 wire | 0.44 | 0.05 | 2.21 | – | 0.54 | m | **2.75** |
| 1400 mm high; 3 wire | 0.44 | 0.05 | 2.21 | – | 0.54 | m | **2.75** |
| 1800 mm high; 3 wire | 0.44 | 0.05 | 2.21 | – | 0.54 | m | **2.75** |

## 36 FENCING

| Item<br>Excluding site overheads and profit | PC<br>£ | Labour<br>hours | Labour<br>£ | Plant<br>£ | Material<br>£ | Unit | Total<br>rate £ |
|---|---|---|---|---|---|---|---|
| **Fencing only to posts and strained wire** | | | | | | | |
| Chain link fencing; AVS Fencing Supplies Ltd; to strained wire and posts priced separately; 3 mm galvanized wire; 51 mm mesh; galvanized steel components; fixing to line wires threaded through posts and strained with eye-bolts; posts (not included) | | | | | | | |
| 1800 mm high | 7.19 | 0.05 | 2.38 | – | 7.29 | m | **9.67** |
| Chain link fencing; to strained wire and posts priced separately; 3.15 mm plastic coated galvanized wire (wire only 2.5 mm); 51 mm mesh; galvanized steel components; fencing with line wires threaded through posts and strained with eye-bolts; posts (not included) (Note: plastic coated fencing can be cheaper than galvanized finish as wire of a smaller cross-sectional area can be used) | | | | | | | |
| 900 mm high | 2.51 | 0.03 | 1.59 | – | 2.71 | m | **4.30** |
| 1200 mm high | 3.39 | 0.03 | 1.59 | – | 3.59 | m | **5.18** |
| 1800 mm high | 4.92 | 0.06 | 2.97 | – | 5.72 | m | **8.69** |
| Extra over strained wire fencing for cranked arms and galvanized barbed wire | | | | | | | |
| 1 row | 0.56 | 0.01 | 0.40 | – | 0.56 | m | **0.96** |
| 2 row | 0.71 | 0.03 | 1.19 | – | 0.71 | m | **1.90** |
| 3 row | 0.86 | 0.03 | 1.19 | – | 0.86 | m | **2.05** |
| Field fencing; AVS Fencing Supplies Ltd; woven wire mesh; fixed to posts and straining wires measured separately | | | | | | | |
| cattle fence; 1200 m high; 114 × 300 mm at bottom to 230 × 300 mm at top | 0.92 | 0.05 | 2.38 | – | 1.12 | m | **3.50** |
| sheep fence; 800 mm high; 140 × 300 mm at bottom to 230 × 300 mm at top | 0.74 | 0.05 | 2.38 | – | 0.94 | m | **3.32** |
| deer fence; 2.0 m high; 89 × 150 mm at bottom to 267 × 300 mm at top | 1.88 | 0.06 | 2.97 | – | 2.08 | m | **5.05** |
| Extra for concreting in posts | – | 0.25 | 11.88 | – | 4.81 | nr | **16.69** |
| Extra for straining post | 14.32 | 0.38 | 17.81 | – | 14.32 | nr | **32.13** |

## 36 FENCING

| Item<br>Excluding site overheads and profit | PC<br>£ | Labour<br>hours | Labour<br>£ | Plant<br>£ | Material<br>£ | Unit | Total<br>rate £ |
|---|---|---|---|---|---|---|---|
| **STRAINED WIRE FENCING/<br>FIELDS FENCING – cont** | | | | | | | |
| **Boundary fencing; strained wire and wire mesh; Jacksons Fencing**<br>Tubular chain link fencing; galvanized; plastic coated; 60.3 mm diameter posts at 3.0 m centres; setting 700 mm into ground; choice of 10 mesh colours; including excavating holes, backfilling and removing surplus soil; with top rail only | | | | | | | |
| 900 mm high | – | – | – | – | – | m | **37.04** |
| 1200 mm high | – | – | – | – | – | m | **39.35** |
| 1800 mm high | – | – | – | – | – | m | **43.85** |
| 2000 mm high | – | – | – | – | – | m | **46.31** |
| Tubular chain link fencing; galvanized; plastic coated; 60.3 mm diameter posts at 3.0 m centres; cranked arms and three lines barbed wire; setting 700 mm into ground; including excavating holes; backfilling and removing surplus soil; with top rail only | | | | | | | |
| 2000 mm high | – | – | – | – | – | m | **46.95** |
| 1800 mm high | – | – | – | – | – | m | **45.62** |
| **RABBIT NETTING** | | | | | | | |
| **Rabbit netting; AVS Fencing Supplies Ltd**<br>Timber stakes; peeled kiln dried pressure treated; pointed; 1.8 m posts driven 750 mm into ground at 3 m centres (line wires and netting priced separately) | | | | | | | |
| 75–100 mm stakes | 4.28 | 0.13 | 5.94 | – | 4.28 | m | **10.22** |
| Corner posts or straining posts 150 mm diameter × 2.4 m high set in concrete; centres to suit local conditions or changes of direction | | | | | | | |
| 1 strut | 17.37 | 0.50 | 23.75 | 4.08 | 24.79 | nr | **52.62** |
| 2 strut | 21.03 | 0.50 | 23.75 | 4.08 | 28.45 | nr | **56.28** |
| Strained wire; to posts (posts not included); 3 mm galvanized wire; fixing with galvanized stirrups | | | | | | | |
| 900 mm high; 2 wire | 0.65 | 0.02 | 0.79 | – | 0.75 | m | **1.54** |
| 1200 mm high; 3 wire | 0.97 | 0.02 | 1.11 | – | 1.07 | m | **2.18** |

## 36 FENCING

| Item<br>Excluding site overheads and profit | PC<br>£ | Labour<br>hours | Labour<br>£ | Plant<br>£ | Material<br>£ | Unit | Total<br>rate £ |
|---|---|---|---|---|---|---|---|
| Rabbit netting; 31 mm 19 gauge 1050 mm high netting fixed to posts line wires and straining posts or corner posts all priced separately | | | | | | | |
|   900 mm high; turned in | 0.89 | 0.02 | 0.95 | – | 0.89 | m | 1.84 |
|   900 mm high; buried 150 mm in trench | 0.89 | 0.04 | 1.98 | – | 0.89 | m | 2.87 |
| **TUBULAR FENCING** | | | | | | | |
| **Boundary fencing;** | | | | | | | |
| **Steelway-Fensecure Ltd** | | | | | | | |
| Classic two rail tubular fencing; top and bottom with stretcher bars and straining wires in between; comprising 60.3 mm tubular posts at 3.00 m centres; setting in concrete; 35 mm top rail tied with aluminium and steel fittings; 50 × 50 × 355/2.5 mm PVC coated chain link; all components galvanized and coated in green nylon | | | | | | | |
|   964 mm high | – | – | – | – | – | m | 39.14 |
|   1269 mm high | – | – | – | – | – | m | 43.09 |
|   1574 mm high | – | – | – | – | – | m | 48.99 |
|   1878 mm high | – | – | – | – | – | m | 51.44 |
|   2188 mm high | – | – | – | – | – | m | 54.83 |
|   2458 mm high | – | – | – | – | – | m | 61.32 |
|   2948 mm high | – | – | – | – | – | m | 77.71 |
|   3562 mm high | – | – | – | – | – | m | 77.51 |
| End posts; Classic range; 60.3 mm diameter; setting in concrete | | | | | | | |
|   964 mm high | – | – | – | – | – | nr | 57.77 |
|   1269 mm high | – | – | – | – | – | nr | 57.77 |
|   1574 mm high | – | – | – | – | – | nr | 64.33 |
|   1878 mm high | – | – | – | – | – | nr | 70.28 |
|   2188 mm high | – | – | – | – | – | nr | 76.43 |
|   2458 mm high | – | – | – | – | – | nr | 87.83 |
|   2948 mm high | – | – | – | – | – | nr | 100.42 |
|   3562 mm high | – | – | – | – | – | nr | 114.82 |
| Corner posts; 60.3 mm diameter; setting in concrete | | | | | | | |
|   964 mm high | – | – | – | – | – | nr | 43.93 |
|   1269 mm high | – | – | – | – | – | nr | 26.33 |
|   1574 mm high | – | – | – | – | – | nr | 51.39 |
|   1878 mm high | – | – | – | – | – | nr | 53.84 |
|   2188 mm high | – | – | – | – | – | nr | 62.62 |
|   2458 mm high | – | – | – | – | – | nr | 66.71 |
|   2948 mm high | – | – | – | – | – | nr | 80.11 |
|   3562 mm high | – | – | – | – | – | nr | 85.29 |

## 36 FENCING

| Item<br>Excluding site overheads and profit | PC<br>£ | Labour<br>hours | Labour<br>£ | Plant<br>£ | Material<br>£ | Unit | Total<br>rate £ |
|---|---|---|---|---|---|---|---|
| **WELDED MESH FENCING** | | | | | | | |
| **Boundary fencing** | | | | | | | |
| Paladin welded mesh colour coated green fencing; fixing to metal posts at 2.975 m centres with manufacturer's fixings; setting 600 mm deep in firm ground; including excavating holes; backfilling and removing surplus excavated material; includes 10 year product guarantee | | | | | | | |
| 1800 mm high | 45.39 | 0.33 | 15.68 | – | 45.39 | m | **61.07** |
| 2000 mm high | 42.05 | 0.33 | 15.68 | – | 50.15 | m | **65.83** |
| 2400 mm high | 48.03 | 0.33 | 15.68 | – | 57.47 | m | **73.15** |
| Extra over welded galvanized plastic coated mesh fencing for concreting in posts | 9.62 | 0.06 | 2.64 | – | 9.62 | m | **12.26** |
| **Boundary fencing; Lang & Fulton** | | | | | | | |
| Orsogril rectangular steel bar mesh fence panels; pleione pattern; bolting to 60 × 8 mm uprights at 2 m centres; mesh 62 × 66 mm; setting in concrete | | | | | | | |
| 930 mm high panels | 49.00 | 0.08 | 5.94 | – | 63.83 | m | **69.77** |
| 1326 mm high panels | 67.00 | 0.10 | 7.13 | – | 81.83 | m | **88.96** |
| 1722 mm high panels | 83.00 | 0.17 | 11.87 | – | 97.83 | m | **109.70** |
| **SECURITY FENCING** | | | | | | | |
| **Security fencing; Jacksons Fencing** | | | | | | | |
| Barbican galvanized steel paling fencing; on 60 × 60 mm posts at 3 m centres; setting in concrete | | | | | | | |
| 1250 mm high | – | – | – | – | – | m | **43.34** |
| 1500 mm high | – | – | – | – | – | m | **48.11** |
| 2000 mm high | – | – | – | – | – | m | **58.13** |
| 2500 mm high | – | – | – | – | – | m | **74.04** |
| Gates; to match Barbican galvanized steel paling fencing | | | | | | | |
| width 1 m | – | – | – | – | – | nr | **499.40** |
| width 2 m | – | – | – | – | – | nr | **520.30** |
| width 3 m | – | – | – | – | – | nr | **555.50** |
| width 4 m | – | – | – | – | – | nr | **592.90** |
| width 8 m | – | – | – | – | – | pair | **1263.21** |
| width 9 m | – | – | – | – | – | pair | **1676.40** |
| width 10 m | – | – | – | – | – | pair | **1997.61** |

## 36 FENCING

| Item<br>Excluding site overheads and profit | PC<br>£ | Labour<br>hours | Labour<br>£ | Plant<br>£ | Material<br>£ | Unit | Total<br>rate £ |
|---|---|---|---|---|---|---|---|
| **Security fencing; Jacksons Fencing; intruder guards** | | | | | | | |
| Viper Spike Intruder Guards; to existing structures; including fixing bolts | | | | | | | |
| Viper 1; 40 × 5 mm × 1.1 m long; with base plate | 40.38 | 1.00 | 47.50 | – | 40.38 | nr | **87.88** |
| Viper 3; 160 × 190 mm wide; U shape; to prevent intruders climbing pipes | 43.15 | 0.50 | 23.75 | – | 43.15 | nr | **66.90** |
| **Security fencing; Jacksons Fencing** | | | | | | | |
| Razor Barb Concertina; spiral wire security barriers; fixing to 600 mm steel ground stakes | | | | | | | |
| Ref 3275; 450 mm diameter roll; medium barb; galvanized | 3.14 | 0.05 | 2.38 | – | 3.36 | m | **5.74** |
| Ref 3276; 730 mm diameter roll; medium barb; galvanized | 4.68 | 0.05 | 2.38 | – | 4.90 | m | **7.28** |
| Ref 3277; 950 mm diameter roll; medium barb; galvanized | 5.72 | 0.05 | 2.38 | – | 5.94 | m | **8.32** |
| Three lines Barbed Tape; medium barb; on 50 × 50 mm mild steel angle posts; setting in concrete | | | | | | | |
| Ref 3283; barbed tape; medium barb; galvanized | 1.25 | 0.21 | 10.00 | – | 13.39 | m | **23.39** |
| Five lines Barbed Tape; medium barb; on 50 × 50 mm mild steel angle posts; setting in concrete | | | | | | | |
| Ref 3283; barbed tape; medium barb; galvanized | 2.08 | 0.17 | 7.92 | – | 14.22 | m | **22.14** |
| **TRIP RAILS** | | | | | | | |
| **Trip rails; Birdsmouth; Jacksons Fencing** | | | | | | | |
| Diamond rail fencing; Birdsmouth; posts 100 × 100 mm softwood planed at 1.35 m centres set in 1:3:6 concrete; rail 75 × 75 mm nominal secured with galvanized straps nailed to posts | | | | | | | |
| posts 900 mm (600 mm above ground); at 1.35 m centres | 10.74 | 0.25 | 11.88 | 0.27 | 12.74 | m | **24.89** |
| posts 1.20 m (900 mm above ground); at 1.35 m centres | 12.44 | 0.25 | 11.88 | 0.27 | 14.45 | m | **26.60** |
| posts 900 mm (600 mm above ground); at 1.80 m centres | 9.00 | 0.19 | 8.93 | 0.20 | 10.52 | m | **19.65** |
| posts 1.20 m (900 mm above ground); at 1.80 m centres | 10.28 | 0.19 | 8.93 | 0.20 | 11.78 | m | **20.91** |

## 36 FENCING

| Item<br>Excluding site overheads and profit | PC<br>£ | Labour<br>hours | Labour<br>£ | Plant<br>£ | Material<br>£ | Unit | Total<br>rate £ |
|---|---|---|---|---|---|---|---|
| **TRIP RAILS – cont** | | | | | | | |
| **Trip rails; Townscape Products Ltd**<br>Hollow steel section knee rails; galvanized; 500 mm high; setting in concrete | | | | | | | |
| 1000 mm bays | 116.85 | 1.00 | 47.50 | – | 126.19 | m | **173.69** |
| 1200 mm bays | 24.62 | 1.00 | 47.50 | – | 109.73 | m | **157.23** |
| **WINDBREAK FENCING** | | | | | | | |
| **Windbreak fencing**<br>Fencing; English Woodlands; Shade and Shelter Netting windbreak fencing; green; to 100 mm diameter treated softwood posts; setting 450 mm into ground; fixing with 50 × 25 mm treated softwood battens nailed to posts; including excavating and backfilling into firm ground; setting in concrete at 3 m centres | | | | | | | |
| 1200 mm high | 1.25 | 0.08 | 3.70 | – | 6.04 | m | **9.74** |
| 1800 mm high | 1.87 | 0.08 | 3.70 | – | 6.66 | m | **10.36** |
| **BALLSTOP FENCING** | | | | | | | |
| **Ball stop fencing;**<br>**Steelway-Fensecure Ltd**<br>Ball stop net; 30 × 30 mm netting fixed to 60.3 mm diameter 12 mm solid bar lattice; galvanized; dual posts; top, middle and bottom rails | | | | | | | |
| 4.5 m high | – | – | – | – | – | m | **108.76** |
| 5.0 m high | – | – | – | – | – | m | **137.31** |
| 6.0 m high | – | – | – | – | – | m | **152.99** |
| 7.0 m high | – | – | – | – | – | m | **168.69** |
| 8.0 m high | – | – | – | – | – | m | **184.34** |
| 9.0 m high | – | – | – | – | – | m | **200.04** |
| 10.0 m high | – | – | – | – | – | m | **182.77** |
| Ball stop net; corner posts | | | | | | | |
| 4.5 m high | – | – | – | – | – | m | **74.63** |
| 5.0 m high | – | – | – | – | – | m | **81.68** |
| 6.0 m high | – | – | – | – | – | m | **126.89** |
| 7.0 m high | – | – | – | – | – | m | **113.04** |
| 8.0 m high | – | – | – | – | – | m | **128.73** |
| 9.0 m high | – | – | – | – | – | m | **133.80** |
| 10 m high | – | – | – | – | – | m | **160.09** |

## 36 FENCING

| Item<br>Excluding site overheads and profit | PC<br>£ | Labour<br>hours | Labour<br>£ | Plant<br>£ | Material<br>£ | Unit | Total<br>rate £ |
|---|---|---|---|---|---|---|---|
| Ball stop net; end posts | | | | | | | |
|   4.5 m high | – | – | – | – | – | m | 62.08 |
|   5.0 m high | – | – | – | – | – | m | 69.14 |
|   6.0 m high | – | – | – | – | – | m | 84.82 |
|   7.0 m high | – | – | – | – | – | m | 100.52 |
|   8.0 m high | – | – | – | – | – | m | 116.17 |
|   9.0 m high | – | – | – | – | – | m | 131.88 |
|   10 m high | – | – | – | – | – | m | 133.80 |
| **RAILINGS** | | | | | | | |
| **Railings; Steelway-Fensecure Ltd** | | | | | | | |
| Mild steel bar railings of balusters at 115 mm centres welded to flat rail top and bottom; bays 2.00 m long; bolting to 51 × 51 mm hollow square section posts; setting in concrete | | | | | | | |
|   galvanized; 900 mm high | – | – | – | – | – | m | 82.28 |
|   galvanized; 1200 mm high | – | – | – | – | – | m | 99.66 |
|   galvanized; 1500 mm high | – | – | – | – | – | m | 110.76 |
|   galvanized; 1800 mm high | – | – | – | – | – | m | 132.87 |
|   primed; 900 mm high | – | – | – | – | – | m | 77.55 |
|   primed; 1200 mm high | – | – | – | – | – | m | 94.16 |
|   primed; 1500 mm high | – | – | – | – | – | m | 104.09 |
|   primed; 1800 mm high | – | – | – | – | – | m | 124.06 |
| Mild steel blunt top railings of 19 mm balusters at 130 mm centres welded to bottom rail; passing through holes in top rail and welded; top and bottom rails 40 × 10 mm; bolting to 51 × 51 mm hollow square section posts; setting in concrete | | | | | | | |
|   galvanized; 800 mm high | – | – | – | – | – | m | 88.59 |
|   galvanized; 1000 mm high | – | – | – | – | – | m | 93.04 |
|   galvanized; 1300 mm high | – | – | – | – | – | m | 110.76 |
|   galvanized; 1500 mm high | – | – | – | – | – | m | 119.61 |
|   primed; 800 mm high | – | – | – | – | – | m | 85.28 |
|   primed; 1000 mm high | – | – | – | – | – | m | 88.59 |
|   primed; 1300 mm high | – | – | – | – | – | m | 105.22 |
|   primed; 1500 mm high | – | – | – | – | – | m | 112.99 |
| Railings; traditional pattern; 16 mm diameter verticals at 127 mm intervals with horizontal bars near top and bottom; balusters with spiked tops; 51 × 20 mm standards; including setting 520 mm into concrete at 2.75 m centres | | | | | | | |
|   primed; 1200 mm high | – | – | – | – | – | m | 99.66 |
|   primed; 1500 mm high | – | – | – | – | – | m | 112.99 |
|   primed; 1800 mm high | – | – | – | – | – | m | 121.83 |

## 36 FENCING

| Item<br>Excluding site overheads and profit | PC<br>£ | Labour<br>hours | Labour<br>£ | Plant<br>£ | Material<br>£ | Unit | Total<br>rate £ |
|---|---|---|---|---|---|---|---|
| **RAILINGS – cont** | | | | | | | |
| **Railings – cont** | | | | | | | |
| Interlaced bow-top mild steel railings; traditional park type; 16 mm diameter verticals at 80 mm intervals; welded at bottom to 50 × 10 mm flat and slotted through 38 × 8 mm top rail to form hooped top profile; 50 × 10 mm standards; setting 560 mm into concrete at 2.75 m centres | | | | | | | |
| galvanized; 900 mm high | – | – | – | – | – | m | 150.63 |
| galvanized; 1200 mm high | – | – | – | – | – | m | 183.87 |
| galvanized; 1500 mm high | – | – | – | – | – | m | 219.30 |
| galvanized; 1800 mm high | – | – | – | – | – | m | 254.74 |
| primed; 900 mm high | – | – | – | – | – | m | 146.19 |
| primed; 1200 mm high | – | – | – | – | – | m | 177.21 |
| primed; 1500 mm high | – | – | – | – | – | m | 210.45 |
| primed; 1800 mm high | – | – | – | – | – | m | 243.65 |
| **ESTATE FENCING/HORIZONTAL BAR FENCING** | | | | | | | |
| **Metal estate and parkland fencing; Stonebank Ironcraft Ltd; continously welded, mild steel, 40 × 10 mm uprights manually driven into normal soil to a minimum depth of 600 mm at 1.00 m centres; 40 × 10 mm flat steel supports welded to uprights beneath topsoil at typically every third upright; horizontal rails threaded through holes within uprights; 20 mm solid steel top rail; 4 nr 25 × 8 mm × 16 mm diameter solid steel lower rails; painted fencing shot-blasted, primed and finished with two coats of paint** | | | | | | | |
| Bare metal; 1200 mm high | | | | | | | |
| 10– 50 m | – | – | – | – | – | m | 65.33 |
| 51–100 m | – | – | – | – | – | m | 55.17 |
| 10–500 m | – | – | – | – | – | m | 51.51 |
| above 500 m | – | – | – | – | – | m | 49.50 |
| Primed and painted | | | | | | | |
| 10–50 m | – | – | – | – | – | m | 78.04 |
| 51–100 m | – | – | – | – | – | m | 66.08 |
| 101–500 m | – | – | – | – | – | m | 61.77 |
| above 500 m | – | – | – | – | – | m | 59.41 |

## 36 FENCING

| Item<br>Excluding site overheads and profit | PC<br>£ | Labour<br>hours | Labour<br>£ | Plant<br>£ | Material<br>£ | Unit | Total<br>rate £ |
|---|---|---|---|---|---|---|---|
| Galvanized and painted | | | | | | | |
| 10–50 m | – | – | – | – | – | m | 105.35 |
| 51–100 m | – | – | – | – | – | m | 89.21 |
| 101–500 m | – | – | – | – | – | m | 83.39 |
| above 500 m | – | – | – | – | – | m | 80.20 |
| End posts and corner posts; 40 × 40 mm; standard hollow section; end or corner posts with capped tops set into concrete | | | | | | | |
| bare metal | – | – | – | – | – | nr | 65.00 |
| primed and painted | – | – | – | – | – | nr | 85.00 |
| galvanized and painted | – | – | – | – | – | nr | 100.00 |
| **Gates for metal estate fencing; Stonebank Ironcraft Ltd; fully welded gates to match fence; scroll at hinge end; turnover at latch end; posts topped with finials and set into concrete** | | | | | | | |
| Pedestrian gates; 1.0–1.5 m wide | | | | | | | |
| bare metal | – | – | – | – | – | nr | 450.00 |
| galvanized and painted | – | – | – | – | – | nr | 600.00 |
| Field gates; 1.50–4.0 m wide | | | | | | | |
| bare metal | – | – | – | – | – | nr | 600.00 |
| galvanized and painted | – | – | – | – | – | nr | 800.00 |
| Kissing gates | | | | | | | |
| bare metal | – | – | – | – | – | nr | 550.00 |
| galvanized and painted | – | – | – | – | – | nr | 745.00 |
| Quadrant; 4 nr uprights with rails welded to fence | | | | | | | |
| bare metal | – | – | – | – | – | nr | 230.00 |
| primed and painted | – | – | – | – | – | nr | 300.00 |
| Step stile; 2 treads | | | | | | | |
| bare metal | – | – | – | – | – | nr | 140.00 |
| primed and painted | – | – | – | – | – | nr | 200.00 |

### GATES, KISSING GATES AND STILES

### Gates – General

Preamble: Gates in fences; see specification for fencing as gates in traditional or proprietary fencing systems are usually constructed of the same materials and finished as the fencing itself.

## 36 FENCING

| Item<br>Excluding site overheads and profit | PC<br>£ | Labour<br>hours | Labour<br>£ | Plant<br>£ | Material<br>£ | Unit | Total<br>rate £ |
|---|---|---|---|---|---|---|---|
| **GATES, KISSING GATES AND STILES** – cont | | | | | | | |
| **Gates; hardwood; AVS Fencing Supplies Ltd** | | | | | | | |
| Hardwood entrance gate; five bar diamond braced; curved hanging stile; planed iroko; fixed to 150 × 150 mm softwood posts; inclusive of hinges and furniture | | | | | | | |
| Ref 1100 040; 0.9 m wide | 205.00 | 2.50 | 118.75 | – | 274.40 | nr | **393.15** |
| Ref 1100 041; 1.2 m wide | 238.00 | 2.50 | 118.75 | – | 307.40 | nr | **426.15** |
| Ref 1100 042; 1.5 m wide | 277.00 | 2.50 | 118.75 | – | 346.40 | nr | **465.15** |
| Ref 1100 043; 1.8 m wide | 300.00 | 2.50 | 118.75 | – | 369.40 | nr | **488.15** |
| Ref 1100 044; 2.1 m wide | 323.00 | 2.50 | 118.75 | – | 392.40 | nr | **511.15** |
| Ref 1100 045; 2.4 m wide | 345.00 | 2.50 | 118.75 | – | 414.40 | nr | **533.15** |
| Ref 1100 047; 3.0 m wide | 365.00 | 2.50 | 118.75 | – | 434.40 | nr | **553.15** |
| Ref 1100 048; 3.3 m wide | 378.00 | 2.50 | 118.75 | – | 447.40 | nr | **566.15** |
| Ref 1100 049; 3.6 m wide | 395.00 | 2.50 | 118.75 | – | 464.40 | nr | **583.15** |
| Hardwood field gate; five bar diamond braced; planed iroko; fixed to 150 × 150 mm softwood posts; inclusive of hinges and furniture | | | | | | | |
| Ref 1100 100; 0.9 m wide | 150.00 | 2.00 | 95.00 | – | 219.40 | nr | **314.40** |
| Ref 1100 101; 1.2 m wide | 199.00 | 2.00 | 95.00 | – | 268.40 | nr | **363.40** |
| Ref 1100 102; 1.5 m wide | 228.00 | 2.00 | 95.00 | – | 297.40 | nr | **392.40** |
| Ref 1100 103; 1.8 m wide | 248.00 | 2.00 | 95.00 | – | 317.40 | nr | **412.40** |
| Ref 1100 104; 2.1 m wide | 273.00 | 2.00 | 95.00 | – | 342.40 | nr | **437.40** |
| Ref 1100 105; 2.4 m wide | 306.00 | 2.00 | 95.00 | – | 375.40 | nr | **470.40** |
| Ref 1100 106; 2.7 m wide | 307.00 | 2.00 | 95.00 | – | 376.40 | nr | **471.40** |
| Ref 1100 108; 3.3 m wide | 335.00 | 2.00 | 95.00 | – | 404.40 | nr | **499.40** |
| **Gates; treated softwood; Jacksons Fencing** | | | | | | | |
| Timber field gates; including wrought iron ironmongery; five bar type; diamond braced; 1.80 m high; to 200 × 200 mm posts; setting 750 mm into firm ground | | | | | | | |
| width 2400 mm | 117.39 | 3.00 | 142.50 | – | 267.01 | nr | **409.51** |
| width 2700 mm | 126.58 | 3.00 | 142.50 | – | 276.20 | nr | **418.70** |
| width 3000 mm | 135.71 | 3.50 | 166.25 | – | 285.33 | nr | **451.58** |
| width 3300 mm | 145.48 | 3.75 | 178.13 | – | 295.10 | nr | **473.23** |
| Featherboard garden gates; including ironmongery; to 100 × 120 mm posts; one diagonal brace | | | | | | | |
| 1.0 × 1.2 m high | 33.10 | 1.50 | 71.25 | – | 144.66 | nr | **215.91** |
| 1.0 × 1.5 m high | 42.64 | 1.50 | 71.25 | – | 163.70 | nr | **234.95** |
| 1.0 × 1.8 m high | 121.22 | 1.50 | 71.25 | – | 178.39 | nr | **249.64** |

## 36 FENCING

| Item<br>Excluding site overheads and profit | PC<br>£ | Labour<br>hours | Labour<br>£ | Plant<br>£ | Material<br>£ | Unit | Total<br>rate £ |
|---|---|---|---|---|---|---|---|
| **Picket garden gates; including ironmongery; to match picket fence; width 1000 mm; to 100 × 120 mm posts; one diagonal brace** | | | | | | | |
| 950 mm high | 79.58 | 1.50 | 71.25 | – | 122.07 | nr | 193.32 |
| 1200 mm high | 83.88 | 1.50 | 71.25 | – | 126.37 | nr | 197.62 |
| 1800 mm high | 98.43 | 1.50 | 71.25 | – | 155.60 | nr | 226.85 |
| **Gates; tubular mild steel; Jacksons Fencing** | | | | | | | |
| Field gates; galvanized; including ironmongery; diamond braced; 1.80 m high; to tubular steel posts; setting in concrete | | | | | | | |
| width 3000 mm | 133.34 | 2.50 | 118.75 | – | 248.65 | nr | 367.40 |
| width 3300 mm | 145.62 | 2.50 | 118.75 | – | 260.94 | nr | 379.69 |
| width 3600 mm | 149.58 | 2.50 | 118.75 | – | 264.90 | nr | 383.65 |
| width 4200 mm | 165.96 | 2.50 | 118.75 | – | 281.28 | nr | 400.03 |
| **Gates; sliding; Jacksons Fencing** | | | | | | | |
| Sliding Gate; including all galvanized rails and vertical rail infill panels; special guide and shutting frame posts (Note: foundations installed by suppliers) | | | | | | | |
| access width 4.00 m; 1.5 m high gates | – | – | – | – | – | nr | 3434.00 |
| access width 4.00 m; 2.0 m high gates | – | – | – | – | – | nr | 3554.00 |
| access width 4.00 m; 2.5 m high gates | – | – | – | – | – | nr | 3674.00 |
| access width 6.00 m; 1.5 m high gates | – | – | – | – | – | nr | 4207.00 |
| access width 6.00 m; 2.0 m high gates | – | – | – | – | – | nr | 4362.00 |
| access width 6.00 m; 2.5 m high gates | – | – | – | – | – | nr | 4518.00 |
| access width 8.00 m; 1.5 m high gates | – | – | – | – | – | nr | 4928.00 |
| access width 8.00 m; 2.0 m high gates | – | – | – | – | – | nr | 5121.01 |
| access width 8.00 m; 2.5 m high gates | – | – | – | – | – | nr | 5313.01 |
| access width 10.00 m; 2.0 m high gates | – | – | – | – | – | nr | 6464.01 |
| access width 10.00 m; 2.5 m high gates | – | – | – | – | – | nr | 6702.01 |

## 36 FENCING

| Item<br>Excluding site overheads and profit | PC<br>£ | Labour<br>hours | Labour<br>£ | Plant<br>£ | Material<br>£ | Unit | Total<br>rate £ |
|---|---|---|---|---|---|---|---|
| **GATES, KISSING GATES AND STILES – cont** | | | | | | | |
| **Kissing gates** | | | | | | | |
| Kissing gates; Jacksons Fencing; in galvanized metal bar; fixing to fencing posts (posts not included); | | | | | | | |
| 1.65 × 1.30 × 1.00 m high | 283.50 | 2.50 | 118.75 | – | 283.50 | nr | **402.25** |
| **Stiles** | | | | | | | |
| Stiles; two posts; setting into firm ground; three rails; two treads | 84.00 | 1.50 | 71.25 | – | 96.82 | nr | **168.07** |
| **GUARD RAILS** | | | | | | | |
| **Pedestrian guard rails and barriers** | | | | | | | |
| Mild steel pedestrian guard rails; Broxap Street Furniture; 1.00 m high with 150 mm toe space; to posts at 2.00 m centres; galvanized finish | | | | | | | |
|    vertical staggered in-line bar infill panel | 115.00 | 1.00 | 47.50 | – | 122.42 | m | **169.92** |
|    vertical staggered bar infill with 230 mm visibility gap at top | 115.00 | 1.00 | 47.50 | – | 122.42 | m | **169.92** |

## 37 SOFT LANDSCAPING

| Item<br>Excluding site overheads and profit | PC<br>£ | Labour<br>hours | Labour<br>£ | Plant<br>£ | Material<br>£ | Unit | Total<br>rate £ |
|---|---|---|---|---|---|---|---|
| **GENERALLY** | | | | | | | |
| **Seeding/turfing – General**<br>Preamble: The following market prices generally reflect the manufacturer's recommended retail prices. Trade and bulk discounts are often available on the prices shown. The manufacturers of these products generally recommend application rates. Note: the following rates reflect the average rate for each product. | | | | | | | |
| **CLARIFICATION NOTES ON LABOUR COSTS IN THIS SECTION** | | | | | | | |
| **General landscape team**<br>Generally a three man team is used in this section; The column Labour hours reports team hours. The column Labour £ reports the total cost of the team for the unit of work shown | | | | | | | |
| 3 man team | – | 1.00 | 71.25 | – | – | hr | 71.25 |
| **MARKET PRICES OF SEEDING MATERIALS** | | | | | | | |
| **Market prices of pre-seeding materials**<br>Rigby Taylor Ltd | | | | | | | |
| turf fertilizer; Mascot Outfield 16+6+6 | – | – | – | – | 2.42 | 100 m² | 2.42 |
| Boughton Loam | | | | | | | |
| screened topsoil; 100 mm | – | – | – | – | 59.40 | m³ | 59.40 |
| screened Kettering loam; 3 mm | – | – | – | – | 117.00 | m³ | 117.00 |
| screened Kettering loam; sterilized; 3 mm | – | – | – | – | 120.60 | m³ | 120.60 |
| top dressing; sand soil mixtures; 90/10 to 50/50 | – | – | – | – | 93.60 | m³ | 93.60 |
| **Market prices of turf fertilizers; recommended average application rates**<br>Everris; turf fertilizers | | | | | | | |
| grass fertilizer; slow release; Sierraform GT Preseeder 18+22+05 | – | – | – | – | 4.68 | 100 m² | 4.68 |
| grass fertilizer; slow release; Sierraform GT All Season 18+06+18 | – | – | – | – | 3.90 | 100 m² | 3.90 |
| grass fertilizer; slow release; Sierraform GT Anti-Stress 15+00+26 | – | – | – | – | 4.68 | 100 m² | 4.68 |

## 37 SOFT LANDSCAPING

| Item<br>Excluding site overheads and profit | PC<br>£ | Labour<br>hours | Labour<br>£ | Plant<br>£ | Material<br>£ | Unit | Total<br>rate £ |
|---|---|---|---|---|---|---|---|
| **MARKET PRICES OF SEEDING MATERIALS – cont** | | | | | | | |
| **Market prices of turf fertilizers – cont** | | | | | | | |
| Everris – cont | | | | | | | |
| grass fertilizer; slow release; Sierraform GT Momentum 22+05+11 | – | – | – | – | 3.12 | 100 m$^2$ | **3.12** |
| grass fertilizer; slow release; Sierraform GT Spring Start 16+00+16 | – | – | – | – | 3.90 | 100 m$^2$ | **3.90** |
| grass fertilizer; controlled release; Sierrablen 27+05+05 (8–9 months) | – | – | – | – | 5.43 | 100 m$^2$ | **5.43** |
| grass fertilizer; controlled release; Sierrablen 28+05+05 (5–6 months) | – | – | – | – | 5.37 | 100 m$^2$ | **5.37** |
| grass fertilizer; controlled release; Sierrablen 15+00+22 (5–6 months) | – | – | – | – | 4.87 | 100 m$^2$ | **4.87** |
| grass fertilizer; Sierrablen Fine 38+00+00 (2–3 months) | – | – | – | – | 2.00 | 100 m$^2$ | **2.00** |
| grass fertilizer; Sierrablen Mini 00+00+37 (5–6 months) | – | – | – | – | 3.34 | 100 m$^2$ | **3.34** |
| grass fertilizer; Greenmaster Pro-Lite Turf Tonic 08+00+00 | – | – | – | – | 2.31 | 100 m$^2$ | **2.31** |
| grass fertilizer; Greenmaster Pro-Lite Spring & Summer 14+05+10 | – | – | – | – | 3.30 | 100 m$^2$ | **3.30** |
| grass fertilizer; Greenmaster Pro-Lite Zero Phosphate 14+00+10 | – | – | – | – | 3.34 | 100 m$^2$ | **3.34** |
| grass fertilizer; Greenmaster Pro-Lite Mosskiller 14+00+00 | – | – | – | – | 3.68 | 100 m$^2$ | **3.68** |
| grass fertilizer; Greenmaster Pro-Lite Invigorator 04+00+08 | – | – | – | – | 2.50 | 100 m$^2$ | **2.50** |
| grass fertilizer; Greenmaster Pro-Lite Autumn 06+05+10 | – | – | – | – | 3.56 | 100 m$^2$ | **3.56** |
| grass fertilizer; Greenmaster Pro-Lite Double K 07+00+14 | – | – | – | – | 3.56 | 100 m$^2$ | **3.56** |
| grass fertilizer; Greenmaster Pro-Lite NK 12+00+12 | – | – | – | – | 3.56 | 100 m$^2$ | **3.56** |
| outfield turf fertilizer; Sportsmaster Standard Spring & Summer 09+07+07 | – | – | – | – | 2.89 | 100 m$^2$ | **2.89** |
| outfield turf fertilizer; Sportsmaster Standard Autumn 04+12+12 | – | – | – | – | 3.15 | 100 m$^2$ | **3.15** |
| outfield turf fertilizer; Sportsmaster Standard Zero Phosphate 12+00+09 | – | – | – | – | 3.30 | 100 m$^2$ | **3.30** |
| TPMC; tree planting and mulching compost | – | – | – | – | 2.60 | 75 l | **2.60** |
| Rigby Taylor Ltd; 35 g/m$^2$ | | | | | | | |
| grass fertilizer; Mascot Microfine 12+0+10 + 2% Mg + 2% Fe | – | – | – | – | 5.99 | 100 m$^2$ | **5.99** |
| grass fertilizer; Mascot Microfine 8+0+6 + 2% Mg + 4% Fe | – | – | – | – | 5.04 | 100 m$^2$ | **5.04** |
| grass fertilizer; Mascot Microfine Organic OC1 8+0+0 + 2% Fe | – | – | – | – | 5.21 | 100 m$^2$ | **5.21** |

## 37 SOFT LANDSCAPING

| Item<br>Excluding site overheads and profit | PC<br>£ | Labour<br>hours | Labour<br>£ | Plant<br>£ | Material<br>£ | Unit | Total<br>rate £ |
|---|---|---|---|---|---|---|---|
| grass fertilizer; Mascot Microfine Organic OC2 5+2+10 | – | – | – | – | 5.64 | 100 m$^2$ | **5.64** |
| grass fertilizer; Mascot Delta Sport 12+4+8 + 0.5% Fe | – | – | – | – | 4.64 | 100 m$^2$ | **4.64** |
| grass fertilizer; Mascot Fine Turf 12+0+9 + 1% Mg + 1% Fe + seaweed | – | – | – | – | 5.46 | 100 m$^2$ | **5.46** |
| outfield fertilizer; Mascot Outfield 16+6+6 | – | – | – | – | 3.38 | 100 m$^2$ | **3.38** |
| outfield fertilizer; Mascot Outfield 4+10+10 | – | – | – | – | 3.46 | 100 m$^2$ | **3.46** |
| outfield fertilizer; Mascot Outfield 9+5+5 | – | – | – | – | 3.27 | 100 m$^2$ | **3.27** |
| outfield fertilizer; Mascot Outfield 12+4+4 | – | – | – | – | 3.32 | 100 m$^2$ | **3.32** |
| liquid fertilizer; Mascot Microflow-C 26+0+0 + trace elements; 200–1400 ml/100 m$^2$ | – | – | – | – | 6.67 | 100 m$^2$ | **6.67** |
| liquid fertilizer; Mascot Microflow-CX 6+2+18 + trace elements; 200–1400 ml/100 m$^2$ | – | – | – | – | 6.42 | 100 m$^2$ | **6.42** |
| liquid fertilizer; Mascot Microflow-CX 14+0+7 + trace elements; 200–1400 ml/100 m$^2$ | – | – | – | – | 5.66 | 100 m$^2$ | **5.66** |
| liquid fertilizer; Mascot Microflow-CX 17+2+5 + trace elements; 200–1400 ml/100 m$^2$ | – | – | – | – | 6.37 | 100 m$^2$ | **6.37** |
| **Market prices of grass seed** | | | | | | | |
| Preamble: The prices shown are for supply only at one number 20 kg or 25 kg bag purchase price unless otherwise stated. Rates shown are based on the manufacturer's maximum recommendation for each seed type. Trade and bulk discounts are often available on the prices shown for quantities of more than one bag. | | | | | | | |
| Bowling greens; fine lawns; ornamental turf; croquet lawns | | | | | | | |
| Germinal GB; A1 Green; 35 g/m$^2$ | – | – | – | – | 26.60 | 100 m$^2$ | **26.60** |
| DLF Seeds Ltd; J Green; 34–50 g/m$^2$ | – | – | – | – | 43.50 | 100 m$^2$ | **43.50** |
| DLF Seeds Ltd; J Premier Green; 34–50 g/m$^2$ | – | – | – | – | 69.80 | 100 m$^2$ | **69.80** |
| DLF Seeds Ltd; Promaster 20; 35–50 g/m$^2$ | – | – | – | – | 23.15 | 100 m$^2$ | **23.15** |
| DLF Seeds Ltd; Promaster 10; 35–50 g/m$^2$ | – | – | – | – | 37.45 | 100 m$^2$ | **37.45** |
| Tennis courts; cricket squares | | | | | | | |
| Germinal GB; A2 Lawns; 35 g/m$^2$ | – | – | – | – | 19.84 | 100 m$^2$ | **19.84** |
| Germinal GB; A5 Cricket; 35 g/m$^2$ | – | – | – | – | 30.19 | 100 m$^2$ | **30.19** |
| DLF Seeds Ltd; J Premier Green; 34–50 g/m$^2$ | – | – | – | – | 69.80 | 100 m$^2$ | **69.80** |

## 37 SOFT LANDSCAPING

| Item<br>Excluding site overheads and profit | PC<br>£ | Labour<br>hours | Labour<br>£ | Plant<br>£ | Material<br>£ | Unit | Total<br>rate £ |
|---|---|---|---|---|---|---|---|
| **MARKET PRICES OF SEEDING MATERIALS – cont** | | | | | | | |
| **Market prices of grass seed – cont** | | | | | | | |
| Tennis courts – cont | | | | | | | |
|   DLF Seeds Ltd; J Intense; 18–25 g/m² | – | – | – | – | 12.22 | 100 m² | **12.22** |
|   DLF Seeds Ltd; Promaster 35; 35 g/m² | – | – | – | – | 15.19 | 100 m² | **15.19** |
| Amenity grassed areas; general purpose lawns | | | | | | | |
|   Germinal GB; A3 Banks; 25–50 g/m² | – | – | – | – | 24.38 | 100 m² | **24.38** |
|   DLF Seeds Ltd; J Rye Fairway; 18–25 g/m² | – | – | – | – | 12.85 | 100 m² | **12.85** |
|   DLF Seeds Ltd; J Intense; 18–25 g/m² | – | – | – | – | 12.22 | 100 m² | **12.22** |
|   DLF Seeds Ltd; Promaster 50; 25–35 g/m² | – | – | – | – | 16.34 | 100 m² | **16.34** |
|   DLF Seeds Ltd; Promaster 120; 25–35 g/m² | – | – | – | – | 15.40 | 100 m² | **15.40** |
| Conservation; country parks; slopes and banks | | | | | | | |
|   Germinal GB; A4 Parkland; 17–35 g/m² | – | – | – | – | 18.24 | 100 m² | **18.24** |
|   Germinal GB; A16 Country Park; 8–19 g/m² | – | – | – | – | 13.00 | 100 m² | **13.00** |
|   Germinal GB; A17 Legume; 2 g/m² | – | – | – | – | 25.20 | 100 m² | **25.20** |
| Shaded areas | | | | | | | |
|   Germinal GB; A6 Shade; 50 g/m² | – | – | – | – | 36.56 | 100 m² | **36.56** |
|   Germinal GB; J Green; 34–50 g/m² | – | – | – | – | 43.50 | 100 m² | **43.50** |
|   DLF Seeds Ltd; Promaster 60; 35–50 g/m² | – | – | – | – | 24.05 | 100 m² | **24.05** |
| Sports pitches; rugby; soccer pitches | | | | | | | |
|   Germinal GB; A7 Sportsground; 20 g/m² | – | – | – | – | 13.53 | 100 m² | **13.53** |
|   DLF Seeds Ltd; J Pitch; 18–30 g/m² | – | – | – | – | 13.80 | 100 m² | **13.80** |
|   DLF Seeds Ltd; Promaster 70; 15–35 g/m² | – | – | – | – | 14.00 | 100 m² | **14.00** |
|   DLF Seeds Ltd; Promaster 75; 15–35 g/m² | – | – | – | – | 16.91 | 100 m² | **16.91** |
|   DLF Seeds Ltd; Promaster 80; 17–35 g/m² | – | – | – | – | – | 100 m² | **–** |
|   Rigby Taylor; Mascot R11 Football & Rugby; 35 g/m² | – | – | – | – | 15.73 | 100 m² | **15.73** |
|   Rigby Taylor; Mascot R13 General Playing Fields; 35 g/m² | – | – | – | – | 13.54 | 100 m² | **13.54** |
| Outfields | | | | | | | |
|   Germinal GB; A7 Sportsground; 20 g/m² | – | – | – | – | 13.53 | 100 m² | **13.53** |
|   Germinal GB; A9 Outfield; 17–35 g/m² | – | – | – | – | 19.86 | 100 m² | **19.86** |
|   DLF Seeds Ltd; J Fairway; 12–25 g/m² | – | – | – | – | 12.93 | 100 m² | **12.93** |

## 37 SOFT LANDSCAPING

| Item<br>Excluding site overheads and profit | PC<br>£ | Labour<br>hours | Labour<br>£ | Plant<br>£ | Material<br>£ | Unit | Total<br>rate £ |
|---|---|---|---|---|---|---|---|
| DLF Seeds Ltd; Promaster 40; 35 g/m$^2$ | – | – | – | – | 16.84 | 100 m$^2$ | 16.84 |
| DLF Seeds Ltd; Promaster 70;<br>15–35 g/m$^2$ | – | – | – | – | 14.00 | 100 m$^2$ | 14.00 |
| Rigby Taylor; R4 Species Rich<br>Fairways & Outfields; 35 g/m$^2$ | – | – | – | – | 19.84 | 100 m$^2$ | 19.84 |
| Hockey pitches | | | | | | | |
| DLF Seeds Ltd; J Fairway;<br>12–25 g/m$^2$ | – | – | – | – | 12.93 | 100 m$^2$ | 12.93 |
| DLF Seeds Ltd; Promaster 70;<br>15–35 g/m$^2$ | – | – | – | – | 14.00 | 100 m$^2$ | 14.00 |
| Parks | | | | | | | |
| Germinal GB; A7 Sportsground 20 g/m$^2$ | – | – | – | – | 13.53 | 100 m$^2$ | 13.53 |
| Germinal GB; A9 Outfield;<br>17–35 g/m$^2$ | – | – | – | – | 19.86 | 100 m$^2$ | 19.86 |
| DLF Seeds Ltd Promaster 120;<br>25–35 g/m$^2$ | – | – | – | – | 15.40 | 100 m$^2$ | 15.40 |
| Informal playing fields | | | | | | | |
| DLF Seeds Ltd; J Rye Fairway;<br>18–25 g/m$^2$ | – | – | – | – | 12.85 | 100 m$^2$ | 12.85 |
| DLF Seeds Ltd; Promaster 45;<br>35 g/m$^2$ | – | – | – | – | 16.55 | 100 m$^2$ | 16.55 |
| Caravan sites | | | | | | | |
| Germinal GB; A9 Outfield;<br>17–35 g/m$^2$ | – | – | – | – | 19.86 | 100 m$^2$ | 19.86 |
| Sports pitch re-seeding and repair | | | | | | | |
| Germinal GB; A8 Pitch Renovator;<br>20–35 g/m$^2$ | – | – | – | – | 18.73 | 100 m$^2$ | 18.73 |
| Germinal GB; A20 Ryesport;<br>20–35 g/m$^2$ | – | – | – | – | 21.44 | 100 m$^2$ | 21.44 |
| DLF Seeds Ltd; J Pitch; 18–30 g/m$^2$ | – | – | – | – | 13.80 | 100 m$^2$ | 13.80 |
| DLF Seeds Ltd; Promaster 80;<br>17–35 g/m$^2$ | – | – | – | – | 15.86 | 100 m$^2$ | 15.86 |
| DLF Seeds Ltd; Promaster 81;<br>17–35 g/m$^2$ | – | – | – | – | 16.34 | 100 m$^2$ | 16.34 |
| Rigby Taylor; R14 Sports Pitches<br>100% premium perennial ryegrass;<br>35 g/m$^2$ | – | – | – | – | 17.26 | 100 m$^2$ | 17.26 |
| Racecourses; gallops; polo grounds;<br>horse rides | | | | | | | |
| Germinal GB; Racecourse;<br>25–30 g/m$^2$ | – | – | – | – | 18.30 | 100 m$^2$ | 18.30 |
| DLF Seeds Ltd; J Intense;<br>18–25 g/m$^2$ | – | – | – | – | 12.22 | 100 m$^2$ | 12.22 |
| DLF Seeds Ltd; Promaster 65;<br>17–35 g/m$^2$ | – | – | – | – | 17.50 | 100 m$^2$ | 17.50 |
| Motorway and road verges | | | | | | | |
| Germinal GB; A18 Road Verge;<br>6-15 g/m$^2$ | – | – | – | – | 9.38 | 100 m$^2$ | 9.38 |
| DLF Seeds Ltd; Promaster 85; 10 g/m$^2$ | – | – | – | – | 4.89 | 100 m$^2$ | 4.89 |
| DLF Seeds Ltd; Promaster 120;<br>25-35 g/m$^2$ | – | – | – | – | 15.40 | 100 m$^2$ | 15.40 |

## 37 SOFT LANDSCAPING

| Item<br>Excluding site overheads and profit | PC<br>£ | Labour<br>hours | Labour<br>£ | Plant<br>£ | Material<br>£ | Unit | Total<br>rate £ |
|---|---|---|---|---|---|---|---|
| **MARKET PRICES OF SEEDING MATERIALS – cont** | | | | | | | |
| **Market prices of grass seed – cont** | | | | | | | |
| Golf courses; tees | | | | | | | |
|    Germinal GB; A10 Tees, 35–50 g/m$^2$ | – | – | – | – | 34.38 | 100 m$^2$ | **34.38** |
|    DLF Seeds Ltd; J Premier Fairway; 18–30 g/m$^2$ | – | – | – | – | 17.67 | 100 m$^2$ | **17.67** |
|    DLF Seeds Ltd; J Intense; 18–25 g/m$^2$ | – | – | – | – | 12.22 | 100 m$^2$ | **12.22** |
|    DLF Seeds Ltd; Promaster 40; 35 g/m$^2$ | – | – | – | – | 16.84 | 100 m$^2$ | **16.84** |
|    DLF Seeds Ltd; Promaster 45; 35 g/m$^2$ | – | – | – | – | 16.55 | 100 m$^2$ | **16.55** |
| Golf courses; greens | | | | | | | |
|    Germinal GB; A11 Golf Greens; 35 g/m$^2$ | – | – | – | – | 27.13 | 100 m$^2$ | **27.13** |
|    Germinal GB; A13 Roughs; 8 g/m$^2$ | – | – | – | – | 4.20 | 100 m$^2$ | **4.20** |
|    DLF Seeds Ltd; J All Bent; 8 g/m$^2$ | – | – | – | – | 16.08 | 100 m$^2$ | **16.08** |
|    DLF Seeds Ltd; J Green; 34–50 g/m$^2$ | – | – | – | – | 43.50 | 100 m$^2$ | **43.50** |
|    DLF Seeds Ltd; Promaster 5; 35–50 g/m$^2$ | – | – | – | – | 28.00 | 100 m$^2$ | **28.00** |
|    Rigby Taylor; R1 Golf & Bowls Fescue/Bent; 35 g/m$^2$ | – | – | – | – | 32.65 | 100 m$^2$ | **32.65** |
| Golf courses; fairways | | | | | | | |
|    Germinal GB; A12 Fairways; 15–25 g/m$^2$ | – | – | – | – | 14.06 | 100 m$^2$ | **14.06** |
|    DLF Seeds Ltd; J Rye Fairway; 18–30 g/m$^2$ | – | – | – | – | 15.42 | 100 m$^2$ | **15.42** |
|    DLF Seeds Ltd; J Premier Fairway; 18–30 g/m$^2$ | – | – | – | – | 17.67 | 100 m$^2$ | **17.67** |
|    DLF Seeds Ltd; Promaster 40; 35 g/m$^2$ | – | – | – | – | 16.84 | 100 m$^2$ | **16.84** |
|    DLF Seeds Ltd; Promaster 45; 35 g/m$^2$ | – | – | – | – | 16.55 | 100 m$^2$ | **16.55** |
|    Rigby Taylor; Mascot R116 Fescue Rye Fairway; 35 g/m$^2$ | – | – | – | – | 47.91 | 100 m$^2$ | **47.91** |
| Golf courses; roughs | | | | | | | |
|    DLF Seeds Ltd; Promaster 25; 17–35 g/m$^2$ | – | – | – | – | 16.84 | 100 m$^2$ | **16.84** |
|    Rigby Taylor; Mascot R5 Golf Links & Rough; 35 g/m$^2$ | – | – | – | – | 4.69 | 100 m$^2$ | **4.69** |
| Waste land; spoil heaps; quarries | | | | | | | |
|    Germinal GB; A15 Reclamation; 15–20 g/m$^2$ | – | – | – | – | 13.00 | 100 m$^2$ | **13.00** |
|    DLF Seeds Ltd; Promaster 95; 12–35 g/m$^2$ | – | – | – | – | 21.39 | 100 m$^2$ | **21.39** |
|    DLF Seeds Ltd; Promaster 105; 5 g/m$^2$ | – | – | – | – | 5.43 | 100 m$^2$ | **5.43** |
| Low maintenance; housing estates; amenity grassed areas | | | | | | | |
|    Germinal GB; A19 Housing; 25–35 g/m$^2$ | – | – | – | – | 18.38 | 100 m$^2$ | **18.38** |
|    Germinal GB; A22 Low Maintenance; 25–35 g/m$^2$ | – | – | – | – | 22.75 | 100 m$^2$ | **22.75** |
|    DLF Seeds Ltd; Promaster 120; 25–35 g/m$^2$ | – | – | – | – | 15.40 | 100 m$^2$ | **15.40** |

## 37 SOFT LANDSCAPING

| Item<br>Excluding site overheads and profit | PC<br>£ | Labour<br>hours | Labour<br>£ | Plant<br>£ | Material<br>£ | Unit | Total<br>rate £ |
|---|---|---|---|---|---|---|---|
| Saline coastal; roadside areas | | | | | | | |
| Germinal GB; A21 Saline; 15–20 g/m² | – | – | – | – | 11.25 | 100 m² | 11.25 |
| DLF Seeds Ltd; Promaster 90; 15–35 g/m² | – | – | – | – | 17.99 | 100 m² | 17.99 |
| Turf production | | | | | | | |
| Germinal GB; A25 Meadow Ley; 160 kg/ha | – | – | – | – | 1216.00 | ha | 1216.00 |
| Germinal GB; A24 Wear & Tear; 185 kg/ha | – | – | – | – | 1019.81 | ha | 1019.81 |
| **Market prices of wild flora seed mixtures** | | | | | | | |
| Acid soils | | | | | | | |
| Germinal GB; WF1 (Annual Flowering); 1–2 g/m² | – | – | – | – | 22.13 | 100 m² | 22.13 |
| Neutral soils | | | | | | | |
| Germinal GB; WF3 (Neutral Soils); 0.5–1 g/m² | – | – | – | – | 11.06 | 100 m² | 11.06 |
| **Market prices of wild flora and grass seed mixtures** | | | | | | | |
| General purpose | | | | | | | |
| DLF Seeds Ltd; Pro Flora 8 Old English Country Meadow Mix; 5 g/m² | – | – | – | – | 24.00 | 100 m² | 24.00 |
| DLF Seeds Ltd; Pro Flora 9 General purpose; 5 g/m² | – | – | – | – | 10.00 | 100 m² | 10.00 |
| Acid soils | | | | | | | |
| Germinal GB; WFG2 (Annual Meadow); 5 g/m² | – | – | – | – | 27.50 | 100 m² | 27.50 |
| DLF Seeds Ltd; Pro Flora 2 Acidic soils; 5 g/m² | – | – | – | – | 12.00 | 100 m² | 12.00 |
| Neutral soils | | | | | | | |
| Germinal GB; WFG4 (Neutral Meadow); 5 g/m² | – | – | – | – | 29.70 | 100 m² | 29.70 |
| Germinal GB; WFG13 (Scotland); 5 g/m² | – | – | – | – | 26.80 | 100 m² | 26.80 |
| DLF Seeds Ltd; Pro Flora 3 Damp loamy soils; 5 g/m² | – | – | – | – | 16.10 | 100 m² | 16.10 |
| Calcareous soils | | | | | | | |
| Germinal GB; WFG5 (Calcareous Soils); 5 g/m² | – | – | – | – | 28.10 | 100 m² | 28.10 |
| DLF Seeds Ltd; Pro Flora 4 Calcareous soils; 5 g/m² | – | – | – | – | 14.40 | 100 m² | 14.40 |
| Heavy clay soils | | | | | | | |
| Germinal GB; WFG6 (Clay Soils); 5 g/m² | – | – | – | – | 32.75 | 100 m² | 32.75 |
| Germinal GB; WFG12 (Ireland); 5 g/m² | – | – | – | – | 26.88 | 100 m² | 26.88 |
| DLF Seeds Ltd; Pro Flora 5 Wet loamy soils; 5 g/m² | – | – | – | – | 28.60 | 100 m² | 28.60 |

## 37 SOFT LANDSCAPING

| Item<br>Excluding site overheads and profit | PC<br>£ | Labour<br>hours | Labour<br>£ | Plant<br>£ | Material<br>£ | Unit | Total<br>rate £ |
|---|---|---|---|---|---|---|---|
| **MARKET PRICES OF SEEDING MATERIALS – cont** | | | | | | | |
| **Market prices of wild flora and grass seed mixtures – cont** | | | | | | | |
| Sandy soils | | | | | | | |
| Germinal GB; WFG7 (Free Draining Soils); 5 g/m$^2$ | – | – | – | – | 37.50 | 100 m$^2$ | **37.50** |
| Germinal GB; WFG11 (Ireland); 5 g/m$^2$ | – | – | – | – | 28.00 | 100 m$^2$ | **28.00** |
| Germinal GB; WFG14 (Scotland); 5 g/m$^2$ | – | – | – | – | 30.25 | 100 m$^2$ | **30.25** |
| DLF Seeds Ltd; Pro Flora 6 Dry free draining loamy soils; 5 g/m$^2$ | – | – | – | – | 29.25 | 100 m$^2$ | **29.25** |
| Shaded areas | | | | | | | |
| Germinal GB; WFG8 (Woodland and Hedgerow); 5 g/m$^2$ | – | – | – | – | 29.25 | 100 m$^2$ | **29.25** |
| DLF Seeds Ltd; Pro Flora 7 Hedgerow and light shade; 5 g/m$^2$ | – | – | – | – | 36.00 | 100 m$^2$ | **36.00** |
| Educational | | | | | | | |
| Germinal GB; WFG15 (Schools and Colleges); 5 g/m$^2$ | – | – | – | – | 39.00 | 100 m$^2$ | **39.00** |
| Wetlands | | | | | | | |
| Germinal GB; WFG9 (Wetlands and Ponds); 5 g/m$^2$ | – | – | – | – | 35.00 | 100 m$^2$ | **35.00** |
| DLF Seeds Ltd; Pro Flora 5 Wet loamy soils; 5 g/m$^2$ | – | – | – | – | 28.60 | 100 m$^2$ | **28.60** |
| Scrub and moorland | | | | | | | |
| Germinal GB; WFG10 (Cornfield Annuals); 5 g/m$^2$ | – | – | – | – | 38.05 | 100 m$^2$ | **38.05** |
| Hedgerow | | | | | | | |
| DLF Seeds Ltd; Pro Flora 7 Hedgerow and light shade; 5 g/m$^2$ | – | – | – | – | 36.00 | 100 m$^2$ | **36.00** |
| Vacant sites | | | | | | | |
| DLF Seeds Ltd; Pro Flora 1 Cornfield annuals; 5 g/m$^2$ | – | – | – | – | 30.00 | 100 m$^2$ | **30.00** |
| Regional Environmental mixes | | | | | | | |
| Germinal GB; RE1 (Traditional Hay); 5 g/m$^2$ | – | – | – | – | 29.95 | 100 m$^2$ | **29.95** |
| Germinal GB; RE2 (Lowland Meadow); 5 g/m$^2$ | – | – | – | – | 37.75 | 100 m$^2$ | **37.75** |
| Germinal GB; RE3 (Riverflood Plain/Water Meadow); 5 g/m$^2$ | – | – | – | – | 37.25 | 100 m$^2$ | **37.25** |
| Germinal GB; RE4 (Lowland Limestone); 5 g/m$^2$ | – | – | – | – | 36.75 | 100 m$^2$ | **36.75** |
| Germinal GB; RE5 (Calcareous Sub-mountain Restoration); 5 g/m$^2$ | – | – | – | – | 41.10 | 100 m$^2$ | **41.10** |
| Germinal GB; RE6 (Upland Limestone); 5 g/m$^2$ | – | – | – | – | 51.45 | 100 m$^2$ | **51.45** |
| Germinal GB; RE7 (Acid Sub-mountain Restoration); 5 g/m$^2$ | – | – | – | – | 36.25 | 100 m$^2$ | **36.25** |

## 37 SOFT LANDSCAPING

| Item<br>Excluding site overheads and profit | PC<br>£ | Labour<br>hours | Labour<br>£ | Plant<br>£ | Material<br>£ | Unit | Total<br>rate £ |
|---|---|---|---|---|---|---|---|
| Germinal GB; RE8 (Coastal Reclamation); 5 g/m² | – | – | – | – | 45.45 | 100 m² | 45.45 |
| Germinal GB; RE9 (Farmland Mixture); 5 g/m² | – | – | – | – | 32.85 | 100 m² | 32.85 |
| Germinal GB; RE10 (Marginal Land); 5 g/m² | – | – | – | – | 46.75 | 100 m² | 46.75 |
| Germinal GB; RE11 (Heath Scrubland); 5 g/m² | – | – | – | – | 17.80 | 100 m² | 17.80 |
| Germinal GB; RE12 (Drought Land); 5 g/m² | – | – | – | – | 44.50 | 100 m² | 44.50 |
| **CULTIVATION AND SOIL PREPARATION** | | | | | | | |
| **Cultivation by tractor; Agripower Ltd** | | | | | | | |
| Ripping up subsoil; using approved subsoiling machine; minimum depth 250 mm below topsoil; at 1.20 m centres; in | | | | | | | |
| gravel or sandy clay | – | – | – | 2.20 | – | 100 m² | 2.20 |
| soil compacted by machines | – | – | – | 2.44 | – | 100 m² | 2.44 |
| clay | – | – | – | 3.14 | – | 100 m² | 3.14 |
| chalk or other soft rock | – | – | – | 3.67 | – | 100 m² | 3.67 |
| Extra for subsoiling at 1 m centres | – | – | – | 0.63 | – | 100 m² | 0.63 |
| Breaking up existing ground; using tractor drawn Blec ground preparation equipment in one operation | | | | | | | |
| Single pass | | | | | | | |
| 100 mm deep | – | – | – | 15.60 | – | 100 m² | 15.60 |
| 150 mm deep | – | – | – | 19.50 | – | 100 m² | 19.50 |
| 200 mm deep | – | – | – | 22.28 | – | 100 m² | 22.28 |
| Cultivating ploughed ground; using disc, drag or chain harrow | | | | | | | |
| 4 passes | – | – | – | 17.73 | – | 100 m² | 17.73 |
| Rolling cultivated ground lightly; using self-propelled agricultural roller | – | 0.07 | 1.58 | 0.74 | – | 100 m² | 2.32 |
| **Cultivation by pedestrian operated rotavator** | | | | | | | |
| Breaking up existing ground; tine cultivator or rotavator | | | | | | | |
| 100 mm deep | – | 0.28 | 6.53 | 0.99 | – | 100 m² | 7.52 |
| 150 mm deep | – | 0.31 | 7.46 | 1.14 | – | 100 m² | 8.60 |
| 200 mm deep | – | 0.40 | 9.50 | 1.44 | – | 100 m² | 10.94 |
| As above but in heavy clay or wet soils | | | | | | | |
| 100 mm deep | – | 0.44 | 10.45 | 1.59 | – | 100 m² | 12.04 |
| 150 mm deep | – | 0.66 | 15.68 | 2.38 | – | 100 m² | 18.06 |
| 200 mm deep | – | 0.82 | 19.59 | 2.98 | – | 100 m² | 22.57 |

## 37 SOFT LANDSCAPING

| Item<br>Excluding site overheads and profit | PC<br>£ | Labour<br>hours | Labour<br>£ | Plant<br>£ | Material<br>£ | Unit | Total<br>rate £ |
|---|---|---|---|---|---|---|---|
| **CULTIVATION AND SOIL PREPARATION – cont** | | | | | | | |
| **Cultivation by pedestrian operated rotavator – cont** | | | | | | | |
| Importing and storing selected and approved topsoil; inclusive of settlement | | | | | | | |
| small quantities (less than 15 m$^3$) | 78.00 | – | – | – | 78.00 | m$^3$ | 78.00 |
| over 15 m$^3$ | 39.60 | – | – | – | 39.60 | m$^3$ | 39.60 |
| Spreading and lightly consolidating approved topsoil (imported or from spoil heaps); in layers not exceeding 150 mm; travel distance from spoil heaps not exceeding 100 m; by machine (imported topsoil not included) | | | | | | | |
| minimum depth 100 mm | – | 1.55 | 36.81 | 55.60 | – | 100 m$^2$ | 92.41 |
| minimum depth 150 mm | – | 2.33 | 55.42 | 83.68 | – | 100 m$^2$ | 139.10 |
| minimum depth 300 mm | – | 4.67 | 110.83 | 167.35 | – | 100 m$^2$ | 278.18 |
| minimum depth 450 mm | – | 6.99 | 166.01 | 250.75 | – | 100 m$^2$ | 416.76 |
| Spreading and lightly consolidating approved topsoil (imported or from spoil heaps); in layers not exceeding 150 mm; travel distance from spoil heaps not exceeding 100 m; by hand (imported topsoil not included) | | | | | | | |
| minimum depth 100 mm | – | 20.00 | 475.10 | – | – | 100 m$^2$ | 475.10 |
| minimum depth 150 mm | – | 30.01 | 712.64 | – | – | 100 m$^2$ | 712.64 |
| minimum depth 300 mm | – | 60.01 | 1425.29 | – | – | 100 m$^2$ | 1425.29 |
| minimum depth 450 mm | – | 90.02 | 2137.93 | – | – | 100 m$^2$ | 2137.93 |
| Extra over for spreading topsoil to slopes 15–30°; by machine or hand | – | – | – | – | – | 10% | – |
| Extra over for spreading topsoil to slopes over 30°; by machine or hand | – | – | – | – | – | 25% | – |
| Extra over for spreading topsoil from spoil heaps; travel exceeding 100 m; by machine | | | | | | | |
| 100–150 m | – | 0.01 | 0.29 | 0.08 | – | m$^3$ | 0.37 |
| 150–200 m | – | 0.02 | 0.44 | 0.12 | – | m$^3$ | 0.56 |
| 200–300 m | – | 0.03 | 0.66 | 0.19 | – | m$^3$ | 0.85 |
| Extra over spreading topsoil for travel exceeding 100 m; by hand | | | | | | | |
| 100 m | – | 1.20 | 28.50 | – | – | m$^3$ | 28.50 |
| 200 m | – | 1.67 | 39.59 | – | – | m$^3$ | 39.59 |
| 300 m | – | 2.50 | 59.38 | – | – | m$^3$ | 59.38 |
| Evenly grading; to general surfaces to bring to finished levels | | | | | | | |
| by machine (tractor mounted rotavator) | – | – | – | 0.07 | – | m$^2$ | 0.07 |
| by pedestrian operated rotavator | – | – | 0.10 | 0.02 | – | m$^2$ | 0.12 |
| by hand | – | 0.01 | 0.24 | – | – | m$^2$ | 0.24 |

## 37 SOFT LANDSCAPING

| Item<br>Excluding site overheads and profit | PC<br>£ | Labour<br>hours | Labour<br>£ | Plant<br>£ | Material<br>£ | Unit | Total<br>rate £ |
|---|---|---|---|---|---|---|---|
| Extra over grading for slopes 15–30°; by machine or hand | – | – | – | – | – | 10% | – |
| Extra over grading for slopes over 30°; by machine or hand | – | – | – | – | – | 25% | – |
| Apply screened topdressing to grass surfaces; spread using Tru-Lute | | | | | | | |
| sand soil mixes 90/10 to 50/50 | – | – | 0.05 | 0.03 | 0.14 | m² | **0.22** |
| Spread only existing cultivated soil to final levels using Tru-Lute | | | | | | | |
| cultivated soil | – | – | 0.05 | 0.03 | – | m² | **0.08** |
| Clearing stones; disposing off site | | | | | | | |
| by hand; stones not exceeding 50 mm in any direction; loading to skip 4.6 m³ | – | 0.01 | 0.24 | 0.05 | – | m² | **0.29** |
| by mechanical stone rake; stones not exceeding 50 mm in any direction; loading to 15 m³ truck by mechanical loader | – | – | 0.05 | 0.19 | 0.01 | m² | **0.25** |
| Lightly cultivating; weeding; to fallow areas; disposing debris off site | | | | | | | |
| by hand | – | 0.01 | 0.34 | – | 0.15 | m² | **0.49** |

## 37 SOFT LANDSCAPING

| Item<br>Excluding site overheads and profit | PC<br>£ | Labour<br>hours | Labour<br>£ | Plant<br>£ | Material<br>£ | Unit | Total<br>rate £ |
|---|---|---|---|---|---|---|---|
| **SOIL ADDITIVES AND IMPROVEMENTS** | | | | | | | |
| **Surface applications; soil additives; pre-seeding; material delivered to a maximum of 25 m from area of application; applied; by machine** | | | | | | | |
| Soil conditioners; to cultivated ground; mushroom compost; delivered in 25 m³ loads; including turning in | | | | | | | |
| 1 m³ per 40 m² = 25 mm thick | 0.60 | 0.02 | 0.39 | 0.18 | 0.60 | m² | **1.17** |
| 1 m³ per 20 m² = 50 mm thick | 1.21 | 0.03 | 0.73 | 0.26 | 1.21 | m² | **2.20** |
| 1 m³ per 13.33 m² = 75 mm thick | 1.81 | 0.04 | 0.95 | 0.35 | 1.81 | m² | **3.11** |
| 1 m³ per 10 m² = 100 mm thick | 2.42 | 0.05 | 1.19 | 0.43 | 2.42 | m² | **4.04** |
| Soil conditioners; to cultivated ground; mushroom compost; delivered in 35 m³ loads; including turning in | | | | | | | |
| 1 m³ per 40 m² = 25 mm thick | 0.47 | 0.02 | 0.39 | 0.18 | 0.47 | m² | **1.04** |
| 1 m³ per 20 m² = 50 mm thick | 0.95 | 0.03 | 0.73 | 0.26 | 0.95 | m² | **1.94** |
| 1 m³ per 13.33 m² = 75 mm thick | 1.43 | 0.04 | 0.95 | 0.35 | 1.43 | m² | **2.73** |
| 1 m³ per 10 m² = 100 mm thick | 1.90 | 0.05 | 1.19 | 0.43 | 1.90 | m² | **3.52** |
| **Surface applications and soil additives; pre-seeding; material delivered to a maximum of 25 m from area of application; applied; by hand** | | | | | | | |
| Soil conditioners; to cultivated ground; mushroom compost; delivered in 25 m³ loads; including turning in | | | | | | | |
| 1 m³ per 40 m² = 25 mm thick | 0.60 | 0.02 | 0.53 | – | 0.60 | m² | **1.13** |
| 1 m³ per 20 m² = 50 mm thick | 1.21 | 0.04 | 1.05 | – | 1.21 | m² | **2.26** |
| 1 m³ per 13.33 m² = 75 mm thick | 1.81 | 0.07 | 1.58 | – | 1.81 | m² | **3.39** |
| 1 m³ per 10 m² = 100 mm thick | 2.42 | 0.08 | 1.90 | – | 2.42 | m² | **4.32** |
| Soil conditioners; to cultivated ground; mushroom compost; delivered in 60 m³ loads; including turning in | | | | | | | |
| 1 m³ per 40 m² = 25 mm thick | 0.31 | 0.02 | 0.53 | – | 0.31 | m² | **0.84** |
| 1 m³ per 20 m² = 50 mm thick | 0.56 | 0.04 | 1.05 | – | 0.56 | m² | **1.61** |
| 1 m³ per 13.33 m² = 75 mm thick | 0.94 | 0.07 | 1.58 | – | 0.94 | m² | **2.52** |
| 1 m³ per 10 m² = 100 mm thick | 1.25 | 0.08 | 1.90 | – | 1.25 | m² | **3.15** |
| **PREPARATION OF SEED BEDS** | | | | | | | |
| **Preparation of seedbeds – General**<br>Preamble: For preliminary operations see Cultivation section. | | | | | | | |
| **Preparation of seedbeds; soil preparation** | | | | | | | |
| Lifting selected and approved topsoil from spoil heaps; passing through 6 mm screen; removing debris | – | 0.08 | 1.98 | 4.95 | 0.04 | m³ | **6.97** |

## 37 SOFT LANDSCAPING

| Item Excluding site overheads and profit | PC £ | Labour hours | Labour £ | Plant £ | Material £ | Unit | Total rate £ |
|---|---|---|---|---|---|---|---|
| Topsoil; supply only; PC £33.00/m³; allowing for 20% settlement | | | | | | | |
| 25 mm | – | – | – | – | 0.99 | m² | 0.99 |
| 50 mm | – | – | – | – | 1.98 | m² | 1.98 |
| 100 mm | – | – | – | – | 3.96 | m² | 3.96 |
| 150 mm | – | – | – | – | 5.94 | m² | 5.94 |
| 200 mm | – | – | – | – | 7.92 | m² | 7.92 |
| 250 mm | – | – | – | – | 9.90 | m² | 9.90 |
| 300 mm | – | – | – | – | 11.88 | m² | 11.88 |
| 400 mm | – | – | – | – | 15.84 | m² | 15.84 |
| 450 mm | – | – | – | – | 17.82 | m² | 17.82 |
| Spreading topsoil to form seedbeds (topsoil not included); by machine | | | | | | | |
| 25 mm deep | – | – | 0.06 | 0.15 | – | m² | 0.21 |
| 50 mm deep | – | – | 0.08 | 0.19 | – | m² | 0.27 |
| 75 mm deep | – | – | 0.09 | 0.22 | – | m² | 0.31 |
| 100 mm deep | – | 0.01 | 0.12 | 0.29 | – | m² | 0.41 |
| 150 mm deep | – | – | – | – | – | m² | – |
| Spreading only topsoil to form seedbeds (topsoil not included); by hand | | | | | | | |
| 25 mm deep | – | 0.03 | 0.59 | – | – | m² | 0.59 |
| 50 mm deep | – | 0.03 | 0.79 | – | – | m² | 0.79 |
| 75 mm deep | – | 0.04 | 1.02 | – | – | m² | 1.02 |
| 100 mm deep | – | 0.05 | 1.19 | – | – | m² | 1.19 |
| 150 mm deep | – | 0.08 | 1.78 | – | – | m² | 1.78 |
| Bringing existing topsoil to a fine tilth for seeding; by raking or harrowing; stones not to exceed 6 mm; by machine | – | – | 0.10 | 0.11 | – | m² | 0.21 |
| Bringing existing topsoil to a fine tilth for seeding; by raking; stones not to exceed 6 mm; by hand | – | 0.01 | 0.34 | – | – | m² | 0.34 |
| **Preparation of seedbeds; soil treatments** | | | | | | | |
| For the following operations add or subtract the following amounts for every £0.10 difference in the material cost price | | | | | | | |
| 35 g/m² | – | – | – | – | 0.35 | 100 m² | 0.35 |
| 50 g/m² | – | – | – | – | 0.50 | 100 m² | 0.50 |
| 70 g/m² | – | – | – | – | 0.70 | 100 m² | 0.70 |
| 100 g/m² | – | – | – | – | 1.00 | 100 m² | 1.00 |
| 125 kg/ha | – | – | – | – | 12.50 | ha | 12.50 |
| 150 kg/ha | – | – | – | – | 15.00 | ha | 15.00 |
| 175 kg/ha | – | – | – | – | 17.50 | ha | 17.50 |
| 200 kg/ha | – | – | – | – | 20.00 | ha | 20.00 |
| 225 kg/ha | – | – | – | – | 22.50 | ha | 22.50 |
| 250 kg/ha | – | – | – | – | 25.00 | ha | 25.00 |
| 300 kg/ha | – | – | – | – | 30.00 | ha | 30.00 |

## 37 SOFT LANDSCAPING

| Item<br>Excluding site overheads and profit | PC<br>£ | Labour<br>hours | Labour<br>£ | Plant<br>£ | Material<br>£ | Unit | Total<br>rate £ |
|---|---|---|---|---|---|---|---|
| **PREPARATION OF SEED BEDS – cont** | | | | | | | |
| **Preparation of seedbeds – cont** | | | | | | | |
| For the following operations add or subtract the following amounts for every £0.10 difference in the material cost price – cont | | | | | | | |
| 350 kg/ha | – | – | – | – | 35.00 | ha | **35.00** |
| 400 kg/ha | – | – | – | – | 40.00 | ha | **40.00** |
| 500 kg/ha | – | – | – | – | 50.00 | ha | **50.00** |
| 700 kg/ha | – | – | – | – | 70.00 | ha | **70.00** |
| 1000 kg/ha | – | – | – | – | 100.00 | ha | **100.00** |
| 1250 kg/ha | – | – | – | – | 125.00 | ha | **125.00** |
| Pre-seeding fertilizers (12+00+09); PC £0.63/kg; to seedbeds; by machine | | | | | | | |
| 35 g/m$^2$ | 2.20 | – | – | 0.19 | 2.20 | 100 m$^2$ | **2.39** |
| 50 g/m$^2$ | 3.14 | – | – | 0.19 | 3.14 | 100 m$^2$ | **3.33** |
| 70 g/m$^2$ | 4.40 | – | – | 0.19 | 4.40 | 100 m$^2$ | **4.59** |
| 100 g/m$^2$ | 6.29 | – | – | 0.19 | 6.29 | 100 m$^2$ | **6.48** |
| Pre-seeding fertilizers (12+00+09); PC £0.63/kg; to seedbeds; by hand | | | | | | | |
| 35 g/m$^2$ | 2.20 | 0.17 | 3.96 | – | 2.20 | 100 m$^2$ | **6.16** |
| 50 g/m$^2$ | 3.14 | 0.17 | 3.96 | – | 3.14 | 100 m$^2$ | **7.10** |
| 70 g/m$^2$ | 4.40 | 0.17 | 3.96 | – | 4.40 | 100 m$^2$ | **8.36** |
| 100 g/m$^2$ | 6.29 | 0.20 | 4.75 | – | 6.29 | 100 m$^2$ | **11.04** |
| **SEEDING** | | | | | | | |
| **Seeding** | | | | | | | |
| Seeding labours only in two operations; by machine (for seed prices see above) | | | | | | | |
| 35 g/m$^2$ | – | – | – | 0.52 | – | 100 m$^2$ | **0.52** |
| Grass seed; spreading in two operations; PC £4.50/kg (for changes in material prices please refer to table above); by machine | | | | | | | |
| 35 g/m$^2$ | – | – | – | 0.52 | 15.75 | 100 m$^2$ | **16.27** |
| 50 g/m$^2$ | – | – | – | 0.52 | 22.50 | 100 m$^2$ | **23.02** |
| 70 g/m$^2$ | – | – | – | 0.52 | 31.50 | 100 m$^2$ | **32.02** |
| 100 g/m$^2$ | – | – | – | 0.52 | 45.00 | 100 m$^2$ | **45.52** |
| 125 kg/ha | – | – | – | 51.99 | 562.50 | ha | **614.49** |
| 150 kg/ha | – | – | – | 51.99 | 675.00 | ha | **726.99** |
| 200 kg/ha | – | – | – | 51.99 | 900.00 | ha | **951.99** |
| 250 kg/ha | – | – | – | 51.99 | 1125.00 | ha | **1176.99** |
| 300 kg/ha | – | – | – | 51.99 | 1350.00 | ha | **1401.99** |
| 350 kg/ha | – | – | – | 51.99 | 1575.00 | ha | **1626.99** |
| 400 kg/ha | – | – | – | 51.99 | 1800.00 | ha | **1851.99** |
| 500 kg/ha | – | – | – | 51.99 | 2250.00 | ha | **2301.99** |
| 700 kg/ha | – | – | – | 51.99 | 3150.00 | ha | **3201.99** |

## 37 SOFT LANDSCAPING

| Item<br>Excluding site overheads and profit | PC<br>£ | Labour<br>hours | Labour<br>£ | Plant<br>£ | Material<br>£ | Unit | Total<br>rate £ |
|---|---|---|---|---|---|---|---|
| **Extra over seeding by machine for slopes over 30° (allowing for the actual area but measured in plan)** | | | | | | | |
| 35 g/m² | – | – | – | 0.08 | 2.36 | 100 m² | 2.44 |
| 50 g/m² | – | – | – | 0.08 | 3.38 | 100 m² | 3.46 |
| 70 g/m² | – | – | – | 0.08 | 4.72 | 100 m² | 4.80 |
| 100 g/m² | – | – | – | 0.08 | 6.75 | 100 m² | 6.83 |
| 125 kg/ha | – | – | – | 7.80 | 84.38 | ha | 92.18 |
| 150 kg/ha | – | – | – | 7.80 | 101.25 | ha | 109.05 |
| 200 kg/ha | – | – | – | 7.80 | 135.00 | ha | 142.80 |
| 250 kg/ha | – | – | – | 7.80 | 168.75 | ha | 176.55 |
| 300 kg/ha | – | – | – | 7.80 | 202.50 | ha | 210.30 |
| 350 kg/ha | – | – | – | – | – | ha | – |
| 400 kg/ha | – | – | – | 7.80 | 270.00 | ha | 277.80 |
| 500 kg/ha | – | – | – | 7.80 | 337.50 | ha | 345.30 |
| 700 kg/ha | – | – | – | 7.80 | 472.50 | ha | 480.30 |
| **Seeding labours only in two operations; by hand (for seed prices see above)** | | | | | | | |
| 35 g/m² | – | 0.17 | 3.96 | – | – | 100 m² | 3.96 |
| **Grass seed; spreading in two operations; PC £4.50/kg (for changes in material prices please refer to table above); by hand** | | | | | | | |
| 35 g/m² | – | 0.17 | 3.96 | – | 15.75 | 100 m² | 19.71 |
| 50 g/m² | – | 0.17 | 3.96 | – | 22.50 | 100 m² | 26.46 |
| 70 g/m² | – | 0.17 | 3.96 | – | 31.50 | 100 m² | 35.46 |
| 100 g/m² | – | 0.20 | 4.75 | – | 45.00 | 100 m² | 49.75 |
| 125 g/m² | – | 0.20 | 4.75 | – | 56.25 | 100 m² | 61.00 |
| **Extra over seeding by hand for slopes over 30° (allowing for the actual area but measured in plan)** | | | | | | | |
| 35 g/m² | 2.34 | – | 0.09 | – | 2.34 | 100 m² | 2.43 |
| 50 g/m² | 3.38 | – | 0.09 | – | 3.38 | 100 m² | 3.47 |
| 70 g/m² | 4.72 | – | 0.09 | – | 4.72 | 100 m² | 4.81 |
| 100 g/m² | 6.75 | – | 0.10 | – | 6.75 | 100 m² | 6.85 |
| 125 g/m² | 8.41 | – | 0.10 | – | 8.41 | 100 m² | 8.51 |
| Harrowing seeded areas; light chain harrow | – | – | – | 0.09 | – | 100 m² | 0.09 |
| **Raking over seeded areas** | | | | | | | |
| by mechanical stone rake | – | – | – | 2.23 | – | 100 m² | 2.23 |
| by hand | – | 0.80 | 19.00 | – | – | 100 m² | 19.00 |
| **Rolling seeded areas; light roller** | | | | | | | |
| by tractor drawn roller | – | – | – | 0.55 | – | 100 m² | 0.55 |
| by pedestrian operated mechanical roller | – | 0.08 | 1.98 | 0.41 | – | 100 m² | 2.39 |
| by hand drawn roller | – | 0.17 | 3.96 | – | – | 100 m² | 3.96 |

## 37 SOFT LANDSCAPING

| Item<br>Excluding site overheads and profit | PC<br>£ | Labour<br>hours | Labour<br>£ | Plant<br>£ | Material<br>£ | Unit | Total<br>rate £ |
|---|---|---|---|---|---|---|---|
| **SEEDING – cont** | | | | | | | |
| **Seeding labours only in two operations – cont** | | | | | | | |
| Extra over harrowing, raking or rolling seeded areas for slopes over 30°; by machine or hand | – | – | – | – | – | 25% | – |
| Turf edging; to seeded areas; 300 mm wide | – | 0.05 | 1.13 | – | 2.05 | m² | 3.18 |
| **PREPARATION OF TURF BEDS** | | | | | | | |
| **Preparation of turf beds** | | | | | | | |
| Rolling turf to be lifted; lifting by hand or mechanical turf stripper; stacks to be not more than 1 m high | | | | | | | |
| cutting only preparing to lift; pedestrian turf cutter | – | 0.75 | 17.81 | 7.08 | – | 100 m² | 24.89 |
| lifting and stacking; by hand | – | 8.33 | 197.92 | – | – | 100 m² | 197.92 |
| Rolling up; moving to stacks | | | | | | | |
| distance not exceeding 100 m | – | 2.50 | 59.38 | – | – | 100 m² | 59.38 |
| extra over rolling and moving turf to stacks to transport per additional 100 m | – | 0.83 | 19.79 | – | – | 100 m² | 19.79 |
| Lifting selected and approved topsoil from spoil heaps | | | | | | | |
| passing through 6 mm screen; removing debris | – | 0.17 | 3.96 | 9.90 | – | m³ | 13.86 |
| Extra over lifting topsoil and passing through screen for imported topsoil; plus 20% allowance for settlement | – | – | – | – | 39.60 | m³ | 39.60 |
| Topsoil; PC £33.00/m³; plus 20% allowance for settlement | | | | | | | |
| 25 mm deep | – | – | – | – | 0.99 | m² | 0.99 |
| 50 mm deep | – | – | – | – | 1.98 | m² | 1.98 |
| 100 mm deep | – | – | – | – | 3.96 | m² | 3.96 |
| 150 mm deep | – | – | – | – | 5.94 | m² | 5.94 |
| 200 mm deep | – | – | – | – | 7.92 | m² | 7.92 |
| 250 mm deep | – | – | – | – | 9.90 | m² | 9.90 |
| 300 mm deep | – | – | – | – | 11.88 | m² | 11.88 |
| 400 mm deep | – | – | – | – | 15.84 | m² | 15.84 |
| 450 mm deep | – | – | – | – | 17.82 | m² | 17.82 |
| Spreading topsoil to form turfbeds (topsoil not included); by machine | | | | | | | |
| 25 mm deep | – | – | 0.06 | 0.15 | – | m² | 0.21 |
| 50 mm deep | – | – | 0.08 | 0.19 | – | m² | 0.27 |
| 75 mm deep | – | – | 0.09 | 0.22 | – | m² | 0.31 |
| 100 mm deep | – | 0.01 | 0.12 | 0.29 | – | m² | 0.41 |
| 150 mm deep | – | 0.01 | 0.18 | 0.43 | – | m² | 0.61 |

## 37 SOFT LANDSCAPING

| Item Excluding site overheads and profit | PC £ | Labour hours | Labour £ | Plant £ | Material £ | Unit | Total rate £ |
|---|---|---|---|---|---|---|---|
| **Spreading topsoil to form turfbeds** (topsoil not included); by hand | | | | | | | |
| 25 mm deep | – | 0.03 | 0.59 | – | – | m² | **0.59** |
| 50 mm deep | – | 0.03 | 0.79 | – | – | m² | **0.79** |
| 75 mm deep | – | 0.04 | 1.02 | – | – | m² | **1.02** |
| 100 mm deep | – | 0.05 | 1.19 | – | – | m² | **1.19** |
| 150 mm deep | – | 0.08 | 1.78 | – | – | m² | **1.78** |
| **Cultivation by pedestrian operated rotavator** | | | | | | | |
| Breaking up existing ground; tine cultivator or rotavator | | | | | | | |
| 100 mm deep | – | 0.28 | 6.53 | 0.99 | – | 100 m² | **7.52** |
| 150 mm deep | – | 0.31 | 7.46 | 1.14 | – | 100 m² | **8.60** |
| 200 mm deep | – | 0.40 | 9.50 | 1.44 | – | 100 m² | **10.94** |
| As above but in heavy clay or wet soils | | | | | | | |
| 100 mm deep | – | 0.44 | 10.45 | 1.59 | – | 100 m² | **12.04** |
| 150 mm deep | – | 0.66 | 15.68 | 2.38 | – | 100 m² | **18.06** |
| 200 mm deep | – | 0.82 | 19.59 | 2.98 | – | 100 m² | **22.57** |
| Bringing existing topsoil to a fine tilth for turfing by raking or harrowing; stones not to exceed 6 mm; by machine | – | – | 0.10 | 0.11 | – | m² | **0.21** |
| Bringing existing topsoil to a fine tilth for turfing by raking; stones not to exceed 6 mm; by hand | | | | | | | |
| general commercial or amenity turf | – | 0.01 | 0.30 | – | – | m² | **0.30** |
| higher grade fine culivation | – | 0.02 | 0.47 | – | – | m² | **0.47** |
| **TURFING** | | | | | | | |
| **Turfing (see also Preparation of turf beds – above)** | | | | | | | |
| Hand load and transport and unload turf from delivery area to laying area | | | | | | | |
| by small site dumper up to 50 m | – | 0.01 | 0.47 | 0.04 | – | m² | **0.51** |
| by hand 25 m | – | 0.03 | 0.59 | – | – | m² | **0.59** |
| by hand 50 m | – | 0.03 | 0.68 | – | – | m² | **0.68** |
| Turfing; laying only; to stretcher bond; butt joints; including providing and working from barrow plank runs where necessary to surfaces not exceeding 30° from horizontal | | | | | | | |
| specially selected lawn turves from previously lifted stockpile | – | 0.08 | 1.78 | – | – | m² | **1.78** |
| cultivated lawn turves; to large open areas | – | 0.06 | 1.38 | – | – | m² | **1.38** |
| cultivated lawn turves; to domestic or garden areas | – | 0.08 | 1.85 | – | – | m² | **1.85** |
| road verge quality turf | – | 0.04 | 0.95 | – | – | m² | **0.95** |

## 37 SOFT LANDSCAPING

| Item<br>Excluding site overheads and profit | PC<br>£ | Labour<br>hours | Labour<br>£ | Plant<br>£ | Material<br>£ | Unit | Total<br>rate £ |
|---|---|---|---|---|---|---|---|
| **TURFING – cont**<br>**Industrially grown turf; PC prices listed represent the general range of industrial turf prices for sportsfields and amenity purposes; prices will vary with quantity and site location**<br>Rolawn | | | | | | | |
| RB Medallion; sports fields, domestic lawns, general landscape; full loads 1720 m$^2$ | 1.87 | 0.08 | 1.98 | – | 1.87 | m$^2$ | **3.85** |
| RB Medallion; sports fields, domestic lawns, general landscape; part loads | 2.05 | 0.08 | 1.98 | – | 2.05 | m$^2$ | **4.03** |
| Tensar Ltd | | | | | | | |
| Tensar Mat 400 reinforced turf for embankments; laid to embankments 3.3 m$^2$ turves | 3.85 | 0.11 | 2.61 | – | 3.85 | m$^2$ | **6.46** |
| Inturf | | | | | | | |
| Inturf Masters; formal lawns, golf greens, bowling greens and low maintenance areas | 5.83 | 0.10 | 2.27 | – | 5.83 | m$^2$ | **8.10** |
| Inturf Classic; golf tees, surrounds, lawns, parks, general purpose landscaping areas and winter sports | 4.83 | 0.05 | 1.25 | – | 4.83 | m$^2$ | **6.08** |
| Inturf Ornamental; medium fine turf for ornamental lawns, fairways and green surrounds | 5.33 | 0.05 | 1.29 | – | 5.33 | m$^2$ | **6.62** |
| Inturf Custom Grown Turf; tailor made for any turfgrass specification | 10.33 | 0.08 | 1.90 | – | 10.33 | m$^2$ | **12.23** |
| RTF Rhizomatous Tall Fescue; deep rooted self repairing for our changing climate | 5.33 | 0.08 | 1.90 | – | 5.33 | m$^2$ | **7.23** |
| Turf laid in Big roll; Inturf | | | | | | | |
| Inturf Classic; golf tees, surrounds, lawns, parks, general purpose landscaping areas and winter sports | 4.00 | 0.01 | 0.57 | – | 5.13 | m$^2$ | **5.70** |
| **Specialized turf products; Lindum Turf Ltd; Laid to prepared and cultivated soil (not included)**<br>Rhizomatous Tall Fescue (RTF), new generation tall fescue which produces rhizomes that send a shoot up to the soil surface while extending new roots downwards | | | | | | | |
| shaded areas or drought tolerant turf | 4.10 | 0.08 | 1.98 | – | 5.97 | m$^2$ | **7.95** |
| Wildflower Meadow Turf; British wildflowers and grasses 50:50; P | | | | | | | |
| Flowering height 40–60 cm | 14.35 | 0.08 | 1.98 | – | 15.85 | m$^2$ | **17.83** |

## 37 SOFT LANDSCAPING

| Item<br>Excluding site overheads and profit | PC<br>£ | Labour<br>hours | Labour<br>£ | Plant<br>£ | Material<br>£ | Unit | Total<br>rate £ |
|---|---|---|---|---|---|---|---|
| **Wildflower & Perennial Turf;** wildflowers, herbs and perennials. For dry or exposed gardens, roof gardens or other situations requiring drought tolerant planting | | | | | | | |
| Flowering height 20–30 cm | 22.00 | 0.08 | 1.98 | – | 22.00 | m² | **23.98** |
| **Instant use turf; LT6 Lindum turf; grown in an imported sand/soil rootzone, harvested up to 40 mm thickness and installed using specialist machinery. Compatibility with existing rootzones** | | | | | | | |
| For sports pitches or high wear areas where instant use is required inclusive of removal of existing rootzone | | | | | | | |
| minimum area 150 m² | 8.00 | – | – | – | 24.50 | m² | **24.50** |
| **Reinforced turf; AGB Advanced Turf; blended mesh elements incorporated into root zone; root zone spread and levelled over cultivated, prepared and reduced and levelled ground (not included); compacted with vibratory roller** | | | | | | | |
| ABG ATS 400/B with selected turf and fertilizer 100 mm thick | | | | | | | |
| 100–500 m² | 25.00 | 0.03 | 0.79 | 1.93 | 25.00 | m² | **27.72** |
| over 500 m² | 23.50 | 0.03 | 0.79 | 1.93 | 23.50 | m² | **26.22** |
| ABG ATS 400/B with selected turf and fertilizer 150 mm thick | | | | | | | |
| 100–500 m² | 30.00 | 0.04 | 0.95 | 2.25 | 30.00 | m² | **33.20** |
| over 500 m² | 28.00 | 0.04 | 0.95 | 2.25 | 28.00 | m² | **31.20** |
| ABG ATS 400/B with selected turf and fertilizer 200 mm thick | | | | | | | |
| 100–500 m² | 36.00 | 0.05 | 1.19 | 2.73 | 36.00 | m² | **39.92** |
| over 500 m² | 34.00 | 0.05 | 1.19 | 2.73 | 34.00 | m² | **37.92** |
| Firming turves with wooden beater | – | 0.01 | 0.24 | – | – | m² | **0.24** |
| Rolling turfed areas; light roller | | | | | | | |
| by tractor with turf tyres and roller | – | – | – | 0.55 | – | 100 m² | **0.55** |
| by pedestrian operated mechanical roller | – | 0.08 | 1.98 | 0.41 | – | 100 m² | **2.39** |
| by hand drawn roller | – | 0.17 | 3.96 | – | – | 100 m² | **3.96** |
| Dressing with finely sifted topsoil; brushing into joints | 0.05 | 0.05 | 1.19 | – | 0.05 | m² | **1.24** |
| Turfing; laying only | | | | | | | |
| to slopes over 30°; to diagonal bond (measured as plan area – add 15% to these rates for the incline area of 30° slopes) | – | 0.12 | 2.85 | – | – | m² | **2.85** |

## 37 SOFT LANDSCAPING

| Item<br>Excluding site overheads and profit | PC<br>£ | Labour<br>hours | Labour<br>£ | Plant<br>£ | Material<br>£ | Unit | Total<br>rate £ |
|---|---|---|---|---|---|---|---|
| **TURFING – cont** | | | | | | | |
| **Reinforced turf – cont** | | | | | | | |
| Extra over laying turfing for pegging down turves | | | | | | | |
| wooden or galvanized wire pegs; 200 mm long; 2 pegs per 0.50 m² | 1.63 | 0.01 | 0.32 | – | 1.63 | m² | **1.95** |
| **WILD FLOWER MEADOWS** | | | | | | | |
| Note for this section; Prices exclude for delivery and collection of specialized machinery where used to the site. | | | | | | | |
| **Topsoil stripping for wild flower planting** | | | | | | | |
| Stripping to subsoil layer 300 deep | | | | | | | |
| moving to stockpile | – | – | – | 1.33 | – | m² | **1.33** |
| spreading locally | – | – | – | 1.92 | – | m² | **1.92** |
| **Soil testing for wildflower meadows** | | | | | | | |
| Collection of samples and submission to soil laboratory for analysis and recommendation for wildflower meadow establishment | | | | | | | |
| 3 nr random samples | – | – | – | – | – | nr | **800.00** |
| **Cultivation and preparation of seedbeds for wildflower meadows** | | | | | | | |
| Cultivate stripped, filled or existing surface prior to application of herbicides and leave fallow for seedbank eradication period | | | | | | | |
| by tractor drawn implements including rough grading to medium tilth | – | – | – | 1.56 | – | 100 m² | **1.56** |
| by hand drawn rotavator including hand raking, and hand grading, to medium tilth | – | 0.72 | 17.10 | 0.79 | – | 100 m² | **17.89** |

## 37 SOFT LANDSCAPING

| Item<br>Excluding site overheads and profit | PC<br>£ | Labour<br>hours | Labour<br>£ | Plant<br>£ | Material<br>£ | Unit | Total<br>rate £ |
|---|---|---|---|---|---|---|---|
| **Seedbank eradication for wildflower meadows** | | | | | | | |
| Treat stripped intended meadow area to eradicate inherent seedbank; allow optimum 2 years of repeat eradication by herbicide returning regularly to reapply herbicides | | | | | | | |
| Existing or reclaimed brownfield area by tractor drawn application; 4 applications | – | 0.04 | 0.95 | 0.61 | 0.25 | 100 m$^2$ | **1.81** |
| Existing or reclaimed brownfield area by tractor drawn application; 3 applications | – | 0.03 | 0.71 | 0.46 | 0.19 | 100 m$^2$ | **1.36** |
| Existing paddocks or agricultural area; knapsack; 4 applications | – | 2.12 | 50.35 | – | 0.25 | 100 m$^2$ | **50.60** |
| Existing or reclaimed brownfield area by backpack spray application; 3 applications | – | 1.59 | 37.76 | – | 0.19 | 100 m$^2$ | **37.95** |
| **Subsoil filling to wildflower meadows** | | | | | | | |
| Average thickness 300 mm; filling of poor nutrient value soil to form seedbed for wildflower area; grading to level and preparing to medium tilth | | | | | | | |
| from stockpile on site | – | – | – | 22.17 | – | 100 m$^2$ | **22.17** |
| imported subsoil fill | – | – | – | 22.17 | 0.72 | 100 m$^2$ | **22.89** |
| **Seeding to wildflower meadows; Spreading of selected wildflower seed to manufacturer's specifications; by tractor drawn seeder** | | | | | | | |
| General areas | | | | | | | |
| DLF Seeds Ltd; Pro Flora 8 Old English Country Meadow Mix; 5 g/m$^2$ | – | – | – | 1.23 | 24.00 | 100 m$^2$ | **25.23** |
| Acid soils | | | | | | | |
| Germinal GB; WFG2 (Annual Meadow); 5 g/m$^2$ | – | – | – | 1.23 | 27.50 | 100 m$^2$ | **28.73** |
| Neutral soils | | | | | | | |
| Germinal GB; WFG4 (Neutral Meadow); 5 g/m$^2$ | – | – | – | 1.23 | 29.70 | 100 m$^2$ | **30.93** |
| DLF Seeds Ltd; Pro Flora 3 Damp loamy soils; 5 g/m$^2$ | – | – | – | 1.23 | 16.10 | 100 m$^2$ | **17.33** |
| Heavy clay soils | | | | | | | |
| Germinal GB; WFG6 (Clay Soils); 5 g/m$^2$ | – | – | – | 1.23 | 32.75 | 100 m$^2$ | **33.98** |

## 37 SOFT LANDSCAPING

| Item<br>Excluding site overheads and profit | PC<br>£ | Labour<br>hours | Labour<br>£ | Plant<br>£ | Material<br>£ | Unit | Total<br>rate £ |
|---|---|---|---|---|---|---|---|
| **WILD FLOWER MEADOWS – cont** | | | | | | | |
| **Seeding to wildflower meadows; Spreading of selected wildflower seed to manufacturer's specifications; by pedestrian drawn seeder** | | | | | | | |
| General areas | | | | | | | |
| DLF Seeds Ltd; Pro Flora 8 Old English Country Meadow Mix; 5 g/m$^2$ | – | 0.25 | 5.94 | – | 24.00 | 100 m$^2$ | **29.94** |
| Acid soils | | | | | | | |
| Germinal GB; WFG2 (Annual Meadow); 5 g/m$^2$ | – | 0.25 | 5.94 | – | 27.50 | 100 m$^2$ | **33.44** |
| Neutral soils | | | | | | | |
| Germinal GB; WFG4 (Neutral Meadow); 5 g/m$^2$ | – | 0.25 | 5.94 | – | 29.70 | 100 m$^2$ | **35.64** |
| DLF Seeds Ltd; Pro Flora 3 Damp loamy soils; 5 g/m$^2$ | – | 0.25 | 5.94 | – | 16.10 | 100 m$^2$ | **22.04** |
| Heavy clay soils | | | | | | | |
| Germinal GB; WFG6 (Clay Soils); 5 g/m$^2$ | – | 0.25 | 5.94 | – | 32.75 | 100 m$^2$ | **38.69** |
| **ARTIFICIAL TURF** | | | | | | | |
| **Artificial grass; Trulawn Limited**<br>Artificial lawns laid to cleared and levelled area (not included); laid to 50 mm Type 1 drainage layer on 25 mm sharp sand bed | | | | | | | |
| Trulawn Value; 16 mm roof; terraces, balconies, general use | – | – | – | – | – | m$^2$ | **35.95** |
| Trulawn Play; 20 mm; entry level; lawn and play areas | – | – | – | – | – | m$^2$ | **37.95** |
| Trulawn Continental; 20 mm realistic appearance; lawns and patios | – | – | – | – | – | m$^2$ | **33.85** |
| Trulawn Optimum; 26 mm; softest pile; lush green; lawns and patios | – | – | – | – | – | m$^2$ | **45.50** |
| Trulawn Luxury; 30 mm; deep pile; realistic lawn | – | – | – | – | – | m$^2$ | **47.50** |

## 37 SOFT LANDSCAPING

| Item Excluding site overheads and profit | PC £ | Labour hours | Labour £ | Plant £ | Material £ | Unit | Total rate £ |
|---|---|---|---|---|---|---|---|
| **MAINTENANCE OF SEEDED/ TURFED AREAS** | | | | | | | |
| **Maintenance operations (Note: the following rates apply to aftercare maintenance executed as part of a landscaping contract only)** | | | | | | | |
| Initial cutting; to turfed areas | | | | | | | |
| 20 mm high; using pedestrian guided power driven cylinder mower; including boxing off cuttings (stone picking and rolling not included) | – | 0.18 | 4.28 | 0.31 | – | 100 m² | **4.59** |
| Repairing damaged grass areas scraping out; removing slurry; from ruts and holes; average 100 mm deep | – | 0.13 | 3.17 | – | – | m² | **3.17** |
| 100 mm topsoil | – | 0.13 | 3.17 | – | 3.96 | m² | **7.13** |
| Repairing damaged grass areas; sowing grass seed to match existing or as specified; to individually prepared worn patches | | | | | | | |
| 35 g/m² | 0.18 | 0.01 | 0.24 | – | 0.18 | m² | **0.42** |
| 50 g/m² | 0.26 | 0.01 | 0.24 | – | 0.26 | m² | **0.50** |
| Sweeping leaves; disposing off site; motorized vacuum sweeper or rotary brush sweeper | | | | | | | |
| areas of maximum 2500 m² with occasional large tree and established boundary planting; 4.6 m³ (one skip of material to be removed) | – | 0.40 | 9.50 | 3.64 | – | 100 m² | **13.14** |
| Leaf clearance; clearing grassed area of leaves and other extraneous debris Using equipment towed by tractor | | | | | | | |
| large grassed areas with perimeters of mature trees such as sports fields and amenity areas | – | 0.01 | 0.30 | 0.07 | – | 100 m² | **0.37** |
| large grassed areas containing ornamental trees and shrub beds | – | 0.03 | 0.59 | 0.09 | – | 100 m² | **0.68** |
| Using pedestrian operated mechanical equipment and blowers | | | | | | | |
| grassed areas with perimeters of mature trees such as sports fields and amenity areas | – | 0.04 | 0.95 | 0.03 | – | 100 m² | **0.98** |
| grassed areas containing ornamental trees and shrub beds | – | 0.10 | 2.38 | 0.07 | – | 100 m² | **2.45** |
| verges | – | 0.07 | 1.58 | 0.04 | – | 100 m² | **1.62** |

## 37 SOFT LANDSCAPING

| Item<br>Excluding site overheads and profit | PC<br>£ | Labour<br>hours | Labour<br>£ | Plant<br>£ | Material<br>£ | Unit | Total<br>rate £ |
|---|---|---|---|---|---|---|---|
| **MAINTENANCE OF SEEDED/ TURFED AREAS – cont** | | | | | | | |
| By hand | | | | | | | |
| grassed areas with perimeters of mature trees such as sports fields and amenity areas | – | 0.05 | 1.19 | 0.10 | – | 100 m² | **1.29** |
| grassed areas containing ornamental trees and shrub beds | – | 0.08 | 1.98 | 0.17 | – | 100 m² | **2.15** |
| verges | – | 1.00 | 23.75 | 1.99 | – | 100 m² | **25.74** |
| Removal of arisings | | | | | | | |
| areas with perimeters of mature trees | – | 0.01 | 0.13 | 0.09 | 0.10 | 100 m² | **0.32** |
| areas containing ornamental trees and shrub beds | – | 0.02 | 0.38 | 0.34 | 0.26 | 100 m² | **0.98** |
| Cutting grass to specified height; per cut | | | | | | | |
| multi-unit gang mower | – | 0.59 | 13.97 | 19.35 | – | ha | **33.32** |
| ride-on triple cylinder mower | – | 0.01 | 0.33 | 0.14 | – | 100 m² | **0.47** |
| ride-on triple rotary mower | – | 0.01 | 0.33 | – | – | 100 m² | **0.33** |
| pedestrian mower | – | 0.18 | 4.28 | 0.73 | – | 100 m² | **5.01** |
| Cutting grass to banks; per cut | | | | | | | |
| side arm cutter bar mower | – | 0.02 | 0.55 | 0.36 | – | 100 m² | **0.91** |
| Cutting rough grass; per cut | | | | | | | |
| power flail or scythe cutter | – | 0.04 | 0.83 | – | – | 100 m² | **0.83** |
| Extra over cutting grass for slopes not exceeding 30° | – | – | – | – | – | 10% | – |
| Extra over cutting grass for slopes exceeding 30° | – | – | – | – | – | 40% | – |
| Cutting fine sward | | | | | | | |
| pedestrian operated seven-blade cylinder lawn mower | – | 0.14 | 3.33 | 0.23 | – | 100 m² | **3.56** |
| Extra over cutting fine sward for boxing off cuttings | | | | | | | |
| pedestrian mower | – | 0.03 | 0.67 | 0.05 | – | 100 m² | **0.72** |
| Cutting areas of rough grass | | | | | | | |
| scythe | – | 1.00 | 23.75 | – | – | 100 m² | **23.75** |
| sickle | – | 2.00 | 47.50 | – | – | 100 m² | **47.50** |
| petrol operated strimmer | – | 0.30 | 7.13 | 0.43 | – | 100 m² | **7.56** |
| Cutting areas of rough grass which contain trees or whips | | | | | | | |
| petrol operated strimmer | – | 0.40 | 9.50 | 0.57 | – | 100 m² | **10.07** |
| Extra over cutting rough grass for on site raking up and dumping | – | 0.33 | 7.92 | – | – | 100 m² | **7.92** |
| Trimming edge of grass areas; edging tool | | | | | | | |
| with petrol powered strimmer | – | 0.13 | 3.17 | 0.19 | – | 100 m | **3.36** |
| by hand | – | 0.67 | 15.83 | – | – | 100 m | **15.83** |

## 37 SOFT LANDSCAPING

| Item<br>Excluding site overheads and profit | PC<br>£ | Labour<br>hours | Labour<br>£ | Plant<br>£ | Material<br>£ | Unit | Total<br>rate £ |
|---|---|---|---|---|---|---|---|
| Marking out pitches using approved line marking compound; including initial setting out and marking | | | | | | | |
|   discus, hammer, javelin or shot putt area | 4.68 | 2.00 | 47.50 | – | 4.68 | nr | 52.18 |
|   cricket square | 3.12 | 2.00 | 47.50 | – | 3.12 | nr | 50.62 |
|   cricket boundary | 10.92 | 8.00 | 190.00 | – | 10.92 | nr | 200.92 |
|   grass tennis court | 4.68 | 4.00 | 95.00 | – | 4.68 | nr | 99.68 |
|   hockey pitch | 15.60 | 8.00 | 190.00 | – | 15.60 | nr | 205.60 |
|   football pitch | 15.60 | 8.00 | 190.00 | – | 15.60 | nr | 205.60 |
|   rugby pitch | 15.60 | 8.00 | 190.00 | – | 15.60 | nr | 205.60 |
|   eight lane running track; 400 m | 31.20 | 16.00 | 380.00 | – | 31.20 | nr | 411.20 |
| Re-marking out pitches using approved line marking compound | | | | | | | |
|   discus, hammer, javelin or shot putt area | 3.12 | 0.50 | 11.88 | – | 3.12 | nr | 15.00 |
|   cricket square | 3.12 | 0.50 | 11.88 | – | 3.12 | nr | 15.00 |
|   grass tennis court | 10.92 | 1.00 | 23.75 | – | 10.92 | nr | 34.67 |
|   hockey pitch | 10.92 | 1.00 | 23.75 | – | 10.92 | nr | 34.67 |
|   football pitch | 10.92 | 1.00 | 23.75 | – | 10.92 | nr | 34.67 |
|   rugby pitch | 10.92 | 1.00 | 23.75 | – | 10.92 | nr | 34.67 |
|   eight lane running track; 400 m | 31.20 | 2.50 | 59.38 | – | 31.20 | nr | 90.58 |
| Rolling grass areas; light roller | | | | | | | |
|   by tractor drawn roller | – | – | – | 0.55 | – | 100 m$^2$ | 0.55 |
|   by pedestrian operated mechanical roller | – | 0.08 | 1.98 | 0.41 | – | 100 m$^2$ | 2.39 |
|   by hand drawn roller | – | 0.17 | 3.96 | – | – | 100 m$^2$ | 3.96 |
| Aerating grass areas; to a depth of 100 mm | | | | | | | |
|   using tractor drawn aerator | – | 0.06 | 1.38 | 5.04 | – | 100 m$^2$ | 6.42 |
|   using pedestrian guided motor powered solid or slitting tine turf aerator | – | 0.18 | 4.16 | 1.86 | – | 100 m$^2$ | 6.02 |
|   using hollow tine aerator; including sweeping up and dumping corings | – | 0.50 | 11.88 | 3.73 | – | 100 m$^2$ | 15.61 |
|   using hand aerator or fork | – | 1.67 | 39.58 | – | – | 100 m$^2$ | 39.58 |
| Extra over aerating grass areas for on site sweeping up and dumping corings | – | 0.17 | 3.96 | – | – | 100 m$^2$ | 3.96 |
| Switching off dew; from fine turf areas | – | 0.20 | 4.75 | – | – | 100 m$^2$ | 4.75 |
| Scarifying grass areas to break up thatch; removing dead grass | | | | | | | |
|   using tractor drawn scarifier | – | 0.07 | 1.66 | 0.31 | – | 100 m$^2$ | 1.97 |
|   using self-propelled scarifier; including removing and disposing of grass on site | – | 0.33 | 7.92 | 0.12 | – | 100 m$^2$ | 8.04 |
| Harrowing grass areas | | | | | | | |
|   using drag mat | – | 0.03 | 0.67 | 0.31 | – | 100 m$^2$ | 0.98 |
|   using chain harrow | – | 0.04 | 0.83 | 0.39 | – | 100 m$^2$ | 1.22 |
|   using drag mat | – | 2.80 | 66.51 | 30.86 | – | ha | 97.37 |
|   using chain harrow | – | 3.50 | 83.13 | 38.58 | – | ha | 121.71 |

## 37 SOFT LANDSCAPING

| Item<br>Excluding site overheads and profit | PC<br>£ | Labour<br>hours | Labour<br>£ | Plant<br>£ | Material<br>£ | Unit | Total<br>rate £ |
|---|---|---|---|---|---|---|---|
| **MAINTENANCE OF SEEDED/ TURFED AREAS – cont** | | | | | | | |
| Extra for scarifying and harrowing grass areas for disposing excavated material off site; to tip; loading by machine | | | | | | | |
| slightly contaminated | – | – | – | 2.52 | 48.33 | m³ | 50.85 |
| rubbish | – | – | – | 2.52 | 48.33 | m³ | 50.85 |
| inert material | – | – | – | 1.68 | 30.81 | m³ | 32.49 |
| For the following topsoil improvement and seeding operations add or subtract the following amounts for every £0.10 difference in the material cost price | | | | | | | |
| 35 g/m² | – | – | – | – | 0.35 | 100 m² | 0.35 |
| 50 g/m² | – | – | – | – | 0.50 | 100 m² | 0.50 |
| 70 g/m² | – | – | – | – | 0.70 | 100 m² | 0.70 |
| 100 g/m² | – | – | – | – | 1.00 | 100 m² | 1.00 |
| 125 kg/ha | – | – | – | – | 12.50 | ha | 12.50 |
| 150 kg/ha | – | – | – | – | 15.00 | ha | 15.00 |
| 175 kg/ha | – | – | – | – | 17.50 | ha | 17.50 |
| 200 kg/ha | – | – | – | – | 20.00 | ha | 20.00 |
| 225 kg/ha | – | – | – | – | 22.50 | ha | 22.50 |
| 250 kg/ha | – | – | – | – | 25.00 | ha | 25.00 |
| 300 kg/ha | – | – | – | – | 30.00 | ha | 30.00 |
| 350 kg/ ha | – | – | – | – | 35.00 | ha | 35.00 |
| 400 kg/ha | – | – | – | – | 40.00 | ha | 40.00 |
| 500 kg/ha | – | – | – | – | 50.00 | ha | 50.00 |
| 700 kg/ha | – | – | – | – | 70.00 | ha | 70.00 |
| 1000 kg/ha | – | – | – | – | 100.00 | ha | 100.00 |
| 1250 kg/ha | – | – | – | – | 125.00 | ha | 125.00 |
| Top dressing fertilizers (7+7+7); PC £1.11/kg; to seedbeds; by machine | | | | | | | |
| 35 g/m² | 3.91 | – | – | 0.19 | 3.91 | 100 m² | 4.10 |
| 50 g/m² | 5.58 | – | – | 0.19 | 5.58 | 100 m² | 5.77 |
| 300 kg/ha | 334.80 | – | – | 19.14 | 334.80 | ha | 353.94 |
| 350 kg/ha | 390.60 | – | – | 19.14 | 390.60 | ha | 409.74 |
| 400 kg/ha | 446.40 | – | – | 19.14 | 446.40 | ha | 465.54 |
| 500 kg/ha | 558.00 | – | – | 30.62 | 558.00 | ha | 588.62 |
| Top dressing fertilizers (7+7+7); PC £1.11/kg; to seedbeds; by hand | | | | | | | |
| 35 g/m² | 3.91 | 0.17 | 3.96 | – | 3.91 | 100 m² | 7.87 |
| 50 g/m² | 5.58 | 0.17 | 3.96 | – | 5.58 | 100 m² | 9.54 |
| 70 g/m² | 7.81 | 0.17 | 3.96 | – | 7.81 | 100 m² | 11.77 |
| Watering turf; evenly; at a rate of 5 l/m² using movable spray lines powering 3 nr sprinkler heads with a radius of 15 m and allowing for 60% overlap (irrigation machinery costs not included) | – | 0.02 | 0.37 | – | – | 100 m² | 0.37 |
| using sprinkler equipment and with sufficient water pressure to run 1 nr 15 m radius sprinkler | – | 0.02 | 0.47 | – | – | 100 m² | 0.47 |
| using hand-held watering equipment | – | 0.25 | 5.94 | – | – | 100 m² | 5.94 |

## 37 SOFT LANDSCAPING

| Item Excluding site overheads and profit | PC £ | Labour hours | Labour £ | Plant £ | Material £ | Unit | Total rate £ |
|---|---|---|---|---|---|---|---|
| **PLANTING GENERAL** | | | | | | | |
| **Planting – General** | | | | | | | |
| Preamble: Prices for all planting work are deemed to include carrying out planting in accordance with all good horticultural practice. | | | | | | | |
| **MARKET PRICES OF MATERIALS** | | | | | | | |
| Market prices of planting materials (Note: the rates shown generally reflect the manufacturer's recommended retail prices; trade and bulk discounts are often available on the prices shown) | | | | | | | |
| Imported topsoil; market prices; British Sugar Topsoil; sustainably sourced graded topsoil to British standards and with independent certification; rates shown allow for 20% settlement | | | | | | | |
| Landscape 20; BS3882:2007; sandy loam | | | | | | | |
| 8 wheel delivery; 20 tonne loads (approximately 13.33 m$^3$) | – | – | – | – | 37.50 | m$^3$ | 37.50 |
| articulated delivery | – | – | – | – | 30.00 | m$^3$ | 30.00 |
| Sports 10; for topdressing to sports pitches; application rate 40–80 t/6000 m$^2$ pitch | | | | | | | |
| 8 wheel delivery; 20 tonne loads (approximately 13.33 m$^3$) | – | – | – | – | 45.00 | m$^3$ | 45.00 |
| articulated delivery | – | – | – | – | 37.50 | m$^3$ | 37.50 |
| **Topsoil prices; Average locally sourced prices** | | | | | | | |
| Topsoil | | | | | | | |
| general purpose | – | – | – | – | 20.00 | m$^3$ | 20.00 |
| Topsoil enhanced with organic matter | – | – | – | – | 20.00 | m$^3$ | 20.00 |
| **Market prices of mulching materials** | | | | | | | |
| Melcourt Industries Ltd; 25 m$^3$ loads (items labelled FSC are Forest Stewardship Council certified, items marked FT are certified fire tested) | | | | | | | |
| Ornamental Bark Mulch FT | – | – | – | – | 58.90 | m$^3$ | 58.90 |
| Bark Nuggets® FT | – | – | – | – | 55.20 | m$^3$ | 55.20 |
| Amenity Bark Mulch FSC FT | – | – | – | – | 40.05 | m$^3$ | 40.05 |

## 37 SOFT LANDSCAPING

| Item<br>Excluding site overheads and profit | PC<br>£ | Labour<br>hours | Labour<br>£ | Plant<br>£ | Material<br>£ | Unit | Total<br>rate £ |
|---|---|---|---|---|---|---|---|
| **MARKET PRICES OF MATERIALS – cont** | | | | | | | |
| **Market prices of mulching materials – cont** | | | | | | | |
| Melcourt Industries Ltd – cont | | | | | | | |
| Contract Bark Mulch FSC | – | – | – | – | 37.40 | m$^3$ | 37.40 |
| Spruce Ornamental FSC FT | – | – | – | – | 40.90 | m$^3$ | 40.90 |
| Decorative Biomulch® | – | – | – | – | 40.05 | m$^3$ | 40.05 |
| Rustic Biomulch® | – | – | – | – | 43.80 | m$^3$ | 43.80 |
| Mulch 2000 | – | – | – | – | 30.20 | m$^3$ | 30.20 |
| Forest BioMulch® | – | – | – | – | 37.55 | m$^3$ | 37.55 |
| Melcourt Industries Ltd; 50 m$^3$ loads | | | | | | | |
| Mulch 2000 | – | – | – | – | 19.45 | m$^3$ | 19.45 |
| Melcourt Industries Ltd; 70 m$^3$ loads | | | | | | | |
| Contract Bark Mulch FSC | – | – | – | – | 25.00 | m$^3$ | 25.00 |
| Melcourt Industries Ltd; 80 m$^3$ loads | | | | | | | |
| Ornamental Bark Mulch FT | – | – | – | – | 45.90 | m$^3$ | 45.90 |
| Bark Nuggets® FT | – | – | – | – | 42.20 | m$^3$ | 42.20 |
| Amenity Bark Mulch FSC | – | – | – | – | 27.15 | m$^3$ | 27.15 |
| Spruce Ornamental FSC FT | – | – | – | – | 27.90 | m$^3$ | 27.90 |
| Decorative Biomulch® | – | – | – | – | 27.05 | m$^3$ | 27.05 |
| Rustic Biomulch® | – | – | – | – | 30.80 | m$^3$ | 30.80 |
| Forest BioMulch® | – | – | – | – | 24.55 | m$^3$ | 24.55 |
| Mulch; Melcourt Industries Ltd; 25 m$^3$ loads | | | | | | | |
| Composted Fine Bark FSC | – | – | – | – | 37.15 | m$^3$ | 37.15 |
| Humus 2000 | – | – | – | – | 30.85 | m$^3$ | 30.85 |
| Spent Mushroom Compost | – | – | – | – | 24.20 | m$^3$ | 24.20 |
| Topgrow | – | – | – | – | 36.55 | m$^3$ | 36.55 |
| Mulch; Melcourt Industries Ltd; 50 m$^3$ loads | | | | | | | |
| Humus 2000 | – | – | – | – | 19.35 | m$^3$ | 19.35 |
| Mulch; Melcourt Industries Ltd; 60 m$^3$ loads | | | | | | | |
| Spent Mushroom Compost | – | – | – | – | 12.50 | m$^3$ | 12.50 |
| Mulch; Melcourt Industries Ltd; 65 m$^3$ loads | | | | | | | |
| Composted Fine Bark | – | – | – | – | 24.65 | m$^3$ | 24.65 |
| Topgrow | – | – | – | – | 28.55 | m$^3$ | 28.55 |
| **Market prices of fertilizers** | | | | | | | |
| Fertilizers; Germinal GB | | | | | | | |
| BSH1; 6+9+6 | – | – | – | – | 30.50 | kg | 30.50 |
| BSH4; 11+5+5 | – | – | – | – | 30.00 | kg | 30.00 |
| BSH6; 20+10+10 | – | – | – | – | 33.00 | kg | 33.00 |
| BSH9; 12+2+9+Fe | – | – | – | – | 31.35 | kg | 31.35 |
| Fertilizers; Everris | | | | | | | |
| Enmag; controlled release fertilizer (8–9 months); 11+22+09; 70 g/m$^2$ | – | – | – | – | 14.00 | 100 m$^2$ | 14.00 |

## 37 SOFT LANDSCAPING

| Item<br>Excluding site overheads and profit | PC<br>£ | Labour<br>hours | Labour<br>£ | Plant<br>£ | Material<br>£ | Unit | Total<br>rate £ |
|---|---|---|---|---|---|---|---|
| Fertilizers; Everris; granular Osmocote Flora; 15+09+11; controlled release fertilizer; costs for recommended application rates | | | | | | | |
| transplant | – | – | – | – | 0.12 | nr | 0.12 |
| whip | – | – | – | – | 0.18 | nr | 0.18 |
| feathered | – | – | – | – | 0.24 | nr | 0.24 |
| light standard | – | – | – | – | 0.24 | nr | 0.24 |
| standard | – | – | – | – | 0.30 | nr | 0.30 |
| selected standard | – | – | – | – | 0.42 | nr | 0.42 |
| heavy standard | – | – | – | – | 0.48 | nr | 0.48 |
| extra heavy standard | – | – | – | – | 0.60 | nr | 0.60 |
| 16–18 cm girth | – | – | – | – | 0.66 | nr | 0.66 |
| 18–20 cm girth | – | – | – | – | 0.72 | nr | 0.72 |
| 20–22 cm girth | – | – | – | – | 0.84 | nr | 0.84 |
| 22–24 cm girth | – | – | – | – | 0.90 | nr | 0.90 |
| 24–26 cm girth | – | – | – | – | 0.96 | nr | 0.96 |
| Fertilizers; Farmura Environmental Ltd; Seanure Soilbuilder | | | | | | | |
| soil amelioration; 70 g/m$^2$ | – | – | – | – | 12.68 | 100 m$^2$ | 12.68 |
| to plant pits; 300 × 300 × 300 mm | – | – | – | – | 0.05 | nr | 0.05 |
| to plant pits; 600 × 600 × 600 mm | – | – | – | – | 0.59 | nr | 0.59 |
| to tree pits; 1.00 × 1.00 × 1.00 | – | – | – | – | 2.72 | nr | 2.72 |
| Fertilizers; Everris | | | | | | | |
| TPMC tree planting and mulching compost | – | – | – | – | 2.60 | bag | 2.60 |
| Fertilizers; Rigby Taylor Ltd; fertilizer application rates 35 g/m$^2$ unless otherwise shown | | | | | | | |
| liquid fertilizer; Mascot Microflow-C; 25+0+0 + trace elements; 200–1400 ml/100 m$^2$ | – | – | – | – | 6.67 | 100 m$^2$ | 6.67 |
| liquid fertilizer; Mascot Microflow-CX; 4+2+18 + trace elements; 200–1400 ml/100 m$^2$ | – | – | – | – | 6.42 | 100 m$^2$ | 6.42 |
| liquid fertilizer; Mascot Microflow-CX; 12+0+8 + trace elements; 200–1400 ml/100 m$^2$ | – | – | – | – | 5.66 | 100 m$^2$ | 5.66 |
| liquid fertilizer; Mascot Microflow-CX; 17+2+5 + trace elements; 200–1400 ml/100 m$^2$ | – | – | – | – | 6.37 | 100 m$^2$ | 6.37 |
| Wetting agents; Rigby Taylor Ltd wetting agent; Breaker Advance Granules; 20 kg | – | – | – | – | 16.00 | 100 m$^2$ | 16.00 |

## 37 SOFT LANDSCAPING

| Item<br>Excluding site overheads and profit | PC<br>£ | Labour<br>hours | Labour<br>£ | Plant<br>£ | Material<br>£ | Unit | Total<br>rate £ |
|---|---|---|---|---|---|---|---|
| **PLANTING PREPARATION AND CULTIVATION** | | | | | | | |
| **Site protection; temporary protective fencing** | | | | | | | |
| Cleft chestnut rolled fencing; to 100 mm diameter chestnut posts; driving into firm ground at 3 m centres; pales at 50 mm centres | | | | | | | |
| 900 mm high | 5.16 | 0.11 | 2.53 | – | 11.69 | m | 14.22 |
| 1100 mm high | 6.30 | 0.11 | 2.53 | – | 12.84 | m | 15.37 |
| 1500 mm high | 8.29 | 0.11 | 2.53 | – | 14.82 | m | 17.35 |
| Extra over temporary protective fencing for removing and making good (no allowance for reuse of material) | – | 0.07 | 1.58 | 0.20 | – | m | 1.78 |
| **Cultivation by tractor; Agripower Ltd; Specialist subcontract prepared turf and sports area preparation** | | | | | | | |
| Ripping up subsoil; using approved subsoiling machine; minimum depth 250 mm below topsoil; at 1.20 m centres; in | | | | | | | |
| gravel or sandy clay | – | – | – | 2.20 | – | 100 m$^2$ | 2.20 |
| soil compacted by machines | – | – | – | 2.44 | – | 100 m$^2$ | 2.44 |
| clay | – | – | – | 3.14 | – | 100 m$^2$ | 3.14 |
| chalk or other soft rock | – | – | – | 3.67 | – | 100 m$^2$ | 3.67 |
| Extra for subsoiling at 1 m centres | – | – | – | 0.63 | – | 100 m$^2$ | 0.63 |
| **Breaking up existing ground; using tractor drawn tine cultivator or rotavator with stone burier** | | | | | | | |
| Single pass | | | | | | | |
| 100 mm deep | – | – | – | 1.56 | – | 100 m$^2$ | 1.56 |
| 150 mm deep | – | – | – | 1.95 | – | 100 m$^2$ | 1.95 |
| 200 mm deep | – | – | – | 2.59 | – | 100 m$^2$ | 2.59 |
| Cultivating ploughed ground; using disc, drag or chain harrow | | | | | | | |
| 4 passes | – | – | – | – | – | 100 m$^2$ | – |
| Rolling cultivated ground lightly; using self-propelled pedestrian agricultural roller | – | 0.10 | 2.38 | 0.46 | – | 100 m$^2$ | 2.84 |
| **Cultivation; pedestrian operated rotavator** | | | | | | | |
| Breaking up existing ground; tine cultivator or rotavator | | | | | | | |
| 100 mm deep | – | 0.22 | 5.22 | 0.79 | – | 100 m$^2$ | 6.01 |
| 150 mm deep | – | 0.28 | 6.53 | 0.99 | – | 100 m$^2$ | 7.52 |
| 200 mm deep | – | 0.37 | 8.71 | 1.32 | – | 100 m$^2$ | 10.03 |

## 37 SOFT LANDSCAPING

| Item<br>Excluding site overheads and profit | PC<br>£ | Labour<br>hours | Labour<br>£ | Plant<br>£ | Material<br>£ | Unit | Total<br>rate £ |
|---|---|---|---|---|---|---|---|
| As above but in heavy clay or wet soils | | | | | | | |
| 100 mm deep | – | 0.44 | 10.45 | 1.59 | – | 100 m² | 12.04 |
| 150 mm deep | – | 0.66 | 15.68 | 2.38 | – | 100 m² | 18.06 |
| 200 mm deep | – | 0.82 | 19.59 | 2.98 | – | 100 m² | 22.57 |
| Clearing stones; disposing off site | | | | | | | |
| by hand; stones not exceeding 50 mm in any direction; loading to skip 4.6 m³ | – | 0.01 | 0.24 | 0.05 | – | m² | 0.29 |
| by mechanical stone rake; stones not exceeding 50 mm in any direction; loading to 15 m³ truck by mechanical loader | – | – | 0.05 | 0.19 | 0.01 | m² | 0.25 |
| Lightly cultivating; weeding; to fallow areas; disposing debris off site | | | | | | | |
| by hand | – | 0.01 | 0.34 | – | 0.15 | m² | 0.49 |
| **TOPSOIL FILLING** | | | | | | | |
| **Topsoil filling** | | | | | | | |
| Market prices of topsoil; price shown includes a 20% settlement factor | | | | | | | |
| topsoil to BS3882 | 51.00 | – | – | – | 51.00 | m³ | 51.00 |
| small quantities (less than 15 m³) | 78.00 | – | – | – | 78.00 | m³ | 78.00 |
| Spreading and lightly consolidating approved topsoil (imported or from spoil heaps); in layers not exceeding 150 mm; travel distance from spoil heaps not exceeding 100 m; by machine (imported topsoil not included) | | | | | | | |
| minimum depth 100 mm | – | 1.55 | 36.81 | 55.60 | – | 100 m² | 92.41 |
| minimum depth 150 mm | – | 2.33 | 55.42 | 83.68 | – | 100 m² | 139.10 |
| minimum depth 300 mm | – | 4.67 | 110.83 | 167.35 | – | 100 m² | 278.18 |
| minimum depth 450 mm | – | 6.99 | 166.01 | 250.75 | – | 100 m² | 416.76 |
| Spreading and lightly consolidating approved topsoil (imported or from spoil heaps); in layers not exceeding 150 mm; travel distance from spoil heaps not exceeding 100 m; by hand (imported topsoil not included) | | | | | | | |
| minimum depth 100 mm | – | 20.00 | 475.10 | – | – | 100 m² | 475.10 |
| minimum depth 150 mm | – | 30.01 | 712.64 | – | – | 100 m² | 712.64 |
| minimum depth 300 mm | – | 60.01 | 1425.29 | – | – | 100 m² | 1425.29 |
| minimum depth 450 mm | – | 90.02 | 2137.93 | – | – | 100 m² | 2137.93 |
| Spreading topsoil; from dump not exceeding 100 m distance; by machine (topsoil not included) | | | | | | | |
| at 1 m³ per 13 m²; 75 mm thick | – | 0.56 | 13.20 | 41.58 | – | 100 m² | 54.78 |
| at 1 m³ per 10 m²; 100 mm thick | – | 0.89 | 21.11 | 42.36 | – | 100 m² | 63.47 |
| at 1 m³ per 6.50 m²; 150 mm thick | – | 1.11 | 26.36 | 74.67 | – | 100 m² | 101.03 |
| at 1 m³ per 5 m²; 200 mm thick | – | 1.48 | 35.15 | 99.26 | – | 100 m² | 134.41 |

## 37 SOFT LANDSCAPING

| Item<br>Excluding site overheads and profit | PC<br>£ | Labour<br>hours | Labour<br>£ | Plant<br>£ | Material<br>£ | Unit | Total<br>rate £ |
|---|---|---|---|---|---|---|---|
| **TOPSOIL FILLING – cont** | | | | | | | |
| **Topsoil filling – cont** | | | | | | | |
| Spreading topsoil; from dump not exceeding 100 m distance; by hand (topsoil not included) | | | | | | | |
|   at 1 m³ per 13 m²; 75 mm thick | – | 15.00 | 356.32 | – | – | 100 m² | **356.32** |
|   at 1 m³ per 10 m²; 100 mm thick | – | 20.00 | 475.10 | – | – | 100 m² | **475.10** |
|   at 1 m³ per 6.50 m²; 150 mm thick | – | 30.01 | 712.64 | – | – | 100 m² | **712.64** |
|   at 1 m³ per 5 m²; 200 mm thick | – | 36.67 | 871.01 | – | – | 100 m² | **871.01** |
| Imported topsoil; tipped 100 m from area of application; by machine | | | | | | | |
|   at 1 m³ per 13 m²; 75 mm thick | 297.00 | 0.56 | 13.20 | 41.58 | 297.00 | 100 m² | **351.78** |
|   at 1 m³ per 10 m²; 100 mm thick | 396.00 | 0.89 | 21.11 | 42.36 | 396.00 | 100 m² | **459.47** |
|   at 1 m³ per 6.50 m²; 150 mm thick | 594.00 | 1.11 | 26.36 | 74.67 | 594.00 | 100 m² | **695.03** |
|   at 1 m³ per 5 m²; 200 mm thick | 792.00 | 1.48 | 35.15 | 99.26 | 792.00 | 100 m² | **926.41** |
| Imported topsoil; tipped 100 m from area of application; by hand | | | | | | | |
|   at 1 m³ per 13 m²; 75 mm thick | 297.00 | 15.00 | 356.32 | – | 297.00 | 100 m² | **653.32** |
|   at 1 m³ per 10 m²; 100 mm thick | 396.00 | 20.00 | 475.10 | – | 396.00 | 100 m² | **871.10** |
|   at 1 m³ per 6.50 m²; 150 mm thick | 594.00 | 30.01 | 712.64 | – | 594.00 | 100 m² | **1306.64** |
|   at 1 m³ per 5 m²; 200 mm thick | 792.00 | 36.67 | 871.01 | – | 792.00 | 100 m² | **1663.01** |
| Extra over for spreading topsoil to slopes 15–30°; by machine or hand | – | – | – | – | – | 10% | – |
| Extra over for spreading topsoil to slopes over 30°; by machine or hand | – | – | – | – | – | 25% | – |
| Extra over for spreading topsoil from spoil heaps; travel exceeding 100 m; by machine | | | | | | | |
|   100–150 m | – | 0.01 | 0.29 | 0.08 | – | m³ | **0.37** |
|   150–200 m | – | 0.02 | 0.44 | 0.12 | – | m³ | **0.56** |
|   200–300 m | – | 0.03 | 0.66 | 0.19 | – | m³ | **0.85** |
| Extra over spreading topsoil for travel exceeding 100 m; by hand | | | | | | | |
|   100 m | – | 0.83 | 19.79 | – | – | m³ | **19.79** |
|   200 m | – | 1.67 | 39.59 | – | – | m³ | **39.59** |
|   300 m | – | 2.50 | 59.38 | – | – | m³ | **59.38** |
| Evenly grading; to general surfaces to bring to finished levels | | | | | | | |
|   by machine (tractor mounted rotavator) | – | – | – | 0.08 | – | m² | **0.08** |
|   by pedestrian operated rotavator | – | – | 0.10 | 0.02 | – | m² | **0.12** |
|   by hand | – | 0.01 | 0.24 | – | – | m² | **0.24** |
| Extra over grading for slopes 15–30°; by machine or hand | – | – | – | – | – | 10% | – |
| Extra over grading for slopes over 30°; by machine or hand | – | – | – | – | – | 25% | – |
| Apply screened topdressing to grass surfaces; spread using Tru-Lute | | | | | | | |
|   sand soil mixes 90/10 to 50/50 | – | – | 0.05 | 0.03 | 0.14 | m² | **0.22** |

## 37 SOFT LANDSCAPING

| Item<br>Excluding site overheads and profit | PC<br>£ | Labour<br>hours | Labour<br>£ | Plant<br>£ | Material<br>£ | Unit | Total<br>rate £ |
|---|---|---|---|---|---|---|---|
| Spread only existing cultivated soil to final levels using Tru-Lute | | | | | | | |
| cultivated soil | – | – | 0.05 | 0.03 | – | m² | 0.08 |
| **SOIL IMPROVEMENTS** | | | | | | | |
| **Preparation of planting operations** | | | | | | | |
| For the following topsoil improvement and planting operations add or subtract the following amounts for every £0.10 difference in the material cost price | | | | | | | |
| 35 g/m² | – | – | – | – | 0.35 | 100 m² | 0.35 |
| 50 g/m² | – | – | – | – | 0.50 | 100 m² | 0.50 |
| 70 g/m² | – | – | – | – | 0.70 | 100 m² | 0.70 |
| 100 g/m² | – | – | – | – | 1.00 | 100 m² | 1.00 |
| 150 kg/ha | – | – | – | – | 15.00 | ha | 15.00 |
| 200 kg/ha | – | – | – | – | 20.00 | ha | 20.00 |
| 250 kg/ha | – | – | – | – | 25.00 | ha | 25.00 |
| 300 kg/ha | – | – | – | – | 30.00 | ha | 30.00 |
| 400 kg/ha | – | – | – | – | 40.00 | ha | 40.00 |
| 500 kg/ha | – | – | – | – | 50.00 | ha | 50.00 |
| 700 kg/ha | – | – | – | – | 70.00 | ha | 70.00 |
| 1000 kg/ha | – | – | – | – | 100.00 | ha | 100.00 |
| 1250 kg/ha | – | – | – | – | 125.00 | ha | 125.00 |
| Fertilizers; in top 150 mm of topsoil; at 35 g/m² | | | | | | | |
| Mascot Microfine; controlled release turf fertilizer; 8+0+6 + 2% Mg + 4% Fe | 0.05 | – | 0.03 | – | 0.05 | m² | 0.08 |
| Enmag; controlled release fertilizer (8–9 months); 11+22+09 | 0.07 | – | 0.03 | – | 0.07 | m² | 0.10 |
| Mascot Outfield; turf fertilizer; 4+10+10 | 0.03 | – | 0.03 | – | 0.03 | m² | 0.06 |
| Mascot Outfield; turf fertilizer; 9+5+5 | 0.03 | – | 0.03 | – | 0.03 | m² | 0.06 |
| Fertilizers; in top 150 mm of topsoil at 70 g/m² | | | | | | | |
| Mascot Microfine; controlled release turf fertilizer; 8+0+6 + 2% Mg + 4% Fe | 0.10 | – | 0.03 | – | 0.10 | m² | 0.13 |
| Enmag; controlled release fertilizer (8–9 months); 11+22+09 | 0.15 | – | 0.03 | – | 0.15 | m² | 0.18 |
| Mascot Outfield; turf fertilizer; 4+10+10 | 0.07 | – | 0.03 | – | 0.07 | m² | 0.10 |
| Mascot Outfield; turf fertilizer; 9+5+5 | 0.07 | – | 0.03 | – | 0.07 | m² | 0.10 |
| Topgrow; Melcourt; peat free tree and shrub compost (25 m³ loads); placing on beds by mechanical loader; spreading and rotavating into topsoil by tractor drawn machine | | | | | | | |
| 50 mm thick | 182.75 | – | – | 16.86 | 182.75 | 100 m² | 199.61 |
| 100 mm thick | 383.77 | – | – | 23.35 | 383.77 | 100 m² | 407.12 |
| 150 mm thick | 575.66 | – | – | 30.10 | 575.66 | 100 m² | 605.76 |
| 200 mm thick | 767.55 | – | – | 36.73 | 767.55 | 100 m² | 804.28 |

## 37 SOFT LANDSCAPING

| Item<br>Excluding site overheads and profit | PC<br>£ | Labour<br>hours | Labour<br>£ | Plant<br>£ | Material<br>£ | Unit | Total<br>rate £ |
|---|---|---|---|---|---|---|---|
| **SOIL IMPROVEMENTS – cont** | | | | | | | |
| Topgrow; Melcourt; peat free tree and shrub compost (65 m³ loads); placing on beds by mechanical loader; spreading and rotavating into topsoil by tractor drawn machine | | | | | | | |
| 50 mm thick | 142.75 | – | – | 16.86 | 142.75 | 100 m² | **159.61** |
| 100 mm thick | 299.77 | – | – | 23.35 | 299.77 | 100 m² | **323.12** |
| 150 mm thick | 449.66 | – | – | 30.10 | 449.66 | 100 m² | **479.76** |
| 200 mm thick | 599.55 | – | – | 36.73 | 599.55 | 100 m² | **636.28** |
| Mushroom compost; Melcourt; (25 m³ loads); from not further than 25 m from location; cultivating into topsoil by pedestrian drawn machine | | | | | | | |
| 50 mm thick | 121.00 | 2.86 | 67.86 | 0.66 | 121.00 | 100 m² | **189.52** |
| 100 mm thick | 242.00 | 6.05 | 143.66 | 0.66 | 242.00 | 100 m² | **386.32** |
| 150 mm thick | 363.00 | 8.90 | 211.49 | 0.66 | 363.00 | 100 m² | **575.15** |
| 200 mm thick | 484.00 | 12.90 | 306.49 | 0.66 | 484.00 | 100 m² | **791.15** |
| Mushroom compost; Melcourt (25 m³ loads); placing on beds by mechanical loader; spreading and rotavating into topsoil by tractor drawn cultivator | | | | | | | |
| 50 mm thick | 121.00 | – | – | 16.86 | 121.00 | 100 m² | **137.86** |
| 100 mm thick | 242.00 | – | – | 23.35 | 242.00 | 100 m² | **265.35** |
| 150 mm thick | 363.00 | – | – | 30.10 | 363.00 | 100 m² | **393.10** |
| 200 mm thick | 484.00 | – | – | 36.73 | 484.00 | 100 m² | **520.73** |
| Manure (60 m³ loads); from not further than 25 m from location; cultivating into topsoil by pedestrian drawn machine | | | | | | | |
| 50 mm thick | 114.90 | 2.86 | 67.86 | 0.66 | 114.90 | 100 m² | **183.42** |
| 100 mm thick | 241.29 | 6.05 | 143.66 | 0.66 | 241.29 | 100 m² | **385.61** |
| 150 mm thick | 361.94 | 8.90 | 211.49 | 0.66 | 361.94 | 100 m² | **574.09** |
| 200 mm thick | 482.58 | 12.90 | 306.49 | 0.66 | 482.58 | 100 m² | **789.73** |

## 37 SOFT LANDSCAPING

| Item<br>Excluding site overheads and profit | PC<br>£ | Labour<br>hours | Labour<br>£ | Plant<br>£ | Material<br>£ | Unit | Total<br>rate £ |
|---|---|---|---|---|---|---|---|
| **Surface applications and soil additives; pre-planting; from not further than 25 m from location; by machine** | | | | | | | |
| Medium bark soil conditioner; Melcourt Ltd; including turning in to cultivated ground; delivered in 25 m³ loads | | | | | | | |
| 1 m³ per 40 m² = 25 mm thick | 0.93 | – | – | 0.18 | 0.93 | m² | **1.11** |
| 1 m³ per 20 m² = 50 mm thick | 1.86 | – | – | 0.26 | 1.86 | m² | **2.12** |
| 1 m³ per 13.33 m² = 75 mm thick | 2.79 | – | – | 0.35 | 2.79 | m² | **3.14** |
| 1 m³ per 10 m² = 100 mm thick | 3.71 | – | – | 0.43 | 3.71 | m² | **4.14** |
| **Works by hand; surface applications and soil additives; pre-planting; from not further than 25 m from location; by hand** | | | | | | | |
| Mushroom compost soil conditioner; Melcourt Industries Ltd; including turning in to cultivated ground; delivered in 25 m³ loads | | | | | | | |
| 1 m³ per 40 m² = 25 mm thick | 0.60 | 0.02 | 0.53 | – | 0.60 | m² | **1.13** |
| 1 m³ per 20 m² = 50 mm thick | 1.21 | 0.04 | 1.05 | – | 1.21 | m² | **2.26** |
| 1 m³ per 13.33 m² = 75 mm thick | 1.81 | 0.07 | 1.58 | – | 1.81 | m² | **3.39** |
| 1 m³ per 10 m² = 100 mm thick | 2.42 | 0.08 | 1.90 | – | 2.42 | m² | **4.32** |
| Medium bark soil conditioner; Melcourt Industries Ltd; including turning in to cultivated ground; delivered in 25 m³ loads | | | | | | | |
| 1 m³ per 40 m² = 25 mm thick | 0.93 | 0.02 | 0.53 | – | 0.93 | m² | **1.46** |
| 1 m³ per 20 m² = 50 mm thick | 1.86 | 0.04 | 1.05 | – | 1.86 | m² | **2.91** |
| 1 m³ per 13.33 m² = 75 mm thick | 2.79 | 0.07 | 1.58 | – | 2.79 | m² | **4.37** |
| 1 m³ per 10 m² = 100 mm thick | 3.71 | 0.08 | 1.90 | – | 3.71 | m² | **5.61** |
| Manure; 20 m³ loads; delivered not further than 25 m from location; cultivating into topsoil by pedestrian operated machine | | | | | | | |
| 50 mm thick | 225.00 | 2.86 | 67.86 | 0.66 | 225.00 | 100 m² | **293.52** |
| 100 mm thick | 241.29 | 6.05 | 143.66 | 0.66 | 241.29 | 100 m² | **385.61** |
| 150 mm thick | 361.94 | 8.90 | 211.49 | 0.66 | 361.94 | 100 m² | **574.09** |
| 200 mm thick | 482.58 | 12.90 | 306.49 | 0.66 | 482.58 | 100 m² | **789.73** |
| Fertilizer (7+7+7); PC £1.11/kg; to beds; by hand | | | | | | | |
| 35 g/m² | 3.91 | 0.17 | 3.96 | – | 3.91 | 100 m² | **7.87** |
| 50 g/m² | 5.58 | 0.17 | 3.96 | – | 5.58 | 100 m² | **9.54** |
| 70 g/m² | 7.81 | 0.17 | 3.96 | – | 7.81 | 100 m² | **11.77** |
| Fertilizers; Enmag; PC £2.00/kg; controlled release fertilizer; to beds; by hand | | | | | | | |
| 35 g/m² | 7.00 | 0.17 | 3.96 | – | 7.00 | 100 m² | **10.96** |
| 50 g/m² | 10.00 | 0.17 | 3.96 | – | 10.00 | 100 m² | **13.96** |
| 70 g/m² | 15.00 | 0.17 | 3.96 | – | 15.00 | 100 m² | **18.96** |

## 37 SOFT LANDSCAPING

| Item<br>Excluding site overheads and profit | PC<br>£ | Labour<br>hours | Labour<br>£ | Plant<br>£ | Material<br>£ | Unit | Total<br>rate £ |
|---|---|---|---|---|---|---|---|
| **SOIL IMPROVEMENTS – cont** | | | | | | | |
| **Works by hand – cont** | | | | | | | |
| Note: For machine incorporation of fertilizers and soil conditioners see 'Cultivation'. | | | | | | | |
| **TREE PLANTING** | | | | | | | |
| **Tree planting; pre-planting operations** | | | | | | | |
| Excavating tree pits; depositing soil alongside pits; by machine | | | | | | | |
| 600 × 600 × 600 mm deep | – | 0.15 | 3.50 | 0.69 | – | nr | 4.19 |
| 900 × 900 × 600 mm deep | – | 0.33 | 7.84 | 1.55 | – | nr | 9.39 |
| 1.00 m × 1.00 m × 600 mm deep | – | 0.61 | 14.57 | 1.92 | – | nr | 16.49 |
| 1.25 m × 1.25 m × 600 mm deep | – | 0.96 | 22.81 | 3.00 | – | nr | 25.81 |
| 1.00 × 1.00 × 1.00 m deep | – | 1.02 | 24.29 | 3.20 | – | nr | 27.49 |
| 1.50 m × 1.50 m × 750 mm deep | – | 1.73 | 40.99 | 5.39 | – | nr | 46.38 |
| 1.50 × 1.50 × 1.00 m deep | – | 2.30 | 54.51 | 7.17 | – | nr | 61.68 |
| 1.75 × 1.75 × 1.00 m deep | – | 3.13 | 74.39 | 9.79 | – | nr | 84.18 |
| 2.00 × 2.00 × 1.00 m deep | – | 4.09 | 97.15 | 12.78 | – | nr | 109.93 |
| Excavating tree pits; depositing soil alongside pits; by hand | | | | | | | |
| 600 × 600 × 600 mm deep | – | 0.44 | 10.45 | – | – | nr | 10.45 |
| 900 × 900 × 600 mm deep | – | 1.00 | 23.75 | – | – | nr | 23.75 |
| 1.00 m × 1.00x 600 mm deep | – | 1.13 | 26.72 | – | – | nr | 26.72 |
| 1.25 m × 1.25 mx 600 mm deep | – | 1.93 | 45.84 | – | – | nr | 45.84 |
| 1.00 × 1.00 × 1.00 m deep | – | 2.06 | 48.92 | – | – | nr | 48.92 |
| 1.50 m × 1.50 m × 750 mm deep | – | 3.47 | 82.41 | – | – | nr | 82.41 |
| 1.75 m × 1.50 m × 750 mm deep | – | 4.05 | 96.19 | – | – | nr | 96.19 |
| 1.50 × 1.50 × 1.00 m deep | – | 4.63 | 109.96 | – | – | nr | 109.96 |
| 2.00 m × 2.00 m × 750 mm deep | – | 6.17 | 146.54 | – | – | nr | 146.54 |
| 2.00 × 2.00 × 1.00 m deep | – | 8.23 | 195.47 | – | – | nr | 195.47 |
| Breaking up subsoil in tree pits; to a depth of 200 mm | – | 0.03 | 0.79 | – | – | m² | 0.79 |
| Spreading and lightly consolidating approved topsoil (imported or from spoil heaps); in layers not exceeding 150 mm; distance from spoil heaps not exceeding 100 m (imported topsoil not included); by machine | | | | | | | |
| minimum depth 100 mm | – | 1.55 | 36.81 | 55.60 | – | 100 m² | 92.41 |
| minimum depth 150 mm | – | 2.33 | 55.42 | 83.68 | – | 100 m² | 139.10 |
| minimum depth 300 mm | – | 4.67 | 110.83 | 167.35 | – | 100 m² | 278.18 |
| minimum depth 450 mm | – | 6.99 | 166.01 | 250.75 | – | 100 m² | 416.76 |

## 37 SOFT LANDSCAPING

| Item<br>Excluding site overheads and profit | PC<br>£ | Labour<br>hours | Labour<br>£ | Plant<br>£ | Material<br>£ | Unit | Total<br>rate £ |
|---|---|---|---|---|---|---|---|
| Spreading and lightly consolidating approved topsoil (imported or from spoil heaps); in layers not exceeding 150 mm; distance from spoil heaps not exceeding 100 m (imported topsoil not included); by hand | | | | | | | |
| minimum depth 100 mm | – | 20.00 | 475.10 | – | – | 100 m² | 475.10 |
| minimum depth 150 mm | – | 30.01 | 712.64 | – | – | 100 m² | 712.64 |
| minimum depth 300 mm | – | 60.01 | 1425.29 | – | – | 100 m² | 1425.29 |
| minimum depth 450 mm | – | 90.02 | 2137.93 | – | – | 100 m² | 2137.93 |
| Extra for filling tree pits with imported topsoil; PC £33.00/m³; plus allowance for 20% settlement | | | | | | | |
| depth 100 mm | – | – | – | – | 4.75 | m² | 4.75 |
| depth 150 mm | – | – | – | – | 7.13 | m² | 7.13 |
| depth 200 mm | – | – | – | – | 9.50 | m² | 9.50 |
| depth 300 mm | – | – | – | – | 14.26 | m² | 14.26 |
| depth 400 mm | – | – | – | – | 19.01 | m² | 19.01 |
| depth 450 mm | – | – | – | – | 21.38 | m² | 21.38 |
| depth 500 mm | – | – | – | – | 23.76 | m² | 23.76 |
| depth 600 mm | – | – | – | – | 28.51 | m² | 28.51 |
| depth 750 mm | – | – | – | – | 29.70 | m² | 29.70 |
| depth 1.00 m | – | – | – | – | 39.60 | m² | 39.60 |
| Backfilling of tree pits with imported topsoil; compacting in layers | | | | | | | |
| depth 500 mm | – | – | – | 4.86 | 19.80 | m² | 24.66 |
| depth 600 mm | – | – | – | 5.56 | 23.76 | m² | 29.32 |
| depth 750 mm | – | – | – | 6.48 | 29.70 | m² | 36.18 |
| depth 1.00 m | – | – | – | 9.72 | 39.60 | m² | 49.32 |
| Add or deduct the following amounts for every £1.00 change in the material price of topsoil | | | | | | | |
| depth 100 mm | – | – | – | – | 0.12 | m² | 0.12 |
| depth 150 mm | – | – | – | – | 0.18 | m² | 0.18 |
| depth 200 mm | – | – | – | – | 0.24 | m² | 0.24 |
| depth 300 mm | – | – | – | – | 0.36 | m² | 0.36 |
| depth 400 mm | – | – | – | – | 0.48 | m² | 0.48 |
| depth 450 mm | – | – | – | – | 0.54 | m² | 0.54 |
| depth 500 mm | – | – | – | – | 0.60 | m² | 0.60 |
| depth 600 mm | – | – | – | – | 0.72 | m² | 0.72 |

## 37 SOFT LANDSCAPING

| Item<br>Excluding site overheads and profit | PC<br>£ | Labour<br>hours | Labour<br>£ | Plant<br>£ | Material<br>£ | Unit | Total<br>rate £ |
|---|---|---|---|---|---|---|---|
| **TREE PLANTING – cont** | | | | | | | |
| **Structural soils (Amsterdam tree sand); Heicom; non-compressive soil mixture for tree planting in areas to receive compressive surface treatments** | | | | | | | |
| Backfilling and lightly compacting in layers; excavation, disposal, moving of material from delivery position and surface treatments not included; by machine | | | | | | | |
| individual tree pits (29 t loads) | 49.50 | 0.50 | 11.88 | 2.34 | 49.50 | m³ | **63.72** |
| individual tree pits (20 t loads) | 57.75 | 0.50 | 11.88 | 2.34 | 57.75 | m³ | **71.97** |
| in trenches (29 t loads) | 49.50 | 0.42 | 9.90 | 1.95 | 49.50 | m³ | **61.35** |
| Backfilling and lightly compacting in layers; excavation, disposal, moving of material from delivery position and surface treatments not included; by hand | | | | | | | |
| individual tree pits (29 t loads) | 49.50 | 1.60 | 38.00 | – | 49.50 | m³ | **87.50** |
| individual tree pits (20 t loads) | 57.75 | 1.60 | 38.00 | – | 57.75 | m³ | **95.75** |
| in trenches | 49.50 | 1.33 | 31.59 | – | 49.50 | m³ | **81.09** |
| **Tree staking** | | | | | | | |
| J Toms Ltd; extra over trees for tree stake(s); driving 500 mm into firm ground; trimming to approved height; including two tree ties to approved pattern | | | | | | | |
| one stake; 1.50 m long × 32 mm × 32 mm | 1.54 | 0.20 | 4.75 | – | 1.54 | nr | **6.29** |
| two stakes; 900 mm long × 25 mm × 25 mm | 1.50 | 0.30 | 7.13 | – | 1.50 | nr | **8.63** |
| two stakes; 1.50 m long × 32 mm × 32 mm | 2.28 | 0.30 | 7.13 | – | 2.28 | nr | **9.41** |
| three stakes; 1.50 m long × 32 mm × 32 mm | 3.42 | 0.36 | 8.55 | – | 3.42 | nr | **11.97** |
| **Tree anchors** | | | | | | | |
| Platipus Anchors Ltd; extra over trees for tree anchors | | | | | | | |
| RF1P rootball kit; for 75–220 mm girth × 2–4.5 m high; inclusive of Plati-Mat PM1 | 29.67 | 1.00 | 23.75 | – | 29.67 | nr | **53.42** |
| RF2P; rootball kit; for 220–450 mm girth × 4.5–7.5 m high; inclusive of Plati-Mat PM2 | 50.51 | 1.33 | 31.59 | – | 50.51 | nr | **82.10** |
| RF3P; rootball kit; for 450–750 mm girth × 7.5 to 12 m high; inclusive of Plati-Mat PM3 | 106.92 | 1.50 | 35.63 | – | 106.92 | nr | **142.55** |

## 37 SOFT LANDSCAPING

| Item<br>Excluding site overheads and profit | PC<br>£ | Labour<br>hours | Labour<br>£ | Plant<br>£ | Material<br>£ | Unit | Total<br>rate £ |
|---|---|---|---|---|---|---|---|
| CG1; guy fixing kit; 75–220 mm girth × 2–4.5 m high | 19.08 | 1.67 | 39.58 | – | 19.08 | nr | 58.66 |
| CG2; guy fixing kit; 220–450 mm girth × 4.5–7.5 m high | 33.56 | 2.00 | 47.50 | – | 33.56 | nr | 81.06 |
| installation tools; drive rod for RF1/CG1 kits | – | – | – | – | 61.85 | nr | 61.85 |
| installation tools; drive rod for RF2/CG2 kits | – | – | – | – | 93.20 | nr | 93.20 |
| Extra over trees for land drain to tree pits; 100 mm diameter perforated flexible agricultural drain; including excavating drain trench; laying pipe; backfilling | 1.49 | 1.00 | 23.75 | – | 1.49 | m | 25.24 |
| **Platipus anchors; deadman anchoring** | | | | | | | |
| Deadman kits; anchoring to concrete kerbs placed in base of tree pit | | | | | | | |
| trees 12–25 cm girth; 2.5–4.0 m high | 26.62 | 0.75 | 17.81 | – | 48.00 | nr | 65.81 |
| trees 25–45 cm girth; 4.0–7.5 m high | 45.82 | 1.50 | 35.63 | – | 67.20 | nr | 102.83 |
| trees 45–75 cm girth; 7.5–12.0 m high | 100.23 | 2.00 | 47.50 | – | 121.61 | nr | 169.11 |
| **Tree planting; tree pit additives** | | | | | | | |
| Melcourt Industries Ltd; Topgrow; incorporating into topsoil at 1 part Topgrow to 3 parts excavated topsoil; supplied in 75 l bags; pit size | | | | | | | |
| 600 × 600 × 600 mm | 2.12 | 0.02 | 0.47 | – | 2.12 | nr | 2.59 |
| 900 × 900 × 900 mm | 7.17 | 0.06 | 1.43 | – | 7.17 | nr | 8.60 |
| 1.00 × 1.00 × 1.00 m | 9.82 | 0.24 | 5.70 | – | 9.82 | nr | 15.52 |
| 1.25 × 1.25 × 1.25 m | 19.20 | 0.40 | 9.50 | – | 19.20 | nr | 28.70 |
| 1.50 × 1.50 × 1.50 m | 33.19 | 0.90 | 21.38 | – | 33.19 | nr | 54.57 |
| Melcourt Industries Ltd; Topgrow; incorporating into topsoil at 1 part Topgrow to 3 parts excavated topsoil; supplied in 60 m³ loose loads; pit size | | | | | | | |
| 600 × 600 × 600 mm | 1.54 | 0.02 | 0.40 | – | 1.54 | nr | 1.94 |
| 900 × 900 × 900 mm | 5.20 | 0.05 | 1.19 | – | 5.20 | nr | 6.39 |
| 1.00 × 1.00 × 1.00 m | 7.14 | 0.20 | 4.75 | – | 7.14 | nr | 11.89 |
| 1.25 × 1.25 × 1.25 m | 13.94 | 0.33 | 7.92 | – | 13.94 | nr | 21.86 |
| 1.50 × 1.50 × 1.50 m | 24.09 | 0.75 | 17.81 | – | 24.09 | nr | 41.90 |

## 37 SOFT LANDSCAPING

| Item<br>Excluding site overheads and profit | PC<br>£ | Labour<br>hours | Labour<br>£ | Plant<br>£ | Material<br>£ | Unit | Total<br>rate £ |
|---|---|---|---|---|---|---|---|
| **TREE PLANTING – cont** | | | | | | | |
| **Tree planting; root barriers** | | | | | | | |
| GreenBlue Urban; Root Director; one-piece root control planters for installation at time of planting to divert root growth down away from pavements and out for anchorage; excavation measured separately | | | | | | | |
| RD 1050; 1050 × 1050 mm to 1300 × 1300 mm at base | 104.81 | 0.25 | 5.94 | – | 104.81 | nr | 110.75 |
| RD 640; 640 × 640 mm to 870 × 870 mm at base | 68.07 | 0.25 | 5.94 | – | 68.07 | nr | 74.01 |
| GreenBlue Urban; linear root deflection barriers; installed to trench measured separately | | | | | | | |
| Re-Root 2000; 2.0 mm thick × 2 m wide | 13.48 | 0.05 | 1.19 | – | 13.48 | m | 14.67 |
| Re-Root 2000; 1.0 mm thick × 1 m wide | 4.70 | 0.05 | 1.19 | – | 4.70 | m | 5.89 |
| Re-Root 600; 1.0 mm thick | 7.70 | 0.05 | 1.19 | – | 7.70 | m | 8.89 |
| **GreenBlue Urban; irrigation systems; Root Rain tree pit irrigation systems** | | | | | | | |
| Metro; small; 35 mm pipe diameter × 1.25 m long; for specimen shrubs and standard trees | | | | | | | |
| plastic | 4.02 | 0.25 | 5.94 | – | 4.02 | nr | 9.96 |
| plastic; with chain | 7.53 | 0.25 | 5.94 | – | 7.53 | nr | 13.47 |
| metal; with chain | 10.59 | 0.25 | 5.94 | – | 10.59 | nr | 16.53 |
| Metro; medium; 35 mm pipe diameter × 1.75 m long; for standard and selected standard trees | | | | | | | |
| plastic | 4.74 | 0.29 | 6.79 | – | 4.74 | nr | 11.53 |
| plastic; with chain | 8.37 | 0.29 | 6.79 | – | 8.37 | nr | 15.16 |
| metal; with chain | 11.43 | 0.29 | 6.79 | – | 11.43 | nr | 18.22 |
| Metro; large; 35 mm pipe diameter × 2.50 m long; for selected standards and extra heavy standards | | | | | | | |
| plastic | 5.82 | 0.33 | 7.92 | – | 5.82 | nr | 13.74 |
| plastic; with chain | 9.45 | 0.33 | 7.92 | – | 9.45 | nr | 17.37 |
| metal; with chain | 12.51 | 0.33 | 7.92 | – | 12.51 | nr | 20.43 |
| Urban; for large capacity general purpose irrigation to parkland and street verge planting | | | | | | | |
| RRUrb1; 3.0 m pipe | – | 0.29 | 6.79 | – | 7.11 | nr | 13.90 |
| RRUrb2; 5.0 m pipe | 8.98 | 0.33 | 7.92 | – | 8.98 | nr | 16.90 |
| RRUrb3; 8.0 m pipe | 11.77 | 0.40 | 9.50 | – | 11.77 | nr | 21.27 |

## 37 SOFT LANDSCAPING

| Item<br>Excluding site overheads and profit | PC<br>£ | Labour<br>hours | Labour<br>£ | Plant<br>£ | Material<br>£ | Unit | Total<br>rate £ |
|---|---|---|---|---|---|---|---|
| **Civic; heavy cast aluminium inlet; for heavily trafficked locations** | | | | | | | |
| 5.0 m pipe | 32.19 | 0.33 | 7.92 | – | 32.19 | nr | 40.11 |
| 8.0 m pipe | 36.16 | 0.40 | 9.50 | – | 36.16 | nr | 45.66 |
| **Tree irrigation kits; Platipus; direct water delivery to the rootball area of trees** | | | | | | | |
| Piddler tree irrigation system; permeable membrane delivering targeted irrigation to rootball surround | | | | | | | |
| rootballs up to 550 mm diameter; Irrigation Kit PID0 | 6.30 | 0.25 | 5.94 | – | 6.30 | nr | 12.24 |
| rootballs up to 900 mm diameter; Irrigation Kit PID1 | 11.25 | 0.30 | 7.13 | – | 11.25 | nr | 18.38 |
| rootballs up to 1.55 m diameter; Irrigation Kit PID2 | 14.85 | 0.40 | 9.50 | – | 14.85 | nr | 24.35 |
| rootballs up to 2.40 m diameter; Irrigation Kit PID3 | 18.45 | 0.45 | 10.69 | – | 18.45 | nr | 29.14 |
| rootballs up to 3.10 m diameter; Irrigation Kit PID4 | 25.50 | 0.50 | 11.88 | – | 25.50 | nr | 37.38 |
| **GreenBlue Urban; Arboresin porous tree pit surfacing resin to galvanized tree pit frames** | | | | | | | |
| Two part support frame system | | | | | | | |
| support frame 1.00 m × 1.00 m | 127.92 | 1.00 | 47.50 | – | 305.57 | nr | 353.07 |
| support frame 1.20 m × 1.20 m | 184.20 | 1.20 | 57.00 | – | 410.14 | nr | 467.14 |
| support frame 1.50 m × 1.50 m | 287.82 | 1.40 | 66.50 | – | 570.24 | nr | 636.74 |
| support frame 2.00 m × 2.00 m | 575.64 | 1.70 | 80.75 | – | 1042.68 | nr | 1123.43 |
| support frame 1.00 m diameter | 336.31 | 1.00 | 47.50 | – | 336.31 | nr | 383.81 |
| support frame 1.20 m diameter | 144.55 | 1.10 | 52.25 | – | 426.06 | nr | 478.31 |
| support frame 1.50 m diameter | 225.78 | 1.40 | 66.50 | – | 507.78 | nr | 574.28 |
| support frame 2.00 m diameter | 401.67 | 1.70 | 80.75 | – | 932.39 | nr | 1013.14 |
| **Resin bound mulch or surface to tree pits; Dekorgrip Ltd** | | | | | | | |
| resin bound system with a typical makeup of 10 mm aggregate to create voids to allow and promote the free flow of air and water around tree beds to surface of tree pit | | | | | | | |
| Dekortree permeable resin tree pits 50 mm uV stabilized 10–25 m² | – | 0.75 | 35.63 | – | 80.00 | m² | 115.63 |
| Dekortree permeable resin tree pits 50 mm uV stabilized over 25 m² | – | 0.75 | 35.63 | – | 70.00 | m² | 105.63 |
| Dekortree 50 mm; Standard resin 10–25 m² | – | 0.75 | 35.63 | – | 62.00 | m² | 97.63 |
| Dekortree 50 mm uV Standard over 25 m² | – | 0.75 | 35.63 | – | 52.00 | m² | 87.63 |

## 37 SOFT LANDSCAPING

| Item<br>Excluding site overheads and profit | PC<br>£ | Labour<br>hours | Labour<br>£ | Plant<br>£ | Material<br>£ | Unit | Total<br>rate £ |
|---|---|---|---|---|---|---|---|
| **TREE PLANTING – cont** | | | | | | | |
| **Mulching of tree pits; Melcourt Industries Ltd; FSC (Forest Stewardship Council) and FT (fire tested) certified** | | | | | | | |
| Spreading mulch; to individual trees; maximum distance 25 m (mulch not included) | | | | | | | |
| 50 mm thick | – | 0.05 | 1.15 | – | – | m² | **1.15** |
| 75 mm thick | – | 0.07 | 1.73 | – | – | m² | **1.73** |
| 100 mm thick | – | 0.10 | 2.31 | – | – | m² | **2.31** |
| Mulch; Bark Nuggets®; to individual trees; delivered in 80 m³ loads; maximum distance 25 m | | | | | | | |
| 50 mm thick | 2.22 | 0.05 | 1.15 | – | 2.22 | m² | **3.37** |
| 75 mm thick | 3.33 | 0.07 | 1.73 | – | 3.33 | m² | **5.06** |
| 100 mm thick | 4.43 | 0.10 | 2.31 | – | 4.43 | m² | **6.74** |
| Mulch; Bark Nuggets®; to individual trees; delivered in 25 m³ loads; maximum distance 25 m | | | | | | | |
| 50 mm thick | 2.90 | 0.05 | 1.15 | – | 2.90 | m² | **4.05** |
| 75 mm thick | 4.35 | 0.05 | 1.19 | – | 4.35 | m² | **5.54** |
| 100 mm thick | 5.80 | 0.07 | 1.58 | – | 5.80 | m² | **7.38** |
| Mulch; Amenity Bark Mulch; to individual trees; delivered in 80 m³ loads; maximum distance 25 m | | | | | | | |
| 50 mm thick | 1.43 | 0.05 | 1.15 | – | 1.43 | m² | **2.58** |
| 75 mm thick | 2.14 | 0.07 | 1.73 | – | 2.14 | m² | **3.87** |
| 100 mm thick | 2.85 | 0.10 | 2.31 | – | 2.85 | m² | **5.16** |
| Mulch; Amenity Bark Mulch; to individual trees; delivered in 25 m³ loads; maximum distance 25 m | | | | | | | |
| 50 mm thick | 2.10 | 0.05 | 1.15 | – | 2.10 | m² | **3.25** |
| 75 mm thick | 3.16 | 0.07 | 1.73 | – | 3.16 | m² | **4.89** |
| 100 mm thick | 4.21 | 0.07 | 1.58 | – | 4.21 | m² | **5.79** |
| **Trees; planting labours only** | | | | | | | |
| Bare root trees; including backfilling with previously excavated material (all other operations and materials not included) | | | | | | | |
| light standard; 6–8 cm girth | – | 0.35 | 8.31 | – | – | nr | **8.31** |
| standard; 8–10 cm girth | – | 0.40 | 9.50 | – | – | nr | **9.50** |
| selected standard; 10–12 cm girth | – | 0.58 | 13.78 | – | – | nr | **13.78** |
| heavy standard; 12–14 cm girth | – | 0.83 | 19.79 | – | – | nr | **19.79** |
| extra heavy standard; 14–16 cm girth | – | 1.00 | 23.75 | – | – | nr | **23.75** |

## 37 SOFT LANDSCAPING

| Item<br>Excluding site overheads and profit | PC<br>£ | Labour<br>hours | Labour<br>£ | Plant<br>£ | Material<br>£ | Unit | Total<br>rate £ |
|---|---|---|---|---|---|---|---|
| **Rootballed trees; including backfilling with previously excavated material (all other operations and materials not included)** | | | | | | | |
| standard; 8–10 cm girth | – | 0.50 | 11.88 | – | – | nr | **11.88** |
| selected standard; 10–12 cm girth | – | 0.60 | 14.25 | – | – | nr | **14.25** |
| heavy standard; 12–14 cm girth | – | 0.80 | 19.00 | – | – | nr | **19.00** |
| extra heavy standard; 14–16 cm girth | – | 1.50 | 35.63 | – | – | nr | **35.63** |
| 16–18 cm girth | – | 1.30 | 30.78 | 32.68 | – | nr | **63.46** |
| 18–20 cm girth | – | 1.60 | 38.00 | 40.35 | – | nr | **78.35** |
| 20–25 cm girth | – | 4.50 | 106.88 | 124.45 | – | nr | **231.33** |
| 25–30 cm girth | – | 6.00 | 142.50 | 164.47 | – | nr | **306.97** |
| 30–35 cm girth | – | 11.00 | 261.25 | 293.85 | – | nr | **555.10** |
| **Tree pits for urban planting** | | | | | | | |
| Excavating tree pits; depositing soil alongside pits; by machine | | | | | | | |
| 1.75 × 1.75 × 1.00 m deep | – | 2.09 | 49.59 | 40.60 | – | nr | **90.19** |
| 2.00 × 2.00 × 1.00 m deep | – | 2.73 | 64.77 | 52.88 | – | nr | **117.65** |
| 2.50 × 2.50 × 1.50 m deep | – | 3.28 | 77.90 | 127.53 | – | nr | **205.43** |
| **Earthwork retention to tree pits** | | | | | | | |
| Bracing of deep tree pits by plywood boards braced by timber stays | | | | | | | |
| tree pit 1.00 m × 1.00 m × 1.00 m | – | 0.75 | 35.63 | – | 17.19 | nr | **52.82** |
| tree pit 1.50 m × 1.50 m × 1.00 m | – | 0.90 | 42.75 | – | 22.84 | nr | **65.59** |
| tree pit 2.00 m × 2.00 m × 1.50 m | – | 1.00 | 47.50 | – | 28.49 | nr | **75.99** |
| Excavate tree break-out zone for tree roots to develop in restricted and compacted urban environments; depositing soil alongside pits by machine | | | | | | | |
| trenches 1.50 m wide × 1.50 m deep average | – | 0.17 | 3.96 | 6.48 | – | m³ | **10.44** |
| Bases to tree pits; levelling base breaking up base by hand and placing of granular drainage material 200 mm thick | | | | | | | |
| tree pit 1.00 m × 1.00 m | – | 0.25 | 5.94 | 9.72 | 8.19 | nr | **23.85** |
| tree pit 1.50 m × 1.50 m | – | 0.45 | 10.69 | 17.50 | 18.43 | nr | **46.62** |
| tree pit 2.00 m × 2.00 m | – | 0.75 | 17.81 | 29.16 | 32.76 | nr | **79.73** |
| tree pit 2.50 m × 2.50 m | – | 0.75 | 17.81 | 29.16 | 51.19 | nr | **98.16** |

## 37 SOFT LANDSCAPING

| Item<br>Excluding site overheads and profit | PC<br>£ | Labour<br>hours | Labour<br>£ | Plant<br>£ | Material<br>£ | Unit | Total<br>rate £ |
|---|---|---|---|---|---|---|---|
| **TREE PLANTING – cont** | | | | | | | |
| **Gt ArborRaft System; Greentech Ltd; Strong geocellular system replaces traditional subbase layer in urban tree planting enabling shallow tree planting in a nutrient rich uncompacted environment that better reflects natural environment encouraging tree establishment and growth** | | | | | | | |
| 70 t Loading capacity; single layer installation to fully graded and levelled partially backfilled tree pit | | | | | | | |
| ArborRaft 150 mm unit (708 mm × 354 mm × 150 mm) | 65.00 | 0.17 | 7.92 | – | 65.00 | $m^2$ | **72.92** |
| ArborRaft 85 mm unit (708 mm × 354 mm × 85 mm) | 60.00 | 0.14 | 6.79 | – | 60.00 | $m^2$ | **66.79** |
| tree planting soil with nutrients compaction resistance and water retention qualities; structure remains open for water, oxygen and nutrients to reach developing root system of the tree | | | | | | | |
| ArborRaft tree pit soil 29 t loads; by machine | 27.50 | – | – | 11.86 | 27.50 | $m^3$ | **39.36** |
| ArborRaft tree pit soil 20 t loads; by machine | 31.25 | – | – | 11.86 | 31.25 | $m^3$ | **43.11** |
| ArborRaft tree pit soil bagged; by machine | 67.50 | – | – | 19.77 | 67.50 | $m^3$ | **87.27** |
| ArborRaft tree pit soil bagged; by hand | 67.50 | 0.50 | 23.75 | – | 67.50 | $m^3$ | **91.25** |
| **StrataCell; GreenBlue Urban; Tree planting accessories for Urban tree planting; Load bearing tree planting modules for load support in tree environments; Placed to excavated tree pit** | | | | | | | |
| single layer; 500 mm × 500 mm × 250 mm deep; interlocking load bearing cells; backfilling with imported topsoil | | | | | | | |
| 7.5 tonne load limit | 55.92 | 0.25 | 5.94 | 2.40 | 67.17 | $m^2$ | **75.51** |
| 40 tonne load limit | 79.52 | 0.25 | 5.94 | 2.40 | 90.77 | $m^2$ | **99.11** |
| double layer; 500 mm × 500 mm × 500 mm deep; interlocking load bearing cells; backfilling with imported topsoil | | | | | | | |
| 7.5 tonne load limit | 111.84 | 0.50 | 11.88 | 4.80 | 134.34 | $m^2$ | **151.02** |
| 40 tonne load limit | 159.04 | 0.50 | 11.88 | 4.80 | 181.54 | $m^2$ | **198.22** |

## 37 SOFT LANDSCAPING

| Item Excluding site overheads and profit | PC £ | Labour hours | Labour £ | Plant £ | Material £ | Unit | Total rate £ |
|---|---|---|---|---|---|---|---|
| **Root director; GreenBlue Urban; preformed root barrier system with integral root deflecting ribs** | | | | | | | |
| Placed over tree rootball; under pavements; overall size | | | | | | | |
| 510 × 595 × 310 mm | 44.94 | 0.25 | 11.88 | – | 44.94 | nr | **56.82** |
| 650 × 855 × 455 mm | 74.12 | 0.33 | 15.68 | – | 74.12 | nr | **89.80** |
| 975 × 1370 × 545 mm | 114.14 | 0.45 | 21.38 | – | 114.14 | nr | **135.52** |
| 1300 × 1805 × 500 mm | 195.20 | 0.60 | 28.50 | – | 195.20 | nr | **223.70** |
| **Tree planting; containerized trees; nursery stock; James Coles & Sons (Nurseries) Ltd** | | | | | | | |
| Acer platanoides 'Emerald Queen'; including backfillling with excavated material (other operations not included) | | | | | | | |
| standard; 8–10 cm girth | 49.00 | 0.48 | 11.40 | – | 49.00 | nr | **60.40** |
| selected standard; 10–12 cm girth | 56.00 | 0.56 | 13.31 | – | 56.00 | nr | **69.31** |
| heavy standard; 12–14 cm girth | 70.00 | 0.76 | 18.15 | – | 70.00 | nr | **88.15** |
| extra heavy standard; 14–16 cm girth | 77.00 | 1.20 | 28.50 | – | 77.00 | nr | **105.50** |
| Carpinus betulus; including backfillling with excavated material (other operations not included) | | | | | | | |
| standard; 8–10 cm girth | 19.50 | 0.48 | 11.40 | – | 19.50 | nr | **30.90** |
| selected standard; 10–12 cm girth | 70.00 | 0.56 | 13.31 | – | 70.00 | nr | **83.31** |
| heavy standard; 12–14 cm girth | 91.00 | 0.76 | 18.15 | – | 91.00 | nr | **109.15** |
| extra heavy standard; 14–16 cm girth | 112.00 | 1.20 | 28.50 | – | 112.00 | nr | **140.50** |
| Fagus sylvatica; including backfillling with excavated material (other operations not included) | | | | | | | |
| standard; 8–10 cm girth | 49.00 | 0.48 | 11.40 | – | 49.00 | nr | **60.40** |
| selected standard; 10–12 cm girth | 105.00 | 0.56 | 13.31 | – | 105.00 | nr | **118.31** |
| heavy standard; 12–14 cm girth | 119.00 | 0.76 | 18.15 | – | 119.00 | nr | **137.15** |
| extra heavy standard; 14–16 cm girth | 133.00 | 1.20 | 28.50 | – | 133.00 | nr | **161.50** |
| Prunus avium; including backfillling with excavated material (other operations not included) | | | | | | | |
| standard; 8–10 cm girth | 49.00 | 0.40 | 9.50 | – | 49.00 | nr | **58.50** |
| selected standard; 10–12 cm girth | 70.00 | 0.56 | 13.31 | – | 70.00 | nr | **83.31** |
| heavy standard; 12–14 cm girth | 77.00 | 0.76 | 18.15 | – | 77.00 | nr | **95.15** |
| extra heavy standard; 14–16 cm girth | 91.00 | 1.20 | 28.50 | – | 91.00 | nr | **119.50** |
| Quercus robur; including backfillling with excavated material (other operations not included) | | | | | | | |
| standard; 8–10 cm girth | 49.00 | 0.48 | 11.40 | – | 49.00 | nr | **60.40** |
| selected standard; 10–12 cm girth | 77.00 | 0.56 | 13.31 | – | 77.00 | nr | **90.31** |
| heavy standard; 12–14 cm girth | 105.00 | 0.76 | 18.15 | – | 105.00 | nr | **123.15** |
| extra heavy standard; 14–16 cm girth | 119.00 | 1.20 | 28.50 | – | 119.00 | nr | **147.50** |

## 37 SOFT LANDSCAPING

| Item<br>Excluding site overheads and profit | PC<br>£ | Labour<br>hours | Labour<br>£ | Plant<br>£ | Material<br>£ | Unit | Total<br>rate £ |
|---|---|---|---|---|---|---|---|
| **TREE PLANTING – cont** | | | | | | | |
| **Tree planting – cont** | | | | | | | |
| Betula utilis jaquemontii; multi-stemmed; including backfillling with excavated material (other operations not included) | | | | | | | |
| 175/200 mm high | 18.50 | 0.48 | 11.40 | – | 18.50 | nr | **29.90** |
| 200/250 mm high | 98.00 | 0.56 | 13.31 | – | 98.00 | nr | **111.31** |
| 250/300 mm high | 104.00 | 0.76 | 18.15 | – | 104.00 | nr | **122.15** |
| 300/350 mm high | 196.00 | 1.20 | 28.50 | – | 196.00 | nr | **224.50** |
| **Tree planting; rootballed trees; advanced nursery stock and semi-mature – General** | | | | | | | |
| Preamble: The cost of planting semi-mature trees will depend on the size and species and on the access to the site for tree handling machines. Prices should be obtained for individual trees and planting. | | | | | | | |
| **Tree planting; bare root trees; nursery stock; James Coles & Sons (Nurseries) Ltd** | | | | | | | |
| Acer platanoides; including backfillling with excavated material (other operations not included) | | | | | | | |
| light standard; 6–8 cm girth | 13.25 | 0.35 | 8.31 | – | 13.25 | nr | **21.56** |
| standard; 8–10 cm girth | 17.50 | 0.40 | 9.50 | – | 17.50 | nr | **27.00** |
| selected standard; 10–12 cm girth | 19.50 | 0.58 | 13.78 | – | 19.50 | nr | **33.28** |
| heavy standard; 12–14 cm girth | 49.00 | 0.83 | 19.79 | – | 49.00 | nr | **68.79** |
| extra heavy standard; 14–16 cm girth | 68.50 | 1.00 | 23.75 | – | 68.50 | nr | **92.25** |
| Carpinus betulus; including backfillling with excavated material (other operations not included) | | | | | | | |
| light standard; 6–8 cm girth | 13.25 | 0.35 | 8.31 | – | 13.25 | nr | **21.56** |
| standard; 8–10 cm girth | 19.50 | 0.40 | 9.50 | – | 19.50 | nr | **29.00** |
| selected standard; 10–12 cm girth | 28.00 | 0.58 | 13.78 | – | 28.00 | nr | **41.78** |
| heavy standard; 12–14 cm girth | 70.00 | 0.83 | 19.80 | – | 70.00 | nr | **89.80** |
| extra heavy standard; 14–16 cm girth | 42.00 | 1.00 | 23.75 | – | 42.00 | nr | **65.75** |
| Fagus sylvatica; including backfillling with excavated material (other operations not included) | | | | | | | |
| light standard; 6–8 cm girth | 18.95 | 0.35 | 8.31 | – | 18.95 | nr | **27.26** |
| standard; 8–10 cm girth | 22.75 | 0.40 | 9.50 | – | 22.75 | nr | **32.25** |
| selected standard; 10–12 cm girth | 39.25 | 0.58 | 13.78 | – | 39.25 | nr | **53.03** |
| heavy standard; 12–14 cm girth | 70.00 | 0.83 | 19.71 | – | 70.00 | nr | **89.71** |
| extra heavy standard; 14–16 cm girth | 105.00 | 1.00 | 23.75 | – | 105.00 | nr | **128.75** |

## 37 SOFT LANDSCAPING

| Item<br>Excluding site overheads and profit | PC<br>£ | Labour<br>hours | Labour<br>£ | Plant<br>£ | Material<br>£ | Unit | Total<br>rate £ |
|---|---|---|---|---|---|---|---|
| Prunus avium; including backfillling with excavated material (other operations not included) | | | | | | | |
|   light standard; 6–8 cm girth | 18.95 | 0.36 | 8.64 | – | 18.95 | nr | 27.59 |
|   standard; 8–10 cm girth | 22.75 | 0.40 | 9.50 | – | 22.75 | nr | 32.25 |
|   selected standard; 10–12 cm girth | 21.00 | 0.58 | 13.78 | – | 21.00 | nr | 34.78 |
|   heavy standard; 12–14 cm girth | 35.00 | 0.83 | 19.79 | – | 35.00 | nr | 54.79 |
|   extra heavy standard; 14–16 cm girth | 44.75 | 1.00 | 23.75 | – | 44.75 | nr | 68.50 |
| Quercus robur; including backfillling with excavated material (other operations not included) | | | | | | | |
|   light standard; 6–8 cm girth | 18.25 | 0.35 | 8.31 | – | 18.25 | nr | 26.56 |
|   standard; 8–10 cm girth | 30.75 | 0.40 | 9.50 | – | 30.75 | nr | 40.25 |
|   selected standard; 10–12 cm girth | 42.00 | 0.58 | 13.78 | – | 42.00 | nr | 55.78 |
|   heavy standard; 12–14 cm girth | 67.25 | 0.83 | 19.79 | – | 67.25 | nr | 87.04 |
| Robinia pseudoacacia 'Frisia'; including backfillling with excavated material (other operations not included) | | | | | | | |
|   light standard; 6–8 cm girth | 21.00 | 0.35 | 8.31 | – | 21.00 | nr | 29.31 |
|   standard; 8–10 cm girth | 28.00 | 0.40 | 9.50 | – | 28.00 | nr | 37.50 |
|   selected standard; 10–12 cm girth | 44.75 | 0.58 | 13.78 | – | 44.75 | nr | 58.53 |
| **Tree planting; rootballed trees; nursery stock; James Coles & Sons (Nurseries) Ltd** | | | | | | | |
| Acer platanoides; including backfillling with excavated material (other operations not included) | | | | | | | |
|   standard; 8–10 cm girth | 26.00 | 0.48 | 11.40 | – | 26.00 | nr | 37.40 |
|   selected standard; 10–12 cm girth | 56.00 | 0.56 | 13.31 | – | 56.00 | nr | 69.31 |
|   heavy standard; 12–14 cm girth | 70.00 | 0.76 | 18.15 | – | 70.00 | nr | 88.15 |
|   extra heavy standard; 14–16 cm girth | 77.00 | 1.20 | 28.50 | – | 77.00 | nr | 105.50 |
| Carpinus betulus; including backfillling with excavated material (other operations not included) | | | | | | | |
|   standard; 8–10 cm girth | 49.00 | 0.48 | 11.40 | – | 49.00 | nr | 60.40 |
|   selected standard; 10–12 cm girth | 56.00 | 0.56 | 13.31 | – | 56.00 | nr | 69.31 |
|   heavy standard; 12–14 cm girth | 91.00 | 0.76 | 18.15 | – | 91.00 | nr | 109.15 |
|   extra heavy standard; 14–16 cm girth | 112.00 | 1.20 | 28.50 | – | 112.00 | nr | 140.50 |

## 37 SOFT LANDSCAPING

| Item<br>Excluding site overheads and profit | PC<br>£ | Labour<br>hours | Labour<br>£ | Plant<br>£ | Material<br>£ | Unit | Total<br>rate £ |
|---|---|---|---|---|---|---|---|
| **TREE PLANTING – cont** | | | | | | | |
| | | | | | | | |
| **Tree planting – cont** | | | | | | | |
| Fagus sylvatica; including backfillling with excavated material (other operations not included) | | | | | | | |
| standard; 8–10 cm girth | 31.25 | 0.48 | 11.40 | – | 31.25 | nr | **42.65** |
| selected standard; 10–12 cm girth | 51.75 | 0.56 | 13.31 | – | 51.75 | nr | **65.06** |
| heavy standard; 12–14 cm girth | 85.50 | 0.76 | 18.15 | – | 85.50 | nr | **103.65** |
| extra heavy standard; 14–16 cm girth | 122.15 | 1.20 | 28.50 | – | 122.15 | nr | **150.65** |
| Prunus avium; including backfillling with excavated material (other operations not included) | | | | | | | |
| standard; 8–10 cm girth | 8.50 | 0.40 | 9.50 | – | 8.50 | nr | **18.00** |
| selected standard; 10–12 cm girth | 33.50 | 0.56 | 13.31 | – | 33.50 | nr | **46.81** |
| heavy standard; 12–14 cm girth | 50.50 | 0.76 | 18.15 | – | 50.50 | nr | **68.65** |
| extra heavy standard; 14–16 cm girth | 17.15 | 1.20 | 28.50 | – | 61.90 | nr | **90.40** |
| Quercus robur; including backfillling with excavated material (other operations not included) | | | | | | | |
| standard; 8–10 cm girth | 30.75 | 0.48 | 11.40 | – | 30.75 | nr | **42.15** |
| selected standard; 10–12 cm girth | 42.00 | 0.56 | 13.31 | – | 42.00 | nr | **55.31** |
| heavy standard; 12–14 cm girth | 67.25 | 0.76 | 18.15 | – | 67.25 | nr | **85.40** |
| extra heavy standard; 14–16 cm girth | 94.50 | 1.20 | 28.50 | – | 94.50 | nr | **123.00** |
| Robinia pseudoacacia 'Frisia'; including backfillling with excavated material (other operations not included) | | | | | | | |
| standard; 8–10 cm girth | 40.00 | 0.48 | 11.40 | – | 40.00 | nr | **51.40** |
| selected standard; 10–12 cm girth | 62.25 | 0.56 | 13.31 | – | 62.25 | nr | **75.56** |
| heavy standard; 12–14 cm girth | 112.00 | 0.76 | 18.15 | – | 112.00 | nr | **130.15** |
| **Tree planting; Airpot container grown trees; advanced nursery stock and semi-mature; Hillier Nurseries** | | | | | | | |
| Acer platanoides 'Emerald Queen'; including backfilling with excavated material (other operations not included) | | | | | | | |
| 16–18 cm girth | 100.00 | 1.98 | 47.02 | 45.38 | 100.00 | nr | **192.40** |
| 18–20 cm girth | 118.00 | 2.18 | 51.73 | 48.83 | 118.00 | nr | **218.56** |
| 20–25 cm girth | 182.00 | 2.38 | 56.43 | 54.46 | 182.00 | nr | **292.89** |
| 25–30 cm girth | 225.00 | 2.97 | 70.54 | 83.79 | 225.00 | nr | **379.33** |
| 30–35 cm girth | 307.00 | 3.96 | 94.05 | 90.77 | 307.00 | nr | **491.82** |
| Prunus avium 'Flora Plena'; including backfilling with excavated material (other operations not included) | | | | | | | |
| 16–18 cm girth | 95.00 | 1.98 | 47.02 | 45.38 | 95.00 | nr | **187.40** |
| 18–20 cm girth | 118.00 | 1.60 | 38.00 | 48.83 | 118.00 | nr | **204.83** |
| 20–25 cm girth | 182.00 | 2.38 | 56.43 | 54.46 | 182.00 | nr | **292.89** |
| 25–30 cm girth | 225.00 | 2.97 | 70.54 | 83.79 | 225.00 | nr | **379.33** |
| 30–35 cm girth | 350.00 | 3.96 | 94.05 | 90.77 | 350.00 | nr | **534.82** |

## 37 SOFT LANDSCAPING

| Item<br>Excluding site overheads and profit | PC<br>£ | Labour<br>hours | Labour<br>£ | Plant<br>£ | Material<br>£ | Unit | Total<br>rate £ |
|---|---|---|---|---|---|---|---|
| Quercus palustris 'Pin Oak'; including backfilling with excavated material (other operations not included) | | | | | | | |
| 16–18 cm girth | 108.00 | 1.98 | 47.02 | 45.38 | 108.00 | nr | 200.40 |
| 18–20 cm girth | 165.00 | 1.60 | 38.00 | 48.83 | 165.00 | nr | 251.83 |
| 20–25 cm girth | 245.00 | 2.38 | 56.43 | 54.46 | 245.00 | nr | 355.89 |
| 25–30 cm girth | 375.00 | 2.97 | 70.54 | 83.79 | 375.00 | nr | 529.33 |
| 30–35 cm girth | 575.00 | 3.96 | 94.05 | 90.77 | 575.00 | nr | 759.82 |
| **Tree planting; Multi-stem trees; Hillier Nurseries; Airpot container grown trees; semi-mature;** | | | | | | | |
| Betula Pendula; including backfilling with excavated material (other operations not included) | | | | | | | |
| Betula pendula; 300–400 cm | 135.00 | 3.00 | 213.75 | 83.79 | 135.00 | nr | 432.54 |
| Betula pendula; 400–500 cm | 250.00 | 3.25 | 231.56 | 83.79 | 250.00 | nr | 565.35 |
| Betula pendula; 500–600 cm | 375.00 | 3.50 | 249.38 | 88.58 | 375.00 | nr | 712.96 |
| Betula pendula; 600–700 cm | 850.00 | 3.75 | 267.19 | 91.87 | 850.00 | nr | 1209.06 |
| Betula pendula; 700–800 cm | 1500.00 | 4.00 | 285.00 | 93.51 | 1500.00 | nr | 1878.51 |
| Betula pendula; 800–900 cm | 1900.00 | 4.50 | 320.63 | 102.90 | 1900.00 | nr | 2323.53 |
| Betula jacquemontii; including backfilling with excavated material (other operations not included) | | | | | | | |
| Betula jacquemontii; 300–400 cm | 300.00 | 3.00 | 213.75 | 83.79 | 300.00 | nr | 597.54 |
| Betula jacquemontii; 400–500 cm | 450.00 | 3.25 | 231.56 | 83.79 | 450.00 | nr | 765.35 |
| Betula jacquemontii; 500–600 cm | 750.00 | 3.50 | 249.38 | 88.58 | 750.00 | nr | 1087.96 |
| Betula jacquemontii; 600–700 cm | 1500.00 | 4.00 | 285.00 | 93.51 | 1500.00 | nr | 1878.51 |
| Betula jacquemontii; 700–800 cm | 1950.00 | 4.50 | 320.63 | 102.90 | 1950.00 | nr | 2373.53 |
| **Tree planting; Airpot container grown trees; semi-mature and mature trees; Hillier Nurseries; planting and back filling; planted by telehandler or by crane; delivery included; all other operations priced separately** | | | | | | | |
| Semi-mature trees; indicative prices | | | | | | | |
| 40–45 cm girth | 550.00 | 4.00 | 95.00 | 57.26 | 550.00 | nr | 702.26 |
| 45–50 cm girth | 750.00 | 4.00 | 95.00 | 57.26 | 750.00 | nr | 902.26 |
| 55–60 cm girth | 1350.00 | 6.00 | 142.50 | 57.26 | 1350.00 | nr | 1549.76 |
| 60–70 cm girth | 2500.00 | 7.00 | 166.25 | 77.20 | 2500.00 | nr | 2743.45 |
| 70–80 cm girth | 3500.00 | 7.50 | 178.13 | 97.14 | 3500.00 | nr | 3775.27 |
| 80–90 cm girth | 4500.00 | 8.00 | 190.00 | 114.53 | 4500.00 | nr | 4804.53 |

## 37 SOFT LANDSCAPING

| Item<br>Excluding site overheads and profit | PC<br>£ | Labour<br>hours | Labour<br>£ | Plant<br>£ | Material<br>£ | Unit | Total<br>rate £ |
|---|---|---|---|---|---|---|---|
| **TREE PLANTING – cont** | | | | | | | |
| | | | | | | | |
| **Tree planting; rootballed trees; advanced nursery stock and semi-mature; Lorenz von Ehren** | | | | | | | |
| Acer platanoides 'Emerald Queen'; including backfilling with excavated material (other operations not included) | | | | | | | |
| 16–18 cm girth | 85.00 | 1.30 | 30.78 | 3.91 | 85.00 | nr | **119.69** |
| 18–20 cm girth | 105.00 | 1.60 | 38.00 | 3.91 | 105.00 | nr | **146.91** |
| 20–25 cm girth | 130.00 | 4.50 | 106.88 | 16.83 | 130.00 | nr | **253.71** |
| 25–30 cm girth | 170.00 | 6.00 | 142.50 | 20.98 | 170.00 | nr | **333.48** |
| 30–35 cm girth | 310.00 | 11.00 | 261.25 | 28.18 | 310.00 | nr | **599.43** |
| Aesculus carnea 'Briotti'; including backfilling with excavated material (other operations not included) | | | | | | | |
| 16–18 cm girth | 180.00 | 1.30 | 30.78 | 3.91 | 180.00 | nr | **214.69** |
| 18–20 cm girth | 230.00 | 1.60 | 38.00 | 3.91 | 230.00 | nr | **271.91** |
| 20–25 cm girth | 280.00 | 4.50 | 106.88 | 16.83 | 280.00 | nr | **403.71** |
| 25–30 cm girth | 360.00 | 6.00 | 142.50 | 20.98 | 360.00 | nr | **523.48** |
| 30–35 cm girth | 480.00 | 11.00 | 261.25 | 28.18 | 480.00 | nr | **769.43** |
| Prunus avium 'Plena'; including backfilling with excavated material (other operations not included) | | | | | | | |
| 16–18 cm girth | 115.00 | 1.30 | 30.78 | 3.91 | 115.00 | nr | **149.69** |
| 18–20 cm girth | 130.00 | 1.60 | 38.00 | 3.91 | 130.00 | nr | **171.91** |
| 20–25 cm girth | 150.00 | 4.50 | 106.88 | 16.83 | 150.00 | nr | **273.71** |
| 25–30 cm girth | 200.00 | 6.00 | 142.50 | 20.98 | 200.00 | nr | **363.48** |
| 30–35 cm girth | 290.00 | 11.00 | 261.25 | 24.88 | 290.00 | nr | **576.13** |
| Quercus palustris 'Pin Oak'; including backfilling with excavated material (other operations not included) | | | | | | | |
| 16–18 cm girth | 140.00 | 1.30 | 30.78 | 3.91 | 140.00 | nr | **174.69** |
| 18–20 cm girth | 170.00 | 1.60 | 38.00 | 3.91 | 170.00 | nr | **211.91** |
| 20–25 cm girth | 210.00 | 4.50 | 106.88 | 16.83 | 210.00 | nr | **333.71** |
| 25–30 cm girth | 300.00 | 6.00 | 142.50 | 20.98 | 300.00 | nr | **463.48** |
| 30–35 cm girth | 400.00 | 11.00 | 261.25 | 28.18 | 400.00 | nr | **689.43** |
| Tilia cordata 'Greenspire'; including backfilling with excavated material (other operations not included) | | | | | | | |
| 16–18 cm girth | 95.00 | 1.30 | 30.78 | 3.91 | 95.00 | nr | **129.69** |
| 18–20 cm girth | 115.00 | 1.60 | 38.00 | 3.91 | 115.00 | nr | **156.91** |
| 20–25 cm girth | 140.00 | 4.50 | 106.88 | 16.83 | 140.00 | nr | **263.71** |
| 25–30 cm girth; 5 × transplanted; 4.0–5.0 m tall | 170.00 | 6.00 | 142.50 | 20.98 | 170.00 | nr | **333.48** |
| 30–35 cm girth; 5 × transplanted; 5.0–7.0 m tall | 360.00 | 11.00 | 261.25 | 28.18 | 360.00 | nr | **649.43** |

## 37 SOFT LANDSCAPING

| Item<br>Excluding site overheads and profit | PC<br>£ | Labour<br>hours | Labour<br>£ | Plant<br>£ | Material<br>£ | Unit | Total<br>rate £ |
|---|---|---|---|---|---|---|---|
| Betula pendula (3 stems); including backfilling with excavated material (other operations not included) | | | | | | | |
| 3.0–3.5 m high | 90.00 | 1.98 | 47.02 | 45.38 | 90.00 | nr | **182.40** |
| 3.5–4.0 m high | 160.00 | 1.60 | 38.00 | 48.83 | 160.00 | nr | **246.83** |
| 4.0–4.5 m high | 135.00 | 2.38 | 56.43 | 54.46 | 135.00 | nr | **245.89** |
| 4.5–5.0 m high | 155.00 | 2.97 | 70.54 | 83.79 | 155.00 | nr | **309.33** |
| 5.0–6.0 m high | 260.00 | 3.96 | 94.05 | 90.77 | 260.00 | nr | **444.82** |
| 6.0–7.0 m high | 360.00 | 4.50 | 106.88 | 105.91 | 360.00 | nr | **572.79** |
| Pinus sylvestris; including backfilling with excavated material (other operations not included) | | | | | | | |
| 3.0–3.5 m high | 410.00 | 1.98 | 47.02 | 45.38 | 410.00 | nr | **502.40** |
| 3.5–4.0 m high | 510.00 | 1.60 | 38.00 | 48.83 | 510.00 | nr | **596.83** |
| 4.0–4.5 m high | 700.00 | 2.38 | 56.43 | 54.46 | 700.00 | nr | **810.89** |
| 4.5–5.0 m high | 960.00 | 2.97 | 70.54 | 83.79 | 960.00 | nr | **1114.33** |
| 5.0–6.0 m high | 1520.00 | 3.96 | 94.05 | 90.77 | 1520.00 | nr | **1704.82** |
| 6.0–7.0 m high | 2540.00 | 4.50 | 106.88 | 109.39 | 2540.00 | nr | **2756.27** |
| **Pleached trees; Carpinus betulus; specimen trees; Lorenz Von Ehren; supply and planting only; excavation, support and tree pit additives not included** | | | | | | | |
| Box shaped trees to provide 'floating hedge' or screen effect; planting distance 1 tree per m run; trunk and box alignment; wire rootballs | | | | | | | |
| 4 × transplanted; 20–25 cm; 1.00 m centres | 320.00 | 5.00 | 118.75 | 16.83 | 320.00 | m | **455.58** |
| 5 × transplanted; 25–30 cm; 1.00 m centres | 450.00 | 6.50 | 154.38 | 20.98 | 450.00 | m | **625.36** |
| 5 × transplanted; 30–35 cm; 1.20 m centres | 720.00 | 11.00 | 261.25 | 28.18 | 720.00 | m | **1009.43** |
| **Pleached trees; Tilia europaea 'Pallida'; specimen trees; Lorenz Von Ehren; supply and planting only; excavation, support and tree pit additives not included** | | | | | | | |
| Box pleached trees to provide 'floating hedge' or screen effect; wire rootballs | | | | | | | |
| 4 × transplanted; 20–25 cm; 1.00 m centres | 320.00 | 5.00 | 118.75 | 16.83 | 320.00 | m | **455.58** |
| 4 × transplanted; 25–30 cm; 1.00 m centres | 380.00 | 5.00 | 118.75 | 16.83 | 380.00 | m | **515.58** |
| 5 × transplanted; 30–35 cm; 1.20 m centres | 374.99 | 5.00 | 118.75 | 16.83 | 374.99 | m | **510.57** |

## 37 SOFT LANDSCAPING

| Item<br>Excluding site overheads and profit | PC<br>£ | Labour<br>hours | Labour<br>£ | Plant<br>£ | Material<br>£ | Unit | Total<br>rate £ |
|---|---|---|---|---|---|---|---|
| **TREE PLANTING – cont** | | | | | | | |
| **Pleached trees – cont** | | | | | | | |
| Box pleached trees to provide 'floating hedge' or screen effect – cont | | | | | | | |
| 5 × transplanted; 35–40 cm; 1.50 m centres | 393.35 | 5.00 | 118.75 | 16.83 | 393.35 | m | **528.93** |
| 6 × transplanted; 40–45 cm; 1.80 m centres | 455.59 | 5.00 | 118.75 | 16.83 | 455.59 | m | **591.17** |
| 6 × transplanted; 45–50 cm; 1.80 m centres | 583.38 | 5.00 | 118.75 | 16.83 | 583.38 | m | **718.96** |
| 6 × transplanted; 50–60 cm; 2.00 m centres | 700.00 | 5.00 | 118.75 | 16.83 | 700.00 | m | **835.58** |
| **Espalier pleached trees; specimen trees; Lorenz Von Ehren; supply and planting only; excavation, support and tree pit additives not included** | | | | | | | |
| Frame pleached trees; to provide floating screen effects; wire rootballs | | | | | | | |
| Carpinus betulus; 20–25 cm; 1.00 m centres | 450.00 | 5.00 | 118.75 | 16.83 | 450.00 | m | **585.58** |
| Carpinus betulus; 25–30 cm; 1.00 m centres | 600.00 | 5.00 | 118.75 | 16.83 | 600.00 | m | **735.58** |
| Tilia europaea 'Pallida'; 20–25 cm; 1.00 m centres | 400.00 | 5.00 | 118.75 | 16.83 | 400.00 | m | **535.58** |
| Tilia europaea 'Pallida'; 25–30 cm; 1.00 m centres | 550.00 | 5.00 | 118.75 | 16.83 | 550.00 | m | **685.58** |
| Tilia europaea 'Pallida'; 30–35 cm; 1.20 m centres | 574.98 | 5.00 | 118.75 | 16.83 | 574.98 | m | **710.56** |
| **Umbrella shaped pleaches** | | | | | | | |
| Umbrella or roof pleached trees to provide umbrella effect; wire rootballs | | | | | | | |
| Tilia euchlora; specimen; 4 × transplanted; 20–25 cm | 390.00 | 5.00 | 118.75 | 16.83 | 390.00 | nr | **525.58** |
| Tilia euchlora; specimen; 5 × transplanted; 25–30 cm | 520.00 | 5.00 | 118.75 | 16.83 | 520.00 | nr | **655.58** |
| Platanus acerifolia; clear stem; 4 × transplanted; 20–25 cm | 390.00 | 5.00 | 118.75 | 16.83 | 390.00 | nr | **525.58** |
| Platanus acerifolia; specimen; 5 × transplanted; 30–35 cm | 650.00 | 5.00 | 118.75 | 16.83 | 650.00 | nr | **785.58** |
| Tilia europaea; specimen; 5 × transplanted; 30–35 cm | 650.00 | 5.00 | 118.75 | 16.83 | 650.00 | nr | **785.58** |
| Platanus acerifolia; specimen; 5 × transplanted; 25–30 cm | 520.00 | 5.00 | 118.75 | 16.83 | 520.00 | nr | **655.58** |

## 37 SOFT LANDSCAPING

| Item<br>Excluding site overheads and profit | PC<br>£ | Labour<br>hours | Labour<br>£ | Plant<br>£ | Material<br>£ | Unit | Total<br>rate £ |
|---|---|---|---|---|---|---|---|
| **Tree planting; containerized trees; Lorenz von Ehren; to the tree prices above add for Airpot containerization only** | | | | | | | |
| Tree size | | | | | | | |
| 20–25 cm | – | – | – | – | 80.00 | nr | 80.00 |
| 25–30 cm | – | – | – | – | 120.00 | nr | 120.00 |
| 30–35 cm | – | – | – | – | 120.00 | nr | 120.00 |
| 35–40 cm | – | – | – | – | 200.00 | nr | 200.00 |
| 40–45 cm | – | – | – | – | 200.00 | nr | 200.00 |
| 45–50 cm | – | – | – | – | 280.00 | nr | 280.00 |
| 50–60 cm | – | – | – | – | 400.00 | nr | 400.00 |
| 60–70 cm | – | – | – | – | 540.00 | nr | 540.00 |
| 70–80 cm | – | – | – | – | 700.00 | nr | 700.00 |
| 80–90 cm | – | – | – | – | 940.00 | nr | 940.00 |
| **Tree planting; rootballed trees; semi-mature and mature trees; Lorenz von Ehren; planting and back filling; planted by telehandler or by crane; delivery included; all other operations priced separately** | | | | | | | |
| Semi-mature trees | | | | | | | |
| 40–45 cm girth | 600.00 | 8.00 | 190.00 | 68.56 | 600.00 | nr | 858.56 |
| 45–50 cm girth | 850.00 | 8.00 | 190.00 | 87.71 | 850.00 | nr | 1127.71 |
| 50–60 cm girth | 1200.00 | 10.00 | 237.50 | 107.37 | 1200.00 | nr | 1544.87 |
| 60–70 cm girth | 2200.00 | 15.00 | 356.25 | 261.31 | 2200.00 | nr | 2817.56 |
| 70–80 cm girth | 3200.00 | 18.00 | 427.50 | 235.87 | 3200.00 | nr | 3863.37 |
| 80–90 cm girth | 4800.00 | 18.00 | 427.50 | 235.87 | 4800.00 | nr | 5463.37 |
| 90–100 cm girth | 5900.00 | 18.00 | 427.50 | 235.87 | 5900.00 | nr | 6563.37 |
| **SHRUB PLANTING** | | | | | | | |
| **Shrub planting – General**<br>Preamble: For preparation of planting areas see 'Cultivation' at the beginning of the section on planting. | | | | | | | |
| **Market prices of Shrubs**<br>The prices for shrubs are so varied that it is not possible to provide a list in this publication. For general budgeting purposes however the following average prices may be used with deciduous shrubs being slightly less and evergreeens slighty more | | | | | | | |
| 1 litre shrub | – | – | – | – | 1.50 | nr | 1.50 |
| 2 litre shrub | – | – | – | – | 2.80 | nr | 2.80 |
| 3 litre shrub | – | – | – | – | 3.20 | nr | 3.20 |
| 5 litre shrub | – | – | – | – | 5.50 | nr | 5.50 |
| 10 litre specimen shrub | – | – | – | – | 10.00 | nr | 10.00 |
| 25 litre specimen shrub | – | – | – | – | 35.00 | nr | 35.00 |

## 37 SOFT LANDSCAPING

| Item<br>Excluding site overheads and profit | PC<br>£ | Labour<br>hours | Labour<br>£ | Plant<br>£ | Material<br>£ | Unit | Total<br>rate £ |
|---|---|---|---|---|---|---|---|
| **SHRUB PLANTING** – cont | | | | | | | |
| **Market prices of Shrubs** – cont | | | | | | | |
| Market prices of high impact specimen shrubs; Griffin Nurseries; Prices are varied and the prices below reflect the price ranges for plants as described; container sizes | | | | | | | |
| 25 litre container | 40.00 | – | – | – | 40.00 | nr | **40.00** |
| 25 litre container; specialist feature plant specimen | 54.00 | – | – | – | 54.00 | nr | **54.00** |
| 40 litre container | 68.00 | – | – | – | 68.00 | nr | **68.00** |
| 40 litre container; specialist feature plant specimen | 105.00 | – | – | – | 105.00 | nr | **105.00** |
| 80 litre container | 95.00 | – | – | – | 95.00 | nr | **95.00** |
| 80 litre container; specialist feature plant specimen | 195.00 | – | – | – | 195.00 | nr | **195.00** |
| 130 litre container; specialist feature plant specimen | 295.00 | – | – | – | 295.00 | nr | **295.00** |
| **Shrub planting** | | | | | | | |
| Setting out; selecting planting from holding area; loading to wheelbarrows; planting as plan or as directed; distance from holding area maximum 50 m; plants 2–3 litre containers | | | | | | | |
| single plants not grouped | – | 0.04 | 0.95 | – | – | nr | **0.95** |
| plants in groups of 3–5 nr | – | 0.03 | 0.59 | – | – | nr | **0.59** |
| plants in groups of 10–100 nr | – | 0.02 | 0.39 | – | – | nr | **0.39** |
| plants in groups of 100 nr minimum | – | 0.01 | 0.28 | – | – | nr | **0.28** |
| Forming planting holes; in cultivated ground (cultivating not included); by mechanical auger; trimming holes by hand; depositing excavated material alongside holes | | | | | | | |
| 250 mm diameter | – | 0.03 | 0.79 | 0.05 | – | nr | **0.84** |
| 250 × 250 mm | – | 0.04 | 0.95 | 0.08 | – | nr | **1.03** |
| 300 × 300 mm | – | 0.08 | 1.78 | 0.10 | – | nr | **1.88** |
| Planting shrubs; hand excavation; forming planting holes; in cultivated ground (cultivating not included); depositing excavated material alongside holes; Placing plants previously set out alongside and backfilling (plants not included) | | | | | | | |
| 100 × 100 × 100 mm deep; with mattock or hoe | – | 0.01 | 0.26 | – | – | nr | **0.26** |
| 250 × 250 × 300 mm deep | – | 0.01 | 0.32 | – | – | nr | **0.32** |
| 300 × 300 × 300 mm deep | – | 0.02 | 0.40 | – | – | nr | **0.40** |

## 37 SOFT LANDSCAPING

| Item<br>Excluding site overheads and profit | PC<br>£ | Labour<br>hours | Labour<br>£ | Plant<br>£ | Material<br>£ | Unit | Total<br>rate £ |
|---|---|---|---|---|---|---|---|
| 400 × 400 × 400 mm deep | – | 0.08 | 1.98 | – | – | nr | 1.98 |
| 500 × 500 × 500 mm deep | – | 0.17 | 3.96 | – | – | nr | 3.96 |
| 600 × 600 × 600 mm deep | – | 0.25 | 5.94 | – | – | nr | 5.94 |
| 900 × 900 × 600 mm deep | – | 0.50 | 11.88 | – | – | nr | 11.88 |
| 1.00 m × 1.00 m × 600 mm deep | – | 1.00 | 23.75 | – | – | nr | 23.75 |
| 1.25 m × 1.25 m × 600 mm deep | – | 1.33 | 31.67 | – | – | nr | 31.67 |
| Hand excavation; forming planting holes; in uncultivated ground; depositing excavated material alongside holes | | | | | | | |
| 100 × 100 × 100 mm deep; with mattock or hoe | – | 0.01 | 0.32 | – | – | nr | 0.32 |
| 250 × 250 × 250 mm deep | – | 0.03 | 0.59 | – | – | nr | 0.59 |
| 300 × 300 × 300 mm deep | – | 0.06 | 1.48 | – | – | nr | 1.48 |
| 400 × 400 × 400 mm deep | – | 0.25 | 5.94 | – | – | nr | 5.94 |
| 500 × 500 × 500 mm deep | – | 0.33 | 7.73 | – | – | nr | 7.73 |
| 600 × 600 × 600 mm deep | – | 0.55 | 13.06 | – | – | nr | 13.06 |
| 900 × 900 × 600 mm deep | – | 1.25 | 29.69 | – | – | nr | 29.69 |
| 1.00 m × 1.00 m × 600 mm deep | – | 1.54 | 36.52 | – | – | nr | 36.52 |
| 1.25 m × 1.25 m × 600 mm deep | – | 2.41 | 57.30 | – | – | nr | 57.30 |
| Bare root planting; to planting holes (forming holes not included); including backfilling with excavated material (bare root plants not included) | | | | | | | |
| bare root 1+1; 30–90 mm high | – | 0.02 | 0.40 | – | – | nr | 0.40 |
| bare root 1+2; 90–20 mm high | – | 0.02 | 0.40 | – | – | nr | 0.40 |
| **Shrub planting where plants are set out at equal off set centres where the spacings are always the specified distance between the plant centres. This introduces more uniform but slightly higher distances than rectangular centres** | | | | | | | |
| 2 litre containerized plants; in cultivated ground (cultivating not included); PC £2.80/nr | | | | | | | |
| 1.00 m centres; 1.43 plants/m² | – | 0.02 | 0.45 | – | 4.00 | m² | 4.45 |
| 900 mm centres; 1.69 plants/m² | – | 0.02 | 0.53 | – | 4.73 | m² | 5.26 |
| 750 mm centres; 2.6 plants/m² | – | 0.04 | 0.85 | – | 7.28 | m² | 8.13 |
| 600 mm centres; 3.04 plants/m² | – | 0.04 | 0.96 | – | 8.51 | m² | 9.47 |
| 500 mm centres; 4.5 plants/m² | – | 0.06 | 1.42 | – | 12.60 | m² | 14.02 |
| 450 mm centres; 6.75 plants/m² | – | 0.09 | 2.14 | – | 18.90 | m² | 21.04 |
| 400 mm centres; 7.50 plants/m² | – | 0.17 | 3.99 | – | 28.06 | m² | 32.05 |
| 300 mm centres; 12.7 plants/m² | – | 0.17 | 4.01 | – | 35.76 | m² | 39.77 |

## 37 SOFT LANDSCAPING

| Item<br>Excluding site overheads and profit | PC<br>£ | Labour<br>hours | Labour<br>£ | Plant<br>£ | Material<br>£ | Unit | Total<br>rate £ |
|---|---|---|---|---|---|---|---|
| **SHRUB PLANTING – cont** | | | | | | | |
| **Shrub planting where plants are set out at equal off set centres where the spacings are always the specified distance between the plant centres. – cont** | | | | | | | |
| Shrub planting; 3 litre containerised plants; in cultivated ground (cultivating not included); PC £3.20/nr | | | | | | | |
| 1.00 m centres; 1.43 plants/m$^2$ | – | 0.02 | 0.45 | – | 4.58 | m$^2$ | **5.03** |
| 900 mm centres; 1.69 plants/m$^2$ | – | 0.02 | 0.53 | – | 5.41 | m$^2$ | **5.94** |
| 750 mm centres; 2.6 plants/m$^2$ | – | 0.04 | 0.85 | – | 8.32 | m$^2$ | **9.17** |
| 600 mm centres; 3.04 plants/m$^2$ | – | 0.04 | 0.96 | – | 9.73 | m$^2$ | **10.69** |
| 500 mm centres; 4.5 plants/m$^2$ | – | 0.06 | 1.42 | – | 14.40 | m$^2$ | **15.82** |
| 450 mm centres; 6.75 plants/m$^2$ | – | 0.09 | 2.14 | – | 21.60 | m$^2$ | **23.74** |
| 400 mm centres; 7.50 plants/m$^2$ | – | 0.17 | 3.99 | – | 32.06 | m$^2$ | **36.05** |
| 300 mm centres; 12.7 plants/m$^2$ | – | 0.17 | 4.01 | – | 40.86 | m$^2$ | **44.87** |
| Extra over shrubs for stakes | 0.84 | 0.02 | 0.50 | – | 0.84 | nr | **1.34** |
| **HEDGE PLANTING** | | | | | | | |
| **Hedges** | | | | | | | |
| Excavating trench for hedges; depositing soil alongside trench; by machine | | | | | | | |
| 300 mm deep × 300 mm wide | – | 0.03 | 1.44 | 1.07 | – | m | **2.51** |
| 300 mm deep × 600 mm wide | – | 0.05 | 2.16 | 1.62 | – | m | **3.78** |
| 300 mm deep × 450 mm wide | – | 0.05 | 2.14 | 1.60 | – | m | **3.74** |
| 500 mm deep × 500 mm wide | – | 0.06 | 2.98 | 2.23 | – | m | **5.21** |
| 500 mm deep × 700 mm wide | – | 0.09 | 4.17 | 3.12 | – | m | **7.29** |
| 600 mm deep × 750 mm wide | – | 0.11 | 5.35 | 4.01 | – | m | **9.36** |
| 750 mm deep × 900 mm wide | – | 0.17 | 8.04 | 6.02 | – | m | **14.06** |
| Excavating trench for hedges; depositing soil alongside trench; by hand | | | | | | | |
| 300 mm deep × 300 mm wide | – | 0.08 | 5.94 | – | – | m | **5.94** |
| 300 mm deep × 450 mm wide | – | 0.13 | 8.91 | – | – | m | **8.91** |
| 300 mm deep × 600 mm wide | – | 0.17 | 11.88 | – | – | m | **11.88** |
| 500 mm deep × 500 mm wide | – | 0.23 | 16.11 | – | – | m | **16.11** |
| 500 mm deep × 700 mm wide | – | 0.32 | 22.59 | – | – | m | **22.59** |
| 600 mm deep × 750 mm wide | – | 0.41 | 29.04 | – | – | m | **29.04** |
| 750 mm deep × 900 mm wide | – | 0.61 | 43.48 | – | – | m | **43.48** |

## 37 SOFT LANDSCAPING

| Item Excluding site overheads and profit | PC £ | Labour hours | Labour £ | Plant £ | Material £ | Unit | Total rate £ |
|---|---|---|---|---|---|---|---|
| Setting out; notching out; excavating trench; breaking up subsoil to minimum depth 300 mm | | | | | | | |
| minimum 400 mm deep | – | 0.25 | 5.94 | – | – | m | 5.94 |
| **Disposal of excavated soil to stockpile** | | | | | | | |
| Up to 25 m distant | | | | | | | |
| By hand | – | 2.40 | 57.00 | – | – | m³ | 57.00 |
| By machine | – | – | – | 2.44 | – | m³ | 2.44 |
| **Instant Hedge planting; Readyhedge Ltd; continuous strips of instant hedges; delivered in 2.50 m lengths; planting and backfilling with excavated material of hedging plants to trenches (not included); including minor trimming and shaping after planting** | | | | | | | |
| Beech (Fagus) hedges | | | | | | | |
| height 1.10/1.20 m × 400/450 mm wide | 100.00 | 0.20 | 9.50 | 6.58 | 100.00 | m | 116.08 |
| height 1.80 m × 500/600 mm wide | 100.00 | 0.25 | 11.88 | 8.23 | 100.00 | m | 120.11 |
| Purple Beech (Fagus) hedges | | | | | | | |
| height 1.80 m × 500/600 mm wide | 95.00 | 0.25 | 11.88 | 8.23 | 95.00 | m | 115.11 |
| Yew (Taxus) hedges | | | | | | | |
| height 1.0/1.2 m × 400/450 mm wide | 75.00 | 0.20 | 9.50 | 6.58 | 75.00 | m | 91.08 |
| Prunus lauruscerasus hedges | | | | | | | |
| height 1.80/2.00 m × 500 mm wide | 145.00 | 0.25 | 11.88 | 8.23 | 145.00 | m | 165.11 |
| Prunus lucitanica (Portuguese laurel) hedges | | | | | | | |
| height 1.10/1.20 m × 400/450 mm wide | 82.00 | 0.20 | 9.50 | 6.58 | 82.00 | m | 98.08 |
| height 1.60/1.80 m × 400/450 mm wide | 157.00 | 0.25 | 11.88 | 8.23 | 157.00 | m | 177.11 |
| Carpinus (Hornbeam) hedges | | | | | | | |
| height 1.10/1.20 m × 400/450 mm wide | 82.00 | 0.20 | 9.50 | 6.58 | 82.00 | m | 98.08 |
| height 1.60/1.80 m × 400/450 mm wide | 157.00 | 0.25 | 11.88 | 8.23 | 157.00 | m | 177.11 |
| Box (Buxus) hedges | | | | | | | |
| height 0.80/1.00 m × 300 mm wide | 123.00 | 0.10 | 4.75 | 3.29 | 123.00 | m | 131.04 |
| **Instant Hedge planting; Readyhedge Ltd; individual pre-clipped hedge plants to form mature hedge; planting and backfilling of hedging plants to trenches excavated separately; including minor trimming and shaping after planting** | | | | | | | |
| Beech and Hornbeam hedges | | | | | | | |
| 1.50 m high × 500 mm wide; 500 mm centres | 190.00 | 0.50 | 35.63 | 3.07 | 190.00 | m | 228.70 |
| 1.50 m high × 500 mm wide; 750 mm centres | 126.35 | 0.19 | 13.39 | 3.07 | 126.35 | m | 142.81 |
| 2.00 m high × 500 mm wide; 500 mm centres | 280.00 | 0.57 | 40.71 | 3.07 | 280.00 | m | 323.78 |

## 37 SOFT LANDSCAPING

| Item<br>Excluding site overheads and profit | PC<br>£ | Labour<br>hours | Labour<br>£ | Plant<br>£ | Material<br>£ | Unit | Total<br>rate £ |
|---|---|---|---|---|---|---|---|
| **HEDGE PLANTING – cont** | | | | | | | |
| **Instant Hedge planting – cont** | | | | | | | |
| Beech and Hornbeam hedges – cont | | | | | | | |
| 2.00 m high × 500 mm wide; 750 mm centres | 186.20 | 0.57 | 40.71 | 3.07 | 186.20 | m | **229.98** |
| Yew (Taxus) hedges | | | | | | | |
| 1.50 m high × 500 mm wide; 500 mm centres | 300.00 | 0.50 | 35.63 | 3.07 | 300.00 | m | **338.70** |
| 1.50 m high × 500 mm wide; 750 mm centres | 199.50 | 0.19 | 13.39 | 3.07 | 199.50 | m | **215.96** |
| 1.75 m high × 500 mm wide; 500 mm centres | 340.00 | 0.57 | 40.71 | 3.07 | 340.00 | m | **383.78** |
| 1.75 m high × 500 mm wide; 750 mm centres | 226.10 | 0.21 | 15.00 | 3.07 | 226.10 | m | **244.17** |
| 2.00 m high × 500 mm wide; 500 mm centres | 400.00 | 0.67 | 47.50 | 3.07 | 400.00 | m | **450.57** |
| 2.00 m high × 500 mm wide; 750 mm centres | 266.00 | 0.57 | 40.71 | 3.07 | 266.00 | m | **309.78** |
| 2.00 m high × 500 mm wide; 900 mm centres | 222.00 | 0.27 | 19.26 | 3.07 | 222.00 | m | **244.33** |
| Feathered hedges; Beech or Hornbeam | | | | | | | |
| 1.75 m high × 300 mm wide; 400 mm centres | 75.00 | 0.20 | 14.25 | 1.53 | 75.00 | m | **90.78** |
| 1.75 m high × 300 mm wide; 500 mm centres | 60.00 | 0.16 | 11.40 | 1.53 | 60.00 | m | **72.93** |
| 1.75 m high × 300 mm wide; 750 mm centres | 42.60 | 0.11 | 8.14 | 1.53 | 42.60 | m | **52.27** |
| 1.75 m high × 300 mm wide; 900 mm centres | 33.30 | 0.09 | 6.33 | 1.53 | 33.30 | m | **41.16** |
| 1.75–2.00 m high × 300 mm wide; 600 mm centres | 50.00 | 0.33 | 23.75 | 15.01 | 50.00 | m | **88.76** |
| 1.75–2.00 m high × 300 mm wide; 900 mm centres | 33.30 | 0.29 | 20.36 | 13.16 | 33.30 | m | **66.82** |
| 2.00–2.25 m high 300 wide; 600 mm centres | 61.13 | 0.40 | 28.50 | 18.11 | 61.13 | m | **107.74** |
| 2.00–2.25 m high × 300 mm wide; 900 mm centres | 40.34 | 0.33 | 23.75 | 15.01 | 40.34 | m | **79.10** |
| 2.00–2.25 m high × 300 mm wide; 1.20 m centres | 30.56 | 0.25 | 17.81 | 11.77 | 30.56 | m | **60.14** |

## 37 SOFT LANDSCAPING

| Item<br>Excluding site overheads and profit | PC<br>£ | Labour<br>hours | Labour<br>£ | Plant<br>£ | Material<br>£ | Unit | Total<br>rate £ |
|---|---|---|---|---|---|---|---|
| **Hedge planting; Griffin Nurseries Ltd; feathered hedge plants to form mature hedge; planting and backfilling of hedging plants to trenches excavated separately; including minor trimming and shaping after planting** | | | | | | | |
| (Yew) Taxus hedging; rootballed | | | | | | | |
| 600–800 mm high × 300 mm wide; 250 mm centres | 40.00 | 0.20 | 14.25 | 1.53 | 40.00 | m | **55.78** |
| 600–800 mm high × 300 mm wide; 500 mm centres | 20.00 | 0.11 | 7.92 | 1.53 | 20.00 | m | **29.45** |
| 800 mm–1.00 m high × 300 mm wide; 250 mm centres | 47.00 | 0.20 | 14.25 | 1.53 | 47.00 | m | **62.78** |
| 800 mm–1.00 m high × 300 mm wide; 500 mm centres | 23.50 | 0.11 | 7.92 | 1.53 | 23.50 | m | **32.95** |
| 1.00–1.20 m high × 300/350 mm wide; 3 nr/m | 61.60 | 0.17 | 11.88 | 1.53 | 61.60 | m | **75.01** |
| 1.00–1.20 m high × 300/350 mm wide; 2 nr/m | 37.00 | 0.13 | 8.91 | 1.53 | 37.00 | m | **47.44** |
| 1.25–1.50 m high × 300/350 mm wide; 3 nr/m | 3.33 | 0.17 | 11.88 | 1.53 | 3.33 | m | **16.74** |
| 1.25–1.50 m high × 300/350 mm wide; 2 nr/m | 2.00 | 0.13 | 9.50 | 1.53 | 2.00 | m | **13.03** |
| 1.50–1.75 m high × 400 mm wide; 450 mm centres | 130.98 | 0.33 | 23.75 | 11.26 | 130.98 | m | **165.99** |
| 1.50–1.75 m high × 400 mm wide; 600 mm centres | 98.34 | 0.22 | 15.83 | 10.17 | 98.34 | m | **124.34** |
| 1.50–1.75 m high × 400 mm wide; 900 mm centres | 65.49 | 0.20 | 14.25 | 9.31 | 65.49 | m | **89.05** |
| 1.75–2.00 m high × 500 mm wide; 600 mm centres | 165.00 | 0.33 | 23.75 | 15.01 | 165.00 | m | **203.76** |
| 1.75–2.00 m high × 500 mm wide; 900 mm centres | 135.42 | 0.29 | 20.36 | 13.16 | 245.31 | m | **278.83** |
| 1.75–2.00 m high × 500 mm wide; 1.20 m centres | 82.50 | 0.25 | 17.81 | 11.77 | 82.50 | m | **112.08** |
| 2.00–2.25 m high × 600 mm wide; 600 mm centres | 203.34 | 0.40 | 28.50 | 18.11 | 203.34 | m | **249.95** |
| 2.00–2.25 m high × 600 mm wide; 900 mm centres | 135.42 | 0.33 | 23.75 | 15.01 | 135.42 | m | **174.18** |
| 2.00–2.25 m high × 600 mm wide; 1.20 m centres | 101.66 | 0.31 | 21.92 | 14.01 | 101.66 | m | **137.59** |
| (Yew) Taxus hedging; wire rootballed | | | | | | | |
| 1.50–1.75 m high × 400 mm wide; 450 mm centres | 333.00 | 0.33 | 23.75 | 11.26 | 333.00 | m | **368.01** |
| 1.50–1.75 m high × 400 mm wide; 600 mm centres | 250.00 | 0.22 | 15.83 | 10.17 | 250.00 | m | **276.00** |
| 1.50–1.75 m high × 400 mm wide; 900 mm centres | 165.00 | 0.20 | 14.25 | 9.31 | 165.00 | m | **188.56** |

## 37 SOFT LANDSCAPING

| Item<br>Excluding site overheads and profit | PC<br>£ | Labour<br>hours | Labour<br>£ | Plant<br>£ | Material<br>£ | Unit | Total<br>rate £ |
|---|---|---|---|---|---|---|---|
| **HEDGE PLANTING – cont** | | | | | | | |
| **Hedge planting – cont** | | | | | | | |
| (Yew) Taxus hedging – cont | | | | | | | |
| 1.75–2.00 m high × 600 mm wide; 600 mm centres | 163.37 | 0.33 | 23.75 | 15.01 | 163.37 | m | **202.13** |
| 1.75–2.00 m high × 600 mm wide; 900 mm centres | 108.78 | 0.29 | 20.36 | 13.16 | 108.78 | m | **142.30** |
| 1.75–2.00 m high × 600 mm wide; 1.20 m centres | 81.66 | 0.25 | 17.81 | 11.77 | 81.66 | m | **111.24** |
| 2.00–2.25 m high × 650 mm wide; 700 mm centres | 227.20 | 0.44 | 31.66 | 19.84 | 227.20 | m | **278.70** |
| 2.00–2.25 m high × 600 mm wide; 900 mm centres | 177.60 | 0.36 | 25.91 | 16.18 | 177.60 | m | **219.69** |
| 2.00–2.25 m high × 600 mm wide; 1.20 m centres | 101.66 | 0.31 | 21.92 | 14.01 | 101.66 | m | **137.59** |
| Buxus (Box) hedges; rootballed | | | | | | | |
| 400–500 mm high; 200 mm rootball; 200 mm centres | 37.00 | 0.09 | 2.11 | – | 37.00 | m | **39.11** |
| 400–500 mm high; 200 mm rootball; 300 mm centres | 24.64 | 0.07 | 1.59 | – | 24.64 | m | **26.23** |
| 400–500 mm high; 200 mm rootball; 500 mm centres | 14.80 | 0.05 | 1.19 | – | 14.80 | m | **15.99** |
| 500–600 mm high; 200 mm rootball; 200 mm centres | 50.50 | 0.09 | 2.11 | – | 50.50 | m | **52.61** |
| 500–600 mm high; 200 mm rootball; 300 mm centres | 33.63 | 0.07 | 1.59 | – | 33.63 | m | **35.22** |
| 500–600 mm high; 200 mm rootball; 500 mm centres | 20.20 | 0.08 | 1.83 | – | 20.20 | m | **22.03** |
| 600–800 mm high; 250 mm rootball; 250 mm centres | 55.00 | 0.09 | 2.16 | – | 55.00 | m | **57.16** |
| 600–800 mm high; 250 mm rootball; 400 mm centres | 34.38 | 0.08 | 1.98 | – | 34.38 | m | **36.36** |
| 600–800 mm high; 250 mm rootball; 600 mm centres | 16.83 | 0.08 | 1.83 | – | 16.83 | m | **18.66** |
| 800 mm–1.00 m high; 300 mm rootball; 300 mm centres | 71.93 | 0.10 | 2.38 | – | 71.93 | m | **74.31** |
| 800 mm–1.00 m high; 300 mm rootball; 500 mm centres | 43.20 | 0.11 | 2.64 | – | 43.20 | m | **45.84** |
| 800 mm–1.00 m high; 300 mm rootball; 750 mm centres | 28.73 | 0.13 | 2.97 | – | 28.73 | m | **31.70** |
| 1.00 m–1.25 m high; 350 mm rootball; 400 mm centres | 82.75 | 0.13 | 2.97 | – | 82.75 | m | **85.72** |
| 1.00 m–1.25 m high; 350 mm rootball; 600 mm centres | 55.18 | 0.11 | 2.64 | – | 55.18 | m | **57.82** |
| 1.00 m–1.25 m high; 350 mm rootball; 900 mm centres | 36.74 | 0.10 | 2.38 | – | 36.74 | m | **39.12** |

## 37 SOFT LANDSCAPING

| Item<br>Excluding site overheads and profit | PC<br>£ | Labour<br>hours | Labour<br>£ | Plant<br>£ | Material<br>£ | Unit | Total<br>rate £ |
|---|---|---|---|---|---|---|---|
| **Native species and bare root hedge planting** | | | | | | | |
| Hedge planting operations only (excavation of trenches priced separately) including backfill with excavated topsoil; PC £0.50/nr; 4060 cm | | | | | | | |
| single row; 200 mm centres | 2.50 | 0.06 | 1.48 | – | 2.50 | m | **3.98** |
| single row; 300 mm centres | 1.67 | 0.06 | 1.32 | – | 1.67 | m | **2.99** |
| single row; 400 mm centres | 1.25 | 0.04 | 0.99 | – | 1.25 | m | **2.24** |
| single row; 500 mm centres | 1.00 | 0.03 | 0.79 | – | 1.00 | m | **1.79** |
| double row; 200 mm centres | 5.00 | 0.17 | 3.96 | – | 5.00 | m | **8.96** |
| double row; 300 mm centres | 3.33 | 0.13 | 3.17 | – | 3.33 | m | **6.50** |
| double row; 400 mm centres | 2.50 | 0.08 | 1.98 | – | 2.50 | m | **4.48** |
| double row; 500 mm centres | 2.00 | 0.07 | 1.58 | – | 2.00 | m | **3.58** |
| Extra over hedges for incorporating manure; at 1 m$^3$ per 30 m | 0.77 | 0.03 | 0.59 | – | 0.77 | m | **1.36** |

## 37 SOFT LANDSCAPING

| Item<br>Excluding site overheads and profit | PC<br>£ | Labour<br>hours | Labour<br>£ | Plant<br>£ | Material<br>£ | Unit | Total<br>rate £ |
|---|---|---|---|---|---|---|---|
| **TOPIARY** | | | | | | | |
| **Clipped topiary; Lorenz von Ehren; German field grown clipped and transplanted as detailed; planted to plant pit; including backfilling with excavated material and TPMC** | | | | | | | |
| Buxus sempervirens (box); balls | | | | | | | |
| 300 mm diameter; 3 × transplanted; container grown or rootballed | 16.00 | 0.25 | 5.94 | – | 25.40 | nr | **31.34** |
| 500 mm diameter; 4 × transplanted; wire rootballed | 50.00 | 1.50 | 35.63 | 4.77 | 122.60 | nr | **163.00** |
| 900 mm diameter; 5 × transplanted; wire rootballed | 210.00 | 2.75 | 65.31 | 4.77 | 330.63 | nr | **400.71** |
| 1300 mm diameter; 6 × transplanted; wire rootballed | 660.00 | 3.10 | 73.63 | 5.72 | 779.33 | nr | **858.68** |
| Buxus sempervirens (box); pyramids | | | | | | | |
| 500 mm high; 3 × transplanted; container grown or rootballed | 35.00 | 1.50 | 35.63 | 4.77 | 107.60 | nr | **148.00** |
| 900 mm high; 4 × transplanted; wire rootballed | 130.00 | 2.75 | 65.31 | 4.77 | 250.63 | nr | **320.71** |
| 1300 mm high; 5 × transplanted; wire rootballed | 380.00 | 3.10 | 73.63 | 5.72 | 499.33 | nr | **578.68** |
| Buxus sempervirens (box); truncated pyramids | | | | | | | |
| 500 mm high; 3 × transplanted; container grown or rootballed | 85.00 | 1.50 | 35.63 | 4.77 | 157.60 | nr | **198.00** |
| 900 mm high; 4 × transplanted; wire rootballed | 310.00 | 2.75 | 65.31 | 4.77 | 430.63 | nr | **500.71** |
| Buxus sempervirens (box); truncated cone | | | | | | | |
| 1300 mm high; 5 × transplanted; wire rootballed | 940.00 | 3.10 | 73.63 | 5.72 | 1059.33 | nr | **1138.68** |
| Buxus sempervirens (box); cubes | | | | | | | |
| 500 mm square; 4 × transplanted; rootballed | 85.00 | 1.50 | 35.63 | 4.77 | 157.60 | nr | **198.00** |
| 900 mm square; 5 × transplanted; wire rootballed | 320.00 | 2.75 | 65.31 | 4.77 | 440.63 | nr | **510.71** |
| Taxus baccata (yew); balls | | | | | | | |
| 500 mm diameter; 4 × transplanted; rootballed | 60.00 | 1.50 | 35.63 | 4.77 | 132.60 | nr | **173.00** |
| 900 mm diameter; 5 × transplanted; wire rootballed | 185.00 | 2.75 | 65.31 | 4.77 | 305.63 | nr | **375.71** |
| 1300 mm diameter; 6 × transplanted; wire rootballed | 570.00 | – | – | – | – | nr | **–** |
| Taxus baccata (yew); cones | | | | | | | |
| 800 mm high; 4 × transplanted; wire rootballed | 60.00 | 1.50 | 35.63 | 4.77 | 132.60 | nr | **173.00** |
| 1500 mm high; 5 × transplanted; wire rootballed | 145.00 | 2.00 | 47.50 | 4.77 | 235.10 | nr | **287.37** |
| 2500 mm high; 6 × transplanted; wire rootballed | 480.00 | 3.10 | 73.63 | 5.72 | 599.33 | nr | **678.68** |

## 37 SOFT LANDSCAPING

| Item<br>Excluding site overheads and profit | PC<br>£ | Labour<br>hours | Labour<br>£ | Plant<br>£ | Material<br>£ | Unit | Total<br>rate £ |
|---|---|---|---|---|---|---|---|
| Taxus baccata (yew); cubes | | | | | | | |
| 500 mm square; 4 × transplanted; rootballed | 70.00 | 1.50 | 35.63 | 4.77 | 142.60 | nr | **183.00** |
| 900 mm square; 5 × transplanted; wire rootballed | 245.00 | 2.75 | 65.31 | 4.77 | 365.63 | nr | **435.71** |
| Taxus baccata (yew); pyramids | | | | | | | |
| 900 mm high; 4 × transplanted; wire rootballed | 125.00 | 1.50 | 35.63 | 4.77 | 197.60 | nr | **238.00** |
| 1500 mm high; 5 × transplanted; wire rootballed | 250.00 | 3.10 | 73.63 | 5.72 | 369.33 | nr | **448.68** |
| 2500 mm high; 6 × transplanted; wire rootballed | 700.00 | 4.00 | 95.00 | 9.53 | 850.98 | nr | **955.51** |
| Carpinus betulus (common hornbeam); columns; round base | | | | | | | |
| 800 mm wide × 2000 mm high; 4 × transplanted; wire rootballed | 155.00 | 3.10 | 73.63 | 5.72 | 274.33 | nr | **353.68** |
| 800 mm wide × 2750 mm high; 4 × transplanted; wire rootballed | 305.00 | 4.00 | 95.00 | 9.53 | 455.98 | nr | **560.51** |
| Carpinus betulus (common hornbeam); cones | | | | | | | |
| 3 m high; 5 × transplanted; wire rootballed | 275.00 | 4.00 | 95.00 | 9.53 | 425.98 | nr | **530.51** |
| 4 m high; 5 × transplanted; wire rootballed | 570.00 | 5.00 | 118.75 | 9.53 | 750.20 | nr | **878.48** |
| Carpinus betulus 'Fastigiata'; pyramids | | | | | | | |
| 4 m high; 6 × transplanted; wire rootballed | 970.00 | 5.00 | 118.75 | 9.53 | 1150.20 | nr | **1278.48** |
| 7 m high; 7 × transplanted; wire rootballed | 2650.00 | 7.50 | 178.13 | 14.30 | 2873.95 | nr | **3066.38** |
| **LIVING WALLS** | | | | | | | |
| **Design to living walls; Scotscape Living Walls Ltd** | | | | | | | |
| Fabric Living Wall Design | | | | | | | |
| 10–30 m² | – | – | – | – | – | nr | **300.00** |
| 30–70 m² | – | – | – | – | – | nr | **400.00** |
| 70–100 m² | – | – | – | – | – | nr | **450.00** |
| 100–150 m² | – | – | – | – | – | nr | **500.00** |
| over 150 m² | – | – | – | – | – | nr | **600.00** |

## 37 SOFT LANDSCAPING

| Item<br>Excluding site overheads and profit | PC<br>£ | Labour<br>hours | Labour<br>£ | Plant<br>£ | Material<br>£ | Unit | Total<br>rate £ |
|---|---|---|---|---|---|---|---|
| **LIVING WALLS – cont** | | | | | | | |
| **Living Wall; fabric and soil based system; Scotscape Living Walls; self contained, irrigated vertical planting system; Fabric modules, fully-established upon installation; saturated weight 40 kg per m$^2$; access costs excluded** System units 250 × 500 × 100 mm overall; fixed to horizontal hanging rails on softwood batten and DPM; planting depth 150 mm at 30° angle to vertical surface; inclusive of automatic dripline irrigation system and guttering with running outlets to drain (not included); Area of installation | | | | | | | |
| 5–10 m$^2$ | – | – | – | – | – | m$^2$ | 415.00 |
| 10–20 m$^2$ | – | – | – | – | – | m$^2$ | 400.00 |
| 20–30 m$^2$ | – | – | – | – | – | m$^2$ | 380.00 |
| 30–70 m$^2$ | – | – | – | – | – | m$^2$ | 360.00 |
| 70–100 m$^2$ | – | – | – | – | – | m$^2$ | 350.00 |
| 100–150 m$^2$ | – | – | – | – | – | m$^2$ | 345.00 |
| over 150 m$^2$ | – | – | – | – | – | m$^2$ | 340.00 |
| Fully automatic irrigation systems to the above; comprising breaktank, station controller, solenoid valves, plant feeders and dripline emitters | | | | | | | |
| 10–30 m$^2$ | – | – | – | – | – | nr | 2750.00 |
| 30–70 m$^2$ | – | – | – | – | – | nr | 3250.00 |
| 70–100 m$^2$ | – | – | – | – | – | nr | 4000.00 |
| 100–150 m$^2$ | – | – | – | – | – | nr | 6250.00 |
| over 150 m$^2$ | – | – | – | – | – | nr | 7750.00 |
| Rainwater Harvesting System | – | – | – | – | – | nr | 2250.00 |
| **Preliminary costs for all living walls** | | | | | | | |
| General preliminaries | – | – | – | – | – | nr | 350.00 |
| Design costs; site visits, consultations and presentations | – | – | – | – | – | nr | 345.00 |
| **Maintenance to living walls** Annual maintenance to living walls; regular visits to maintain planting and systems; inclusive of feeding; pest control and calibration of irrigation systems | | | | | | | |
| 10–30 m$^2$ | – | – | – | – | – | annum | 2250.00 |
| 30–70 m$^2$ | – | – | – | – | – | annum | 3500.00 |
| 70–100 m$^2$ | – | – | – | – | – | annum | 4250.00 |
| 100–150 m$^2$ | – | – | – | – | – | annum | 5500.00 |
| over 150 m$^2$ | – | – | – | – | – | annum | 7500.00 |

## 37 SOFT LANDSCAPING

| Item<br>Excluding site overheads and profit | PC<br>£ | Labour<br>hours | Labour<br>£ | Plant<br>£ | Material<br>£ | Unit | Total<br>rate £ |
|---|---|---|---|---|---|---|---|
| **Accessories to living walls** | | | | | | | |
| planters to base of living wall in GRP; black; inclusive of drainage and planting | – | – | – | – | – | m | 308.75 |
| **Green Screen; Mobilane Ltd** | | | | | | | |
| Fully installed vertical green screen of Hedera hibernica; 65 plants per linear m; all on 5 mm galvanized steel weldmesh; attached to existing wall fence or hoarding | | | | | | | |
| screen 1.80 high | – | – | – | – | 95.00 | m | 95.00 |
| screen 2.20 m high | – | – | – | – | 170.00 | m | 170.00 |
| screen 4.00 m high | – | – | – | – | 300.00 | m | 300.00 |
| **HERBACEOUS/GROUNDCOVER/ BULB/BEDDING PLANTING** | | | | | | | |
| **Groundcover plants** | | | | | | | |
| Groundcover plants in massed areas; PC £1.50/nr; including forming planting holes in cultivated ground (cultivating not included); backfilling with excavated material; 1 litre containers | | | | | | | |
| average 4 plants per m$^2$; 500 mm centres | 6.00 | 0.09 | 2.22 | – | 6.00 | m$^2$ | 8.22 |
| average 6 plants per m$^2$; 408 mm centres | 9.00 | 0.14 | 3.33 | – | 9.00 | m$^2$ | 12.33 |
| average 8 plants per m$^2$; 354 mm centres | 12.00 | 0.19 | 4.44 | – | 12.00 | m$^2$ | 16.44 |
| Groundcover plants in massed areas; PC £2.80/nr; including forming planting holes in cultivated ground (cultivating not included); backfilling with excavated material; 1 litre containers | | | | | | | |
| average 4 plants per m$^2$; 500 mm centres | 11.20 | 0.09 | 2.22 | – | 11.20 | m$^2$ | 13.42 |
| average 6 plants per m$^2$; 408 mm centres | 16.80 | 0.14 | 3.33 | – | 16.80 | m$^2$ | 20.13 |
| average 8 plants per m$^2$; 354 mm centres | 22.40 | 0.19 | 4.44 | – | 22.40 | m$^2$ | 26.84 |

## 37 SOFT LANDSCAPING

| Item<br>Excluding site overheads and profit | PC<br>£ | Labour<br>hours | Labour<br>£ | Plant<br>£ | Material<br>£ | Unit | Total<br>rate £ |
|---|---|---|---|---|---|---|---|
| **HERBACEOUS/GROUNDCOVER/ BULB/BEDDING PLANTING – cont** | | | | | | | |
| **Herbaceous planting** | | | | | | | |
| Herbaceous plants; PC £1.50/nr; including forming planting holes in cultivated ground (cultivating not included); backfilling with excavated material; 1 litre containers | | | | | | | |
| average 4 plants per m²; 500 mm centres | 6.00 | 0.09 | 2.22 | – | 6.00 | m² | **8.22** |
| average 6 plants per m²; 408 mm centres | 9.00 | 0.14 | 3.33 | – | 9.00 | m² | **12.33** |
| average 8 plants per m²; 354 mm centres | 12.00 | 0.19 | 4.44 | – | 12.00 | m² | **16.44** |
| Note: For machine incorporation of fertilizers and soil conditioners see 'Cultivation'. | | | | | | | |
| **Bulb planting** | | | | | | | |
| Bulbs; including forming planting holes in cultivated area (cultivating not included); backfilling with excavated material | | | | | | | |
| small | 13.00 | 0.83 | 19.79 | – | 13.00 | 100 nr | **32.79** |
| medium | 22.00 | 0.83 | 19.79 | – | 22.00 | 100 nr | **41.79** |
| large | 25.00 | 0.91 | 21.59 | – | 25.00 | 100 nr | **46.59** |
| Bulbs; in grassed area; using bulb planter; including backfilling with screened topsoil or peat and cut turf plug | | | | | | | |
| small | 13.00 | 1.67 | 39.58 | – | 13.00 | 100 nr | **52.58** |
| medium | 22.00 | 1.67 | 39.58 | – | 22.00 | 100 nr | **61.58** |
| large | 25.00 | 2.00 | 47.50 | – | 25.00 | 100 nr | **72.50** |
| **Aquatic planting** | | | | | | | |
| Aquatic plants; Planting only in prepared growing medium in pool; plant size 2–3 litre containerized (plants not included) | – | 0.04 | 0.95 | – | – | nr | **0.95** |

## 37 SOFT LANDSCAPING

| Item Excluding site overheads and profit | PC £ | Labour hours | Labour £ | Plant £ | Material £ | Unit | Total rate £ |
|---|---|---|---|---|---|---|---|
| **MARKET PRICES OF NATIVE SPECIES** | | | | | | | |
| The following prices are typical of some of the more popular species used in native plantations; prices vary between species; readers should check the catalogue of the supplier | | | | | | | |
| 1+1; 1 year seedling transplanted and grown for a year | | | | | | | |
| 30–40 cm | – | – | – | – | 0.48 | nr | 0.48 |
| 40–60 cm | – | – | – | – | 0.55 | nr | 0.55 |
| 1+2; 1 year seedling transplanted and grown for 2 years | | | | | | | |
| 60–80 cm | – | – | – | – | 0.65 | nr | 0.65 |
| Whips | | | | | | | |
| 80–100 cm | – | – | – | – | 0.90 | nr | 0.90 |
| 100–125 cm | – | – | – | – | 1.40 | nr | 1.40 |
| Feathered trees | | | | | | | |
| 125–150 cm | – | – | – | – | 1.70 | nr | 1.70 |
| 150–180 cm | – | – | – | – | 2.40 | nr | 2.40 |
| 180–200 cm | – | – | – | – | 4.20 | nr | 4.20 |
| 200–250 cm | – | – | – | – | 5.60 | nr | 5.60 |
| 250–300 cm | – | – | – | – | 6.20 | nr | 6.20 |
| **Native species planting** | | | | | | | |
| Forming planting holes; in cultivated ground (cultivating not included); by mechanical auger; trimming holes by hand; depositing excavated material alongside holes | | | | | | | |
| 250 mm diameter | – | 0.03 | 0.79 | 0.05 | – | nr | 0.84 |
| 250 × 250 mm | – | 0.04 | 0.95 | 0.08 | – | nr | 1.03 |
| 300 × 300 mm | – | 0.08 | 1.78 | 0.10 | – | nr | 1.88 |
| Hand excavation; forming planting holes; in cultivated ground (cultivating not included); depositing excavated material alongside holes | | | | | | | |
| 100 × 100 × 100 mm deep; with mattock or hoe | – | 0.01 | 0.16 | – | – | nr | 0.16 |
| 250 × 250 × 300 mm deep | – | 0.04 | 0.95 | – | – | nr | 0.95 |
| 300 × 300 × 300 mm deep | – | 0.06 | 1.32 | – | – | nr | 1.32 |

## 37 SOFT LANDSCAPING

| Item<br>Excluding site overheads and profit | PC<br>£ | Labour<br>hours | Labour<br>£ | Plant<br>£ | Material<br>£ | Unit | Total<br>rate £ |
|---|---|---|---|---|---|---|---|
| **TREE/SHRUB PROTECTION** | | | | | | | |
| **Tree planting; tree protection – General** | | | | | | | |
| Preamble: Care must be taken to ensure that tree grids and guards are removed when trees grow beyond the specified diameter of guard. | | | | | | | |
| **Tree planting; tree protection** | | | | | | | |
| Tree tube; olive green | | | | | | | |
| 1200 mm high × 80 × 80 mm | 1.74 | 0.07 | 1.58 | – | 1.74 | nr | **3.32** |
| stakes; 1500 mm high for Crowders Tree Tube; driving into ground | 0.84 | 0.05 | 1.19 | – | 0.84 | nr | **2.03** |
| Expandable plastic tree guards; including 25 mm softwood stakes | | | | | | | |
| 500 mm high | 0.64 | 0.17 | 3.96 | – | 0.64 | nr | **4.60** |
| 1.00 m high | 0.85 | 0.17 | 3.96 | – | 0.85 | nr | **4.81** |
| English Woodlands; Weldmesh tree guards; nailing to tree stakes (tree stakes not included) | | | | | | | |
| 1800 mm high × 300 mm diameter | 19.00 | 0.25 | 5.94 | – | 19.00 | nr | **24.94** |
| J. Toms Ltd; spiral rabbit guards; clear or brown | | | | | | | |
| 450 × 38 mm | 0.21 | 0.03 | 0.79 | – | 0.21 | nr | **1.00** |
| 610 × 38 mm | 0.23 | 0.03 | 0.79 | – | 0.23 | nr | **1.02** |
| English Woodlands; Plastic Mesh Tree Guards; black; supplied in 50 m rolls | | | | | | | |
| 13 × 13 mm small mesh; roll width 60 cm | 0.98 | 0.04 | 0.85 | – | 1.36 | nr | **2.21** |
| 13 × 13 mm small mesh; roll width 120 cm | 2.17 | 0.06 | 1.40 | – | 2.55 | nr | **3.95** |
| Tree guards of 3 nr 2.40 m × 100 mm stakes; driving 600 mm into firm ground; bracing with timber braces at top and bottom; including 3 strands barbed wire | 0.53 | 1.00 | 23.75 | – | 0.53 | nr | **24.28** |
| English Woodlands; strimmer guard in heavy duty black plastic; 225 mm high | 3.15 | 0.07 | 1.58 | – | 3.15 | nr | **4.73** |
| Standard Treeshelter inclusive of 25 mm stake; prices shown for quantities of 500 nr | | | | | | | |
| 0.6 m high | 1.06 | 0.05 | 1.19 | – | 1.36 | nr | **2.55** |
| 0.75 m high | 1.25 | 0.05 | 1.19 | – | 1.67 | nr | **2.86** |
| 1.2 m high | 1.74 | 0.05 | 1.19 | – | 2.23 | nr | **3.42** |
| 1.5 m high | 2.18 | 0.05 | 1.19 | – | 2.77 | nr | **3.96** |
| Tubex Ltd; Shrubshelter inclusive of 25 mm stake; prices shown for quantities of 500 nr | | | | | | | |
| Ecostart shelter for forestry transplants and seedlings | 1.32 | 0.07 | 1.58 | – | 1.32 | nr | **2.90** |
| 0.6 m high | 1.80 | 0.07 | 1.58 | – | 2.10 | nr | **3.68** |
| 0.75 m high | 1.94 | 0.07 | 1.58 | – | 2.30 | nr | **3.88** |

## 37 SOFT LANDSCAPING

| Item Excluding site overheads and profit | PC £ | Labour hours | Labour £ | Plant £ | Material £ | Unit | Total rate £ |
|---|---|---|---|---|---|---|---|
| Extra over trees for spraying with anti-desiccant spray; Wiltpruf | | | | | | | |
|   selected standards; standards; light standards | 2.62 | 0.20 | 4.75 | – | 2.62 | nr | **7.37** |
|   standards; heavy standards | 4.36 | 0.25 | 5.94 | – | 4.36 | nr | **10.30** |
| **FORESTRY PLANTING** | | | | | | | |
| **Forestry planting** | | | | | | | |
| Deep ploughing rough ground to form planting ridges at | | | | | | | |
|   2.00 m centres | – | 0.63 | 14.84 | 13.75 | – | 100 m² | **28.59** |
|   3.00 m centres | – | 0.59 | 13.97 | 12.94 | – | 100 m² | **26.91** |
|   4.00 m centres | – | 0.40 | 9.50 | 8.80 | – | 100 m² | **18.30** |
| Notching plant forestry seedlings; T or L notch | 42.00 | 0.75 | 17.81 | – | 42.00 | 100 nr | **59.81** |
| Turf planting forestry seedlings | 42.00 | 2.00 | 47.50 | – | 42.00 | 100 nr | **89.50** |
| Tree tubes; to young trees | 258.00 | 0.30 | 7.13 | – | 258.00 | 100 nr | **265.13** |
| Cleaning and weeding around seedlings; once | – | 0.50 | 11.88 | – | – | 100 nr | **11.88** |
| Treading in and firming ground around seedlings planted; at 2500 per ha after frost or other ground disturbance; once | – | 0.33 | 7.92 | – | – | 100 nr | **7.92** |
| Beating up initial planting; once (including supply of replacement seedlings at 10% of original planting) | 4.20 | 0.25 | 5.94 | – | 4.20 | 100 nr | **10.14** |
| **OPERATIONS AFTER PLANTING** | | | | | | | |
| **Operations after planting** | | | | | | | |
| Initial cutting back to shrubs and hedge plants; including disposal of all cuttings | – | 1.00 | 23.75 | – | – | 100 m² | **23.75** |
| Mulch; Melcourt Industries Ltd; Bark Nuggets®; to plant beds; delivered in 80 m³ loads; maximum distance 25 m | | | | | | | |
|   50 mm thick | 2.22 | 0.04 | 1.05 | – | 2.22 | m² | **3.27** |
|   75 mm thick | 3.33 | 0.07 | 1.58 | – | 3.33 | m² | **4.91** |
|   100 mm thick | 4.43 | 0.09 | 2.11 | – | 4.43 | m² | **6.54** |
| Mulch; Melcourt Industries Ltd; Bark Nuggets®; to plant beds; delivered in 25 m³ loads; maximum distance 25 m | | | | | | | |
|   50 mm thick | 2.57 | 0.03 | 0.69 | – | 2.57 | m² | **3.26** |
|   75 mm thick | 4.35 | 0.07 | 1.58 | – | 4.35 | m² | **5.93** |
|   100 mm thick | 5.80 | 0.09 | 2.11 | – | 5.80 | m² | **7.91** |
| Mulch; Melcourt Industries Ltd; Amenity Bark Mulch FSC; to plant beds; delivered in 80 m³ loads; maximum distance 25 m | | | | | | | |
|   50 mm thick | 1.43 | 0.04 | 1.05 | – | 1.43 | m² | **2.48** |
|   75 mm thick | 2.14 | 0.07 | 1.58 | – | 2.14 | m² | **3.72** |
|   100 mm thick | 2.85 | 0.09 | 2.11 | – | 2.85 | m² | **4.96** |

## 37 SOFT LANDSCAPING

| Item<br>Excluding site overheads and profit | PC<br>£ | Labour<br>hours | Labour<br>£ | Plant<br>£ | Material<br>£ | Unit | Total<br>rate £ |
|---|---|---|---|---|---|---|---|
| **OPERATIONS AFTER PLANTING – cont** | | | | | | | |
| **Operations after planting – cont** | | | | | | | |
| Mulch; Melcourt Industries Ltd; Amenity Bark Mulch FSC; to plant beds; delivered in 25 m³ loads; maximum distance 25 m | | | | | | | |
|   50 mm thick | 2.10 | 0.04 | 1.05 | – | 2.10 | m² | **3.15** |
|   75 mm thick | 3.16 | 0.07 | 1.58 | – | 3.16 | m² | **4.74** |
|   100 mm thick | 4.21 | 0.09 | 2.11 | – | 4.21 | m² | **6.32** |
| **Mulch mats; English Woodlands; lay mulch mat to planted area or plant station; mat secured with metal J pins 240 × 3 mm** | | | | | | | |
| Mats to individual plants and plant stations | | | | | | | |
|   Woven Polypropylene; 50 × 50 cm; square mat | 0.32 | 0.05 | 1.19 | – | 0.84 | nr | **2.03** |
|   Woven Polypropylene; 1 × 1 m; square mat | 0.87 | 0.07 | 1.58 | – | 1.39 | nr | **2.97** |
| Hedging mats | | | | | | | |
|   Woven Polypropylene; 1 × 100 m roll; hedge planting | 0.47 | 0.01 | 0.30 | – | 0.73 | m² | **1.03** |
| General planting areas | | | | | | | |
|   Plantex membrane; 1 × 14 m roll; weed control | 1.39 | 0.01 | 0.24 | – | 1.65 | m² | **1.89** |
| **Fertilizers; application during aftercare period** | | | | | | | |
| Fertilizers; in top 150 mm of topsoil at 35 g/m² | | | | | | | |
|   Mascot Microfine; turf fertilizer; 8+0+6 + 2% Mg + 4% Fe | 5.04 | 0.12 | 2.91 | – | 5.04 | 100 m² | **7.95** |
|   Enmag; controlled release fertilizer (8-9 months); 11+22+09 | 7.35 | 0.12 | 2.91 | – | 7.35 | 100 m² | **10.26** |
|   Mascot Outfield; turf fertilizer; 8+12+8 | 3.91 | 0.12 | 2.91 | – | 3.91 | 100 m² | **6.82** |
|   Mascot Outfield; turf fertilizer; 9+5+5 | 3.27 | 0.12 | 2.91 | – | 3.27 | 100 m² | **6.18** |
| Fertilizers; in top 150 mm of topsoil at 70 g/m² | | | | | | | |
|   Mascot Microfine; turf fertilizer; 8+0+6 + 2% Mg + 4% Fe | 10.07 | 0.12 | 2.91 | – | 10.07 | 100 m² | **12.98** |
|   Enmag; controlled release fertilizer (8-9 months); 11+22+09 | 14.70 | 0.12 | 2.91 | – | 14.70 | 100 m² | **17.61** |
|   Mascot Outfield; turf fertilizer; 8+12+8 | 7.81 | 0.12 | 2.91 | – | 7.81 | 100 m² | **10.72** |
|   Mascot Outfield; turf fertilizer; 9+5+5 | 6.53 | 0.12 | 2.91 | – | 6.53 | 100 m² | **9.44** |

## 37 SOFT LANDSCAPING

| Item<br>Excluding site overheads and profit | PC<br>£ | Labour<br>hours | Labour<br>£ | Plant<br>£ | Material<br>£ | Unit | Total<br>rate £ |
|---|---|---|---|---|---|---|---|
| **AFTERCARE AS PART OF A LANDSCAPE CONTRACT** | | | | | | | |
| Maintenance operations (Note: the following rates apply to aftercare maintenance executed as part of a landscaping contract only) | | | | | | | |
| Weeding and hand forking planted areas; including disposing weeds and debris on site; areas maintained weekly | – | – | 0.10 | – | – | m² | **0.10** |
| Weeding and hand forking planted areas; including disposing weeds and debris on site; areas maintained monthly | – | 0.01 | 0.24 | – | – | m² | **0.24** |
| Extra over weeding and hand forking planted areas for disposing excavated material off site; loaded to maintenance vehicel and transported to the contractor's yard | | | | | | | |
| green waste for composting | – | – | – | – | – | m³ | **2.60** |
| green waste to contractor's yard deposited in skips | – | – | – | 45.65 | 2.60 | m³ | **48.25** |
| **Edging to plant beds; maintaing edge against turfed surround or boundary; edging by half moon tool** | | | | | | | |
| Planting beds | | | | | | | |
| maintaining edges | – | 0.04 | 0.95 | – | – | m | **0.95** |
| reforming edges; removing arisings | – | 0.07 | 1.58 | – | – | m | **1.58** |
| Tree pit edges | | | | | | | |
| maintaining edges | – | 0.10 | 2.38 | – | – | nr | **2.38** |
| reforming edges; removing arisings | – | 0.20 | 4.75 | – | – | nr | **4.75** |
| **Sweeping and cleaning pathways and surfaces** | | | | | | | |
| Sweep paths or surfaces to high footfall public or commercial areas | | | | | | | |
| weekly visits | – | – | – | – | – | m² | – |
| sweeping snow | – | 0.03 | 0.59 | – | – | m² | **0.59** |
| gritting and clearing ice | – | 0.03 | 0.59 | – | 0.03 | m² | **0.62** |
| **Top up mulch at the end of the 12 month maintenance period to the full thickness as shown below** | | | | | | | |
| Mulch; Melcourt Industries Ltd; Bark Nuggets®; to plant beds; delivered in 80 m³ loads; maximum distance 25 m | | | | | | | |
| 50 mm thick | 0.44 | 0.03 | 0.68 | – | 0.44 | m² | **1.12** |
| 75 mm thick | 0.67 | 0.01 | 0.12 | – | 0.67 | m² | **0.79** |

## 37 SOFT LANDSCAPING

| Item<br>Excluding site overheads and profit | PC<br>£ | Labour<br>hours | Labour<br>£ | Plant<br>£ | Material<br>£ | Unit | Total<br>rate £ |
|---|---|---|---|---|---|---|---|
| **AFTERCARE AS PART OF A LANDSCAPE CONTRACT – cont** | | | | | | | |
| **Top up mulch at the end of the 12 month maintenance period to the full thickness as shown below – cont** | | | | | | | |
| Mulch; Melcourt Industries Ltd; Bark Nuggets®; to plant beds; delivered in 25 m³ loads; maximum distance 25 m | | | | | | | |
| 50 mm thick | 0.58 | 0.03 | 0.68 | – | 0.58 | m² | **1.26** |
| 75 mm thick | 0.87 | 0.03 | 0.79 | – | 0.87 | m² | **1.66** |
| Mulch; Melcourt Industries Ltd; Amenity Bark Mulch FSC; to plant beds; delivered in 80 m³ loads; maximum distance 25 m | | | | | | | |
| 50 mm thick | 0.71 | 0.03 | 0.68 | – | 0.71 | m² | **1.39** |
| 75 mm thick | 1.00 | 0.03 | 0.79 | – | 1.00 | m² | **1.79** |
| Mulch; Melcourt Industries Ltd; Amenity Bark Mulch FSC; to plant beds; delivered in 25 m³ loads; maximum distance 25 m | | | | | | | |
| 50 mm thick | 1.05 | 0.03 | 0.59 | – | 1.05 | m² | **1.64** |
| 75 mm thick | 1.37 | 0.03 | 0.68 | – | 1.37 | m² | **2.05** |
| Fertilizers; at 35 g/m² | | | | | | | |
| Mascot Microfine; turf fertilizer; 8+0 +6 + 2% Mg + 4% Fe | 5.04 | 0.12 | 2.91 | – | 5.04 | 100 m² | **7.95** |
| Enmag; controlled release fertilizer (8–9 months); 11+22+09 | 7.35 | 0.12 | 2.91 | – | 7.35 | 100 m² | **10.26** |
| Mascot Outfield; turf fertilizer; 8+12 +8 | 3.91 | 0.12 | 2.91 | – | 3.91 | 100 m² | **6.82** |
| Mascot Outfield; turf fertilizer; 9+5+5 | 3.27 | 0.12 | 2.91 | – | 3.27 | 100 m² | **6.18** |
| Fertilizers; at 70 g/m² | | | | | | | |
| Mascot Microfine; turf fertilizer; 8+0 +6 + 2% Mg + 4% Fe | 10.07 | 0.12 | 2.91 | – | 10.07 | 100 m² | **12.98** |
| Enmag; controlled release fertilizer (8–9 months); 11+22+09 | 14.70 | 0.12 | 2.91 | – | 14.70 | 100 m² | **17.61** |
| Mascot Outfield; turf fertilizer; 8+12 +8 | 7.81 | 0.12 | 2.91 | – | 7.81 | 100 m² | **10.72** |
| Mascot Outfield; turf fertilizer; 9+5+5 | 6.53 | 0.12 | 2.91 | – | 6.53 | 100 m² | **9.44** |
| Watering planting; evenly; at a rate of 5 l/m² | | | | | | | |
| using hand-held watering equipment | – | 0.25 | 5.94 | – | – | 100 m² | **5.94** |
| using sprinkler equipment and with sufficient water pressure to run one 15 m radius sprinkler | – | 0.14 | 3.30 | – | – | 100 m² | **3.30** |
| using movable spray lines powering three sprinkler heads with a radius of 15 m and allowing for 60% overlap (irrigation machinery costs not included) | – | 0.02 | 0.37 | – | – | 100 m² | **0.37** |

## 37 SOFT LANDSCAPING

| Item<br>Excluding site overheads and profit | PC<br>£ | Labour<br>hours | Labour<br>£ | Plant<br>£ | Material<br>£ | Unit | Total<br>rate £ |
|---|---|---|---|---|---|---|---|
| **LONG-TERM MAINTENANCE** | | | | | | | |
| **Preamble: Long-term landscape maintenance** | | | | | | | |
| Maintenance on long-term contracts differs in cost from that of maintenance as part of a landscape contract. In this section the contract period is generally 3–5 years. Staff are generally allocated to a single project only and therefore productivity is higher whilst overhead costs are lower. Labour costs in this section are lower than the costs used in other parts of the book. Machinery is assumed to be leased over a 5 year period and written off over the same period. The costs of maintenance and consumables for the various machinery types have been included in the information that follows. Finance costs for the machinery have not been allowed for. | | | | | | | |
| The rates shown below are for machines working in unconfined contiguous areas. Users should adjust the times and rates if working in smaller spaces or spaces with obstructions. | | | | | | | |
| **MARKET PRICES OF LANDSCAPE CHEMICALS** | | | | | | | |
| The following table provides the areas of spraying in each planting scenario for plants which require 1.00 m diameter weed free circles | | | | | | | |
| Area of weed free circles required in each planting density | | | | | | | |
| plants at 500 mm centres; 400 nr/100 m² | – | – | – | – | 400.00 | m² | **400.00** |
| plants at 600 m centres; 278 nr/100 m² | – | – | – | – | 278.00 | m² | **278.00** |
| plants at 750 mm centres; 178 nr/100 m² | – | – | – | – | 178.00 | m² | **178.00** |
| plants at 1.00 m centres; 100 nr/100 m² | – | – | – | – | 100.00 | m² | **100.00** |
| plants at 1.50 m centres; 4444 nr/ha | – | – | – | – | 3490.00 | m² | **3490.00** |
| plants at 1.75 m centres; 3265 nr/ha | – | – | – | – | 2564.00 | m² | **2564.00** |
| plants at 2.00 m centres; 2500 nr/ha | – | – | – | – | 1963.00 | m² | **1963.00** |

## 37 SOFT LANDSCAPING

| Item<br>Excluding site overheads and profit | PC<br>£ | Labour<br>hours | Labour<br>£ | Plant<br>£ | Material<br>£ | Unit | Total<br>rate £ |
|---|---|---|---|---|---|---|---|
| **MARKET PRICES OF LANDSCAPE CHEMICALS AT SUGGESTED APPLICATION RATES** | | | | | | | |
| Note: All chemicals are standard knapsack or backpack applied unless specifically stated as CDA or TDA. | | | | | | | |
| **Total herbicides** | | | | | | | |
| **Herbicide applications; CDA; chemical application via low pressure specialized wands to landscape planting; application to maintain 1.00 m diameter clear circles (0.79 m$^2$) around new planting** | | | | | | | |
| Vanquish Biactive; ALS Ltd; Bayer; enhanced movement glyphosate; application rate 15 litre/ha | | | | | | | |
| plants at 1.50 m centres; 4444 nr/ha | – | – | – | – | 59.46 | ha | **59.46** |
| plants at 1.75 m centres; 3265 nr/ha | – | – | – | – | 43.88 | ha | **43.88** |
| plants at 2.00 m centres; 2500 nr/ha | – | – | – | – | 33.65 | ha | **33.65** |
| mass spraying | – | – | – | – | 170.53 | ha | **170.53** |
| spot spraying | – | – | – | – | 1.71 | ha | **1.71** |
| **Total herbicides; CDA; low pressure sustainable applications** | | | | | | | |
| Hilite; Nomix Enviro Ltd; glyphosate oil-emulsion; Total Droplet Control (TDC formerly CDA); low volume sustainable chemical applications; formulation controls annual and perennial grasses and non-woody broad-leaved weeds in amenity and industrial areas, parks, highways, paved areas and factory sites; also effective for clearing vegetation prior to planting of ornamental species, around the bases of trees, shrubs and roses, as well as in forestry areas | | | | | | | |
| 10 l/ha | – | – | – | – | 1.90 | 100 m$^2$ | **1.90** |
| 7.5 l/ ha | – | – | – | – | 1.43 | 100 m$^2$ | **1.43** |

## 37 SOFT LANDSCAPING

| Item<br>Excluding site overheads and profit | PC<br>£ | Labour<br>hours | Labour<br>£ | Plant<br>£ | Material<br>£ | Unit | Total<br>rate £ |
|---|---|---|---|---|---|---|---|
| Tribute Plus; Nomix Enviro Ltd; TDA; MCPA/dicamba/mecaprop-p; TDC; for selective control of broad-leaved weeds in turf | | | | | | | |
|   6 l/ha; white clover, plantains, creeping buttercup, thistle | – | – | – | – | 2.15 | 100 m² | **2.15** |
|   8 l/ha; yarrow, ragwort, parsley, piert, cinquefoil, mouse eared chickweed, selfheal, pearlwort, sorrel, black meddick, dandelion and related weeds, bulbous buttercup | – | – | – | – | 2.86 | 100 m² | **2.86** |
| Stirrup; Rigby Taylor Ltd; glyphosate; CDA; 192 g | | | | | | | |
|   10 l/ha | – | – | – | – | 2.12 | 100 m² | **2.12** |
|   7.5 l/ha | – | – | – | – | 2.12 | 100 m² | **2.12** |
| Gallup Biograde Amenity; Rigby Taylor Ltd; glyphosate 360 g/l formulation | | | | | | | |
|   general use; woody weeds; ash, beech, bracken, bramble; 3 l/ha | – | – | – | – | 0.33 | 100 m² | **0.33** |
|   annual and perennial grasses; heather (peat soils); 4 l/ha | – | – | – | – | 0.44 | 100 m² | **0.44** |
|   pre-planting; general clearance; 5 l/ha | – | – | – | – | 0.55 | 100 m² | **0.55** |
|   heather; mineral soils; 6 l/ha | – | – | – | – | 0.66 | 100 m² | **0.66** |
|   rhododendron; 10 l/ha | – | – | – | – | 1.10 | 100 m² | **1.10** |
| Gallup Hi-aktiv Amenity; Rigby Taylor Ltd; 490 g/l formulation | | | | | | | |
|   general use; woody weeds; ash, beech, bracken, bramble; 2.2 l/ha | – | – | – | – | 0.26 | 100 m² | **0.26** |
|   annual and perennial grasses; heather (peat soils); 2.9 l/ha | – | – | – | – | 0.35 | 100 m² | **0.35** |
|   pre-planting; general clearance; 3.7 l/ha | – | – | – | – | 0.44 | 100 m² | **0.44** |
|   heather; mineral soils; 4.4 l/ha | – | – | – | – | 0.53 | 100 m² | **0.53** |
|   rhododendron; 7.3 l/ha | – | – | – | – | 0.88 | 100 m² | **0.88** |
| Discman CDA Biograde; Rigby Taylor Ltd | | | | | | | |
|   application rate; 7 l/ha; light vegetation and annual weeds | – | – | – | – | 1.22 | 100 m² | **1.22** |
|   application rate; 8.5 l/ha; forestry/ woodlands; before planting, to control broad-leaved and grass weeds | – | – | – | – | 1.48 | 100 m² | **1.48** |
|   application rate; 10 l/ha; established deep-rooted annuals, perennial grasses and broad-leaved weeds | – | – | – | – | 1.74 | 100 m² | **1.74** |
| **Roundup Pro Biactive 360; Everris** | | | | | | | |
| application rate; 5 l/ha; land not intended for cropping | – | – | – | – | 0.32 | 100 m² | **0.32** |
| application rate; 10 l/ha; forestry; weed control | – | – | – | – | 0.64 | 100 m² | **0.64** |

## 37 SOFT LANDSCAPING

| Item<br>Excluding site overheads and profit | PC<br>£ | Labour<br>hours | Labour<br>£ | Plant<br>£ | Material<br>£ | Unit | Total<br>rate £ |
|---|---|---|---|---|---|---|---|
| **MARKET PRICES OF LANDSCAPE CHEMICALS AT SUGGESTED APPLICATION RATES – cont** | | | | | | | |
| **Roundup Pro Biactive 450; Everris; 450 g/litre glyphosate** | | | | | | | |
| Application rate 4 l/ha; natural surfaces not intended to bear vegetation; amenity vegetation control; permeable surfaces overlaying soil; hard surfaces | – | – | – | – | 0.30 | 100 m$^2$ | 0.30 |
| Application rate 8 l/ha; forest; forest nursery weed control | – | – | – | – | 0.60 | 100 m$^2$ | 0.60 |
| Application rate 16 mm/l; stump application | – | – | – | – | 0.12 | litre | 0.12 |
| **Roundup Pro Biactive 450; Everris; enhanced movement glyphosate; application rate 4 litre/ha** | | | | | | | |
| Plants at 1.50 m centres; 4444 nr/ha | – | – | – | – | 43.74 | ha | 43.74 |
| Plants at 1.75 m centres; 3265 nr/ha | – | – | – | – | 32.17 | ha | 32.17 |
| Plants at 2.00 m centres; 2500 nr/ha | – | – | – | – | 24.62 | ha | 24.62 |
| Mass spraying | – | – | – | – | 33.26 | ha | 33.26 |
| Spot spraying | – | – | – | – | 1.25 | 100 m$^2$ | 1.25 |
| **Dual; Nomix Enviro Ltd; glyphosate/ sulfosulfuron; TDC; low volume sustainable chemical applications; residual effect lasting up to 6 months; shrub beds, gravelled areas, tree bases, parks, industrial areas, hard areas and around obstacles** | | | | | | | |
| Application rate 9 l/ha | – | – | – | – | 2.97 | 100 m$^2$ | 2.97 |
| **Glyfos Dakar Pro; Headland Amenity Ltd; 68% glyphosate as water dispersible granules; broad-leaved weeds and grasses** | | | | | | | |
| Application rate 2.5 kg/ha | – | – | – | – | 0.34 | 100 m$^2$ | 0.34 |

## 37 SOFT LANDSCAPING

| Item Excluding site overheads and profit | PC £ | Labour hours | Labour £ | Plant £ | Material £ | Unit | Total rate £ |
|---|---|---|---|---|---|---|---|
| **Total herbicides with residual pre-emergent action** | | | | | | | |
| **Pistol; ALS; Bayer;** barrier prevention for emergence control and glyphosate for control of emerged weeds at a single application per season; application rate 4.5 litre/ha; 40 g/l diflufenican and 250 g/l glyphosate | | | | | | | |
| Plants at 1.50 m centres; 4444 nr/ha | – | – | – | – | 45.13 | ha | **45.13** |
| Plants at 1.75 m centres; 3265 nr/ha | – | – | – | – | 33.05 | ha | **33.05** |
| Plants at 2.00 m centres; 2500 nr/ha | – | – | – | – | 32.34 | ha | **32.34** |
| **Residual herbicides** | | | | | | | |
| **Chikara; Rigby Taylor Ltd; Belchim Crop Protection;** flazasulfuron; systemic pre-emergent and early post-emergent herbicide for the control of annual and perennial weeds on natural surfaces not intended to bear vegetation and permeable surfaces overlying soil | | | | | | | |
| 150 g/ha | – | – | – | – | 2.25 | 100 m² | **2.25** |
| **Contact herbicides** | | | | | | | |
| **Jewel; Everris;** carfentrazone and mecoprop-p; selective weed and mosskiller | | | | | | | |
| application rate; 1.5 kg/ha | – | – | – | – | 1.14 | 100 m² | **1.14** |
| **Natural Weed and Moss Spray; Headland Amenity Ltd;** acetic acid | | | | | | | |
| application rate; 100 l/ha | – | – | – | – | 4.50 | 100 m² | **4.50** |
| **Selective herbicides CDA** | | | | | | | |
| **Hilite; Nomix Enviro;** glyphosate oil-emulsion; TD; low volume sustainable chemical applications; formulation controls annual and perennial grasses and non-woody broad-leaved weeds in amenity and industrial areas, parks, highways, paved areas and factory sites; also effective for clearing vegetation prior to planting of ornamental species, around the bases of trees, shrubs and roses, as well as in forestry areas | | | | | | | |
| 10 l/ha | – | – | – | – | 1.90 | 100 m² | **1.90** |
| 7.5 l/ha | – | – | – | – | 1.43 | 100 m² | **1.43** |

## 37 SOFT LANDSCAPING

| Item<br>Excluding site overheads and profit | PC<br>£ | Labour<br>hours | Labour<br>£ | Plant<br>£ | Material<br>£ | Unit | Total<br>rate £ |
|---|---|---|---|---|---|---|---|
| **MARKET PRICES OF LANDSCAPE CHEMICALS AT SUGGESTED APPLICATION RATES – cont** | | | | | | | |
| **Selective herbicides CDA – cont** | | | | | | | |
| Tribute Plus; Nomix Enviro; MCPA/ dicamba/mecaprop-p; TDC; for selective control of broad-leaved weeds in turf | | | | | | | |
| 6 l/ha; white clover, plantains, creeping buttercup, thistle | – | – | – | – | 2.15 | 100 m$^2$ | **2.15** |
| 8 l/ha; yarrow, ragwort, parsley, piert, cinquefoil, mouse eared chickweed, selfheal, pearlwort, sorrel, black meddick, dandelion and related weeds, bulbous buttercup | – | – | – | – | 2.86 | 100 m$^2$ | **2.86** |
| **Contact and systemic combined herbicides** | | | | | | | |
| Jewel; Everris; carfentrazone and mecoprop-p; selective weed and mosskiller | | | | | | | |
| application rate; 1.5 kg/ha | – | – | – | – | 1.14 | 100 m$^2$ | **1.14** |
| Finale; ALS; Bayer; glufosinate-ammonium; controls grasses and broad-leaved weeds; it can also be used for weed control around non-edible ornamental crops including trees and shrubs and in tree nurseries; may be used to 'burn off' weeds in shrub beds prior to treatment with a residual herbicide; is also ideal for preparing sports turf for line marking | | | | | | | |
| application rate; 5 l/ha; seedlings of all species; established annual weeds and grasses as specified | – | – | – | – | 1.75 | 100 m$^2$ | **1.75** |
| application rate; 8 l/ha; established annual and perennial weeds and grasses as specified in the product guide | – | – | – | – | 2.80 | 100 m$^2$ | **2.80** |

## 37 SOFT LANDSCAPING

| Item<br>Excluding site overheads and profit | PC<br>£ | Labour<br>hours | Labour<br>£ | Plant<br>£ | Material<br>£ | Unit | Total<br>rate £ |
|---|---|---|---|---|---|---|---|
| **Selective herbicides; power or knapsack sprayer** | | | | | | | |
| Junction; Rigby Taylor Ltd; selective post-emergence turf herbicide containing 6.25 g/litre florasulam plus 452 g/litre 2,4-D; specially designed for use by contractors for maintenance treatments of a wide range of broad-leaved weeds whilst giving excellent safety to turfgrass | | | | | | | |
| application rate; 1.2 l/ha | – | – | – | – | 0.54 | 100 m² | 0.54 |
| Greenor; Rigby Taylor Ltd; fluroxypyr, clopyralid and MCPA; systemic selective treatment of weeds in turfgrass | | | | | | | |
| application rate; 4 l/ha | – | – | – | – | 1.09 | 100 m² | 1.09 |
| Crossbar Herbicide; Rigby Taylor Ltd; fluroxypyr, 2,4-D and dicamba for control of weeds in amenity turf including fine turf | | | | | | | |
| application rate; 2 l/ha | – | – | – | – | 0.85 | 100 m² | 0.85 |
| Re-Act; Everris; MCPA, mecoprop-p and dicamba; selective herbicide; established managed amenity turf and newly seeded grass; controls many annual and perennial weeds; will not vaporize in hot conditions | | | | | | | |
| application rate; 9090–14,285 m²/5 l | – | – | – | – | 0.32 | 100 m² | 0.32 |
| Praxys; Everris; fluroxypyr-meptyl, clopyralid and florasulam; selective, systemic post-emergence herbicide for managed amenity turf, lawns and amenity grassland with high selectivity to established and young turf | | | | | | | |
| application rate; 9090–14,285 m²/5 l | – | – | – | – | 1.15 | 100 m² | 1.15 |
| Asulox; ALS; Bayer; post-emergence translocated herbicide for the control of bracken and docks | | | | | | | |
| application rate; docks; 17857 m²/5 l | – | – | – | – | 0.34 | 100 m² | 0.34 |
| application rate; bracken; 4545 m²/5 l | – | – | – | – | 1.35 | 100 m² | 1.35 |
| Cabadex; Headland Amenity Ltd; fluroxypyr and florasulam; newly sown and established turf (particularly speedwell, yarrow and yellow suckling clover) | | | | | | | |
| application rate 2 l/ha | – | – | – | – | 0.70 | 100 m² | 0.70 |
| Blaster; Headland Amenity Ltd; triclopyr clopyralid; nettles, docks, thistles, brambles, woody weeds in amenity grassland | | | | | | | |
| application rate; 4 l/ha | – | – | – | – | 0.50 | 100 m² | 0.50 |

## 37 SOFT LANDSCAPING

| Item<br>Excluding site overheads and profit | PC<br>£ | Labour<br>hours | Labour<br>£ | Plant<br>£ | Material<br>£ | Unit | Total<br>rate £ |
|---|---|---|---|---|---|---|---|
| **MARKET PRICES OF LANDSCAPE CHEMICALS AT SUGGESTED APPLICATION RATES – cont** | | | | | | | |
| **Aquatic herbicides; backpack or power sprayers** | | | | | | | |
| Roundup Pro Biactive 360; Everris application rate; 6 l/ha; aquatic weed control | – | – | – | – | 0.38 | 100 m$^2$ | **0.38** |
| Roundup Pro Biactive 450; Everris; 450 g/litre glyphosate application rate 4.8 l/ha; enclosed waters, open waters, land immediately adjacent to aquatic areas | – | – | – | – | 0.36 | 100 m$^2$ | **0.36** |
| **Aquatic herbicides; TDC** | | | | | | | |
| Conqueror; Nomix Enviro; low volume sustainable chemical applications glyphosate with aquatic approval | | | | | | | |
| 10 l/ha; established annuals, deep-rooted perennials and broad-leaved weeds | – | – | – | – | 0.23 | 100 m$^2$ | **0.23** |
| 7.5 l/ha; light vegetation and small annual weeds, aquatic situations | – | – | – | – | 1.72 | 100 m$^2$ | **1.72** |
| 12.5 l/ha; aquatic situations, emergent weeds e.g. reed, grasses and watercress | – | – | – | – | 0.29 | 100 m$^2$ | **0.29** |
| 15 l/ha; aquatic situations, floating weeds such as water lilies | – | – | – | – | 0.34 | 100 m$^2$ | **0.34** |
| **Fungicides** | | | | | | | |
| Defender; Rigby Taylor Ltd; broad spectrum fungicide displaying protective and curative activity against fusarium patch and red thread | | | | | | | |
| application rate; 700 g/ha | – | – | – | – | 6.86 | 100 m$^2$ | **6.86** |
| Fusion; Rigby Taylor Ltd; contact and systemic fungicide with curative and preventative activity against major turf diseases such as fusarium, red thread, dollar spot, anthracnose and rust | | | | | | | |
| application rate; 700 g/ha | – | – | – | – | 7.19 | 100 m$^2$ | **7.19** |
| Rayzor; Rigby Taylor Ltd; broad spectrum contact fungicide with both protectant and eradicant activity for the control of fusarium patch and red thread in managed amenity turf | | | | | | | |
| application rate; 5 l/2000 m$^2$ | – | – | – | – | 7.92 | 100 m$^2$ | **7.92** |

## 37 SOFT LANDSCAPING

| Item<br>Excluding site overheads and profit | PC<br>£ | Labour<br>hours | Labour<br>£ | Plant<br>£ | Material<br>£ | Unit | Total<br>rate £ |
|---|---|---|---|---|---|---|---|
| **Moss control**<br>Jewel; Everris; carfentrazone and mecoprop-p; selective weed and moss killer | | | | | | | |
| application rate; 1.5 kg/ha | – | – | – | – | 1.14 | 100 m² | 1.14 |
| **Insect control**<br>Merit Turf; Bayer; insecticide for the control of chafer grubs and leatherjackets | | | | | | | |
| application rate; 10 kg/3333 m² | – | – | – | – | 6.41 | 100 m² | 6.41 |
| **LANDSCAPE CHEMICAL APPLICATION** | | | | | | | |
| **Chemical applications labours only**<br>Controlled droplet application (CDA) | | | | | | | |
| plants at 1.50 m centres; 4444 nr/ha | – | 9.26 | 395.87 | – | – | ha | 395.87 |
| plants at 1.50 m centres; 44.44 nr/100 m² | – | 0.93 | 39.59 | – | – | 100 m² | 39.59 |
| plants at 1.75 m centres; 3265 nr/ha | – | 6.80 | 290.70 | – | – | ha | 290.70 |
| plants at 1.75 m centres; 3265 nr/ha | – | 0.68 | 29.07 | – | – | 100 m² | 29.07 |
| plants at 2.00 m centres; 2500 nr/ha | – | 5.21 | 222.73 | – | – | ha | 222.73 |
| plants at 2.00 m centres; 2500 nr/ha | – | 0.52 | 22.27 | – | – | 100 m² | 22.27 |
| spot spraying to shrub beds | – | 0.20 | 4.75 | – | – | 100 m² | 4.75 |
| spot spraying to hard surfaces | – | 0.13 | 2.97 | – | – | 100 m² | 2.97 |
| mass spraying | – | 2.50 | 106.88 | – | – | ha | 106.88 |
| mass spraying | – | 0.25 | 10.69 | – | – | 100 m² | 10.69 |
| Power spraying<br>mass spraying to areas containing vegetation | – | 2.00 | 47.50 | 61.25 | – | ha | 108.75 |
| mass spraying to hard surfaces | – | – | 0.10 | 0.13 | – | 100 m² | 0.23 |
| spraying to kerbs and channels along roadways 300 mm wide | – | 1.25 | 29.69 | 38.28 | – | 1 km | 67.97 |
| General herbicides; in accordance with manufacturer's instructions; Knapsack spray application; see 'Market rates of chemicals' for costs of specific applications | | | | | | | |
| knapsack sprayer; selective spraying around bases of plants | – | 0.18 | 4.17 | – | – | 100 m² | 4.17 |
| knapsack sprayer; spot spraying | – | 0.25 | 5.94 | – | – | 100 m² | 5.94 |
| knapsack sprayer; mass spraying | – | 0.40 | 9.50 | – | – | 100 m² | 9.50 |
| granular distribution by hand or hand applicator | – | 0.20 | 4.75 | – | – | 100 m² | 4.75 |
| Backpack spraying; keeping weed free at bases of plants | | | | | | | |
| plants at 1.50 m centres; 4444 nr/ha | – | 13.89 | 593.80 | – | – | ha | 593.80 |
| plants at 1.75 m centres; 3265 nr/ha | – | 10.20 | 436.05 | – | – | ha | 436.05 |
| plants at 2.00 m centres; 2500 nr/ha | – | 7.82 | 334.09 | – | – | ha | 334.09 |
| mass spraying | – | 7.50 | 320.63 | – | – | ha | 320.63 |

## 37 SOFT LANDSCAPING

| Item<br>Excluding site overheads and profit | PC<br>£ | Labour<br>hours | Labour<br>£ | Plant<br>£ | Material<br>£ | Unit | Total<br>rate £ |
|---|---|---|---|---|---|---|---|
| **LANDSCAPE CHEMICAL APPLICATION – cont** | | | | | | | |
| **Herbicide applications; standard backpack spray applicators; application to maintain 1.00 m diameter clear circles (0.79 m$^2$) around new planting** | | | | | | | |
| Everris; Roundup Pro Biactive 360; glyphosate; enhanced movement glyphosate; application rate 5 litre/ha | | | | | | | |
| plants at 1.50 m centres; 4444 nr/ha | 12.23 | 12.35 | 527.75 | – | 12.23 | ha | **539.98** |
| plants at 1.75 m centres; 3265 nr/ha | 8.94 | 9.07 | 387.74 | – | 8.94 | ha | **396.68** |
| plants at 2.00 m centres; 2500 nr/ha | 6.90 | 6.95 | 296.90 | – | 6.90 | ha | **303.80** |
| mass spraying | – | – | 0.03 | – | – | m$^2$ | **0.03** |
| **GRASS CUTTING** | | | | | | | |
| **Grass cutting only; ride-on or tractor drawn equipment; Norris & Gardiner Ltd; works carried out on one site only such as large fields or amenity areas where labour and machinery is present for a full day; grass cutting only; strimming not included** | | | | | | | |
| Using multiple-gang mower with cylindrical cutters; contiguous areas such as playing fields and the like larger than 3000 m$^2$ | | | | | | | |
| 3 gang; 2.13 m cutting width | – | 0.01 | 0.22 | 0.12 | – | 100 m$^2$ | **0.34** |
| 5 gang; 3.40 m cutting width | – | – | 0.20 | 0.14 | – | 100 m$^2$ | **0.34** |
| 7 gang; 4.65 m cutting width | – | – | 0.12 | 0.11 | – | 100 m$^2$ | **0.23** |
| Using multiple-gang mower with cylindrical cutters; non-contiguous areas such as verges and general turf areas | | | | | | | |
| 3 gang | – | 0.01 | 0.44 | 0.12 | – | 100 m$^2$ | **0.56** |
| 5 gang | – | 0.01 | 0.29 | 0.14 | – | 100 m$^2$ | **0.43** |
| Using multiple-rotary mower with vertical drive shaft and horizontally rotating bar or disc cutters; contiguous areas larger than 3000 m$^2$ | | | | | | | |
| cutting grass, overgrowth or the like using flail mower or reaper | – | 0.01 | 0.43 | 0.30 | – | 100 m$^2$ | **0.73** |

## 37 SOFT LANDSCAPING

| Item<br>Excluding site overheads and profit | PC<br>£ | Labour<br>hours | Labour<br>£ | Plant<br>£ | Material<br>£ | Unit | Total<br>rate £ |
|---|---|---|---|---|---|---|---|
| **Grass cutting with strimming; ride-on or tractor drawn equipment; Norris & Gardiner Ltd; works carried out on one site only such as large fields or amenity areas where labour and machinery is present for a full day; includes accompanying strimmer operative** | | | | | | | |
| Using multiple-gang mower with cylindrical cutters; contiguous areas such as playing fields and the like larger than 3000 m² | | | | | | | |
| 3 gang; 2.13 m cutting width | – | 0.01 | 0.44 | 0.13 | – | 100 m² | **0.57** |
| 5 gang; 3.40 m cutting width | – | 0.01 | 0.41 | 0.15 | – | 100 m² | **0.56** |
| 7 gang; 4.65 m cutting width | – | 0.01 | 0.25 | 0.14 | – | 100 m² | **0.39** |
| Using multiple-gang mower with cylindrical cutters; non-contiguous areas such as verges and general turf areas | | | | | | | |
| 3 gang | – | 0.02 | 0.89 | 0.18 | – | 100 m² | **1.07** |
| 5 gang | – | 0.01 | 0.57 | 0.16 | – | 100 m² | **0.73** |
| Using multiple-rotary mower with vertical drive shaft and horizontally rotating bar or disc cutters; contiguous areas larger than 3000 m² | | | | | | | |
| cutting grass, overgrowth or the like using flail mower or reaper | – | 0.02 | 0.85 | 0.36 | – | 100 m² | **1.21** |
| **Grass cutting only; ride-on or tractor drawn equipment; Norris & Gardiner Ltd; works carried out on multiple sites; machinery moved by trailer between sites; strimming not included** | | | | | | | |
| Using multiple-gang mower with cylindrical cutters; contiguous areas such as playing fields and the like larger than 3000 m² | | | | | | | |
| 3 gang; 2.13 m cutting width | – | 0.01 | 0.36 | 0.12 | – | 100 m² | **0.48** |
| 5 gang; 3.40 m cutting width | – | 0.01 | 0.33 | 0.14 | – | 100 m² | **0.47** |
| 7 gang; 4.65 m cutting width | – | – | 0.21 | 0.11 | – | 100 m² | **0.32** |
| Using multiple-gang mower with cylindrical cutters; non-contiguous areas such as verges and general turf areas | | | | | | | |
| 3 gang | – | 0.02 | 0.72 | 0.12 | – | 100 m² | **0.84** |
| 5 gang | – | 0.01 | 0.46 | 0.14 | – | 100 m² | **0.60** |

## 37 SOFT LANDSCAPING

| Item<br>Excluding site overheads and profit | PC<br>£ | Labour<br>hours | Labour<br>£ | Plant<br>£ | Material<br>£ | Unit | Total<br>rate £ |
|---|---|---|---|---|---|---|---|
| **GRASS CUTTING – cont** | | | | | | | |
| **Grass cutting only – cont** | | | | | | | |
| Using multiple-rotary mower with vertical drive shaft and horizontally rotating bar or disc cutters; contiguous areas larger than 3000 m$^2$ | | | | | | | |
|    cutting grass, overgrowth or the like using flail mower or reaper | – | 0.02 | 0.70 | 0.30 | – | 100 m$^2$ | **1.00** |
| Cutting grass, overgrowth or the like; using tractor-mounted side-arm flail mower; in areas inaccessible to alternative machine; on surface | | | | | | | |
|    not exceeding 30° from horizontal | – | 0.01 | 0.50 | 0.34 | – | 100 m$^2$ | **0.84** |
|    30–50° from horizontal | – | 0.02 | 1.00 | 0.34 | – | 100 m$^2$ | **1.34** |
| **Grass cutting and strimming; ride-on or tractor drawn equipment; Norris & Gardiner Ltd; works carried out on multiple sites; machinery moved by trailer between sites; includes for accompanying strimmer operative** | | | | | | | |
| Using multiple-gang mower with cylindrical cutters; contiguous areas such as playing fields and the like larger than 3000 m$^2$ | | | | | | | |
|    3 gang; 2.13 m cutting width | – | 0.02 | 0.73 | 0.14 | – | 100 m$^2$ | **0.87** |
|    5 gang; 3.40 m cutting width | – | 0.02 | 0.66 | 0.16 | – | 100 m$^2$ | **0.82** |
|    7 gang; 4.65 m cutting width | – | 0.01 | 0.41 | 0.12 | – | 100 m$^2$ | **0.53** |
| Using multiple-gang mower with cylindrical cutters; non-contiguous areas such as verges and general turf areas | | | | | | | |
|    3 gang | – | 0.03 | 1.44 | 0.17 | – | 100 m$^2$ | **1.61** |
|    5 gang | – | 0.02 | 0.93 | 0.20 | – | 100 m$^2$ | **1.13** |
| Using multiple-rotary mower with vertical drive shaft and horizontally rotating bar or disc cutters; contiguous areas larger than 3000 m$^2$ | | | | | | | |
|    cutting grass, overgrowth or the like using flail mower or reaper | – | 0.03 | 1.39 | 0.35 | – | 100 m$^2$ | **1.74** |
| Cutting grass, overgrowth or the like; using tractor-mounted side-arm flail mower; in areas inaccessible to alternative machine; on surface | | | | | | | |
|    not exceeding 30° from horizontal | – | 0.02 | 1.02 | 0.37 | – | 100 m$^2$ | **1.39** |
|    30°–50° from horizontal | – | 0.05 | 2.00 | 0.41 | – | 100 m$^2$ | **2.41** |

## 37 SOFT LANDSCAPING

| Item<br>Excluding site overheads and profit | PC<br>£ | Labour<br>hours | Labour<br>£ | Plant<br>£ | Material<br>£ | Unit | Total<br>rate £ |
|---|---|---|---|---|---|---|---|
| **Grass cutting; pedestrian operated equipment; Norris & Gardiner Ltd; the following rates are for grass cutting only; there is no allowance for accompanying strimming operations** | | | | | | | |
| Using cylinder lawn mower fitted with not less than five cutting blades, front and rear rollers; on surface not exceeding 30° from horizontal; arisings let fly; width of cut | | | | | | | |
| 51 cm | – | 0.03 | 1.30 | 0.16 | – | 100 m | **1.46** |
| 61 cm | – | 0.03 | 1.09 | 0.18 | – | 100 m | **1.27** |
| 71 cm | – | 0.02 | 0.94 | 0.19 | – | 100 m | **1.13** |
| 91 cm | – | 0.02 | 0.73 | 0.20 | – | 100 m | **0.93** |
| Using rotary self-propelled mower; width of cut | | | | | | | |
| 45 cm | – | 0.02 | 0.85 | 0.08 | – | 100 m² | **0.93** |
| 81 cm | – | 0.01 | 0.47 | 0.08 | – | 100 m² | **0.55** |
| 120 cm | – | 0.01 | 0.32 | 0.36 | – | 100 m² | **0.68** |
| Add for using grass box for collecting and depositing arisings | | | | | | | |
| removing and depositing arisings | – | 0.03 | 1.07 | – | – | 100 m² | **1.07** |

## 37 SOFT LANDSCAPING

| Item<br>Excluding site overheads and profit | PC<br>£ | Labour<br>hours | Labour<br>£ | Plant<br>£ | Material<br>£ | Unit | Total<br>rate £ |
|---|---|---|---|---|---|---|---|
| **GRASS CUTTING – cont** | | | | | | | |
| **Grass cutting; pedestrian operated equipment; Norris & Gardiner Ltd; the following rates are for grass cutting with accompanying strimming operations** | | | | | | | |
| Using cylinder lawn mower fitted with not less than five cutting blades, front and rear rollers; on surface not exceeding 30° from horizontal; arisings let fly; width of cut | | | | | | | |
| 51 cm | – | 0.06 | 2.60 | 0.25 | – | 100 m | **2.85** |
| 61 cm | – | 0.05 | 2.18 | 0.25 | – | 100 m | **2.43** |
| 71 cm | – | 0.04 | 1.87 | 0.25 | – | 100 m | **2.12** |
| 91 cm | – | 0.03 | 1.46 | 0.25 | – | 100 m | **1.71** |
| Using rotary self-propelled mower; width of cut | | | | | | | |
| 45 cm | – | 0.04 | 1.69 | 0.14 | – | 100 m$^2$ | **1.83** |
| 81 cm | – | 0.02 | 0.94 | 0.11 | – | 100 m$^2$ | **1.05** |
| 91 cm | – | 0.02 | 0.83 | 0.14 | – | 100 m$^2$ | **0.97** |
| 120 cm | – | 0.01 | 0.63 | 0.11 | – | 100 m$^2$ | **0.74** |
| Add for using grass box for collecting and depositing arisings | | | | | | | |
| removing and depositing arisings | – | 0.03 | 1.07 | – | – | 100 m$^2$ | **1.07** |
| Add for 30–50° from horizontal | – | – | – | – | – | 33% | – |
| Add for slopes exceeding 50° | – | – | – | – | – | 100% | – |
| Cutting grass or light woody undergrowth; using trimmer with nylon cord or metal disc cutter; on surface | | | | | | | |
| not exceeding 30° from horizontal | – | 0.10 | 4.28 | 0.29 | – | 100 m$^2$ | **4.57** |
| 30–50° from horizontal | – | 0.20 | 8.55 | 0.57 | – | 100 m$^2$ | **9.12** |
| exceeding 50° from horizontal | – | 0.25 | 10.69 | 0.71 | – | 100 m$^2$ | **11.40** |
| **Grass cutting; collecting arisings; Norris & Gardener Ltd** | | | | | | | |
| Extra over for tractor drawn and self-propelled machinery using attached grass boxes; depositing arisings | | | | | | | |
| 22 cuts per year | – | 0.03 | 1.07 | – | – | 100 m$^2$ | **1.07** |
| 18 cuts per year | – | 0.04 | 1.60 | – | – | 100 m$^2$ | **1.60** |
| 12 cuts per year | – | 0.05 | 2.14 | – | – | 100 m$^2$ | **2.14** |
| 4 cuts per year | – | 0.13 | 5.34 | – | – | 100 m$^2$ | **5.34** |
| Arisings collected by trailed sweepers | | | | | | | |
| 22 cuts per year | – | 0.01 | 0.36 | 0.26 | – | 100 m$^2$ | **0.62** |
| 18 cuts per year | – | 0.01 | 0.47 | 0.26 | – | 100 m$^2$ | **0.73** |
| 12 cuts per year | – | 0.02 | 0.71 | 0.26 | – | 100 m$^2$ | **0.97** |
| 4 cuts per year | – | 0.03 | 1.43 | 0.26 | – | 100 m$^2$ | **1.69** |
| Disposing arisings on site; 100 m distance maximum | | | | | | | |
| 22 cuts per year | – | – | 0.17 | 0.02 | – | 100 m$^2$ | **0.19** |
| 18 cuts per year | – | – | 0.21 | 0.03 | – | 100 m$^2$ | **0.24** |
| 12 cuts per year | – | 0.01 | 0.31 | 0.04 | – | 100 m$^2$ | **0.35** |

## 37 SOFT LANDSCAPING

| Item<br>Excluding site overheads and profit | PC<br>£ | Labour<br>hours | Labour<br>£ | Plant<br>£ | Material<br>£ | Unit | Total<br>rate £ |
|---|---|---|---|---|---|---|---|
| 4 cuts per year | – | 0.02 | 0.93 | 0.13 | – | 100 m² | 1.06 |
| Disposal of arisings off site | | | | | | | |
| 22 cuts per year | – | – | 0.18 | 0.04 | 0.07 | 100 m² | 0.29 |
| 18 cuts per year | – | 0.01 | 0.21 | 0.05 | 0.09 | 100 m² | 0.35 |
| 12 cuts per year | – | – | 0.14 | 0.14 | 0.16 | 100 m² | 0.44 |
| 4 cuts per year | – | 0.01 | 0.43 | 0.82 | 0.26 | 100 m² | 1.51 |
| **GROUNDCARE OPERATIONS** | | | | | | | |
| **Harrowing** | | | | | | | |
| Harrowing grassed area with | | | | | | | |
| drag harrow | – | 0.01 | 0.27 | 0.15 | – | 100 m² | 0.42 |
| chain or light flexible spiked harrow | – | 0.01 | 0.36 | 0.15 | – | 100 m² | 0.51 |
| **Scarifying** | | | | | | | |
| Mechanical | | | | | | | |
| Sisis ARP4; including grass collection box; towed by tractor; area scarified annually | – | 0.01 | 0.50 | 0.10 | – | 100 m² | 0.60 |
| Sisis ARP4; including grass collection box; towed by tractor; area scarified 2 years previously | – | 0.01 | 0.60 | 0.12 | – | 100 m² | 0.72 |
| pedestrian operated self-powered equipment | – | 0.04 | 1.50 | 0.18 | – | 100 m² | 1.68 |
| add for disposal of arisings | – | 0.01 | 0.53 | 1.02 | 1.30 | 100 m² | 2.85 |
| By hand | | | | | | | |
| hand implement | – | 0.25 | 10.69 | – | – | 100 m² | 10.69 |
| add for disposal of arisings | – | 0.01 | 0.53 | 1.02 | 1.30 | 100 m² | 2.85 |
| **Rolling** | | | | | | | |
| Rolling grassed area; equipment towed by tractor; once over; using | | | | | | | |
| smooth roller | – | 0.01 | 0.31 | 0.16 | – | 100 m² | 0.47 |
| **Turf aeration** | | | | | | | |
| By machine | | | | | | | |
| Vertidrain turf aeration equipment towed by tractor to effect a minimum penetration of 100–250 mm at 100 mm centres | – | 0.02 | 0.83 | 1.31 | – | 100 m² | 2.14 |
| Ryan GA 30; self-propelled turf aerating equipment; to effect a minimum penetration of 100 mm at varying centres | – | 0.02 | 0.83 | 1.04 | – | 100 m² | 1.87 |
| Groundsman; pedestrian operated; self-powered solid or slitting tine turf aerating equipment to effect a minimum penetration of 100 mm | – | 0.07 | 3.00 | 0.95 | – | 100 m² | 3.95 |
| Cushman core harvester; self-propelled; for collection of arisings | – | 0.01 | 0.41 | 0.76 | – | 100 m² | 1.17 |

## 37 SOFT LANDSCAPING

| Item<br>Excluding site overheads and profit | PC<br>£ | Labour<br>hours | Labour<br>£ | Plant<br>£ | Material<br>£ | Unit | Total<br>rate £ |
|---|---|---|---|---|---|---|---|
| **GROUNDCARE OPERATIONS – cont** | | | | | | | |
| **Turf aeration – cont** | | | | | | | |
| By hand | | | | | | | |
| hand fork; to effect a minimum penetration of 100 mm and spaced 150 mm apart | – | 0.67 | 28.50 | – | – | 100 m² | **28.50** |
| hollow tine hand implement; to effect a minimum penetration of 100 mm and spaced 150 mm apart | – | 1.00 | 42.75 | – | – | 100 m² | **42.75** |
| collection of arisings by hand | – | 1.50 | 64.13 | – | – | 100 m² | **64.13** |
| **Turf areas; surface treatments and top dressing; British sugar topsoil** | | | | | | | |
| Apply screened topdressing to grass surfaces; spread using Tru-Lute | | | | | | | |
| Sand soil mixes 90/10 to 50/50 | 0.18 | – | 0.04 | 0.03 | 0.18 | m² | **0.25** |
| Apply screened soil 3 mm Kettering loam to goal mouths and worn areas | | | | | | | |
| 20 mm thick | 1.08 | 0.01 | 0.21 | – | 1.08 | m² | **1.29** |
| 10 mm thick | 0.54 | 0.01 | 0.21 | – | 0.54 | m² | **0.75** |
| **Leaf clearance; clearing grassed area of leaves and other extraneous debris** | | | | | | | |
| Using equipment towed by tractor | | | | | | | |
| large grassed areas with perimeters of mature trees such as sports fields and amenity areas | – | 0.01 | 0.27 | 0.07 | – | 100 m² | **0.34** |
| large grassed areas containing ornamental trees and shrub beds | – | – | 0.09 | 0.98 | – | 100 m² | **1.07** |
| Using pedestrian operated mechanical equipment and blowers | | | | | | | |
| grassed areas with perimeters of mature trees such as sports fields and amenity areas | – | 0.01 | 0.36 | 0.79 | – | 100 m² | **1.15** |
| grassed areas containing ornamental trees and shrub beds | – | 0.05 | 2.14 | 0.34 | – | 100 m² | **2.48** |
| verges | – | 0.03 | 1.43 | 0.04 | – | 100 m² | **1.47** |
| By hand | | | | | | | |
| grassed areas with perimeters of mature trees such as sports fields and amenity areas | – | 0.05 | 2.14 | 0.20 | – | 100 m² | **2.34** |
| grassed areas containing ornamental trees and shrub beds | – | 0.08 | 3.56 | 0.33 | – | 100 m² | **3.89** |
| verges | – | 0.17 | 7.12 | 0.66 | – | 100 m² | **7.78** |
| Removal of arisings | | | | | | | |
| areas with perimeters of mature trees | – | – | 0.15 | 0.09 | 0.10 | 100 m² | **0.34** |
| areas containing ornamental trees and shrub beds | – | 0.01 | 0.43 | 0.34 | 0.26 | 100 m² | **1.03** |

## 37 SOFT LANDSCAPING

| Item<br>Excluding site overheads and profit | PC<br>£ | Labour<br>hours | Labour<br>£ | Plant<br>£ | Material<br>£ | Unit | Total<br>rate £ |
|---|---|---|---|---|---|---|---|
| **Litter clearance** | | | | | | | |
| Collection and disposal of litter from grassed area | – | 0.01 | 0.21 | – | 0.16 | 100 m² | **0.37** |
| Collection and disposal of litter from isolated grassed area not exceeding 1000 m² | – | 0.02 | 0.85 | – | 0.16 | 100 m² | **1.01** |
| **Tree guards, stakes and ties** | | | | | | | |
| Adjusting existing tree tie | – | 0.02 | 0.71 | – | – | nr | **0.71** |
| Taking up single or double tree stake and ties; removing and disposing | – | 0.03 | 0.59 | – | – | nr | **0.59** |
| **SHRUB BED MAINTENANCE** | | | | | | | |
| **Edge maintenance** | | | | | | | |
| Maintain edges where lawn abuts pathway or hard surface using | | | | | | | |
|   strimmer | – | – | 0.11 | 0.01 | – | m | **0.12** |
|   shears | – | 0.01 | 0.34 | – | – | m | **0.34** |
| Maintain edges where lawn abuts plant bed using | | | | | | | |
|   mechanical edging tool | – | – | 0.14 | 0.03 | – | m | **0.17** |
|   shears | – | – | 0.23 | – | – | m | **0.23** |
|   half moon edging tool | – | 0.01 | 0.41 | – | – | m | **0.41** |
| **Pruning shrubs** | | | | | | | |
| Trimming ground cover planting | | | | | | | |
|   soft groundcover; vinca ivy and the like | – | 0.33 | 20.38 | – | – | 100 m² | **20.38** |
|   woody groundcover; cotoneaster and the like | – | 0.50 | 30.57 | – | – | 100 m² | **30.57** |
| Pruning massed shrub border (measure ground area) | | | | | | | |
|   shrub beds pruned annually | – | – | 0.20 | – | – | m² | **0.20** |
|   shrub beds pruned hard every 3 years | – | 0.01 | 0.57 | – | – | m² | **0.57** |
| Cutting off dead heads | | | | | | | |
|   bush or standard rose | – | 0.02 | 1.02 | – | – | nr | **1.02** |
|   climbing rose | – | 0.03 | 1.70 | – | – | nr | **1.70** |
| Pruning roses | | | | | | | |
|   bush or standard rose | – | 0.02 | 1.02 | – | – | nr | **1.02** |
|   climbing rose or rambling rose; tying in as required | – | 0.02 | 1.36 | – | – | nr | **1.36** |
| Pruning ornamental shrub; height before pruning (increase these rates by 50% if pruning work has not been executed during the previous 2 years) | | | | | | | |
|   not exceeding 1 m | – | 0.01 | 0.82 | – | – | nr | **0.82** |
|   1 to 2 m | – | 0.02 | 1.13 | – | – | nr | **1.13** |
|   exceeding 2 m | – | 0.04 | 2.55 | – | – | nr | **2.55** |

## 37 SOFT LANDSCAPING

| Item<br>Excluding site overheads and profit | PC<br>£ | Labour<br>hours | Labour<br>£ | Plant<br>£ | Material<br>£ | Unit | Total<br>rate £ |
|---|---|---|---|---|---|---|---|
| **SHRUB BED MAINTENANCE – cont** | | | | | | | |
| **Pruning shrubs – cont** | | | | | | | |
| Removing excess growth etc. from face of building etc.; height before pruning | | | | | | | |
| not exceeding 2 m | – | 0.01 | 0.58 | – | – | nr | **0.58** |
| 2 to 4 m | – | 0.02 | 1.02 | – | – | nr | **1.02** |
| 4 to 6 m | – | 0.03 | 1.70 | – | – | nr | **1.70** |
| 6 to 8 m | – | 0.04 | 2.55 | – | – | nr | **2.55** |
| 8 to 10 m | – | 0.05 | 2.91 | – | – | nr | **2.91** |
| Removing epicormic growth from base of shrub or trunk and base of tree (any height, any diameter); number of growths | | | | | | | |
| not exceeding 10 | – | 0.02 | 1.02 | – | – | nr | **1.02** |
| 10 to 20 | – | 0.02 | 1.36 | – | – | nr | **1.36** |
| **Shrub beds, borders and planters** | | | | | | | |
| Lifting | | | | | | | |
| bulbs | – | 0.17 | 10.19 | – | – | 100 nr | **10.19** |
| tubers or corms | – | 0.13 | 8.15 | – | – | 100 nr | **8.15** |
| established herbaceous plants; hoeing and depositing for replanting | – | 0.66 | 40.76 | – | – | 100 nr | **40.76** |
| Temporary staking and tying in herbaceous plant | – | 0.01 | 0.68 | – | 0.17 | nr | **0.85** |
| Cutting down spent growth of herbaceous plant; clearing arisings | | | | | | | |
| unstaked | – | 0.01 | 0.41 | – | – | nr | **0.41** |
| staked; not exceeding four stakes per plant; removing stakes and putting into store | – | 0.01 | 0.51 | – | – | nr | **0.51** |
| Hand weeding; regular visits; costs per occasion | | | | | | | |
| newly planted areas; mulched beds; | – | 0.03 | 1.70 | – | – | 100 m² | **1.70** |
| newly planted areas; unmulched | – | 0.04 | 2.72 | – | – | 100 m² | **2.72** |
| established areas | – | 0.01 | 0.68 | – | – | 100 m² | **0.68** |
| Removing grasses from groundcover areas | – | 0.02 | 1.36 | – | – | 100 m² | **1.36** |
| Hand digging with fork between shrubs; not exceeding 150 mm deep; breaking down lumps; leaving surface with a medium tilth | – | 0.44 | 27.17 | – | – | 100 m² | **27.17** |
| Hand digging with fork or spade to an average depth of 230 mm; breaking down lumps; leaving surface with a medium tilth | – | 0.66 | 40.76 | – | – | 100 m² | **40.76** |

## 37 SOFT LANDSCAPING

| Item<br>Excluding site overheads and profit | PC<br>£ | Labour<br>hours | Labour<br>£ | Plant<br>£ | Material<br>£ | Unit | Total<br>rate £ |
|---|---|---|---|---|---|---|---|
| Hand hoeing; not exceeding 50 mm deep; leaving surface with a medium tilth | – | 0.13 | 8.15 | – | – | 100 m$^2$ | **8.15** |
| Hand raking to remove stones etc.; breaking down lumps: leaving surface with a fine tilth prior to planting | – | 0.22 | 13.59 | – | – | 100 m$^2$ | **13.59** |
| Hand weeding; planter, window box; not exceeding 1.00 m$^2$ | | | | | | | |
|   ground level box | – | 0.02 | 1.02 | – | – | nr | **1.02** |
|   box accessed by stepladder | – | 0.03 | 1.70 | – | – | nr | **1.70** |
| Spreading only compost, mulch or processed bark to a depth of 75 mm | | | | | | | |
|   on shrub bed with existing mature planting | – | 0.03 | 1.85 | – | – | m$^2$ | **1.85** |
|   recently planted areas | – | 0.02 | 1.36 | – | – | m$^2$ | **1.36** |
|   groundcover and herbaceous areas | – | 0.02 | 1.53 | – | – | m$^2$ | **1.53** |
| Clearing cultivated area of litter and other extraneous debris; using hand implement (exludes winter leaf clearance | | | | | | | |
|   weekly maintenance; private areas | – | 0.01 | 0.34 | – | – | 100 m$^2$ | **0.34** |
|   weekly maintenance; public areas | – | 0.01 | 0.68 | – | – | 100 m$^2$ | **0.68** |
|   daily maintenance public areas | – | 0.01 | 0.34 | – | – | 100 m$^2$ | **0.34** |
| Clearing cultivated area of winter leaf fall, extraneous debris; using hand implement; winter leaf; removal of cleared material to stockpile on site | | | | | | | |
|   per occasion; occasional trees and deciduous shrubs within 100 m$^2$ | – | 0.01 | 0.85 | – | – | 100 m$^2$ | **0.85** |
|   per occasion; dense tall trees and deciduous shrubs in close proximity | – | 0.03 | 1.70 | – | – | 100 m$^2$ | **1.70** |
| **BEDDING MAINTENANCE** | | | | | | | |
| **Works to established bedding areas** | | | | | | | |
| Lifting | | | | | | | |
|   bedding plants; hoeing and depositing for disposal | – | 0.90 | 55.42 | – | – | 100 m$^2$ | **55.42** |
| Hand digging with fork; not exceeding 150 mm deep: breaking down lumps; leaving surface with a medium tilth | – | 0.24 | 14.74 | – | – | 100 m$^2$ | **14.74** |
| Hand weeding | | | | | | | |
|   newly planted areas | – | 0.64 | 39.30 | – | – | 100 m$^2$ | **39.30** |
|   established areas | – | 0.16 | 9.82 | – | – | 100 m$^2$ | **9.82** |
| Hand digging with fork or spade to an average depth of 230 mm; breaking down lumps; leaving surface with a medium tilth | – | 0.16 | 9.82 | – | – | 100 m$^2$ | **9.82** |

## 37 SOFT LANDSCAPING

| Item<br>Excluding site overheads and profit | PC<br>£ | Labour<br>hours | Labour<br>£ | Plant<br>£ | Material<br>£ | Unit | Total<br>rate £ |
|---|---|---|---|---|---|---|---|
| **BEDDING MAINTENANCE – cont** | | | | | | | |
| **Works to established bedding areas – cont** | | | | | | | |
| Hand hoeing: not exceeding 50 mm deep; leaving surface with a medium tilth | – | 0.13 | 7.86 | – | – | 100 m$^2$ | **7.86** |
| Hand raking to remove stones etc.; breaking down lumps; leaving surface with a fine tilth prior to planting | – | 0.21 | 13.10 | – | – | 100 m$^2$ | **13.10** |
| Hand weeding; planter, window box; not exceeding 1.00 m$^2$ | | | | | | | |
|   ground level box | – | 0.02 | 0.99 | – | – | nr | **0.99** |
|   box accessed by stepladder | – | 0.03 | 1.64 | – | – | nr | **1.64** |
| Spreading only; compost, mulch or processed bark to a depth of 75 mm | | | | | | | |
|   on shrub bed with existing mature planting | – | 0.03 | 1.78 | – | – | m$^2$ | **1.78** |
|   recently planted areas | – | 0.02 | 1.32 | – | – | m$^2$ | **1.32** |
|   groundcover and herbaceous areas | – | 0.02 | 1.47 | – | – | m$^2$ | **1.47** |
| Collecting bedding from nursery | – | 0.95 | 58.95 | 13.64 | – | 100 m$^2$ | **72.59** |
| Setting out | | | | | | | |
|   mass planting single variety | – | 0.04 | 2.46 | – | – | m$^2$ | **2.46** |
|   pattern | – | 0.11 | 6.55 | – | – | m$^2$ | **6.55** |
| Planting only | | | | | | | |
|   massed bedding plants | – | 0.06 | 3.93 | – | – | m$^2$ | **3.93** |
| Clearing cultivated area of leaves, litter and other extraneous debris; using hand implement | | | | | | | |
|   weekly maintenance | – | 0.04 | 2.46 | – | – | 100 m$^2$ | **2.46** |
|   daily maintenance | – | – | – | – | – | 100 m$^2$ | **–** |
| **IRRIGATION AND WATERING** | | | | | | | |
| **Irrigation and watering** | | | | | | | |
| Hand held hosepipe; flow rate 25 litres per minute; irrigation requirement | | | | | | | |
|   10 l/m$^2$ | – | 0.74 | 14.00 | – | – | 100 m$^2$ | **14.00** |
|   15 l/m$^2$ | – | 1.10 | 20.90 | – | – | 100 m$^2$ | **20.90** |
|   20 l/m$^2$ | – | 1.46 | 27.80 | – | – | 100 m$^2$ | **27.80** |
|   25 l/m$^2$ | – | 1.84 | 34.90 | – | – | 100 m$^2$ | **34.90** |
| Hand held hosepipe; flow rate 40 litres per minute; irrigation requirement | | | | | | | |
|   10 l/m$^2$ | – | 0.46 | 8.78 | – | – | 100 m$^2$ | **8.78** |
|   15 l/m$^2$ | – | 0.69 | 13.17 | – | – | 100 m$^2$ | **13.17** |
|   20 l/m$^2$ | – | 0.91 | 17.35 | – | – | 100 m$^2$ | **17.35** |
|   25 l/m$^2$ | – | 1.15 | 21.78 | – | – | 100 m$^2$ | **21.78** |

## 37 SOFT LANDSCAPING

| Item Excluding site overheads and profit | PC £ | Labour hours | Labour £ | Plant £ | Material £ | Unit | Total rate £ |
|---|---|---|---|---|---|---|---|
| **HEDGE MAINTENANCE** | | | | | | | |
| **Hedge cutting; field hedges cut once or twice annually** | | | | | | | |
| Trimming sides and top using hand tool or hand held mechanical tools | | | | | | | |
| not exceeding 2 m high | – | 0.05 | 2.14 | 0.14 | – | 10 m² | **2.28** |
| 2–4 m high | – | 0.17 | 7.12 | 0.48 | – | 10 m² | **7.60** |
| **Hedge cutting; ornamental** | | | | | | | |
| Trimming sides and top using hand tool or hand held mechanical tools | | | | | | | |
| not exceeding 2 m high | – | 0.06 | 2.67 | 0.18 | – | 10 m² | **2.85** |
| 2–4 m high | – | 0.25 | 10.69 | 0.71 | – | 10 m² | **11.40** |
| **Hedge cutting; reducing width; hand tool or hand held mechanical tools** | | | | | | | |
| Not exceeding 2 m high | | | | | | | |
| average depth of cut not exceeding 300 mm | – | 0.01 | 0.53 | 0.04 | – | m² | **0.57** |
| average depth of cut 300–600 mm | – | 0.03 | 1.19 | 0.08 | – | m² | **1.27** |
| average depth of cut 600–900 mm | – | 0.03 | 1.34 | 0.09 | – | m² | **1.43** |
| 2–4 m high | | | | | | | |
| average depth of cut not exceeding 300 mm | – | 0.02 | 0.85 | 0.06 | – | m² | **0.91** |
| average depth of cut 300–600 mm | – | 0.03 | 1.34 | 0.09 | – | m² | **1.43** |
| average depth of cut 600–900 mm | – | 0.05 | 2.14 | 0.14 | – | m² | **2.28** |
| 4–6 m high | | | | | | | |
| average depth of cut not exceeding 300 mm | – | 0.05 | 2.14 | 0.14 | – | m² | **2.28** |
| average depth of cut 300–600 mm | – | 0.08 | 3.56 | 0.24 | – | m² | **3.80** |
| average depth of cut 600–900 mm | – | 0.25 | 10.69 | 0.71 | – | m² | **11.40** |
| **Hedge cutting; reducing width; tractor mounted hedge cutting equipment** | | | | | | | |
| Not exceeding 2 m high | | | | | | | |
| average depth of cut not exceeding 300 mm | – | 0.02 | 0.85 | 0.67 | – | 10 m² | **1.52** |
| average depth of cut 300–600 mm | – | 0.03 | 1.07 | 0.83 | – | 10 m² | **1.90** |
| average depth of cut 600–900 mm | – | 0.10 | 4.28 | 3.33 | – | 10 m² | **7.61** |
| 2–4 m high | | | | | | | |
| average depth of cut not exceeding 300 mm | – | 0.01 | 0.27 | 0.21 | – | 10 m² | **0.48** |
| average depth of cut 300–600 mm | – | 0.01 | 0.53 | 0.42 | – | 10 m² | **0.95** |
| average depth of cut 600–900 mm | – | 0.01 | 0.43 | 0.33 | – | 10 m² | **0.76** |

## 37 SOFT LANDSCAPING

| Item<br>Excluding site overheads and profit | PC<br>£ | Labour<br>hours | Labour<br>£ | Plant<br>£ | Material<br>£ | Unit | Total<br>rate £ |
|---|---|---|---|---|---|---|---|
| **HEDGE MAINTENANCE – cont** | | | | | | | |
| **Hedge cutting; reducing height; hand tool or hand held mechanical tools** | | | | | | | |
| Not exceeding 2 m high | | | | | | | |
| average depth of cut not exceeding 300 mm | – | 0.03 | 1.43 | 0.10 | – | 10 m² | **1.53** |
| average depth of cut 300–600 mm | – | 0.07 | 2.85 | 0.19 | – | 10 m² | **3.04** |
| average depth of cut 600–900 mm | – | 0.20 | 8.55 | 0.57 | – | 10 m² | **9.12** |
| 2–4 m high | | | | | | | |
| average depth of cut not exceeding 300 mm | – | 0.02 | 0.71 | 0.05 | – | m² | **0.76** |
| average depth of cut 300–600 mm | – | 0.03 | 1.43 | 0.10 | – | m² | **1.53** |
| average depth of cut 600–900 mm | – | 0.10 | 4.28 | 0.29 | – | m² | **4.57** |
| 4–6 m high | | | | | | | |
| average depth of cut not exceeding 300 mm | – | 0.03 | 1.43 | 0.10 | – | m² | **1.53** |
| average depth of cut 300–600 mm | – | 0.06 | 2.67 | 0.18 | – | m² | **2.85** |
| average depth of cut 600–900 mm | – | 0.13 | 5.34 | 0.36 | – | m² | **5.70** |
| **Hedge cutting; removal and disposal of arisings** | | | | | | | |
| Sweeping up and depositing arisings | | | | | | | |
| 300 mm cut | – | 0.03 | 1.07 | – | – | 10 m² | **1.07** |
| 600 mm cut | – | 0.10 | 4.28 | – | – | 10 m² | **4.28** |
| 900 mm cut | – | 0.20 | 8.55 | – | – | 10 m² | **8.55** |
| Chipping arisings | | | | | | | |
| 300 mm cut | – | 0.01 | 0.43 | 0.11 | – | 10 m² | **0.54** |
| 600 mm cut | – | 0.04 | 1.78 | 0.44 | – | 10 m² | **2.22** |
| 900 mm cut | – | 0.10 | 4.28 | 1.07 | – | 10 m² | **5.35** |
| Disposal of unchipped arisings | | | | | | | |
| 300 mm cut | – | 0.01 | 0.36 | 0.51 | 0.17 | 10 m² | **1.04** |
| 600 mm cut | – | 0.02 | 0.71 | 1.02 | 0.26 | 10 m² | **1.99** |
| 900 mm cut | – | 0.04 | 1.78 | 2.56 | 0.65 | 10 m² | **4.99** |
| Disposal of chipped arisings | | | | | | | |
| 300 mm cut | – | – | 0.07 | 0.17 | 0.26 | 10 m² | **0.50** |
| 600 mm cut | – | 0.01 | 0.36 | 0.17 | 0.52 | 10 m² | **1.05** |
| 900 mm cut | – | 0.02 | 0.71 | 0.34 | 1.30 | 10 m² | **2.35** |

## 37 SOFT LANDSCAPING

| Item<br>Excluding site overheads and profit | PC £ | Labour hours | Labour £ | Plant £ | Material £ | Unit | Total rate £ |
|---|---|---|---|---|---|---|---|
| **CYCLIC MAINTENANCE OPERATIONS** | | | | | | | |
| Note: These items for landscape maintenance may be featured for maintenance in the section Aftercare as Part of a Landscape Contract. | | | | | | | |
| **Path or hard surface maintenance** | | | | | | | |
| Cyclic maintenance to surfaces based on annual requirements for long-term maintenance contractors | | | | | | | |
| Check or clear daily for litter or dog faeces | – | 12.00 | 228.00 | – | – | 100 m² | 228.00 |
| Sweeping monthly | – | 1.00 | 19.00 | – | – | 100 m² | 19.00 |
| Jet washing bi-annually | – | 0.33 | 6.27 | – | – | 100 m² | 6.27 |
| Surface grit application; six occasions per annum | – | 3.00 | 57.00 | – | 24.00 | 100 m² | 81.00 |
| Snow clearance; per occasion | – | 1.25 | 23.75 | – | – | 100 m² | 23.75 |
| Childrens play area maintenance | | | | | | | |
| Visit daily to inspect for safety (Play area as part of larger maintained staffed area) | – | 0.10 | 1.90 | – | – | nr | 1.90 |
| Visit daily to inspect for safety (Play area as isolated play area ) | – | 0.50 | 9.50 | – | – | nr | 9.50 |
| Monthly inspections of play equipment integrity and safety | – | 1.00 | 19.00 | – | – | nr | 19.00 |
| Annual repairs and maintenance to equipment in play areas; high density populations | – | 4.00 | 76.00 | – | – | nr | 76.00 |
| Annual repairs and maintenance to equipment in play areas; low density populations | – | 1.00 | 19.00 | – | – | nr | 19.00 |
| Street furniture maintenance – annual cost | | | | | | | |
| Clean benches weekly | – | 2.60 | 61.75 | – | – | nr | 61.75 |
| Sand timber bench bi-annually; remove splinters | – | 0.50 | 11.88 | – | – | nr | 11.88 |
| Maintain lighting bollard inclusive of cleaning and bulb replacement | – | 0.50 | 11.88 | – | 4.51 | nr | 16.39 |
| Maintain lighting standard inclusive of cleaning and bulb replacement | – | 0.75 | 17.81 | – | 4.51 | nr | 22.32 |
| Maintain clean exterior to litter and dog bins | – | 3.00 | 57.00 | – | – | nr | 57.00 |
| Litter maintenance | | | | | | | |
| Check and clear litter bins daily (high density) based on 240 days per annum | – | 20.00 | 380.00 | 319.55 | 12.00 | nr | 711.55 |
| Check and clear litter bins weekly | – | 4.33 | 82.33 | 213.64 | 2.60 | nr | 298.57 |
| Clear dog litter bins daily (in conjunction with bin clearance on site) | – | 6.00 | 114.00 | 240.00 | 3.00 | nr | 357.00 |

## 37 SOFT LANDSCAPING

| Item<br>Excluding site overheads and profit | PC<br>£ | Labour<br>hours | Labour<br>£ | Plant<br>£ | Material<br>£ | Unit | Total<br>rate £ |
|---|---|---|---|---|---|---|---|
| **CYCLIC MAINTENANCE OPERATIONS – cont** | | | | | | | |
| **Path or hard surface maintenance – cont** | | | | | | | |
| Litter maintenance – cont | | | | | | | |
| Clear dog litter bins weekly (in conjunction with bin clearance on site) | – | 1.00 | 19.00 | 52.00 | 12.00 | nr | **83.00** |
| Sports maintenance | | | | | | | |
| Take football posts from store and erect | – | 2.00 | 95.00 | – | – | set | **95.00** |
| Remove football posts and remove to store | – | 0.75 | 35.63 | – | – | set | **35.63** |
| Mark football/rugby field initial mark | 19.50 | 5.00 | 95.00 | – | 19.50 | nr | **114.50** |
| Mark football/rugby field re-mark per occasion | 4.50 | 1.00 | 19.00 | – | 4.50 | nr | **23.50** |
| Graffiti maintenance per occasion based on term graffiti removal contract | | | | | | | |
| Remove graffiti from street furniture | – | – | – | – | 45.00 | nr | **45.00** |
| Remove graffiti from masonry | – | – | – | – | 45.00 | nr | **45.00** |
| **JAPANESE KNOTWEED** | | | | | | | |
| **Japanese Knotweed Management Plan; PBA Solutions Ltd** | | | | | | | |
| Formulate Japanese Knotweed Management Plan (KMP) as recommended by the Environment Agency Code of Practice 2006 (EA CoP) | – | – | – | – | – | nr | **1200.00** |
| **Treatment of Japanese Knotweed; PBA Solutions Ltd; removal and/or treatment in accordance with The Knotweed Code of Practice published by the Environment Agency 2006 (EA CoP); Japanese knotweed must be treated by qualified practitioners** | | | | | | | |
| Commercial chemical treatments based on areas less than 50 m² knotweed area; allows for the provision of contractual arrangements including method statements and spraying records; also allows for the provision of risk, environment and COSHH assessments; rates for 3–5 year programme, incl. management plan and post-treatment monitoring | | | | | | | |
| rate per visit | – | – | – | – | – | nr | **291.00** |
| rate for complete treatment programme lasting 3 years | – | – | – | – | – | nr | **2620.00** |

## 37 SOFT LANDSCAPING

| Item<br>Excluding site overheads and profit | PC<br>£ | Labour<br>hours | Labour<br>£ | Plant<br>£ | Material<br>£ | Unit | Total<br>rate £ |
|---|---|---|---|---|---|---|---|
| Commercial chemical treatments based on areas between 50–300 m² knotweed area; allows for the provision of contractual arrangements including method statements and spraying records; also allows for the provision of risk, environment and COSHH assessments; rates for 3–5 year programme, incl. management plan and post treatment monitoring | | | | | | | |
| rate per visit | – | – | – | – | – | nr | **361.00** |
| rate for complete treatment programme lasting 3 years | – | – | – | – | – | nr | **3250.00** |
| Commercial chemical treatments based on areas greater than 900 m² knotweed area; allows for the provision of contractual arrangements including method statements and spraying records; also allows for the provision of risk, environment and COSHH assessments; rates for 3–5 year programme, incl. management plan and post treatment monitoring | | | | | | | |
| Rate per visit | – | – | – | – | – | m² | **0.79** |
| Rate per year (3–5 year) | – | – | – | – | – | m² | **7.16** |
| Extra over minimum rate above; for areas over 900 m² | | | | | | | |
| rate per visit | – | – | – | – | – | m² | **0.79** |
| rate for complete treatment programme lasting 3 years | – | – | – | – | – | m² | **7.16** |

Note: Taking waste off site will require pretreatment. Landfill operators are required by the Environment Agency to obtain evidence of pretreatment to show that the waste has been treated to either reduce its volume or hazardousness nature, or facilitate its handling, or enhance its recovery by the waste produce (site of origin) prior being taken to landfill. Generally pretreatment of Japanese Knotweed can include chemical treatment, cutting or burning. The cost to implement this operation would be similar to other minimum treatment costs above. To enable waste to go to landfill soil testing would need to be completed and consultation with a landfill operator or invasive weed specialist is recommended in the first instance.

## 37 SOFT LANDSCAPING

| Item<br>Excluding site overheads and profit | PC<br>£ | Labour<br>hours | Labour<br>£ | Plant<br>£ | Material<br>£ | Unit | Total<br>rate £ |
|---|---|---|---|---|---|---|---|
| **JAPANESE KNOTWEED – cont** | | | | | | | |
| **Japanese Knotweed contaminated waste removal; PBA Solutions Ltd (see note above)** | | | | | | | |
| Removal off site to a licensed landfill site using 20 tonne gross vehicle weight lorries; price is for haulage and tipping only; the mileage stated below is the distance from point of collection to licensed landfill site; price excludes landfill tax and loading of material into lorries; rates will fluctuate between landfill operators and the figures given are for initial guidance only | | | | | | | |
| less than 20 miles | – | – | – | – | – | load | **800.00** |
| 20–40 miles | – | – | – | – | – | load | **850.00** |
| 40–60 miles | – | – | – | – | – | load | **1150.00** |

## 37 SOFT LANDSCAPING

| Item<br>Excluding site overheads and profit | PC<br>£ | Labour<br>hours | Labour<br>£ | Plant<br>£ | Material<br>£ | Unit | Total<br>rate £ |
|---|---|---|---|---|---|---|---|
| **Root barriers; PBA Solutions Ltd; used in conjunction with on site control methods such as bund treatment, cell burial, vertical boundary protection and capping. The following barriers have been reliably used in connection with Japanese Knotweed control and as such are effective when correctly installed as part of a control programme which would usually include for chemical control.** | | | | | | | |
| Linear root deflection barriers; installed to trench measured separately | | | | | | | |
|   Flexiroot non permeable; reinforced LDPE coated polypropylene geomembrane | – | – | – | – | – | m | **4.50** |
|   Bioroot-barrier X permeable root barrier; mechanically bonded geocomposite consisting of a copper-foil bonded between 2 layers of geotextile | – | – | – | – | – | m$^2$ | **12.00** |
| **Landfill tax rates; based on waste being within landfill operator thresholds for controlled waste and with Japanese Knotweed being the only form of contamination; the rates below are based on visual inspections of the load at the gate of the tip** | | | | | | | |
| Rates to April 2016 | | | | | | | |
|   standard rate from April 2014 (where knotweed is visible at more than 5% on inspection at landfill) | – | – | – | – | 84.40 | tonne | **84.40** |
|   lower rate (where knotweed is visible at less than 5% on inspection at landfill) | – | – | – | – | 2.65 | tonne | **2.65** |

## 39 ELECTRICAL SERVICES

| Item<br>Excluding site overheads and profit | PC<br>£ | Labour<br>hours | Labour<br>£ | Plant<br>£ | Material<br>£ | Unit | Total<br>rate £ |
|---|---|---|---|---|---|---|---|
| **CLARIFICATION NOTES ON LABOUR COSTS IN THIS SECTION** | | | | | | | |
| **General groundworks team** | | | | | | | |
| Generally a two man team is used in this section; The column Labour hours reports team hours. The column Labour £ reports the total cost of the team for the unit of work shown | | | | | | | |
| 2 man expert team | – | 1.00 | 47.50 | – | – | hr | **47.50** |
| craftman | – | 1.00 | 23.75 | – | – | hr | **23.75** |
| subcontract electrician | – | – | – | – | – | hr | **35.00** |
| **CABLES AND SWITCHING** | | | | | | | |
| **General groundworks team** | | | | | | | |
| Generally a three man team is used in this section; The column Labour hours reports team hours. The column Labour £ reports the total cost of the team for the unit of work shown | | | | | | | |
| 3 man team | – | 1.00 | 71.25 | – | – | hr | **71.25** |
| **Trenching for electrical services** | | | | | | | |
| 3 tonne excavator (bucket volume 0.13 m$^3$); arisings laid alongside | | | | | | | |
| 600 mm deep | – | – | – | 1.69 | – | m | **1.69** |
| 800 mm deep | – | – | – | 1.98 | – | m | **1.98** |
| 1.00 m deep | – | – | – | 2.37 | – | m | **2.37** |
| Extra over any types of excavating irrespective of depth for breaking out existing materials; heavy duty 110 volt breaker tool | | | | | | | |
| hard rock | – | 5.00 | 118.75 | 25.64 | – | m$^3$ | **144.39** |
| concrete | – | 3.00 | 71.25 | 15.38 | – | m$^3$ | **86.63** |
| reinforced concrete | – | 4.00 | 95.00 | 25.64 | – | m$^3$ | **120.64** |
| brickwork, blockwork or stonework | – | 1.50 | 35.63 | 7.69 | – | m$^3$ | **43.32** |
| By hand | | | | | | | |
| 600 mm deep | – | 0.24 | 5.71 | – | – | m | **5.71** |
| 800 mm deep | – | 0.43 | 10.15 | – | – | m | **10.15** |
| 1.00 m deep | – | 0.67 | 15.83 | – | – | m | **15.83** |
| **Backfilling of trenches; including laying of electrical marker tape; compacting lightly as work proceeds** | | | | | | | |
| To trenches containing armoured cable | | | | | | | |
| by machine | 0.16 | – | – | 2.37 | 0.16 | m | **2.53** |
| by hand | 0.16 | 0.20 | 4.75 | – | 0.16 | m | **4.91** |

## 39 ELECTRICAL SERVICES

| Item<br>Excluding site overheads and profit | PC<br>£ | Labour<br>hours | Labour<br>£ | Plant<br>£ | Material<br>£ | Unit | Total<br>rate £ |
|---|---|---|---|---|---|---|---|
| **To trenches containing ducted cable; including 150 mm sharp sand over the duct** | | | | | | | |
| by machine | 2.62 | – | – | 2.97 | 2.62 | m | **5.59** |
| by hand | 2.62 | 0.25 | 5.94 | – | 2.62 | m | **8.56** |
| **Cable to trenches (trenching operations not included); laying only cable** | | | | | | | |
| Twin core steel wire armoured cable; 50 m drums | | | | | | | |
| 1.5 mm core | 0.69 | – | – | – | 0.69 | m | **0.69** |
| 2.5 mm core | 0.95 | – | – | – | 0.95 | m | **0.95** |
| Twin core steel wire armoured cable; lengths less than 50 m runs | | | | | | | |
| 1.5 mm core | 0.72 | – | – | – | 0.72 | m | **0.72** |
| 2.5 mm core | 1.02 | – | – | – | 1.02 | m | **1.02** |
| Three core steel wire armoured cable; 50 m drums | | | | | | | |
| 1.5 mm core | 0.84 | – | – | – | 0.84 | m | **0.84** |
| 2.5 mm core | 1.19 | – | – | – | 1.19 | m | **1.19** |
| 4.0 mm core | 1.64 | – | – | – | 1.64 | m | **1.64** |
| 6.0 mm core | 2.21 | – | – | – | 2.21 | m | **2.21** |
| 10.0 mm core | 3.53 | – | – | – | 3.53 | m | **3.53** |
| Three core steel wire armoured cable; lengths less than 50 m runs | | | | | | | |
| 1.5 mm core | 0.90 | – | – | – | 0.90 | m | **0.90** |
| 2.5 mm core | 1.27 | – | – | – | 1.27 | m | **1.27** |
| **Ducts to trenches; twin wall flexible cable ducts with drawstrings; laid on 150 mm clean sharp sand** | | | | | | | |
| Twin wall duct; laying to trenches | | | | | | | |
| 63 mm × 50 m coils | 1.00 | 0.02 | 0.47 | – | 1.00 | m | **1.47** |
| 110 mm × 50 m coils | 1.50 | 0.02 | 0.47 | – | 1.50 | m | **1.97** |
| Cable drawing through ducts | | | | | | | |
| straight runs up to 50 m lengths | – | 1.00 | 23.75 | – | – | nr | **23.75** |
| **Terminations to armoured cables; cutting coiling and taping length of cable to receive connection to light fitting, transformer and the like; fixing to temporary post** | | | | | | | |
| Armoured cable | – | 0.33 | 7.91 | – | – | m | **7.91** |
| Plain ducted cable | – | 0.13 | 2.97 | – | – | m | **2.97** |
| **Junction boxes; fixing to cables to receive connections to light fittings or transformers; to IP68** | | | | | | | |

## 39 ELECTRICAL SERVICES

| Item<br>Excluding site overheads and profit | PC<br>£ | Labour<br>hours | Labour<br>£ | Plant<br>£ | Material<br>£ | Unit | Total<br>rate £ |
|---|---|---|---|---|---|---|---|
| **CABLES AND SWITCHING – cont** | | | | | | | |
| Underground jointing box | | | | | | | |
| 2 way | 32.80 | 0.33 | 9.99 | – | 32.80 | m | 42.79 |
| 3 way | 26.45 | 0.57 | 17.14 | – | 26.45 | m | 43.59 |
| Above-ground jointing box | | | | | | | |
| 2 way | 11.15 | 0.33 | 9.99 | – | 11.15 | m | 21.14 |
| 3 way | 14.95 | 0.40 | 12.00 | – | 14.95 | m | 26.95 |
| **Electrical connections** | | | | | | | |
| Connections to existing distribution boards | | | | | | | |
| per switching circuit located within 1 m of the distribution board; chasing of cables to walls not included | – | – | – | – | – | nr | 46.68 |
| Connections to switches; external power cable circuits; connections to internal wall switches; drilling through walls and making good; chasing of cables to walls not included | | | | | | | |
| price for the first circuit | 4.50 | 1.00 | 23.75 | – | 51.18 | nr | 74.93 |
| additional circuits | 4.50 | – | – | – | 15.70 | nr | 15.70 |
| **LOW VOLTAGE** | | | | | | | |
| **Low voltage lights; connecting to transformers (transformers shown separately); fixing as described** | | | | | | | |
| Hunza Spike adjustable spotlight; 63.5 mm diameter × 75 mm long; c/w 20/35/50 W lamp | | | | | | | |
| stainless steel | 161.98 | – | – | – | 206.43 | nr | 206.43 |
| copper | 135.54 | – | – | – | 147.19 | nr | 147.19 |
| black | 108.18 | – | – | – | 119.83 | nr | 119.83 |
| Hunza Spike spotlight;; 63.5 mm diameter × 75 mm long; c/w 20/35/50 W lamp | | | | | | | |
| stainless steel | 186.97 | – | – | – | 198.62 | nr | 198.62 |
| copper | 146.96 | – | – | – | 158.61 | nr | 158.61 |
| black | 77.33 | – | – | – | 88.98 | nr | 88.98 |
| Hunza pole lights; single; fixing to concrete base 150 × 150 × 150 mm | | | | | | | |
| stainless steel | 204.35 | 0.75 | 17.81 | – | 216.43 | nr | 234.24 |
| copper | 150.46 | 0.75 | 17.81 | – | 162.52 | nr | 180.33 |
| black | 104.05 | 0.75 | 17.81 | – | 116.12 | nr | 133.93 |
| Hunza pole lights; double; fixing to concrete base 150 × 150 × 150 mm | | | | | | | |
| stainless steel | 345.97 | 0.75 | 17.81 | – | 358.05 | nr | 375.86 |
| copper | 203.74 | 0.75 | 17.81 | – | 215.82 | nr | 233.63 |
| black | 345.97 | 0.75 | 17.81 | – | 358.05 | nr | 375.86 |

## 39 ELECTRICAL SERVICES

| Item<br>Excluding site overheads and profit | PC<br>£ | Labour<br>hours | Labour<br>£ | Plant<br>£ | Material<br>£ | Unit | Total<br>rate £ |
|---|---|---|---|---|---|---|---|
| Recessed wall lights; eyelid steplights; fixing to brick or stone walls; inclusive of core drilling | | | | | | | |
| stainless steel | 32.80 | 0.75 | 17.81 | – | 192.84 | nr | 210.65 |
| copper | 32.80 | 0.75 | 17.81 | – | 188.01 | nr | 205.82 |
| black | 32.80 | 0.75 | 17.81 | – | 188.01 | nr | 205.82 |
| Hunza; recessed deck path or lawn lights; inclusive of core drilling excavation and all making good | | | | | | | |
| stainless steel light installed to deck | 191.16 | 0.50 | 11.88 | – | 202.81 | nr | 214.69 |
| stainless steel light installed to path | 191.16 | 0.75 | 17.81 | – | 202.81 | nr | 220.62 |
| stainless steel light installed to lawn | 191.16 | 0.40 | 9.50 | – | 202.81 | nr | 212.31 |
| stainless steel driveway light | 191.16 | 0.50 | 11.88 | – | 202.81 | nr | 214.69 |
| **Exterior Lighting Transformers** | | | | | | | |
| **Outdoor transformers for 12 volt exterior lighting; connecting to junction box (not included); IP67** | | | | | | | |
| Single light transformers; 100 × 68 × 72 mm; 50 vA | | | | | | | |
| fused | 22.82 | – | – | – | 57.82 | nr | 57.82 |
| unfused | 15.86 | – | – | – | 50.86 | nr | 50.86 |
| Two light transformer; 130 × 85 × 85 mm; 100 vA | | | | | | | |
| fused | 38.90 | – | – | – | 73.90 | nr | 73.90 |
| unfused | 35.50 | – | – | – | 70.50 | nr | 70.50 |
| Three light transformer; 130 × 85 × 85 mm; 150 vA | | | | | | | |
| fused | 40.38 | – | – | – | 75.38 | nr | 75.38 |
| unfused | – | – | – | – | 74.90 | nr | 74.90 |
| Four light transformer; 130 × 85 × 100 mm; 200 vA | | | | | | | |
| fused | 51.13 | – | – | – | 86.13 | nr | 86.13 |
| unfused | – | – | – | – | 80.20 | nr | 80.20 |

## 39 ELECTRICAL SERVICES

| Item<br>Excluding site overheads and profit | PC<br>£ | Labour<br>hours | Labour<br>£ | Plant<br>£ | Material<br>£ | Unit | Total<br>rate £ |
|---|---|---|---|---|---|---|---|
| **GENERAL** | | | | | | | |
| **Street area floodlighting – General**<br>Preamble: There are an enormous number of luminaires available which are designed for small scale urban and garden projects. The designs are continually changing and the landscape designer is advised to consult the manufacturer's latest catalogue. Most manufacturers supply light fittings suitable for column, bracket, bulkhead, wall or soffit mounting. Highway lamps and columns for trafficked roads are not included in this section as the design of highway lighting is a very specialized subject outside the scope of most landscape contracts. The IP reference number refers to the waterproof properties of the fitting; the higher the number the more waterproof the fitting. Most items can be fitted with time clocks or PIR controls. | | | | | | | |
| **MARKET PRICES OF LAMPS** | | | | | | | |
| **Market prices of lamps**<br>Lamps | | | | | | | |
| 70 W HQIT-S | – | – | – | – | 9.84 | nr | 9.84 |
| 70 W HQIT | – | – | – | – | 19.98 | nr | 19.98 |
| 70 W SON | – | – | – | – | 8.28 | nr | 8.28 |
| 70 W SONT | – | – | – | – | 9.03 | nr | 9.03 |
| 100 W SONT | – | – | – | – | 12.26 | nr | 12.26 |
| 150 W SONT | – | – | – | – | 12.26 | nr | 12.26 |
| 28 W 2D | – | – | – | – | 4.74 | nr | 4.74 |
| 100 W GLS\E27 | – | – | – | – | 1.26 | nr | 1.26 |
| **BULKHEAD LIGHTING** | | | | | | | |
| **Bulkhead and canopy fittings; including fixing to wall and light fitting (lamp, final painting, electric wiring, connections or related fixtures such as switch gear and time clock mechanisms not included unless otherwise indicated)**<br>Bulkhead and canopy fittings | | | | | | | |
| Die cast aluminium body, powder painted in titan grey with anodizing pretreatment | 320.00 | 0.25 | 11.88 | – | 320.60 | nr | 332.48 |

## 39 ELECTRICAL SERVICES

| Item<br>Excluding site overheads and profit | PC<br>£ | Labour<br>hours | Labour<br>£ | Plant<br>£ | Material<br>£ | Unit | Total<br>rate £ |
|---|---|---|---|---|---|---|---|
| **FLOODLIGHTS** | | | | | | | |
| **Floodlighting; ground, wall or pole mounted; including fixing (lamp, final painting, electric wiring, connections or related fixtures such as switch gear and time clock mechanisms not included unless otherwise indicated)** | | | | | | | |
| Floodlighting | | | | | | | |
|   multi-purpose ground/spike mounted wide angle floodlight; 70 W HIT or SON; to IP65 | 375.00 | 0.50 | 23.75 | – | 378.73 | nr | **402.48** |
| Floodlight accessories | | | | | | | |
|   earth spike | 39.05 | – | – | – | 39.05 | nr | **39.05** |
|   louvre | 52.80 | – | – | – | 52.80 | nr | **52.80** |
|   cowl | 52.80 | – | – | – | 52.80 | nr | **52.80** |
|   barn doors | 123.20 | – | – | – | 123.20 | nr | **123.20** |
| Large area/pitch floodlighting; | | | | | | | |
|   1000 W SON-T; floodlights with lamp and loose gear; narrow asymmetric beam | 330.00 | – | – | – | 365.00 | nr | **365.00** |
|   2.0 kw MBIOS; floodlight with lamp and loose gear; projector beam | 519.00 | – | – | – | 554.00 | nr | **554.00** |
| Large area floodlighting; | | | | | | | |
|   floodlight with lamp and integral gear | 110.86 | – | – | – | 145.86 | nr | **145.86** |
|   floodlight with lamp and integral gear | 126.00 | – | – | – | 161.00 | nr | **161.00** |
| **SPOTLIGHTS** | | | | | | | |
| **Spotlights for uplighting and for illuminating signs and notice boards, statuary and other features; ground, wall or pole mounted; including fixing, light fitting and priming (lamp, final painting, electric wiring, connections or related fixtures such as switch gear and time clock mechanisms not included); all mains voltage (240 V) unless otherwise stated** | | | | | | | |
| Spotlights | | | | | | | |
|   LED spot. IP67; 21 W; Professional compact projector for halogen, metal halide lamps, white and rgb LED | 160.00 | 0.50 | 11.88 | – | 198.73 | nr | **210.61** |

## 39 ELECTRICAL SERVICES

| Item<br>Excluding site overheads and profit | PC<br>£ | Labour<br>hours | Labour<br>£ | Plant<br>£ | Material<br>£ | Unit | Total<br>rate £ |
|---|---|---|---|---|---|---|---|
| **SPOTLIGHTS – cont** | | | | | | | |
| **Spotlights for uplighting and for illuminating signs and notice boards, statuary and other features; ground, wall or pole mounted – cont** | | | | | | | |
| Spotlighters, uplighters and cowl lighting – cont | | | | | | | |
| Spotlighters, uplighters and cowl lighting; Havells Sylvania; TECHNO – SHORT ARM; head adjustable 130°; rotation 350°; projection 215 mm on 210 mm base plate; 355 mm high; integral gear; PG16 cable gland; black or aluminium | | | | | | | |
| 70 W CDMT/HQIT | 766.09 | – | – | – | 801.09 | nr | **801.09** |
| 150 W CDMT/HQIT | 805.29 | – | – | – | 840.29 | nr | **840.29** |
| **RECESSED LIGHTING** | | | | | | | |
| **Recessed uplighting; including walk/ drive over fully recessed uplighting; excavating, ground fixing, concreting in and making good surfaces (electric wiring, connections or related fixtures such as switch gear and time clock mechanisms not included unless otherwise stated) (Note: transformers will power multiple lights dependent on the distance between the light units); all mains voltage (240 v) unless otherwise stated** | | | | | | | |
| Recessed uplighting; 266 mm diameter; diecast aluminium with stainless steel top plate 10 mm toughened safety glass; 2000 kg drive over; integral control gear; to IP67 | | | | | | | |
| metal halide; white light; spot or flood or wall wash distribution | 655.00 | 0.50 | 11.88 | – | 693.73 | nr | **705.61** |
| compact fluorescent; white light; low power consumption; flood distribution | 620.00 | 0.50 | 11.88 | – | 658.73 | nr | **670.61** |
| Accessories for uplighters | | | | | | | |
| rockguard | 169.95 | – | – | – | 169.95 | nr | **169.95** |
| stainless steel installation sleeve | 54.45 | – | – | – | 54.45 | nr | **54.45** |
| anti-glare louvre (internal tilt) | 108.35 | – | – | – | 108.35 | nr | **108.35** |
| Accessories for uplighters | | | | | | | |
| IS installation sleeve | 62.70 | – | – | – | 62.70 | nr | **62.70** |

## 39 ELECTRICAL SERVICES

| Item Excluding site overheads and profit | PC £ | Labour hours | Labour £ | Plant £ | Material £ | Unit | Total rate £ |
|---|---|---|---|---|---|---|---|
| Wall recessed; Havells Sylvania; EOS range; integral gear; asymmetric reflector; toughened reeded glass | | | | | | | |
| 145 × 90 mm; black or aluminium; 20 W QT9 lamp | 147.51 | – | – | – | 191.26 | nr | **191.26** |
| 270 × 145 mm; black or aluminium; 20 W QT9 lamp | 213.62 | – | – | – | 285.27 | nr | **285.27** |
| **LIGHTED BOLLARDS** | | | | | | | |
| Lighted bollards; including excavating, ground fixing, concreting in and making good surfaces (lamp, final painting, electric wiring, connections or related fixtures such as switch gear and time clock mechanisms not included) (Note: all illuminated bollards must be earthed); heights given are from ground level to top of bollards | | | | | | | |
| Lighted bollards | | | | | | | |
| radial emission bollard for fluorescent, metal halide lamps and white LED. Extruded aluminium body. Vandal-proof anti-UV clear polycarbonate diffuser, 40 joule shock resistant | 650.16 | 2.50 | 118.75 | – | 683.83 | nr | **802.58** |
| radial emission bollard; tapered, anodized aluminium body and die-cast anodized head, powder painted with a titanium gray finish; cusp reflector made of matt finish aluminium | 541.80 | 2.50 | 118.75 | – | 575.47 | nr | **694.22** |
| path bollard; powder coated in titanium grey finish anodized extruded aluminium body and die-cast aluminium optical unit; high luminous efficiency pure aluminium reflector; ultra flat transparent methacrylate protective screen | 518.24 | 2.50 | 118.75 | – | 551.91 | nr | **670.66** |
| illuminated bollard; in 'very durable hardwood'; integral die cast aluminium lighting unit; 165 × 165 mm × 1.00 m high | 352.20 | 2.50 | 118.75 | – | 392.93 | nr | **511.68** |

## 39 ELECTRICAL SERVICES

| Item<br>Excluding site overheads and profit | PC<br>£ | Labour<br>hours | Labour<br>£ | Plant<br>£ | Material<br>£ | Unit | **Total<br>rate £** |
|---|---|---|---|---|---|---|---|
| **STREET AND PRECINCT LIGHTING** | | | | | | | |
| **Precinct lighting lanterns; ground, wall or pole mounted; including fixing and light fitting (lamps, poles, brackets, final painting, electric wiring, connections or related fixtures such as switch gear and time clock mechanisms not included)** | | | | | | | |
| Precinct lighting lanterns; | | | | | | | |
| side entry 90° curved arm mounted; steel rings over domed top; housing cast silumin sandblasted with integral gear; protected by UV stabilized clear polycarbonate diffuser; IP55; PC £800.00 | 795.00 | 0.50 | 23.75 | – | 795.00 | nr | **818.75** |
| bottom entry pole-top; shot blasted or lacquered Hanover design award; hinged diffuser for simple maintenance; indirect lighting technique ensures low glare; IP55; PC £1000.00 | 1000.00 | 0.50 | 23.75 | – | 1000.00 | nr | **1023.75** |
| Precinct lighting lanterns; IP66 | | | | | | | |
| molded GRP body; graphite finish | 560.00 | 0.50 | 23.75 | – | 560.00 | nr | **583.75** |
| molded GRP body; graphite finish | 560.00 | 0.50 | 23.75 | – | 560.00 | nr | **583.75** |
| cast aluminium body; silver and blue finish | 560.00 | 0.50 | 23.75 | – | 560.00 | nr | **583.75** |

# PART 8

# Tables and Memoranda

This part contains the following sections:

|  | page |
|---|---|
| Conversion Tables | 491 |
| Geometry | 495 |
| Formulae | 499 |
| Typical Thermal Conductivity of Building Materials | 500 |
| Earthwork | 501 |
| Concrete Work | 507 |
| Reinforcement | 512 |
| Formwork | 514 |
| Masonry | 515 |
| Timber | 518 |
| Roofing | 523 |
| Glazing | 527 |
| Metal | 528 |
| Kerbs, Paving, etc. | 546 |
| Seeding/Turfing and Planting | 553 |
| Fencing and Gates | 556 |
| Drainage | 561 |
| Electrical Supply/Power/Lighting Systems | 568 |
| Rail Tracks | 569 |
| Fractions, Decimals and Millimetre Equivalents | 572 |
| Imperial Standard Wire Gauge (SWG) | 573 |
| Pipes, Water, Storage, Insulation | 574 |

# ESSENTIAL READING FROM TAYLOR AND FRANCIS

## Living Roofs in Integrated Urban Water Systems
### ROEHR & FASSMAN-BECK

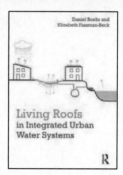

With the infrastructure to manage storm water threats in cities becoming increasingly expensive to build or repair, the design community needs to look at alternative approaches. Living roofs present an opportunity to compliment ground-level storm water control measures, contributing to a holistic, integrated urban water management system.

This book offers tools to plan and design living roofs, in the context of effectively mitigating storm water. Quantitative tools for engineering calculations and qualitative discussion of potential influences and interactions of the design team and assembly elements are addressed.

February 2015; 246 x 174 mm; 192 pp
Hb: 978-0-415-53553-3; £49.99

To Order: Tel: +44 (0) 1235 400524  Fax: +44 (0) 1235 400525
or Post: Taylor and Francis Customer Services,
Bookpoint Ltd, Unit T1, 200 Milton Park, Abingdon, Oxon, OX14 4TA UK
Email: book.orders@tandf.co.uk

For a complete listing of all our titles visit:
www.tandf.co.uk

## CONVERSION TABLES

## CONVERSION TABLES

| Length | Unit | Conversion factors | | | | |
|---|---|---|---|---|---|---|
| Millimetre | mm | 1 in | = 25.4 mm | 1 mm | = 0.0394 in |
| Centimetre | cm | 1 in | = 2.54 cm | 1 cm | = 0.3937 in |
| Metre | m | 1 ft | = 0.3048 m | 1 m | = 3.2808 ft |
| | | 1 yd | = 0.9144 m | | = 1.0936 yd |
| Kilometre | km | 1 mile | = 1.6093 km | 1 km | = 0.6214 mile |

Note: 1 cm = 10 mm  1 ft = 12 in
1 m = 1 000 mm  1 yd = 3 ft
1 km = 1 000 m  1 mile = 1 760 yd

| Area | Unit | Conversion factors | | | | |
|---|---|---|---|---|---|---|
| Square Millimetre | $mm^2$ | 1 $in^2$ | = 645.2 $mm^2$ | 1 $mm^2$ | = 0.0016 $in^2$ |
| Square Centimetre | $cm^2$ | 1 $in^2$ | = 6.4516 $cm^2$ | 1 $cm^2$ | = 1.1550 $in^2$ |
| Square Metre | $m^2$ | 1 $ft^2$ | = 0.0929 $m^2$ | 1 $m^2$ | = 10.764 $ft^2$ |
| | | 1 $yd^2$ | = 0.8361 $m^2$ | 1 $m^2$ | = 1.1960 $yd^2$ |
| Square Kilometre | $km^2$ | 1 $mile^2$ | = 2.590 $km^2$ | 1 $km^2$ | = 0.3861 $mile^2$ |

Note: 1 $cm^2$ = 100 $mm^2$  1 $ft^2$ = 144 $in^2$
1 $m^2$ = 10 000 $cm^2$  1 $yd^2$ = 9 $ft^2$
1 $km^2$ = 100 hectares  1 acre = 4 840 $yd^2$
1 $mile^2$ = 640 acres

| Volume | Unit | Conversion factors | | | | |
|---|---|---|---|---|---|---|
| Cubic Centimetre | $cm^3$ | 1 $cm^3$ | = 0.0610 $in^3$ | 1 $in^3$ | = 16.387 $cm^3$ |
| Cubic Decimetre | $dm^3$ | 1 $dm^3$ | = 0.0353 $ft^3$ | 1 $ft^3$ | = 28.329 $dm^3$ |
| Cubic Metre | $m^3$ | 1 $m^3$ | = 35.3147 $ft^3$ | 1 $ft^3$ | = 0.0283 $m^3$ |
| | | 1 $m^3$ | = 1.3080 $yd^3$ | 1 $yd^3$ | = 0.7646 $m^3$ |
| Litre | l | 1 l | = 1.76 pint | 1 pint | = 0.5683 l |
| | | | = 2.113 US pt | | = 0.4733 US l |

Note: 1 $dm^3$ = 1 000 $cm^3$  1 $ft^3$ = 1 728 $in^3$  1 pint = 20 fl oz
1 $m^3$ = 1 000 $dm^3$  1 $yd^3$ = 27 $ft^3$  1 gal = 8 pints
1 l = 1 $dm^3$

Neither the Centimetre nor Decimetre are SI units, and as such their use, particularly that of the Decimetre, is not widespread outside educational circles.

| Mass | Unit | Conversion factors | | | | |
|---|---|---|---|---|---|---|
| Milligram | mg | 1 mg | = 0.0154 grain | 1 grain | = 64.935 mg |
| Gram | g | 1 g | = 0.0353 oz | 1 oz | = 28.35 g |
| Kilogram | kg | 1 kg | = 2.2046 lb | 1 lb | = 0.4536 kg |
| Tonne | t | 1 t | = 0.9842 ton | 1 ton | = 1.016 t |

Note: 1 g = 1000 mg  1 oz = 437.5 grains  1 cwt = 112 lb
1 kg = 1000 g  1 lb = 16 oz  1 ton = 20 cwt
1 t = 1000 kg  1 stone = 14 lb

| Force | Unit | Conversion factors | | | | |
|---|---|---|---|---|---|---|
| Newton | N | 1 lbf | = 4.448 N | 1 kgf | = 9.807 N |
| Kilonewton | kN | 1 lbf | = 0.004448 kN | 1 ton f | = 9.964 kN |
| Meganewton | MN | 100 tonf | = 0.9964 MN | | |

## CONVERSION TABLES

| Pressure and stress | Unit | Conversion factors | |
|---|---|---|---|
| Kilonewton per square metre | $kN/m^2$ | 1 $lbf/in^2$ | = 6.895 $kN/m^2$ |
| | | 1 bar | = 100 $kN/m^2$ |
| Meganewton per square metre | $MN/m^2$ | 1 $tonf/ft^2$ | = 107.3 $kN/m^2$ = 0.1073 $MN/m^2$ |
| | | 1 $kgf/cm^2$ | = 98.07 $kN/m^2$ |
| | | 1 $lbf/ft^2$ | = 0.04788 $kN/m^2$ |

| Coefficient of consolidation (Cv) or swelling | Unit | Conversion factors | |
|---|---|---|---|
| Square metre per year | $m^2$/year | 1 $cm^2$/s | = 3 154 $m^2$/year |
| | | 1 $ft^2$/year | = 0.0929 $m^2$/year |

| Coefficient of permeability | Unit | Conversion factors | |
|---|---|---|---|
| Metre per second | m/s | 1 cm/s | = 0.01 m/s |
| Metre per year | m/year | 1 ft/year | = 0.3048 m/year |
| | | | = 0.9651 × $(10)^8$ m/s |

| Temperature | Unit | Conversion factors | |
|---|---|---|---|
| Degree Celsius | °C | °C = 5/9 × (°F − 32) | °F = (9 × °C)/ 5 + 32 |

*Tables and Memoranda*

**CONVERSION TABLES**

## SPEED CONVERSION

| km/h | m/min | mph | fpm |
|---|---|---|---|
| 1 | 16.7 | 0.6 | 54.7 |
| 2 | 33.3 | 1.2 | 109.4 |
| 3 | 50.0 | 1.9 | 164.0 |
| 4 | 66.7 | 2.5 | 218.7 |
| 5 | 83.3 | 3.1 | 273.4 |
| 6 | 100.0 | 3.7 | 328.1 |
| 7 | 116.7 | 4.3 | 382.8 |
| 8 | 133.3 | 5.0 | 437.4 |
| 9 | 150.0 | 5.6 | 492.1 |
| 10 | 166.7 | 6.2 | 546.8 |
| 11 | 183.3 | 6.8 | 601.5 |
| 12 | 200.0 | 7.5 | 656.2 |
| 13 | 216.7 | 8.1 | 710.8 |
| 14 | 233.3 | 8.7 | 765.5 |
| 15 | 250.0 | 9.3 | 820.2 |
| 16 | 266.7 | 9.9 | 874.9 |
| 17 | 283.3 | 10.6 | 929.6 |
| 18 | 300.0 | 11.2 | 984.3 |
| 19 | 316.7 | 11.8 | 1038.9 |
| 20 | 333.3 | 12.4 | 1093.6 |
| 21 | 350.0 | 13.0 | 1148.3 |
| 22 | 366.7 | 13.7 | 1203.0 |
| 23 | 383.3 | 14.3 | 1257.7 |
| 24 | 400.0 | 14.9 | 1312.3 |
| 25 | 416.7 | 15.5 | 1367.0 |
| 26 | 433.3 | 16.2 | 1421.7 |
| 27 | 450.0 | 16.8 | 1476.4 |
| 28 | 466.7 | 17.4 | 1531.1 |
| 29 | 483.3 | 18.0 | 1585.7 |
| 30 | 500.0 | 18.6 | 1640.4 |
| 31 | 516.7 | 19.3 | 1695.1 |
| 32 | 533.3 | 19.9 | 1749.8 |
| 33 | 550.0 | 20.5 | 1804.5 |
| 34 | 566.7 | 21.1 | 1859.1 |
| 35 | 583.3 | 21.7 | 1913.8 |
| 36 | 600.0 | 22.4 | 1968.5 |
| 37 | 616.7 | 23.0 | 2023.2 |
| 38 | 633.3 | 23.6 | 2077.9 |
| 39 | 650.0 | 24.2 | 2132.5 |
| 40 | 666.7 | 24.9 | 2187.2 |
| 41 | 683.3 | 25.5 | 2241.9 |
| 42 | 700.0 | 26.1 | 2296.6 |
| 43 | 716.7 | 26.7 | 2351.3 |
| 44 | 733.3 | 27.3 | 2405.9 |
| 45 | 750.0 | 28.0 | 2460.6 |

## CONVERSION TABLES

| km/h | m/min | mph | fpm |
|---|---|---|---|
| 46 | 766.7 | 28.6 | 2515.3 |
| 47 | 783.3 | 29.2 | 2570.0 |
| 48 | 800.0 | 29.8 | 2624.7 |
| 49 | 816.7 | 30.4 | 2679.4 |
| 50 | 833.3 | 31.1 | 2734.0 |

## GEOMETRY

### Two dimensional figures

| Figure | Diagram of figure | Surface area | Perimeter |
|---|---|---|---|
| Square | | $a^2$ | $4a$ |
| Rectangle | | $ab$ | $2(a+b)$ |
| Triangle | | $\frac{1}{2}ch$ | $a+b+c$ |
| Circle | | $\pi r^2$ <br> $\frac{1}{4}\pi d^2$ <br> where $2r = d$ | $2\pi r$ <br> $\pi d$ |
| Parallelogram | | $ah$ | $2(a+b)$ |
| Trapezium | | $\frac{1}{2}h(a+b)$ | $a+b+c+d$ |
| Ellipse | | Approximately $\pi ab$ | $\pi(a+b)$ |
| Hexagon | | $2.6 \times a^2$ | |

## GEOMETRY

| Figure | Diagram of figure | Surface area | Perimeter |
|---|---|---|---|
| Octagon | | $4.83 \times a^2$ | $6a$ |
| Sector of a circle | | $\frac{1}{2}rb$ or $\frac{q}{360}\pi r^2$  note b = angle $\frac{q}{360} \times \pi 2r$ | |
| Segment of a circle | | $S - T$  where S = area of sector, T = area of triangle | |
| Bellmouth | | $\frac{3}{14} \times r^2$ | |

**GEOMETRY**

## Three dimensional figures

| Figure | Diagram of figure | Surface area | Volume |
|---|---|---|---|
| Cube | | $6a^2$ | $a^3$ |
| Cuboid/rectangular block | | $2(ab + ac + bc)$ | $abc$ |
| Prism/triangular block | | $bd + hc + dc + ad$ | $\frac{1}{2}hcd$ |
| Cylinder | | $2\pi r^2 + 2\pi rh$ | $\pi r^2 h$ |
| Sphere | | $4\pi r^2$ | $\frac{4}{3}\pi r^3$ |
| Segment of sphere | | $2\pi Rh$ | $\frac{1}{6}\pi h(3r^2 + h^2)$ $\frac{1}{3}\pi h^2(3R - H)$ |
| Pyramid | | $(a+b)l + ab$ | $\frac{1}{3}abh$ |

## GEOMETRY

| Figure | Diagram of figure | Surface area | Volume |
|---|---|---|---|
| Frustum of a pyramid | | $l(a+b+c+d) + \sqrt{(ab+cd)}$ [rectangular figure only] | $\frac{h}{3}(ab + cd + \sqrt{abcd})$ |
| Cone | | $\pi r l$ (excluding base) $\pi r l + \pi r^2$ (including base) | $\frac{1}{3}\pi r^2 h$ $\frac{1}{12}\pi d^2 h$ |
| Frustum of a cone | | $\pi r^2 + \pi R^2 + \pi l(R+r)$ | $\frac{1}{3}\pi(R^2 + Rr + r^2)$ |

## FORMULAE

### Formulae

| Formula | Description |
|---|---|
| Pythagoras Theorem | $A^2 = B^2 + C^2$<br>where A is the hypotenuse of a right-angled triangle and B and C are the two adjacent sides |
| Simpsons Rule | The Area is divided into an even number of strips of equal width, and therefore has an odd number of ordinates at the division points<br><br>$$\text{area} = \frac{S(A + 2B + 4C)}{3}$$<br><br>where   S = common interval (strip width)<br>           A = sum of first and last ordinates<br>           B = sum of remaining odd ordinates<br>           C = sum of the even ordinates<br><br>The Volume can be calculated by the same formula, but by substituting the area of each coordinate rather than its length |
| Trapezoidal Rule | A given trench is divided into two equal sections, giving three ordinates, the first, the middle and the last<br><br>$$\text{volume} = \frac{S \times (A + B + 2C)}{2}$$<br><br>where   S = width of the strips<br>           A = area of the first section<br>           B = area of the last section<br>           C = area of the rest of the sections |
| Prismoidal Rule | A given trench is divided into two equal sections, giving three ordinates, the first, the middle and the last<br><br>$$\text{volume} = \frac{L \times (A + 4B + C)}{6}$$<br><br>where   L = total length of trench<br>           A = area of the first section<br>           B = area of the middle section<br>           C = area of the last section |

## TYPICAL THERMAL CONDUCTIVITY OF BUILDING MATERIALS

(Always check manufacturer's details – variation will occur depending on product and nature of materials)

| | Thermal conductivity (W/mK) | | Thermal conductivity (W/mK) |
|---|---|---|---|
| Acoustic plasterboard | 0.25 | Oriented strand board | 0.13 |
| Aerated concrete slab (500 kg/m$^3$) | 0.16 | Outer leaf brick | 0.77 |
| Aluminium | 237 | Plasterboard | 0.22 |
| Asphalt (1700 kg/m$^3$) | 0.5 | Plaster dense (1300 kg/m$^3$) | 0.5 |
| Bitumen-impregnated fibreboard | 0.05 | Plaster lightweight (600 kg/m$^3$) | 0.16 |
| Blocks (standard grade 600 kg/m$^3$) | 0.15 | Plywood (950 kg/m$^3$) | 0.16 |
| Blocks (solar grade 460 kg/m$^3$) | 0.11 | Prefabricated timber wall panels (check manufacturer) | 0.12 |
| Brickwork (outer leaf 1700 kg/m$^3$) | 0.84 | Screed (1200 kg/m$^3$) | 0.41 |
| Brickwork (inner leaf 1700 kg/m$^3$) | 0.62 | Stone chippings (1800 kg/m$^3$) | 0.96 |
| Dense aggregate concrete block 1800 kg/m$^3$ (exposed) | 1.21 | Tile hanging (1900 kg/m$^3$) | 0.84 |
| Dense aggregate concrete block 1800 kg/m$^3$ (protected) | 1.13 | Timber (650 kg/m$^3$) | 0.14 |
| Calcium silicate board (600 kg/m$^3$) | 0.17 | Timber flooring (650 kg/m$^3$) | 0.14 |
| Concrete general | 1.28 | Timber rafters | 0.13 |
| Concrete (heavyweight 2300 kg/m$^3$) | 1.63 | Timber roof or floor joists | 0.13 |
| Concrete (dense 2100 kg/m$^3$ typical floor) | 1.4 | Roof tile (1900 kg/m$^3$) | 0.84 |
| Concrete (dense 2000 kg/m$^3$ typical floor) | 1.13 | Timber blocks (650 kg/m$^3$) | 0.14 |
| Concrete (medium 1400 kg/m$^3$) | 0.51 | Cellular glass | 0.045 |
| Concrete (lightweight 1200 kg/m$^3$) | 0.38 | Expanded polystyrene | 0.034 |
| Concrete (lightweight 600 kg/m$^3$) | 0.19 | Expanded polystyrene slab (25 kg/m$^3$) | 0.035 |
| Concrete slab (aerated 500 kg/m$^3$) | 0.16 | Extruded polystyrene | 0.035 |
| Copper | 390 | Glass mineral wool | 0.04 |
| External render sand/cement finish | 1 | Mineral quilt (12 kg/m$^3$) | 0.04 |
| External render (1300 kg/m$^3$) | 0.5 | Mineral wool slab (25 kg/m$^3$) | 0.035 |
| Felt – Bitumen layers (1700 kg/m$^3$) | 0.5 | Phenolic foam | 0.022 |
| Fibreboard (300 kg/m$^3$) | 0.06 | Polyisocyanurate | 0.025 |
| Glass | 0.93 | Polyurethane | 0.025 |
| Marble | 3 | Rigid polyurethane | 0.025 |
| Metal tray used in wriggly tin concrete floors (7800 kg/m$^3$) | 50 | Rock mineral wool | 0.038 |
| Mortar (1750 kg/m$^3$) | 0.8 | | |

# EARTHWORK

## Weights of Typical Materials Handled by Excavators

The weight of the material is that of the state in its natural bed and includes moisture. Adjustments should be made to allow for loose or compacted states

| Material | Mass (kg/m³) | Mass (lb/cu yd) |
| --- | --- | --- |
| Ashes, dry | 610 | 1028 |
| Ashes, wet | 810 | 1365 |
| Basalt, broken | 1954 | 3293 |
| Basalt, solid | 2933 | 4943 |
| Bauxite, crushed | 1281 | 2159 |
| Borax, fine | 849 | 1431 |
| Caliche | 1440 | 2427 |
| Cement, clinker | 1415 | 2385 |
| Chalk, fine | 1221 | 2058 |
| Chalk, solid | 2406 | 4055 |
| Cinders, coal, ash | 641 | 1080 |
| Cinders, furnace | 913 | 1538 |
| Clay, compacted | 1746 | 2942 |
| Clay, dry | 1073 | 1808 |
| Clay, wet | 1602 | 2700 |
| Coal, anthracite, solid | 1506 | 2538 |
| Coal, bituminous | 1351 | 2277 |
| Coke | 610 | 1028 |
| Dolomite, lumpy | 1522 | 2565 |
| Dolomite, solid | 2886 | 4864 |
| Earth, dense | 2002 | 3374 |
| Earth, dry, loam | 1249 | 2105 |
| Earth, Fullers, raw | 673 | 1134 |
| Earth, moist | 1442 | 2430 |
| Earth, wet | 1602 | 2700 |
| Felsite | 2495 | 4205 |
| Fieldspar, solid | 2613 | 4404 |
| Fluorite | 3093 | 5213 |
| Gabbro | 3093 | 5213 |
| Gneiss | 2696 | 4544 |
| Granite | 2690 | 4534 |
| Gravel, dry ¼ to 2 inch | 1682 | 2835 |
| Gravel, dry, loose | 1522 | 2565 |
| Gravel, wet ¼ to 2 inch | 2002 | 3374 |
| Gypsum, broken | 1450 | 2444 |
| Gypsum, solid | 2787 | 4697 |
| Hardcore (consolidated) | 1928 | 3249 |
| Lignite, dry | 801 | 1350 |
| Limestone, broken | 1554 | 2619 |
| Limestone, solid | 2596 | 4375 |
| Magnesite, magnesium ore | 2993 | 5044 |
| Marble | 2679 | 4515 |
| Marl, wet | 2216 | 3735 |
| Mica, broken | 1602 | 2700 |
| Mica, solid | 2883 | 4859 |
| Peat, dry | 400 | 674 |
| Peat, moist | 700 | 1179 |
| Peat, wet | 1121 | 1889 |

## EARTHWORK

| Material | Mass (kg/m$^3$) | Mass (lb/cu yd) |
|---|---|---|
| Potash | 1281 | 2159 |
| Pumice, stone | 640 | 1078 |
| Quarry waste | 1438 | 2423 |
| Quartz sand | 1201 | 2024 |
| Quartz, solid | 2584 | 4355 |
| Rhyolite | 2400 | 4045 |
| Sand and gravel, dry | 1650 | 2781 |
| Sand and gravel, wet | 2020 | 3404 |
| Sand, dry | 1602 | 2700 |
| Sand, wet | 1831 | 3086 |
| Sandstone, solid | 2412 | 4065 |
| Shale, solid | 2637 | 4444 |
| Slag, broken | 2114 | 3563 |
| Slag, furnace granulated | 961 | 1619 |
| Slate, broken | 1370 | 2309 |
| Slate, solid | 2667 | 4495 |
| Snow, compacted | 481 | 810 |
| Snow, freshly fallen | 160 | 269 |
| Taconite | 2803 | 4724 |
| Trachyte | 2400 | 4045 |
| Trap rock, solid | 2791 | 4704 |
| Turf | 400 | 674 |
| Water | 1000 | 1685 |

## Transport Capacities

| Type of vehicle | Capacity of vehicle | |
|---|---|---|
| | Payload | Heaped capacity |
| Wheelbarrow | 150 | 0.10 |
| 1 tonne dumper | 1250 | 1.00 |
| 2.5 tonne dumper | 4000 | 2.50 |
| Articulated dump truck (Volvo A20 6 × 4) | 18500 | 11.00 |
| Articulated dump truck (Volvo A35 6 × 6) | 32000 | 19.00 |
| Large capacity rear dumper (Euclid R35) | 35000 | 22.00 |
| Large capacity rear dumper (Euclid R85) | 85000 | 50.00 |

**EARTHWORK**

## Machine Volumes for Excavating and Filling

| Machine type | Cycles per minute | Volume per minute (m³) |
|---|---|---|
| 1.5 tonne excavator | 1 | 0.04 |
| | 2 | 0.08 |
| | 3 | 0.12 |
| 3 tonne excavator | 1 | 0.13 |
| | 2 | 0.26 |
| | 3 | 0.39 |
| 5 tonne excavator | 1 | 0.28 |
| | 2 | 0.56 |
| | 3 | 0.84 |
| 7 tonne excavator | 1 | 0.28 |
| | 2 | 0.56 |
| | 3 | 0.84 |
| 21 tonne excavator | 1 | 1.21 |
| | 2 | 2.42 |
| | 3 | 3.63 |
| Backhoe loader JCB3CX excavator Rear bucket capacity 0.28 m³ | 1 | 0.28 |
| | 2 | 0.56 |
| | 3 | 0.84 |
| Backhoe loader JCB3CX loading Front bucket capacity 1.00 m³ | 1 | 1.00 |
| | 2 | 2.00 |

## Machine Volumes for Excavating and Filling

| Machine type | Loads per hour | Volume per hour (m³) |
|---|---|---|
| 1 tonne high tip skip loader Volume 0.485 m³ | 5 | 2.43 |
| | 7 | 3.40 |
| | 10 | 4.85 |
| 3 tonne dumper Max volume 2.40 m³ Available volume 1.9 m³ | 4 | 7.60 |
| | 5 | 9.50 |
| | 7 | 13.30 |
| | 10 | 19.00 |
| 6 tonne dumper Max volume 3.40 m³ Available volume 3.77 m³ | 4 | 15.08 |
| | 5 | 18.85 |
| | 7 | 26.39 |
| | 10 | 37.70 |

# EARTHWORK

## Bulkage of Soils (after excavation)

| Type of soil | Approximate bulking of 1 m³ after excavation |
|---|---|
| Vegetable soil and loam | 25–30% |
| Soft clay | 30–40% |
| Stiff clay | 10–20% |
| Gravel | 20–25% |
| Sand | 40–50% |
| Chalk | 40–50% |
| Rock, weathered | 30–40% |
| Rock, unweathered | 50–60% |

## Shrinkage of Materials (on being deposited)

| Type of soil | Approximate bulking of 1 m³ after excavation |
|---|---|
| Clay | 10% |
| Gravel | 8% |
| Gravel and sand | 9% |
| Loam and light sandy soils | 12% |
| Loose vegetable soils | 15% |

## Voids in Material Used as Subbases or Beddings

| Material | m³ of voids/m³ |
|---|---|
| Alluvium | 0.37 |
| River grit | 0.29 |
| Quarry sand | 0.24 |
| Shingle | 0.37 |
| Gravel | 0.39 |
| Broken stone | 0.45 |
| Broken bricks | 0.42 |

## Angles of Repose

| Type of soil | | Degrees |
|---|---|---|
| Clay | – dry | 30 |
| | – damp, well drained | 45 |
| | – wet | 15–20 |
| Earth | – dry | 30 |
| | – damp | 45 |
| Gravel | – moist | 48 |
| Sand | – dry or moist | 35 |
| | – wet | 25 |
| Loam | | 40 |

# EARTHWORK

## Slopes and Angles

| Ratio of base to height | Angle in degrees |
|---|---|
| 5:1 | 11 |
| 4:1 | 14 |
| 3:1 | 18 |
| 2:1 | 27 |
| 1½:1 | 34 |
| 1:1 | 45 |
| 1:1½ | 56 |
| 1:2 | 63 |
| 1:3 | 72 |
| 1:4 | 76 |
| 1:5 | 79 |

## Grades (in Degrees and Percents)

| Degrees | Percent | Degrees | Percent |
|---|---|---|---|
| 1 | 1.8 | 24 | 44.5 |
| 2 | 3.5 | 25 | 46.6 |
| 3 | 5.2 | 26 | 48.8 |
| 4 | 7.0 | 27 | 51.0 |
| 5 | 8.8 | 28 | 53.2 |
| 6 | 10.5 | 29 | 55.4 |
| 7 | 12.3 | 30 | 57.7 |
| 8 | 14.0 | 31 | 60.0 |
| 9 | 15.8 | 32 | 62.5 |
| 10 | 17.6 | 33 | 64.9 |
| 11 | 19.4 | 34 | 67.4 |
| 12 | 21.3 | 35 | 70.0 |
| 13 | 23.1 | 36 | 72.7 |
| 14 | 24.9 | 37 | 75.4 |
| 15 | 26.8 | 38 | 78.1 |
| 16 | 28.7 | 39 | 81.0 |
| 17 | 30.6 | 40 | 83.9 |
| 18 | 32.5 | 41 | 86.9 |
| 19 | 34.4 | 42 | 90.0 |
| 20 | 36.4 | 43 | 93.3 |
| 21 | 38.4 | 44 | 96.6 |
| 22 | 40.4 | 45 | 100.0 |
| 23 | 42.4 | | |

**EARTHWORK**

## Bearing Powers

| Ground conditions | | Bearing power | | |
|---|---|---|---|---|
| | | kg/m² | lb/in² | Metric t/m² |
| Rock, | broken | 483 | 70 | 50 |
| | solid | 2415 | 350 | 240 |
| Clay, | dry or hard | 380 | 55 | 40 |
| | medium dry | 190 | 27 | 20 |
| | soft or wet | 100 | 14 | 10 |
| Gravel, | cemented | 760 | 110 | 80 |
| Sand, | compacted | 380 | 55 | 40 |
| | clean dry | 190 | 27 | 20 |
| Swamp and alluvial soils | | 48 | 7 | 5 |

## Earthwork Support

### Maximum depth of excavation in various soils without the use of earthwork support

| Ground conditions | Feet (ft) | Metres (m) |
|---|---|---|
| Compact soil | 12 | 3.66 |
| Drained loam | 6 | 1.83 |
| Dry sand | 1 | 0.3 |
| Gravelly earth | 2 | 0.61 |
| Ordinary earth | 3 | 0.91 |
| Stiff clay | 10 | 3.05 |

It is important to note that the above table should only be used as a guide. Each case must be taken on its merits and, as the limited distances given above are approached, careful watch must be kept for the slightest signs of caving in

## CONCRETE WORK

### Weights of Concrete and Concrete Elements

| Type of material | | kg/m³ | lb/cu ft |
|---|---|---|---|
| **Ordinary concrete (dense aggregates)** | | | |
| Non-reinforced plain or mass concrete | | | |
| Nominal weight | | 2305 | 144 |
| Aggregate | – limestone | 2162 to 2407 | 135 to 150 |
| | – gravel | 2244 to 2407 | 140 to 150 |
| | – broken brick | 2000 (av) | 125 (av) |
| | – other crushed stone | 2326 to 2489 | 145 to 155 |
| Reinforced concrete | | | |
| Nominal weight | | 2407 | 150 |
| Reinforcement | – 1% | 2305 to 2468 | 144 to 154 |
| | – 2% | 2356 to 2519 | 147 to 157 |
| | – 4% | 2448 to 2703 | 153 to 163 |
| **Special concretes** | | | |
| Heavy concrete | | | |
| Aggregates | – barytes, magnetite | 3210 (min) | 200 (min) |
| | – steel shot, punchings | 5280 | 330 |
| Lean mixes | | | |
| Dry-lean (gravel aggregate) | | 2244 | 140 |
| Soil-cement (normal mix) | | 1601 | 100 |

## CONCRETE WORK

| Type of material | | kg/m² per mm thick | lb/sq ft per inch thick |
|---|---|---|---|
| **Ordinary concrete (dense aggregates)** | | | |
| Solid slabs (floors, walls etc.) | | | |
| Thickness: | 75 mm or 3 in | 184 | 37.5 |
| | 100 mm or 4 in | 245 | 50 |
| | 150 mm or 6 in | 378 | 75 |
| | 250 mm or 10 in | 612 | 125 |
| | 300 mm or 12 in | 734 | 150 |
| Ribbed slabs | | | |
| Thickness: | 125 mm or 5 in | 204 | 42 |
| | 150 mm or 6 in | 219 | 45 |
| | 225 mm or 9 in | 281 | 57 |
| | 300 mm or 12 in | 342 | 70 |
| **Special concretes** | | | |
| Finishes etc. | | | |
| | Rendering, screed etc. Granolithic, terrazzo | 1928 to 2401 | 10 to 12.5 |
| | Glass-block (hollow) concrete | 1734 (approx) | 9 (approx) |
| Prestressed concrete | | Weights as for reinforced concrete (upper limits) | |
| Air-entrained concrete | | Weights as for plain or reinforced concrete | |

# CONCRETE WORK

## Average Weight of Aggregates

| Materials | Voids % | Weight kg/m³ |
|---|---|---|
| Sand | 39 | 1660 |
| Gravel 10–20 mm | 45 | 1440 |
| Gravel 35–75 mm | 42 | 1555 |
| Crushed stone | 50 | 1330 |
| Crushed granite (over 15 mm) | 50 | 1345 |
| (n.e. 15 mm) | 47 | 1440 |
| 'All-in' ballast | 32 | 1800–2000 |

| Material | kg/m³ | lb/cu yd |
|---|---|---|
| Vermiculite (aggregate) | 64–80 | 108–135 |
| All-in aggregate | 1999 | 125 |

## Applications and Mix Design

### Site mixed concrete

| Recommended mix | Class of work suitable for | Cement (kg) | Sand (kg) | Coarse aggregate (kg) | Nr 25 kg bags cement per m³ of combined aggregate |
|---|---|---|---|---|---|
| 1:3:6 | Roughest type of mass concrete such as footings, road haunching over 300 mm thick | 208 | 905 | 1509 | 8.30 |
| 1:2.5:5 | Mass concrete of better class than 1:3:6 such as bases for machinery, walls below ground etc. | 249 | 881 | 1474 | 10.00 |
| 1:2:4 | Most ordinary uses of concrete, such as mass walls above ground, road slabs etc. and general reinforced concrete work | 304 | 889 | 1431 | 12.20 |
| 1:1.5:3 | Watertight floors, pavements and walls, tanks, pits, steps, paths, surface of 2 course roads, reinforced concrete where extra strength is required | 371 | 801 | 1336 | 14.90 |
| 1:1:2 | Works of thin section such as fence posts and small precast work | 511 | 720 | 1206 | 20.40 |

## CONCRETE WORK

### Ready mixed concrete

| Application | Designated concrete | Standardized prescribed concrete | Recommended consistence (nominal slump class) |
|---|---|---|---|
| **Foundations** | | | |
| Mass concrete fill or blinding | GEN 1 | ST2 | S3 |
| Strip footings | GEN 1 | ST2 | S3 |
| **Mass concrete foundations** | | | |
| Single storey buildings | GEN 1 | ST2 | S3 |
| Double storey buildings | GEN 3 | ST4 | S3 |
| **Trench fill foundations** | | | |
| Single storey buildings | GEN 1 | ST2 | S4 |
| Double storey buildings | GEN 3 | ST4 | S4 |
| **General applications** | | | |
| Kerb bedding and haunching | GEN 0 | ST1 | S1 |
| Drainage works – immediate support | GEN 1 | ST2 | S1 |
| Other drainage works | GEN 1 | ST2 | S3 |
| Oversite below suspended slabs | GEN 1 | ST2 | S3 |
| **Floors** | | | |
| Garage and house floors with no embedded steel | GEN 3 | ST4 | S2 |
| Wearing surface: Light foot and trolley traffic | RC30 | ST4 | S2 |
| Wearing surface: General industrial | RC40 | N/A | S2 |
| Wearing surface: Heavy industrial | RC50 | N/A | S2 |
| **Paving** | | | |
| House drives, domestic parking and external parking | PAV 1 | N/A | S2 |
| Heavy-duty external paving | PAV 2 | N/A | S2 |

# CONCRETE WORK

## Prescribed Mixes for Ordinary Structural Concrete

Weights of cement and total dry aggregates in kg to produce approximately one cubic metre of fully compacted concrete together with the percentages by weight of fine aggregate in total dry aggregates

| Conc. grade | Nominal max size of aggregate (mm) | 40 | | 20 | | 14 | | 10 | |
|---|---|---|---|---|---|---|---|---|---|
| | Workability | Med. | High | Med. | High | Med. | High | Med. | High |
| | Limits to slump that may be expected (mm) | 50–100 | 100–150 | 25–75 | 75–125 | 10–50 | 50–100 | 10–25 | 25–50 |
| 7 | Cement (kg) | 180 | 200 | 210 | 230 | – | – | – | – |
| | Total aggregate (kg) | 1950 | 1850 | 1900 | 1800 | – | – | – | – |
| | Fine aggregate (%) | 30–45 | 30–45 | 35–50 | 35–50 | – | – | – | – |
| 10 | Cement (kg) | 210 | 230 | 240 | 260 | – | – | – | – |
| | Total aggregate (kg) | 1900 | 1850 | 1850 | 1800 | – | – | – | – |
| | Fine aggregate (%) | 30–45 | 30–45 | 35–50 | 35–50 | – | – | – | – |
| 15 | Cement (kg) | 250 | 270 | 280 | 310 | – | – | – | – |
| | Total aggregate (kg) | 1850 | 1800 | 1800 | 1750 | – | – | – | – |
| | Fine aggregate (%) | 30–45 | 30–45 | 35–50 | 35–50 | – | – | – | – |
| 20 | Cement (kg) | 300 | 320 | 320 | 350 | 340 | 380 | 360 | 410 |
| | Total aggregate (kg) | 1850 | 1750 | 1800 | 1750 | 1750 | 1700 | 1750 | 1650 |
| | Sand | | | | | | | | |
| | Zone 1 (%) | 35 | 40 | 40 | 45 | 45 | 50 | 50 | 55 |
| | Zone 2 (%) | 30 | 35 | 35 | 40 | 40 | 45 | 45 | 50 |
| | Zone 3 (%) | 30 | 30 | 30 | 35 | 35 | 40 | 40 | 45 |
| 25 | Cement (kg) | 340 | 360 | 360 | 390 | 380 | 420 | 400 | 450 |
| | Total aggregate (kg) | 1800 | 1750 | 1750 | 1700 | 1700 | 1650 | 1700 | 1600 |
| | Sand | | | | | | | | |
| | Zone 1 (%) | 35 | 40 | 40 | 45 | 45 | 50 | 50 | 55 |
| | Zone 2 (%) | 30 | 35 | 35 | 40 | 40 | 45 | 45 | 50 |
| | Zone 3 (%) | 30 | 30 | 30 | 35 | 35 | 40 | 40 | 45 |
| 30 | Cement (kg) | 370 | 390 | 400 | 430 | 430 | 470 | 460 | 510 |
| | Total aggregate (kg) | 1750 | 1700 | 1700 | 1650 | 1700 | 1600 | 1650 | 1550 |
| | Sand | | | | | | | | |
| | Zone 1 (%) | 35 | 40 | 40 | 45 | 45 | 50 | 50 | 55 |
| | Zone 2 (%) | 30 | 35 | 35 | 40 | 40 | 45 | 45 | 50 |
| | Zone 3 (%) | 30 | 30 | 30 | 35 | 35 | 40 | 40 | 45 |

## REINFORCEMENT

### REINFORCEMENT

#### Weights of Bar Reinforcement

| Nominal sizes (mm) | Cross-sectional area (mm²) | Mass (kg/m) | Length of bar (m/tonne) |
|---|---|---|---|
| 6 | 28.27 | 0.222 | 4505 |
| 8 | 50.27 | 0.395 | 2534 |
| 10 | 78.54 | 0.617 | 1622 |
| 12 | 113.10 | 0.888 | 1126 |
| 16 | 201.06 | 1.578 | 634 |
| 20 | 314.16 | 2.466 | 405 |
| 25 | 490.87 | 3.853 | 260 |
| 32 | 804.25 | 6.313 | 158 |
| 40 | 1265.64 | 9.865 | 101 |
| 50 | 1963.50 | 15.413 | 65 |

#### Weights of Bars (at specific spacings)

#### Weights of metric bars in kilogrammes per square metre

| Size (mm) | Spacing of bars in millimetres | | | | | | | | | |
|---|---|---|---|---|---|---|---|---|---|---|
| | 75 | 100 | 125 | 150 | 175 | 200 | 225 | 250 | 275 | 300 |
| 6 | 2.96 | 2.220 | 1.776 | 1.480 | 1.27 | 1.110 | 0.99 | 0.89 | 0.81 | 0.74 |
| 8 | 5.26 | 3.95 | 3.16 | 2.63 | 2.26 | 1.97 | 1.75 | 1.58 | 1.44 | 1.32 |
| 10 | 8.22 | 6.17 | 4.93 | 4.11 | 3.52 | 3.08 | 2.74 | 2.47 | 2.24 | 2.06 |
| 12 | 11.84 | 8.88 | 7.10 | 5.92 | 5.07 | 4.44 | 3.95 | 3.55 | 3.23 | 2.96 |
| 16 | 21.04 | 15.78 | 12.63 | 10.52 | 9.02 | 7.89 | 7.02 | 6.31 | 5.74 | 5.26 |
| 20 | 32.88 | 24.66 | 19.73 | 16.44 | 14.09 | 12.33 | 10.96 | 9.87 | 8.97 | 8.22 |
| 25 | 51.38 | 38.53 | 30.83 | 25.69 | 22.02 | 19.27 | 17.13 | 15.41 | 14.01 | 12.84 |
| 32 | 84.18 | 63.13 | 50.51 | 42.09 | 36.08 | 31.57 | 28.06 | 25.25 | 22.96 | 21.04 |
| 40 | 131.53 | 98.65 | 78.92 | 65.76 | 56.37 | 49.32 | 43.84 | 39.46 | 35.87 | 32.88 |
| 50 | 205.51 | 154.13 | 123.31 | 102.76 | 88.08 | 77.07 | 68.50 | 61.65 | 56.05 | 51.38 |

Basic weight of steelwork taken as 7850 kg/m³
Basic weight of bar reinforcement per metre run = 0.00785 kg/mm²
The value of $\pi$ has been taken as 3.141592654

## REINFORCEMENT

### Fabric Reinforcement

**Preferred range of designated fabric types and stock sheet sizes**

| Fabric reference | Longitudinal wires | | | Cross wires | | | Mass |
|---|---|---|---|---|---|---|---|
| | Nominal wire size (mm) | Pitch (mm) | Area (mm²/m) | Nominal wire size (mm) | Pitch (mm) | Area (mm²/m) | (kg/m²) |
| Square mesh | | | | | | | |
| A393 | 10 | 200 | 393 | 10 | 200 | 393 | 6.16 |
| A252 | 8 | 200 | 252 | 8 | 200 | 252 | 3.95 |
| A193 | 7 | 200 | 193 | 7 | 200 | 193 | 3.02 |
| A142 | 6 | 200 | 142 | 6 | 200 | 142 | 2.22 |
| A98 | 5 | 200 | 98 | 5 | 200 | 98 | 1.54 |
| Structural mesh | | | | | | | |
| B1131 | 12 | 100 | 1131 | 8 | 200 | 252 | 10.90 |
| B785 | 10 | 100 | 785 | 8 | 200 | 252 | 8.14 |
| B503 | 8 | 100 | 503 | 8 | 200 | 252 | 5.93 |
| B385 | 7 | 100 | 385 | 7 | 200 | 193 | 4.53 |
| B283 | 6 | 100 | 283 | 7 | 200 | 193 | 3.73 |
| B196 | 5 | 100 | 196 | 7 | 200 | 193 | 3.05 |
| Long mesh | | | | | | | |
| C785 | 10 | 100 | 785 | 6 | 400 | 70.8 | 6.72 |
| C636 | 9 | 100 | 636 | 6 | 400 | 70.8 | 5.55 |
| C503 | 8 | 100 | 503 | 5 | 400 | 49.0 | 4.34 |
| C385 | 7 | 100 | 385 | 5 | 400 | 49.0 | 3.41 |
| C283 | 6 | 100 | 283 | 5 | 400 | 49.0 | 2.61 |
| Wrapping mesh | | | | | | | |
| D98 | 5 | 200 | 98 | 5 | 200 | 98 | 1.54 |
| D49 | 2.5 | 100 | 49 | 2.5 | 100 | 49 | 0.77 |

Stock sheet size 4.8 m × 2.4 m, Area 11.52 m²

### Average weight kg/m³ of steelwork reinforcement in concrete for various building elements

| Substructure | kg/m³ concrete | Substructure | kg/m³ concrete |
|---|---|---|---|
| Pile caps | 110–150 | Plate slab | 150–220 |
| Tie beams | 130–170 | Cant slab | 145–210 |
| Ground beams | 230–330 | Ribbed floors | 130–200 |
| Bases | 125–180 | Topping to block floor | 30–40 |
| Footings | 100–150 | Columns | 210–310 |
| Retaining walls | 150–210 | Beams | 250–350 |
| Raft | 60–70 | Stairs | 130–170 |
| Slabs – one way | 120–200 | Walls – normal | 40–100 |
| Slabs – two way | 110–220 | Walls – wind | 70–125 |

Note: For exposed elements add the following %:
Walls 50%, Beams 100%, Columns 15%

## FORMWORK

### Formwork Stripping Times – Normal Curing Periods

| Conditions under which concrete is maturing | Minimum periods of protection for different types of cement | | | | | |
|---|---|---|---|---|---|---|
| | Number of days (where the average surface temperature of the concrete exceeds 10°C during the whole period) | | | Equivalent maturity (degree hours) calculated as the age of the concrete in hours multiplied by the number of degrees Celsius by which the average surface temperature of the concrete exceeds 10°C | | |
| | Other | SRPC | OPC or RHPC | Other | SRPC | OPC or RHPC |
| 1. Hot weather or drying winds | 7 | 4 | 3 | 3500 | 2000 | 1500 |
| 2. Conditions not covered by 1 | 4 | 3 | 2 | 2000 | 1500 | 1000 |

**KEY**
OPC – Ordinary Portland Cement
RHPC – Rapid-hardening Portland Cement
SRPC – Sulphate-resisting Portland Cement

### Minimum Period before Striking Formwork

| | Minimum period before striking | | |
|---|---|---|---|
| | Surface temperature of concrete | | |
| | 16°C | 17°C | t°C (0–25) |
| Vertical formwork to columns, walls and large beams | 12 hours | 18 hours | 300 hours t+10 |
| Soffit formwork to slabs | 4 days | 6 days | 100 days t+10 |
| Props to slabs | 10 days | 15 days | 250 days t+10 |
| Soffit formwork to beams | 9 days | 14 days | 230 days t+10 |
| Props to beams | 14 days | 21 days | 360 days t+10 |

## MASONRY

### Number of Bricks Required for Various Types of Work per m² of Walling

| Description | Brick size | |
|---|---|---|
| | 215 × 102.5 × 50 mm | 215 × 102.5 × 65 mm |
| **Half brick thick** | | |
| Stretcher bond | 74 | 59 |
| English bond | 108 | 86 |
| English garden wall bond | 90 | 72 |
| Flemish bond | 96 | 79 |
| Flemish garden wall bond | 83 | 66 |
| | | |
| **One brick thick and cavity wall of two half brick skins** | | |
| Stretcher bond | 148 | 119 |

### Quantities of Bricks and Mortar Required per m² of Walling

| | Unit | No of bricks required | Mortar required (cubic metres) | | |
|---|---|---|---|---|---|
| | | | No frogs | Single frogs | Double frogs |
| **Standard bricks** | | | | | |
| Brick size 215 × 102.5 × 50 mm | | | | | |
| half brick wall (103 mm) | m² | 72 | 0.022 | 0.027 | 0.032 |
| 2 × half brick cavity wall (270 mm) | m² | 144 | 0.044 | 0.054 | 0.064 |
| one brick wall (215 mm) | m² | 144 | 0.052 | 0.064 | 0.076 |
| one and a half brick wall (322 mm) | m² | 216 | 0.073 | 0.091 | 0.108 |
| Mass brickwork | m³ | 576 | 0.347 | 0.413 | 0.480 |
| | | | | | |
| Brick size 215 × 102.5 × 65 mm | | | | | |
| half brick wall (103 mm) | m² | 58 | 0.019 | 0.022 | 0.026 |
| 2 × half brick cavity wall (270 mm) | m² | 116 | 0.038 | 0.045 | 0.055 |
| one brick wall (215 mm) | m² | 116 | 0.046 | 0.055 | 0.064 |
| one and a half brick wall (322 mm) | m² | 174 | 0.063 | 0.074 | 0.088 |
| Mass brickwork | m³ | 464 | 0.307 | 0.360 | 0.413 |
| **Metric modular bricks** | | | Perforated | | |
| Brick size 200 × 100 × 75 mm | | | | | |
| 90 mm thick | m² | 67 | 0.016 | 0.019 | |
| 190 mm thick | m² | 133 | 0.042 | 0.048 | |
| 290 mm thick | m² | 200 | 0.068 | 0.078 | |
| | | | | | |
| Brick size 200 × 100 × 100 mm | | | | | |
| 90 mm thick | m² | 50 | 0.013 | 0.016 | |
| 190 mm thick | m² | 100 | 0.036 | 0.041 | |
| 290 mm thick | m² | 150 | 0.059 | 0.067 | |
| | | | | | |
| Brick size 300 × 100 × 75 mm | | | | | |
| 90 mm thick | m² | 33 | – | 0.015 | |
| | | | | | |
| Brick size 300 × 100 × 100 mm | | | | | |
| 90 mm thick | m² | 44 | 0.015 | 0.018 | |

Note: Assuming 10 mm thick joints

## MASONRY

### Mortar Required per m² Blockwork (9.88 blocks/m²)

| Wall thickness | 75 | 90 | 100 | 125 | 140 | 190 | 215 |
|---|---|---|---|---|---|---|---|
| Mortar m³/m² | 0.005 | 0.006 | 0.007 | 0.008 | 0.009 | 0.013 | 0.014 |

| Mortar Group | Cement: lime: sand | Masonry cement: sand | Cement: sand with plasticizer |
|---|---|---|---|
| 1 | 1:0–0.25:3 | | |
| 2 | 1:0.5:4–4.5 | 1:2.5-3.5 | 1:3–4 |
| 3 | 1:1:5–6 | 1:4–5 | 1:5–6 |
| 4 | 1:2:8–9 | 1:5.5–6.5 | 1:7–8 |
| 5 | 1:3:10–12 | 1:6.5–7 | 1:8 |

Group 1: strong inflexible mortar
Group 5: weak but flexible

All mixes within a group are of approximately similar strength
Frost resistance increases with the use of plasticizers
Cement: lime: sand mixes give the strongest bond and greatest resistance to rain penetration
Masonry cement equals ordinary Portland cement plus a fine neutral mineral filler and an air entraining agent

### Calcium Silicate Bricks

| Type | Strength | Location |
|---|---|---|
| Class 2 crushing strength | 14.0 N/mm² | not suitable for walls |
| Class 3 | 20.5 N/mm² | walls above dpc |
| Class 4 | 27.5 N/mm² | cappings and copings |
| Class 5 | 34.5 N/mm² | retaining walls |
| Class 6 | 41.5 N/mm² | walls below ground |
| Class 7 | 48.5 N/mm² | walls below ground |

The Class 7 calcium silicate bricks are therefore equal in strength to Class B bricks
Calcium silicate bricks are not suitable for DPCs

| Durability of Bricks | |
|---|---|
| FL | Frost resistant with low salt content |
| FN | Frost resistant with normal salt content |
| ML | Moderately frost resistant with low salt content |
| MN | Moderately frost resistant with normal salt content |

## MASONRY

### Brickwork Dimensions

| No. of horizontal bricks | Dimensions (mm) | No. of vertical courses | Height of vertical courses (mm) |
|---|---|---|---|
| ½ | 112.5 | 1 | 75 |
| 1 | 225.0 | 2 | 150 |
| 1½ | 337.5 | 3 | 225 |
| 2 | 450.0 | 4 | 300 |
| 2½ | 562.5 | 5 | 375 |
| 3 | 675.0 | 6 | 450 |
| 3½ | 787.5 | 7 | 525 |
| 4 | 900.0 | 8 | 600 |
| 4½ | 1012.5 | 9 | 675 |
| 5 | 1125.0 | 10 | 750 |
| 5½ | 1237.5 | 11 | 825 |
| 6 | 1350.0 | 12 | 900 |
| 6½ | 1462.5 | 13 | 975 |
| 7 | 1575.0 | 14 | 1050 |
| 7½ | 1687.5 | 15 | 1125 |
| 8 | 1800.0 | 16 | 1200 |
| 8½ | 1912.5 | 17 | 1275 |
| 9 | 2025.0 | 18 | 1350 |
| 9½ | 2137.5 | 19 | 1425 |
| 10 | 2250.0 | 20 | 1500 |
| 20 | 4500.0 | 24 | 1575 |
| 40 | 9000.0 | 28 | 2100 |
| 50 | 11250.0 | 32 | 2400 |
| 60 | 13500.0 | 36 | 2700 |
| 75 | 16875.0 | 40 | 3000 |

## TIMBER

# TIMBER

## Weights of Timber

| Material | kg/m³ | lb/cu ft |
|---|---|---|
| *General* | 806 (avg) | 50 (avg) |
| Douglas fir | 479 | 30 |
| Yellow pine, spruce | 479 | 30 |
| Pitch pine | 673 | 42 |
| Larch, elm | 561 | 35 |
| Oak (English) | 724 to 959 | 45 to 60 |
| Teak | 643 to 877 | 40 to 55 |
| Jarrah | 959 | 60 |
| Greenheart | 1040 to 1204 | 65 to 75 |
| Quebracho | 1285 | 80 |
| **Material** | kg/m² per mm thickness | lb/sq ft per inch thickness |
| *Wooden boarding and blocks* | | |
| Softwood | 0.48 | 2.5 |
| Hardwood | 0.76 | 4 |
| Hardboard | 1.06 | 5.5 |
| Chipboard | 0.76 | 4 |
| Plywood | 0.62 | 3.25 |
| Blockboard | 0.48 | 2.5 |
| Fibreboard | 0.29 | 1.5 |
| Wood-wool | 0.58 | 3 |
| Plasterboard | 0.96 | 5 |
| Weather boarding | 0.35 | 1.8 |

# TIMBER

## Conversion Tables (for timber only)

| Inches | Millimetres | Feet | Metres |
|---|---|---|---|
| 1 | 25 | 1 | 0.300 |
| 2 | 50 | 2 | 0.600 |
| 3 | 75 | 3 | 0.900 |
| 4 | 100 | 4 | 1.200 |
| 5 | 125 | 5 | 1.500 |
| 6 | 150 | 6 | 1.800 |
| 7 | 175 | 7 | 2.100 |
| 8 | 200 | 8 | 2.400 |
| 9 | 225 | 9 | 2.700 |
| 10 | 250 | 10 | 3.000 |
| 11 | 275 | 11 | 3.300 |
| 12 | 300 | 12 | 3.600 |
| 13 | 325 | 13 | 3.900 |
| 14 | 350 | 14 | 4.200 |
| 15 | 375 | 15 | 4.500 |
| 16 | 400 | 16 | 4.800 |
| 17 | 425 | 17 | 5.100 |
| 18 | 450 | 18 | 5.400 |
| 19 | 475 | 19 | 5.700 |
| 20 | 500 | 20 | 6.000 |
| 21 | 525 | 21 | 6.300 |
| 22 | 550 | 22 | 6.600 |
| 23 | 575 | 23 | 6.900 |
| 24 | 600 | 24 | 7.200 |

## Planed Softwood

The finished end section size of planed timber is usually 3/16" less than the original size from which it is produced. This however varies slightly depending upon availability of material and origin of the species used.

## Standards (timber) to cubic metres and cubic metres to standards (timber)

| Cubic metres | Cubic metres standards | Standards |
|---|---|---|
| 4.672 | 1 | 0.214 |
| 9.344 | 2 | 0.428 |
| 14.017 | 3 | 0.642 |
| 18.689 | 4 | 0.856 |
| 23.361 | 5 | 1.070 |
| 28.033 | 6 | 1.284 |
| 32.706 | 7 | 1.498 |
| 37.378 | 8 | 1.712 |
| 42.050 | 9 | 1.926 |
| 46.722 | 10 | 2.140 |
| 93.445 | 20 | 4.281 |
| 140.167 | 30 | 6.421 |
| 186.890 | 40 | 8.561 |
| 233.612 | 50 | 10.702 |
| 280.335 | 60 | 12.842 |
| 327.057 | 70 | 14.982 |
| 373.779 | 80 | 17.122 |

# TIMBER

1 cu metre = 35.3148 cu ft = 0.21403 std

1 cu ft = 0.028317 cu metres

1 std = 4.67227 cu metres

## Basic sizes of sawn softwood available (cross-sectional areas)

| Thickness (mm) | Width (mm) | | | | | | | | |
|---|---|---|---|---|---|---|---|---|---|
| | 75 | 100 | 125 | 150 | 175 | 200 | 225 | 250 | 300 |
| 16 | X | X | X | X | | | | | |
| 19 | X | X | X | X | | | | | |
| 22 | X | X | X | X | | | | | |
| 25 | X | X | X | X | X | X | X | X | X |
| 32 | X | X | X | X | X | X | X | X | X |
| 36 | X | X | X | X | | | | | |
| 38 | X | X | X | X | X | X | X | | |
| 44 | X | X | X | X | X | X | X | X | X |
| 47* | X | X | X | X | X | X | X | X | X |
| 50 | X | X | X | X | X | X | X | X | X |
| 63 | X | X | X | X | X | X | X | | |
| 75 | X | X | X | X | X | X | X | X | |
| 100 | | X | | X | | X | | X | X |
| 150 | | | | X | | X | | | X |
| 200 | | | | | | X | | | |
| 250 | | | | | | | | X | |
| 300 | | | | | | | | | X |

* This range of widths for 47 mm thickness will usually be found to be available in construction quality only

**Note:** The smaller sizes below 100 mm thick and 250 mm width are normally but not exclusively of European origin. Sizes beyond this are usually of North and South American origin

## Basic lengths of sawn softwood available (metres)

| | | | | | | |
|---|---|---|---|---|---|---|
| 1.80 | 2.10 | 3.00 | 4.20 | 5.10 | 6.00 | 7.20 |
| | 2.40 | 3.30 | 4.50 | 5.40 | 6.30 | |
| | 2.70 | 3.60 | 4.80 | 5.70 | 6.60 | |
| | | 3.90 | | | 6.90 | |

**Note:** Lengths of 6.00 m and over will generally only be available from North American species and may have to be recut from larger sizes

## TIMBER

### Reductions from basic size to finished size by planning of two opposed faces

| Purpose | | Reductions from basic sizes for timber | | | |
|---|---|---|---|---|---|
| | | 15–35 mm | 36–100 mm | 101–150 mm | over 150 mm |
| a) | Constructional timber | 3 mm | 3 mm | 5 mm | 6 mm |
| b) | Matching interlocking boards | 4 mm | 4 mm | 6 mm | 6 mm |
| c) | Wood trim not specified in BS 584 | 5 mm | 7 mm | 7 mm | 9 mm |
| d) | Joinery and cabinet work | 7 mm | 9 mm | 11 mm | 13 mm |

**Note:** The reduction of width or depth is overall the extreme size and is exclusive of any reduction of the face by the machining of a tongue or lap joints

### Maximum Spans for Various Roof Trusses

### Maximum permissible spans for rafters for Fink trussed rafters

| Basic size (mm) | Actual size (mm) | Pitch (degrees) | | | | | | | | |
|---|---|---|---|---|---|---|---|---|---|---|
| | | 15 (m) | 17.5 (m) | 20 (m) | 22.5 (m) | 25 (m) | 27.5 (m) | 30 (m) | 32.5 (m) | 35 (m) |
| 38 × 75 | 35 × 72 | 6.03 | 6.16 | 6.29 | 6.41 | 6.51 | 6.60 | 6.70 | 6.80 | 6.90 |
| 38 × 100 | 35 × 97 | 7.48 | 7.67 | 7.83 | 7.97 | 8.10 | 8.22 | 8.34 | 8.47 | 8.61 |
| 38 × 125 | 35 × 120 | 8.80 | 9.00 | 9.20 | 9.37 | 9.54 | 9.68 | 9.82 | 9.98 | 10.16 |
| 44 × 75 | 41 × 72 | 6.45 | 6.59 | 6.71 | 6.83 | 6.93 | 7.03 | 7.14 | 7.24 | 7.35 |
| 44 × 100 | 41 × 97 | 8.05 | 8.23 | 8.40 | 8.55 | 8.68 | 8.81 | 8.93 | 9.09 | 9.22 |
| 44 × 125 | 41 × 120 | 9.38 | 9.60 | 9.81 | 9.99 | 10.15 | 10.31 | 10.45 | 10.64 | 10.81 |
| 50 × 75 | 47 × 72 | 6.87 | 7.01 | 7.13 | 7.25 | 7.35 | 7.45 | 7.53 | 7.67 | 7.78 |
| 50 × 100 | 47 × 97 | 8.62 | 8.80 | 8.97 | 9.12 | 9.25 | 9.38 | 9.50 | 9.66 | 9.80 |
| 50 × 125 | 47 × 120 | 10.01 | 10.24 | 10.44 | 10.62 | 10.77 | 10.94 | 11.00 | 11.00 | 11.00 |

**TIMBER**

## Sizes of Internal and External Doorsets

| Description | Internal size (mm) | Permissible deviation | External size (mm) | Permissible deviation |
|---|---|---|---|---|
| Coordinating dimension: height of door leaf height sets | 2100 | | 2100 | |
| Coordinating dimension: height of ceiling height set | 2300<br>2350<br>2400<br>2700<br>3000 | | 2300<br>2350<br>2400<br>2700<br>3000 | |
| Coordinating dimension: width of all doorsets<br><br>S = Single leaf set<br>D = Double leaf set | 600 S<br>700 S<br>800 S&D<br>900 S&D<br>1000 S&D<br>1200 D<br>1500 D<br>1800 D<br>2100 D | | 900 S<br>1000 S<br>1200 D<br><br>1800 D<br>2100 D | |
| Work size: height of door leaf height set | 2090 | ± 2.0 | 2095 | ± 2.0 |
| Work size: height of ceiling height set | 2285<br>2335<br>2385<br>2685<br>2985 | ± 2.0 | 2295<br>2345<br>2395<br>2695<br>2995 | ± 2.0 |
| Work size: width of all doorsets<br><br>S = Single leaf set<br>D = Double leaf set | 590 S<br>690 S<br>790 S&D<br>890 S&D<br>990 S&D<br>1190 D<br>1490 D<br>1790 D<br>2090 D | ± 2.0 | 895 S<br>995 S<br>1195 D<br>1495 D<br>1795 D<br>2095 D | ± 2.0 |
| Width of door leaf in single leaf sets<br>F = Flush leaf<br>P = Panel leaf | 526 F<br>626 F<br>726 F&P<br>826 F&P<br>926 F&P | ± 1.5 | 806 F&P<br>906 F&P | ± 1.5 |
| Width of door leaf in double leaf sets<br>F = Flush leaf<br>P = Panel leaf | 362 F<br>412 F<br>426 F<br>562 F&P<br>712 F&P<br>826 F&P<br>1012 F&P | ± 1.5 | 552 F&P<br>702 F&P<br>852 F&P<br>1002 F&P | ± 1.5 |
| Door leaf height for all doorsets | 2040 | ± 1.5 | 1994 | ± 1.5 |

**ROOFING**

## ROOFING

### Total Roof Loadings for Various Types of Tiles/Slates

|  |  | Roof load (slope) kg/m² | | |
|---|---|---|---|---|
|  |  | Slate/Tile | Roofing underlay and battens² | Total dead load kg/m |
| Asbestos cement slate (600 × 300) |  | 21.50 | 3.14 | 24.64 |
| Clay tile | interlocking | 67.00 | 5.50 | 72.50 |
|  | plain | 43.50 | 2.87 | 46.37 |
| Concrete tile | interlocking | 47.20 | 2.69 | 49.89 |
|  | plain | 78.20 | 5.50 | 83.70 |
| Natural slate (18" × 10") |  | 35.40 | 3.40 | 38.80 |
|  |  | Roof load (plan) kg/m² | | |
| Asbestos cement slate (600 × 300) |  | 28.45 | 76.50 | 104.95 |
| Clay tile | interlocking | 53.54 | 76.50 | 130.04 |
|  | plain | 83.71 | 76.50 | 60.21 |
| Concrete tile | interlocking | 57.60 | 76.50 | 134.10 |
|  | plain | 96.64 | 76.50 | 173.14 |

## ROOFING

**Tiling Data**

| Product | | Lap (mm) | Gauge of battens | No. slates per m$^2$ | Battens (m/m$^2$) | Weight as laid (kg/m$^2$) |
|---|---|---|---|---|---|---|
| **CEMENT SLATES** | | | | | | |
| Eternit slates | 600 × 300 mm | 100 | 250 | 13.4 | 4.00 | 19.50 |
| (Duracem) | | 90 | 255 | 13.1 | 3.92 | 19.20 |
| | | 80 | 260 | 12.9 | 3.85 | 19.00 |
| | | 70 | 265 | 12.7 | 3.77 | 18.60 |
| | 600 × 350 mm | 100 | 250 | 11.5 | 4.00 | 19.50 |
| | | 90 | 255 | 11.2 | 3.92 | 19.20 |
| | 500 × 250 mm | 100 | 200 | 20.0 | 5.00 | 20.00 |
| | | 90 | 205 | 19.5 | 4.88 | 19.50 |
| | | 80 | 210 | 19.1 | 4.76 | 19.00 |
| | | 70 | 215 | 18.6 | 4.65 | 18.60 |
| | 400 × 200 mm | 90 | 155 | 32.3 | 6.45 | 20.80 |
| | | 80 | 160 | 31.3 | 6.25 | 20.20 |
| | | 70 | 165 | 30.3 | 6.06 | 19.60 |
| **CONCRETE TILES/SLATES** | | | | | | |
| Redland Roofing | | | | | | |
| Stonewold slate | 430 × 380 mm | 75 | 355 | 8.2 | 2.82 | 51.20 |
| Double Roman tile | 418 × 330 mm | 75 | 355 | 8.2 | 2.91 | 45.50 |
| Grovebury pantile | 418 × 332 mm | 75 | 343 | 9.7 | 2.91 | 47.90 |
| Norfolk pantile | 381 × 227 mm | 75 | 306 | 16.3 | 3.26 | 44.01 |
| | | 100 | 281 | 17.8 | 3.56 | 48.06 |
| Renown interlocking tile | 418 × 330 mm | 75 | 343 | 9.7 | 2.91 | 46.40 |
| '49' tile | 381 × 227 mm | 75 | 306 | 16.3 | 3.26 | 44.80 |
| | | 100 | 281 | 17.8 | 3.56 | 48.95 |
| Plain, vertical tiling | 265 × 165 mm | 35 | 115 | 52.7 | 8.70 | 62.20 |
| | | | | | | |
| Marley Roofing | | | | | | |
| Bold roll tile | 420 × 330 mm | 75 | 344 | 9.7 | 2.90 | 47.00 |
| | | 100 | – | 10.5 | 3.20 | 51.00 |
| Modern roof tile | 420 × 330 mm | 75 | 338 | 10.2 | 3.00 | 54.00 |
| | | 100 | – | 11.0 | 3.20 | 58.00 |
| Ludlow major | 420 × 330 mm | 75 | 338 | 10.2 | 3.00 | 45.00 |
| | | 100 | – | 11.0 | 3.20 | 49.00 |
| Ludlow plus | 387 × 229 mm | 75 | 305 | 16.1 | 3.30 | 47.00 |
| | | 100 | – | 17.5 | 3.60 | 51.00 |
| Mendip tile | 420 × 330 mm | 75 | 338 | 10.2 | 3.00 | 47.00 |
| | | 100 | – | 11.0 | 3.20 | 51.00 |
| Wessex | 413 × 330 mm | 75 | 338 | 10.2 | 3.00 | 54.00 |
| | | 100 | – | 11.0 | 3.20 | 58.00 |
| Plain tile | 267 × 165 mm | 65 | 100 | 60.0 | 10.00 | 76.00 |
| | | 75 | 95 | 64.0 | 10.50 | 81.00 |
| | | 85 | 90 | 68.0 | 11.30 | 86.00 |
| Plain vertical tiles (feature) | 267 × 165 mm | 35 | 110 | 53.0 | 8.70 | 67.00 |
| | | 34 | 115 | 56.0 | 9.10 | 71.00 |

## ROOFING

### Slate Nails, Quantity per Kilogram

| Length | Type | | | |
|---|---|---|---|---|
| | Plain wire | Galvanized wire | Copper nail | Zinc nail |
| 28.5 mm | 325 | 305 | 325 | 415 |
| 34.4 mm | 286 | 256 | 254 | 292 |
| 50.8 mm | 242 | 224 | 194 | 200 |

## Metal Sheet Coverings

### Thicknesses and weights of sheet metal coverings

**Lead to BS 1178**

| BS Code No | 3 | 4 | 5 | 6 | 7 | 8 | | |
|---|---|---|---|---|---|---|---|---|
| Colour code | Green | Blue | Red | Black | White | Orange | | |
| Thickness (mm) | 1.25 | 1.80 | 2.24 | 2.50 | 3.15 | 3.55 | | |
| Density kg/m$^2$ | 14.18 | 20.41 | 25.40 | 30.05 | 35.72 | 40.26 | | |

**Copper to BS 2870**

| | | | | | | | | |
|---|---|---|---|---|---|---|---|---|
| Thickness (mm) | | 0.60 | 0.70 | | | | | |
| Bay width | | | | | | | | |
|   Roll (mm) | | 500 | 650 | | | | | |
|   Seam (mm) | | 525 | 600 | | | | | |
| Standard width to form bay | 600 | 750 | | | | | | |
| Normal length of sheet | 1.80 | 1.80 | | | | | | |

**Zinc to BS 849**

| Zinc Gauge (Nr) | 9 | 10 | 11 | 12 | 13 | 14 | 15 | 16 |
|---|---|---|---|---|---|---|---|---|
| Thickness (mm) | 0.43 | 0.48 | 0.56 | 0.64 | 0.71 | 0.79 | 0.91 | 1.04 |
| Density (kg/m$^2$) | 3.1 | 3.2 | 3.8 | 4.3 | 4.8 | 5.3 | 6.2 | 7.0 |

**Aluminium to BS 4868**

| Thickness (mm) | 0.5 | 0.6 | 0.7 | 0.8 | 0.9 | 1.0 | 1.2 | |
|---|---|---|---|---|---|---|---|---|
| Density (kg/m$^2$) | 12.8 | 15.4 | 17.9 | 20.5 | 23.0 | 25.6 | 30.7 | |

**ROOFING**

| Type of felt | Nominal mass per unit area (kg/10 m) | Nominal mass per unit area of fibre base (g/m$^2$) | Nominal length of roll (m) |
|---|---|---|---|
| **Class 1** | | | |
| 1B fine granule | 14 | 220 | 10 or 20 |
| surfaced bitumen | 18 | 330 | 10 or 20 |
| | 25 | 470 | 10 |
| 1E mineral surfaced bitumen | 38 | 470 | 10 |
| 1F reinforced bitumen | 15 | 160 (fibre) 110 (hessian) | 15 |
| 1F reinforced bitumen, aluminium faced | 13 | 160 (fibre) 110 (hessian) | 15 |
| **Class 2** | | | |
| 2B fine granule surfaced bitumen asbestos | 18 | 500 | 10 or 20 |
| 2E mineral surfaced bitumen asbestos | 38 | 600 | 10 |
| **Class 3** | | | |
| 3B fine granule surfaced bitumen glass fibre | 18 | 60 | 20 |
| 3E mineral surfaced bitumen glass fibre | 28 | 60 | 10 |
| 3E venting base layer bitumen glass fibre | 32 | 60* | 10 |
| 3H venting base layer bitumen glass fibre | 17 | 60* | 20 |

* Excluding effect of perforations

## GLAZING

| Nominal thickness (mm) | Tolerance on thickness (mm) | Approximate weight (kg/m²) | Normal maximum size (mm) |
|---|---|---|---|
| **Float and polished plate glass** | | | |
| 3 | + 0.2 | 7.50 | 2140 × 1220 |
| 4 | + 0.2 | 10.00 | 2760 × 1220 |
| 5 | + 0.2 | 12.50 | 3180 × 2100 |
| 6 | + 0.2 | 15.00 | 4600 × 3180 |
| 10 | + 0.3 | 25.00 ) | 6000 × 3300 |
| 12 | + 0.3 | 30.00 ) | |
| 15 | + 0.5 | 37.50 | 3050 × 3000 |
| 19 | + 1.0 | 47.50 ) | 3000 × 2900 |
| 25 | + 1.0 | 63.50 ) | |
| **Clear sheet glass** | | | |
| 2 * | + 0.2 | 5.00 | 1920 × 1220 |
| 3 | + 0.3 | 7.50 | 2130 × 1320 |
| 4 | + 0.3 | 10.00 | 2760 × 1220 |
| 5 * | + 0.3 | 12.50 ) | 2130 × 2400 |
| 6 * | + 0.3 | 15.00 ) | |
| **Cast glass** | | | |
| 3 | + 0.4 / − 0.2 | 6.00 ) | 2140 × 1280 |
| 4 | + 0.5 | 7.50 ) | |
| 5 | + 0.5 | 9.50 | 2140 × 1320 |
| 6 | + 0.5 | 11.50 ) | 3700 × 1280 |
| 10 | + 0.8 | 21.50 ) | |
| **Wired glass** (Cast wired glass) | | | |
| 6 | + 0.3 / − 0.7 | − ) | 3700 × 1840 |
| 7 | + 0.7 | ) − ) | |
| (Polished wire glass) | | | |
| 6 | + 1.0 | − | 330 × 1830 |

* The 5 mm and 6 mm thickness are known as *thick drawn sheet*. Although 2 mm sheet glass is available it is not recommended for general glazing purposes

## METAL

### Weights of Metals

| Material | kg/m³ | lb/cu ft |
|---|---|---|
| **Metals, steel construction, etc.** | | |
| **Iron** | | |
| – cast | 7207 | 450 |
| – wrought | 7687 | 480 |
| – ore – general | 2407 | 150 |
| – (crushed) Swedish | 3682 | 230 |
| **Steel** | 7854 | 490 |
| **Copper** | | |
| – cast | 8731 | 545 |
| – wrought | 8945 | 558 |
| **Brass** | 8497 | 530 |
| **Bronze** | 8945 | 558 |
| **Aluminium** | 2774 | 173 |
| **Lead** | 11322 | 707 |
| **Zinc (rolled)** | 7140 | 446 |
| | g/mm² per metre | lb/sq ft per foot |
| **Steel bars** | 7.85 | 3.4 |
| **Structural steelwork** | Net weight of member @ 7854 kg/m³ | |
| riveted | + 10% for cleats, rivets, bolts, etc. | |
| welded | + 1.25% to 2.5% for welds, etc. | |
| **Rolled sections** | | |
| beams | + 2.5% | |
| stanchions | + 5% (extra for caps and bases) | |
| **Plate** | | |
| web girders | + 10% for rivets or welds, stiffeners, etc. | |
| | kg/m | lb/ft |
| **Steel stairs**: industrial type | | |
| 1 m or 3 ft wide | 84 | 56 |
| **Steel tubes** | | |
| 50 mm or 2 in bore | 5 to 6 | 3 to 4 |
| **Gas piping** | | |
| 20 mm or ¾ in | 2 | 1¼ |

METAL

## Universal Beams BS 4: Part 1: 2005

| Designation | Mass (kg/m) | Depth of section (mm) | Width of section (mm) | Thickness Web (mm) | Thickness Flange (mm) | Surface area (m²/m) |
|---|---|---|---|---|---|---|
| 1016 × 305 × 487 | 487.0 | 1036.1 | 308.5 | 30.0 | 54.1 | 3.20 |
| 1016 × 305 × 438 | 438.0 | 1025.9 | 305.4 | 26.9 | 49.0 | 3.17 |
| 1016 × 305 × 393 | 393.0 | 1016.0 | 303.0 | 24.4 | 43.9 | 3.15 |
| 1016 × 305 × 349 | 349.0 | 1008.1 | 302.0 | 21.1 | 40.0 | 3.13 |
| 1016 × 305 × 314 | 314.0 | 1000.0 | 300.0 | 19.1 | 35.9 | 3.11 |
| 1016 × 305 × 272 | 272.0 | 990.1 | 300.0 | 16.5 | 31.0 | 3.10 |
| 1016 × 305 × 249 | 249.0 | 980.2 | 300.0 | 16.5 | 26.0 | 3.08 |
| 1016 × 305 × 222 | 222.0 | 970.3 | 300.0 | 16.0 | 21.1 | 3.06 |
| 914 × 419 × 388 | 388.0 | 921.0 | 420.5 | 21.4 | 36.6 | 3.44 |
| 914 × 419 × 343 | 343.3 | 911.8 | 418.5 | 19.4 | 32.0 | 3.42 |
| 914 × 305 × 289 | 289.1 | 926.6 | 307.7 | 19.5 | 32.0 | 3.01 |
| 914 × 305 × 253 | 253.4 | 918.4 | 305.5 | 17.3 | 27.9 | 2.99 |
| 914 × 305 × 224 | 224.2 | 910.4 | 304.1 | 15.9 | 23.9 | 2.97 |
| 914 × 305 × 201 | 200.9 | 903.0 | 303.3 | 15.1 | 20.2 | 2.96 |
| 838 × 292 × 226 | 226.5 | 850.9 | 293.8 | 16.1 | 26.8 | 2.81 |
| 838 × 292 × 194 | 193.8 | 840.7 | 292.4 | 14.7 | 21.7 | 2.79 |
| 838 × 292 × 176 | 175.9 | 834.9 | 291.7 | 14.0 | 18.8 | 2.78 |
| 762 × 267 × 197 | 196.8 | 769.8 | 268.0 | 15.6 | 25.4 | 2.55 |
| 762 × 267 × 173 | 173.0 | 762.2 | 266.7 | 14.3 | 21.6 | 2.53 |
| 762 × 267 × 147 | 146.9 | 754.0 | 265.2 | 12.8 | 17.5 | 2.51 |
| 762 × 267 × 134 | 133.9 | 750.0 | 264.4 | 12.0 | 15.5 | 2.51 |
| 686 × 254 × 170 | 170.2 | 692.9 | 255.8 | 14.5 | 23.7 | 2.35 |
| 686 × 254 × 152 | 152.4 | 687.5 | 254.5 | 13.2 | 21.0 | 2.34 |
| 686 × 254 × 140 | 140.1 | 383.5 | 253.7 | 12.4 | 19.0 | 2.33 |
| 686 × 254 × 125 | 125.2 | 677.9 | 253.0 | 11.7 | 16.2 | 2.32 |
| 610 × 305 × 238 | 238.1 | 635.8 | 311.4 | 18.4 | 31.4 | 2.45 |
| 610 × 305 × 179 | 179.0 | 620.2 | 307.1 | 14.1 | 23.6 | 2.41 |
| 610 × 305 × 149 | 149.1 | 612.4 | 304.8 | 11.8 | 19.7 | 2.39 |
| 610 × 229 × 140 | 139.9 | 617.2 | 230.2 | 13.1 | 22.1 | 2.11 |
| 610 × 229 × 125 | 125.1 | 612.2 | 229.0 | 11.9 | 19.6 | 2.09 |
| 610 × 229 × 113 | 113.0 | 607.6 | 228.2 | 11.1 | 17.3 | 2.08 |
| 610 × 229 × 101 | 101.2 | 602.6 | 227.6 | 10.5 | 14.8 | 2.07 |
| 533 × 210 × 122 | 122.0 | 544.5 | 211.9 | 12.7 | 21.3 | 1.89 |
| 533 × 210 × 109 | 109.0 | 539.5 | 210.8 | 11.6 | 18.8 | 1.88 |
| 533 × 210 × 101 | 101.0 | 536.7 | 210.0 | 10.8 | 17.4 | 1.87 |
| 533 × 210 × 92 | 92.1 | 533.1 | 209.3 | 10.1 | 15.6 | 1.86 |
| 533 × 210 × 82 | 82.2 | 528.3 | 208.8 | 9.6 | 13.2 | 1.85 |
| 457 × 191 × 98 | 98.3 | 467.2 | 192.8 | 11.4 | 19.6 | 1.67 |
| 457 × 191 × 89 | 89.3 | 463.4 | 191.9 | 10.5 | 17.7 | 1.66 |
| 457 × 191 × 82 | 82.0 | 460.0 | 191.3 | 9.9 | 16.0 | 1.65 |
| 457 × 191 × 74 | 74.3 | 457.0 | 190.4 | 9.0 | 14.5 | 1.64 |
| 457 × 191 × 67 | 67.1 | 453.4 | 189.9 | 8.5 | 12.7 | 1.63 |
| 457 × 152 × 82 | 82.1 | 465.8 | 155.3 | 10.5 | 18.9 | 1.51 |
| 457 × 152 × 74 | 74.2 | 462.0 | 154.4 | 9.6 | 17.0 | 1.50 |
| 457 × 152 × 67 | 67.2 | 458.0 | 153.8 | 9.0 | 15.0 | 1.50 |
| 457 × 152 × 60 | 59.8 | 454.6 | 152.9 | 8.1 | 13.3 | 1.50 |
| 457 × 152 × 52 | 52.3 | 449.8 | 152.4 | 7.6 | 10.9 | 1.48 |
| 406 × 178 × 74 | 74.2 | 412.8 | 179.5 | 9.5 | 16.0 | 1.51 |
| 406 × 178 × 67 | 67.1 | 409.4 | 178.8 | 8.8 | 14.3 | 1.50 |
| 406 × 178 × 60 | 60.1 | 406.4 | 177.9 | 7.9 | 12.8 | 1.49 |

## METAL

| Designation | Mass (kg/m) | Depth of section (mm) | Width of section (mm) | Thickness Web (mm) | Thickness Flange (mm) | Surface area (m²/m) |
|---|---|---|---|---|---|---|
| 406 × 178 × 50 | 54.1 | 402.6 | 177.7 | 7.7 | 10.9 | 1.48 |
| 406 × 140 × 46 | 46.0 | 403.2 | 142.2 | 6.8 | 11.2 | 1.34 |
| 406 × 140 × 39 | 39.0 | 398.0 | 141.8 | 6.4 | 8.6 | 1.33 |
| 356 × 171 × 67 | 67.1 | 363.4 | 173.2 | 9.1 | 15.7 | 1.38 |
| 356 × 171 × 57 | 57.0 | 358.0 | 172.2 | 8.1 | 13.0 | 1.37 |
| 356 × 171 × 51 | 51.0 | 355.0 | 171.5 | 7.4 | 11.5 | 1.36 |
| 356 × 171 × 45 | 45.0 | 351.4 | 171.1 | 7.0 | 9.7 | 1.36 |
| 356 × 127 × 39 | 39.1 | 353.4 | 126.0 | 6.6 | 10.7 | 1.18 |
| 356 × 127 × 33 | 33.1 | 349.0 | 125.4 | 6.0 | 8.5 | 1.17 |
| 305 × 165 × 54 | 54.0 | 310.4 | 166.9 | 7.9 | 13.7 | 1.26 |
| 305 × 165 × 46 | 46.1 | 306.6 | 165.7 | 6.7 | 11.8 | 1.25 |
| 305 × 165 × 40 | 40.3 | 303.4 | 165.0 | 6.0 | 10.2 | 1.24 |
| 305 × 127 × 48 | 48.1 | 311.0 | 125.3 | 9.0 | 14.0 | 1.09 |
| 305 × 127 × 42 | 41.9 | 307.2 | 124.3 | 8.0 | 12.1 | 1.08 |
| 305 × 127 × 37 | 37.0 | 304.4 | 123.3 | 7.1 | 10.7 | 1.07 |
| 305 × 102 × 33 | 32.8 | 312.7 | 102.4 | 6.6 | 10.8 | 1.01 |
| 305 × 102 × 28 | 28.2 | 308.7 | 101.8 | 6.0 | 8.8 | 1.00 |
| 305 × 102 × 25 | 24.8 | 305.1 | 101.6 | 5.8 | 7.0 | 0.992 |
| 254 × 146 × 43 | 43.0 | 259.6 | 147.3 | 7.2 | 12.7 | 1.08 |
| 254 × 146 × 37 | 37.0 | 256.0 | 146.4 | 6.3 | 10.9 | 1.07 |
| 254 × 146 × 31 | 31.1 | 251.4 | 146.1 | 6.0 | 8.6 | 1.06 |
| 254 × 102 × 28 | 28.3 | 260.4 | 102.2 | 6.3 | 10.0 | 0.904 |
| 254 × 102 × 25 | 25.2 | 257.2 | 101.9 | 6.0 | 8.4 | 0.897 |
| 254 × 102 × 22 | 22.0 | 254.0 | 101.6 | 5.7 | 6.8 | 0.890 |
| 203 × 133 × 30 | 30.0 | 206.8 | 133.9 | 6.4 | 9.6 | 0.923 |
| 203 × 133 × 25 | 25.1 | 203.2 | 133.2 | 5.7 | 7.8 | 0.915 |
| 203 × 102 × 23 | 23.1 | 203.2 | 101.8 | 5.4 | 9.3 | 0.790 |
| 178 × 102 × 19 | 19.0 | 177.8 | 101.2 | 4.8 | 7.9 | 0.738 |
| 152 × 89 × 16 | 16.0 | 152.4 | 88.7 | 4.5 | 7.7 | 0.638 |
| 127 × 76 × 13 | 13.0 | 127.0 | 76.0 | 4.0 | 7.6 | 0.537 |

**METAL**

## Universal Columns BS 4: Part 1: 2005

| Designation | Mass (kg/m) | Depth of section (mm) | Width of section (mm) | Thickness | | Surface area (m²/m) |
|---|---|---|---|---|---|---|
| | | | | Web (mm) | Flange (mm) | |
| 356 × 406 × 634 | 633.9 | 474.7 | 424.0 | 47.6 | 77.0 | 2.52 |
| 356 × 406 × 551 | 551.0 | 455.6 | 418.5 | 42.1 | 67.5 | 2.47 |
| 356 × 406 × 467 | 467.0 | 436.6 | 412.2 | 35.8 | 58.0 | 2.42 |
| 356 × 406 × 393 | 393.0 | 419.0 | 407.0 | 30.6 | 49.2 | 2.38 |
| 356 × 406 × 340 | 339.9 | 406.4 | 403.0 | 26.6 | 42.9 | 2.35 |
| 356 × 406 × 287 | 287.1 | 393.6 | 399.0 | 22.6 | 36.5 | 2.31 |
| 356 × 406 × 235 | 235.1 | 381.0 | 384.8 | 18.4 | 30.2 | 2.28 |
| 356 × 368 × 202 | 201.9 | 374.6 | 374.7 | 16.5 | 27.0 | 2.19 |
| 356 × 368 × 177 | 177.0 | 368.2 | 372.6 | 14.4 | 23.8 | 2.17 |
| 356 × 368 × 153 | 152.9 | 362.0 | 370.5 | 12.3 | 20.7 | 2.16 |
| 356 × 368 × 129 | 129.0 | 355.6 | 368.6 | 10.4 | 17.5 | 2.14 |
| 305 × 305 × 283 | 282.9 | 365.3 | 322.2 | 26.8 | 44.1 | 1.94 |
| 305 × 305 × 240 | 240.0 | 352.5 | 318.4 | 23.0 | 37.7 | 1.91 |
| 305 × 305 × 198 | 198.1 | 339.9 | 314.5 | 19.1 | 31.4 | 1.87 |
| 305 × 305 × 158 | 158.1 | 327.1 | 311.2 | 15.8 | 25.0 | 1.84 |
| 305 × 305 × 137 | 136.9 | 320.5 | 309.2 | 13.8 | 21.7 | 1.82 |
| 305 × 305 × 118 | 117.9 | 314.5 | 307.4 | 12.0 | 18.7 | 1.81 |
| 305 × 305 × 97 | 96.9 | 307.9 | 305.3 | 9.9 | 15.4 | 1.79 |
| 254 × 254 × 167 | 167.1 | 289.1 | 265.2 | 19.2 | 31.7 | 1.58 |
| 254 × 254 × 132 | 132.0 | 276.3 | 261.3 | 15.3 | 25.3 | 1.55 |
| 254 × 254 × 107 | 107.1 | 266.7 | 258.8 | 12.8 | 20.5 | 1.52 |
| 254 × 254 × 89 | 88.9 | 260.3 | 256.3 | 10.3 | 17.3 | 1.50 |
| 254 × 254 × 73 | 73.1 | 254.1 | 254.6 | 8.6 | 14.2 | 1.49 |
| 203 × 203 × 86 | 86.1 | 222.2 | 209.1 | 12.7 | 20.5 | 1.24 |
| 203 × 203 × 71 | 71.0 | 215.8 | 206.4 | 10.0 | 17.3 | 1.22 |
| 203 × 203 × 60 | 60.0 | 209.6 | 205.8 | 9.4 | 14.2 | 1.21 |
| 203 × 203 × 52 | 52.0 | 206.2 | 204.3 | 7.9 | 12.5 | 1.20 |
| 203 × 203 × 46 | 46.1 | 203.2 | 203.6 | 7.2 | 11.0 | 1.19 |
| 152 × 152 × 37 | 37.0 | 161.8 | 154.4 | 8.0 | 11.5 | 0.912 |
| 152 × 152 × 30 | 30.0 | 157.6 | 152.9 | 6.5 | 9.4 | 0.901 |
| 152 × 152 × 23 | 23.0 | 152.4 | 152.2 | 5.8 | 6.8 | 0.889 |

**METAL**

## Joists BS 4: Part 1: 2005 (retained for reference, Corus have ceased manufacture in UK)

| Designation | Mass (kg/m) | Depth of section (mm) | Width of section (mm) | Thickness Web (mm) | Thickness Flange (mm) | Surface area (m²/m) |
|---|---|---|---|---|---|---|
| 254 × 203 × 82 | 82.0 | 254.0 | 203.2 | 10.2 | 19.9 | 1.210 |
| 203 × 152 × 52 | 52.3 | 203.2 | 152.4 | 8.9 | 16.5 | 0.932 |
| 152 × 127 × 37 | 37.3 | 152.4 | 127.0 | 10.4 | 13.2 | 0.737 |
| 127 × 114 × 29 | 29.3 | 127.0 | 114.3 | 10.2 | 11.5 | 0.646 |
| 127 × 114 × 27 | 26.9 | 127.0 | 114.3 | 7.4 | 11.4 | 0.650 |
| 102 × 102 × 23 | 23.0 | 101.6 | 101.6 | 9.5 | 10.3 | 0.549 |
| 102 × 44 × 7 | 7.5 | 101.6 | 44.5 | 4.3 | 6.1 | 0.350 |
| 89 × 89 × 19 | 19.5 | 88.9 | 88.9 | 9.5 | 9.9 | 0.476 |
| 76 × 76 × 13 | 12.8 | 76.2 | 76.2 | 5.1 | 8.4 | 0.411 |

## Parallel Flange Channels

| Designation | Mass (kg/m) | Depth of section (mm) | Width of section (mm) | Thickness Web (mm) | Thickness Flange (mm) | Surface area (m²/m) |
|---|---|---|---|---|---|---|
| 430 × 100 × 64 | 64.4 | 430 | 100 | 11.0 | 19.0 | 1.23 |
| 380 × 100 × 54 | 54.0 | 380 | 100 | 9.5 | 17.5 | 1.13 |
| 300 × 100 × 46 | 45.5 | 300 | 100 | 9.0 | 16.5 | 0.969 |
| 300 × 90 × 41 | 41.4 | 300 | 90 | 9.0 | 15.5 | 0.932 |
| 260 × 90 × 35 | 34.8 | 260 | 90 | 8.0 | 14.0 | 0.854 |
| 260 × 75 × 28 | 27.6 | 260 | 75 | 7.0 | 12.0 | 0.79 |
| 230 × 90 × 32 | 32.2 | 230 | 90 | 7.5 | 14.0 | 0.795 |
| 230 × 75 × 26 | 25.7 | 230 | 75 | 6.5 | 12.5 | 0.737 |
| 200 × 90 × 30 | 29.7 | 200 | 90 | 7.0 | 14.0 | 0.736 |
| 200 × 75 × 23 | 23.4 | 200 | 75 | 6.0 | 12.5 | 0.678 |
| 180 × 90 × 26 | 26.1 | 180 | 90 | 6.5 | 12.5 | 0.697 |
| 180 × 75 × 20 | 20.3 | 180 | 75 | 6.0 | 10.5 | 0.638 |
| 150 × 90 × 24 | 23.9 | 150 | 90 | 6.5 | 12.0 | 0.637 |
| 150 × 75 × 18 | 17.9 | 150 | 75 | 5.5 | 10.0 | 0.579 |
| 125 × 65 × 15 | 14.8 | 125 | 65 | 5.5 | 9.5 | 0.489 |
| 100 × 50 × 10 | 10.2 | 100 | 50 | 5.0 | 8.5 | 0.382 |

**METAL**

## Equal Angles BS EN 10056-1

| Designation | Mass (kg/m) | Surface area (m²/m) |
|---|---|---|
| 200 × 200 × 24 | 71.1 | 0.790 |
| 200 × 200 × 20 | 59.9 | 0.790 |
| 200 × 200 × 18 | 54.2 | 0.790 |
| 200 × 200 × 16 | 48.5 | 0.790 |
| 150 × 150 × 18 | 40.1 | 0.59 |
| 150 × 150 × 15 | 33.8 | 0.59 |
| 150 × 150 × 12 | 27.3 | 0.59 |
| 150 × 150 × 10 | 23.0 | 0.59 |
| 120 × 120 × 15 | 26.6 | 0.47 |
| 120 × 120 × 12 | 21.6 | 0.47 |
| 120 × 120 × 10 | 18.2 | 0.47 |
| 120 × 120 × 8 | 14.7 | 0.47 |
| 100 × 100 × 15 | 21.9 | 0.39 |
| 100 × 100 × 12 | 17.8 | 0.39 |
| 100 × 100 × 10 | 15.0 | 0.39 |
| 100 × 100 × 8 | 12.2 | 0.39 |
| 90 × 90 × 12 | 15.9 | 0.35 |
| 90 × 90 × 10 | 13.4 | 0.35 |
| 90 × 90 × 8 | 10.9 | 0.35 |
| 90 × 90 × 7 | 9.61 | 0.35 |
| 90 × 90 × 6 | 8.30 | 0.35 |

## Unequal Angles BS EN 10056-1

| Designation | Mass (kg/m) | Surface area (m²/m) |
|---|---|---|
| 200 × 150 × 18 | 47.1 | 0.69 |
| 200 × 150 × 15 | 39.6 | 0.69 |
| 200 × 150 × 12 | 32.0 | 0.69 |
| 200 × 100 × 15 | 33.7 | 0.59 |
| 200 × 100 × 12 | 27.3 | 0.59 |
| 200 × 100 × 10 | 23.0 | 0.59 |
| 150 × 90 × 15 | 26.6 | 0.47 |
| 150 × 90 × 12 | 21.6 | 0.47 |
| 150 × 90 × 10 | 18.2 | 0.47 |
| 150 × 75 × 15 | 24.8 | 0.44 |
| 150 × 75 × 12 | 20.2 | 0.44 |
| 150 × 75 × 10 | 17.0 | 0.44 |
| 125 × 75 × 12 | 17.8 | 0.40 |
| 125 × 75 × 10 | 15.0 | 0.40 |
| 125 × 75 × 8 | 12.2 | 0.40 |
| 100 × 75 × 12 | 15.4 | 0.34 |
| 100 × 75 × 10 | 13.0 | 0.34 |
| 100 × 75 × 8 | 10.6 | 0.34 |
| 100 × 65 × 10 | 12.3 | 0.32 |
| 100 × 65 × 8 | 9.94 | 0.32 |
| 100 × 65 × 7 | 8.77 | 0.32 |

**METAL**

## Structural Tees Split from Universal Beams BS 4: Part 1: 2005

| Designation | Mass (kg/m) | Surface area (m²/m) |
|---|---|---|
| 305 × 305 × 90 | 89.5 | 1.22 |
| 305 × 305 × 75 | 74.6 | 1.22 |
| 254 × 343 × 63 | 62.6 | 1.19 |
| 229 × 305 × 70 | 69.9 | 1.07 |
| 229 × 305 × 63 | 62.5 | 1.07 |
| 229 × 305 × 57 | 56.5 | 1.07 |
| 229 × 305 × 51 | 50.6 | 1.07 |
| 210 × 267 × 61 | 61.0 | 0.95 |
| 210 × 267 × 55 | 54.5 | 0.95 |
| 210 × 267 × 51 | 50.5 | 0.95 |
| 210 × 267 × 46 | 46.1 | 0.95 |
| 210 × 267 × 41 | 41.1 | 0.95 |
| 191 × 229 × 49 | 49.2 | 0.84 |
| 191 × 229 × 45 | 44.6 | 0.84 |
| 191 × 229 × 41 | 41.0 | 0.84 |
| 191 × 229 × 37 | 37.1 | 0.84 |
| 191 × 229 × 34 | 33.6 | 0.84 |
| 152 × 229 × 41 | 41.0 | 0.76 |
| 152 × 229 × 37 | 37.1 | 0.76 |
| 152 × 229 × 34 | 33.6 | 0.76 |
| 152 × 229 × 30 | 29.9 | 0.76 |
| 152 × 229 × 26 | 26.2 | 0.76 |

## Universal Bearing Piles BS 4: Part 1: 2005

| Designation | Mass (kg/m) | Depth of Section (mm) | Width of section (mm) | Thickness Web (mm) | Thickness Flange (mm) |
|---|---|---|---|---|---|
| 356 × 368 × 174 | 173.9 | 361.4 | 378.5 | 20.3 | 20.4 |
| 356 × 368 × 152 | 152.0 | 356.4 | 376.0 | 17.8 | 17.9 |
| 356 × 368 × 133 | 133.0 | 352.0 | 373.8 | 15.6 | 15.7 |
| 356 × 368 × 109 | 108.9 | 346.4 | 371.0 | 12.8 | 12.9 |
| 305 × 305 × 223 | 222.9 | 337.9 | 325.7 | 30.3 | 30.4 |
| 305 × 305 × 186 | 186.0 | 328.3 | 320.9 | 25.5 | 25.6 |
| 305 × 305 × 149 | 149.1 | 318.5 | 316.0 | 20.6 | 20.7 |
| 305 × 305 × 126 | 126.1 | 312.3 | 312.9 | 17.5 | 17.6 |
| 305 × 305 × 110 | 110.0 | 307.9 | 310.7 | 15.3 | 15.4 |
| 305 × 305 × 95 | 94.9 | 303.7 | 308.7 | 13.3 | 13.3 |
| 305 × 305 × 88 | 88.0 | 301.7 | 307.8 | 12.4 | 12.3 |
| 305 × 305 × 79 | 78.9 | 299.3 | 306.4 | 11.0 | 11.1 |
| 254 × 254 × 85 | 85.1 | 254.3 | 260.4 | 14.4 | 14.3 |
| 254 × 254 × 71 | 71.0 | 249.7 | 258.0 | 12.0 | 12.0 |
| 254 × 254 × 63 | 63.0 | 247.1 | 256.6 | 10.6 | 10.7 |
| 203 × 203 × 54 | 53.9 | 204.0 | 207.7 | 11.3 | 11.4 |
| 203 × 203 × 45 | 44.9 | 200.2 | 205.9 | 9.5 | 9.5 |

**METAL**

## Hot Formed Square Hollow Sections EN 10210 S275J2H & S355J2H

| Size (mm) | Wall thickness (mm) | Mass (kg/m) | Superficial area (m²/m) |
|---|---|---|---|
| 40 × 40 | 2.5 | 2.89 | 0.154 |
| | 3.0 | 3.41 | 0.152 |
| | 3.2 | 3.61 | 0.152 |
| | 3.6 | 4.01 | 0.151 |
| | 4.0 | 4.39 | 0.150 |
| | 5.0 | 5.28 | 0.147 |
| 50 × 50 | 2.5 | 3.68 | 0.194 |
| | 3.0 | 4.35 | 0.192 |
| | 3.2 | 4.62 | 0.192 |
| | 3.6 | 5.14 | 0.191 |
| | 4.0 | 5.64 | 0.190 |
| | 5.0 | 6.85 | 0.187 |
| | 6.0 | 7.99 | 0.185 |
| | 6.3 | 8.31 | 0.184 |
| 60 × 60 | 3.0 | 5.29 | 0.232 |
| | 3.2 | 5.62 | 0.232 |
| | 3.6 | 6.27 | 0.231 |
| | 4.0 | 6.90 | 0.230 |
| | 5.0 | 8.42 | 0.227 |
| | 6.0 | 9.87 | 0.225 |
| | 6.3 | 10.30 | 0.224 |
| | 8.0 | 12.50 | 0.219 |
| 70 × 70 | 3.0 | 6.24 | 0.272 |
| | 3.2 | 6.63 | 0.272 |
| | 3.6 | 7.40 | 0.271 |
| | 4.0 | 8.15 | 0.270 |
| | 5.0 | 9.99 | 0.267 |
| | 6.0 | 11.80 | 0.265 |
| | 6.3 | 12.30 | 0.264 |
| | 8.0 | 15.00 | 0.259 |
| 80 × 80 | 3.2 | 7.63 | 0.312 |
| | 3.6 | 8.53 | 0.311 |
| | 4.0 | 9.41 | 0.310 |
| | 5.0 | 11.60 | 0.307 |
| | 6.0 | 13.60 | 0.305 |
| | 6.3 | 14.20 | 0.304 |
| | 8.0 | 17.50 | 0.299 |
| 90 × 90 | 3.6 | 9.66 | 0.351 |
| | 4.0 | 10.70 | 0.350 |
| | 5.0 | 13.10 | 0.347 |
| | 6.0 | 15.50 | 0.345 |
| | 6.3 | 16.20 | 0.344 |
| | 8.0 | 20.10 | 0.339 |
| 100 × 100 | 3.6 | 10.80 | 0.391 |
| | 4.0 | 11.90 | 0.390 |
| | 5.0 | 14.70 | 0.387 |
| | 6.0 | 17.40 | 0.385 |
| | 6.3 | 18.20 | 0.384 |
| | 8.0 | 22.60 | 0.379 |
| | 10.0 | 27.40 | 0.374 |
| 120 × 120 | 4.0 | 14.40 | 0.470 |
| | 5.0 | 17.80 | 0.467 |
| | 6.0 | 21.20 | 0.465 |

**METAL**

| Size (mm) | Wall thickness (mm) | Mass (kg/m) | Superficial area (m²/m) |
|---|---|---|---|
| | 6.3 | 22.20 | 0.464 |
| | 8.0 | 27.60 | 0.459 |
| | 10.0 | 33.70 | 0.454 |
| | 12.0 | 39.50 | 0.449 |
| | 12.5 | 40.90 | 0.448 |
| 140 × 140 | 5.0 | 21.00 | 0.547 |
| | 6.0 | 24.90 | 0.545 |
| | 6.3 | 26.10 | 0.544 |
| | 8.0 | 32.60 | 0.539 |
| | 10.0 | 40.00 | 0.534 |
| | 12.0 | 47.00 | 0.529 |
| | 12.5 | 48.70 | 0.528 |
| 150 × 150 | 5.0 | 22.60 | 0.587 |
| | 6.0 | 26.80 | 0.585 |
| | 6.3 | 28.10 | 0.584 |
| | 8.0 | 35.10 | 0.579 |
| | 10.0 | 43.10 | 0.574 |
| | 12.0 | 50.80 | 0.569 |
| | 12.5 | 52.70 | 0.568 |
| Hot formed from seamless hollow | 16.0 | 65.2 | 0.559 |
| 160 × 160 | 5.0 | 24.10 | 0.627 |
| | 6.0 | 28.70 | 0.625 |
| | 6.3 | 30.10 | 0.624 |
| | 8.0 | 37.60 | 0.619 |
| | 10.0 | 46.30 | 0.614 |
| | 12.0 | 54.60 | 0.609 |
| | 12.5 | 56.60 | 0.608 |
| | 16.0 | 70.20 | 0.599 |
| 180 × 180 | 5.0 | 27.30 | 0.707 |
| | 6.0 | 32.50 | 0.705 |
| | 6.3 | 34.00 | 0.704 |
| | 8.0 | 42.70 | 0.699 |
| | 10.0 | 52.50 | 0.694 |
| | 12.0 | 62.10 | 0.689 |
| | 12.5 | 64.40 | 0.688 |
| | 16.0 | 80.20 | 0.679 |
| 200 × 200 | 5.0 | 30.40 | 0.787 |
| | 6.0 | 36.20 | 0.785 |
| | 6.3 | 38.00 | 0.784 |
| | 8.0 | 47.70 | 0.779 |
| | 10.0 | 58.80 | 0.774 |
| | 12.0 | 69.60 | 0.769 |
| | 12.5 | 72.30 | 0.768 |
| | 16.0 | 90.30 | 0.759 |
| 250 × 250 | 5.0 | 38.30 | 0.987 |
| | 6.0 | 45.70 | 0.985 |
| | 6.3 | 47.90 | 0.984 |
| | 8.0 | 60.30 | 0.979 |
| | 10.0 | 74.50 | 0.974 |
| | 12.0 | 88.50 | 0.969 |
| | 12.5 | 91.90 | 0.968 |
| | 16.0 | 115.00 | 0.959 |

**METAL**

| Size (mm) | Wall thickness (mm) | Mass (kg/m) | Superficial area (m²/m) |
|---|---|---|---|
| 300 × 300 | 6.0 | 55.10 | 1.18 |
| | 6.3 | 57.80 | 1.18 |
| | 8.0 | 72.80 | 1.18 |
| | 10.0 | 90.20 | 1.17 |
| | 12.0 | 107.00 | 1.17 |
| | 12.5 | 112.00 | 1.17 |
| | 16.0 | 141.00 | 1.16 |
| 350 × 350 | 8.0 | 85.40 | 1.38 |
| | 10.0 | 106.00 | 1.37 |
| | 12.0 | 126.00 | 1.37 |
| | 12.5 | 131.00 | 1.37 |
| | 16.0 | 166.00 | 1.36 |
| 400 × 400 | 8.0 | 97.90 | 1.58 |
| | 10.0 | 122.00 | 1.57 |
| | 12.0 | 145.00 | 1.57 |
| | 12.5 | 151.00 | 1.57 |
| | 16.0 | 191.00 | 1.56 |
| (Grade S355J2H only) | 20.00* | 235.00 | 1.55 |

Note: * SAW process

**METAL**

## Hot Formed Square Hollow Sections JUMBO RHS: JIS G3136

| Size (mm) | Wall thickness (mm) | Mass (kg/m) | Superficial area (m²/m) |
|---|---|---|---|
| 350 × 350 | 19.0 | 190.00 | 1.33 |
|  | 22.0 | 217.00 | 1.32 |
|  | 25.0 | 242.00 | 1.31 |
| 400 × 400 | 22.0 | 251.00 | 1.52 |
|  | 25.0 | 282.00 | 1.51 |
| 450 × 450 | 12.0 | 162.00 | 1.76 |
|  | 16.0 | 213.00 | 1.75 |
|  | 19.0 | 250.00 | 1.73 |
|  | 22.0 | 286.00 | 1.72 |
|  | 25.0 | 321.00 | 1.71 |
|  | 28.0 * | 355.00 | 1.70 |
|  | 32.0 * | 399.00 | 1.69 |
| 500 × 500 | 12.0 | 181.00 | 1.96 |
|  | 16.0 | 238.00 | 1.95 |
|  | 19.0 | 280.00 | 1.93 |
|  | 22.0 | 320.00 | 1.92 |
|  | 25.0 | 360.00 | 1.91 |
|  | 28.0 * | 399.00 | 1.90 |
|  | 32.0 * | 450.00 | 1.89 |
|  | 36.0 * | 498.00 | 1.88 |
| 550 × 550 | 16.0 | 263.00 | 2.15 |
|  | 19.0 | 309.00 | 2.13 |
|  | 22.0 | 355.00 | 2.12 |
|  | 25.0 | 399.00 | 2.11 |
|  | 28.0 * | 443.00 | 2.10 |
|  | 32.0 * | 500.00 | 2.09 |
|  | 36.0 * | 555.00 | 2.08 |
|  | 40.0 * | 608.00 | 2.06 |
| 600 × 600 | 25.0 * | 439.00 | 2.31 |
|  | 28.0 * | 487.00 | 2.30 |
|  | 32.0 * | 550.00 | 2.29 |
|  | 36.0 * | 611.00 | 2.28 |
|  | 40.0 * | 671.00 | 2.26 |
| 700 × 700 | 25.0 * | 517.00 | 2.71 |
|  | 28.0 * | 575.00 | 2.70 |
|  | 32.0 * | 651.00 | 2.69 |
|  | 36.0 * | 724.00 | 2.68 |
|  | 40.0 * | 797.00 | 2.68 |

Note: * SAW process

**METAL**

## Hot Formed Rectangular Hollow Sections: EN10210 S275J2h & S355J2H

| Size (mm) | Wall thickness (mm) | Mass (kg/m) | Superficial area (m²/m) |
|---|---|---|---|
| 50 × 30 | 2.5 | 2.89 | 0.154 |
| | 3.0 | 3.41 | 0.152 |
| | 3.2 | 3.61 | 0.152 |
| | 3.6 | 4.01 | 0.151 |
| | 4.0 | 4.39 | 0.150 |
| | 5.0 | 5.28 | 0.147 |
| 60 × 40 | 2.5 | 3.68 | 0.194 |
| | 3.0 | 4.35 | 0.192 |
| | 3.2 | 4.62 | 0.192 |
| | 3.6 | 5.14 | 0.191 |
| | 4.0 | 5.64 | 0.190 |
| | 5.0 | 6.85 | 0.187 |
| | 6.0 | 7.99 | 0.185 |
| | 6.3 | 8.31 | 0.184 |
| 80 × 40 | 3.0 | 5.29 | 0.232 |
| | 3.2 | 5.62 | 0.232 |
| | 3.6 | 6.27 | 0.231 |
| | 4.0 | 6.90 | 0.230 |
| | 5.0 | 8.42 | 0.227 |
| | 6.0 | 9.87 | 0.225 |
| | 6.3 | 10.30 | 0.224 |
| | 8.0 | 12.50 | 0.219 |
| 76.2 × 50.8 | 3.0 | 5.62 | 0.246 |
| | 3.2 | 5.97 | 0.246 |
| | 3.6 | 6.66 | 0.245 |
| | 4.0 | 7.34 | 0.244 |
| | 5.0 | 8.97 | 0.241 |
| | 6.0 | 10.50 | 0.239 |
| | 6.3 | 11.00 | 0.238 |
| | 8.0 | 13.40 | 0.233 |
| 90 × 50 | 3.0 | 6.24 | 0.272 |
| | 3.2 | 6.63 | 0.272 |
| | 3.6 | 7.40 | 0.271 |
| | 4.0 | 8.15 | 0.270 |
| | 5.0 | 9.99 | 0.267 |
| | 6.0 | 11.80 | 0.265 |
| | 6.3 | 12.30 | 0.264 |
| | 8.0 | 15.00 | 0.259 |
| 100 × 50 | 3.0 | 6.71 | 0.292 |
| | 3.2 | 7.13 | 0.292 |
| | 3.6 | 7.96 | 0.291 |
| | 4.0 | 8.78 | 0.290 |
| | 5.0 | 10.80 | 0.287 |
| | 6.0 | 12.70 | 0.285 |
| | 6.3 | 13.30 | 0.284 |
| | 8.0 | 16.30 | 0.279 |

## METAL

| Size (mm) | Wall thickness (mm) | Mass (kg/m) | Superficial area (m²/m) |
|---|---|---|---|
| 100 × 60 | 3.0 | 7.18 | 0.312 |
| | 3.2 | 7.63 | 0.312 |
| | 3.6 | 8.53 | 0.311 |
| | 4.0 | 9.41 | 0.310 |
| | 5.0 | 11.60 | 0.307 |
| | 6.0 | 13.60 | 0.305 |
| | 6.3 | 14.20 | 0.304 |
| | 8.0 | 17.50 | 0.299 |
| 120 × 60 | 3.6 | 9.70 | 0.351 |
| | 4.0 | 10.70 | 0.350 |
| | 5.0 | 13.10 | 0.347 |
| | 6.0 | 15.50 | 0.345 |
| | 6.3 | 16.20 | 0.344 |
| | 8.0 | 20.10 | 0.339 |
| 120 × 80 | 3.6 | 10.80 | 0.391 |
| | 4.0 | 11.90 | 0.390 |
| | 5.0 | 14.70 | 0.387 |
| | 6.0 | 17.40 | 0.385 |
| | 6.3 | 18.20 | 0.384 |
| | 8.0 | 22.60 | 0.379 |
| | 10.0 | 27.40 | 0.374 |
| 150 × 100 | 4.0 | 15.10 | 0.490 |
| | 5.0 | 18.60 | 0.487 |
| | 6.0 | 22.10 | 0.485 |
| | 6.3 | 23.10 | 0.484 |
| | 8.0 | 28.90 | 0.479 |
| | 10.0 | 35.30 | 0.474 |
| | 12.0 | 41.40 | 0.469 |
| | 12.5 | 42.80 | 0.468 |
| 160 × 80 | 4.0 | 14.40 | 0.470 |
| | 5.0 | 17.80 | 0.467 |
| | 6.0 | 21.20 | 0.465 |
| | 6.3 | 22.20 | 0.464 |
| | 8.0 | 27.60 | 0.459 |
| | 10.0 | 33.70 | 0.454 |
| | 12.0 | 39.50 | 0.449 |
| | 12.5 | 40.90 | 0.448 |
| 200 × 100 | 5.0 | 22.60 | 0.587 |
| | 6.0 | 26.80 | 0.585 |
| | 6.3 | 28.10 | 0.584 |
| | 8.0 | 35.10 | 0.579 |
| | 10.0 | 43.10 | 0.574 |
| | 12.0 | 50.80 | 0.569 |
| | 12.5 | 52.70 | 0.568 |
| | 16.0 | 65.20 | 0.559 |
| 250 × 150 | 5.0 | 30.40 | 0.787 |
| | 6.0 | 36.20 | 0.785 |
| | 6.3 | 38.00 | 0.784 |
| | 8.0 | 47.70 | 0.779 |
| | 10.0 | 58.80 | 0.774 |
| | 12.0 | 69.60 | 0.769 |
| | 12.5 | 72.30 | 0.768 |
| | 16.0 | 90.30 | 0.759 |

**METAL**

| Size (mm) | Wall thickness (mm) | Mass (kg/m) | Superficial area (m²/m) |
|---|---|---|---|
| 300 × 200 | 5.0 | 38.30 | 0.987 |
| | 6.0 | 45.70 | 0.985 |
| | 6.3 | 47.90 | 0.984 |
| | 8.0 | 60.30 | 0.979 |
| | 10.0 | 74.50 | 0.974 |
| | 12.0 | 88.50 | 0.969 |
| | 12.5 | 91.90 | 0.968 |
| | 16.0 | 115.00 | 0.959 |
| 400 × 200 | 6.0 | 55.10 | 1.18 |
| | 6.3 | 57.80 | 1.18 |
| | 8.0 | 72.80 | 1.18 |
| | 10.0 | 90.20 | 1.17 |
| | 12.0 | 107.00 | 1.17 |
| | 12.5 | 112.00 | 1.17 |
| | 16.0 | 141.00 | 1.16 |
| 450 × 250 | 8.0 | 85.40 | 1.38 |
| | 10.0 | 106.00 | 1.37 |
| | 12.0 | 126.00 | 1.37 |
| | 12.5 | 131.00 | 1.37 |
| | 16.0 | 166.00 | 1.36 |
| 500 × 300 | 8.0 | 98.00 | 1.58 |
| | 10.0 | 122.00 | 1.57 |
| | 12.0 | 145.00 | 1.57 |
| | 12.5 | 151.00 | 1.57 |
| | 16.0 | 191.00 | 1.56 |
| | 20.0 | 235.00 | 1.55 |

**METAL**

## Hot Formed Circular Hollow Sections EN 10210 S275J2H & S355J2H

| Outside diameter (mm) | Wall thickness (mm) | Mass (kg/m) | Superficial area (m²/m) |
|---|---|---|---|
| 21.3 | 3.2 | 1.43 | 0.067 |
| 26.9 | 3.2 | 1.87 | 0.085 |
| 33.7 | 3.0 | 2.27 | 0.106 |
| | 3.2 | 2.41 | 0.106 |
| | 3.6 | 2.67 | 0.106 |
| | 4.0 | 2.93 | 0.106 |
| 42.4 | 3.0 | 2.91 | 0.133 |
| | 3.2 | 3.09 | 0.133 |
| | 3.6 | 3.44 | 0.133 |
| | 4.0 | 3.79 | 0.133 |
| 48.3 | 2.5 | 2.82 | 0.152 |
| | 3.0 | 3.35 | 0.152 |
| | 3.2 | 3.56 | 0.152 |
| | 3.6 | 3.97 | 0.152 |
| | 4.0 | 4.37 | 0.152 |
| | 5.0 | 5.34 | 0.152 |
| 60.3 | 2.5 | 3.56 | 0.189 |
| | 3.0 | 4.24 | 0.189 |
| | 3.2 | 4.51 | 0.189 |
| | 3.6 | 5.03 | 0.189 |
| | 4.0 | 5.55 | 0.189 |
| | 5.0 | 6.82 | 0.189 |
| 76.1 | 2.5 | 4.54 | 0.239 |
| | 3.0 | 5.41 | 0.239 |
| | 3.2 | 5.75 | 0.239 |
| | 3.6 | 6.44 | 0.239 |
| | 4.0 | 7.11 | 0.239 |
| | 5.0 | 8.77 | 0.239 |
| | 6.0 | 10.40 | 0.239 |
| | 6.3 | 10.80 | 0.239 |
| 88.9 | 2.5 | 5.33 | 0.279 |
| | 3.0 | 6.36 | 0.279 |
| | 3.2 | 6.76 | 0.27 |
| | 3.6 | 7.57 | 0.279 |
| | 4.0 | 8.38 | 0.279 |
| | 5.0 | 10.30 | 0.279 |
| | 6.0 | 12.30 | 0.279 |
| | 6.3 | 12.80 | 0.279 |
| 114.3 | 3.0 | 8.23 | 0.359 |
| | 3.2 | 8.77 | 0.359 |
| | 3.6 | 9.83 | 0.359 |
| | 4.0 | 10.09 | 0.359 |
| | 5.0 | 13.50 | 0.359 |
| | 6.0 | 16.00 | 0.359 |
| | 6.3 | 16.80 | 0.359 |

**METAL**

| Outside diameter (mm) | Wall thickness (mm) | Mass (kg/m) | Superficial area (m²/m) |
|---|---|---|---|
| 139.7 | 3.2 | 10.80 | 0.439 |
| | 3.6 | 12.10 | 0.439 |
| | 4.0 | 13.40 | 0.439 |
| | 5.0 | 16.60 | 0.439 |
| | 6.0 | 19.80 | 0.439 |
| | 6.3 | 20.70 | 0.439 |
| | 8.0 | 26.00 | 0.439 |
| | 10.0 | 32.00 | 0.439 |
| 168.3 | 3.2 | 13.00 | 0.529 |
| | 3.6 | 14.60 | 0.529 |
| | 4.0 | 16.20 | 0.529 |
| | 5.0 | 20.10 | 0.529 |
| | 6.0 | 24.00 | 0.529 |
| | 6.3 | 25.20 | 0.529 |
| | 8.0 | 31.60 | 0.529 |
| | 10.0 | 39.00 | 0.529 |
| | 12.0 | 46.30 | 0.529 |
| | 12.5 | 48.00 | 0.529 |
| 193.7 | 5.0 | 23.30 | 0.609 |
| | 6.0 | 27.80 | 0.609 |
| | 6.3 | 29.10 | 0.609 |
| | 8.0 | 36.60 | 0.609 |
| | 10.0 | 45.30 | 0.609 |
| | 12.0 | 53.80 | 0.609 |
| | 12.5 | 55.90 | 0.609 |
| 219.1 | 5.0 | 26.40 | 0.688 |
| | 6.0 | 31.50 | 0.688 |
| | 6.3 | 33.10 | 0.688 |
| | 8.0 | 41.60 | 0.688 |
| | 10.0 | 51.60 | 0.688 |
| | 12.0 | 61.30 | 0.688 |
| | 12.5 | 63.70 | 0.688 |
| | 16.0 | 80.10 | 0.688 |
| 244.5 | 5.0 | 29.50 | 0.768 |
| | 6.0 | 35.30 | 0.768 |
| | 6.3 | 37.00 | 0.768 |
| | 8.0 | 46.70 | 0.768 |
| | 10.0 | 57.80 | 0.768 |
| | 12.0 | 68.80 | 0.768 |
| | 12.5 | 71.50 | 0.768 |
| | 16.0 | 90.20 | 0.768 |
| 273.0 | 5.0 | 33.00 | 0.858 |
| | 6.0 | 39.50 | 0.858 |
| | 6.3 | 41.40 | 0.858 |
| | 8.0 | 52.30 | 0.858 |
| | 10.0 | 64.90 | 0.858 |
| | 12.0 | 77.20 | 0.858 |
| | 12.5 | 80.30 | 0.858 |
| | 16.0 | 101.00 | 0.858 |

**METAL**

| Outside diameter (mm) | Wall thickness (mm) | Mass (kg/m) | Superficial area (m²/m) |
|---|---|---|---|
| 323.9 | 5.0 | 39.30 | 1.02 |
| | 6.0 | 47.00 | 1.02 |
| | 6.3 | 49.30 | 1.02 |
| | 8.0 | 62.30 | 1.02 |
| | 10.0 | 77.40 | 1.02 |
| | 12.0 | 92.30 | 1.02 |
| | 12.5 | 96.00 | 1.02 |
| | 16.0 | 121.00 | 1.02 |
| 355.6 | 6.3 | 54.30 | 1.12 |
| | 8.0 | 68.60 | 1.12 |
| | 10.0 | 85.30 | 1.12 |
| | 12.0 | 102.00 | 1.12 |
| | 12.5 | 106.00 | 1.12 |
| | 16.0 | 134.00 | 1.12 |
| 406.4 | 6.3 | 62.20 | 1.28 |
| | 8.0 | 79.60 | 1.28 |
| | 10.0 | 97.80 | 1.28 |
| | 12.0 | 117.00 | 1.28 |
| | 12.5 | 121.00 | 1.28 |
| | 16.0 | 154.00 | 1.28 |
| 457.0 | 6.3 | 70.00 | 1.44 |
| | 8.0 | 88.60 | 1.44 |
| | 10.0 | 110.00 | 1.44 |
| | 12.0 | 132.00 | 1.44 |
| | 12.5 | 137.00 | 1.44 |
| | 16.0 | 174.00 | 1.44 |
| 508.0 | 6.3 | 77.90 | 1.60 |
| | 8.0 | 98.60 | 1.60 |
| | 10.0 | 123.00 | 1.60 |
| | 12.0 | 147.00 | 1.60 |
| | 12.5 | 153.00 | 1.60 |
| | 16.0 | 194.00 | 1.60 |

## METAL

### Spacing of Holes in Angles

| Nominal leg length (mm) | Spacing of holes | | | | | | Maximum diameter of bolt or rivet | | |
|---|---|---|---|---|---|---|---|---|---|
| | A | B | C | D | E | F | A | B and C | D, E and F |
| 200 |    |    | 75 | 75 | 55 | 55 | 55 |    | 30 | 20 |
| 150 |    |    | 55 | 55 |    |    |    |    | 20 |    |
| 125 |    |    | 45 | 60 |    |    |    |    | 20 |    |
| 120 |    |    |    |    |    |    |    |    |    |    |
| 100 | 55 |    |    |    |    |    | 24 |    |    |    |
| 90  | 50 |    |    |    |    |    | 24 |    |    |    |
| 80  | 45 |    |    |    |    |    | 20 |    |    |    |
| 75  | 45 |    |    |    |    |    | 20 |    |    |    |
| 70  | 40 |    |    |    |    |    | 20 |    |    |    |
| 65  | 35 |    |    |    |    |    | 20 |    |    |    |
| 60  | 35 |    |    |    |    |    | 16 |    |    |    |
| 50  | 28 |    |    |    |    |    | 12 |    |    |    |
| 45  | 25 |    |    |    |    |    |    |    |    |    |
| 40  | 23 |    |    |    |    |    |    |    |    |    |
| 30  | 20 |    |    |    |    |    |    |    |    |    |
| 25  | 15 |    |    |    |    |    |    |    |    |    |

**KERBS, PAVING, ETC.**

## KERBS/EDGINGS/CHANNELS

### Precast Concrete Kerbs to BS 7263
### Straight kerb units: length from 450 to 915 mm

| | | |
|---|---|---|
| 150 mm high × 125 mm thick | | |
|   bullnosed | type BN | |
|   half battered | type HB3 | |
| | | |
| 255 mm high × 125 mm thick | | |
|   45° splayed | type SP | |
|   half battered | type HB2 | |
| | | |
| 305 mm high × 150 mm thick | | |
|   half battered | type HB1 | |
| | | |
| Quadrant kerb units | | |
|   150 mm high × 305 and 455 mm radius to match | type BN | type QBN |
|   150 mm high × 305 and 455 mm radius to match | type HB2, HB3 | type QHB |
|   150 mm high × 305 and 455 mm radius to match | type SP | type QSP |
|   255 mm high × 305 and 455 mm radius to match | type BN | type QBN |
|   255 mm high × 305 and 455 mm radius to match | type HB2, HB3 | type QHB |
|   225 mm high × 305 and 455 mm radius to match | type SP | type QSP |
| | | |
| Angle kerb units | | |
|   305 × 305 × 225 mm high × 125 mm thick | | |
|   bullnosed external angle | type XA | |
|   splayed external angle to match type SP | type XA | |
|   bullnosed internal angle | type IA | |
|   splayed internal angle to match type SP | type IA | |
| | | |
| Channels | | |
|   255 mm wide × 125 mm high flat | type CS1 | |
|   150 mm wide × 125 mm high flat type | CS2 | |
|   255 mm wide × 125 mm high dished | type CD | |

**KERBS, PAVING, ETC.**

| Transition kerb units | | | |
|---|---|---|---|
| from kerb type SP to HB | left handed | type TL | |
| | right handed | type TR | |
| from kerb type BN to HB | left handed | type DL1 | |
| | right handed | type DR1 | |
| from kerb type BN to SP | left handed | type DL2 | |
| | right handed | type DR2 | |

## Number of kerbs required per quarter circle (780 mm kerb lengths)

| Radius (m) | Number in quarter circle |
|---|---|
| 12 | 24 |
| 10 | 20 |
| 8 | 16 |
| 6 | 12 |
| 5 | 10 |
| 4 | 8 |
| 3 | 6 |
| 2 | 4 |
| 1 | 2 |

## Precast Concrete Edgings

| Round top type ER | Flat top type EF | Bullnosed top type EBN |
|---|---|---|
| 150 × 50 mm | 150 × 50 mm | 150 × 50 mm |
| 200 × 50 mm | 200 × 50 mm | 200 × 50 mm |
| 250 × 50 mm | 250 × 50 mm | 250 × 50 mm |

## KERBS, PAVING, ETC.

# BASES

## Cement Bound Material for Bases and Subbases

| CBM1: | very carefully graded aggregate from 37.5–75 mm, with a 7-day strength of 4.5 N/mm² |
| --- | --- |
| CBM2: | same range of aggregate as CBM1 but with more tolerance in each size of aggregate with a 7-day strength of 7.0 N/mm² |
| CBM3: | crushed natural aggregate or blast furnace slag, graded from 37.5–150 mm for 40 mm aggregate, and from 20–75 mm for 20 mm aggregate, with a 7-day strength of 10 N/mm² |
| CBM4: | crushed natural aggregate or blast furnace slag, graded from 37.5–150 mm for 40 mm aggregate, and from 20–75 mm for 20 mm aggregate, with a 7-day strength of 15 N/mm² |

# INTERLOCKING BRICK/BLOCK ROADS/PAVINGS

## Sizes of Precast Concrete Paving Blocks

**Type R blocks**
200 × 100 × 60 mm
200 × 100 × 65 mm
200 × 100 × 80 mm
200 × 100 × 100 mm

**Type S**
Any shape within a 295 mm space

**Sizes of clay brick pavers**
200 × 100 × 50 mm
200 × 100 × 65 mm
210 × 105 × 50 mm
210 × 105 × 65 mm
215 × 102.5 × 50 mm
215 × 102.5 × 65 mm

**Type PA: 3 kN**
Footpaths and pedestrian areas, private driveways, car parks, light vehicle traffic and over-run

**Type PB: 7 kN**
Residential roads, lorry parks, factory yards, docks, petrol station forecourts, hardstandings, bus stations

**KERBS, PAVING, ETC.**

## PAVING AND SURFACING

### Weights and Sizes of Paving and Surfacing

| Description of item | Size | Quantity per tonne |
|---|---|---|
| Paving 50 mm thick | 900 × 600 mm | 15 |
| Paving 50 mm thick | 750 × 600 mm | 18 |
| Paving 50 mm thick | 600 × 600 mm | 23 |
| Paving 50 mm thick | 450 × 600 mm | 30 |
| Paving 38 mm thick | 600 × 600 mm | 30 |
| Path edging | 914 × 50 × 150 mm | 60 |
| Kerb (including radius and tapers) | 125 × 254 × 914 mm | 15 |
| Kerb (including radius and tapers) | 125 × 150 × 914 mm | 25 |
| Square channel | 125 × 254 × 914 mm | 15 |
| Dished channel | 125 × 254 × 914 mm | 15 |
| Quadrants | 300 × 300 × 254 mm | 19 |
| Quadrants | 450 × 450 × 254 mm | 12 |
| Quadrants | 300 × 300 × 150 mm | 30 |
| Internal angles | 300 × 300 × 254 mm | 30 |
| Fluted pavement channel | 255 × 75 × 914 mm | 25 |
| Corner stones | 300 × 300 mm | 80 |
| Corner stones | 360 × 360 mm | 60 |
| Cable covers | 914 × 175 mm | 55 |
| Gulley kerbs | 220 × 220 × 150 mm | 60 |
| Gulley kerbs | 220 × 200 × 75 mm | 120 |

## KERBS, PAVING, ETC.

### Weights and Sizes of Paving and Surfacing

| Material | kg/m³ | lb/cu yd |
|---|---|---|
| Tarmacadam | 2306 | 3891 |
| Macadam (waterbound) | 2563 | 4325 |
| Vermiculite (aggregate) | 64–80 | 108–135 |
| Terracotta | 2114 | 3568 |
| Cork – compressed | 388 | 24 |
| | kg/m² | lb/sq ft |
| Clay floor tiles, 12.7 mm | 27.3 | 5.6 |
| Pavement lights | 122 | 25 |
| Damp-proof course | 5 | 1 |
| | kg/m² per mm thickness | lb/sq ft per inch thickness |
| Paving slabs (stone) | 2.3 | 12 |
| Granite setts | 2.88 | 15 |
| Asphalt | 2.30 | 12 |
| Rubber flooring | 1.68 | 9 |
| Polyvinyl chloride | 1.94 (avg) | 10 (avg) |

### Coverage (m²) Per Cubic Metre of Materials Used as Subbases or Capping Layers

| Consolidated thickness laid in (mm) | Square metre coverage | | |
|---|---|---|---|
| | Gravel | Sand | Hardcore |
| 50 | 15.80 | 16.50 | – |
| 75 | 10.50 | 11.00 | – |
| 100 | 7.92 | 8.20 | 7.42 |
| 125 | 6.34 | 6.60 | 5.90 |
| 150 | 5.28 | 5.50 | 4.95 |
| 175 | – | – | 4.23 |
| 200 | – | – | 3.71 |
| 225 | – | – | 3.30 |
| 300 | – | – | 2.47 |

## KERBS, PAVING, ETC.

## Approximate Rate of Spreads

| Average thickness of course (mm) | Description | Approximate rate of spread | | | |
|---|---|---|---|---|---|
| | | Open Textured | | Dense, Medium & Fine Textured | |
| | | (kg/m²) | (m²/t) | (kg/m²) | (m²/t) |
| 35 | 14 mm open textured or dense wearing course | 60–75 | 13–17 | 70–85 | 12–14 |
| 40 | 20 mm open textured or dense base course | 70–85 | 12–14 | 80–100 | 10–12 |
| 45 | 20 mm open textured or dense base course | 80–100 | 10–12 | 95–100 | 9–10 |
| 50 | 20 mm open textured or dense, or 28 mm dense base course | 85–110 | 9–12 | 110–120 | 8–9 |
| 60 | 28 mm dense base course, 40 mm open textured of dense base course or 40 mm single course as base course | | 8–10 | 130–150 | 7–8 |
| 65 | 28 mm dense base course, 40 mm open textured or dense base course or 40 mm single course | 100–135 | 7–10 | 140–160 | 6–7 |
| 75 | 40 mm single course, 40 mm open textured or dense base course, 40 mm dense roadbase | 120–150 | 7–8 | 165–185 | 5–6 |
| 100 | 40 mm dense base course or roadbase | – | – | 220–240 | 4–4.5 |

## KERBS, PAVING, ETC.

### Surface Dressing Roads: Coverage (m²) per Tonne of Material

| Size in mm | Sand | Granite chips | Gravel | Limestone chips |
|---|---|---|---|---|
| Sand | 168 | – | – | – |
| 3 | – | 148 | 152 | 165 |
| 6 | – | 130 | 133 | 144 |
| 9 | – | 111 | 114 | 123 |
| 13 | – | 85 | 87 | 95 |
| 19 | – | 68 | 71 | 78 |

### Sizes of Flags

| Reference | Nominal size (mm) | Thickness (mm) |
|---|---|---|
| A | 600 × 450 | 50 and 63 |
| B | 600 × 600 | 50 and 63 |
| C | 600 × 750 | 50 and 63 |
| D | 600 × 900 | 50 and 63 |
| E | 450 × 450 | 50 and 70 chamfered top surface |
| F | 400 × 400 | 50 and 65 chamfered top surface |
| G | 300 × 300 | 50 and 60 chamfered top surface |

### Sizes of Natural Stone Setts

| Width (mm) | | Length (mm) | | Depth (mm) |
|---|---|---|---|---|
| 100 | × | 100 | × | 100 |
| 75 | × | 150 to 250 | × | 125 |
| 75 | × | 150 to 250 | × | 150 |
| 100 | × | 150 to 250 | × | 100 |
| 100 | × | 150 to 250 | × | 150 |

## SEEDING/TURFING AND PLANTING

### Topsoil Quality

| Topsoil grade | Properties |
|---|---|
| Premium | Natural topsoil, high fertility, loamy texture, good soil structure, suitable for intensive cultivation. |
| General purpose | Natural or manufactured topsoil of lesser quality than Premium, suitable for agriculture or amenity landscape, may need fertilizer or soil structure improvement. |
| Economy | Selected subsoil, natural mineral deposit such as river silt or greensand. The grade comprises two subgrades; 'Low clay' and 'High clay' which is more liable to compaction in handling. This grade is suitable for low-production agricultural land and amenity woodland or conservation planting areas. |

### Forms of Trees

| | |
|---|---|
| Standards: | Shall be clear with substantially straight stems. Grafted and budded trees shall have no more than a slight bend at the union. Standards shall be designated as Half, Extra light, Light, Standard, Selected standard, Heavy, and Extra heavy. |
| Sizes of Standards | |
| Heavy standard | 12–14 cm girth × 3.50 to 5.00 m high |
| Extra Heavy standard | 14–16 cm girth × 4.25 to 5.00 m high |
| Extra Heavy standard | 16–18 cm girth × 4.25 to 6.00 m high |
| Extra Heavy standard | 18–20 cm girth × 5.00 to 6.00 m high |
| | |
| Semi-mature trees: | Between 6.0 m and 12.0 m tall with a girth of 20 to 75 cm at 1.0 m above ground. |
| Feathered trees: | Shall have a defined upright central leader, with stem furnished with evenly spread and balanced lateral shoots down to or near the ground. |
| Whips: | Shall be without significant feather growth as determined by visual inspection. |
| Multi-stemmed trees: | Shall have two or more main stems at, near, above or below ground. |

Seedlings grown from seed and not transplanted shall be specified when ordered for sale as:

| | |
|---|---|
| 1+0 | one year old seedling |
| 2+0 | two year old seedling |
| 1+1 | one year seed bed, one year transplanted = two year old seedling |
| 1+2 | one year seed bed, two years transplanted = three year old seedling |
| 2+1 | two years seed bed, one year transplanted = three year old seedling |
| 1u1 | two years seed bed, undercut after 1 year = two year old seedling |
| 2u2 | four years seed bed, undercut after 2 years = four year old seedling |

## SEEDING/TURFING AND PLANTING

## Cuttings

The age of cuttings (plants grown from shoots, stems, or roots of the mother plant) shall be specified when ordered for sale. The height of transplants and undercut seedlings/cuttings (which have been transplanted or undercut at least once) shall be stated in centimetres. The number of growing seasons before and after transplanting or undercutting shall be stated.

| | |
|---|---|
| 0 + 1 | one year cutting |
| 0 + 2 | two year cutting |
| 0 + 1 + 1 | one year cutting bed, one year transplanted = two year old seedling |
| 0 + 1 + 2 | one year cutting bed, two years transplanted = three year old seedling |

## Grass Cutting Capacities in $m^2$ per hour

| Speed mph | Width of cut in metres | | | | | | | | | | | | |
|---|---|---|---|---|---|---|---|---|---|---|---|---|---|
| | 0.5 | 0.7 | 1.0 | 1.2 | 1.5 | 1.7 | 2.0 | 2.0 | 2.1 | 2.5 | 2.8 | 3.0 | 3.4 |
| 1.0 | 724 | 1127 | 1529 | 1931 | 2334 | 2736 | 3138 | 3219 | 3380 | 4023 | 4506 | 4828 | 5472 |
| 1.5 | 1086 | 1690 | 2293 | 2897 | 3500 | 4104 | 4707 | 4828 | 5069 | 6035 | 6759 | 7242 | 8208 |
| 2.0 | 1448 | 2253 | 3058 | 3862 | 4667 | 5472 | 6276 | 6437 | 6759 | 8047 | 9012 | 9656 | 10944 |
| 2.5 | 1811 | 2816 | 3822 | 4828 | 5834 | 6840 | 7846 | 8047 | 8449 | 10058 | 11265 | 12070 | 13679 |
| 3.0 | 2173 | 3380 | 4587 | 5794 | 7001 | 8208 | 9415 | 9656 | 10139 | 12070 | 13518 | 14484 | 16415 |
| 3.5 | 2535 | 3943 | 5351 | 6759 | 8167 | 9576 | 10984 | 11265 | 11829 | 14082 | 15772 | 16898 | 19151 |
| 4.0 | 2897 | 4506 | 6115 | 7725 | 9334 | 10944 | 12553 | 12875 | 13518 | 16093 | 18025 | 19312 | 21887 |
| 4.5 | 3259 | 5069 | 6880 | 8690 | 10501 | 12311 | 14122 | 14484 | 15208 | 18105 | 20278 | 21726 | 24623 |
| 5.0 | 3621 | 5633 | 7644 | 9656 | 11668 | 13679 | 15691 | 16093 | 16898 | 20117 | 22531 | 24140 | 27359 |
| 5.5 | 3983 | 6196 | 8409 | 10622 | 12834 | 15047 | 17260 | 17703 | 18588 | 22128 | 24784 | 26554 | 30095 |
| 6.0 | 4345 | 6759 | 9173 | 11587 | 14001 | 16415 | 18829 | 19312 | 20278 | 24140 | 27037 | 28968 | 32831 |
| 6.5 | 4707 | 7322 | 9938 | 12553 | 15168 | 17783 | 20398 | 20921 | 21967 | 26152 | 29290 | 31382 | 35566 |
| 7.0 | 5069 | 7886 | 10702 | 13518 | 16335 | 19151 | 21967 | 22531 | 23657 | 28163 | 31543 | 33796 | 38302 |

## Number of Plants per $m^2$: For Plants Planted on an Evenly Spaced Grid

### Planting distances

| mm | 0.10 | 0.15 | 0.20 | 0.25 | 0.35 | 0.40 | 0.45 | 0.50 | 0.60 | 0.75 | 0.90 | 1.00 | 1.20 | 1.50 |
|---|---|---|---|---|---|---|---|---|---|---|---|---|---|---|
| 0.10 | 100.00 | 66.67 | 50.00 | 40.00 | 28.57 | 25.00 | 22.22 | 20.00 | 16.67 | 13.33 | 11.11 | 10.00 | 8.33 | 6.67 |
| 0.15 | 66.67 | 44.44 | 33.33 | 26.67 | 19.05 | 16.67 | 14.81 | 13.33 | 11.11 | 8.89 | 7.41 | 6.67 | 5.56 | 4.44 |
| 0.20 | 50.00 | 33.33 | 25.00 | 20.00 | 14.29 | 12.50 | 11.11 | 10.00 | 8.33 | 6.67 | 5.56 | 5.00 | 4.17 | 3.33 |
| 0.25 | 40.00 | 26.67 | 20.00 | 16.00 | 11.43 | 10.00 | 8.89 | 8.00 | 6.67 | 5.33 | 4.44 | 4.00 | 3.33 | 2.67 |
| 0.35 | 28.57 | 19.05 | 14.29 | 11.43 | 8.16 | 7.14 | 6.35 | 5.71 | 4.76 | 3.81 | 3.17 | 2.86 | 2.38 | 1.90 |
| 0.40 | 25.00 | 16.67 | 12.50 | 10.00 | 7.14 | 6.25 | 5.56 | 5.00 | 4.17 | 3.33 | 2.78 | 2.50 | 2.08 | 1.67 |
| 0.45 | 22.22 | 14.81 | 11.11 | 8.89 | 6.35 | 5.56 | 4.94 | 4.44 | 3.70 | 2.96 | 2.47 | 2.22 | 1.85 | 1.48 |
| 0.50 | 20.00 | 13.33 | 10.00 | 8.00 | 5.71 | 5.00 | 4.44 | 4.00 | 3.33 | 2.67 | 2.22 | 2.00 | 1.67 | 1.33 |
| 0.60 | 16.67 | 11.11 | 8.33 | 6.67 | 4.76 | 4.17 | 3.70 | 3.33 | 2.78 | 2.22 | 1.85 | 1.67 | 1.39 | 1.11 |
| 0.75 | 13.33 | 8.89 | 6.67 | 5.33 | 3.81 | 3.33 | 2.96 | 2.67 | 2.22 | 1.78 | 1.48 | 1.33 | 1.11 | 0.89 |
| 0.90 | 11.11 | 7.41 | 5.56 | 4.44 | 3.17 | 2.78 | 2.47 | 2.22 | 1.85 | 1.48 | 1.23 | 1.11 | 0.93 | 0.74 |
| 1.00 | 10.00 | 6.67 | 5.00 | 4.00 | 2.86 | 2.50 | 2.22 | 2.00 | 1.67 | 1.33 | 1.11 | 1.00 | 0.83 | 0.67 |
| 1.20 | 8.33 | 5.56 | 4.17 | 3.33 | 2.38 | 2.08 | 1.85 | 1.67 | 1.39 | 1.11 | 0.93 | 0.83 | 0.69 | 0.56 |
| 1.50 | 6.67 | 4.44 | 3.33 | 2.67 | 1.90 | 1.67 | 1.48 | 1.33 | 1.11 | 0.89 | 0.74 | 0.67 | 0.56 | 0.44 |

## SEEDING/TURFING AND PLANTING

### Grass Clippings Wet: Based on 3.5 m³/tonne

| Annual kg/100 m² | | Average 20 cuts kg/100 m² | | m²/tonne | | m²/m³ | |
|---|---|---|---|---|---|---|---|
| 32.0 | | 1.6 | | 61162.1 | | 214067.3 | |

| Nr of cuts | 22 | 20 | 18 | 16 | 12 | 4 |
|---|---|---|---|---|---|---|
| kg/cut | 1.45 | 1.60 | 1.78 | 2.00 | 2.67 | 8.00 |
| **Area capacity of 3 tonne vehicle per load** | | | | | | |
| m² | 206250 | 187500 | 168750 | 150000 | 112500 | 37500 |
| **Load m³** | **100 m² units/m³ of vehicle space** | | | | | |
| 1 | 196.4 | 178.6 | 160.7 | 142.9 | 107.1 | 35.7 |
| 2 | 392.9 | 357.1 | 321.4 | 285.7 | 214.3 | 71.4 |
| 3 | 589.3 | 535.7 | 482.1 | 428.6 | 321.4 | 107.1 |
| 4 | 785.7 | 714.3 | 642.9 | 571.4 | 428.6 | 142.9 |
| 5 | 982.1 | 892.9 | 803.6 | 714.3 | 535.7 | 178.6 |

### Transportation of Trees

To unload large trees a machine with the necessary lifting strength is required. The weight of the trees must therefore be known in advance. The following table gives a rough overview. The additional columns with root ball dimensions and the number of plants per trailer provide additional information, for example about preparing planting holes and calculating unloading times.

| Girth in cm | Rootball diameter in cm | Ball height in cm | Weight in kg | Numbers of trees per trailer |
|---|---|---|---|---|
| 16–18 | 50–60 | 40 | 150 | 100–120 |
| 18–20 | 60–70 | 40–50 | 200 | 80–100 |
| 20–25 | 60–70 | 40–50 | 270 | 50–70 |
| 25–30 | 80 | 50–60 | 350 | 50 |
| 30–35 | 90–100 | 60–70 | 500 | 12–18 |
| 35–40 | 100–110 | 60–70 | 650 | 10–15 |
| 40–45 | 110–120 | 60–70 | 850 | 8–12 |
| 45–50 | 110–120 | 60–70 | 1100 | 5–7 |
| 50–60 | 130–140 | 60–70 | 1600 | 1–3 |
| 60–70 | 150–160 | 60–70 | 2500 | 1 |
| 70–80 | 180–200 | 70 | 4000 | 1 |
| 80–90 | 200–220 | 70–80 | 5500 | 1 |
| 90–100 | 230–250 | 80–90 | 7500 | 1 |
| 100–120 | 250–270 | 80–90 | 9500 | 1 |

Data supplied by Lorenz von Ehren GmbH
The information in the table is approximate; deviations depend on soil type, genus and weather

## FENCING AND GATES

## FENCING AND GATES

### Types of Preservative

| | |
|---|---|
| Creosote (tar oil) can be 'factory' applied | by pressure to BS 144: pts 1&2<br>by immersion to BS 144: pt 1<br>by hot and cold open tank to BS 144: pts 1&2 |
| Copper/chromium/arsenic (CCA) | by full cell process to BS 4072 pts 1&2 |
| Organic solvent (OS) | by double vacuum (vacvac) to BS 5707 pts 1&3<br>by immersion to BS 5057 pts 1&3 |
| Pentachlorophenol (PCP) | by heavy oil double vacuum to BS 5705 pts 2&3 |
| Boron diffusion process (treated with disodium octaborate to BWPA Manual 1986) | |

Note: Boron is used on green timber at source and the timber is supplied dry

### Cleft Chestnut Pale Fences

| Pales | Pale spacing | Wire lines | |
|---|---|---|---|
| 900 mm | 75 mm | 2 | temporary protection |
| 1050 mm | 75 or 100 mm | 2 | light protective fences |
| 1200 mm | 75 mm | 3 | perimeter fences |
| 1350 mm | 75 mm | 3 | perimeter fences |
| 1500 mm | 50 mm | 3 | narrow perimeter fences |
| 1800 mm | 50 mm | 3 | light security fences |

### Close-Boarded Fences

Close-boarded fences 1.05 to 1.8 m high
Type BCR (recessed) or BCM (morticed) with concrete posts 140 × 115 mm tapered and Type BW with timber posts

### Palisade Fences

Wooden palisade fences
Type WPC with concrete posts 140 × 115 mm tapered and Type WPW with timber posts

For both types of fence:
Height of fence 1050 mm: two rails
Height of fence 1200 mm: two rails
Height of fence 1500 mm: three rails
Height of fence 1650 mm: three rails
Height of fence 1800 mm: three rails

## FENCING AND GATES

### Post and Rail Fences

Wooden post and rail fences
Type MPR 11/3 morticed rails and Type SPR 11/3 nailed rails
Height to top of rail 1100 mm
Rails: three rails 87 mm, 38 mm

Type MPR 11/4 morticed rails and Type SPR 11/4 nailed rails
Height to top of rail 1100 mm
Rails: four rails 87 mm, 38 mm

Type MPR 13/4 morticed rails and Type SPR 13/4 nailed rails
Height to top of rail 1300 mm
Rail spacing 250 mm, 250 mm, and 225 mm from top
Rails: four rails 87 mm, 38 mm

### Steel Posts

#### Rolled steel angle iron posts for chain link fencing

| Posts | Fence height | Strut | Straining post |
|---|---|---|---|
| 1500 × 40 × 40 × 5 mm | 900 mm | 1500 × 40 × 40 × 5 mm | 1500 × 50 × 50 × 6 mm |
| 1800 × 40 × 40 × 5 mm | 1200 mm | 1800 × 40 × 40 × 5 mm | 1800 × 50 × 50 × 6 mm |
| 2000 × 45 × 45 × 5 mm | 1400 mm | 2000 × 45 × 45 × 5 mm | 2000 × 60 × 60 × 6 mm |
| 2600 × 45 × 45 × 5 mm | 1800 mm | 2600 × 45 × 45 × 5 mm | 2600 × 60 × 60 × 6 mm |
| 3000 × 50 × 50 × 6 mm with arms | 1800 mm | 2600 × 45 × 45 × 5 mm | 3000 × 60 × 60 × 6 mm |

### Concrete Posts

#### Concrete posts for chain link fencing

| Posts and straining posts | Fence height | Strut |
|---|---|---|
| 1570 mm 100 × 100 mm | 900 mm | 1500 mm × 75 × 75 mm |
| 1870 mm 125 × 125 mm | 1200 mm | 1830 mm × 100 × 75 mm |
| 2070 mm 125 × 125 mm | 1400 mm | 1980 mm × 100 × 75 mm |
| 2620 mm 125 × 125 mm | 1800 mm | 2590 mm × 100 × 85 mm |
| 3040 mm 125 × 125 mm | 1800 mm | 2590 mm × 100 × 85 mm (with arms) |

**FENCING AND GATES**

## Rolled Steel Angle Posts

### Rolled steel angle posts for rectangular wire mesh (field) fencing

| Posts | Fence height | Strut | Straining post |
|---|---|---|---|
| 1200 × 40 × 40 × 5 mm | 600 mm | 1200 × 75 × 75 mm | 1350 × 100 × 100 mm |
| 1400 × 40 × 40 × 5 mm | 800 mm | 1400 × 75 × 75 mm | 1550 × 100 × 100 mm |
| 1500 × 40 × 40 × 5 mm | 900 mm | 1500 × 75 × 75 mm | 1650 × 100 × 100 mm |
| 1600 × 40 × 40 × 5 mm | 1000 mm | 1600 × 75 × 75 mm | 1750 × 100 × 100 mm |
| 1750 × 40 × 40 × 5 mm | 1150 mm | 1750 × 75 × 100 mm | 1900 × 125 × 125 mm |

## Concrete Posts

### Concrete posts for rectangular wire mesh (field) fencing

| Posts | Fence height | Strut | Straining post |
|---|---|---|---|
| 1270 × 100 × 100 mm | 600 mm | 1200 × 75 × 75 mm | 1420 × 100 × 100 mm |
| 1470 × 100 × 100 mm | 800 mm | 1350 × 75 × 75 mm | 1620 × 100 × 100 mm |
| 1570 × 100 × 100 mm | 900 mm | 1500 × 75 × 75 mm | 1720 × 100 × 100 mm |
| 1670 × 100 × 100 mm | 600 mm | 1650 × 75 × 75 mm | 1820 × 100 × 100 mm |
| 1820 × 125 × 125 mm | 1150 mm | 1830 × 75 × 100 mm | 1970 × 125 × 125 mm |

## Cleft Chestnut Pale Fences

### Timber Posts

### Timber posts for wire mesh and hexagonal wire netting fences

Round timber for general fences

| Posts | Fence height | Strut | Straining post |
|---|---|---|---|
| 1300 × 65 mm dia. | 600 mm | 1200 × 80 mm dia. | 1450 × 100 mm dia. |
| 1500 × 65 mm dia. | 800 mm | 1400 × 80 mm dia. | 1650 × 100 mm dia. |
| 1600 × 65 mm dia. | 900 mm | 1500 × 80 mm dia. | 1750 × 100 mm dia. |
| 1700 × 65 mm dia. | 1050 mm | 1600 × 80 mm dia. | 1850 × 100 mm dia. |
| 1800 × 65 mm dia. | 1150 mm | 1750 × 80 mm dia. | 2000 × 120 mm dia. |

Squared timber for general fences

| Posts | Fence height | Strut | Straining post |
|---|---|---|---|
| 1300 × 75 × 75 mm | 600 mm | 1200 × 75 × 75 mm | 1450 × 100 × 100 mm |
| 1500 × 75 × 75 mm | 800 mm | 1400 × 75 × 75 mm | 1650 × 100 × 100 mm |
| 1600 × 75 × 75 mm | 900 mm | 1500 × 75 × 75 mm | 1750 × 100 × 100 mm |
| 1700 × 75 × 75 mm | 1050 mm | 1600 × 75 × 75 mm | 1850 × 100 × 100 mm |
| 1800 × 75 × 75 mm | 1150 mm | 1750 × 75 × 75 mm | 2000 × 125 × 100 mm |

## FENCING AND GATES

### Steel Fences to BS 1722: Part 9: 1992

|  | Fence height | Top/bottom rails and flat posts | Vertical bars |
|---|---|---|---|
| Light | 1000 mm | 40 × 10 mm 450 mm in ground | 12 mm dia. at 115 mm cs |
|  | 1200 mm | 40 × 10 mm 550 mm in ground | 12 mm dia. at 115 mm cs |
|  | 1400 mm | 40 × 10 mm 550 mm in ground | 12 mm dia. at 115 mm cs |
| Light | 1000 mm | 40 × 10 mm 450 mm in ground | 16 mm dia. at 120 mm cs |
|  | 1200 mm | 40 × 10 mm 550 mm in ground | 16 mm dia. at 120 mm cs |
|  | 1400 mm | 40 × 10 mm 550 mm in ground | 16 mm dia. at 120 mm cs |
| Medium | 1200 mm | 50 × 10 mm 550 mm in ground | 20 mm dia. at 125 mm cs |
|  | 1400 mm | 50 × 10 mm 550 mm in ground | 20 mm dia. at 125 mm cs |
|  | 1600 mm | 50 × 10 mm 600 mm in ground | 22 mm dia. at 145 mm cs |
|  | 1800 mm | 50 × 10 mm 600 mm in ground | 22 mm dia. at 145 mm cs |
| Heavy | 1600 mm | 50 × 10 mm 600 mm in ground | 22 mm dia. at 145 mm cs |
|  | 1800 mm | 50 × 10 mm 600 mm in ground | 22 mm dia. at 145 mm cs |
|  | 2000 mm | 50 × 10 mm 600 mm in ground | 22 mm dia. at 145 mm cs |
|  | 2200 mm | 50 × 10 mm 600 mm in ground | 22 mm dia. at 145 mm cs |

Notes: Mild steel fences: round or square verticals; flat standards and horizontals. Tops of vertical bars may be bow-top, blunt, or pointed. Round or square bar railings

### Timber Field Gates to BS 3470: 1975

Gates made to this standard are designed to open one way only
All timber gates are 1100 mm high
Width over stiles 2400, 2700, 3000, 3300, 3600, and 4200 mm
Gates over 4200 mm should be made in two leaves

### Steel Field Gates to BS 3470: 1975

All steel gates are 1100 mm high
Heavy duty:   width over stiles 2400, 3000, 3600 and 4500 mm
Light duty:   width over stiles 2400, 3000, and 3600 mm

## FENCING AND GATES

### Domestic Front Entrance Gates to BS 4092: Part 1: 1966

| Metal gates: | Single gates are 900 mm high minimum, 900 mm, 1000 mm and 1100 mm wide |
|---|---|

### Domestic Front Entrance Gates to BS 4092: Part 2: 1966

| Wooden gates: | All rails shall be tenoned into the stiles<br>Single gates are 840 mm high minimum, 801 mm and 1020 mm wide<br>Double gates are 840 mm high minimum, 2130, 2340 and 2640 mm wide |
|---|---|

### Timber Bridle Gates to BS 5709:1979 (Horse or Hunting Gates)

| Gates open one way only<br>Minimum width between posts<br>Minimum height | 1525 mm<br>1100 mm |
|---|---|

### Timber Kissing Gates to BS 5709:1979

| Minimum width | 700 mm |
|---|---|
| Minimum height | 1000 mm |
| Minimum distance between shutting posts | 600 mm |
| Minimum clearance at mid-point | 600 mm |

### Metal Kissing Gates to BS 5709:1979

Sizes are the same as those for timber kissing gates
Maximum gaps between rails 120 mm

### Categories of Pedestrian Guard Rail to BS 3049:1976

Class A for normal use
Class B where vandalism is expected
Class C where crowd pressure is likely

# DRAINAGE

## Width Required for Trenches for Various Diameters of Pipes

| Pipe diameter (mm) | Trench n.e. 1.50 m deep | Trench over 1.50 m deep |
| --- | --- | --- |
| n.e. 100 mm | 450 mm | 600 mm |
| 100–150 mm | 500 mm | 650 mm |
| 150–225 mm | 600 mm | 750 mm |
| 225–300 mm | 650 mm | 800 mm |
| 300–400 mm | 750 mm | 900 mm |
| 400–450 mm | 900 mm | 1050 mm |
| 450–600 mm | 1100 mm | 1300 mm |

## Weights and Dimensions – Vitrified Clay Pipes

| Product | Nominal diameter (mm) | Effective length (mm) | BS 65 limits of tolerance min (mm) | BS 65 limits of tolerance max (mm) | Crushing strength (kN/m) | Weight (kg/pipe) | Weight (kg/m) |
| --- | --- | --- | --- | --- | --- | --- | --- |
| Supersleve | 100 | 1600 | 96 | 105 | 35.00 | 14.71 | 9.19 |
|  | 150 | 1750 | 146 | 158 | 35.00 | 29.24 | 16.71 |
| Hepsleve | 225 | 1850 | 221 | 236 | 28.00 | 84.03 | 45.42 |
|  | 300 | 2500 | 295 | 313 | 34.00 | 193.05 | 77.22 |
|  | 150 | 1500 | 146 | 158 | 22.00 | 37.04 | 24.69 |
| Hepseal | 225 | 1750 | 221 | 236 | 28.00 | 85.47 | 48.84 |
|  | 300 | 2500 | 295 | 313 | 34.00 | 204.08 | 81.63 |
|  | 400 | 2500 | 394 | 414 | 44.00 | 357.14 | 142.86 |
|  | 450 | 2500 | 444 | 464 | 44.00 | 454.55 | 181.63 |
|  | 500 | 2500 | 494 | 514 | 48.00 | 555.56 | 222.22 |
|  | 600 | 2500 | 591 | 615 | 57.00 | 796.23 | 307.69 |
|  | 700 | 3000 | 689 | 719 | 67.00 | 1111.11 | 370.45 |
|  | 800 | 3000 | 788 | 822 | 72.00 | 1351.35 | 450.45 |
| Hepline | 100 | 1600 | 95 | 107 | 22.00 | 14.71 | 9.19 |
|  | 150 | 1750 | 145 | 160 | 22.00 | 29.24 | 16.71 |
|  | 225 | 1850 | 219 | 239 | 28.00 | 84.03 | 45.42 |
|  | 300 | 1850 | 292 | 317 | 34.00 | 142.86 | 77.22 |
| Hepduct (conduit) | 90 | 1500 | – | – | 28.00 | 12.05 | 8.03 |
|  | 100 | 1600 | – | – | 28.00 | 14.71 | 9.19 |
|  | 125 | 1750 | – | – | 28.00 | 20.73 | 11.84 |
|  | 150 | 1750 | – | – | 28.00 | 29.24 | 16.71 |
|  | 225 | 1850 | – | – | 28.00 | 84.03 | 45.42 |
|  | 300 | 1850 | – | – | 34.00 | 142.86 | 77.22 |

**DRAINAGE**

## Weights and Dimensions – Vitrified Clay Pipes

| Nominal internal diameter (mm) | Nominal wall thickness (mm) | Approximate weight (kg/m) |
|---|---|---|
| 150 | 25 | 45 |
| 225 | 29 | 71 |
| 300 | 32 | 122 |
| 375 | 35 | 162 |
| 450 | 38 | 191 |
| 600 | 48 | 317 |
| 750 | 54 | 454 |
| 900 | 60 | 616 |
| 1200 | 76 | 912 |
| 1500 | 89 | 1458 |
| 1800 | 102 | 1884 |
| 2100 | 127 | 2619 |

Wall thickness, weights and pipe lengths vary, depending on type of pipe required
The particulars shown above represent a selection of available diameters and are applicable to strength class 1 pipes with flexible rubber ring joints
Tubes with Ogee joints are also available

## DRAINAGE

## Weights and Dimensions – u-PVC Pipes

| | Nominal size | Mean outside diameter (mm) | | Wall thickness | Weight |
|---|---|---|---|---|---|
| | | min | max | (mm) | (kg/m) |
| Standard pipes | 82.4 | 82.4 | 82.7 | 3.2 | 1.2 |
| | 110.0 | 110.0 | 110.4 | 3.2 | 1.6 |
| | 160.0 | 160.0 | 160.6 | 4.1 | 3.0 |
| | 200.0 | 200.0 | 200.6 | 4.9 | 4.6 |
| | 250.0 | 250.0 | 250.7 | 6.1 | 7.2 |
| Perforated pipes heavy grade | As above | As above | As above | As above | As above |
| thin wall | 82.4 | 82.4 | 82.7 | 1.7 | – |
| | 110.0 | 110.0 | 110.4 | 2.2 | – |
| | 160.0 | 160.0 | 160.6 | 3.2 | – |

## Width of Trenches Required for Various Diameters of Pipes

| Pipe diameter (mm) | Trench n.e. 1.5 m deep (mm) | Trench over 1.5 m deep (mm) |
|---|---|---|
| n.e. 100 | 450 | 600 |
| 100–150 | 500 | 650 |
| 150–225 | 600 | 750 |
| 225–300 | 650 | 800 |
| 300–400 | 750 | 900 |
| 400–450 | 900 | 1050 |
| 450–600 | 1100 | 1300 |

## DRAINAGE

## DRAINAGE BELOW GROUND AND LAND DRAINAGE

### Flow of Water Which Can Be Carried by Various Sizes of Pipe

#### Clay or concrete pipes

| | Gradient of pipeline | | | | | | | |
|---|---|---|---|---|---|---|---|---|
| | 1:10 | 1:20 | 1:30 | 1:40 | 1:50 | 1:60 | 1:80 | 1:100 |
| Pipe size | Flow in litres per second | | | | | | | |
| DN 100 15.0 | 8.5 | 6.8 | 5.8 | 5.2 | 4.7 | 4.0 | 3.5 | |
| DN 150 28.0 | 19.0 | 16.0 | 14.0 | 12.0 | 11.0 | 9.1 | 8.0 | |
| DN 225 140.0 | 95.0 | 76.0 | 66.0 | 58.0 | 53.0 | 46.0 | 40.0 | |

#### Plastic pipes

| | Gradient of pipeline | | | | | | | |
|---|---|---|---|---|---|---|---|---|
| | 1:10 | 1:20 | 1:30 | 1:40 | 1:50 | 1:60 | 1:80 | 1:100 |
| Pipe size | Flow in litres per second | | | | | | | |
| 82.4 mm i/dia. | 12.0 | 8.5 | 6.8 | 5.8 | 5.2 | 4.7 | 4.0 | 3.5 |
| 110 mm i/dia. | 28.0 | 19.0 | 16.0 | 14.0 | 12.0 | 11.0 | 9.1 | 8.0 |
| 160 mm i/dia. | 76.0 | 53.0 | 43.0 | 37.0 | 33.0 | 29.0 | 25.0 | 22.0 |
| 200 mm i/dia. | 140.0 | 95.0 | 76.0 | 66.0 | 58.0 | 53.0 | 46.0 | 40.0 |

### Vitrified (Perforated) Clay Pipes and Fittings to BS En 295-5 1994

| Length not specified | | |
|---|---|---|
| 75 mm bore | 250 mm bore | 600 mm bore |
| 100 | 300 | 700 |
| 125 | 350 | 800 |
| 150 | 400 | 1000 |
| 200 | 450 | 1200 |
| 225 | 500 | |

### Precast Concrete Pipes: Prestressed Non-pressure Pipes and Fittings: Flexible Joints to BS 5911: Pt. 103: 1994

| Rationalized metric nominal sizes: 450, 500 | |
|---|---|
| Length: | 500–1000 by 100 increments |
| | 1000–2200 by 200 increments |
| | 2200–2800 by 300 increments |
| Angles: length: | 450–600 angles 45, 22.5, 11.25° |
| | 600 or more angles 22.5, 11.25° |

## DRAINAGE

**Precast Concrete Pipes: Unreinforced and Circular Manholes and Soakaways to BS 5911: Pt. 200: 1994**

| Nominal sizes: | |
|---|---|
| Shafts: | 675, 900 mm |
| Chambers: | 900, 1050, 1200, 1350, 1500, 1800, 2100, 2400, 2700, 3000 mm |
| Large chambers: | To have either tapered reducing rings or a flat reducing slab in order to accept the standard cover |
| Ring depths: | 1. 300–1200 mm by 300 mm increments except for bottom slab and rings below cover slab, these are by 150 mm increments |
| | 2. 250–1000 mm by 250 mm increments except for bottom slab and rings below cover slab, these are by 125 mm increments |
| Access hole: | 750 × 750 mm for DN 1050 chamber |
| | 1200 × 675 mm for DN 1350 chamber |

## Calculation of Soakaway Depth

The following formula determines the depth of concrete ring soakaway that would be required for draining given amounts of water.

$$h = \frac{4ar}{3\pi D^2}$$

h = depth of the chamber below the invert pipe
a = the area to be drained
r = the hourly rate of rainfall (50 mm per hour)
$\pi$ = pi
D = internal diameter of the soakaway

This table shows the depth of chambers in each ring size which would be required to contain the volume of water specified. These allow a recommended storage capacity of ⅓ (one third of the hourly rainfall figure).

### Table Showing Required Depth of Concrete Ring Chambers in Metres

| Area m² | 50 | 100 | 150 | 200 | 300 | 400 | 500 |
|---|---|---|---|---|---|---|---|
| Ring size | | | | | | | |
| 0.9 | 1.31 | 2.62 | 3.93 | 5.24 | 7.86 | 10.48 | 13.10 |
| 1.1 | 0.96 | 1.92 | 2.89 | 3.85 | 5.77 | 7.70 | 9.62 |
| 1.2 | 0.74 | 1.47 | 2.21 | 2.95 | 4.42 | 5.89 | 7.37 |
| 1.4 | 0.58 | 1.16 | 1.75 | 2.33 | 3.49 | 4.66 | 5.82 |
| 1.5 | 0.47 | 0.94 | 1.41 | 1.89 | 2.83 | 3.77 | 4.72 |
| 1.8 | 0.33 | 0.65 | 0.98 | 1.31 | 1.96 | 2.62 | 3.27 |
| 2.1 | 0.24 | 0.48 | 0.72 | 0.96 | 1.44 | 1.92 | 2.41 |
| 2.4 | 0.18 | 0.37 | 0.55 | 0.74 | 1.11 | 1.47 | 1.84 |
| 2.7 | 0.15 | 0.29 | 0.44 | 0.58 | 0.87 | 1.16 | 1.46 |
| 3.0 | 0.12 | 0.24 | 0.35 | 0.47 | 0.71 | 0.94 | 1.18 |

## DRAINAGE

### Precast Concrete Inspection Chambers and Gullies to BS 5911: Part 230: 1994

| | |
|---|---|
| Nominal sizes: | 375 diameter, 750, 900 mm deep |
| | 450 diameter, 750, 900, 1050, 1200 mm deep |
| Depths: | from the top for trapped or untrapped units: |
| |     centre of outlet 300 mm |
| |     invert (bottom) of the outlet pipe 400 mm |
| Depth of water seal for trapped gullies: | |
| | 85 mm, rodding eye int. dia. 100 mm |
| Cover slab: | 65 mm min |

### Bedding Flexible Pipes: u-PVC Or Ductile Iron

| | |
|---|---|
| Type 1 = | 100 mm fill below pipe, 300 mm above pipe: single size material |
| Type 2 = | 100 mm fill below pipe, 300 mm above pipe: single size or graded material |
| Type 3 = | 100 mm fill below pipe, 75 mm above pipe with concrete protective slab over |
| Type 4 = | 100 mm fill below pipe, fill laid level with top of pipe |
| Type 5 = | 200 mm fill below pipe, fill laid level with top of pipe |
| Concrete = | 25 mm sand blinding to bottom of trench, pipe supported on chocks, 100 mm concrete under the pipe, 150 mm concrete over the pipe |

## DRAINAGE

### Bedding Rigid Pipes: Clay or Concrete
(for vitrified clay pipes the manufacturer should be consulted)

| | |
|---|---|
| Class D: | Pipe laid on natural ground with cut-outs for joints, soil screened to remove stones over 40 mm and returned over pipe to 150 m min depth. Suitable for firm ground with trenches trimmed by hand. |
| Class N: | Pipe laid on 50 mm granular material of graded aggregate to Table 4 of BS 882, or 10 mm aggregate to Table 6 of BS 882, or as dug light soil (not clay) screened to remove stones over 10 mm. Suitable for machine dug trenches. |
| Class B: | As Class N, but with granular bedding extending half way up the pipe diameter. |
| Class F: | Pipe laid on 100 mm granular fill to BS 882 below pipe, minimum 150 mm granular fill above pipe: single size material. Suitable for machine dug trenches. |
| Class A: | Concrete 100 mm thick under the pipe extending half way up the pipe, backfilled with the appropriate class of fill. Used where there is only a very shallow fall to the drain. Class A bedding allows the pipes to be laid to an exact gradient. |
| Concrete surround: | 25 mm sand blinding to bottom of trench, pipe supported on chocks, 100 mm concrete under the pipe, 150 mm concrete over the pipe. It is preferable to bed pipes under slabs or wall in granular material. |

## PIPED SUPPLY SYSTEMS

### Identification of Service Tubes From Utility to Dwellings

| Utility | Colour | Size | Depth |
|---|---|---|---|
| British Telecom | grey | 54 mm od | 450 mm |
| Electricity | black | 38 mm od | 450 mm |
| Gas | yellow | 42 mm od rigid 60 mm od convoluted | 450 mm |
| Water | may be blue | (normally untubed) | 750 mm |

## ELECTRICAL SUPPLY/POWER/LIGHTING SYSTEMS

### Electrical Insulation Class En 60.598 BS 4533

Class 1: luminaires comply with class 1 (I) earthed electrical requirements
Class 2: luminaires comply with class 2 (II) double insulated electrical requirements
Class 3: luminaires comply with class 3 (III) electrical requirements

## Protection to Light Fittings

BS EN 60529:1992 Classification for degrees of protection provided by enclosures.
(IP Code – International or ingress Protection)

1st characteristic: against ingress of solid foreign objects

| The figure | | |
|---|---|---|
| | 2 | indicates that fingers cannot enter |
| | 3 | that a 2.5 mm diameter probe cannot enter |
| | 4 | that a 1.0 mm diameter probe cannot enter |
| | 5 | the fitting is dust proof (no dust around live parts) |
| | 6 | the fitting is dust tight (no dust entry) |

2nd characteristic: ingress of water with harmful effects

| The figure | | |
|---|---|---|
| | 0 | indicates unprotected |
| | 1 | vertically dripping water cannot enter |
| | 2 | water dripping 15° (tilt) cannot enter |
| | 3 | spraying water cannot enter |
| | 4 | splashing water cannot enter |
| | 5 | jetting water cannot enter |
| | 6 | powerful jetting water cannot enter |
| | 7 | proof against temporary immersion |
| | 8 | proof against continuous immersion |

Optional additional codes: A–D protects against access to hazardous parts

| | H | high voltage apparatus |
|---|---|---|
| | M | fitting was in motion during water test |
| | S | fitting was static during water test |
| | W | protects against weather |

Marking code arrangement: (example) IPX5S = IP (International or Ingress Protection)
X (denotes omission of first characteristic)
5 = jetting
S = static during water test

## RAIL TRACKS

|  | kg/m of track | lb/ft of track |
|---|---|---|
| **Standard gauge** <br> Bull-head rails, chairs, transverse timber (softwood) sleepers etc. | 245 | 165 |
| **Main lines** <br> Flat-bottom rails, transverse prestressed concrete sleepers, etc. <br> Add for electric third rail <br> Add for crushed stone ballast | 418 <br> 51 <br> 2600 | 280 <br> 35 <br> 1750 |
|  | kg/m² | lb/sq ft |
| Overall average weight – rails connections, sleepers, ballast, etc. | 733 | 150 |
|  | kg/m of track | lb/ft of track |
| Bridge rails, longitudinal timber sleepers, etc. | 112 | 75 |

**RAIL TRACKS**

## Heavy Rails

| British Standard Section No. | Rail height (mm) | Foot width (mm) | Head width (mm) | Min web thickness (mm) | Section weight (kg/m) |
|---|---|---|---|---|---|
| Flat Bottom Rails | | | | | |
| 60 A | 114.30 | 109.54 | 57.15 | 11.11 | 30.62 |
| 70 A | 123.82 | 111.12 | 60.32 | 12.30 | 34.81 |
| 75 A | 128.59 | 114.30 | 61.91 | 12.70 | 37.45 |
| 80 A | 133.35 | 117.47 | 63.50 | 13.10 | 39.76 |
| 90 A | 142.88 | 127.00 | 66.67 | 13.89 | 45.10 |
| 95 A | 147.64 | 130.17 | 69.85 | 14.68 | 47.31 |
| 100 A | 152.40 | 133.35 | 69.85 | 15.08 | 50.18 |
| 110 A | 158.75 | 139.70 | 69.85 | 15.87 | 54.52 |
| 113 A | 158.75 | 139.70 | 69.85 | 20.00 | 56.22 |
| 50 'O' | 100.01 | 100.01 | 52.39 | 10.32 | 24.82 |
| 80 'O' | 127.00 | 127.00 | 63.50 | 13.89 | 39.74 |
| 60R | 114.30 | 109.54 | 57.15 | 11.11 | 29.85 |
| 75R | 128.59 | 122.24 | 61.91 | 13.10 | 37.09 |
| 80R | 133.35 | 127.00 | 63.50 | 13.49 | 39.72 |
| 90R | 142.88 | 136.53 | 66.67 | 13.89 | 44.58 |
| 95R | 147.64 | 141.29 | 68.26 | 14.29 | 47.21 |
| 100R | 152.40 | 146.05 | 69.85 | 14.29 | 49.60 |
| 95N | 147.64 | 139.70 | 69.85 | 13.89 | 47.27 |
| Bull Head Rails | | | | | |
| 95R BH | 145.26 | 69.85 | 69.85 | 19.05 | 47.07 |

## Light Rails

| British Standard Section No. | Rail height (mm) | Foot width (mm) | Head width (mm) | Min web thickness (mm) | Section weight (kg/m) |
|---|---|---|---|---|---|
| Flat Bottom Rails | | | | | |
| 20M | 65.09 | 55.56 | 30.96 | 6.75 | 9.88 |
| 30M | 75.41 | 69.85 | 38.10 | 9.13 | 14.79 |
| 35M | 80.96 | 76.20 | 42.86 | 9.13 | 17.39 |
| 35R | 85.73 | 82.55 | 44.45 | 8.33 | 17.40 |
| 40 | 88.11 | 80.57 | 45.64 | 12.3 | 19.89 |
| Bridge Rails | | | | | |
| 13 | 48.00 | 92 | 36.00 | 18.0 | 13.31 |
| 16 | 54.00 | 108 | 44.50 | 16.0 | 16.06 |
| 20 | 55.50 | 127 | 50.00 | 20.5 | 19.86 |
| 28 | 67.00 | 152 | 50.00 | 31.0 | 28.62 |
| 35 | 76.00 | 160 | 58.00 | 34.5 | 35.38 |
| 50 | 76.00 | 165 | 58.50 | – | 50.18 |
| Crane Rails | | | | | |
| A65 | 75.00 | 175.00 | 65.00 | 38.0 | 43.10 |
| A75 | 85.00 | 200.00 | 75.00 | 45.0 | 56.20 |
| A100 | 95.00 | 200.00 | 100.00 | 60.0 | 74.30 |
| A120 | 105.00 | 220.00 | 120.00 | 72.0 | 100.00 |
| 175CR | 152.40 | 152.40 | 107.95 | 38.1 | 86.92 |

## RAIL TRACKS

### Fish Plates

| British Standard Section No. | Overall plate length | | Hole diameter | Finished weight per pair | |
|---|---|---|---|---|---|
| | 4 Hole (mm) | 6 Hole (mm) | (mm) | 4 Hole (kg/pair) | 6 Hole (kg/pair) |
| For British Standard Heavy Rails: Flat Bottom Rails ||||||
| 60 A | 406.40 | 609.60 | 20.64 | 9.87 | 14.76 |
| 70 A | 406.40 | 609.60 | 22.22 | 11.15 | 16.65 |
| 75 A | 406.40 | – | 23.81 | 11.82 | 17.73 |
| 80 A | 406.40 | 609.60 | 23.81 | 13.15 | 19.72 |
| 90 A | 457.20 | 685.80 | 25.40 | 17.49 | 26.23 |
| 100 A | 508.00 | – | pear | 25.02 | – |
| 110 A (shallow) | 507.00 | – | 27.00 | 30.11 | 54.64 |
| 113 A (heavy) | 507.00 | – | 27.00 | 30.11 | 54.64 |
| 50 'O' (shallow) | 406.40 | – | – | 6.68 | 10.14 |
| 80 'O' (shallow) | 495.30 | – | 23.81 | 14.72 | 22.69 |
| 60R (shallow) | 406.40 | 609.60 | 20.64 | 8.76 | 13.13 |
| 60R (angled) | 406.40 | 609.60 | 20.64 | 11.27 | 16.90 |
| 75R (shallow) | 406.40 | – | 23.81 | 10.94 | 16.42 |
| 75R (angled) | 406.40 | – | 23.81 | 13.67 | – |
| 80R (shallow) | 406.40 | 609.60 | 23.81 | 11.93 | 17.89 |
| 80R (angled) | 406.40 | 609.60 | 23.81 | 14.90 | 22.33 |
| For British Standard Heavy Rails: Bull head rails ||||||
| 95R BH (shallow) | – | 457.20 | 27.00 | 14.59 | 14.61 |
| For British Standard Light Rails: Flat Bottom Rails ||||||
| 30M | 355.6 | – | – | – | 2.72 |
| 35M | 355.6 | – | – | – | 2.83 |
| 40 | 355.6 | – | – | 3.76 | – |

## FRACTIONS, DECIMALS AND MILLIMETRE EQUIVALENTS

| Fractions | Decimals | (mm) | Fractions | Decimals | (mm) |
|---|---|---|---|---|---|
| 1/64 | 0.015625 | 0.396875 | 33/64 | 0.515625 | 13.096875 |
| 1/32 | 0.03125 | 0.79375 | 17/32 | 0.53125 | 13.49375 |
| 3/64 | 0.046875 | 1.190625 | 35/64 | 0.546875 | 13.890625 |
| 1/16 | 0.0625 | 1.5875 | 9/16 | 0.5625 | 14.2875 |
| 5/64 | 0.078125 | 1.984375 | 37/64 | 0.578125 | 14.684375 |
| 3/32 | 0.09375 | 2.38125 | 19/32 | 0.59375 | 15.08125 |
| 7/64 | 0.109375 | 2.778125 | 39/64 | 0.609375 | 15.478125 |
| 1/8 | 0.125 | 3.175 | 5/8 | 0.625 | 15.875 |
| 9/64 | 0.140625 | 3.571875 | 41/64 | 0.640625 | 16.271875 |
| 5/32 | 0.15625 | 3.96875 | 21/32 | 0.65625 | 16.66875 |
| 11/64 | 0.171875 | 4.365625 | 43/64 | 0.671875 | 17.065625 |
| 3/16 | 0.1875 | 4.7625 | 11/16 | 0.6875 | 17.4625 |
| 13/64 | 0.203125 | 5.159375 | 45/64 | 0.703125 | 17.859375 |
| 7/32 | 0.21875 | 5.55625 | 23/32 | 0.71875 | 18.25625 |
| 15/64 | 0.234375 | 5.953125 | 47/64 | 0.734375 | 18.653125 |
| 1/4 | 0.25 | 6.35 | 3/4 | 0.75 | 19.05 |
| 17/64 | 0.265625 | 6.746875 | 49/64 | 0.765625 | 19.446875 |
| 9/32 | 0.28125 | 7.14375 | 25/32 | 0.78125 | 19.84375 |
| 19/64 | 0.296875 | 7.540625 | 51/64 | 0.796875 | 20.240625 |
| 5/16 | 0.3125 | 7.9375 | 13/16 | 0.8125 | 20.6375 |
| 21/64 | 0.328125 | 8.334375 | 53/64 | 0.828125 | 21.034375 |
| 11/32 | 0.34375 | 8.73125 | 27/32 | 0.84375 | 21.43125 |
| 23/64 | 0.359375 | 9.128125 | 55/64 | 0.859375 | 21.828125 |
| 3/8 | 0.375 | 9.525 | 7/8 | 0.875 | 22.225 |
| 25/64 | 0.390625 | 9.921875 | 57/64 | 0.890625 | 22.621875 |
| 13/32 | 0.40625 | 10.31875 | 29/32 | 0.90625 | 23.01875 |
| 27/64 | 0.421875 | 10.71563 | 59/64 | 0.921875 | 23.415625 |
| 7/16 | 0.4375 | 11.1125 | 15/16 | 0.9375 | 23.8125 |
| 29/64 | 0.453125 | 11.50938 | 61/64 | 0.953125 | 24.209375 |
| 15/32 | 0.46875 | 11.90625 | 31/32 | 0.96875 | 24.60625 |
| 31/64 | 0.484375 | 12.30313 | 63/64 | 0.984375 | 25.003125 |
| 1/2 | 0.5 | 12.7 | 1.0 | 1 | 25.4 |

## IMPERIAL STANDARD WIRE GAUGE (SWG)

| SWG No. | Diameter (inches) | Diameter (mm) | SWG No. | Diameter (inches) | Diameter (mm) |
|---|---|---|---|---|---|
| 7/0 | 0.5 | 12.7 | 23 | 0.024 | 0.61 |
| 6/0 | 0.464 | 11.79 | 24 | 0.022 | 0.559 |
| 5/0 | 0.432 | 10.97 | 25 | 0.02 | 0.508 |
| 4/0 | 0.4 | 10.16 | 26 | 0.018 | 0.457 |
| 3/0 | 0.372 | 9.45 | 27 | 0.0164 | 0.417 |
| 2/0 | 0.348 | 8.84 | 28 | 0.0148 | 0.376 |
| 1/0 | 0.324 | 8.23 | 29 | 0.0136 | 0.345 |
| 1 | 0.3 | 7.62 | 30 | 0.0124 | 0.315 |
| 2 | 0.276 | 7.01 | 31 | 0.0116 | 0.295 |
| 3 | 0.252 | 6.4 | 32 | 0.0108 | 0.274 |
| 4 | 0.232 | 5.89 | 33 | 0.01 | 0.254 |
| 5 | 0.212 | 5.38 | 34 | 0.009 | 0.234 |
| 6 | 0.192 | 4.88 | 35 | 0.008 | 0.213 |
| 7 | 0.176 | 4.47 | 36 | 0.008 | 0.193 |
| 8 | 0.16 | 4.06 | 37 | 0.007 | 0.173 |
| 9 | 0.144 | 3.66 | 38 | 0.006 | 0.152 |
| 10 | 0.128 | 3.25 | 39 | 0.005 | 0.132 |
| 11 | 0.116 | 2.95 | 40 | 0.005 | 0.122 |
| 12 | 0.104 | 2.64 | 41 | 0.004 | 0.112 |
| 13 | 0.092 | 2.34 | 42 | 0.004 | 0.102 |
| 14 | 0.08 | 2.03 | 43 | 0.004 | 0.091 |
| 15 | 0.072 | 1.83 | 44 | 0.003 | 0.081 |
| 16 | 0.064 | 1.63 | 45 | 0.003 | 0.071 |
| 17 | 0.056 | 1.42 | 46 | 0.002 | 0.061 |
| 18 | 0.048 | 1.22 | 47 | 0.002 | 0.051 |
| 19 | 0.04 | 1.016 | 48 | 0.002 | 0.041 |
| 20 | 0.036 | 0.914 | 49 | 0.001 | 0.031 |
| 21 | 0.032 | 0.813 | 50 | 0.001 | 0.025 |
| 22 | 0.028 | 0.711 | | | |

## PIPES, WATER, STORAGE, INSULATION

## WATER PRESSURE DUE TO HEIGHT

### Imperial

| Head (Feet) | Pressure (lb/in$^2$) | | Head (Feet) | Pressure (lb/in$^2$) |
|---|---|---|---|---|
| 1 | 0.43 | | 70 | 30.35 |
| 5 | 2.17 | | 75 | 32.51 |
| 10 | 4.34 | | 80 | 34.68 |
| 15 | 6.5 | | 85 | 36.85 |
| 20 | 8.67 | | 90 | 39.02 |
| 25 | 10.84 | | 95 | 41.18 |
| 30 | 13.01 | | 100 | 43.35 |
| 35 | 15.17 | | 105 | 45.52 |
| 40 | 17.34 | | 110 | 47.69 |
| 45 | 19.51 | | 120 | 52.02 |
| 50 | 21.68 | | 130 | 56.36 |
| 55 | 23.84 | | 140 | 60.69 |
| 60 | 26.01 | | 150 | 65.03 |
| 65 | 28.18 | | | |

### Metric

| Head (m) | Pressure (bar) | | Head (m) | Pressure (bar) |
|---|---|---|---|---|
| 0.5 | 0.049 | | 18.0 | 1.766 |
| 1.0 | 0.098 | | 19.0 | 1.864 |
| 1.5 | 0.147 | | 20.0 | 1.962 |
| 2.0 | 0.196 | | 21.0 | 2.06 |
| 3.0 | 0.294 | | 22.0 | 2.158 |
| 4.0 | 0.392 | | 23.0 | 2.256 |
| 5.0 | 0.491 | | 24.0 | 2.354 |
| 6.0 | 0.589 | | 25.0 | 2.453 |
| 7.0 | 0.687 | | 26.0 | 2.551 |
| 8.0 | 0.785 | | 27.0 | 2.649 |
| 9.0 | 0.883 | | 28.0 | 2.747 |
| 10.0 | 0.981 | | 29.0 | 2.845 |
| 11.0 | 1.079 | | 30.0 | 2.943 |
| 12.0 | 1.177 | | 32.5 | 3.188 |
| 13.0 | 1.275 | | 35.0 | 3.434 |
| 14.0 | 1.373 | | 37.5 | 3.679 |
| 15.0 | 1.472 | | 40.0 | 3.924 |
| 16.0 | 1.57 | | 42.5 | 4.169 |
| 17.0 | 1.668 | | 45.0 | 4.415 |

| | | | |
|---|---|---|---|
| 1 bar | = | 14.5038 lbf/in$^2$ |
| 1 lbf/in$^2$ | = | 0.06895 bar |
| 1 metre | = | 3.2808 ft or 39.3701 in |
| 1 foot | = | 0.3048 metres |
| 1 in wg | = | 2.5 mbar (249.1 N/m$^2$) |

## PIPES, WATER, STORAGE, INSULATION

### Dimensions and Weights of Copper Pipes to BSEN 1057, BSEN 12499, BSEN 14251

| Outside diameter (mm) | Internal diameter (mm) | Weight per metre (kg) | Internal diameter (mm) | Weight per metre (kg) | Internal siameter (mm) | Weight per metre (kg) |
|---|---|---|---|---|---|---|
| | Formerly Table X | | Formerly Table Y | | Formerly Table Z | |
| 6 | 4.80 | 0.0911 | 4.40 | 0.1170 | 5.00 | 0.0774 |
| 8 | 6.80 | 0.1246 | 6.40 | 0.1617 | 7.00 | 0.1054 |
| 10 | 8.80 | 0.1580 | 8.40 | 0.2064 | 9.00 | 0.1334 |
| 12 | 10.80 | 0.1914 | 10.40 | 0.2511 | 11.00 | 0.1612 |
| 15 | 13.60 | 0.2796 | 13.00 | 0.3923 | 14.00 | 0.2031 |
| 18 | 16.40 | 0.3852 | 16.00 | 0.4760 | 16.80 | 0.2918 |
| 22 | 20.22 | 0.5308 | 19.62 | 0.6974 | 20.82 | 0.3589 |
| 28 | 26.22 | 0.6814 | 25.62 | 0.8985 | 26.82 | 0.4594 |
| 35 | 32.63 | 1.1334 | 32.03 | 1.4085 | 33.63 | 0.6701 |
| 42 | 39.63 | 1.3675 | 39.03 | 1.6996 | 40.43 | 0.9216 |
| 54 | 51.63 | 1.7691 | 50.03 | 2.9052 | 52.23 | 1.3343 |
| 76.1 | 73.22 | 3.1287 | 72.22 | 4.1437 | 73.82 | 2.5131 |
| 108 | 105.12 | 4.4666 | 103.12 | 7.3745 | 105.72 | 3.5834 |
| 133 | 130.38 | 5.5151 | – | – | 130.38 | 5.5151 |
| 159 | 155.38 | 8.7795 | – | – | 156.38 | 6.6056 |

### Dimensions of Stainless Steel Pipes to BS 4127

| Outside siameter (mm) | Maximum outside siameter (mm) | Minimum outside diameter (mm) | Wall thickness (mm) | Working pressure (bar) |
|---|---|---|---|---|
| 6 | 6.045 | 5.940 | 0.6 | 330 |
| 8 | 8.045 | 7.940 | 0.6 | 260 |
| 10 | 10.045 | 9.940 | 0.6 | 210 |
| 12 | 12.045 | 11.940 | 0.6 | 170 |
| 15 | 15.045 | 14.940 | 0.6 | 140 |
| 18 | 18.045 | 17.940 | 0.7 | 135 |
| 22 | 22.055 | 21.950 | 0.7 | 110 |
| 28 | 28.055 | 27.950 | 0.8 | 121 |
| 35 | 35.070 | 34.965 | 1.0 | 100 |
| 42 | 42.070 | 41.965 | 1.1 | 91 |
| 54 | 54.090 | 53.940 | 1.2 | 77 |

**PIPES, WATER, STORAGE, INSULATION**

## Dimensions of Steel Pipes to BS 1387

| Nominal Size (mm) | Approx. Outside Diameter (mm) | Outside diameter | | | | Thickness | | |
|---|---|---|---|---|---|---|---|---|
| | | Light | | Medium & Heavy | | Light | Medium | Heavy |
| | | Max (mm) | Min (mm) | Max (mm) | Min (mm) | (mm) | (mm) | (mm) |
| 6 | 10.20 | 10.10 | 9.70 | 10.40 | 9.80 | 1.80 | 2.00 | 2.65 |
| 8 | 13.50 | 13.60 | 13.20 | 13.90 | 13.30 | 1.80 | 2.35 | 2.90 |
| 10 | 17.20 | 17.10 | 16.70 | 17.40 | 16.80 | 1.80 | 2.35 | 2.90 |
| 15 | 21.30 | 21.40 | 21.00 | 21.70 | 21.10 | 2.00 | 2.65 | 3.25 |
| 20 | 26.90 | 26.90 | 26.40 | 27.20 | 26.60 | 2.35 | 2.65 | 3.25 |
| 25 | 33.70 | 33.80 | 33.20 | 34.20 | 33.40 | 2.65 | 3.25 | 4.05 |
| 32 | 42.40 | 42.50 | 41.90 | 42.90 | 42.10 | 2.65 | 3.25 | 4.05 |
| 40 | 48.30 | 48.40 | 47.80 | 48.80 | 48.00 | 2.90 | 3.25 | 4.05 |
| 50 | 60.30 | 60.20 | 59.60 | 60.80 | 59.80 | 2.90 | 3.65 | 4.50 |
| 65 | 76.10 | 76.00 | 75.20 | 76.60 | 75.40 | 3.25 | 3.65 | 4.50 |
| 80 | 88.90 | 88.70 | 87.90 | 89.50 | 88.10 | 3.25 | 4.05 | 4.85 |
| 100 | 114.30 | 113.90 | 113.00 | 114.90 | 113.30 | 3.65 | 4.50 | 5.40 |
| 125 | 139.70 | – | – | 140.60 | 138.70 | – | 4.85 | 5.40 |
| 150 | 165.1* | – | – | 166.10 | 164.10 | – | 4.85 | 5.40 |

* 165.1 mm (6.5in) outside diameter is not generally recommended except where screwing to BS 21 is necessary
All dimensions are in accordance with ISO R65 except approximate outside diameters which are in accordance with ISO R64
Light quality is equivalent to ISO R65 Light Series II

## Approximate Metres Per Tonne of Tubes to BS 1387

| Nom. size (mm) | BLACK | | | | | | GALVANIZED | | | | | |
|---|---|---|---|---|---|---|---|---|---|---|---|---|
| | Plain/screwed ends | | | Screwed & socketed | | | Plain/screwed ends | | | Screwed & socketed | | |
| | L (m) | M (m) | H (m) | L (m) | M (m) | H (m) | L (m) | M (m) | H (m) | L (m) | M (m) | H (m) |
| 6 | 2765 | 2461 | 2030 | 2743 | 2443 | 2018 | 2604 | 2333 | 1948 | 2584 | 2317 | 1937 |
| 8 | 1936 | 1538 | 1300 | 1920 | 1527 | 1292 | 1826 | 1467 | 1254 | 1811 | 1458 | 1247 |
| 10 | 1483 | 1173 | 979 | 1471 | 1165 | 974 | 1400 | 1120 | 944 | 1386 | 1113 | 939 |
| 15 | 1050 | 817 | 688 | 1040 | 811 | 684 | 996 | 785 | 665 | 987 | 779 | 661 |
| 20 | 712 | 634 | 529 | 704 | 628 | 525 | 679 | 609 | 512 | 673 | 603 | 508 |
| 25 | 498 | 410 | 336 | 494 | 407 | 334 | 478 | 396 | 327 | 474 | 394 | 325 |
| 32 | 388 | 319 | 260 | 384 | 316 | 259 | 373 | 308 | 254 | 369 | 305 | 252 |
| 40 | 307 | 277 | 226 | 303 | 273 | 223 | 296 | 268 | 220 | 292 | 264 | 217 |
| 50 | 244 | 196 | 162 | 239 | 194 | 160 | 235 | 191 | 158 | 231 | 188 | 157 |
| 65 | 172 | 153 | 127 | 169 | 151 | 125 | 167 | 149 | 124 | 163 | 146 | 122 |
| 80 | 147 | 118 | 99 | 143 | 116 | 98 | 142 | 115 | 97 | 139 | 113 | 96 |
| 100 | 101 | 82 | 69 | 98 | 81 | 68 | 98 | 81 | 68 | 95 | 79 | 67 |
| 125 | – | 62 | 56 | – | 60 | 55 | – | 60 | 55 | – | 59 | 54 |
| 150 | – | 52 | 47 | – | 50 | 46 | – | 51 | 46 | – | 49 | 45 |

The figures for 'plain or screwed ends' apply also to tubes to BS 1775 of equivalent size and thickness
Key:
L – Light
M – Medium
H – Heavy

## PIPES, WATER, STORAGE, INSULATION

## Flange Dimension Chart to BS 4504 & BS 10

### Normal Pressure Rating (PN 6) 6 Bar

| Nom. size | Flange outside dia. | Table 6/2 Forged Welding Neck | Table 6/3 Plate Slip on | Table 6/4 Forged Bossed Screwed | Table 6/5 Forged Bossed Slip on | Table 6/8 Plate Blank | Raised face Dia. | Raised face T'ness | Nr. bolt hole | Size of bolt |
|---|---|---|---|---|---|---|---|---|---|---|
| 15 | 80 | 12 | 12 | 12 | 12 | 12 | 40 | 2 | 4 | M10 × 40 |
| 20 | 90 | 14 | 14 | 14 | 14 | 14 | 50 | 2 | 4 | M10 × 45 |
| 25 | 100 | 14 | 14 | 14 | 14 | 14 | 60 | 2 | 4 | M10 × 45 |
| 32 | 120 | 14 | 16 | 14 | 14 | 14 | 70 | 2 | 4 | M12 × 45 |
| 40 | 130 | 14 | 16 | 14 | 14 | 14 | 80 | 3 | 4 | M12 × 45 |
| 50 | 140 | 14 | 16 | 14 | 14 | 14 | 90 | 3 | 4 | M12 × 45 |
| 65 | 160 | 14 | 16 | 14 | 14 | 14 | 110 | 3 | 4 | M12 × 45 |
| 80 | 190 | 16 | 18 | 16 | 16 | 16 | 128 | 3 | 4 | M16 × 55 |
| 100 | 210 | 16 | 18 | 16 | 16 | 16 | 148 | 3 | 4 | M16 × 55 |
| 125 | 240 | 18 | 20 | 18 | 18 | 18 | 178 | 3 | 8 | M16 × 60 |
| 150 | 265 | 18 | 20 | 18 | 18 | 18 | 202 | 3 | 8 | M16 × 60 |
| 200 | 320 | 20 | 22 | – | 20 | 20 | 258 | 3 | 8 | M16 × 60 |
| 250 | 375 | 22 | 24 | – | 22 | 22 | 312 | 3 | 12 | M16 × 65 |
| 300 | 440 | 22 | 24 | – | 22 | 22 | 365 | 4 | 12 | M20 × 70 |

### Normal Pressure Rating (PN 16) 16 Bar

| Nom. size | Flange outside dia. | Table 6/2 Forged Welding Neck | Table 6/3 Plate Slip on | Table 6/4 Forged Bossed Screwed | Table 6/5 Forged Bossed Slip on | Table 6/8 Plate Blank | Raised face Dia. | Raised face T'ness | Nr. bolt hole | Size of bolt |
|---|---|---|---|---|---|---|---|---|---|---|
| 15 | 95 | 14 | 14 | 14 | 14 | 14 | 45 | 2 | 4 | M12 × 45 |
| 20 | 105 | 16 | 16 | 16 | 16 | 16 | 58 | 2 | 4 | M12 × 50 |
| 25 | 115 | 16 | 16 | 16 | 16 | 16 | 68 | 2 | 4 | M12 × 50 |
| 32 | 140 | 16 | 16 | 16 | 16 | 16 | 78 | 2 | 4 | M16 × 55 |
| 40 | 150 | 16 | 16 | 16 | 16 | 16 | 88 | 3 | 4 | M16 × 55 |
| 50 | 165 | 18 | 18 | 18 | 18 | 18 | 102 | 3 | 4 | M16 × 60 |
| 65 | 185 | 18 | 18 | 18 | 18 | 18 | 122 | 3 | 4 | M16 × 60 |
| 80 | 200 | 20 | 20 | 20 | 20 | 20 | 138 | 3 | 8 | M16 × 60 |
| 100 | 220 | 20 | 20 | 20 | 20 | 20 | 158 | 3 | 8 | M16 × 65 |
| 125 | 250 | 22 | 22 | 22 | 22 | 22 | 188 | 3 | 8 | M16 × 70 |
| 150 | 285 | 22 | 22 | 22 | 22 | 22 | 212 | 3 | 8 | M20 × 70 |
| 200 | 340 | 24 | 24 | – | 24 | 24 | 268 | 3 | 12 | M20 × 75 |
| 250 | 405 | 26 | 26 | – | 26 | 26 | 320 | 3 | 12 | M24 × 90 |
| 300 | 460 | 28 | 28 | – | 28 | 28 | 378 | 4 | 12 | M24 × 90 |

**PIPES, WATER, STORAGE, INSULATION**

## Minimum Distances Between Supports/Fixings

| Material | BS Nominal pipe size | | Pipes – Vertical | Pipes – Horizontal on to low gradients |
|---|---|---|---|---|
| | (inch) | (mm) | Support distance in metres | Support distance in metres |
| Copper | 0.50 | 15.00 | 1.90 | 1.30 |
| | 0.75 | 22.00 | 2.50 | 1.90 |
| | 1.00 | 28.00 | 2.50 | 1.90 |
| | 1.25 | 35.00 | 2.80 | 2.50 |
| | 1.50 | 42.00 | 2.80 | 2.50 |
| | 2.00 | 54.00 | 3.90 | 2.50 |
| | 2.50 | 67.00 | 3.90 | 2.80 |
| | 3.00 | 76.10 | 3.90 | 2.80 |
| | 4.00 | 108.00 | 3.90 | 2.80 |
| | 5.00 | 133.00 | 3.90 | 2.80 |
| | 6.00 | 159.00 | 3.90 | 2.80 |
| muPVC | 1.25 | 32.00 | 1.20 | 0.50 |
| | 1.50 | 40.00 | 1.20 | 0.50 |
| | 2.00 | 50.00 | 1.20 | 0.60 |
| Polypropylene | 1.25 | 32.00 | 1.20 | 0.50 |
| | 1.50 | 40.00 | 1.20 | 0.50 |
| uPVC | – | 82.40 | 1.20 | 0.50 |
| | – | 110.00 | 1.80 | 0.90 |
| | – | 160.00 | 1.80 | 1.20 |
| Steel | 0.50 | 15.00 | 2.40 | 1.80 |
| | 0.75 | 20.00 | 3.00 | 2.40 |
| | 1.00 | 25.00 | 3.00 | 2.40 |
| | 1.25 | 32.00 | 3.00 | 2.40 |
| | 1.50 | 40.00 | 3.70 | 2.40 |
| | 2.00 | 50.00 | 3.70 | 2.40 |
| | 2.50 | 65.00 | 4.60 | 3.00 |
| | 3.00 | 80.40 | 4.60 | 3.00 |
| | 4.00 | 100.00 | 4.60 | 3.00 |
| | 5.00 | 125.00 | 5.50 | 3.70 |
| | 6.00 | 150.00 | 5.50 | 4.50 |
| | 8.00 | 200.00 | 8.50 | 6.00 |
| | 10.00 | 250.00 | 9.00 | 6.50 |
| | 12.00 | 300.00 | 10.00 | 7.00 |
| | 16.00 | 400.00 | 10.00 | 8.25 |

## PIPES, WATER, STORAGE, INSULATION

### Litres of Water Storage Required Per Person Per Building Type

| Type of building | Storage (litres) |
| --- | --- |
| Houses and flats (up to 4 bedrooms) | 120/bedroom |
| Houses and flats (more than 4 bedrooms) | 100/bedroom |
| Hostels | 90/bed |
| Hotels | 200/bed |
| Nurses homes and medical quarters | 120/bed |
| Offices with canteen | 45/person |
| Offices without canteen | 40/person |
| Restaurants | 7/meal |
| Boarding schools | 90/person |
| Day schools – Primary | 15/person |
| Day schools – Secondary | 20/person |

### Recommended Air Conditioning Design Loads

| Building type | Design loading |
| --- | --- |
| Computer rooms | 500 W/m² of floor area |
| Restaurants | 150 W/m² of floor area |
| Banks (main area) | 100 W/m² of floor area |
| Supermarkets | 25 W/m² of floor area |
| Large office block (exterior zone) | 100 W/m² of floor area |
| Large office block (interior zone) | 80 W/m² of floor area |
| Small office block (interior zone) | 80 W/m² of floor area |

PIPES, WATER, STORAGE, INSULATION

## Capacity and Dimensions of Galvanized Mild Steel Cisterns – BS 417

| Capacity (litres) | BS type (SCM) | Dimensions | | |
|---|---|---|---|---|
| | | Length (mm) | Width (mm) | Depth (mm) |
| 18 | 45 | 457 | 305 | 305 |
| 36 | 70 | 610 | 305 | 371 |
| 54 | 90 | 610 | 406 | 371 |
| 68 | 110 | 610 | 432 | 432 |
| 86 | 135 | 610 | 457 | 482 |
| 114 | 180 | 686 | 508 | 508 |
| 159 | 230 | 736 | 559 | 559 |
| 191 | 270 | 762 | 584 | 610 |
| 227 | 320 | 914 | 610 | 584 |
| 264 | 360 | 914 | 660 | 610 |
| 327 | 450/1 | 1220 | 610 | 610 |
| 336 | 450/2 | 965 | 686 | 686 |
| 423 | 570 | 965 | 762 | 787 |
| 491 | 680 | 1090 | 864 | 736 |
| 709 | 910 | 1070 | 889 | 889 |

## Capacity of Cold Water Polypropylene Storage Cisterns – BS 4213

| Capacity (litres) | BS type (PC) | Maximum height (mm) |
|---|---|---|
| 18 | 4 | 310 |
| 36 | 8 | 380 |
| 68 | 15 | 430 |
| 91 | 20 | 510 |
| 114 | 25 | 530 |
| 182 | 40 | 610 |
| 227 | 50 | 660 |
| 273 | 60 | 660 |
| 318 | 70 | 660 |
| 455 | 100 | 760 |

## PIPES, WATER, STORAGE, INSULATION

### Minimum Insulation Thickness to Protect Against Freezing for Domestic Cold Water Systems (8 Hour Evaluation Period)

| Pipe size (mm) | Insulation thickness (mm) | | | | | |
|---|---|---|---|---|---|---|
| | Condition 1 | | | Condition 2 | | |
| | $\lambda = 0.020$ | $\lambda = 0.030$ | $\lambda = 0.040$ | $\lambda = 0.020$ | $\lambda = 0.030$ | $\lambda = 0.040$ |
| Copper pipes | | | | | | |
| 15 | 11 | 20 | 34 | 12 | 23 | 41 |
| 22 | 6 | 9 | 13 | 6 | 10 | 15 |
| 28 | 4 | 6 | 9 | 4 | 7 | 10 |
| 35 | 3 | 5 | 7 | 4 | 5 | 7 |
| 42 | 3 | 4 | 5 | 8 | 4 | 6 |
| 54 | 2 | 3 | 4 | 2 | 3 | 4 |
| 76 | 2 | 2 | 3 | 2 | 2 | 3 |
| Steel pipes | | | | | | |
| 15 | 9 | 15 | 24 | 10 | 18 | 29 |
| 20 | 6 | 9 | 13 | 6 | 10 | 15 |
| 25 | 4 | 7 | 9 | 5 | 7 | 10 |
| 32 | 3 | 5 | 6 | 3 | 5 | 7 |
| 40 | 3 | 4 | 5 | 3 | 4 | 6 |
| 50 | 2 | 3 | 4 | 2 | 3 | 4 |
| 65 | 2 | 2 | 3 | 2 | 3 | 3 |

Condition 1: water temperature 7°C; ambient temperature −6°C; evaluation period 8 h; permitted ice formation 50%; normal installation, i.e. inside the building and inside the envelope of the structural insulation
Condition 2: water temperature 2°C; ambient temperature −6°C; evaluation period 8 h; permitted ice formation 50%; extreme installation, i.e. inside the building but outside the envelope of the structural insulation
$\lambda$ = thermal conductivity [W/(mK)]

### Insulation Thickness for Chilled And Cold Water Supplies to Prevent Condensation

#### On a Low Emissivity Outer Surface (0.05, i.e. Bright Reinforced Aluminium Foil) with an Ambient Temperature of +25°C and a Relative Humidity of 80%

| Steel pipe size (mm) | t = +10 | | | t = +5 | | | t = 0 | | |
|---|---|---|---|---|---|---|---|---|---|
| | Insulation thickness (mm) | | | Insulation thickness (mm) | | | Insulation thickness (mm) | | |
| | $\lambda = 0.030$ | $\lambda = 0.040$ | $\lambda = 0.050$ | $\lambda = 0.030$ | $\lambda = 0.040$ | $\lambda = 0.050$ | $\lambda = 0.030$ | $\lambda = 0.040$ | $\lambda = 0.050$ |
| 15 | 16 | 20 | 25 | 22 | 28 | 34 | 28 | 36 | 43 |
| 25 | 18 | 24 | 29 | 25 | 32 | 39 | 32 | 41 | 50 |
| 50 | 22 | 28 | 34 | 30 | 39 | 47 | 38 | 49 | 60 |
| 100 | 26 | 34 | 41 | 36 | 47 | 57 | 46 | 60 | 73 |
| 150 | 29 | 38 | 46 | 40 | 52 | 64 | 51 | 67 | 82 |
| 250 | 33 | 43 | 53 | 46 | 60 | 74 | 59 | 77 | 94 |
| Flat surfaces | 39 | 52 | 65 | 56 | 75 | 93 | 73 | 97 | 122 |

$t$ = temperature of contents (°C)
$\lambda$ = thermal conductivity at mean temperature of insulation [W/(mK)]

## PIPES, WATER, STORAGE, INSULATION

## Insulation Thickness for Non-domestic Heating Installations to Control Heat Loss

| Steel pipe size (mm) | t = 75 | | | t = 100 | | | t = 150 | | |
|---|---|---|---|---|---|---|---|---|---|
| | Insulation thickness (mm) | | | Insulation thickness (mm) | | | Insulation thickness (mm) | | |
| | $\lambda = 0.030$ | $\lambda = 0.040$ | $\lambda = 0.050$ | $\lambda = 0.030$ | $\lambda = 0.040$ | $\lambda = 0.050$ | $\lambda = 0.030$ | $\lambda = 0.040$ | $\lambda = 0.050$ |
| 10 | 18 | 32 | 55 | 20 | 36 | 62 | 23 | 44 | 77 |
| 15 | 19 | 34 | 56 | 21 | 38 | 64 | 26 | 47 | 80 |
| 20 | 21 | 36 | 57 | 23 | 40 | 65 | 28 | 50 | 83 |
| 25 | 23 | 38 | 58 | 26 | 43 | 68 | 31 | 53 | 85 |
| 32 | 24 | 39 | 59 | 28 | 45 | 69 | 33 | 55 | 87 |
| 40 | 25 | 40 | 60 | 29 | 47 | 70 | 35 | 57 | 88 |
| 50 | 27 | 42 | 61 | 31 | 49 | 72 | 37 | 59 | 90 |
| 65 | 29 | 43 | 62 | 33 | 51 | 74 | 40 | 63 | 92 |
| 80 | 30 | 44 | 62 | 35 | 52 | 75 | 42 | 65 | 94 |
| 100 | 31 | 46 | 63 | 37 | 54 | 76 | 45 | 68 | 96 |
| 150 | 33 | 48 | 64 | 40 | 57 | 77 | 50 | 73 | 100 |
| 200 | 35 | 49 | 65 | 42 | 59 | 79 | 53 | 76 | 103 |
| 250 | 36 | 50 | 66 | 43 | 61 | 80 | 55 | 78 | 105 |

$t$ = hot face temperature (°C)
$\lambda$ = thermal conductivity at mean temperature of insulation [W/(mK)]

# Index

Access chambers 115–116
Access covers 233–234
  load classes 232
  recessed 233
  solid top 232–233
Aeration of turf 81
Aftercare 451–452
  clearing pathways 451
  fertilisers 452
  maintaining edgings to plant beds 451
  maintenance operations 451
  mulching 451–452
  watering plants 452
  weeding 451
Agricultural fencing 98–99
  deer fencing 99
  forestry 99
  rabbit fencing 98
Airport container grown trees 428–429
Aluminium edgings 275–276
Anchoring systems 182–183
  retaining walls 182–183
Annual bedding 91–92
Anti-skate devices, seating 330
Aquatic herbicides, market prices 460
Aquatic planting 446
Asbestos survey 146
Automatic irrigation 96

Backfilling, soakaways 262
Backfilling materials, market prices 259
Backhoe loaders 162
Ball stop fencing, net 374–375
Bank stabilization 45–51
  erosion control mats 51
Banks
  excavation 45–47
  stabilization 51
Bar reinforcement 192–193
Bare root, hedges 441
Bare root trees
  nursery stock 425–531

  planting labours 422–423
Bark
  mulch 450
  paving 68
  playgrounds 73
Barriers 313
  electrically operated 313
  manually operated 313
Bases
  cellular confined bases 266
  earth anchors 314
  hardcore 265
  for paving 55
  street furniture 313–314
  SUDS 289–291
  surface treatments 266
  type 1 265–266
Beach cobble paving 68
Bedding 91–92
  living wall 92
  remulching 94
  watering 92
Bedding maintenance 471–472
  hoeing 472
  lifting 471
  weeding 472
Benches
  stone 326
  tree benches 329
Benches and seating 113
Bins 315–317
  cigarette 317
  dog waste 315
  grit 317
  litter 315–318
Block
  edgings 270
  faced block walling 106
  paving 59–60
  piers 106
  walls 103–104
Block paving 59–60
  car parks 59–60

concrete block permeable paving
  (CBPP) 289
concrete sett 59
edgings 60
rectangular permeable block
  paving 291
Block retaining walls
  concrete 110–111
  hollow 110–111
Block walling
  block piers 205
  concrete 103–104
  gauged mortar 204–205
  retaining walls 103–104
Bollards
  cast iron 319–320
  concrete 318–319
  excavation 318
  lighted 487
  recycled plastic 319
  removable 319
  security 321
  steel 320
  timber 320
  vehicle deterrent 321
  with rails 320
Boreholes 343
  drilling 343
Boundary fencing
  chain link 370
  field fencing 370
  Orsogril 372
  rabbit netting 370
  tubular 371
  tubular chain link 370
  welded mesh 372
Brick
  channels 55
  class B engineering bricks 199
  copings 201
  curved walls 200
  edgings 273–274
  English bond piers 201
  English garden wall bond 199–201
  Flemish bond 199
  mortar mixes 196–197
  movement of materials 198
  painting 230
  pathways 66
  paved paths 58

paving 57–58
pavings 278–281
planters 92–93
price variation 274
quantities of mortar 197–198
retaining walls 182
variation in brick prices 279
walls 104–106
Brick paving
  by hand 59
  by machine 57–58
  clay 279–281
  edge restraints 279
  excavation 57–58
  movement of materials 278
  paths 58
  variation in brick prices 279
Brick piers 107–108
  English bond 204–205
Brick walling 104–106
  brick copings 201
  class B engineering bricks 199
  demolition 155
  english garden wall bond 199–201
  facing bricks 199–200
  mortar mixes 195–202
  with piers 105–106
Bulb planting, forming plant
  holes 446

Cables and switching 480–482
  cable to trenches 481
  ducts to trenches 481
  electrical connections 482
  junction boxes 481–482
  terminations to armoured cables 481
Car parks 59–60
  block paving 59–60
  excavation (Macadam) 62
  grass concrete 63
  macadam 62
  marking 304
Cast iron
  bollards 319–320
  litter bins 316
  tree grilles 332
Cast iron bollards, with rails 320
Cast iron drainage 243–244
  below ground drainage 243–244
  inspection chambers 243–244

Cast stone
  buildings 153
  copings 209
Catchwater/French drains 251–252
Cellular pavings 282–283
  granular 282–283
  polyethylene reinforcement 283
Cellular retaining systems 51
Cellular slope retention 49
Cement
  access covers 232–234
  market prices 195
Channel drains 246–247
  stainless steel 247
Channels
  3 coursed granite setts;
    dished 55
  brick 55
  edgings 274–275
  excavation 55
  foundations 267
  granite sett 274
  vitrified clay 234–235
Chip vegetation 42
Cigarette bins 317
Cladding
  cast stone 213
  natural stone 213
  stainless steel fixings 213
  Yorkstone 213
Clay
  brick paving 279–281
  pipes 250
Clay brick paving 279–281
  basketweave 280
  herringbone 280
Clearance
  fencing/gates 43
  leaf 468
  litter 469
  shrubs 471
  site 42–44
  surfaces 44
  vegetation 42
Cobble, beach cobble paving 68
Cold water 343–344
  hose tap 343–344
  hose union bib taps 343
  standpipes 343–344
  stopcocks 343

Commercial urban development, design
    costs 148–149
Compost
  mushroom compost 414
  peat free 413–414
  spreading (shrub beds) 471
Concrete
  bases (street furniture) 313–314
  block permeable paving 289
  block retaining walls 110–111
  block walling 103–104
  bollards 318–319
  bulk bags 187
  channel foundations 274
  concrete sett paving 59
  copings 209
  cycle blocks 321
  demolition 44
  edge restraint 54
  embankments 183
  foundations 188–190
  foundations (soakaways) 260–262
  grass 283–284
  grass concrete paving 63
  heavy loads 187
  in situ 188–191
  in-situ (walls) 108–109
  inspection chambers 239–240
  jointed pipes 254–255
  kerbs 52–53
  manholes 241
  mix information 186
  mixes 186–187
  modular glass paving 63
  outfalls 255
  pavings formwork 304
  precast concrete block paving 287–288
  precast concrete channels 244–245
  precast concrete drainage 134–136
  precast concrete edgings 270
  precast concrete kerbs 268–269
  precast concrete litter bins 315
  precast concrete retaining walls 124
  precast concrete slab paving 291–295
  reinforced concrete paths 62–63
  reinforced concrete roadbed 56
  reinforced concrete walls (foundations)
    102–103
  revetments 182
  road excavation by hand 56

road gullies 116
roads 56
seating 326–327
self adhesive 212
setts 288
site mixed 109
soakaways 121–122
steps (insitu) 310–311
suspended walkway 194
walls 102–112
Concrete mixes 186–187
bulk bags 187
heavy loads 187
Concrete paving
by hand 63
in situ 305–306
modular grass concrete 63
pedestrian areas 62
trafficked areas 62
Concrete roads 56
Concrete sett paving 59
Concrete sundries 306
Conservation
cast iron bollards 320
historic railings 101
kerbs 268
metal railings 160–161
paving 292–293
Considerate contract scheme (ccs) 149
Contact herbicides, market prices 457–458
Containerised trees 433
nursery stock 425
Contract drawing, project management 147
Contract management 40
Contract/sub contract evaluation 148
Conveyor belts 143
Copings, brick 204
Cost models 35–39
business or science park 36–37
civic environments and streetscape 37–38
domestic 39
housing and house builder 35–36
playgrounds 39
Crane hire 145
Crib wall, timber 112
Crib walling 180
Cultivating
by pedestrian operated rotavator 389–391
by tractor 389
ripping subsoil 178–179

rolling 177–178
Cultivation
bulbs 88
by pedestrian operated rotavator 410–411
by tractor 410
ploughed ground 410
shrubs 88
turf 397
Cultivation and soil preparation 389–391
by pedestrian operated rotavator 389–391
by tractor 389
grading 390
Curriculum vitae of staff 150
Cycle blocks, concrete 322
Cycle holders 321–322
Cycle shelters 322
Cycle stand 113
Cyclic maintenance operations 475–476
litter 457–476
pathways 475–476

Damp proof courses, walls 210
Deadman anchoring 419
Decking
commercial applications 308
composite decking 308
domestic garden applications 308
handrails 309
roof decks 308
surface treatments 309
timber joists 309
Demolition 155–159
brick walls 155
concrete footings 155
disconnection of water 156
electrical disconnection 156
existing free standing walls 43
existing structures 44
existing surfaces 44
fencing 157
free standing buildings 155–156
posts/straining posts 157
Design costs
commercial urban development 148
hospitals and education 148
housing 149
landscape works 148–149
public urban open space 148
reclamation project 149
Deterrent pavings 295

edge restraints 295
  pedestrian 295
Directional signage 330
Disconnection of services
  electricity 156
  water 156
Disposal 168–170
  by hand 169
  by machine 168–169
  by skip 169
  clearing stones 391
  excavated materials 169
  excavated materials (drainage) 235
  excavated materials (land drains) 256–257
  excavated soil (hedge planting) 436–437
  grass cuttings 466–467
  hedge cuttings 474
  hedges 170
  shrubs 170
  soakaways 260
  topsoil for fencing 368
Ditching 253–255
  clearing silt 253
  clearing vegetation 253
  excavation 253–254
  outfalls 255
  piped 254–255
Dog waste bins 315
Drainage 133–136
  access chambers 115–116
  access covers 232–234
  bands 259
  below ground 133–136, 231–232
  below ground, cast iron drainage 243–244
  by hand 230
  cast iron accessories 244
  cast iron inspection chambers 243–244
  cast iron pipes 243
  catchwater/French drains 251–252
  channel drain accessories 117
  channel drains 246
  clay pipes 250
  cover slabs 136
  culverts 133, 263
  ditches 121
  ditching 253–255
  drainlock gratings 247
  earthwork support 232
  excavation 235–236
  foul water 115–118

  fountains 358–359
  galvanised hexagonal slot drains 245–246
  inspection chambers 134, 239–244
  interception traps 236–239
  land drainage 118–122
  manhole covers 234
  outfalls 255
  piped ditching 254–255
  pipes 231
  planters 230
  precast concrete 134–136
  roof decks 230
  roofs 215–226
  sand grooving 259
  sand slitting 259
  slits 259
  slot drains 245–246
  sports 264
  step irons 243
  subsoil 118–121
  SUDS paving 289–291
  surface water 115–118
  swimming pools 360
  trench excavation 231–232
  u-PVC pipes 250–251
Drainage boards
  planters 230
  roof decks 230
Drainlock gratings 247
Drip irrigation 347
Driveways, cellular pavings 282–283
Drystone walling 206
  Cotswold stone 206
  Purbeck 206
  Yorkstone 206
Dumpers 162

Earth anchors
  protective fencing 362
  street furniture 314
Earthwork retention, to tree pits 423
Earthwork support
  drainage 232
  pipe laying 248–249
Ecavation
  manhole boxes 166
  trench boxes 166
Edge restraint
  brick 54
  concrete block 54

Edge restraints 273
Edgings 52–54
　block paving 60
　brick 272–274
　brick/block stretchers 273
　channels 274
　edge restraints 273–274
　excavation 53
　footpaths 56–57
　foundations 267
　granite 54
　granite flag 272
　granite sett 271
　metal 275–276
　natural Yorkstone 272
　precast concrete block 287
　sawn Yorkstone 54
　slate 272
　steel 276
　to paths/pavings 52–54
Electrical connections 482
Electrical disconnection 156
Electrical installation 480–482
　bulkhead lighting 484
　cable to trenches 481
　cables and switching 480–482
　ducts to trenches 481
　electrical connections 482
　exterior lighting transformers 483
　floodlights 485
　junction boxes 481–482
　low voltage lights 482–483
　precinct lighting 488
　recessed lighting 486–487
　spotlights 485–486
　terminations to armoured cables 481
　trenching 480
Electrical lighting, lighted bollards 487
Electrical Services
　electrical connections 482
　junction boxes 481–482
Embankments 180–181
　erosion control mats 181
　excavated materials 180–181
　reinforced concrete 183
　tensar sheets 180–181
Employers requirements
　considerate contract scheme (ccs) 149
　curriculum vitae of staff 150
　health and safety 142–143

　indicative staff resource chart 150
　management of works 150
　provision, content and use of documents 149
　security/protection 143
　tendering 141–143
Erosion control
　cellular retaining systems 51
　mats 51
　wetland establishment 352–353
Excavation
　artificial turf 79–80
　banks 45–46
　bollards 318
　brick paving 57–58
　by hand 165–166
　by machine 163–165
　car parks (Macadam) 63
　cellular confined surfaces 65
　channel 55
　concrete Roads 56
　disposal of materials 169
　ditching 253–254
　earthwork retention 166
　earthwork support 248–249
　edgings 54
　footpaths 56–57
　for concrete foundations 102
　for embankments 180–181
　forming plant holes (bulbs) 446
　forming planting holes 434–435
　foundations (retaining walls) 175
　foundations to kerbs 267
　general 46–47
　granular beds 248
　ha-ha 48
　hand 235
　hedge planting 89–91
　kerbs 52–54
　lake construction 356
　land drains 256–257
　lighted bollards 487
　machine 235
　on-site material 159
　pathways 277–278
　pit works 235
　planting holes 87
　playgrounds 72–73
　rainwater harvesting tanks 344
　rainwater tanks 97
　recessed lighting 486–487

retaining wall 209–210
slab paving 67
soakaways 260
spreading materials 47
steps 71
stone kerbs 54
stone slab 68–69
street furniture 313
swimming pools 360
trenches 48
trenches (for hedge planting) 436–437
trenches (pipe laying) 247–249
urban planting (tree pits) 423
wall foundations 102–112
walls 102–112
water features 113–114
Excavators 161–162
Existing mains and services 151
Expansion joints, paving 307
Extensive green roofs 215–220
Exterior lighting, transformers 483

Faced block walling 106
Fences, clearance 43
Fences and gates, clearance 43
Fencing 98–101
 agricultural 98–99
 barbed 373
 barbed wire 368
 boundary fencing 370
 chain link 369
 chain link and wire 98
 cleft chestnut rolled 362
 cleft rail 365
 closeboarded 363–364
 deer fencing 99
 demolition 157
 feather edged 364
 field fencing 369
 fixing to posts 368
 forestry 99
 galvanised steel 369
 gates 101
 guard rails 380
 hit and miss horizontal rail 365
 horizontal bar 376–377
 metal estate fencing 376–377
 painting 229
 palisade 365
 post and rail 365

 protective 362
 rabbit fencing 98
 rabbit netting 370–371
 railings 375–376
 security 100
 security fences 372–373
 site 41
 strained wire 367–370
 temporary protective fencing for planting 410
 temporary security fence 143
 temporary security fencing 362
 trellis top 365
 tubular 371
 tubular chain link 370
 welded mesh 372
 windbreak 374
Fertilizers
 planting preparation 413
 Pre-seeding 394
 turf (market prices) 381–383
Field gates 378
Field hedges, cutting 473
Filling 166–167
 by hand 167
 by machine 166–167
 manhole boxes 166
 surface treatments 278
 to make up levels 278
 topsoil 411–413
 topsoil (tree pits) 417
Firepaths 63
Flag paving, pedestrian areas 67
Flagpoles 322–323
 ground mounted 322
 wall mounted 323
Flexible sheet materials 184
 erosion control mats 184
 river banks and revetments 184
 Tensar mat 184
Floodlights 485
Flooring
 ceramic tiles 214
 clay tiles 214
 finishes 214–215
 quarry tiles 215
Footings
 demolition 155
 retaining walls 110
Footpath gravels 285–286
 porous loose 285–286

porous self binding  285
Footpaths  56–57
  excavation  56–57
  gravels  285–286
  lay edgings  56–57
  macadam  61
Forestry
  fencing  99
  planting  449
Form landform
  excavated materials  47
  imported material  47
Formwork
  in situ concrete  191
  plain vertical  191
Foul water drainage  115–118
Foundations
  concrete channel  274
  in situ concrete (walls)  102–103
  in-situ concrete  188–190
  kerbs  267
  reinforced concrete walls  102–103
  retaining walls  175
  soakaways  260
  walls  102–112
Fountains  349–350
  fountain kits  359
  surrounds  359
  wall fountains  358–359
Fraise mowing  178
Free standing walls, demolition  43
Fungicides, market prices  460

Gabions  180–185
  filling by hand  180
  mattress revetments  49
  PVC coated  180
  Reno mattress  180
  retaining walls  180
  walls  111
Gates
  clearance  43
  featherboard garden  378
  hardwood field gate  378
  kissing gates  380
  metal estate fencing  377
  picket garden  379
  security fences  372
  sliding  379
  stiles  380

stock gate  101
treated softwood  378–379
tubular mild steel  379
Grading
  cultivation  390
  cultivation and soil preparation  390
  excavating trenches  249
  trenches and banks  45–49
Grading and surface preparation  176–178
  cultivating  178–179
  fraise mowing  178
  grading operations  176–177
  surface grading  178
  surface treatments  179
  trenches  177
Graffiti maintenance  476
Granite edgings  54
Granite sett paving
  pedestrian  68–69
  trafficked areas  68–69
Granular bases
  play surfaces  340
  surface treatments  265
Grass
  aeration  405
  rolling  467
  turf aeration  467–468
Grass concrete  283–284
  car parks  63
  full mortar bedding  284
  paving  63
  revetments  49
  type 1 bases  283–284
Grass cutting
  collecting arisings  466–467
  disposal  466–467
  pedestrian operated equipment  465–466
  ride-on equipment  462–464
  strimming  463–464
  tractor drawn equipment  462–464
Grass seed
  golf courses  386
  market prices  383–389
  sports pitches  384–385
  tennis courts  384
  wild flora and grass seed mixtures  387–389
Gravel
  backfilling trenches  258–259
  for footpaths  285–286
  market prices  255

pathways 65
surroundings to soakaways 261–262
Green roofs
  extensive 215–220
  extensive elements 125–127
  inspection chambers 225
  installation 225
  inverted waterproofing systems 226–227
  landscape options 225–226
  lightweight 218
  preamable 225
  vapour barriers 226
  waterproofing 226
Green screen 445
Grit bins 317
Ground retention, retaining walls 124
Groundcare operations 467–469
  harrowing 467
  leaf clearance 468
  litter clearance 469
  rolling 467
  scarifying 467
  top dressings 468
  tree guards 469
  tree stakes 469
  tree ties 468
  turf aeration 467–468
Groundcover plants 445
Guard rails 380
  pedestrian 380
Gullies
  bottle gullies 238–239
  cast iron 244
  concrete road 116
  road gullies 236
  universal gully trap 238
  yard gullies 237

Ha-ha 48
  ditch excavation 48
Handrails, timber decking 309
Hardcore bases 265
Hardwood
  entrance gate 378
  field gate 378
Harrowing 467
Haunching 274–275
Health and safety, producing health and safety file 142–143
Hedge cutting 473–474

hand held tools 473
reducing height 474
reducing width 473
removal/disposal of arisings 474
Hedge planting
  bare root 441
  beech and hornbeam hedges 437–438
  box 440
  by hand 89
  by machine 89–91
  disposal of excavated soil 437
  feathered hedges 438
  fully preformed 89
  instant hedge planting 437–438
  native species 441
  pre-clipped and preformed 90
  rootballed hedges 439–440
  yew (taxus) 438
Hedges
  bare root planting 441
  beech and hornbeam hedges 437–438
  box 440
  cutting 473–474
  disposal 170
  excavation for planting 89–91
  feathered 438
  instant hedge planting 437–438
  maintenance 473–474
  native species 441
  rootballed planting 439–440
  trench excavation 436–437
  yew (taxus) 438
Herbaceous and groundcover planting 445–446
Herbicide application 461–462
  backpack spraying 462
  power spraying 461
Herbicides
  application 461–462
  aquatic herbicides (market prices) 460
  contact herbicides (market prices) 457–458
  for seeding 77
  market prices 454–456
  residual herbicides (market prices) 457
  selective herbicides (market prices) 457–459
  total herbicides (market prices) 454–456
Horizontal bar fencing 376–377
Hose tap 343
Hosepipe, hand held 472

Hospitals and education, design costs 148
Housing, design costs 149

In situ concrete 188–191
  edgings 270
  formwork 191
  foundations (walls) 102–103
  haunching 274–275
  kerbs 267–268
  paving 304–306
  plain 188–190
  ready mixed 189–191
  reinforcement bars 192–193
  reinforcement fabric 193
  site mixed 188–189
  steps 310–311
  subbase 305
  sundries 306
  swimming pools 360
In situ paving and surfaces 62–70
Indicative staff resource chart 150
Industrially grown turf 398
Insect control, market prices 461
Inspection chambers
  cast iron 243–244
  green roofs 225
  polypropylene 242
  precast concrete 239
Inspection chambers 239–244
Intensive green roofs 220–225
Interception traps 236–239
Inturf big roll 79
Irrigation 346–349
  commissioning and testing 347
  drip irrigation 347
  for plants 472
  hand held hosepipe 472
  header tanks 346–347
  infrastructure 346
  leaky pipe 348–349
  maintenance costs 348
  moisture control 348
  sprinklers 347
  valves 346
  water supply and control 346–347
Irrigation sprinklers 347
  sprayheads 347
Irrigation systems 96–97
  automatic irrigation 96
  leaky pipe irrigation 96

living walls 444
trees 420–421

Japanese knotweed 476–479
  commercial chemical treatments 476–477
  contaminated waste 478
  landfill tax rates 478–479
  removal 478
  root barriers 479
Jogging track 74–75
Junction boxes 481–482

Kerbs 52–54
  bullnosed 269
  foundations 267
  half battered 269
  natural stone 270–271
  stone 53–54
  straight 268

Labour costs 161
  brick/block walling team 195
  concrete team 186
  groundworks team 235
Lakes 354–356
  construction 356
  liners 354–356
Lamps, market prices 484
Land drainage 118–122
  ditches 121
  soakaways 121–122
  subsoil 118–121
Land drains
  backfilling 258–259
  backfilling trenches 258
  calculation table 255
  disposal of excavated material 256–257
  excavation 256–257
  pipe laying 258
Landfill, tax rates (Japanese knotweed) 478–479
Landscape chemicals
  application 461
  market prices 453–461
Landscape design
  commercial urban development 148
  hospitals and education 148
  housing 149
  public urban open space 148
  reclamation project 149

Landscape maintenance 93–95
  defects liability 95
  long term 453
  remulching 94
  rose beds 94
  shrub beds 93–94
  vegetation control 95
Landscape works, design costs 148–149
Landscaping
  grass concrete 284
  intensive green roofs 224–226
  linear drains 245
  macadam 303
  macadam surfacing 299–300
  ornamental shrubs 140
  precast concrete block paving 288
  precast concrete slabs 292
  soft landscaping 139–140
Lanscaping, mature multi stem trees 139
Leaf clearance 468
Leaky pipe irrigation 96
Lifebuoy stations 324
Lighting 484–488
  bulkhead 484
  exterior lighting transformers 483
  floodlights 485
  lamps (market prices) 484
  lighted bollards 487
  low voltage 482–483
  precinct lighting 488
  recessed lighting 486–487
  spotlights 485–486
Linear drainage
  channel drain accessories 117
  mini beany combined kerb 244–245
  pedestrian areas 117
  sensitive areas 116
  vehicular areas 117
Litter bins 315–317
  cast iron 316
  metal 315–316
  plastic 317
  timber faced 317
Litter clearance 469
Living walls
  irrigation systems 444
  maintenance 444–445
  preliminary costs 444
Location factors 25
Low voltage lighting 482–483

Macadam surfaces 61–62
  car parks 62
  footpaths 61
  roadway 61
Macadam surfacing 301–304
  base (roadbase) course 299–300
  dense bitumen 301–304
  play surfaces 340
  resin bound 299–300
  roadbases 301–303
Machine selection 161–162
  backhoe loaders 162
  dumpers 162
  excavators 161–162
Maintenance
  cyclic maintenance operations 475–476
  graffiti 476
  groundcare operations 467–469
  hedges 473–474
  Japanese knotweed 476–479
  litter 475–476
  living walls 444–445
  pathways 475–476
  plant bedding 471–472
  playgrounds 475
  shrub beds 469–471
  sports areas 476
  street furniture 475
Maintenance of grass area 80–81
Maintenance of seeded/turfed areas 403–406
  initial cutting 403
  leaf clearance 403
  repairing grass areas 403
  watering 406
Management of works
  employers requirements 150
  setting-out 150
Manholes 241–242
  brick 117
  concrete 241
  covers 234
  walls 241–242
Manure 414–415
Market prices
  aquatic herbicides 460
  cement 195
  contact herbicides 458
  fertilizers (planting) 408–409
  fungicides 461

grass seed 383–389
gravel 255
herbicides 454–456
imported topsoil 407
insect control 461
lamps 484
landscape chemicals 453–461
moss control 461
mulching materials 407–408
native species 447
plant containers 324–325
pre-seeding materials 381
residual herbicides 457
selective herbicides 457–459
shrubs 433–434
sports/play surfacing materials 333
timber decking 307–308
total herbicides 454–456
turf fertilizers 381–383
wild flora and grass seed mixtures 387–389
wild flora seed mixtures 387–389
Measured works contract 40
Mechanical plant 144–145
Metal edgings 275–276
   aluminium edgings 275–276
   maintenance strip 276
   roads 275
   steel 276
Metal estate fencing 376–377
   bare metal 376
   end posts 377
   galvanised 377
   gates 377
   painted 376
   pedestrian gates 377
Method statements 147
   local authority type projects 142
Mortar mixes 196–198
   lime mortar 203–204
Moss control, market prices 461
Mulch
   aftercare 451–452
   Amenity Bark Mulch (trees) 422
   Bark Mulch 449
   bark nuggets 449
   bark nuggets (trees) 422
   market prices 407–408
   mats 450
Mulching
   mobile mulching machines 159

   resin bound 421
   tree pits 422
Mushroom compost 414

Native species
   market prices 447
   planting 447
Natural stone
   cladding 213
   kerbs 270–271
Natural stone kerbs, dressed 270–271
Natural stone paving
   granite flag 296–298
   granite sett 68–69
   indian sandstone 297
   Yorkstone 296–297
Natural stone slab paving 69
   by hand 70
   by machine 69
Natural stone walling 205–206
   granite 206
   stonework team 205
Nursery stock trees 425–431

Operations after planting 449–450
   fertilizers 450
   Mulch 449–450
   Mulch mats 450
Ornamental buildings 114
   pavilion 114
   pergola 114
   temple 114
Ornamental hedges, cutting 473
Ornamental pools 357–358
   centre pieces 358
   liners 357
   surrounds 358
   wall fountains 358–359
Ornamental stone structures 154
Ornamental water features 113–114
Outdoor seating 326–330
   all steel 328
   anti-skate devices 330
   concrete 326–327
   metal 327–328
   recycled plastic 329
   stone bench 326
   timber 328–329
   tree benches 329
Outfalls, reinforced concrete 255

Painting 228–230
 fences 229
 metal surfaces 228
 new wood 228–229
 railings 228
 sawn surfaces 229
 sheds 229
 surface preparation 228–229
 walls 230
Park signage 330
 miscellaneous 330
Parking 40
 site 152
Paths, brick paved 58
Pathways
 excavation 277–278
 firepaths 63
 gravel 65
 macadam surfacing 299–300
 maintenance 475–476
 preparation 277–278
 reinforced concrete 62–63
 resin bonded 64
 resin bound 64
 slab/brick 66
 stepping stone path 66
 sweeping/clearing 451
 woodchip surfaces 307
Pavilion 114
Paving
 access covers 234
 bark 68
 bases 55
 block and beam 304
 block and beam 137
 brick 57–58
 car parks 59–60
 celestial smooth 294
 Charnwood textured 294
 clay brick 279–281
 concrete block permeable paving (CBPP) 289–291
 concrete cobble paviors 288
 conservation paving 292–293
 edge restraints 279
 edgings 287
 expansion joints 307
 fibre reinforced 293
 flag 67
 granite sett 68–69
 grass concrete 63
 grass concrete 283–284
 macadam roads 301–303
 modular grass concrete 63
 mortar mixes 196–197
 natural stone 68–70
 natural stone slab 69
 pebble paving 292
 pedestrian 68
 pedestrian deterrent 67
 precast concrete slab 291–295
 rectangular permeable block paving 291
 resin bound 299–300
 slab 66
 steps 136–137
 SUDS paving 289–291
 suspended 304
 textured (blind pedestrians) 292
 trafficked areas 69
 vehicular 287
 woodchip surfaces 307
Pavings
 brick 278–281
 cellular 282–283
 cellular confined granular 282–283
 concrete formwork 304
 deterrent pavings 295
 edge restraints 295
 granite 296–298
 granite flag 296–298
 in situ concrete 305–306
 Indian sandstone 297
 natural Yorkstone 296–297
 polyethylene grass reinforcement 283
 precast concrete steps 311–312
 reconstituted stone 298–299
 reconstituted Yorkstone 298–299
 resin bound 299–300
 specialised mortars 277
 stone 296–298
Pebble, paving 292
Pedestrian deterrent paving 67
Picnic tables 324
 recycled plastic 324
Piers 106–108
 block 106
 brick 107–108
Pipe laying
 backfilling 258
 catchwater/French drains 251–252

clay pipes 250
concrete jointed pipes 254–255
ditching 254–255
earthwork support 248–249
land drainage 258
PVC-u 115
PVC-u pipes 250–251
straight jointing pipes 251–252
trench excavation 247–249
vitrified clay 115
Pipes
  cast iron 243
  cold water services 343–344
  excavation 115
  laying 115
  ployethene 343–344
  standpipes 343
Pit works, drainage excavation 235
Plant containers 324–326
  faux lead 325
  market prices 324–325
  precast concrete 325
  timber 325
  window boxes 325
Planters, drainage 230
Planting
  aftercare 451–452
  airport container grown trees 428–429
  aquatic 446
  aquatic (water features) 114
  bare root hedges 441
  bare root shrubs 435
  bedding 91–92
  bedding maintenance 471–472
  beech and hornbeam hedges 437–438
  compost 413–415
  containerised shrubs 435–436
  containerised trees 433
  containerised trees (nursery stock) 425
  cultivation by pedestrian operated
    rotavator 410–411
  cultivation by tractor 410
  excavation of planting holes 87
  feathered hedges 438
  fertilisers (aftercare) 452
  fertilizers (market prices) 408–409
  forestry 449
  forming planting holes 434–435
  fungicides (market prices) 460
  groundcover plants 445

herbaceous 446
instant hedge 437–438
irrigation 472
irrigation (trees) 420–421
Japanese knotweed 476–478
maintaining edgings to plant beds 451
maintenance operations 451
manure 414–415
market prices 407–409
moss control (market prices) 461
mulch 449–450
mulch (market prices) 407–408
mulch mats 449–450
native species 447
native species (hedges) 441
operations after planting 449–450
ornamental shrubs 140
plant containers 324–326
planters 92–93
pleached trees 431–432
ploughed ground 410
post-planting maintenance 95
preparation/cultivation 410–411
protection 448–449
root barriers (trees) 420
rootballed hedges 439–440
rootballed trees (adv. nursery stock/semi
  mature) 430–431
rootballed trees (nursery stock) 427
rootballed trees (semi mature/mature) 433
semi mature trees 85
shrub maintenance 93–94
shrubs 433–436
shrubs and groundcover 87–88
site protection 410
soil additives 415
soil conditioners 415
soil improvements 413–416
spreading topsoil 412–413
structural soils (trees) 418
surface applications 415
surface preparation 82–83
temporary protective fencing 410
top up mulch 451–452
topiary 442–443
topsoil (market price) 407
topsoil filling 411–413
tree pit additives 419
tree staking 418
vegetation control 95

watering 472
watering bedding 92
watering plants 452
window boxes 325
yew (taxus) 438
Planting preparation, fertilizers 413
Play areas
　surfaces 340–341
　surfacing materials 333
Play surfaces 340–341
　granular type 1 base 340
Playground equipment 334–338
　climbing equipment 337–338
　installation 334
　moving equipment 336
　multiple activity play systems 335
　play structures 337–338
　roundabouts 337
　sandpits 338
　slides 336
　sports and social areas 335
　spring equipment 338
　swings 335–336
Playgrounds
　ball games 72
　bark 73
　excavation 72–73
　macadam 72–73
　maintenance 475
　safety surfaced 72–73
　surfaces 72–73
Pleached trees 431–432
　box pleached trees 431–432
　frame pleached trees 432
　umbrella shaped 432
Portable roadway systems 146
Posts/ straining posts, demolition 157
Pre-seeding materials, market prices 381
Precast concrete
　block paving 287–288
　channels 274–275
　cobble paviors 288
　copings 210
　drainage 134–136
　fencing 362
　litter bins 315
　paving for steps 136–137
　plant containers 325
　slab paving 291–295
　steps 311–312

suspended walkway 194
Precast concrete kerbs 52–53
　excavation 52–53
Precast concrete paving 291–295
　celestia smooth 294
　Charnwood textured 294
　coloured 291
　conservation 292–293
　fibre reinforced 293
Precinct lighting 488
　lanterns 488
Prefabricated buildings, stone structures 153–154
Preliminaries 40–41
Preparatory Groundworks 45–51
Protective fencing
　cleft chestnut rolled 362
　earth anchors 362
Provision, content and use of documents 149–150
Pruning, shrubs 469–470
Pruning shrubs 469–470
Public urban open space, design costs 148
PVC-u pipes 250–251

Rabbit netting 370–371
Railings
　guard rails 380
　historic 101
　mild steel blunt top 375
　painting 228
　restoration 228
　traditional pattern 375
　transportation off site 160
Railway sleeper wall
　hardwood 174–175
　softwood 174
　trench excavation 175
Rainwater harvesting
　accessories 345
　backfilling 344
　excavation 344
　low profile 345
　tanks 344–345
　type 1 granular base 344
Rainwater harvesting tanks 97
　pumps 346
Recessed lighting 486–487
Reclamation project, design costs 149
Reconstituted stone

paving 298–299
walls 206–210
Reconstituted stone walls 206–210
  balls and bases 208
  balustrades 207
  bottle balustrades 207
  copings 209
  gate piers 208
  pier caps 207
  pier plinths 206
  pier shafts 207
Recycled plastic
  bollards 319
  outdoor seating 329
  picnic tables 324
Regional variations 25
Reinforced earth 180–185
  anchoring systems 182–183
Reinforced soil 183
  green faced 183
  split face 183
Reinforcement bars 192–193
  rolled plain 192
Reinforcement fabric 193
Residual herbicides, market prices 457
Resin bound surfacing 64
  crushed rock 299
  marble 299
  natural gravel 299
  paving 299–300
  pavings 299–300
  pedestrian 299
  recycled glass 299
  spectrum 299
  tree pits 300
  vehicular 299
Restoration, railings 228
Restoration brickwork 202–205
  brick copings 204
  constructing walls 203–204
  damaged brickwork 202–203
  dismantling copings 202
  dismantling existing wall 202
  facing bricks 204
  lime mortar 203–204
  replacing pointing 203
Retaining walls
  block 110–111
  brick/block 182
  concrete 171–173

concrete block 104
footings 110
gabions 180
grass concrete 172
ground retention 124
in situ concrete 183
keystone 173
precast concrete 111
railway sleeper wall 174–175
reconstituted stone 209–210
reinforced 209
Stepoc interlocking blocks 173
Revetments
  brushwood faggot 175–176
  cellular slope retention 49
  coir 175
  concrete 182
  gabion mattress 49
  grass concrete 49
  willow 175
River bank and water ecology 351–353
  dredging 351
  reclemation/establishment of waterways 351
  wetland establishment 352–353
River banks and water ecology, willow walling 175
Roads
  concrete 56
  concrete excavation by hand 56
  macadam 61
  reinforced concrete roadbed 56
  temporary roadway systems 146
Roads, Paths, Pavings and Surfaces 52–73
Rockery stone 331
Rolling, grassed areas 467
Roof decks, drainage 230
Roof finishings 124–132
  extensive green roof elements 125–127
Roofs
  drainage 225–227
  drainage boards 230
  extensive green roofs 215–220
  inspection chambers 225
  inverted waterproofing systems 226–227
  landscape options 224–226
  vapour barriers 226
  waterproofing 226–227
Root barriers
  Japanese knotweed 479
  root delfecting barrier 435

trees 420
Root director 425
Rootballed trees
  nursery stock 427
  planting labours 423
  semi mature/mature 433

Safety surfacing, playgrounds 72–73
Sand
  banding 259
  grooving 259
  sandpits 338
  slitting 259
Sawn Yorkstone edgings 54
Scarifying
  by hand 467
  mechanical 467
Seating 327–330
  all steel 328
  concrete 326–327
  metal 327–328
  recycled plastic 329
  stone bench 326
  timber 328–329
  tree benches 329
Security
  bollards 321
  guards and kiosk 152
Security fencing
  barbed 373
  barbican galvanised 372
  gates 372
  intruder guards 373
Seedbed preparation 392–394
  pre-seeding fertilisers 394
  soil preparation 392–394
  soil treatments 393–394
  spreading topsoil 393
  wildflower meadows 400
Seeding
  by hand 185
  by machine 185, 394–396
  by pedestrian drawn seeder 402
  by tractor drawn seeder 401
  grass seed 394–395
  grass seeds (market prices) 383–389
  harrowing 395–396
  herbicide 77
  maintenance of seeded areas 403–406
  pre-seeding fertilisers 394

rolling 395
soil additives and improvements 392
soil conditioners 392
soil treatments 393–394
surface preparation 76–78
Topsoil 76–78
wild flora and grass seed mixtures 387–389
wild flora seed mixtures (market prices) 387–389
wildflower meadows 400–402
Seeding to wildflower meadows
  by pedestrian drawn seeder 402
  by tractor drawn seeder 401
Selective herbicides
  market prices 457–459
  power/knapsack sprayer 459
Setts, stone 298
Shrub bed maintenance 469–470
  clearing cultivated area 471
  digging/hoeing 470–471
  edge maintenance 469
  hand weeding 470
  lifting 470
  pruning 469–470
  spreading compost 471
Shrub beds
  digging/hoeing 471
  edge maintenance 469
  hand weeding 470
  lifting 470
  maintenance 469
  pruning 469–470
  spreading compost 471
Shrub planting
  clearing cultivated area 471
  digging/hoeing 471
  excavation of planting holes 87
Shrub protection, shrubshelter 448–449
Shrubs
  bare root 435
  beds in ornamental gardens 93–94
  containerised 435–436
  disposal 170
  edge maintenance 469
  forming planting holes 434–435
  maintenance 93–94
  market prices 433–434
  protection 448–449
  removal 158–159
  shrub bed maintenance 469–471

spreading compost 471
weeding 470
Signage 330
  directional 330
  park 330
Site
  access 152
  administration area 41
  clearance 42–44
  compounds 41
  existing mains and services 151
  fencing 41
  mixed concrete 109
  parking 152
  paving 136–137
  playing surfaces 137–138
  preparation 158–159
  protection (planting) 410
  security kiosk 152
  setup 41
  shrub pruning 158–159
  survey 41
  temporary accommodation 150–151
  Temporary fencing 98
  toilet facilities 151
  visit 152
Site access 152
  security kiosk 152
Site accommodation 150–151
Site clearance
  disposal of excavated materials 169
  mobile mulching machines 159
  shrub removal 158–159
Site compound equipment, weekly
  hire rates 41
Site compounds 41
  equipment 41
  removal 41
Site fencing
  cleft chestnut paling 41
  heras 41
Site preparation 158–159
  site clearance 158–159
  tree felling 158
  turf lifting 158
Site protection, planting 410
Site setup 41
Skips, loading excavated materials 169
Slab paving 66
  precast concrete 291–295

Slab/brick pathways 66
Slot drains 245–246
Smoking shelters 331
Soakaways
  backfilling 261–262
  concrete rings 260–262
  cover slabs 261
  disposal of excavated material 260
  excavation 260
  surroundings 262
Soakaways rings, concrete 260–262
Soakaways
  geofabric 262
  polypropylene 262
Soft landscaping, intensive green roofs
  224–226
Softwood
  featherboard garden gates 378
  field gate 378
  Hit and miss horizontal rail 365
  picket garden gates 379
Soil
  nailing 171
  reinforced 183
  reinforcement 50
  stabilization 49
Soil additives, planting 415
Soil additives and improvements 392
  soil conditioners 392
  surface applications 392
Soil conditioners 415
  manure 415
  medium bark 415
  mushroom compost 415
Soil disposal 168–169
Soil improvements 413–416
  compost 413–415
  fertilizers 413
  manure 415
Soil nailing 183
Soil preparation
  clearing stones 391
  cultivation 389–391
  lifting topsoil 392–393
  playing surfaces 138
  seedbeds 392–393
  spreading 390
Soil prepartion, playing surfaces 138
Soil reinforcement
  backfill excavated bank 50

geogrid 50
green faced 183
split face 183
Soil testing, for wildflower meadows 400
Soil treatments
   pre-seeding fertilisers 394
   seedbed preparation 392–394
Specialised mortars for paving 277
Sport pitches
   drainage 264
   natural 338–339
Sports and amenity drainage 264
Sports areas
   athletics track 342
   equipment 333–334
   grass seed 384–386
   maintenance 476
   marking/remarking pitches 405
   multi-use game areas 335
   natural sport pitches 338–339
   surfacing materials 333
   tennis courts 341
Sports equipment 333–334
   football goals 333
   tennis nets 333
Sports field construction 74–75
Sports pitches 74
   artificial pitches 342
   grass seed (market prices) 384–385
Spotlights 485–486
Spreading
   excavated materials 47
   grass seed 185
   mulch 422
Standing stone 331
Standpipes 343–344
Steel
   bollards 320
   horizontal bar fencing 376–377
   tubular gates 379
Steel edgings 275
Steel trip rail 113
Step irons 243
Stepping stone path 66
Steps 310–312
   excavation 71
   in situ concrete 310–311
   precast concrete 311–312
   stone clad 71
   treads 310

Stone
   kerbs 270–271
   natural stone paving 68–70
   ornamental buildings 114
   paving 296–298
   rockery stone 331
   standing stones 331
   street furniture 331
Stone cladding, steps 71
Stone edgings
   granite flag 272
   granite sett 272
   natural Yorkstone 272
   slate 272
Stone kerbs
   edgings 53–54
   excavation 53
Stone setts 298
   granite 298
Stone structures 153–154
   ornamental 154
Stopcocks 343
Street furniture 113
   bases 313–314
   benches and seating 113
   bins 315–317
   cycle blocks 322
   cycle holders 321–322
   cycle shelters 322
   cycle stand 113
   excavation 313
   flagpoles 322–323
   lifebuoy stations 324
   maintenance 475
   outdoor seating 327–330
   pergolas 323–324
   picnic tables 324
   plant containers 324–326
   rockery stone 331
   signage 330
   smoking shelters 331
   standing stones 331
   stones and boulders 331
   tree grilles 332
Street lighting 488
   bulkhead lighting 484
Street planters 93
Stretchers
   brick paving 279
   brick/block 273

three course 274
Strimming 463–467
Structures, demolition 155–159
Submission of samples 143
Subsoil 47
  drainage 118–121
Subsoil drainage 118–121
  by hand 120–121
  by machine 118–120
Subsoil filling, for wildflower meadows 401
SUDS 289–291
  bases 289–291
  concrete block permeable paving
    (CBPP) 289
  paving 289–291
Surface clearance 44
Surface finishes, painting 228–230
Surface grading 178
  laser controlled 178
  trenches 177
Surface preparation 76–77
  cultivating 178–179
  for planting 82–83
  for seeding 76–78
  for turfing 78–81
  fraise mowing 178
  surface treatments 179
Surface preparation for planting
  by hand 83
  by machine 82–83
  excavating topsoil 82–83
  spreading topsoil 82–83
Surface treatments 178–179
  compacting 179
  granular bases 265
  sand blinding 278
  timber decking 309
  turfed areas 467–469
Surface water drainage 115–118
  access chambers 115–116
  gullies 116
  linear 116–117
  pipe laying 115
Surfaces
  bark playgrounds 73
  cellular confined 65
  maintenance 475–476
  playgrounds 72–73
  playing surfaces 137–138
  resin bonded 64

resin bound 64
safety surfacing 72–73
sports fields 74–75
sports pitches 74
Surfacing
  artificial pitches 342
  macadam 299–300
  pervious surfacing materials 291
  playgrounds 340–341
  resin bonded 300–301
  resin bound 299–300
  sports/play areas 333
  SUDS 289–291
  tennis courts 341
Surveying 41
  asbestos 146
  building fabric 146
  ground penetrating radar 146
  underground services 146
Suspended walkway 194
Swimming pools 359–361
  accessories 360–361
  covers 361
  drainage layer 360
  heating 361
  in situ concrete 360
  natural 359
  refurbishment 361
  underwater lighting 360–361

Tanking and damp proofing 211–212
  cold applied bituminous emulsion 211
  liquid asphalt 212
  mastic asphalt 211
  self adhesive 212
Tanks, rainwater harvesting 344–345
Temporary accommodation 150–151
Temporary fencing
  for planting 410
  precast concrete 362
  protection 362
  security fence hire 143
  security fences 362
Temporary storage 151
Tender and contract documents 147–148
  drawing and plan printing 147
  method statements 147
Tender bid 40
Tender stage programmes 141
Tendering costs

for employed estimator 141
  material costs 141
Tennis courts 341
  grass seed 384
Tensar Mat 180–181
Timber
  bollards 320
  cleft rail fencing 365
  closeboarded fencing 363–364
  crib wall 112
  demolition of timber buildings 156
  feather edged fencing 364
  field gates 378
  hardwood field gate 378
  hit and miss horizontal rail fencing 365
  lap panels 363
  litter bins 317
  outdoor seating 328–329
  palisade fencing 365
  pergolas 323–324
  plant containers 325
  post and rail fencing 365
  railway sleeper wall 112
  trellis 365
  trellis top fencing 365
Timber decking
  commercial applications 308
  composite decking 308
  domestic garden applications 308
  handrails 309
  joists 309
  market prices 307–308
  roof decks 308
  surface treatments 309
Timber fencing 363–365
  cleft rail 365
  closeboarded 363–364
  feather edged 364
  hit and miss horizontal rail 365
  lap panels 363
  palisade 365
  post and rail 365
  trellis top 365
Toilet facilities, temporary; on site 152
Topiary 442–443
  clipped 442–443
  clipped (box) 442
  clipped (yew) 442–443
Topsoil
  cultivation 78–79

filling 411–413
filling (tree pits) 417
lifting (for turfing) 396
lifting for seedbeds 392–393
seeding 76–78
spreading (for turfing) 396–397
spreading (tree pits) 417
spreading for seedbeds 393
stripping (for wildflower meadows) 400
surface preparation 76–77
turfing 78–79
Tree benches 329
Tree felling 42
  site preparation 158
Tree grilles
  cast iron 332
  steel 332
Tree guards 448
Tree irrigation kits 421
Tree pit, excavation 84–86
Tree pits 416–417
  additives 419
  amenity bark mulch 422
  backfilling 417
  bark nugget mulch 422
  earthwork retention 423
  excavation 416
  for urban planting 423
  frames 421
  mulching 422
  resin bound 300
  resin bound mulch 421
  structural soils 418
  surfacing resin 421
Tree planting 416–433
  airport container grown trees 428–429
  anchors 418–419
  backfilling 417
  bare root 422–423
  bare root trees (nursery stock) 426–427
  by hand 84
  by machine 85–86
  containerised trees 433
  containerised trees (nursery stock) 425
  earthwork retention (pits) 423
  filling 417
  forestry 449
  geocellular system 424
  in soft Landscape 85
  irrigation kits 421

irrigation systems 420–421
mulching (tree pits) 422
nursery stock 425–431
pleached trees 431–432
pre-planting operations 416–417
protection 448–449
root director 425
rootballed 423
rootballed (nursery stock) 427
rootballed trees (semi mature/mature) 433
semi mature 85
spreading topsoil 417
staking 418
structural soils 418
trafficked areas 85
tree pit additives 419
tree pit excavation 416
tree pits 416–417
tree tube 448
urban planting (tree pits) 423–424
with mobile cranes 86
Tree protection 448–449
  tree guards 448
  tree tube 448
Trees
  airport container grown 428–429
  anchors 418–419
  bare root (nursery stock) 426–427
  containerised 433
  containerised (nursery stock) 425
  deadman anchoring 419
  earthwork retention (pits) 423
  forestry planting 449
  guards 469
  irrigation kits 421
  irrigation systems 420–421
  mulching (tree pits) 422
  nursery stock 425–431
  pleached 431–432
  pre-planting operations 416–417
  protection 448–449
  root barriers 420
  root director 425
  rootballed (nursery stock) 427
  rootballed (semi mature/mature) 433
  stakes 469
  staking 418
  structural soils 418
  stump removal 158
  ties 469

tree grilles 332
tree guards 448
tree pit additives 418–419
tree pit excavation 416
tree pits 416
urban planting (tree pits) 423
Trellis
  contemporary panels 366–367
  integral arches 367
  traditional panels 366
Trellis 366–367
Trenches
  backfilling 480–481
  electrical services 480–481
  excavation 48
  excavation (pipe laying) 247–249
  excavation (walls and foundations) 102–112
  excavation for drainage 231–232
  excavation for hedges 89–91
  for foundations 48
  for services 48
  grading operations 177
  granular beds 248
Trip rails
  diamond rail fencing 373
  hollow steel section knee rails 374
  steel 113
Tubular fencing 371
Turf
  cultivation and soil preparation 389–391
  cutting 404
  fertilizers (market prices) 381–383
  fraise mowing 178
  harrowing 405–406, 467
  industrially grown 398
  initial cutting 403
  leaf clearance 403
  lifting for disposal/preservation 158
  lifting topsoil 396
  maintenance of turfed areas 403–406
  marking/remarking pitches 405
  preparation of turf beds 396–397
  reinforced 399–400
  repair 403
  soil additives and improvements 392
  soil conditioners 392
  spreading topsoil 396–397
  stripping 42
  surface treatments 467–469
  top dressings 468

transportation 397
watering 406
wildflower meadows 400–402
Turf aeration 467–468
  by hand 468
  by machine 467
Turf lifting 158
Turf stripping
  for disposal 42
  for preservation 42
Turfing
  aeration 81
  inturf big roll 79
  maintenance 80–81
  sports pitches 398
  surface preparation 78–81
  topsoil 78–79
Type 1 base, rainwater harvesting 344
Type 1 bases 265–266
  grass concrete 283–284

Vapour barriers, roofs 226
Variations, regional 25
Vegetation, control 95
Vegetation clearance
  chipping 42
  scrub vegetation 42
  trees 42
Vehicular paving 287
Vitrified clay channels 234–235

Wall fountains 358–359
Walls
  block piers 205
  brick piers 201–202
  cast stone 208–209
  damp proof courses 210
  drystone 206
  English garden wall bond 203
  faced block 106
  finishes 214
  foundations 102–112
  gabion 111
  granite 206
  in situ concrete foundations 102–103
  living wall 92
  mortar mixes 195–202
  natural stone 205–206
  painting 230
  pier caps 210

piers 106–108
precast concrete 111
precast concrete coping 210
reconstituted stone 206–210
reinforced concrete (foundations) 102–103
roughcast/plastered/rendered coatings 214
site mixed concrete 109
timber crib 112
timber retaining 111–112
to manholes 241–242
waterproof rendering 211
Water disconnection 156
Water features 349–350
  aquatic planting 114
  internal 350
  ornamental 113–114
  waterfalls 349
Watering
  hand held hosepipe 472
  plants 452, 472
  rainwater tanks 97
Waterproof rendering 211
  steel 211
  walls 211
Waterproofing
  inverted waterproofing systems 226–227
  rendering (steel trowelled) 211
  vapour barriers 226
Weeding 451
  bedding maintenance 471–472
  Japanese knotweed 476–478
  plant beds 95
  shrubs 470
Wetland establishment 352–353
  revetments 352–353
Wild flora and grass seed mixtures, market prices 387–389
Wildflower meadows
  seedbank eradication 401
  seedbed preparation 400
  seeding 401–402
  soil testing 400
  subsoil filling 401
  topsoil stripping 400
Willow, revetments 175
Windbreak fencing 374
Window boxes 325
Woodchip surfaces 307

# ESSENTIAL READING FROM TAYLOR AND FRANCIS

## Spon's Asia Pacific Construction Costs Handbook, Fifth Edition

**LANGDON & SEAH**

In the last few years, the global economic outlook has continued to be shrouded in uncertainty and volatility following the financial crisis in the Euro zone. While the US and Europe are going through a difficult period, investors are focusing more keenly on Asia. This fifth edition provides overarching construction cost data for 16 countries: Brunei, Cambodia, China, Hong Kong, India, Indonesia, Japan, Malaysia, Myanmar, Philippines, Singapore, South Korea, Sri Lanka, Taiwan, Thailand and Vietnam.

May 2015: 234X156 mm: 452 pp
Hb: 978-1-4822-4358-1: £160.00

---

To Order: Tel: +44 (0) 1235 400524  Fax: +44 (0) 1235 400525
or Post: Taylor and Francis Customer Services,
Bookpoint Ltd, Unit T1, 200 Milton Park, Abingdon, Oxon, OX14 4TA UK
Email: book.orders@tandf.co.uk

For a complete listing of all our titles visit:
www.tandf.co.uk

# ESSENTIAL READING FROM TAYLOR AND FRANCIS

## Professional Practice for Landscape Architects, 3rd Edition
### TENNANT, GARMORY & WINSCH

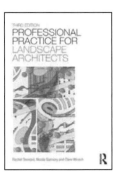

Professional Practice for Landscape Architects third edition deals with the practical issues of being a successful landscape architect professional.

Endorsed by the Landscape Institute, this book is an indispensable guide for licentiate members of the Institute on their Pathway to Chartership. It follows the revised 2013 syllabus covering all aspects of professional judgement, ethics and values, the legal system, organisation and management, legislation and the planning system, environmental policy and control, procurement and implementation. It also serves as a reminder and reference for fully qualified professionals in their everyday practice and for landscape students.

Valuable information is presented in an easy to follow manner with diagrams and schedules, key acts, professional documents and contracts clearly explained and made easy to understand.

A handy list of questions are included to aid with P2C revision, answers of which are found within the text.

September 2015; 234 x 156 mm; 466 pp
Pb: 978-1-138-78597-7; £26.99

To Order: Tel: +44 (0) 1235 400524  Fax: +44 (0) 1235 400525
or Post: Taylor and Francis Customer Services,
Bookpoint Ltd, Unit T1, 200 Milton Park, Abingdon, Oxon, OX14 4TA UK
Email: book.orders@tandf.co.uk

For a complete listing of all our titles visit:
www.tandf.co.uk

# ESSENTIAL READING FROM TAYLOR AND FRANCIS

## NRM1 Cost Management Handbook

David P Benge

The definitive guide to measurement and estimating using NRM1, written by the author of NRM1

The 'RICS New rules of measurement: Order of cost estimating and cost planning of capital building works' (referred to as NRM1) is the cornerstone of good cost management of capital building works projects - enabling more effective and accurate cost advice to be given to clients and other project team members, while facilitating better cost control.

The NRM1 Cost Management Handbook is the essential guide to how to successfully interpret and apply these rules, including explanations of how to:

- quantify building works and prepare order of cost estimates and cost plans
- use the rules as a toolkit for risk management and procurement
- analyse actual costs for the purpose of collecting benchmark data and preparing cost analyses
- capture historical cost data for future order of cost estimates and elemental cost plans
- employ the rules to aid communication
- manage the complete 'cost management cycle'
- use the elemental breakdown and cost structures, together with the coding system developed for NRM1, to effectively integrate cost management with Building Information Modelling (BIM).

March 2014: 246 x 174: 640pp
Pb: 978-0-415-72077-9: £41.99

---

To Order: Tel: +44 (0) 1235 400524  Fax: +44 (0) 1235 400525
or Post: Taylor and Francis Customer Services,
Bookpoint Ltd, Unit T1, 200 Milton Park, Abingdon, Oxon, OX14 4TA UK
Email: book.orders@tandf.co.uk

For a complete listing of all our titles visit:
www.tandf.co.uk

# Ebook Single-User Licence Agreement

We welcome you as a user of this Spon Price Book ebook and hope that you find it a useful and valuable tool. Please read this document carefully. **This is a legal agreement** between you (hereinafter referred to as the "Licensee") and Taylor and Francis Books Ltd. (the "Publisher"), which defines the terms under which you may use the Product. **By accessing and retrieving the access code on the label inside the front cover of this book you agree to these terms and conditions outlined herein. If you do not agree to these terms you must return the Product to your supplier intact, with the seal on the label unbroken and with the access code not accessed.**

1. **Definition of the Product**
   The product which is the subject of this Agreement, (the "Product") consists of online and offline access to the VitalSource ebook edition of *Spon's External Works & Landscape Price Book 2017*.

2. **Commencement and licence**
   2.1 This Agreement commences upon the breaking open of the document containing the access code by the Licensee (the "Commencement Date").
   2.2 This is a licence agreement (the "Agreement") for the use of the Product by the Licensee, and not an agreement for sale.
   2.3 The Publisher licenses the Licensee on a non-exclusive and non-transferable basis to use the Product on condition that the Licensee complies with this Agreement. The Licensee acknowledges that it is only permitted to use the Product in accordance with this Agreement.

3. **Multiple use**
   Use of the Product is not provided or allowed for more than one user or for a wide area network or consortium.

4. **Installation and Use**
   4.1 The Licensee may provide access to the Product for individual study in the following manner: The Licensee may install the Product on a secure local area network on a single site for use by one user.
   4.2 The Licensee shall be responsible for installing the Product and for the effectiveness of such installation.
   4.3 Text from the Product may be incorporated in a coursepack. Such use is only permissible with the express permission of the Publisher in writing and requires the payment of the appropriate fee as specified by the Publisher and signature of a separate licence agreement.
   4.4 The Product is a free addition to the book and the Publisher is under no obligation to provide any technical support.

5. **Permitted Activities**
   5.1 The Licensee shall be entitled to use the Product for its own internal purposes;
   5.2 The Licensee acknowledges that its rights to use the Product are strictly set out in this Agreement, and all other uses (whether expressly mentioned in Clause 6 below or not) are prohibited.

6. **Prohibited Activities**
   The following are prohibited without the express permission of the Publisher:
   6.1 The commercial exploitation of any part of the Product.
   6.2 The rental, loan, (free or for money or money's worth) or hire purchase of this product, save with the express consent of the Publisher.
   6.3 Any activity which raises the reasonable prospect of impeding the Publisher's ability or opportunities to market the Product.
   6.4 Any networking, physical or electronic distribution or dissemination of the product save as expressly permitted by this Agreement.
   6.5 Any reverse engineering, decompilation, disassembly or other alteration of the Product save in accordance with applicable national laws.
   6.6 The right to create any derivative product or service from the Product save as expressly provided for in this Agreement.
   6.7 Any alteration, amendment, modification or deletion from the Product, whether for the purposes of error correction or otherwise.

7. **General Responsibilities of the License**
   7.1 The Licensee will take all reasonable steps to ensure that the Product is used in accordance with the terms and conditions of this Agreement.
   7.2 The Licensee acknowledges that damages may not be a sufficient remedy for the Publisher in the event of breach of this Agreement by the Licensee, and that an injunction may be appropriate.
   7.3 The Licensee undertakes to keep the Product safe and to use its best endeavours to ensure that the product does not fall into the hands of third parties, whether as a result of theft or otherwise.
   7.4 Where information of a confidential nature relating to the product of the business affairs of the Publisher comes into the possession of the Licensee pursuant to this Agreement (or otherwise), the Licensee agrees to use such information solely for the purposes of this Agreement, and under no circumstances to disclose any element of the information to any third party save strictly as permitted under this Agreement. For the avoidance of doubt, the Licensee's obligations under this sub-clause 7.4 shall survive the termination of this Agreement.

8. **Warrant and Liability**
   8.1 The Publisher warrants that it has the authority to enter into this agreement and that it has secured all rights and permissions necessary to enable the Licensee to use the Product in accordance with this Agreement.
   8.2 The Publisher warrants that the Product as supplied on the Commencement Date shall be free of defects in materials and workmanship, and undertakes to replace any defective Product within 28 days of notice of such defect being received provided such notice is received within 30 days of such supply. As an alternative to replacement, the Publisher agrees fully to refund the Licensee in such circumstances, if the Licensee so requests, provided that the Licensee returns this copy of *Spon's External Works & Landscape Price Book 2017* to the Publisher. The provisions of this sub-clause 8.2 do not apply where the defect results from an accident or from misuse of the product by the Licensee.
   8.3 Sub-clause 8.2 sets out the sole and exclusive remedy of the Licensee in relation to defects in the Product.
   8.4 The Publisher and the Licensee acknowledge that the Publisher supplies the Product on an "as is" basis. The Publisher gives no warranties:
   8.4.1 that the Product satisfies the individual requirements of the Licensee; or
   8.4.2 that the Product is otherwise fit for the Licensee's purpose; or
   8.4.3 that the Product is compatible with the Licensee's hardware equipment and software operating environment.
   8.5 The Publisher hereby disclaims all warranties and conditions, express or implied, which are not stated above.
   8.6 Nothing in this Clause 8 limits the Publisher's liability to the Licensee in the event of death or personal injury resulting from the Publisher's negligence.
   8.7 The Publisher hereby excludes liability for loss of revenue, reputation, business, profits, or for indirect or consequential losses, irrespective of whether the Publisher was advised by the Licensee of the potential of such losses.
   8.8 The Licensee acknowledges the merit of independently verifying the price book data prior to taking any decisions of material significance (commercial or otherwise) based on such data. It is agreed that the Publisher shall not be liable for any losses which result from the Licensee placing reliance on the data under any circumstances.
   8.9 Subject to sub-clause 8.6 above, the Publisher's liability under this Agreement shall be limited to the purchase price.

9. **Intellectual Property Rights**
   9.1 Nothing in this Agreement affects the ownership of copyright or other intellectual property rights in the Product.
   9.2 The Licensee agrees to display the Publishers' copyright notice in the manner described in the Product.
   9.3 The Licensee hereby agrees to abide by copyright and similar notice requirements required by the Publisher, details of which are as follows:
   "© 2017 Taylor & Francis. All rights reserved. All materials in *Spon's External Works & Landscape Price Book 2017* are copyright protected. All rights reserved. No such materials may be used, displayed, modified, adapted, distributed, transmitted, transferred, published or otherwise reproduced in any form or by any means now or hereafter developed other than strictly in accordance with the terms of the licence agreement enclosed with *Spon's External Works & Landscape Price Book 2017*. However, text and images may be printed and copied for research and private study within the preset program limitations. Please note the copyright notice above, and that any text or images printed or copied must credit the source."

9.4 This Product contains material proprietary to and copyedited by the Publisher and others. Except for the licence granted herein, all rights, title and interest in the Product, in all languages, formats and media throughout the world, including copyrights therein, are and remain the property of the Publisher or other copyright holders identified in the Product.

## 10. Non-assignment

This Agreement and the licence contained within it may not be assigned to any other person or entity without the written consent of the Publisher.

## 11. Termination and Consequences of Termination.

11.1 The Publisher shall have the right to terminate this Agreement if:

11.1.1 the Licensee is in material breach of this Agreement and fails to remedy such breach (where capable of remedy) within 14 days of a written notice from the Publisher requiring it to do so; or

11.1.2 the Licensee becomes insolvent, becomes subject to receivership, liquidation or similar external administration; or

11.1.3 the Licensee ceases to operate in business.

11.2 The Licensee shall have the right to terminate this Agreement for any reason upon two month's written notice. The Licensee shall not be entitled to any refund for payments made under this Agreement prior to termination under this sub-clause 11.2.

11.3 Termination by either of the parties is without prejudice to any other rights or remedies under the general law to which they may be entitled, or which survive such termination (including rights of the Publisher under sub-clause 7.4 above).

11.4 Upon termination of this Agreement, or expiry of its terms, the Licensee must destroy all copies and any back up copies of the product or part thereof.

## 12. General

12.1 *Compliance with export provisions*

The Publisher hereby agrees to comply fully with all relevant export laws and regulations of the United Kingdom to ensure that the Product is not exported, directly or indirectly, in violation of English law.

12.2 *Force majeure*

The parties accept no responsibility for breaches of this Agreement occurring as a result of circumstances beyond their control.

12.3 *No waiver*

Any failure or delay by either party to exercise or enforce any right conferred by this Agreement shall not be deemed to be a waiver of such right.

12.4 *Entire agreement*

This Agreement represents the entire agreement between the Publisher and the Licensee concerning the Product. The terms of this Agreement supersede all prior purchase orders, written terms and conditions, written or verbal representations, advertising or statements relating in any way to the Product.

12.5 *Severability*

If any provision of this Agreement is found to be invalid or unenforceable by a court of law of competent jurisdiction, such a finding shall not affect the other provisions of this Agreement and all provisions of this Agreement unaffected by such a finding shall remain in full force and effect.

12.6 *Variations*

This agreement may only be varied in writing by means of variation signed in writing by both parties.

12.7 *Notices*

All notices to be delivered to: Spon's Price Books, Taylor & Francis Books Ltd., 3 Park Square, Milton Park, Abingdon, Oxfordshire, OX14 4RN, UK.

12.8 *Governing law*

This Agreement is governed by English law and the parties hereby agree that any dispute arising under this Agreement shall be subject to the jurisdiction of the English courts.

If you have any queries about the terms of this licence, please contact:

Spon's Price Books
Taylor & Francis Books Ltd.
3 Park Square, Milton Park, Abingdon, Oxfordshire, OX14 4RN
Tel: +44 (0) 20 7017 6000
www.sponpress.com

**Ebook options**

Your print copy comes with a free ebook on the VitalSource® Bookshelf platform. Further copies of the ebook are available from https://www.crcpress.com/search/results?kw=spon+2017
Pick your price book and select the ebook under the 'Select Format' option.

To buy Spon's ebooks for five or more users of in your organisation please contact:

Spon's Price Books
eBooks & Online Sales
Taylor & Francis Books Ltd.
3 Park Square, Milton Park, Oxfordshire, OX14 4RN
Tel: (020) 337 73480
onlinesales@informa.com